Formulas for Surface Area

Rectangular solid:
$A = 2lw + 2wh + 2lh$

Sphere: $A = 4\pi r^2$

Right-circular cylinder (including the ends):
$A = 2\pi rh + 2\pi r^2$

Right-circular cone (including the base):
$A = \pi rs + \pi r^2$

Formulas for Volume

Rectangular solid: $V = lwh$

Right prism: $V = Bh$, where B is the area of the base

Sphere: $\frac{4}{3}\pi r^3$

Right-circular cylinder: $V = \pi r^2 h$

Right-circular cone: $V = \frac{1}{3}\pi r^2 h$

Pyramid: $V = \frac{1}{3}Bh$, where B is the area of the base

Business Formulas

Formula relating rate, base, and percentage: $r \cdot b = p$

Formula for simple interest: $I = Prt$

Formula for amount: $A = P + Prt$

Units of Length

12 inches = 1 foot

5280 feet = 1 mile

1 hectometer = 100 meters

1 decimeter = $\frac{1}{10}$ meter

1 millimeter = $\frac{1}{1000}$ meter

3 feet = 1 yard

1 kilometer = 1000 meters

1 dekameter = 10 meters

1 centimeter = $\frac{1}{100}$ meter

Algebraic Formulas

If x, y, and z are real numbers, then

Double-negative rule: $-(-x) = x$

Rule for subtraction: $x - y = x + (-y)$

Absolute value: $\begin{cases} \text{If } x \geq 0, \text{ then } |x| = x. \\ \text{If } x < 0, \text{ then } |x| = -x. \end{cases}$

Closure properties: $x + y$, $x - y$, xy, and $\dfrac{x}{y}$ $(y \neq 0)$ are real numbers

Commutative properties: $x + y = y + x$, $xy = yx$

Associative properties: $(x + y) + z = x + (y + z)$, $(xy)z = x(yz)$

Distributive property: $x(y + z) = xy + xz$

2 ESSENTIAL MATHEMATICS WITH GEOMETRY

2 ESSENTIAL MATHEMATICS WITH GEOMETRY

R. David Gustafson
Rock Valley College

Peter D. Frisk
Rock Valley College

Brooks/Cole Publishing Company
Pacific Grove, California

Brooks/Cole Publishing Company
A Division of Wadsworth, Inc.

© 1994, 1990 by Wadsworth, Inc., Belmont, California 94002.
All rights reserved.
No part of this book may be reproduced, stored in a retrieval system, or transcribed,
in any form or by any means—electronic, mechanical, photocopying, recording,
or otherwise—without the prior written permission of the publisher,
Brooks/Cole Publishing Company, Pacific Grove, California 93950,
a division of Wadsworth, Inc.

Printed in the United States of America
10   9   8   7   6   5   4   3   2   1

Library of Congress Cataloging-in-Publication Data
Gustafson, R. David (Roy David), [date]
    Essential mathematics with geometry / R. David Gustafson, Peter D.
  Frisk. — 2nd ed.
        p.   cm.
    Includes index.
    ISBN 0-534-20268-3
    1. Mathematics.   I. Frisk, Peter D., [date].   II. Title.
QA39.2.G858   1993
510—dc20   92-39254
                                                                           CIP

Sponsoring Editor: *Craig Barth*
Marketing Representative: *Ragu Raghavan*
Editorial Assistant: *Carol Ann Benedict*
Production Editor: *Marjorie Z. Sanders*
Manuscript Editor: *David Hoyt*
Permissions Editor: *Carline Haga*
Interior and Cover Design: *Roy R. Neuhaus*
Cover Photo: *Thomas Craig/FPG International*
Art Coordinator: *Lisa Torri*
Interior Illustration: *Lori Heckelman*
Typesetting: *Integre Technical Publishing Co., Inc.*
Cover Printing: *Phoenix Color Corporation*
Printing and Binding: *Courier/Kendallville*

*Pacific Bell is a registered trademark of Pacific Telesis. Macintosh is a registered trademark
of Apple Computer, Inc. ExamBuilder is a registered trademark of Cooke Publications, Ltd.
EXPTEST is a trademark of Wadsworth, Inc.*

Preface

Educational reform continues to sweep the country. At the heart of the matter is the desire to promote higher student achievement in reading, writing, and mathematics. To pursue this goal, we have written a single-volume text that will prepare college students for college-level mathematics.

Our goal is to write a developmental text that prepares students to succeed in subsequent mathematics courses while holding the attrition rate in the current course to a minimum.

Because *Essential Mathematics with Geometry*, second edition, is more comprehensive than most workbooks and reaches up into the topics of intermediate algebra, it provides the prerequisite background for either a college algebra or a finite mathematics course. It will be effective in either a traditional classroom setting or a self-paced learning laboratory. By picking and choosing topics, the teacher can modify the text to support a one-semester, three-hour course. Most topics can be covered in a one-semester, five-hour course or a sequence of two three-hour courses.

Essential Mathematics with Geometry, second edition, has three parts. The first five chapters cover basic arithmetic, fractions, decimals, percents, reading information from graphs, and some basic statistics.

Chapters 6 and 7 cover topics from basic Euclidean geometry, including the formulas for finding areas and volumes of geometric figures. The concepts of deductive reasoning and problem solving are developed, and many theorems are presented. Proofs are informal and are written in paragraph style.

Chapters 8–16 provide a basic algebra course, including systems of three equations in three variables, complex numbers, and additional topics from geometry. Problem-solving skills are taught throughout the text, and ample opportunity for drill is provided.

Changes for the Second Edition

The overall effect of the changes we have made has been to:

- Make more use of mathematics in the world around us through realistic applications
- Increase the emphasis on visualizing mathematics through the use of graphing and geometry
- Fine-tune the presentation of certain topics

Some of the specific changes we have made are:

- We have interspersed Project Checks throughout the book. These progress checks make the book more interactive and give students instant feedback on whether they understand the concepts and examples. Answers are shown at the bottom of the page.
- A new section has been added to Chapter 5 that deals with the mean, median, and mode.
- Many more realistic applications have been added throughout the text. Each chapter opens with an interesting application problem involving the material of the chapter and ends with a solution of the problem.
- Beginning with the first chapter and continuing throughout the text, students must apply newly acquired skills to solve word problems. Chapter 9 now includes three sections of applications to allow more time to introduce and practice setting up a wide variety of word problems. A section on applications of fractional equations is now included.
- The treatment of exponents has been rewritten and divided into two sections. Section 10.1 now covers the properties of natural-number exponents, and Section 10.2 covers integer exponents.
- The section on factoring by grouping appears earlier in the text in Section 11.2. We now use factoring by grouping as an aid in factoring trinomials.
- More emphasis has been placed on factoring out -1 in the numerators and denominators of fractions.
- Cube roots now receive more emphasis. The section on fractional exponents now appears before the section discussing the distance formula.
- The section on imaginary numbers has been expanded to include the arithmetic of complex numbers.
- Additional worked examples help students apply skills to the more difficult problems in the improved exercise sets. There are over 540 worked examples and over 7300 carefully graded exercises.
- More work is done with function. Domain and range are now presented from a graphical point of view.

Illustrating Our Approach

Our textbooks are noted for their staightforward, no-nonsense approach to teaching mathematics. New material is presented in the context of concrete examples. The ideas are then generalized and written in rule or theorem form. Following each theorem are several examples that illustrate the use of the theorem. (See page 397.)

Each section ends with a set of carefully graded exercises that give students ample practice to develop confidence in their skills, as well as some real challenges. (See page 298.)

More Power Rules for Exponents: If n is a natural number, then

$$(xy)^n = x^n y^n \quad \text{and if } y \neq 0, \text{ then} \quad \left(\frac{x}{y}\right)^n = \frac{x^n}{y^n}$$

EXAMPLE 6

a. $(ab)^4 = a^4 b^4$

b. $(3c)^3 = 3^3 c^3 = 27c^3$

c. $(x^2 y^3)^5 = (x^2)^5 (y^3)^5 = x^{10} y^{15}$

d. $(-2x^3 y)^2 = (-2)^2 (x^3)^2 y^2 = 4x^6 y^2$

e. $\left(\frac{4}{k}\right)^3 = \frac{4^3}{k^3} = \frac{64}{k^3}$

f. $\left(\frac{3x^2}{2y^3}\right)^5 = \frac{3^5 (x^2)^5}{2^5 (y^3)^5} = \frac{243 x^{10}}{32 y^{15}}$

← Students are shown specific examples.

Progress Check 5

Write each expression without using parentheses.

a. $(xy^2)^3$

b. $(-2a^2 b^3)^3$

Work Progress Check 5.

The Quotient Rule for Exponents

To find a rule for dividing exponential expressions we simplify the fraction $\frac{4^5}{4^2}$.

$$\frac{4^5}{4^2} = \frac{4 \cdot 4 \cdot 4 \cdot 4 \cdot 4}{4 \cdot 4}$$

Note that the exponent in the numerator is greater than the exponent in the denominator.

$$= \frac{\cancel{4} \cdot \cancel{4} \cdot 4 \cdot 4 \cdot 4}{\cancel{4} \cdot \cancel{4}}$$

$$= 4^3$$

The result, 4^3, has a base of 4 and an exponent of $5 - 2$, or 3. This suggests that *to divide exponential expressions with the same base, we keep the base and subtract the exponents.*

← A theorem is stated in words.

The Quotient Rule for Exponents: If m and n are natural numbers, $m > n$ and $x \neq 0$, then

$$\frac{x^m}{x^n} = x^{m-n}$$

← A theorem is stated formally.

EXAMPLE 7

If there are no divisions by 0, then

a. $\dfrac{x^4}{x^3} = x^{4-3} = x^1 = x$

b. $\dfrac{8y^2 y^8}{4y^3} = \dfrac{8y^8}{4y^3} = \dfrac{8}{4} y^{8-3} = 2y^5$

← The theorem is illustrated.

5. **a.** $x^3 y^6$ **b.** $-8a^6 b^9$

Section 10.1 Natural-Number Exponents 397

9.3 Exercises

In Exercises 1–30, simplify each expression, if possible.

1. $3x + 17x$
2. $12y - 15y$
3. $8x^2 - 5x^2$
4. $17x^2 + 3x^2$
5. $9x + 3y$
6. $5x + 5y$
7. $3(x + 2) + 4x$
8. $9(y - 3) + 2y$
9. $5(z - 3) + 2z$
10. $4(y + 9) - 6y$
11. $12(x + 11) - 11$
12. $-3(3 + z) + 2z$
13. $-23(x^2 - 2) + x^2$
14. $5(17 + x) - 27$
15. $3x^3 - 2(x^3 + 5)$
16. $-7z^2 - 2z^3 + z$
17. $8x(x + 3) - 3x^2$
18. $2x + x(x + 3)$
19. $8(y + 7) - 2(y - 3)$
20. $9(z + 2) + 5(3 - z)$
21. $2x + 4(y - x) + 3y$
22. $3y - 6(y + z) + y$
23. $9(x + y) - 9(x - y)$
24. $2(x + z) + 3(x - z)$
25. $(x + 2) + (x - y)$
26. $3z + 2(y - z) + y$
27. $2\left(4x + \dfrac{9}{2}\right) - 3\left(x + \dfrac{2}{3}\right)$
28. $7\left(3x - \dfrac{2}{7}\right) - 5\left(2x - \dfrac{3}{5}\right) + x$
29. $2(7x + 2^2) - 3(3x - 3^2)$
30. $2^2 \left(2x - \dfrac{3}{2}\right) + 3^3 \left(\dfrac{x}{3} - \dfrac{2}{3}\right)$

In Exercises 31–88, solve each equation, if possible. Check all solutions.

31. $3x + 2 = 2x$
32. $5x + 7 = 4x$
33. $5x + 3 = 4x$
34. $4x + 3 = 5x$
35. $9y - 3 = 6y$
36. $8y + 4 = 4y$
37. $8y - 7 = y$
38. $9y - 8 = y$
39. $3(a + 2) = 4a$
40. $4(a - 5) = 3a$
41. $5(b + 7) = 6b$
42. $8(b + 2) = 9b$
43. $-9x + 3 = 8x + 20$
44. $-11x + 3 = 4x - 27$
45. $-7a + 5 = 11a - 22$
46. $-6a + 7 = 9a - 8$
47. $4x + 8 + 3(x - 2) = -12$
48. $7x + 2 + 4(x - 3) = 12$

← Exercises are carefully graded.

Topics in *Essential Mathematics with Geometry*, second edition, go beyond what is normally found in a developmental mathematics text. For example, we include exponential expressions with variable exponents, the distance formula, rational exponents, a thorough treatment of cube and higher roots, and complex numbers. We believe it is important for students to recognize these topics when they encounter them in college-level mathematics.

To provide continuous review, we include review exercises at the end of each exercise set (see page 436), chapter review exercises, practice chapter tests, and important formulas listed for reference inside the covers of the text. Each chapter ends with a chapter summary that includes key terms and key ideas. (See page 437.)

In Exercises 31–40, perform each division. If there is a remainder, leave the answer in quotient + $\frac{\text{remainder}}{\text{divisor}}$ form.

31. $\frac{2x^2 + 5x + 2}{2x + 3}$
32. $\frac{3x^2 - 8x + 3}{3x - 2}$
33. $\frac{4x^2 + 6x - 1}{2x + 1}$
34. $\frac{6x^2 - 11x + 2}{3x - 1}$

35. $\frac{x^3 + 3x^2 + 3x + 1}{x + 1}$
36. $\frac{x^3 + 6x^2 + 12x + 8}{x + 2}$

37. $\frac{2x^3 + 7x^2 + 4x + 3}{2x + 3}$
38. $\frac{6x^3 + x^2 + 2x + 1}{3x - 1}$

39. $\frac{2x^3 + 4x^2 - 2x + 3}{x - 2}$
40. $\frac{3y^3 - 4y^2 + 2y + 3}{y + 3}$

In Exercises 41–50, perform each division.

41. $\frac{x^2 - 1}{x - 1}$
42. $\frac{x^2 - 9}{x + 3}$
43. $\frac{4x^2 - 9}{2x + 3}$

45. $\frac{x^3 - 1}{x + 1}$
46. $\frac{x^3 - 8}{x - 2}$
47. $\frac{16a^2 - 9b^2}{4a + 3b}$

49. $3x - 4 \overline{) 15x^3 - 29x^2 + 16}$
50. $2y + 3 \overline{) 23y^2 + 6y}$

Review Exercises

1. List the composite numbers from 20 to 30.
2. Graph the prime on a number line.

In Review Exercises 3–4, let $a = -2$ and $b = -3$. Evaluate each expression.

3. $4a - 2b$
4. $\frac{3a^2 + 2b^2}{3(a - b)}$

In Review Exercises 5–8, write each expression as an equivalent expression without absolute

5. $|25|$
6. $|-32 - 5|$
7. $-|3 - 5|$

In Review Exercises 9–10, simplify each expression.

9. $3(2x^2 - 4x + 5) + 2(x^2 + 3x - 7)$
10. $-2(y^3 + 2y^2 - y)$

11. The area of the ring between two circles of radius r and R (see Illustration 1) is given by
$$A = \frac{22}{7}(R + r)(R - r)$$
If $r = 3$ inches and $R = 17$ inches, find A.

12. Laura bought a color TV set for $502.90, which included a 7% sales tax. What was the selling price of the TV before the tax was added?

436 Chapter 10 Polynomials

✓ Each exercise set ends with 12 review exercises. Most sets of review exercises include word problems.

Mathematics in Medicine

Illustration 1

Recall that the area of a circle is given by the formula $A = \pi r^2$. The bulk of the surface area of the red blood cell in Illustration 1 is contained on its top and bottom. That area is $2\pi r^2$, twice the area of one circle. If there are N discs, their total surface area, T, will be N times the surface area of a single disc: $T = 2N\pi r^2$.

To find the total surface area of the oxygen-carrying red cells, first express the given quantities in scientific notation.

radius = r = 0.00015 in. = 1.5×10^{-4} in.
quantity = N = 25 trillion
 = 2.5×10^{13}

Then substitute these values into the formula for total surface area.

$T = 2N\pi r^2$
$T = 2(2.5 \times 10^{13})(3.14)(1.4 \times 10^{-4})^2$
 $= 2(2.5)(3.14)(1.4)^2 \times 10^{13} \times 10^{-8}$
 $= 31 \times 10^5$
 $= 3.1 \times 10^6$
 $= 3,100,000$

The total surface area of the red blood cells is over 3 million square inches, or 21,500 square feet, almost one-half the area of a football field!

Chapter Summary

Key Words

algebraic terms (10.4)
base (10.1)
binomial (10.4)
conjugate binomial (10.6)
degree of a monomial (10.4)

degree of a polynomial (10.4)
dividend (10.8)
divisor (10.8)
exponent (10.1)
FOIL (10.6)
like terms (10.5)

minuend (10.5)
monomial (10.4)
polynomial (10.4)
power (10.1)
quotient (10.8)
remainder (10.8)
scientific notation (10.3)

special products (10.6)
standard notation (10.3)
subtrahend (10.5)
trinomial (10.4)

Key Ideas

(10.1)–(10.2) **Properties of exponents.** If n is a natural number, then
$$x^n = \overbrace{x \cdot x \cdot x \cdot x \cdots \cdot x}^{n \text{ factors of } x}$$

If m and n are integers and there are no divisions by 0,
$$x^m x^n = x^{m+n}, \quad (x^m)^n = x^{mn}, \quad (xy)^n = x^n y^n$$
$$\left(\frac{x}{y}\right)^n = \frac{x^n}{y^n}, \quad \frac{x^m}{x^n} = x^{m-n}, \quad x^0 = 1, \quad x^{-n} = \frac{1}{x^n}$$

(10.4) If $P(x)$ is a polynomial in x, then $P(r)$ is the value of the polynomial when $x = r$.

(10.3) A number is written in scientific notation if it is written as the product of a number between 1 (including 1) and 10 and an integer power of 10.

(10.5) When adding or subtracting polynomials, combine like terms by adding or subtracting the numerical coefficients and using the same variables and the same exponents.

Chapter Summary 437

Thorough Treatment of Graphing

Although we live in a world of graphing calculators, we believe that students first need to learn the basics by drawing graphs by hand. Thus, our Chapter 13 on graphing begins a discussion of graphing lines. We thoroughly discuss graphing both by plotting points and by the intercept method. We cover graphing vertical and horizontal lines in detail. A complete presentation of the concept of slope and the method of graphing a line using its slope and y-intercept are included.

EXAMPLE 1

Find the equation of the line with slope $-\frac{3}{2}$ and y-intercept 3. Express the equation in general form.

Solution

$y = mx + b$	Use the slope-intercept form.
$y = -\frac{3}{2}x + 3$	Substitute $-\frac{3}{2}$ for m and 3 for b.
$2y = -3x + 6$	Multiply both sides by 2.
$3x + 2y = 6$	Add $3x$ to both sides.

The equation $3x + 2y = 6$ is written in the form $Ax + By = C$, which is the general form of the equation of a line. ∎

Point-Slope Form of the Equation of a Line

Suppose we know the slope of a line, but instead of its y-intercept we know the coordinates of a second point on the line. It is still possible to find the equation of the line. For example, we will find the equation of the line that has a slope of 3 and passes through the point $P(2, 1)$. See Figure 13-17.

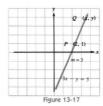

Figure 13-17

We know that $P(2, 1)$ are the coordinates of one point on the line. We can then let $Q(x, y)$ be the coordinates of some other point on the line. To find the slope of the line that passes through the points $P(2, 1)$ and $Q(x, y)$, we use the definition of slope.

$$m = \frac{y_2 - y_1}{x_2 - x_1}$$

1. $m = \dfrac{y - 1}{x - 2}$ Substitute y for y_2, 1 for y_1, x for x_2, and 2 for x_1.

However, we are given that the slope m of the line is 3. Thus, we have

$3 = \dfrac{y - 1}{x - 2}$	Substitute 3 for m in Equation 1.
$3(x - 2) = y - 1$	Multiply both sides by $x - 2$.
$3x - 6 = y - 1$	Use the distributive property to remove parentheses.
$3x - y = 5$	Add $-y$ and 6 to both sides.

The equation of the required line, written in general form, is $3x - y = 5$.

564   Chapter 13 Graphing Linear Equations and Inequalities

← Section 13.3 includes a complete discussion of writing equations of lines with known properties. Section 13.5 provides an introduction to functions.

Progress Checks are included throughout. →

Author's notes explain each step in → the solution process.

EXAMPLE 2

A line has a slope of $\frac{3}{4}$ and passes through the point $(-1, \frac{1}{2})$. Write the equation of the line in general form and graph it.

Solution

We substitute $\frac{3}{4}$ for m, -1 for x_1, and $\frac{1}{2}$ for y_1 in the point-slope form of a linear equation and simplify.

$y - y_1 = m(x - x_1)$	
$y - \dfrac{1}{2} = \dfrac{3}{4}[x - (-1)]$	
$4y - 2 = 3(x + 1)$	Multiply both sides by 4 and simplify.
$4y - 2 = 3x + 3$	Remove parentheses.
$-5 = 3x - 4y$	Add -3 and $-4y$ to both sides.

In general form, the equation of the required line is $3x - 4y = -5$. Work Progress Check 4. ∎

Progress Check 4

Write the equation of a line passing through $(-2, 4)$ with a slope of $\frac{2}{3}$. Write the equation in slope-intercept form.

───────────────
Progress Check Answers

4. $y = \frac{2}{3}x + \frac{16}{3}$

Preface xi

Thorough Treatment of Word Problems

The treatment of word problems is often a distinguishing feature among textbooks. We therefore take great care in presenting word problems. Students begin solving word problems early in Chapter 1. A problem-solving theme continues throughout the text.

We believe the key to solving word problems is the ability to set up word equations. Thus, we include a written analysis before the solution of most algebraic word problems. All of the traditional algebraic word problems are included but many more-current applications, such as break-even analysis, depreciation, and appreciation, are also included.

Supplements for the Student

Conquering Math Anxiety
Cynthia Arem

Conquering Math Anxiety speaks to students about math anxiety and how to recognize it. It includes strategies and practice exercises to cope with it and offers tips on conquering test anxiety.

Student Solutions Manual
Diane Koenig

The Student Solutions Manual contains complete solutions to alternate even-numbered exercises from the text.

Tutorial Software
Teresa Bittner
(IBM and Macintosh)

BCX software. Covers all topics in the text. Provides problems, online prompting, and multiple-choice and free response answers with complete solutions.

The Teaching Package for the Instructor

Transparency Masters
Teresa Bittner

A set of Transparency Masters provides applications and additional thought-stretching problems for reinforcement.

Instructor's Solutions Manual
Michael Welden

Contains complete solutions to all odd-numbered exercises and to the alternate even-numbered exercises that are not in the Student Solutions Manual.

Test Manual
William Hinrichs

Contains four ready-to-use forms of every chapter test.

Classroom Manager

This record-keeping program handles records for up to 1000 students, performs various calculations, imports Scantron data, and exports to most word processors and spreadsheets.

Computerized Test Banks
William Hinrichs

Available with our text is an extensive electronic question bank, in multiple-choice format. The bank contains approximately 1300 test items. The bank is available for IBM and compatible machines and for Macintosh machines.

The testing programs give you all the features of state-of-the-art word processors and more, including the ability to see all technical symbols, fonts, and formatting on the screen just as they will appear when printed. The question bank can be edited.

ExamBuilder® runs on Macintosh computers.

EXPTEST™ runs on IBM and compatible computers.

Accuracy

Gustafson/Frisk textbooks, published by Brooks/Cole, have the reputation of being error-free. This reputation has been earned by the hard work of many people, including very competent production and design departments, and by the diligence of many reviewers and proofreaders. Each exercise has been worked independently by both authors and several other problem checkers.

Acknowledgments

We are grateful to the following people who have reviewed the text at various stages of its development:

Stanley Ball,
University of Texas, El Paso

Theresa Barrie,
Texas Southern University

June Bjercke,
San Jacinto College

Tony Bower,
St. Phillips College

Ginny Crowder,
Brenau College

Ruth Dalrymple,
Texas A & I University

Susan Friedman,
Bernard M. Baruch College

Jerry Gilpin,
University of Texas—Pan American Campus

Roberto Gonzales,
Pan American University

Rebecca Haeberle,
San Francisco State University

Abul Kalam,
Community College of Denver

Paul Kennedy,
South West Texas State University

Josephine Lane,
Eastern Kentucky University

Nenette Loftsgaarder

Nancy Long,
Trinity Valley Community College

James Marcotte,
Cincinatti Technical College

Wayne L. Miller, Jr.,
Lee College

Patricia Morgan,
San Diego State University

Donald Perry,
Lee College

Janet Ritchie,
State University of New York—Old Westbury

Ruth Schaefer,
California State University at Long Beach

Joan Smith,
St. Petersburg Junior College

Patricia M. Stone,
Tomball College

John Watson,
Arkansas Technical University

Jamie Whitehead,
Texarkana College

We are especially thankful to John Wiley Publishing Company for allowing us to adapt material from our text *Elementary Plane Geometry*, third edition.

We also thank the following authors who prepared supplementary materials to accompany the text: Cynthia Arem, William Hinrichs, Diane Koenig, and Michael Weldon.

We wish to thank Diane Koenig, who read the entire manuscript and worked every problem. We also wish to thank Gary Schultz, Bill Hinrichs, James Yarwood, Jerry Frang, David Hinde, Darrell Ropp, Michael Weldon, Jennifer Dollar, and George Mader for their helpful suggestions. We give special thanks to Betty Fernandez for her assistance in the preparation of the manuscript, and Craig Barth, David Hoyt, Roy Neuhaus, Marjorie Sanders, Audra Silverie, and Lisa Torri.

R. David Gustafson
Peter D. Frisk

Contents

1 Whole Numbers 1

- **1.1** Whole Numbers and Place Value; Rounding Whole Numbers 1
- **1.2** Adding and Subtracting Whole Numbers 7
- **1.3** Multiplying and Dividing Whole Numbers 16
- **1.4** Order of Operations; the Distributive Property 28
- **1.5** Prime Numbers, Composite Numbers, and Prime-Factored Form 33
- **1.6** Least Common Multiple and Greatest Common Divisor 38
  - Chapter Summary 43
  - Chapter 1 Review Exercises 44
  - Chapter 1 Test 47

2 Fractions 49

- **2.1** Fractions and Equivalent Fractions 49
- **2.2** Multiplying and Dividing Fractions 55
- **2.3** Adding and Subtracting Fractions 64
- **2.4** Mixed Numbers; Comparing Fractional Expressions 73
  - Chapter Summary 84
  - Chapter 2 Review Exercises 84
  - Chapter 2 Test 87

3 Decimals 89

- **3.1** Decimals and Place Value; Rounding 89
- **3.2** Adding and Subtracting Decimals 94
- **3.3** Multiplying and Dividing Decimals 99
- **3.4** Applications of Decimals 106
  - Chapter Summary 113
  - Chapter 3 Review Exercises 113
  - Chapter 3 Test 115

4 Percent 117

- **4.1** Percents, Fractions, and Decimals 117
- **4.2** Problems Involving Percents 123
- **4.3** Applications of Percents 128
  Chapter Summary 135
  Chapter 4 Review Exercises 136
  Chapter 4 Test 137

5 Measurement, Estimation, and Reading Graphs 139

- **5.1** Measurement 139
- **5.2** Estimation 149
- **5.3** Reading Graphs and Tables 152
- **5.4** Mean, Median, and Mode 166
  Chapter Summary 172
  Chapter 5 Review Exercises 172
  Chapter 5 Test 177

6 Basic Concepts of Geometry 181

- **6.1** Some Undefined Terms and Basic Definitions 181
- **6.2** Special Pairs of Angles 188
- **6.3** Parallel Lines and Transversals 194
- **6.4** Triangles and Other Polygons 200
- **6.5** Congruent Triangles 207
- **6.6** Parallelograms 215
- **6.7** Circles 221
  Chapter Summary 227
  Chapter 6 Review Exercises 229
  Chapter 6 Test 233

7 Proportion, Similar Triangles, Area, and Volume 235

- **7.1** Ratio and Proportion 235
- **7.2** Similar Triangles 240
- **7.3** Areas of Rectangles and Parallelograms 246
- **7.4** Areas of Triangles and Trapezoids 252
- **7.5** Areas of Circles 257
- **7.6** Surface Area and Volume 261
- **7.7** Inductive and Deductive Reasoning 269

Chapter Summary 274
Chapter 7 Review Exercises 275
Chapter 7 Test 279

8 Real Numbers and Their Properties 283

8.1 Sets of Numbers and Their Graphs 283
8.2 Variables and Algebraic Terms 289
8.3 Exponents and Order of Operations 294
8.4 Real Numbers and Their Absolute Values 300
8.5 Adding and Subtracting Real Numbers 307
8.6 Multiplying and Dividing Real Numbers 315
8.7 Properties of Real Numbers 323
Chapter Summary 329
Chapter 8 Review Exercises 330
Chapter 8 Test 333

9 Equations and Inequalities 335

9.1 The Addition and Subtraction Properties of Equality 335
9.2 The Division and Multiplication Properties of Equality 341
9.3 Simplifying Expressions to Solve Equations 349
9.4 Literal Equations 356
9.5 Applications of Equations 362
9.6 More Applications of Equations 367
9.7 Even More Applications of Equations 374
9.8 Inequalities 380
Chapter Summary 388
Chapter 9 Review Exercises 388
Chapter 9 Test 391

10 Polynomials 393

10.1 Natural-Number Exponents 393
10.2 More on Exponents 400
10.3 Scientific Notation 404
10.4 Polynomials 409
10.5 Adding and Subtracting Polynomials 413
10.6 Multiplying Polynomials 418
10.7 Dividing Polynomials by Monomials 426
10.8 Dividing Polynomials by Polynomials 430
Chapter Summary 437
Chapter 10 Review Exercises 438
Chapter 10 Test 441

11 Factoring Polynomials 443

- **11.1** Factoring Out the Greatest Common Factor 443
- **11.2** Factoring by Grouping 448
- **11.3** Factoring the Difference of Two Squares 452
- **11.4** Factoring Trinomials with Lead Coefficients of 1 456
- **11.5** Factoring General Trinomials 463
- **11.6** Factoring the Sum and Difference of Two Cubes 471
- **11.7** Summary of Factoring Techniques 475
- **11.8** Solving Equations by Factoring 477
- **11.9** Applications 482
  Chapter Summary 487
  Chapter 11 Review Exercises 488
  Chapter 11 Test 491

12 Rational Expressions 493

- **12.1** The Basic Properties of Fractions 493
- **12.2** Multiplying Fractions 501
- **12.3** Dividing Fractions 505
- **12.4** Adding and Subtracting Fractions with Like Denominators 510
- **12.5** Adding and Subtracting Fractions with Unlike Denominators 514
- **12.6** Complex Fractions 521
- **12.7** Solving Equations That Contain Fractions 527
- **12.8** Applications of Equations That Contain Fractions 532
  Chapter Summary 538
  Chapter 12 Review Exercises 539
  Chapter 12 Test 541

13 Graphing Linear Equations and Inequalities 543

- **13.1** Graphing Linear Equations 543
- **13.2** The Slope of a Line 554
- **13.3** Writing Equations of Lines 563
- **13.4** Graphing Inequalities 572
- **13.5** Relations and Functions 577
- **13.6** Variation 584
  Chapter Summary 593
  Chapter 13 Review Exercises 594
  Chapter 13 Test 597

14 Solving Systems of Equations and Inequalities 599

- **14.1** Solving Systems of Equations by Graphing 600
- **14.2** Solving Systems of Equations by Substitution 606
- **14.3** Solving Systems of Equations by Addition 611
- **14.4** Applications of Systems of Equations 616
- **14.5** Solving Systems of Inequalities 624
- **14.6** Solving Systems of Three Equations in Three Variables 628
  Chapter Summary 634
  Chapter 14 Review Exercises 634
  Chapter 14 Test 637

15 Roots and Radical Expressions 639

- **15.1** Radicals 639
- **15.2** Simplifying Radical Expressions 645
- **15.3** Adding and Subtracting Radical Expressions 651
- **15.4** Multiplying and Dividing Radical Expressions 656
- **15.5** Solving Equations Containing Radicals 664
- **15.6** Rational Exponents 669
- **15.7** The Distance Formula 674
- **15.8** Special Right Triangles 679
  Chapter Summary 684
  Chapter 15 Review Exercises 686
  Chapter 15 Test 689

16 Quadratic Equations 693

- **16.1** Solving Equations of the Form $x^2 = c$ 693
- **16.2** Completing the Square 698
- **16.3** The Quadratic Formula 703
- **16.4** Complex Numbers 707
- **16.5** Graphing Quadratic Functions 716
  Chapter Summary 723
  Chapter 16 Review Exercises 724
  Chapter 16 Test 725

I   **Roots and Powers**   727

II   **Answers for Selected Exercises**   729

Index   771

1 Whole Numbers

Mathematics and Ecology

Fluorescent lighting provides substantial energy savings over ordinary incandescent lighting, and that is good for the environment. Burned-out fluorescent tubes, however, present an environmental hazard: Poisonous mercury vapor escapes into the air and contaminates soil and water, and glass and metal parts in landfills never decompose.

Several California companies now recycle old fluorescent tubes. The tubes are first crushed, and various screens, filters, and air blowers separate the phosphors from the glass and metal parts and capture the mercury for reuse. The technology is still new, and the companies are processing only 600,000 fluorescent tubes each month.

In this chapter, we introduce the whole numbers and show how the operations of arithmetic can be used to answer such questions as this: How many fluorescent tubes will be processed in the course of a year?

Five thousand years ago, numbers were used by the ancient Babylonians and Egyptians to count their herds and record the changing seasons. However, as travel increased, so did commerce, and with it the need to calculate lengths, weights, and monetary values. Keeping time and developing a calendar required elaborate calculations to predict the positions of the sun, moon, and stars. Because the religious holidays of the Babylonians were dependent on the calendar, most early Babylonian mathematicians were temple priests.

Today, economists use mathematics to analyze the national economy. Ecologists use mathematics to predict the effects of industries on our environment. Computers direct our telephone calls and check our income tax, and computer science relies on mathematics never dreamed of by the ancients. In fact, some knowledge of mathematics is needed by everyone.

We begin our study of mathematics by reviewing the arithmetic of whole numbers.

1.1 Whole Numbers and Place Value; Rounding Whole Numbers

- Standard and Expanded Notation
- Rounding Whole Numbers

The number 0 together with the **natural numbers** (the numbers we count with) are called the **whole numbers**.

> **Definition:** The **whole numbers** are the numbers
> 0, 1, 2, 3, 4, 5, 6, 7, 8, 9, 10, 11, 12, ...

As we move to the right in this list, the numbers become larger. The number 11, for example, is **greater than** 5, because 11 appears after 5 in the list. Since 11 is greater than 5, 5 is **less than** 11. The three dots in the definition indicate that the list of whole numbers continues forever. Thus, there is no largest whole number.

Standard and Expanded Notation

The symbols 0, 1, 2, 3, 4, 5, 6, 7, 8, and 9 are called **digits**. To write the name of a whole number in **standard notation**, several digits may be needed. For example, the height of the World Trade Center is 1352 feet. The meaning of each digit in the numeral 1352 depends on its position, or **place value**. The digit 1 is in the thousands position, the digit 3 is in the hundreds position, the 5 is in the tens position, and the 2 is in the ones, or units, position. The meaning of the numeral 1352 becomes clear when it is written in **expanded notation** as

1 thousand + 3 hundreds + 5 tens + 2 ones

We read 1352 as *one thousand three hundred fifty-two*.

EXAMPLE 1

Expanded notation for 7103 is

7 thousands + 1 hundred + 0 tens + 3 ones

Because 0 tens is zero, it need not be written, and the expanded notation is also

7 thousands + 1 hundred + 3 ones

We read this number as *seven thousand one hundred three*.

EXAMPLE 2

Expanded notation for 37,359 is

3 ten thousands + 7 thousands + 3 hundreds + 5 tens + 9 ones

We read this number as *thirty-seven thousand three hundred fifty-nine*.

EXAMPLE 3

The number *twenty-three thousand forty* is written in standard notation as

23,040

In expanded notation, the number is written as

2 ten thousands + 3 thousands + 4 tens

Work Progress Check 1.

Progress Check 1

a. Write 2002 in expanded notation.

b. Write *seventy-six thousand three* in standard notation.

c. Write 73,952 in words.

---

1. a. 2 thousand + 2 ones
b. 76,003 **c.** seventy-three thousand nine hundred fifty-two

Progress Check Answers

Rounding Whole Numbers

When a teacher with 37 students says, "I have 40 students," she has **rounded** the actual number to the *nearest ten*, because 37 is closer to 40 than it is to 30:

30, 31, 32, 33, 34, 35, 36, 37, 38, 39, 40

If 3,190,268 travelers pass through an airport, a newspaper might report the number to be 3 million. The newspaper has rounded the number to the *nearest million*, because 3,190,268 is closer to 3,000,000 than it is to 4,000,000. If a stereo system costs $1285, and a customer says, "I can't afford $1300," she has rounded the price to the *nearest $100*, because $1285 is closer to $1300 than it is to $1200.

To round a whole number to the nearest 10, we find the digit in the tens position. If the digit to the right of that position (the digit in the units position) is 5 or higher, we **round up**. To do so, we increase the tens digit by one, and place a 0 in the units position. If the digit to the right of the tens position is less than 5, we **round down**. Then, we leave the tens digit unchanged, and place a 0 in the units position.

EXAMPLE 4

Round the number 3464 to the nearest 10.

Solution

We find the digit in the tens position (it is 6)   3464

and then look at the next digit to the right—the digit in the units position. Since that digit is 4, and 4 is less than 5, we round down. We leave the digit in the tens position and replace the digit in the units position with 0. The rounded answer is 3460. ■

EXAMPLE 5

Round the number 12,087 to the nearest 10.

Solution

We find the digit in the tens position (it is 8)   12,087

and then look at the digit in the units position. Because that digit is 7, and 7 is larger than 5, we round up. We increase the digit in the tens position by 1 and replace the digit in the units position with 0. The rounded answer is 12,090. ■

To round a whole number to the nearest 100, we first find the digit in the hundreds position. If the digit to the right of that position (the digit in the tens position) is 5 or higher, we **round up**. To do so, we *increase* the digit in the hundreds position by one, and place 0's in all of the positions to the right. If the digit to the right of the hundreds position is less than 5, we **round down**. To do so, we leave the digit in the hundreds position unchanged and place 0's in every position to the right.

Progress Check 2

a. Round 97,004 to the nearest 10.

b. Round the number 96,906 to the nearest 10.

c. Round 2,345,678 to the nearest 100.

Progress Check 3

Round 81,818,181

a. to the nearest million

b. to the nearest ten thousand

c. to the nearest thousand

d. to the nearest ten

Progress Check 4

Round 98,765,432

a. to the nearest ten million

b. to the nearest hundred million

---

2. **a.** 97,000 **b.** 96,910
   c. 2,345,700

3. **a.** 82,000,000
   b. 81,820,000
   c. 81,818,000
   d. 81,818,180

4. **a.** 100,000,000
   b. 100,000,000

Progress Check Answers

EXAMPLE 6

Round the number 3464 to the nearest 100.

Solution

We find the digit in the hundreds position (it is 4)   3464
                                                          ↑

Then we look at the next digit to the right, the digit in the tens position. Because that digit is 6, and 6 is greater than 5, we round up, increasing the digit in the hundreds position by 1, and replacing the two digits to the right with 0's. The rounded answer is 3500. Work Progress Check 2. ∎

A similar scheme is used to round numbers to the nearest thousand, the nearest ten thousand, and so on.

EXAMPLE 7

Round the number 257,125 **a.** to the nearest thousand **b.** to the nearest ten thousand **c.** to the nearest million.

Solution

a. We find the digit in the thousands position of 257,125. It is the 7. Look at the next digit to the right—the 1. Because 1 is less than 5, we round down. The result is 257,000.

b. We find the digit in the ten thousands position of 257,125. It is 5. We look at the next digit to the right—the 7. Because 7 is greater than 5, we round up. The result is 260,000.

c. We find the digit in the millions position. That digit is missing, but it is understood to be 0. We look at the digit in the next position to the right—the 2 in the hundred thousands position. Because 2 is less than 5, we round down. The result is 0.

Work Progress Check 3. ∎

EXAMPLE 8

Round 7960 to the nearest hundred.

Solution

Find the digit in the hundreds position (it is 9)   7960
                                                       ↑

and look at the next digit to the right, the digit in the tens position. Because 6 is larger than 5, we round up and increase the digit 9 in the hundreds position by 1. Because the 9 in the hundreds position represents 900, increasing 9 by 1 represents increasing 900 to 1000. Therefore, we must replace the 9 with a 0 and add 1 to the 7 in the thousands position. Finally, we replace the two rightmost digits with 0's. The rounded answer is 8000.

We note that 7960 is closer to 8000 than it is to 7900:

7900, 7910, 7920, 7930, 7940, 7950, **7960**, 7970, 7980, 7990, **8000**

Work Progress Check 4. ∎

1.1 Exercises

In Exercises 1–12, write each number in expanded notation and indicate how the number would be read.

1. 245
2. 508
3. 3609
4. 3960
5. 32,500
6. 73,009
7. 104,401
8. 570,003
9. The United Republic of Tanzania covers an area of 363,708 square miles.
10. Estimates of the number of troops that fought in the American Revolution range from 150,000 to 184,000.
11. At the time of the Pilgrims, the estimated population of North America was 9,400,000.
12. The Caspian Sea covers 143,244 square miles.

In Exercises 13–26, write each number in standard notation.

13. 4 hundreds + 2 tens + 5 ones
14. 7 hundreds + 7 tens + 7 ones
15. 2 thousands + 7 hundreds + 3 tens + 6 ones
16. 7 ten thousands + 3 hundreds + 5 tens
17. four hundred fifty-six
18. three thousand seven hundred thirty-seven
19. twenty-seven thousand five hundred ninety-eight
20. seven million four hundred fifty-two thousand eight hundred sixty
21. There are six hundred sixty feet in one furlong.
22. One square foot is approximately nine hundred thirty square centimeters.
23. Hadrian, Emperor of the Roman world, died in the year one hundred thirty-eight.
24. There are eighty-six thousand four hundred seconds in one day.
25. The United States shoreline of Lake Superior stretches eight hundred sixty-three miles.
26. One thousand five hundred three passengers were lost when the *Titanic* hit an iceberg.

In Exercises 27–34, round each number to the nearest ten.

27. 14
28. 37
29. 91
30. 43
31. 98
32. 96
33. 251
34. 989

In Exercises 35–42, round each number to the nearest hundred.

35. 824
36. 277
37. 191
38. 399
39. 51
40. 750
41. 2961
42. 2561

In Exercises 43–50, round each number to the nearest thousand.

43. 2354
44. 7777
45. 500
46. 501
47. 499
48. 79,540
49. 99,215
50. 29,921

In Exercises 51–54, round each number to the nearest ten thousand.

51. 13,234 **52.** 9654 **53.** 54,499 **54.** 1,395,279

In Exercises 55–58, round each number to the nearest hundred thousand.

55. 499,999 **56.** 944,444 **57.** 1,975,115 **58.** 32,110

In Exercises 59–62, round each number to the nearest million.

59. 2,468,354 **60.** 4,444,444 **61.** 510,000 **62.** 490,000

In Exercises 63–66, round 79,593 to the nearest . . .

63. ten **64.** hundred **65.** thousand **66.** ten thousand

In Exercises 67–70, round 5,925,830 to the nearest . . .

67. thousand **68.** ten thousand **69.** hundred thousand **70.** million

71. Write the largest whole number that becomes 45,000 when it is rounded to the nearest 100.

72. Write the smallest whole number that becomes 45,000 when it is rounded to the nearest 100.

73. Write the smallest whole number that becomes 1,234,000 when it is rounded to the nearest 10.

74. Write the largest whole number that becomes 1,234,000 when it is rounded to the nearest 10.

75. Using each of the digits 1 through 9 exactly once, write the largest possible nine-digit number that has a 9 in the units position.

76. Using each of the digits 1 through 9 exactly once, write the largest possible nine-digit number that has 7 in the hundreds position.

77. The speed of light in a vacuum is exactly 299,792,458 meters per second. Round this number to the nearest hundred thousand meters per second.

78. Round the speed of light to the nearest million meters per second. (See Exercise 77.)

Review Exercises

In Review Exercises 1–12, perform each operation. If you are weak on the addition and subtraction facts of whole numbers through 9, learn them now.

1. 7 + 5 = _____   **2.** 9 − 7 = _____   **3.** 8 − 6 = _____   **4.** 7 + 8 = _____

5. 16 − 9 = _____   **6.** 7 + 9 = _____   **7.** 9 + 8 = _____   **8.** 17 − 9 = _____

9. 6 + 7 = _____   **10.** 13 − 8 = _____   **11.** 14 − 9 = _____   **12.** 8 + 5 = _____

1.2 Adding and Subtracting Whole Numbers

- *Addition of Whole Numbers*
- *Finding the Perimeter of a Rectangle*
- *Subtraction of Whole Numbers*
- *The Relationship between Addition and Subtraction*

Addition of Whole Numbers

The addition of whole numbers corresponds to combining sets of objects. If a set of four objects, for example, is combined with a set of five objects, we obtain a total of nine objects.

A set of four objects   A set of five objects   A set of nine objects
★ ★ ★ ★   ★ ★ ★ ★ ★   ★ ★ ★ ★ ★ ★ ★ ★ ★
We combine these two sets   to get this set.

This corresponds to the addition fact

$$4 + 5 = 9$$

In this addition fact, 4 and 5 are called **addends**, and 9 is called the **sum**.
If we combine the sets in the opposite order, we still get the same sum.

A set of five objects   A set of four objects   A set of nine objects
★ ★ ★ ★ ★   ★ ★ ★ ★   ★ ★ ★ ★ ★ ★ ★ ★ ★
We combine these two sets   to get this set.

This corresponds to the addition fact

$$5 + 4 = 9$$

We see that the addition of two numbers can be done in either order, and the same sum will result. This property is called the **commutative property of addition**. Because of the commutative property of addition, $4 + 5 = 5 + 4$.

To add three numbers together, we add two of them together and then add the third. For example, we can add $3 + 4 + 7$ in two ways. The parentheses tell us what to do first.

$(\mathbf{3 + 4}) + 7 = \mathbf{7} + 7 = 14$ Add the 3 and 4 first, and then add 7.
$3 + (\mathbf{4 + 7}) = 3 + \mathbf{11} = 14$ Add 3 to the sum of 4 and 7.

Either way the sum is the same; it does not matter how we group or associate numbers in addition. This property is called the **associative property of addition**. Because of the associative property of addition, $(3 + 4) + 7 = 3 + (4 + 7)$.

Work Progress Check 5.

Whole numbers larger than 10 are often added using a vertical format that adds the corresponding digits with the same place value. Because the additions within each column often exceed the largest possible digit, 9, it is sometimes necessary to **carry** the excess to the next column to the left. For example, to add 27 and 15, we write the numbers with the digits of the same place value aligned vertically.

```
  2 7
+ 1 5
```

We begin by adding the digits in the ones position: $7 + 5 = 12$. Because 12 is 1 ten and 2 ones, we place a 2 in the ones position of the answer and add 1 to the tens column.

Progress Check 5

Find each sum:

a. $8 + 9$

b. $9 + 8$

c. $5 + (7 + 8)$

d. $(5 + 7) + 8$

5. **a.** 17 **b.** 17 **c.** 20 **d.** 20

```
  1
  2 7    Add the digits in the units column: 7 + 5 = 12. Carry 1 to
+ 1 5    the tens column.
    2
```

We continue by adding the digits in the tens column, including the carry digit: 1 + 2 + 1 = 4. We place a 4 in the tens position of the answer.

```
  1
  2 7    Add the digits in the tens column. Place the result, 4, in the
+ 1 5    tens position of the answer.
  4 2
```

Thus, 27 + 15 = 42.

EXAMPLE 1

Find the sum of 9735 and 593.

Solution

To find the sum of 9735 and 593, we write the numbers in vertical format, with their corresponding digits aligned. Then we add the numbers, a column at a time.

```
  9 7 3 5    Write the numerals in a column, with corresponding
+   5 9 3    digits aligned.
```

```
  9 7 3 5    Add the digits in the units column. Place the result
+   5 9 3    in the units position of the answer.
        8
```

```
      1
  9 7 3 5    Add the digits in the tens column. The result, 12,
+   5 9 3    exceeds one digit. Place the 2 in the tens position of the
      2 8    answer and carry 1 to the hundreds column.
```

```
    1 1
  9 7 3 5    Add the digits in the hundreds column. The result, 13,
+   5 9 3    exceeds one digit. Place the 3 in the hundreds position of
    3 2 8    the answer and carry 1 to the thousands column.
```

```
    1 1
  9 7 3 5    Add the digits in the thousands column. That result, 10,
+   5 9 3    requires carrying to the ten thousands position. Write 0
1 0 3 2 8    in the thousands position and 1 in the ten thousands
             position of the answer.
```

Thus, 9735 + 593 = 10,328.

To see that this answer is reasonable, we **estimate** the answer: *9735 is somewhat less than 10,000, and 593 is close to 600. We estimate that the answer is somewhat less than 10,000 + 600, or 10,600. The answer 10,328 is reasonable.* Work Progress Check 6.

In applied problems, words such as *increase, gain, credit, up, forward, rises, in the future,* and *to the right* are used to indicate addition.

Progress Check 6

The United States portion of Lake Superior covers 20,587 square miles. The Canadian portion covers 11,093 square miles. Find the lake's total area.

6. 31,680 sq mi

EXAMPLE 2

On Monday, the average temperature was 31°. By Tuesday, the average temperature had increased 5°, and by Wednesday, the average temperature had risen another 7°. What was the average temperature on Wednesday?

Solution

We find Wednesday's average temperature by addition. To Monday's average temperature, we add the increases that occurred on Tuesday and Wednesday:

$$31 + 5 + 7 = 43$$

The average temperature on Wednesday was 43°.

Finding the Perimeter of a Rectangle

A **rectangle** is a four-sided figure shaped like a dollar bill. The opposite sides of a rectangle are the same length. Either of its longer sides is called the **length** of the rectangle, and either of its shorter sides is called its **width**. The distance around a rectangle is called its **perimeter**. To find the perimeter of a rectangle, we add the lengths of its four sides.

| The perimeter of a rectangle | = | width | + | width | + | length | + | length |

EXAMPLE 3

The width of the rectangle in Figure 1-1 is 47 inches, and its length is 92 inches. Find the perimeter of the rectangle.

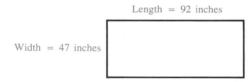

Figure 1-1

Solution

We add the lengths of the four sides of the rectangle.

```
    1
    47
    47
    92
+   92
   ---
   278
```

The perimeter of the rectangle is 278 inches.

To see if the answer is reasonable, we estimate the answer. Because the rectangle is about 50 inches by 100 inches, its perimeter is approximately 50 + 50 + 100 + 100, or 300 inches. An answer of 278 inches is reasonable. Work Progress Check 7.

Progress Check 7

In 1630, the estimated populations of four American colonies were New Hampshire, 500; Plymouth and Massachusetts, 900; New York, 400; Virginia, 2500. Find the total population.

7. 4300

Subtraction of Whole Numbers

The subtraction of whole numbers determines how many objects remain when several objects are removed from a set. For example, if we start with a set of nine objects and take four away, we are left with a set of five objects.

A set of nine objects ★★★★★ ★★★★ We take away four objects

A set of five objects ★★★★★ to get this set.

This corresponds to the subtraction fact

$9 - 4 = 5$

In this subtraction fact, 9 is called the **minuend**, 4 is called the **subtrahend**, and 5 is called the **difference**.

With whole numbers, we cannot subtract in the opposite order and find the difference $4 - 9$, because we cannot take away nine objects from only four objects. Because subtraction cannot be done in either order, subtraction is not commutative.

Subtraction is not associative either, because if we group in different ways, we get different answers.

$(9 - 5) - 1 = 4 - 1 = 3$
$9 - (5 - 1) = 9 - 4 = 5$

Work Progress Check 8.

Whole numbers are often subtracted using a vertical format that subtracts the digits of corresponding columns. Because such subtractions often require subtracting a larger digit from a smaller digit, it is sometimes necessary to **borrow** from the next column to the left. For example, to subtract 15 from 32, we write the minuend, 32, and the subtrahend, 15, in a vertical format with the digits of the same place value aligned.

$\begin{array}{r} 3\ 2 \\ -\ 1\ 5 \\ \hline \end{array}$ Write the numerals in a column, with corresponding digits aligned.

We then attempt to subtract the digits in the ones column. Because five objects cannot be subtracted from only two objects, we must borrow from the tens position of the minuend. The 3 in the tens position represents three tens; if we take one of those tens (leaving two tens behind) and add it to the units position, we will have 12 in the units position of the minuend. Then it is possible to subtract in the units column: $12 - 5 = 7$. We do the calculation as follows:

$\begin{array}{r} \overset{2}{\cancel{3}}_1 2 \\ -\ 1\ 5 \\ \hline 7 \end{array}$ To subtract in the ones column, borrow one ten from the tens position and add 10 to the ones position of the minuend, giving 12. Then subtract: $12 - 5 = 7$.

$\begin{array}{r} \overset{2}{\cancel{3}}_1 2 \\ -\ 1\ 5 \\ \hline 1\ 7 \end{array}$ Subtract in the tens column: $2 - 1 = 1$.

Thus, $32 - 15 = 17$.

EXAMPLE 4

Subtract 576 from 2021.

Progress Check 8

Find each difference:

a. $9 - 3$

b. $8 - 5$

c. $9 - (6 - 3)$

d. $(9 - 6) - 3$

---

8. a. 6 **b.** 3 **c.** 6 **d.** 0

Progress Check Answers

Solution

$$\begin{array}{r} 2\,0\,2\,1 \\ -5\,7\,6 \\ \hline \end{array}$$

Write the numerals in a column, with the digits of the same place value aligned.

$$\begin{array}{r} 2\,0\,2\,\overset{1}{\cancel{1}} \\ -5\,7\,6 \\ \hline 5 \end{array}$$

To subtract in the units column, borrow 1 ten from the tens position and add it to the units digit. Then subtract: $11 - 6 = 5$.

Since we cannot subtract seven objects from one object in the tens column, we must borrow from columns to the left. Because there is a 0 in the hundreds position of the minuend, we cannot borrow from there; we must borrow from the thousands column. The 2 in the thousands position represents 2 thousands. We take one of those thousands (leaving 1 thousand behind) and write it as 10 hundreds: Place a 10 in the hundreds position. From those 10 hundreds, we take 1 hundred (leaving 9 hundreds behind) and write it as 10 tens. Add these 10 tens to the 1 ten that is already in the tens position to get 11 tens. From these 11 tens, we can subtract 7 tens: $11 - 7 = 4$. We place a 4 in the tens position of the answer.

$$\begin{array}{r} \overset{1}{\cancel{2}}\,\overset{9}{\cancel{0}}\,\overset{11}{\cancel{2}}\,1 \\ -\,5\,7\,6 \\ \hline 4\,5 \end{array}$$

To subtract in the tens column, borrow 1000 from the thousands digit and add it to the hundreds digit. Borrow from the hundreds digit and add to the tens digit. Then subtract: $11 - 7 = 4$.

$$\begin{array}{r} \overset{1}{\cancel{2}}\,\overset{9}{\cancel{0}}\,\overset{11}{\cancel{2}}\,1 \\ -\,5\,7\,6 \\ \hline 4\,4\,5 \end{array}$$

Subtract in the hundreds column: $9 - 5 = 4$.

$$\begin{array}{r} \overset{1}{\cancel{2}}\,\overset{9}{\cancel{0}}\,\overset{11}{\cancel{2}}\,1 \\ -\,5\,7\,6 \\ \hline 1\,4\,4\,5 \end{array}$$

Subtract in the thousands column: $1 - 0 = 1$.

Thus, $2021 - 576 = 1445$. Work Progress Check 9.

In applied problems, words such as *decrease*, *loss*, *debit*, *down*, *backward*, *falls*, *reduce*, *in the past*, and *to the left* indicate subtraction.

EXAMPLE 5

A television set sells for $975. To encourage sales, a merchant reduces the price by $100. During a weekend sale, the price is reduced another $40. Find the sale price.

Solution

We can find the sale price with two subtractions. From the original price of $975, we first subtract $100.

$$975 - 100 = 875$$

After the first reduction, the price is $875. To find the sale price, we subtract $40 from $875:

$$875 - 40 = 835$$

The television sells for $835.

Progress Check 9

Subtract 1445 from 2021.

9. 576

The calculations of this example could be written as one expression:

$$975 - 100 - 40$$

The two subtractions are performed in turn, working left to right:

$$(975 - 100) - 40 = 875 - 40$$
$$= 835$$

We could also find the sale price by adding the two price reductions and subtracting that sum from the original price. Since the total reduction was $100 + $40, or $140, the sale price was $835.

$$975 - 140 = 835$$

Additions and subtractions often appear in the same problem. It is important to read the problem carefully, find the useful information, organize it correctly, and do the arithmetic.

EXAMPLE 6

Twenty-seven people are riding the 12:00 o'clock bus on route 47. At Seventh Street, 16 riders get off the bus and 5 get on. How many people are on the bus?

Solution

The route number, street number, and time are not important. When 16 of the original 27 people get off the bus, there are

$$27 - 16 = 11$$

people left. After these 11 are joined by 5 more, there are

$$11 + 5 = 16$$

people on the bus.
The number of riders on the bus can also be described by the expression

$$27 - 16 + 5$$

The calculations are performed left to right:

$$(27 - 16) + 5 = 11 + 5$$
$$= 16$$

Work Progress Check 10.

Progress Check 10

On Monday, a share of ULTRON Corporation stock cost $75. During the next week, the price fell $7 per share. However, it recovered by the end of the month and had risen $13 per share. What was the price at the end of the month?

10. $81

The Relationship between Addition and Subtraction

The operations of addition and subtraction are closely related. For example, $9 - 3$, or 6, is the number that must be added to 3 to produce 9. That is, the subtraction fact

$$9 - 3 = 6$$

is equivalent to the addition fact

$$9 = 3 + 6$$

Likewise, $13 - 8$, or 5, is the number that must be added to 8 to produce 13. That is, the subtraction fact

$$13 - 8 = 5$$

is equivalent to the addition fact

$$13 = 8 + 5$$

In general, if the letters a, b, and x represent numbers for which $a - b = x$, then x is the number that must be added to b to produce a. That is, the subtraction statement

$$a - b = x$$

is equivalent to the addition statement

$$a = b + x$$

Letters that are used to represent numbers are often called **variables**.

EXAMPLE 7

Each subtraction statement in the left column is equivalent to the corresponding addition statement in the right column.

$$15 - 9 = 6 \qquad 15 = 9 + 6$$
$$39 - 21 = 18 \qquad 39 = 21 + 18$$
$$243 - 240 = 3 \qquad 243 = 240 + 3$$
$$x - y = z \qquad x = y + z$$

For any addition fact, there are two related subtraction facts. For example, if 7 is subtracted from 16, the result is 9. Also, if 9 is subtracted from 16, the result is 7. Thus, the addition fact

$$16 = 9 + 7$$

is equivalent to the two subtraction facts:

$$16 - 9 = 7 \quad \text{and} \quad 16 - 7 = 9$$

EXAMPLE 8

Each addition statement in the left column is equivalent to the two corresponding subtraction statements in the right column.

$17 = 12 + 5$	$17 - 12 = 5$	and	$17 - 5 = 12$
$23 = 19 + 4$	$23 - 19 = 4$	and	$23 - 4 = 19$
$157 = 95 + 62$	$157 - 95 = 62$	and	$157 - 62 = 95$
$x = y + z$	$x - y = z$	and	$x - z = y$

Work Progress Check 11.

We have used subtraction to answer the question, *How much is left?* Because of the close relationship between addition and subtraction, we can also use subtraction to answer the question, *How much more is needed?*

EXAMPLE 9

Bob purchased 25 holiday greeting cards but wishes to send cards to 37 friends. How many more cards does he need?

Progress Check 11

Write each subtraction statement as an equivalent addition statement:

a. $23 - 14 = 9$

b. $a - b = c$

11. **a.** $23 = 14 + 9$ **b.** $a = b + c$
Progress Check Answers

Progress Check 12

Nationwide, 930,000 participated in high school football, but only 516,000 played basketball. How many more basketball players must be recruited to make basketball as popular as football?

---

12. 414,000
Progress Check Answers

Solution

The total number of cards Bob plans to send is equal to the sum of the number of cards he has and the number of additional cards he needs.

| The total number of cards Bob will send | = | the number of cards he has | + | the number of extra cards needed |

Bob plans to send 37 cards. He has 25 cards. If the letter x represents the number of additional cards Bob will need, the facts of the problem are represented by the addition statement

$$37 = 25 + x$$

This addition is equivalent to the subtraction statement

$$37 - 25 = x$$

Perform the subtraction to find that x is 12. Bob needs 12 additional cards. Work Progress Check 12.

1.2 Exercises

In Exercises 1–22, perform each addition.

1. $25 + 13 =$ _____
2. $47 + 12 =$ _____
3. $113 + 77 =$ _____
4. $647 + 38 =$ _____
5. $156 + 305 =$ _____
6. $378 + 237 =$ _____
7. $956 + 349 =$ _____
8. $832 + 973 =$ _____
9. $345 + 1789 =$ _____
10. $1865 + 2378 =$ _____

11. 632
   + 347

12. 423
   + 570

13. 1372
   + 613

14. 2477
   + 693

15. 6427
   + 3573

16. 3567
   + 8778

17. 8539
   + 7368

18. 5799
   + 6879

19. 1246
      578
   +   37

20. 4689
     3422
   +   26

21. 3156
     1578
   + 578

22. 2379
     4779
   + 2339

23. During the first year of this century, 18,549 boys and 25,182 girls graduated from American high schools. Find the total number of graduates.

24. In the Civil War, the Union army suffered 140,410 battle casualties and 224,100 deaths for other reasons. Find the total number of war casualties.

25. Walter Schirra's first space flight orbited the earth 6 times and lasted 9 hours. The 16 orbits of his second flight lasted 26 hours. In those two flights, how long was Schirra in space?

26. In 1899–1900, 15,539 college graduates received Bachelor's degrees, 1015 received Master's degrees, and 149 earned Doctor's degrees. How many degrees were awarded?

14 Chapter 1 Whole Numbers

In Exercises 27–32, find the perimeter of a rectangle with the given dimensions.

27. length = 32 feet, width = 12 feet _____

28. length = 127 meters, width = 91 meters _____

29. length = 147 centimeters, width = 122 centimeters _____

30. length = 135 miles, width = 67 miles _____

31. length = 59 feet, width = 51 feet _____

32. length = 2356 centimeters, width = 13 centimeters _____

In Exercises 33–56, perform each subtraction.

33. 17 − 14 = _____

34. 42 − 31 = _____

35. 39 − 14 = _____

36. 45 − 32 = _____

37. 174 − 71 = _____

38. 257 − 155 = _____

39. 134 − 53 = _____

40. 531 − 332 = _____

41. 633 − 598 = _____

42. 600 − 497 = _____

43. 2141 − 34 = _____

44. 3250 − 361 = _____

45. 367 − 343

46. 224 − 122

47. 423 − 305

48. 330 − 270

49. 1537 − 579

50. 2470 − 863

51. 4267 − 2578

52. 7356 − 3578

53. 17246 − 6789

54. 34510 − 27593

55. 15700 − 15397

56. 35021 − 23999

57. Beverly planted 27 seedlings, but only 21 of them survived the first winter. How many plants died?

58. Phil sold 42 toasters on Tuesday, 7 more than he sold on Monday. How many did he sell Monday?

59. Of the 1560 eggs purchased for a fund-raising breakfast, 18 were rotten. The breakfast netted $27,570 for charity. How many eggs were usable?

60. For the 17-mile trip to the airport, Linda paid the taxi driver $23. If $5 was a tip, how much was the fare?

61. Jim's savings account contains $370. If he now makes a deposit of $40 and withdraws $197, find his balance.

62. Nine of the 17 friends Jill invited to her party never arrived. However, 12 other friends did appear, uninvited. How many attended the party?

63. George Washington was inaugurated president in 1789 at the age of 57. Our youngest president was our 26th, Theodore Roosevelt, inaugurated in 1901 at the age of 42. How much older was Washington than Roosevelt, at the time of inauguration?

64. The magazine *TV Guide* recently had an annual circulation of 16,969,260. By what amount did this exceed the circulation of *Reader's Digest*, with 16,566,650 readers?

In Exercises 65–72, write each subtraction as an equivalent addition.

65. $7 - 3 = 4$ _____
66. $9 - 5 = 4$ _____
67. $15 - 7 = 8$ _____
68. $23 - 21 = 2$ _____
69. $35 - 35 = 0$ _____
70. $27 - 0 = 27$ _____
71. $12 - x = 5$ _____
72. $17 - 8 = x$ _____

In Exercises 73–80, write each addition statement as two equivalent subtractions.

73. $8 = 3 + 5$ _____ _____
74. $19 = 9 + 10$ _____ _____
75. $20 = 12 + 8$ _____ _____
76. $34 = 23 + 11$ _____ _____
77. $13 + 4 = 17$ _____ _____
78. $33 + x = 54$ _____ _____
79. $17 + x = 20$ _____ _____
80. $x + 19 = y$ _____ _____

81. At 29,028 feet, Mt. Everest is taller than the 19,340-foot peak of Mt. Kilimanjaro. How much taller is Mt. Everest than Mt. Kilimanjaro?

82. It is 750 miles to Detroit, and Jerry wants to drive there in two days. If he can travel 423 miles on the first day, how many miles would be left for the second day?

83. In 1885, Louis Pasteur used a vaccine to prevent rabies in a small boy. How many years later was it in 1954 when Jonas Salk discovered the polio vaccine?

84. The Wright brothers' first flight was in 1903. In 1969, Neil Armstrong was the first man to walk on the moon. How many years after we left the ground did we leave the earth?

Review Exercises

Multiplication and division of whole numbers is basic in mathematics. If you are weak on multiplication of whole numbers through 9, and related division, spend some time now polishing your skills.

In Review Exercises 1–12, perform each calculation.

1. $7 \cdot 8 =$ _____
2. $7 \cdot 9 =$ _____
3. $45 \div 5 =$ _____
4. $49 \div 7 =$ _____
5. $7 \cdot 5 =$ _____
6. $4 \cdot 7 =$ _____
7. $12 \div 6 =$ _____
8. $27 \div 9 =$ _____
9. $8 \cdot 9 =$ _____
10. $9 \cdot 6 =$ _____
11. $72 \div 9 =$ _____
12. $56 \div 8 =$ _____

1.3 Multiplying and Dividing Whole Numbers

- *Multiplication of Whole Numbers*
- *Finding the Area of a Rectangle*
- *Exponents*
- *Division of Whole Numbers*

Multiplication of Whole Numbers

Numbers that are multiplied together are called **factors**, and the answer is called a **product**. To show multiplication, we use the times sign ×. For example, the product of 4 and 5 is 20. The numbers 4 and 5 are factors of 20, and we write

$$4 \times 5 = 20$$

Multiplication of whole numbers can be thought of as repeated addition. For example, the multiplication 4×5 can be thought of as the sum of four 5's:

$$4 \times 5 = \overbrace{5 + 5 + 5 + 5}^{\text{The sum of four 5's}}$$
$$= 20$$

The multiplication 5×4 can also be thought of as the sum of five 4's:

$$5 \times 4 = \overbrace{4 + 4 + 4 + 4 + 4}^{\text{The sum of five 4's}}$$
$$= 20$$

We see that the sum of four 5's is equal to the sum of five 4's. As with addition, the order in which we multiply makes no difference. This is called the **commutative property of multiplication**.

To multiply three numbers, we multiply two of them together and then multiply that product by the third number. For example, we can multiply $3 \times 4 \times 7$ in two ways. As before, the parentheses tell us what to do first.

$(\mathbf{3 \times 4}) \times 7 = \mathbf{12} \times 7 = 84$ Multiply 3 and 4 first, and then multiply 12 by 7.

$3 \times (\mathbf{4 \times 7}) = 3 \times \mathbf{28} = 84$ Multiply the product of 4 and 7 by 3.

Either way, the product is the same. This shows that it does not matter how we group or associate numbers in multiplication. This property is called the **associative property of multiplication**.

EXAMPLE 1

Jeremy worked for 6 hours at an hourly rate of $9. How much money did he earn?

Solution

For each of the 6 hours, Jeremy earned $9. His earnings were

$$\text{His earnings} = \overbrace{9 + 9 + 9 + 9 + 9 + 9}^{\text{The sum of six 9's}}$$
$$= 54$$

His earnings can be calculated by multiplication. The sum of six 9's is the product of 6 and 9:

$$\text{His earnings} = 6 \times 9 = 54$$

Either way, Jeremy earned $54. ∎

EXAMPLE 2

A car can travel 32 miles on 1 gallon of gasoline. How many miles can the car travel on 3 gallons of gasoline?

Solution

Each of the 3 gallons of gasoline enables the car to go 32 miles. The total distance the car can travel is the sum of three 32's. This can be calculated by multiplication:

$3 \times 32 = 96$

On 3 gallons of gasoline, the car can travel 96 miles. Work Progress Check 13. ■

Progress Check 13

At a rate of $8 per hour, how much will Wendy earn in 7 hours?

To find a product such as 23×435, it is inconvenient to add up twenty-three 435's. Instead, we find the product by multiplying. Because $23 = 20 + 3$, we multiply 435 by 20 and by 3 and add the products.

```
    4 3 5
  ×   2 3
```
Write the factors in a column, with the corresponding digits aligned.

We begin by multiplying 435 by 3:

```
      1
    4 3 5
  ×   2 3
        5
```
Multiply 5 by 3. The product is 15. Place 5 in the ones column and carry 1 to the tens column.

```
    1 1
    4 3 5
  ×   2 3
      0 5
```
Multiply 3 by 3. The product is 9. To the 9, add the carried 1 to get 10. Place the 0 in the tens column and carry the 1 to the hundreds column.

```
    1 1
    4 3 5
  ×   2 3
  1 3 0 5
```
Multiply 4 by 3. The product is 12. Add the 12 to the carried 1 to get 13. Write 13.

We continue by multiplying 435 by 2 tens, or 20:

```
      1
    4 3 5
  ×   2 3
  1 3 0 5
        0
```
Erase the digits that were carried. Multiply 5 by 2 tens. The product is 10 tens. Write 0 in the tens column and carry 1.

```
      1
    4 3 5
  ×   2 3
  1 3 0 5
      7 0
```
Multiply 3 by 2 tens. The product is 6 tens. Add 6 to the carried 1 to get 7 tens. Write the 7. There is no carry.

Progress Check 14

An electronics firm can manufacture 87 television sets in 1 hour. How many can they manufacture in 5 days, working two 8-hour shifts each day?

```
      1
    4 3 5
  ×   2 3
  1 3 0 5
    8 7 0
```
Multiply 4 by 2 tens. The product is 8 tens. There is no carry to add. Write the 8.

```
    4 3 5
  ×   2 3
  1 3 0 5
    8 7 0
 1 0 0 0 5
```
Draw another line beneath the two completed rows. Add the two rows. This sum gives the product of 435 and 23.

The product of 435 and 23 is 10,005. Work Progress Check 14.

13. $56
14. 6960

EXAMPLE 3

Find the product of 257 and 135.

Solution

Because 135 = 100 + 30 + 5, we multiply 257 by 5, by 30, by 100 and then add the three products.

```
      2 5 7
  ×   1 3 5
  ─────────
      1 2 8 5    Multiply 257 by 5.
        7 7 1    Multiply 257 by 3 tens.
      2 5 7      Multiply 257 by 1 hundred.
  ─────────
    3 4 6 9 5    Add.
```

Because the times sign × can be confused with the letter x, multiplications are sometimes indicated by using a dot ·, by writing the factors next to each other, or by enclosing the factors in parentheses. For example, the product of 5 and x can be written as

$5 \cdot x$ or $5x$ or $(5)(x)$ or $5(x)$ or $(5)x$

The product of 23 and 45 cannot be written by placing the factors next to each other, because then 23 × 45 would look like the number 2345. The product of 23 and 45 is written as

23×45 or $23 \cdot 45$ or $(23)(45)$ or $23(45)$ or $(23)45$

We can use multiplication to count objects arranged in rectangular patterns. For example, Figure 1-2 shows a rectangular array consisting of 5 rows of 7 stars. The product 5 × 7, or 35, indicates the total number of stars.

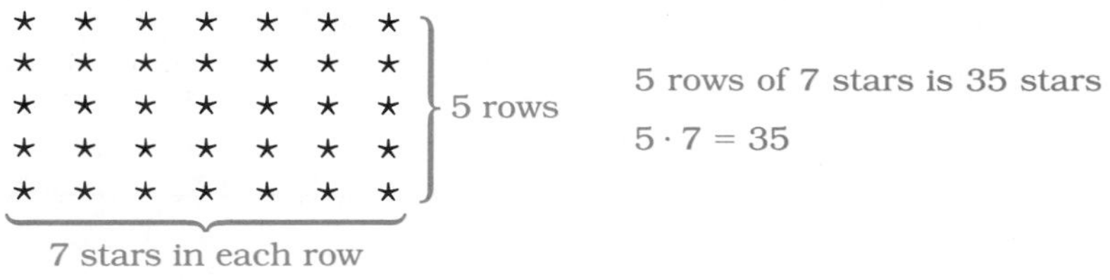

5 rows of 7 stars is 35 stars

$5 \cdot 7 = 35$

Figure 1-2

Because multiplication is commutative, the array consisting of 5 rows of 7 stars will contain the same number of stars as an array with 7 rows of 5 stars:

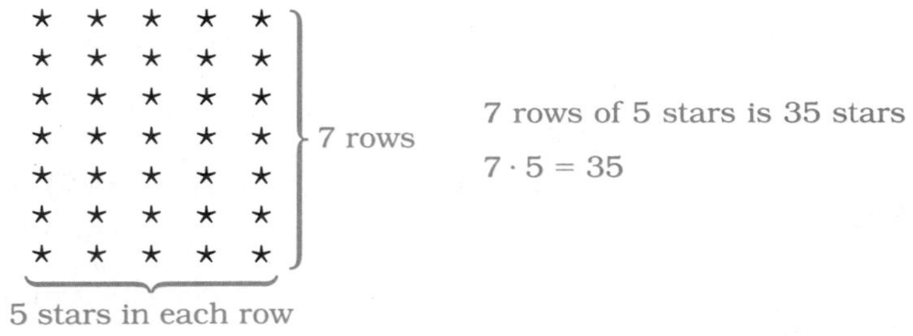

7 rows of 5 stars is 35 stars

$7 \cdot 5 = 35$

EXAMPLE 4

To draw graphics on a computer screen, a computer must control each *pixel* (a tiny dot on the screen). A high-resolution computer graphics image is 800 pixels wide and 600 pixels high. How many pixels must the computer control?

Solution

The graphics image is a rectangular array of pixels. Each of its 600 rows consists of 800 pixels. The total number of pixels is the product of 600 and 800:

$$600 \cdot 800 = 480{,}000$$

The computer must control 480,000 pixels. Work Progress Check 15.

Progress Check 15

On a color monitor, each of the pixels of the previous example can be red, blue, or green. How many colored pixels does the computer control?

Finding the Area of a Rectangle

The rectangle in Figure 1-3 has a length of 5 feet and a width of 3 feet. Each of the small squares is 1 foot long and 1 foot wide, and each is said to cover an **area** of **1 square foot**. The small squares form a rectangular pattern with 3 rows of 5 squares.

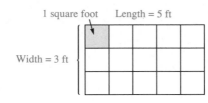

Figure 1-3

Because there are $5 \cdot 3$, or 15, small squares, we say that the **area of the rectangle** is 15 square feet. The area of any rectangle is the product of its length and its width.

> **The Area of a Rectangle:** The area, A, of a rectangle is the product of the rectangle's length, l, and its width, w:
>
> $$A = lw$$

EXAMPLE 5

Find the area of a rectangle that is 21 meters long and 13 meters wide.

Solution

We substitute 21 for l and 13 for w in the formula for the area of a rectangle.

$A = l \cdot w$

$A = 21 \cdot 13$ Replace l with 21 and w with 13.

    $= 273$ Do the multiplication.

The area of the rectangle is 273 square meters. Work Progress Check 16.

Progress Check 16

A package of 9-inch by 12-inch paper contains 500 sheets. How many square inches of paper are in one package?

If one of two numbers is 0, then the product of those numbers is 0. For example, $12 \cdot 0 = 0$ and $0 \cdot 27 = 0$. If one of two numbers is 1, then the product of those numbers is the other number. For example, $12 \cdot 1 = 12$ and $1 \cdot 27 = 27$. These two important properties of multiplication should be memorized:

15. 1,440,000

16. 54,000

If x is a number and x is multiplied by 0, the product is 0:
$$x \cdot 0 = 0 \quad \text{and} \quad 0 \cdot x = 0$$

If x is a number and x is multiplied by 1, the product is x itself:
$$x \cdot 1 = x \quad \text{and} \quad 1 \cdot x = x$$

Exponents

To show that a number appears several times as a factor in a product, we use an **exponent**. For example, we write 5^2 to show the product $5 \cdot 5$. In the expression 5^2, 5 is called the **base**, and 2 is the exponent.

$$5^2 \leftarrow \text{exponent}$$
$$\text{base} \nearrow$$

The number 2 in the notation 5^2 indicates that 5 appears as a factor two times. Likewise, 5^3 indicates that 5 is used as a factor three times: $5^3 = 5 \cdot 5 \cdot 5 = 125$. Expressions such as 5^2 and 5^3 are called **exponential expressions**.

EXAMPLE 6

a. $7^2 = 7 \cdot 7 = 49$ Read 7^2 as "7 squared."

b. $5^2 = 5 \cdot 5 = 25$

c. $2^5 = 2 \cdot 2 \cdot 2 \cdot 2 \cdot 2 = 32$ Read 2^5 as "2 to the fifth power."

d. $10^3 = 10 \cdot 10 \cdot 10 = 1000$ Read 10^3 as "10 cubed."

Work Progress Check 17.

Progress Check 17

Of the numbers 3^5, 4^4, and 5^3,

a. Which is largest?

b. Which is smallest?

Division of Whole Numbers

If $12 is distributed equally among four people, each person would receive $3. This situation illustrates the division of whole numbers:

$$4\overline{)12}^{\,3}$$

In mathematics, division is usually indicated by the division sign ÷, or by a horizontal bar. Thus, the division of 12 by 4 can also be indicated as

$$12 \div 4 = 3 \quad \text{or} \quad \frac{12}{4} = 3$$

The answer to a division is called a **quotient**. The number that we are dividing *by* is called the **divisor**. The number we are dividing *into* is called the **dividend**. In the division problems

$$4\overline{)12}^{\,3}, \quad 12 \div 4 = 3, \quad \text{and} \quad \frac{12}{4} = 3$$

the quotient is 3, the divisor is 4, and the dividend is 12.

17. a. 4^4, or 256 **b.** 5^3, or 125

An expression such as $\frac{12}{4}$ that uses a bar to indicate division is called a **fraction**. The dividend 12 is called the **numerator** of the fraction, and the divisor 4 is called the **denominator**.

Division can be thought of as *repeated subtraction*. To divide 12 by 4 is to ask, *How many 4's can be subtracted from 12?* Exactly three 4's can be subtracted from 12 to get 0:

$$12 \overbrace{- 4 - 4 - 4}^{\text{Three 4's}} = 0$$

Thus, $12 \div 4 = 3$.

Division is also related to multiplication. Twelve divided by 4 is equal to 3 because there are four 3's in 12:

$$12 \div 4 = 3 \quad \text{because} \quad 4 \cdot 3 = 12$$

or

$$\frac{12}{4} = 3 \quad \text{because} \quad 4 \cdot 3 = 12$$

Thus, we have this fact, which should be memorized:

If a, b, and x are any numbers, and b is not equal to 0, then all three of these statements are equivalent: $a \div b = x$, $\frac{a}{b} = x$, and $b \cdot x = a$.

EXAMPLE 7

Each multiplication statement in the left column is equivalent to each of the two corresponding division statements in the right column.

$7 \cdot 5 = 35$	$35 \div 7 = 5$	and	$35 \div 5 = 7$
$9 \cdot 8 = 72$	$72 \div 9 = 8$	and	$72 \div 8 = 9$
$6 \cdot 7 = 42$	$\frac{42}{6} = 7$	and	$\frac{42}{7} = 6$
$11 \cdot 8 = 88$	$\frac{88}{11} = 8$	and	$\frac{88}{8} = 11$
$a \cdot x = y$	$y \div a = x$	and	$y \div x = a$
$a \cdot x = y$	$\frac{y}{a} = x$	and	$\frac{y}{x} = a$

■

It is impossible to divide a number by 0. For example, what number x could be an answer to the division $12 \div 0$? If $12 \div 0$ were equal to x, then $0 \cdot x$ would equal 12. But, $0 \cdot x$ is always equal to 0. Thus, $12 \div 0$ has no answer.

Likewise, what number x could be an answer to the division problem $0 \div 0$? If $0 \div 0$ were equal to x, then $0 \cdot x$ would equal 0. Because $0 \cdot x$ is *always* 0 for *any* number x, *any* answer for $0 \div 0$ is correct! Since neither of these examples of division by 0 provides a *single* correct answer, we have the following fact:

A number cannot be divided by 0. Division by 0 is not defined.

The answer to the division 12 ÷ 1 is the number 12 itself. Likewise, 87 ÷ 1 = 87, and 1234 ÷ 1 = 1234. In general, *any number divided by 1 is the number itself.*

If x is any number, then $x \div 1 = x$.

The answer to the division 12 ÷ 12 is 1. Likewise, 87 ÷ 87 = 1, and 1234 ÷ 1234 = 1. In general, *Any number (except 0) divided by itself is 1.*

If x is any number except 0, then $x \div x = 1$.

If large numbers are to be divided, a process called **long division** can be used. To divide 832 by 23, for example, we proceed as follows:

$$23\overline{)832}$$

Place the divisor and the dividend as indicated. The quotient will appear above the long division symbol.

We will find the quotient one digit at a time. We begin the division process by asking, *How many times does 23 divide 8?* The answer is 0. We continue: *How many times does 23 divide 83?* Now, the answer is *not* 0. We estimate that 83 divided by 23 is approximately 4. Then we proceed as follows:

$$\begin{array}{r} 4 \\ 23\overline{)832} \end{array}$$

Estimate the quotient by asking: *How many 23's in 83?* Because an estimate is 4, place 4 in the tens position of the quotient.

$$\begin{array}{r} 4 \\ 23\overline{)832} \\ 92 \end{array}$$

Multiply 23 · 4. Place that answer, 92, under the 83. Because 92 is larger than 83, our estimate of 4 for the tens position of the quotient is too large.

$$\begin{array}{r} 3 \\ 23\overline{)832} \\ 69\downarrow \\ \overline{142} \end{array}$$

Revise the estimate of the tens position of the quotient to 3. Multiply: 23 · 3 = 69. Place 69 under 83, draw a line, and subtract.
Bring down the 2 in the ones position.

$$\begin{array}{r} 37 \\ 23\overline{)832} \\ 69\downarrow \\ \overline{142} \\ 161 \end{array}$$

Estimate the number of 23's in 142. The answer is approximately 7. Place 7 in the ones position of the quotient. Multiply: 23 · 7 = 161. Place 161 under 142. Because 161 is *larger* than 142, our estimate of 7 is too large.

$$\begin{array}{r} 36 \\ 23\overline{)832} \\ 69\downarrow \\ \overline{142} \\ 138 \\ \overline{4} \end{array}$$

Revise the estimate of the ones position of the quotient to 6. Multiply: 23 · 6 = 138.

Place 138 under 142 and subtract.

The final 4 in this illustration is called the **remainder** of the division. If $832 were distributed equally among 23 people, each person would receive $36, and there would be $4 left over.

To check the result of a division, we multiply the divisor by the quotient and then add the remainder. The result will be the dividend.

$$\text{quotient} \times \text{divisor} + \text{remainder} = \text{dividend}$$
$$36 \times 23 + 4 = 832$$
$$828 + 4 = 832$$
$$832 = 832$$

EXAMPLE 8

Betty sold several calculators for $17 each. If total sales were $1819, how many calculators did she sell?

Solution

The price of one calculator times the number of calculators sold equals the total sales.

$$\boxed{\text{The price of one calculator, in dollars}} \times \boxed{\text{the number of calculators sold}} = \boxed{\text{the total sales, in dollars}}$$

Let the letter x represent the number of calculators sold. Then

$$17 \cdot x = 1819$$

Because of the relation between multiplication and division, we have that

$$17 \cdot x = 1819 \quad \text{is equivalent to} \quad x = 1819 \div 17$$

To find x, we perform the long division:

$$17 \overline{\smash{)}1819} \qquad \text{Place the divisor and the dividend as indicated.}$$

We begin by asking, *How many times does 17 divide into 1?* The answer is 0. We continue: *How many times does 17 divide into 18?* The answer is now 1. We proceed as follows:

$$\begin{array}{r} 1 \\ 17{\overline{\smash{)}1819}} \\ \underline{17{\downarrow}} \\ 11 \end{array}$$

Estimate the hundreds position of the quotient by asking, *How many 17's are there in 18?* The answer is 1. Place a 1 in the hundreds position of the quotient, multiply, subtract, and bring down the 1.

$$\begin{array}{r} 10 \\ 17{\overline{\smash{)}1819}} \\ \underline{17{\downarrow}{\downarrow}} \\ 119 \end{array}$$

Ask, *How many 17's are there in 11?* The answer is 0. Place a 0 in the tens position of the quotient. Don't bother to subtract $17 \cdot 0$, or 0, from 11; the difference would still be 11. Instead, bring down the next digit, 9.

$$\begin{array}{r} 107 \\ 17{\overline{\smash{)}1819}} \\ \underline{17{\downarrow}{\downarrow}} \\ 119 \\ \underline{119} \\ 0 \end{array}$$

Ask, *How many 17's are there in 119?* The answer is 7. Multiply: $17 \cdot 7 = 119$.

Place 119 under 119 and subtract.

Betty sold 107 calculators. To check, we ask the question, *At $17 each, how much would 107 calculators cost?* The answer is $1819, the product of $17 and 107:

$$\$17 \cdot 107 = \$1819$$

The answer checks. ∎

EXAMPLE 9

On Thanksgiving Day, a soup kitchen plans to feed 1990 homeless people. Because of space limitations, only 165 people can be served at one time. How many sittings will be necessary to feed everyone? How many will be served at the last sitting?

Solution

The 1990 people can be fed 165 at a time. We must find how many times 165 can be subtracted from 1990. We do so by performing a long division:

```
        12
165)1990
    165↓
     340
     330
      10
```

Thus, 1990 ÷ 165 = 12, with a remainder of 10. It will require 13 sittings to feed 1990 people: 12 full-capacity sittings and one partial sitting to serve the additional 10. Work Progress Check 18.

Progress Check 18

Each gram of fat in a meal provides 9 calories. A fast-food meal contains 243 calories from fat. How many grams of fat does the meal contain?

Progress Check Answers
18. 27

1.3 Exercises

In Exercises 1–24, perform the indicated multiplication.

1. $4 \times 7 =$ _____
2. $7 \times 9 =$ _____
3. $12 \times 7 =$ _____
4. $15 \times 8 =$ _____
5. $27 \times 12 =$ _____
6. $35 \times 17 =$ _____
7. $47 \cdot 50 =$ _____
8. $75 \cdot 84 =$ _____
9. $135 \cdot 47 =$ _____
10. $273 \cdot 90 =$ _____
11. $(472)(102) =$ _____
12. $(730)(307) =$ _____

13. $\begin{array}{r} 37 \\ \times\ 8 \\ \hline \end{array}$
14. $\begin{array}{r} 48 \\ \times\ 9 \\ \hline \end{array}$
15. $\begin{array}{r} 23 \\ \times 13 \\ \hline \end{array}$
16. $\begin{array}{r} 37 \\ \times 21 \\ \hline \end{array}$

17. $\begin{array}{r} 99 \\ \times 77 \\ \hline \end{array}$
18. $\begin{array}{r} 73 \\ \times 59 \\ \hline \end{array}$
19. $\begin{array}{r} 20 \\ \times 53 \\ \hline \end{array}$
20. $\begin{array}{r} 78 \\ \times 20 \\ \hline \end{array}$

21. $\begin{array}{r} 112 \\ \times\ 23 \\ \hline \end{array}$
22. $\begin{array}{r} 232 \\ \times\ 53 \\ \hline \end{array}$
23. $\begin{array}{r} 207 \\ \times\ 97 \\ \hline \end{array}$
24. $\begin{array}{r} 768 \\ \times\ 70 \\ \hline \end{array}$

In Exercises 25–36, find the value of each exponential expression.

25. 2^2
26. 2^3
27. 3^2
28. 3^3
29. 4^2
30. 4^3
31. 10^3
32. 10^6
33. $(2^2)(3^2)$
34. $5^2 \cdot 7$
35. $2(3^4)$
36. $(3^2)(5^2)$

37. Which is larger, $(2^3)(3^2)$ or $(3^3)(2^2)$?

38. By what number does 3^5 exceed 5^3?

39. Jennifer worked 12 hours at $11 per hour. How much did she earn?

40. Manuel worked 23 hours at $9 per hour. How much did he earn?

41. Robert has 37 quarters. How much is that worth in pennies?

42. Jerald has 17 subway tokens, each worth 75¢. What is their total value in pennies?

43. A small car travels 29 miles on 1 gallon of gasoline. The car's tank holds 14 gallons. Find the number of miles the car can go on a full tank of gas.

44. Becky owns an apartment building with 18 units. Each unit generates a monthly income of $217. Find Becky's total monthly income.

45. A touring music group gives two concerts in each of 37 cities. Approximately 1700 people attend each concert. Find the total number of fans that hear the group on its tour.

46. There are 40 rods in one furlong, 8 furlongs in 1 mile, and 3 miles in 1 league. How many rods are in a league?

47. How many great-great-grandparents does one person have?

48. A breakfast food manufacturer advertises "two cups of raisins in every box of raisin bran." Find the number of cups of raisins in a case of 36 boxes of raisin bran.

49. It takes the juice of 13 oranges to make one can of frozen orange juice. Find the number of oranges used to make a case of 24 cans of frozen orange juice.

50. A checkerboard consists of eight rows with eight squares in each row. How many squares are on a checkerboard?

51. The rulings on a sheet of graph paper are $\frac{1}{4}$ inch apart. How many little squares are on a 9-inch-by-12-inch sheet of graph paper?

52. The mathematics department's lecture hall has 17 rows of 33 seats each. The seats are filled, and there is one instructor in the front of the room. A sign on the wall reads, *Occupancy by more than 570 persons is prohibited.* Is the mathematics department breaking the rule?

53. There are 14 people stuffed into one elevator with a capacity of 2000 pounds. If each person weighs an average of 150 pounds, is the elevator overloaded?

54. The state of Wyoming is a rectangle 360 miles long and 270 miles wide. What is the area of Wyoming?

55. A sheet of poster board is 24 inches by 36 inches. Find its area.

56. Which has the greater area, a room that is 14 feet by 17 feet, or a square room that is 16 feet on each side?

57. A rectangular garden is 27 feet long and 19 feet wide. A path in the garden uses 125 square feet of space. Find the area that remains for planting.

58. A queen-size mattress measures 80 inches by 60 inches, and a full-size mattress is 54 inches by 75 inches. Determine by how much the area of a queen-size mattress exceeds the area of a full-size mattress.

In Exercises 59–82, perform each division.

59. $40 \div 5 =$ _____

60. $40 \div 8 =$ _____

61. $42 \div 14 =$ _____

62. $65 \div 13 =$ _____

63. 132 ÷ 11 = _____

64. 132 ÷ 12 = _____

65. $\dfrac{117}{9}$ = _____

66. $\dfrac{117}{13}$ = _____

67. $\dfrac{221}{17}$ = _____

68. $\dfrac{221}{13}$ = _____

69. $\dfrac{253}{11}$ = _____

70. $\dfrac{253}{23}$ = _____

71. $21\overline{)252}$

72. $12\overline{)252}$

73. $27\overline{)513}$

74. $19\overline{)513}$

75. $13\overline{)949}$

76. $73\overline{)949}$

77. $33\overline{)1353}$

78. $41\overline{)1353}$

79. $39\overline{)7995}$

80. $71\overline{)7313}$

81. $29\overline{)6090}$

82. $13\overline{)7410}$

In Exercises 83–90, perform the indicated division. Then check your work by verifying that divisor × quotient + remainder = dividend.

83. $31\overline{)273}$

84. $25\overline{)290}$

85. $37\overline{)743}$

86. $79\overline{)931}$

87. $42\overline{)1273}$

88. $83\overline{)3280}$

89. $57\overline{)1795}$

90. $99\overline{)9876}$

91. Juan's first-grade class received extra milk today. If 73 half-pint cartons were distributed evenly to his 23 students, how many cartons were left over?

92. How many days more than 3 years is 2000 days? (There are 365 days in one year.)

93. How many inches over 13 feet is 160 inches? (There are 12 inches in 1 foot.)

94. How many feet over two miles is 11,000 feet? (There are 5280 feet in 1 mile.)

95. Hank has been ordered to write 100 times, *I will not chew gum.* What is the 353rd word he will write?

96. The Susquehanna River discharges 38,200 cubic feet of water per second into the Chesapeake Bay. How long does it take the river to discharge 1,719,000 cubic feet of water?

In Exercises 97–102, write each division as an equivalent multiplication.

97. 27 ÷ 3 = 9 _____

98. 91 ÷ 13 = 7 _____

99. 144 ÷ 12 = 12 _____

100. 195 ÷ 13 = 15 _____

101. 29 ÷ x = y _____

102. x ÷ 13 = y _____

In Exercises 103–108, write each multiplication fact as two equivalent expressions involving division.

103. 9 · 5 = 45 _____ _____

104. 12 · 3 = 36 _____ _____

105. 15 · 7 = 105 _____ _____

106. 8 · x = 40 _____ _____

107. x · 7 = 63 _____ _____

108. 8 · 9 = x _____ _____

109. Patrick earned $96 in tips working as a waiter. If each customer tipped $3, how many customers did he serve?

110. A 950,000-gallon swimming pool empties in 16 hours. How many gallons go down the drain each hour?

111. Brian runs 7 miles each day. In how many days will Brian run 371 miles?

112. There are 36 cookies in each package, and 900 children will receive 2 cookies each. How many packages will be needed?

Review Exercises

1. What is the digit in the hundreds position of 372,856?

2. What is the digit in the thousands position of 2,460,621?

3. Round 45,995 to the nearest thousand.

4. Round 45,995 to the nearest hundred.

5. Add the numbers 357, 39, and 476.

6. Add the numbers 289, 567, and 799.

7. Subtract 723 from 961.

8. Subtract 987 from 1010.

9. Write $x + 37 = 52$ as an equivalent subtraction, and find x.

10. Write $x - 149 = 58$ as an equivalent addition, and find x.

11. A radio, originally priced at $97, has been marked down to sell for $75. By how many dollars has the radio been discounted?

12. An automobile, originally priced at $17,550, is being sold at the end of the model year for $13,970. By how many dollars has the price been decreased?

1.4 Order of Operations; The Distributive Property

- Order of Mathematical Operations
- The Arithmetic Mean
- The Distributive Property

Order of Mathematical Operations

To **evaluate** (find the value of) the expression $4 + 6 \div 2$, we can do the division first and obtain

$$4 + 6 \div 2 = 4 + 3 = 7$$

However, if we were to do the addition first, we would get a different result:

$$4 + 6 \div 2 = 10 \div 2 = 5$$

To eliminate the possibility of getting different answers, we will always do multiplications and divisions before additions and subtractions, unless special *grouping symbols* indicate otherwise. Thus, we will agree that the correct calculation is

$$4 + 6 \div 2 = 4 + 3 \quad \text{Do the division first.}$$
$$= 7$$

To indicate that additions are to be done before multiplications, we must indicate that the operations are to be performed in other than the usual order. To do so, we use grouping symbols such as parentheses ( ), brackets [ ], or braces { }. In the expression $(4 + 6) \div 2$, the parentheses indicate that the addition should be done first:

$$(4 + 6) \div 2 = 10 \div 2 \quad \text{Do the addition first.}$$
$$= 5$$

EXAMPLE 1

a. $2 + 3 \cdot 4 = 2 + 12$ Do the multiplication first.
           $= 14$

b. $(2 + 3)4 = 5 \cdot 4$ Because of the parentheses, do the addition first.
             $= 20$

Work Progress Check 19.

Unless grouping symbols indicate otherwise, exponential expressions should be evaluated before multiplications and divisions. Thus the expression $5 + 4 \cdot 3^2$ should be evaluated as follows:

$5 + 4 \cdot 3^2 = 5 + 4 \cdot 9$ Evaluate the exponential expression first.
$ = 5 + 36$ Perform the multiplication.
$ = 41$ Perform the addition.

To guarantee that calculations will have only one correct answer, we use the following set of *priority rules*:

Order of Mathematical Operations: If an expression does not have grouping symbols,

1. Find the values of any exponential expressions.
2. Do all multiplications and divisions as they are encountered while working from left to right.
3. Do all additions and subtractions as they are encountered while working from left to right.

If an expression contains grouping symbols, use the rules above to perform the calculations within each pair of grouping symbols, working from the innermost pair to the outermost pair.

The expression $4 \cdot 3^2$, for example, is not equal to $(4 \cdot 3)^2$. Instead, it is evaluated as follows:

$4 \cdot 3^2 = 4(3 \cdot 3)$ Evaluate the exponential expression first.
$ = 4 \cdot 9$
$ = 36$

However,

$(4 \cdot 3)^2 = (12)^2$ Do the work within the parentheses first.
$ = 12 \cdot 12$
$ = 144$

Work Progress Check 20.

Progress Check 19

Evaluate:

a. $(5 + 7) \cdot 9$

b. $5 + 7 \cdot 9$

Progress Check 20

Evaluate:

a. $3 \cdot 5^2$

b. $(3 \cdot 5)^2$

---

19. a. 108 **b.** 68
20. a. 75   **b.** 225

Progress Check Answers

EXAMPLE 2

Evaluate $5^3 + 2(8 - 3 \cdot 2)$.

Solution

$5^3 + 2(8 - 3 \cdot 2) = 5^3 + 2(8 - 6)$ Within the parentheses, do the multiplication first.

$= 5^3 + 2(2)$ Do the subtraction within the parentheses.

$= 125 + 2(2)$ Find the value of the exponential expression.

$= 125 + 4$ Do the multiplication.

$= 129$ Do the addition.

Because a fraction bar is a grouping symbol, we simplify a fraction by finding the value of the numerator and the value of the denominator separately. Then we perform a division, whenever possible.

EXAMPLE 3

Evaluate $\dfrac{3(3 + 2) + 5}{17 - 3(4)}$.

Solution

Simplify the dividend and the divisor separately. Then perform the division.

$\dfrac{3(3 + 2) + 5}{17 - 3(4)} = \dfrac{3(5) + 5}{17 - 3(4)}$ Do the addition within the parentheses.

$= \dfrac{15 + 5}{17 - 12}$ Do the multiplications.

$= \dfrac{20}{5}$ Do the addition and the subtraction.

$= 4$ Do the division.

Work Progress Check 21.

Progress Check 21

Evaluate $\dfrac{4 + 3(7 - 2)}{1 + 2 \cdot 3^2}$

The Arithmetic Mean

The **arithmetic mean**, or **average**, of several numbers is a value often used as a representative of those numbers. To find the mean of several test scores, for example, we divide the sum of those scores by the number of scores. To find the mean of five numbers, we add the numbers and divide the sum by 5.

Progress Check Answers

21. 1

EXAMPLE 4

Jim earns scores of 72, 83, 66, 88, and 91 on five tests. Find his mean score.

Solution

To find Jim's mean score, we add the five scores and divide by 5:

$$\text{Mean} = \frac{72 + 83 + 66 + 88 + 91}{5}$$

$$= \frac{400}{5} \qquad \text{Do the additions in the numerator.}$$

$$= 80$$

Jim's mean score is 80.

The Distributive Property

The **distributive property** provides a second way of multiplying the sum of two numbers by a third number: We can either add first and then multiply, or multiply first and then add. For example,

$$2(3 + 5) = 2(8) \qquad \text{Add 3 and 5 first.}$$
$$= 16 \qquad \text{Then multiply.}$$

or

$$2(3 + 5) = 2 \cdot 3 + 2 \cdot 5 \qquad \text{Distribute the multiplication by 2 over the addition by multiplying each number within the parentheses by 2.}$$
$$= 6 + 10 \qquad \text{Then add.}$$
$$= 16$$

Either way, the result is 16.

Multiplication also distributes over subtraction. For example,

$$7(8 - 3) = 7(5) \qquad \text{Subtract first.}$$
$$= 35 \qquad \text{Then multiply.}$$

and then subtracting:

$$7(8 - 3) = 7 \cdot 8 - 7 \cdot 3 \qquad \text{Distribute the multiplication by 7 over the subtraction by multiplying each number within the parentheses by 7.}$$
$$= 56 - 21 \qquad \text{Then subtract.}$$
$$= 35$$

Either way, the result is 35. These examples illustrate a general property of arithmetic:

The Distributive Property: For any numbers a, b, and c,

$$a(b + c) = a \cdot b + a \cdot c$$
$$a(b - c) = a \cdot b - a \cdot c$$

Progress Check 22

Suppose that $17(5 + x) = 17 \cdot y + 17 \cdot 13$

a. Find x.

b. Find y.

---

22. a. 13 **b.** 5

Progress Check Answers

EXAMPLE 5

Evaluate $13(27 + 33)$ **a.** directly, and **b.** by using the distributive property.

Solution

a. $13(27 + 33) = 13 \cdot 60$ Do the addition within the parentheses first.
$= 780$ Multiply.

b. $13(27 + 33) = 13 \cdot 27 + 13 \cdot 33$ Distribute the multiplication over the addition.
$= 351 + 429$
$= 780$ Add.

Work Progress Check 22.

1.4 Exercises

In Exercises 1–28, simplify each expression by performing the indicated operations.

1. $2 \cdot 4 - 3$
2. $3 \cdot 5 + 7$
3. $2(4 - 3)$
4. $3(5 + 7)$
5. $2 + 3^2$
6. $(2 + 3)^2$
7. $3 \cdot 5 - 4$
8. $4 \cdot 6 + 5$
9. $3(5 - 4)$
10. $4(6 + 5)$
11. $3 + 5^2$
12. $(4 + 2)^2$
13. $(3 + 5)^2$
14. $4 + 2^2$
15. $(2 + 3) \cdot (5 - 4)$
16. $12 + 2(3) + 2$
17. $64 \div (4 + 4)$
18. $16 \div (2 + 6)$
19. $64 \div 4 + 4$
20. $16 \div 2 + 6$
21. $(2 \cdot 6 - 4)^2$
22. $2(6 - 4)^2$
23. $\dfrac{(2 + 5)^2 - 47}{6 - 2^2}$
24. $\dfrac{3^2 - 2^2}{(3 - 2)^2}$
25. $\dfrac{(3 + 5)^2 + 2}{2(8 - 5)}$
26. $\dfrac{25 - (2 \cdot 3 - 1)}{2 \cdot 9 - 8}$
27. $\dfrac{(5 - 3)^2 + 2}{4^2 - (8 + 2)}$
28. $\dfrac{(4^3 - 2) + 7}{5(2 + 4) - 7}$

In Exercises 29–32, insert parentheses in the expression $3 \cdot 8 + 5 \cdot 3$ to make its value equal to each number.

29. 39
30. 117
31. 87
32. 69

In Exercises 33–36, insert parentheses in the expression $4 + 3 \cdot 5 - 3$ to make its value equal to each number.

33. 14
34. 10
35. 32
36. 16

In Exercises 37–56, evaluate each expression two ways.

37. $3(7 + 9)$
38. $5(3 + 12)$
39. $11(8 - 5)$
40. $15(12 - 9)$
41. $17(12 - 8)$
42. $21(17 - 9)$
43. $19(17 + 7)$
44. $14(11 + 11)$
45. $7(24 - 14)$
46. $4(21 + 13)$
47. $5(5 + 25)$
48. $8(32 - 23)$

49. 11(92 − 57) **50.** 16(34 − 27) **51.** 151(31 + 52) **52.** 122(75 − 67)

53. 32(25 − 10) **54.** 45(24 + 16) **55.** 27(27 + 27) **56.** 59(100 − 31)

57. Bill scored 57, 35, 46, 71, and 51 on his first five exams. Find his mean score.

58. Maria scored 88, 55, 92, 100, 78, and 85 on six exams. Find her mean score.

59. The temperatures last week were 75°, 80°, 83°, 80°, 77°, 72°, and 86°. Find the week's mean temperature.

60. Speeds of six randomly chosen expressway drivers were 65, 65, 72, 62, 60, and 66 mph. Find their mean speed.

61. If $x = 12$ and $y = 9$, find the value of $\dfrac{3 \cdot x}{y}$.

62. If $x = 5$ and $y = 4$, find the value of $\dfrac{3 + x}{y}$.

63. If $x = 14$ and $y = 8$, find the value of $\dfrac{3 \cdot x}{y - 1}$.

64. If $x = 15$ and $y = 13$, find the value of $\dfrac{3 \cdot x + 3}{y + 3}$.

Review Exercises

In Review Exercises 1–10, perform the indicated operations.

1. $2 \cdot 3^2 \cdot 5$ **2.** $3^2 \cdot 5 \cdot 7^2$ **3.** $2^3 \cdot 3^2 \cdot 5$ **4.** $2 \cdot 3^3 \cdot 7$

5. $3^2 \cdot 5^2 \cdot 7^2$ **6.** $7^2 \cdot 11$ **7.** $11^2 \cdot 13$ **8.** $11 \cdot 13^2$

9. $7 \cdot 11 \cdot 13^2$ **10.** $5^2 \cdot 7 \cdot 13$

1.5 Prime Numbers, Composite Numbers, and Prime-Factored Form

- Tests for Divisibility
- Prime and Composite Numbers
- Prime-Factored Form

Since $3 \times 5 = 15$, 3 and 5 are factors of 15. Because 5 is a factor of 15, we say that 5 **divides** 15, that 5 is a **divisor** of 15, and that 15 is **divisible** by 5.

Likewise, 7 divides 21, because 7 is a factor of 21: the product $7 \cdot 3$ is 21. Also, 7 is a divisor of 21, and 21 is divisible by 7.

Definition: If a and b are whole numbers, and $a \cdot b = c$, then a and b **divide** c, and c is **divisible** by both a and b.

EXAMPLE 1

Find the divisors of 36.

Solution

Because $1 \cdot 36 = 36$, both 1 and 36 are divisors of 36.
Because $2 \cdot 18 = 36$, both 2 and 18 are divisors of 36.
Because $3 \cdot 12 = 36$, both 3 and 12 are divisors of 36.

Because 4 · 9 = 36, both 4 and 9 are divisors of 36.
Because 6 · 6 = 36, the number 6 is a divisor of 36.

The divisors of 36 are 1, 2, 3, 4, 6, 9, 12, 18, and 36. Work Progress Check 23. ∎

Progress Check 23

Find all the divisors of 24.

Tests for Divisibility

There are tests to help decide if one given number is divisible by another.

> **Divisibility by 2:** If the last digit of a number is divisible by 2, the number is divisible by 2.
>
> **Divisibility by 4:** If the number formed by the last *two* digits is divisible by 4, the number is divisible by 4.

The number 35,656, for example, is divisible by 2 because the last digit, 6, is divisible by 2. To find the other factor, we divide 35,656 by 2 to obtain 17,828. Thus, 35,656 = **2** · 17,828.

The number 35,656 is also divisible by 4 because the number formed by its last two digits, 56, is divisible by 4. To find the other factor, we divide 35,656 by 4 to obtain 8914. Thus, 35,656 = **4** · 8914.

> **Divisibility by 5:** If the last digit of a number is 0 or 5, the number is divisible by 5.

The number 4365, for example, is divisible by 5 because the last digit is 5. To find the other factor, we divide 4365 by 5 to obtain 873. Thus, 4365 = **5** · 873.

> **Divisibility by 10:** If the last digit of a number is 0, the number is divisible by 10.

For example, the number 57,690,340 is divisible by 10 because its last digit is 0. To find the other factor, we divide 57,690,340 by 10 to obtain 5,769,034.

> **Divisibility by 3:** If the sum of the digits of a number is divisible by 3, the number is divisible by 3.
>
> **Divisibility by 9:** If the sum of the digits of a number is divisible by 9, the number is divisible by 9.

23. 1, 2, 3, 4, 6, 8, 12, 24

Progress Check Answers

The number 3258 is divisible by 3 and by 9, because the sum of its digits is $3 + 2 + 5 + 8 = 18$, and 18 is divisible by 3 and by 9. The number 5721, however, is divisible by 3 but not by 9, because the sum of its digits is $5 + 7 + 2 + 1 = 15$, and 15 is divisible by 3 but not by 9.

There are tests for divisibility by other numbers, but these tests are complicated. To test a number n for divisibility by a number d other than 2, 3, 4, 5, 9, or 10, we simply perform a long division of n by d. If the division leaves no remainder, the number n is divisible by d.

Prime and Composite Numbers

Every whole number greater than 1 has at least two divisors: 1 and the number itself. For example, 17 is divisible by 1 and also by 17. However, some numbers (like 17) have *exactly* two divisors. Such numbers are called **prime numbers**.

> **Definition:** A **prime number** is a whole number greater than 1 that is divisible only by 1 and by itself.

The prime numbers are the numbers

$2, 3, 5, 7, 11, 13, 17, 19, 23, 29, 31, \ldots$

Since the list of prime numbers never ends, there is no largest prime number.

The number 23 is prime because

1. it is a whole number,
2. it is greater than 1, and
3. it can be divided *only* by the numbers 1 and 23.

Although 12 is a whole number greater than 1, it is divisible by numbers other than 1 and 12—the numbers 2, 3, 4, and 6. Thus, 12 is *not* a prime number. Those whole numbers greater than 1 that are not prime numbers are called **composite numbers**. Thus, 12 is a composite number.

> **Definition:** A **composite number** is a whole number greater than 1 that is not a prime number.

The composite numbers from 1 to 20 are

$4, 6, 8, 9, 10, 12, 14, 15, 16, 18, 20$

EXAMPLE 2

Classify **a.** 7, **b.** 6, and **c.** 0 into the categories of prime number or composite number.

Solution

a. The number 7 is a prime number, because its only divisors are 1 and 7.

b. The number 6 is a composite number, because it is divisible by 2 and by 3.

c. The number 0 is neither a prime number nor a composite number, because it is not greater than 1.

Work Progress Check 24. ■

Prime-Factored Form

To **factor** a whole number means to write the number as the product of other whole numbers. If a number is written as the product of prime numbers, it is written in **prime-factored form**. For example, the **prime factorization** of 42 is $2 \cdot 3 \cdot 7$, because $42 = 2 \cdot 3 \cdot 7$, and 2, 3, and 7 are prime numbers. Likewise, the prime factorizations of 60 and 90 are

$$60 = 2 \cdot 2 \cdot 3 \cdot 5 \quad \text{and} \quad 90 = 2 \cdot 3 \cdot 3 \cdot 5$$

By using exponents, we can write these factorizations more compactly as

$$60 = 2^2 \cdot 3 \cdot 5 \quad \text{and} \quad 90 = 2 \cdot 3^2 \cdot 5$$

These examples lead to the following definition.

> **Definition:** A whole number greater than 1 is in **prime-factored form** if it is written as the product of factors that are prime numbers.

The **fundamental theorem of arithmetic** points out that there is *exactly one* prime factorization for each whole number greater than 1. We say that the prime factorization of any whole number is unique.

EXAMPLE 3

Write 2340 in prime-factored form.

Solution

Because the last digit of the number 2340 is 0, the number is divisible by 10. We obtain the other factor by dividing 2340 by 10: $2340 \div 10 = 234$. Thus,

$$2340 = 10 \cdot 234$$

Because $10 = 2 \cdot 5$, we have

$$2340 = 2 \cdot 5 \cdot 234$$

Because the sum of the digits of 234 is $2 + 3 + 4$, or 9, and 9 is divisible by 9, the number 234 is divisible by 9. Perform the division $234 \div 9 = 26$ to find that $234 = 9 \cdot 26$. Thus,

$$2340 = 2 \cdot 5 \cdot 9 \cdot 26$$

Progress Check 24

Is 3,576,233,735 a prime number?

---

24. No. Because the number is divisible by 5.

Progress Check Answers

Progress Check 25

Write 4725 in prime-factored form.

Because $9 = 3 \cdot 3$ and $26 = 2 \cdot 13$, we have

$$2340 = 2 \cdot 5 \cdot 3 \cdot 3 \cdot 2 \cdot 13$$

Because the factors on the right-hand side of this expression are all prime numbers, the number 2340 has been written in prime-factored form. It is customary to arrange the factors in increasing order and use exponents where possible:

$$2340 = 2 \cdot 2 \cdot 3 \cdot 3 \cdot 5 \cdot 13$$
$$= 2^2 \cdot 3^2 \cdot 5 \cdot 13$$

Work Progress Check 25.

25. $3^3 \cdot 5^2 \cdot 7$

Progress Check Answers

1.5 Exercises

In Exercises 1–16, determine if the first number is divisible by the second number.

1. 2526; 2   **2.** 2759; 2   **3.** 11,523; 3   **4.** 51,260; 5

5. 456,327; 9   **6.** 457,250; 10   **7.** 795,203; 5   **8.** 45,261; 3

9. 259; 7   **10.** 523; 11   **11.** 1261; 13   **12.** 4,444,444; 11

13. 1649; 17   **14.** 9797; 97   **15.** 27; 0   **16.** 12,221; 11

In Exercises 17–28, list all the divisors of each number.

17. 42   **18.** 38   **19.** 54   **20.** 100

21. 121   **22.** 142   **23.** 96   **24.** 94

25. 46   **26.** 182   **27.** 23   **28.** 47

In Exercises 29–36, classify each number as a prime number, a composite number, or neither.

29. 23   **30.** 27   **31.** 777   **32.** 101

33. 5,231,475   **34.** 6,782,301   **35.** 0   **36.** 1

In Exercises 37–56, find the prime factorization of each number.

37. 26   **38.** 51   **39.** 171   **40.** 115

41. 513   **42.** 570   **43.** 102   **44.** 530

45. 306   **46.** 228   **47.** 2610   **48.** 1240

49. 1800   **50.** 2400   **51.** 3920   **52.** 1176

53. 1925   **54.** 4725   **55.** 4235   **56.** 3773

57. A whole number is called a **perfect number** if the sum of all of its divisors is equal to 2 times the number. Thus, 6 is a perfect number because its divisors are 1, 2, 3, and 6, and 1 + 2 + 3 + 6 = 12, and 12 is 2 times 6. Show that 28 is a perfect number.

58. Show that 32 is not a perfect number. (See Exercise 57.)

59. Classified information is often transmitted in code. One popular code is difficult to crack because it depends on writing a very large number as the product of two large primes of about 100 digits each. To understand how difficult this factoring is, find the two two-digit prime factors of 7663. (*Hint:* Both primes are greater than 70.)

60. Find the two two-digit prime factors of 8633.

Review Exercises

In Review Exercises 1–8, perform the indicated operations.

1. 341 + 527 = _____

2. 335 + 819 = _____

3. 112 − 76 = _____

4. 215 − 168 = _____

5. 37 × 52 = _____

6. 135 × 37 = _____

7. 555 ÷ 15 = _____

8. 1073 ÷ 37 = _____

9. One case of pencils contains 72 boxes of one dozen pencils (one dozen = 12). Find the number of pencils in three cases.

10. Each of 37 first-grade students brings enough valentines for each of the other students in the class. Find the total number of valentines.

11. When the university marching band lines up in 9 rows of 13 musicians, there are 5 left over. How many band members are there?

12. When the band in Review Exercise 11 lines up in an 11-by-11 square, how many musicians are left over?

1.6 Least Common Multiple and Greatest Common Divisor

- The Least Common Multiple
- The Greatest Common Divisor

The Least Common Multiple

The **multiples** of a whole number x are found by multiplying x successively by 1, 2, 3, 4, and so on. The multiples of 26, for example, are

$$1 \cdot 26, \quad 2 \cdot 26, \quad 3 \cdot 26, \quad 4 \cdot 26, \quad 5 \cdot 26, \quad 6 \cdot 26, \ldots$$

or

26, 52, **78**, 104, 130, 156, ...

Likewise, the multiples of 39 are the numbers

$$1 \cdot 39, \quad 2 \cdot 39, \quad 3 \cdot 39, \quad 4 \cdot 39, \quad 5 \cdot 39, \quad 6 \cdot 39, \ldots$$

or

39, **78**, 117, 156, 195, 234, ...

Because 78 is the smallest number that is a multiple of both 26 and 39, it is called the **least common multiple** of 26 and 39.

> **Definition:** The **least common multiple (LCM)** of several whole numbers a, b, c, ... is the smallest number that is a multiple of each of the numbers a, b, c, ...

We follow these steps to find the least common multiple of several numbers.

> **Finding the Least Common Multiple (LCM) of Several Numbers:**
> 1. Write each of the numbers in prime-factored form.
> 2. Form a product using each of the different prime factors obtained in Step 1. Use each different factor the *greatest* number of times it appears in any single factorization. The product so formed is the least common multiple of the numbers.

EXAMPLE 1

Find the least common multiple of 26 and 39.

Solution

First, we write down each number and find its prime factorization:

$26 = 2 \cdot 13$
$39 = 3 \cdot 13$

Then we form a product with factors of 2, 3, and 13. Use each of these factors the greatest number of times it appears in any single factorization. That is, use each of them once.

$\text{LCM} = 2 \cdot 3 \cdot 13$
$\phantom{\text{LCM}} = 78$

The least common multiple of 26 and 39 is 78. Seventy-eight is the smallest number that is divisible by 26 and 39. ∎

EXAMPLE 2

Find the least common multiple of 24, 18, and 36.

Solution

First, we write down each number and find its prime factorization.

$24 = 2 \cdot 2 \cdot 2 \cdot 3 = 2^3 \cdot 3$
$18 = 2 \cdot 3 \cdot 3 = 2 \cdot 3^2$
$36 = 2 \cdot 2 \cdot 3 \cdot 3 = 2^2 \cdot 3^2$

Then we form a product using the prime factors of 2 and 3. Use each of these the greatest number of times it appears in any single factorization. That is, use the factor 2 three times, because 2 appears three times as a factor of 24. Use the factor 3 twice, because it occurs twice as a factor of 18 and of 36. Thus, the least common multiple of 24, 18, and 36 is

$$\text{LCM} = 2^3 \cdot 3^2$$
$$= 8 \cdot 9$$
$$= 72$$

and 72 is the smallest number divisible by 24, 18, and 36. Work Progress Check 26. ■

Progress Check 26

Find the least common multiple of 6, 10, 12, and 15.

The Greatest Common Divisor

The divisors of 26 are the numbers **1**, 2, **13**, and 26. The divisors of 39 are the numbers **1**, 3, **13**, and 39. The numbers 1 and 13 are common to these two lists. Of these, 13 is the greater, and it is called the **greatest common divisor** of 26 and 39.

> **Definition:** The **greatest common divisor** of several whole numbers is the greatest number that divides each of the numbers.

We follow these steps to find the greatest common divisor of several numbers.

> **Finding the Greatest Common Divisor (GCD) of Several Numbers:**
>
> 1. Write each of the numbers in prime-factored form.
> 2. Form a product using only those prime factors that appear in *all* of the factorizations obtained in Step 1. Use each factor the *smallest* number of times it appears in any single factorization. The product so formed is the greatest common divisor of the numbers.

EXAMPLE 3

Find the greatest common divisor of 45 and 75.

Solution

First, we write down each number and find its prime factorization.

$$45 = 9 \cdot 5 = 3^2 \cdot 5$$
$$75 = 3 \cdot 25 = 3 \cdot 5^2$$

Next we find those primes that appear in *both* factorizations. They are 3 and 5. Finally, we form a product using each of these the *smallest* number of times it appears in any one of the factorizations. Because 3 appears only once in the factorization of 75, we use 3 only once in finding the GCD.

26. 60

Because 5 appears only once in the factorization of 45, we use 5 only once in finding the GCD. Thus, the greatest common divisor of 45 and 75 is

$$\text{GCD} = 3 \cdot 5$$
$$= 15$$

We note that $45 \div 15 = 3$, that $75 \div 15 = 5$, and that there is no number greater than 15 that will divide both 45 and 75. ∎

EXAMPLE 4

Find the greatest common divisor of 48, 72, and 120.

Solution

First, we write down each number and find its prime-factored form.

$$48 = 16 \cdot 3 = 2^4 \cdot 3$$
$$72 = 8 \cdot 9 = 2^3 \cdot 3^2$$
$$120 = 12 \cdot 10 = 3 \cdot 4 \cdot 2 \cdot 5 = 3 \cdot 2 \cdot 2 \cdot 2 \cdot 5 = 2^3 \cdot 3 \cdot 5$$

Next we find the prime factors that appear in all of these factorizations. They are 2 and 3. Finally, we form a product that uses 2 as a factor three times, and 3 as a factor once. That product, $2^3 \cdot 3$, or 24, is the GCD of the numbers. Work Progress Check 27. ∎

Progress Check 27

List *all* of the numbers that divide 28, 42, and 54.

27. 1 and 2
Progress Check Answers

1.6 Exercises

In Exercises 1–4, list the first five multiples of each number, beginning with the number itself.

1. 3 **2.** 7 **3.** 11 **4.** 10

In Exercises 5–24, find the least common multiple of the given numbers.

5. 3, 5 **6.** 7, 11 **7.** 14, 21 **8.** 16, 20

9. 22, 33 **10.** 15, 20 **11.** 15, 30 **12.** 45, 75

13. 100, 120 **14.** 120, 180 **15.** 14, 140 **16.** 15, 300

17. 6, 24, 36 **18.** 6, 10, 18 **19.** 18, 24, 33 **20.** 15, 21, 35

21. 10, 14, 35 **22.** 45, 81, 99 **23.** 35, 45, 50 **24.** 120, 225, 400

In Exercises 25–44, find the greatest common divisor of the given numbers.

25. 3, 5 **26.** 7, 11 **27.** 14, 21 **28.** 16, 20

29. 22, 33 **30.** 15, 20 **31.** 15, 30 **32.** 45, 75

33. 100, 120 **34.** 120, 180 **35.** 14, 140 **36.** 15, 300

37. 6, 24, 36 **38.** 6, 10, 18 **39.** 18, 24, 33 **40.** 15, 21, 35

41. 14, 10, 35 **42.** 45, 81, 99 **43.** 35, 45, 50 **44.** 120, 225, 400

45. Two whole numbers are called **relatively prime** if their greatest common divisor is 1. Verify that 56 and 135 are relatively prime.

46. Verify that 225 and 392 are relatively prime. (See Exercise 45.)

47. Verify that any two of the numbers 15, 21, and 35 are not relatively prime, but that the greatest common divisor of all three numbers is 1. (See Exercise 45.)

48. For any two whole numbers x and y, the product of their greatest common divisor and their least common multiple is equal to the product of x and y. Verify this for the numbers 180 and 280.

Review Exercises

In Review Exercises 1–2, write each addition as an equivalent subtraction, and find the value of x.

1. $x + 95 = 103$ _____

2. $47 + x = 92$ _____

In Review Exercises 3–4, write each subtraction as an equivalent addition, and find the value of x.

3. $x - 12 = 17$ _____

4. $x - 31 = 21$ _____

In Review Exercises 5–6, write each multiplication as an equivalent division, and find the value of x.

5. $x \cdot 5 = 35$ _____

6. $x \cdot 13 = 39$ _____

In Review Exercises 7–8, write each division as an equivalent multiplication, and find the value of x.

7. $x \div 15 = 3$ _____

8. $\dfrac{x}{21} = 5$ _____

In Review Exercises 9–10, find the prime factorization of each number.

9. 1440 _____

10. 38,160 _____

Mathematics and Ecology

Fluorescent tubes are being recycled at the rate of 600,000 per month. To determine the number that are recycled during one year, we multiply 600,000 by 12, the number of months in one year. The work requires neither pencil, paper, nor calculator. Do it mentally, using the properties of whole-number arithmetic:

$$600{,}000 \times 12 = (100{,}000 \times 6) \times 12$$
$$= 100{,}000 \times (6 \times 12) \quad \text{Use the associative property of multiplication.}$$
$$= 100{,}000 \times 72 \quad \text{Multiply } 6 \times 12.$$
$$= 7{,}200{,}000 \quad \text{72 hundred thousand is 7,200,000}$$

These few California corporations can recycle over 7 million fluorescent tubes each year. This industry has considerable room for growth, considering that almost 500 million fluorescent tubes are manufactured each year in the United States.

Chapter Summary

Key Words

addends (1.2)
arithmetic mean (1.4)
associative property
  of addition (1.2)
associative property
  of multiplication (1.3)
average (1.4)
base (1.3)
commutative property
  of addition (1.2)
commutative property
  of multiplication (1.3)
composite number (1.5)
denominator (1.3)
difference (1.2)
digits (1.1)
distributive
  property (1.4)
dividend (1.3)
divisor (1.3)
expanded notation (1.1)
exponent (1.3)
exponential
  expression (1.3)
factor (1.3)
fraction (1.3)
fundamental theorem
  of arithmetic (1.5)
greatest common
  divisor (1.6)
least common
  multiple (1.6)
length (1.2)
long division (1.3)
minuend (1.2)
multiple (1.6)
numerator (1.3)
perimeter (1.2)
place value (1.1)
prime-factored form (1.5)
prime factorization (1.5)
prime number (1.5)
product (1.3)
quotient (1.3)
rectangle (1.2)
remainder (1.3)
standard notation (1.1)
subtrahend (1.2)
sum (1.2)
variable (1.2)
whole number (1.1)
width (1.2)

Key Ideas

(1.1) When precise results are not needed, numbers can be rounded.

(1.2) The subtraction $a - b = x$ is equivalent to the addition $a = b + x$.

The addition $a + b = x$ is equivalent to two subtractions: $x - a = b$ and $x - b = a$.

(1.3) If x is any number, then

$x \cdot 0 = 0$ and $0 \cdot x = 0$
$x \cdot 1 = x$ and $1 \cdot x = x$
$x \div 1 = x$
$x \div x = 1$ (provided x is not 0)

The divisions $a \div b = x$ and $\frac{a}{b} = x$ are equivalent to the multiplication $b \cdot x = a$.

Division by 0 is not defined.

(1.4) **Order of mathematical operations.** If an expression has no grouping symbols,

1. Find the values of any exponential expressions.
2. Do all multiplications and divisions as they are encountered while working from left to right.
3. Do all additions and subtractions as they are encountered while working from left to right.

If an expression has grouping symbols, use these rules to perform the calculations within each pair of grouping symbols, working from the innermost pair to the outermost pair.

The distributive property. For any numbers a, b, and c,

$$a(b + c) = a \cdot b + a \cdot c$$
$$a(b - c) = a \cdot b - a \cdot c$$

(1.5) **Tests for divisibility.** A number is divisible by 2 if its last digit is divisible by 2.

A number is divisible by 3 if the sum of its digits is divisible by 3.

A number is divisible by 4 if the number formed by its last two digits is divisible by 4.

A number is divisible by 5 if its last digit is 0 or 5.

A number is divisible by 9 if the sum of its digits is divisible by 9.

A number is divisible by 10 if its last digit is 0.

(1.6) To find the LCM of two numbers, write each one in prime-factored form. Form a product using each of the different prime factors the *greatest* number of times it appears in any single factorization.

To find the GCD of two numbers, write each one in prime-factored form. Form a product using only those prime factors that appear in *all* of the factorizations, and use each prime factor the *smallest* number of times it appears.

Chapter 1 Review Exercises

[1.1] In Review Exercises 1–2, write each number in expanded notation.

1. 260,206

2. 25,205,025

In Review Exercises 3–4, write each number in standard notation.

3. 3 thousands + 2 hundreds + 7 ones

4. twenty-three million two hundred fifty-three thousand four hundred twelve

[1.2] In Review Exercises 5–8, perform each addition.

5. 135 + 213 = _____

6. 475 + 172 = _____

7. 1247 + 7577 = _____

8. 4447 + 7478 = _____

In Review Exercises 9–12, perform each subtraction.

9. 235 − 218 = _____

10. 723 − 582 = _____

11. 1357 − 1349 = _____

12. 5231 − 5177 = _____

13. Find the perimeter of a rectangle that is 731 feet wide and 642 feet long.

14. A direct flight to San Francisco costs $237. A flight with one stop in Reno costs $192. How much can a traveler save by taking the inexpensive flight?

15. Bob's savings account contains $931. If he deposits $271 and makes withdrawals of $37 and $380, find his final balance.

16. In a shipment of 350 animals, 124 are hogs, 79 are sheep, and the rest are cattle. Find the number of cattle in the shipment.

In Review Exercises 17–20, write each subtraction as an equivalent addition.

17. 13 − 5 = 8 _____

18. 43 − 21 = 22 _____

19. 57 − x = 12 _____

20. 67 − 34 = x _____

In Review Exercises 21–24, write each addition as two equivalent subtractions.

21. 12 = 5 + 7 _____ _____

22. 34 = 14 + 20 _____ _____

23. 21 + x = 57 _____ _____

24. y + 7 = x _____ _____

[1.3] In Review Exercises 25–28, perform each multiplication.

25. 157 · 21 = _____

26. 423 · 78 = _____

27. 1597 · 52 = _____

28. 3723 · 148 = _____

In Review Exercises 29–32, find the value of each exponential expression.

29. 5^4

30. 10^5

31. $7 \cdot 3^4$

32. $3^2 \cdot 5^3$

33. Sarah worked 12 hours at $7 per hour. Her friend, Craig, worked 15 hours at $6 per hour. Who earned more money?

34. There are 12 eggs in one dozen, and 12 dozen in one gross. How many eggs are in a shipment of 5 gross?

In Review Exercises 35–38, perform each division.

35. 357 ÷ 17 = _____

36. 2760 ÷ 23 = _____

37. 4601 ÷ 43 = _____

38. 1443 ÷ 39 = _____

In Review Exercises 39–42, perform each division. Then check your work by verifying that divisor × quotient + remainder = dividend.

39. $21\overline{)405}$

40. $43\overline{)527}$

41. $59\overline{)971}$

42. $21\overline{)1269}$

In Review Exercises 43–46, write each division as an equivalent multiplication.

43. 81 ÷ 3 = 27 _____

44. 153 ÷ 17 = 9 _____

45. 143 ÷ x = 11 _____

46. x ÷ 17 = 5 _____

In Review Exercises 47–50, write each multiplication fact as two equivalent expressions involving division.

47. 8 · 5 = 40 _____ _____

48. 11 · 4 = 44 _____ _____

49. 29 · x = 493 _____ _____

50. y · 4 = 44 _____ _____

51. If 745 candies are divided equally among 45 children, how many will each child receive?

52. In Review Exercise 51, how many candies will be left over?

[1.4] In Review Exercises 53–62, perform the indicated operations.

53. 13 + 12 · 3 = _____

54. 35 − 15 ÷ 5 = _____

55. (13 + 12) · 3 = _____

56. (35 − 15) ÷ 5 = _____

57. 8 · 5 − 4 ÷ 2 = _____

58. 8 · (5 − 4) ÷ 2 = _____

59. $8 \cdot (5 - 4 \div 2) =$ _____

60. $(8 \cdot 5 - 4) \div 2 =$ _____

61. $\dfrac{12 + 3 \cdot 7}{5^2 - 14} =$ _____

62. $\dfrac{3 \cdot 12 - 15}{4^2 - 3^2} =$ _____

In Review Exercises 63–66, evaluate each expression in two ways.

63. $7(11 - 9)$ **64.** $13(17 + 6)$ **65.** $8(12 + 14)$ **66.** $8(16 - 9)$

[1.5] In Review Exercises 67–68, list all the factors of the given number.

67. 81 _____

68. 170 _____

In Review Exercises 69–70, find the prime factorization of each number.

69. 720 _____

70. 1890 _____

[1.6] In Review Exercises 71–74, find the least common multiple of the given numbers.

71. 15, 18 _____

72. 9, 30 _____

73. 12, 25, 30 _____

74. 12, 63, 72 _____

In Review Exercises 75–78, find the greatest common divisor of the given numbers.

75. 15, 18 _____

76. 9, 30 _____

77. 12, 25, 30 _____

78. 12, 63, 72 _____

Name _____   Section _____

Chapter 1 Test

1. List the whole numbers less than 5.

2. Write *five thousand two hundred sixty-six* in expanded notation.

3. Write 7 thousands + 5 hundreds + 7 ones in standard notation.

4. Round 34,752,341 to the nearest million.

5. Add 4521 to 3579.

6. Subtract 3579 from 4521.

7. A rectangle is 327 inches wide and 757 inches long. Find its perimeter.

8. Write $x + 12 = 19$ as an equivalent subtraction, and find the value of x.

9. Write $x - 29 = 17$ as an equivalent addition, and find x.

10. On Tuesday, a share of KBJ Company stock was selling at $73. The price rose $12 on Wednesday and fell $9 on Thursday. Find its price on Thursday.

11. Jenny scored 73, 52, and 70 on three of five exams. She received 0 on the two missing exams. Find Jenny's average.

12. Find the product of 1563 and 927.

13. Find the quotient when 4305 is divided by 35.

14. Find the remainder obtained when 3451 is divided by 74.

15. Find the value of $3 \cdot 4^2 - 2^2$.

16. Find the missing expression: $3(a + b) = 3 \cdot a + \underline{}$ _____

17. Find the missing expression: $x(y + z) = x \cdot y + \underline{}$ _____

18. Write 1260 in prime-factored form. _____

19. Find the least common multiple of 50 and 175. _____

20. Find the least common multiple of 25, 30, and 60. _____

21. Find the greatest common divisor of 50 and 175. _____

22. Find the greatest common divisor of 25, 30, and 60. _____

2 Fractions

Mathematics in Music

The study of mathematics in ancient Greece was a strange mixture of fact and fancy. Pythagoras, born about 590 B.C., avidly studied numbers and geometric figures and discovered many of their important properties.

Other beliefs were less scientific. For example, because there were seven known planets, seven was a sacred number. To reflect the "music of the spheres," a musical scale should contain seven notes. Pythagoras also believed that the number 2 was masculine, that 3 was feminine, and that the notes and harmonies of music should be the "offspring" of the numbers 2 and 3.

Pythagoras used *fractions*—the topic of this chapter—along with his masculine and feminine numbers to create one of the earliest musical scales.

Whole numbers are used primarily for counting: 35 students, 15 minutes, and 500 shares of IBM stock. However, many situations require numbers that are not whole numbers: one-half of a dollar, two-thirds of a pizza, and four-fifths of a mile. Numbers such as one-half, two-thirds, and four-fifths are called **fractions**.

2.1 Fractions and Equivalent Fractions

- Fractions Can Indicate Parts of the Whole
- Fractions Can Indicate Division
- Equal Fractions
- Writing Fractions in Lowest Terms

Fractions Can Indicate Parts of the Whole

Fractions are numbers that represent parts of the whole. For example, there are four sticks of butter in one pound. A recipe that calls for one stick of butter is asking for one stick out of four, or for one-fourth of a pound of butter. The fraction one-fourth is written $\frac{1}{4}$. As another example, there are three feet in one yard. A package tied with two feet of string requires two-thirds, or $\frac{2}{3}$, of a yard of string.

The number above the fraction bar is called the **numerator**, and the number below the bar is called the **denominator**. In the fraction $\frac{5}{12}$, the numerator is 5 and the denominator is 12.

When a fraction is used to indicate a part of the whole, the denominator indicates the total number of equal parts into which the whole is divided. The numerator indicates the number of these equal parts that are being considered. For example, if an airplane has 100 available seats and 93 of them are occupied, the plane is $\frac{93}{100}$ (ninety-three one-hundredths) full.

EXAMPLE 1

In Figure 2-1, what fractional part of each figure is shaded?

(a) 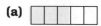 (b)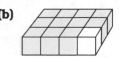

Figure 2-1

Solution

a. Because Figure 2-1**(a)** is divided into 5 equal parts, the denominator of the fraction is 5. Because we are interested in the 3 parts that are shaded, the numerator of the fraction is 3. Thus, $\frac{3}{5}$ of the figure is shaded.

b. Figure 2-1**(b)** is divided into 12 equal parts, so the denominator of the fraction is 12. Because we are interested in the 11 parts that are shaded, the numerator of the fraction is 11. Thus, $\frac{11}{12}$ (eleven-twelfths) of the figure is shaded. ∎

EXAMPLE 2

What fractional part of one inch does the circuit board in Figure 2-2 measure?

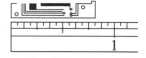

Figure 2-2

Solution

The markings on the ruler divide one inch into 16 equal parts. The circuit board covers 13 of these equal parts. The circuit board is $\frac{13}{16}$ of an inch long. Work Progress Check 1. ∎

Progress Check 1

In a carton of one dozen eggs, five were broken and unusable.

a. How many good eggs were in the carton?

b. What fractional part of the carton is usable?

c. What fractional part of the carton is unusable?

Fractions Can Indicate Division

Every fraction can be interpreted as a division. For example, the fraction $\frac{12}{12}$ indicates the division $12 \div 12$; thus $\frac{12}{12} = 1$. To understand why, we ask the question, *What fractional part of one dozen eggs is one dozen eggs?* We are considering all 12 of the 12 eggs that make up one dozen so 12 is the numerator of the fraction, 12 is the denominator, and we are considering $\frac{12}{12}$ of one dozen—we are considering 1 dozen. Thus, the fraction $\frac{12}{12}$ represents 1, and we can write

$$\frac{12}{12} = 12 \div 12 = 1$$

Likewise, if each of several donuts is cut in half, and Bill eats six halves, Bill has eaten $\frac{6}{2}$ of a donut. Because six half-donuts is really three donuts, we can see that

$$\frac{6}{2} = 6 \div 2 = 3$$

Because division by 0 is impossible, *the denominator of a fraction can never be zero.* To indicate that the denominator of the fraction $\frac{a}{b}$ is not equal to zero, we use the symbol ≠ (read "is not equal to"), and write $b \neq 0$.

1. **a.** 7 **b.** $\frac{7}{12}$ **c.** $\frac{5}{12}$

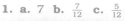

Definition: Any number written in the form $\frac{a}{b}$ and $b \neq 0$ is called a **fraction**. The **numerator** of the fraction is a, and the **denominator** is b.

Since $\frac{a}{b}$ can represent the division $a \div b$, we have the following two facts:

$$\frac{a}{1} = a \div 1 = a$$

$$\frac{a}{a} = a \div a = 1 \quad (a \neq 0)$$

Equal Fractions

Because there are four sticks of butter in one pound, a recipe that calls for two sticks is asking for $\frac{2}{4}$ of a pound of butter. Another recipe that calls for $\frac{1}{2}$ pound of butter is *also* asking for two sticks of butter. Although the fractions $\frac{2}{4}$ and $\frac{1}{2}$ look different, they represent equal amounts of butter, and we write $\frac{2}{4} = \frac{1}{2}$. Two fractions that represent the same number are called **equivalent fractions** or **equal fractions**.

As another example, we consider the identical rectangles in Figure 2-3. In Figure 2-3**(a)**, the shaded part is $\frac{6}{8}$ of the whole. In Figure 2-3**(b)**, it is $\frac{3}{4}$ of the whole. From the figures we see that the *same* fractional parts are shaded. Thus, the fractions $\frac{6}{8}$ and $\frac{3}{4}$ are equal, and we write $\frac{6}{8} = \frac{3}{4}$.

(a)
(b)

$\frac{6}{8}$

$\frac{3}{4}$

Figure 2-3

There is another way of deciding if two fractions are equal. Whenever the product of the numerator of the first fraction and the denominator of the second is equal to the product of the denominator of the first fraction and the numerator of the second, the fractions must be equal. Thus, $\frac{6}{8}$ is equal to $\frac{3}{4}$ because the products $6 \cdot 4$ and $8 \cdot 3$ are equal:

$6 \cdot 4 = 24$

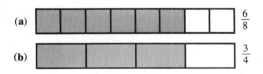

$3 \cdot 8 = 24$

This illustrates a process, called **cross multiplying**, for determining if two fractions are equal: The fractions $\frac{a}{b}$ and $\frac{c}{d}$ are equal if the products ad and bc are equal.

$a \cdot d$

$$\frac{a}{b} = \frac{c}{d}$$

$b \cdot c$

Definition: If a, b, c, and d represent whole numbers, then

$$\frac{a}{b} = \frac{c}{d} \text{ is equivalent to } ad = bc \quad (b \neq 0 \text{ and } d \neq 0)$$

EXAMPLE 3

Show that **a.** $\frac{21}{55} = \frac{105}{275}$ and that **b.** $\frac{3}{64} \neq \frac{5}{107}$.

Solution

a. To show that $\frac{21}{55} = \frac{105}{275}$, we verify that $21 \cdot 275$ is equal to $55 \cdot 105$.

$$21 \cdot 275 = 5775 \qquad 55 \cdot 105 = 5775$$

Because the products are equal, the two fractions are equal.

b. To show that $\frac{3}{64} \neq \frac{5}{107}$, we verify that $3 \cdot 107 \neq 64 \cdot 5$.

$$3 \cdot 107 = 321 \qquad 64 \cdot 5 = 320$$

Because the products are not equal, the fractions are not equal.

Work Progress Check 2.

Progress Check 2

Which two of these fractions are equal: $\frac{6}{10}$, $\frac{14}{22}$, and $\frac{12}{20}$?

Writing Fractions in Lowest Terms

A fraction is written in **lowest terms** if the greatest common divisor of the numerator and denominator is 1. To **simplify** or **reduce** a fraction, we write the fraction in lowest terms. For example, the fraction $\frac{6}{8}$ is not in lowest terms because 2 divides both 6 and 8. To simplify $\frac{6}{8}$, we divide its numerator and denominator by 2. This does not change the value of the fraction:

$$\frac{6}{8} = \frac{6 \div 2}{8 \div 2} = \frac{3}{4}$$

The fraction $\frac{3}{4}$ is written in lowest terms because no number greater than 1 divides both 3 and 4.

The fraction $\frac{15}{18}$ is not in lowest terms but can be simplified by dividing its numerator and denominator by 3:

$$\frac{15}{18} = \frac{15 \div 3}{18 \div 3} = \frac{5}{6}$$

The fraction $\frac{5}{6}$ is in lowest terms because 1 is the greatest common divisor of 5 and 6. Note that $\frac{15}{18} = \frac{5}{6}$ because $15 \cdot 6 = 5 \cdot 18$.

We can take a different approach to simplifying fractions: we write both the numerator and the denominator in factored form, and then divide both numerator and denominator by the factors that appear in both the numerator and denominator. For example, we simplify the fractions $\frac{6}{8}$ and $\frac{15}{18}$ as follows:

$$\frac{6}{8} = \frac{3 \cdot 2}{4 \cdot 2} = \frac{3 \cdot \cancel{2}}{4 \cdot \cancel{2}} = \frac{3}{4} \quad \text{and} \quad \frac{15}{18} = \frac{5 \cdot 3}{6 \cdot 3} = \frac{5 \cdot \cancel{3}}{6 \cdot \cancel{3}} = \frac{5}{6}$$

Progress Check Answers

2. $\frac{6}{10}$ and $\frac{12}{20}$

This discussion suggests a fact known as the **fundamental property of fractions**:

If any factors common to the numerator and the denominator of a fraction are divided out, the value of the fraction is not changed.

> **The Fundamental Property of Fractions:** If $\frac{a}{b}$ is any fraction, $b \neq 0$, and $x \neq 0$, then
> $$\frac{a \cdot x}{b \cdot x} = \frac{a}{b}$$

EXAMPLE 4

a. To simplify the fraction $\frac{6}{30}$, we factor the numerator and denominator and use the fundamental property of fractions to divide out the common factors of 2 and 3:

$$\frac{6}{30} = \frac{2 \cdot 3}{2 \cdot 3 \cdot 5} = \frac{\overset{1}{\cancel{2}} \cdot \overset{1}{\cancel{3}}}{\underset{1}{\cancel{2}} \cdot \underset{1}{\cancel{3}} \cdot 5} = \frac{1}{5}$$

b. To show that $\frac{33}{40}$ is written in lowest terms, we show that the numerator and denominator share no common factors. To do so, we write the numerator and denominator in prime-factored form:

$$\frac{33}{40} = \frac{3 \cdot 11}{2 \cdot 2 \cdot 2 \cdot 5}$$

Because there are no factors common to the numerator and denominator, the fraction $\frac{33}{40}$ is in lowest terms.

Work Progress Check 3.

Progress Check 3

Simplify $\frac{6}{10}$, $\frac{14}{22}$, and $\frac{12}{20}$ and determine which two fractions are equal.

---

3. $\frac{6}{10}$ and $\frac{12}{20}$ both simplify to $\frac{3}{5}$

Progress Check Answers

2.1 Exercises

In Exercises 1–12, determine what fractional part of each figure is shaded. Write the answer in lowest terms.

1.

2.

3.

4.

5.

6.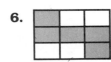

7.

8.

9.   **10.**

11.   **12.**

13. What fractional part of one foot is 5 inches? (There are 12 inches in one foot.)

14. What fractional part of one gallon is one pint? (There are 2 pints in one quart, and 4 quarts in one gallon.)

15. What fractional part of a 24-hour day is 11 hours?

16. What fractional part of January is one week? (January has 31 days.)

17. What fractional part of one pound of butter is three sticks?

18. What fractional part of a leap year is February? (In a leap year, February has one extra day.)

19. What fractional part of one mile is 100 feet? (There are 5280 feet in one mile.)

20. What fractional part of a standard deck of cards are aces?

21. What fractional part of a standard deck of cards are clubs?

22. What fractional part of one mile is one yard? (There are three feet in one yard.)

In Exercises 23–30, perform each division.

23. $\dfrac{12}{6}$ **24.** $\dfrac{18}{9}$ **25.** $\dfrac{163}{163}$ **26.** $\dfrac{72}{8}$

27. $\dfrac{225}{25}$ **28.** $\dfrac{629}{17}$ **29.** $\dfrac{1207}{71}$ **30.** $\dfrac{1248}{48}$

In Exercises 31–38, determine whether the given fractions are equal.

31. $\dfrac{3}{12} \stackrel{?}{=} \dfrac{1}{4}$ **32.** $\dfrac{3}{9} \stackrel{?}{=} \dfrac{1}{3}$ **33.** $\dfrac{3}{15} \stackrel{?}{=} \dfrac{4}{16}$ **34.** $\dfrac{3}{15} \stackrel{?}{=} \dfrac{1}{7}$

35. $\dfrac{5}{20} \stackrel{?}{=} \dfrac{3}{11}$ **36.** $\dfrac{1}{9} \stackrel{?}{=} \dfrac{3}{27}$ **37.** $\dfrac{10}{16} \stackrel{?}{=} \dfrac{15}{24}$ **38.** $\dfrac{27}{35} \stackrel{?}{=} \dfrac{32}{36}$

In Exercises 39–54, write each fraction in lowest terms. If the fraction is already in lowest terms, so indicate.

39. $\dfrac{4}{16}$ **40.** $\dfrac{7}{21}$ **41.** $\dfrac{15}{20}$ **42.** $\dfrac{22}{77}$

43. $\dfrac{72}{64}$ **44.** $\dfrac{26}{21}$ **45.** $\dfrac{45}{49}$ **46.** $\dfrac{45}{30}$

47. $\dfrac{330}{546}$ **48.** $\dfrac{220}{330}$ **49.** $\dfrac{65}{77}$ **50.** $\dfrac{70}{28}$

51. $\dfrac{275}{195}$ **52.** $\dfrac{325}{169}$ **53.** $\dfrac{650}{1300}$ **54.** $\dfrac{175}{441}$

In Exercises 55–58, refer to the diagram of theater seats in Illustration 1 and find the fractional part requested. Simplify the answer, if possible.

55. What fractional part of the whole theater are the orchestra seats?

56. What fractional part of the whole theater are the mezzanine seats?

57. What fractional part of the whole theater are the mezzanine seats to the left of the aisle?

58. What fractional part of the orchestra seats are the seats in the first row of the orchestra?

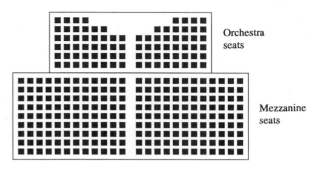

Illustration 1

59. Six pizzas are divided evenly among eight people. What part of a pizza does each person get? Express the answer in lowest terms.

60. Fourteen students drank four gallons of cola. How much cola did each person consume if each drank the same amount? Express the answer in lowest terms.

61. The $266 cost of gasoline for a trip to Florida is shared equally by seven friends. What fractional part of the cost does each person bear? Express the answer as a fraction in lowest terms.

62. In Exercise 61, how much does the trip cost each person?

Review Exercises

In Review Exercises 1–6, find the prime factorization of each number.

1. 315
2. 316
3. 3060
4. 2250
5. 22,500
6. 49,500

In Review Exercises 7–12, perform the indicated operations.

7. $347 \cdot 96 =$ _____

8. $723 \times 51 =$ _____

9. $5529 \div 57 =$ _____

10. $\dfrac{1887}{37} =$ _____

11. $4888 \div 13 =$ _____

12. $\dfrac{4692}{17} =$ _____

2.2 Multiplying and Dividing Fractions

- Multiplying Fractions
- Fractional Parts of a Quantity
- Dividing Fractions

Multiplying Fractions

Fractions can be added, subtracted, multiplied, and divided. To multiply fractions, we multiply their numerators and multiply their denominators. For example, we multiply $\frac{2}{3}$ and $\frac{4}{7}$ as follows:

$$\frac{2}{3} \cdot \frac{4}{7} = \frac{2 \cdot 4}{3 \cdot 7}$$
$$= \frac{8}{21}$$

To justify this process, we consider the square in Figure 2-4. Each side of the square is 1 unit long. Because the area of a square is the product of its length and width, the square in the figure has an area of 1 square unit. If this square is divided into 3 equal parts vertically and 7 equal parts horizontally, it is divided into 21 equal parts, and each of these parts represents $\frac{1}{21}$ of the total area. The area of the shaded rectangle in the square is $\frac{8}{21}$ because it consists of 8 of these 21 parts. Since the shaded rectangle has a width w of $\frac{4}{7}$ and a length l of $\frac{2}{3}$, we have

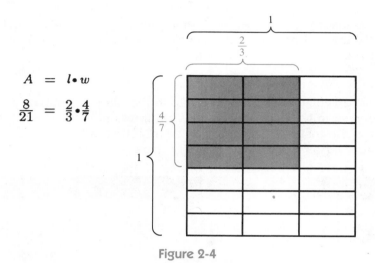

$$A = l \cdot w$$
$$\frac{8}{21} = \frac{2}{3} \cdot \frac{4}{7}$$

Figure 2-4

In general, the following is true:

Multiplying Fractions: The product of the fractions $\frac{a}{b}$ and $\frac{c}{d}$ is found by multiplying the numerators and multiplying the denominators:

$$\frac{a}{b} \cdot \frac{c}{d} = \frac{a \cdot c}{b \cdot d} \quad (b \neq 0 \text{ and } d \neq 0)$$

EXAMPLE 1

a. $\dfrac{3}{7} \cdot \dfrac{13}{5} = \dfrac{3 \cdot 13}{7 \cdot 5}$ Multiply the fractions. Since there are no common factors, the fraction does not simplify.

$\phantom{\dfrac{3}{7} \cdot \dfrac{13}{5}} = \dfrac{39}{35}$ Do the multiplications.

b. $\dfrac{3}{7} \cdot \dfrac{14}{5} = \dfrac{3 \cdot 14}{7 \cdot 5}$ Multiply the numerators and multiply the denominators.

$\phantom{\dfrac{3}{7} \cdot \dfrac{14}{5}} = \dfrac{3 \cdot 2 \cdot 7}{7 \cdot 5}$ To simplify the fraction, factor the numerator.

$\phantom{\dfrac{3}{7} \cdot \dfrac{14}{5}} = \dfrac{3 \cdot 2 \cdot \cancel{7}^{1}}{\cancel{7}_{1} \cdot 5}$ Use the fundamental property of fractions to divide out the common factor of 7.

$\phantom{\dfrac{3}{7} \cdot \dfrac{14}{5}} = \dfrac{6}{5}$ Do the remaining multiplications.

c. $5 \cdot \dfrac{3}{15} = \dfrac{5}{1} \cdot \dfrac{3}{15}$ Write 5 as the fraction $\dfrac{5}{1}$.

$\phantom{5 \cdot \dfrac{3}{15}} = \dfrac{5 \cdot 3}{1 \cdot 15}$ Multiply the fractions.

$\phantom{5 \cdot \dfrac{3}{15}} = \dfrac{5 \cdot 3}{1 \cdot 5 \cdot 3}$ To simplify the fraction, factor the denominator.

$\phantom{5 \cdot \dfrac{3}{15}} = \dfrac{\cancel{5}^{1} \cdot \cancel{3}^{1}}{1 \cdot \cancel{5}_{1} \cdot \cancel{3}_{1}}$ Divide out the common factors.

$\phantom{5 \cdot \dfrac{3}{15}} = 1$

Work Progress Check 4. ■

Progress Check 4

Multiply and simplify: $\dfrac{5}{12} \cdot \dfrac{12}{10}$

Fractional Parts of a Quantity

Multiplication of fractions is used to determine the number of objects that appear in a fractional part of a larger group. When the word *of* follows the name of a fraction in a word problem, the operation of *multiplication* is indicated: For example, two-thirds *of* 51 means $\dfrac{2}{3}$ *times* 51:

$$\text{Two-thirds of } 51 = \dfrac{2}{3} \cdot 51$$

$$= \dfrac{2}{3} \cdot \dfrac{51}{1}$$

$$= \dfrac{2 \cdot 51}{3 \cdot 1}$$

$$= \dfrac{2 \cdot \cancel{3} \cdot 17}{\cancel{3}}$$

$$= 34$$

Two-thirds of 51 is 34.

EXAMPLE 2

A recipe that will make barbecue sauce to serve 200 people calls for 5 gallons of tomato sauce. If the cook cuts back on the recipe to make enough to serve 80, how much tomato sauce should he use?

4. $\dfrac{1}{2}$

Progress Check Answers

Solution

Because the modified recipe is to serve only 80 out of a possible 200, the cook will use $\frac{80}{200}$, or $\frac{2}{5}$, of the 5 gallons of tomato sauce.

$$\frac{2}{5} \text{ of } 5 = \frac{2}{5} \cdot 5$$
$$= \frac{2}{5} \cdot \frac{5}{1}$$
$$= \frac{2 \cdot 5}{5}$$
$$= \frac{2 \cdot \cancel{5}^1}{\cancel{5}_1}$$
$$= 2$$

To prepare 80 servings, the cook will need 2 gallons of tomato sauce. Work Progress Check 5.

Progress Check 5

A doctor warned Jim that during his first week of exercise, his pulse rate should not exceed $\frac{2}{3}$ of his target rate of 120. What should Jim's pulse rate be that first week?

Dividing Fractions

The product of the numbers 2 and $\frac{1}{2}$ is 1, because $2 \cdot \frac{1}{2} = 1$. Likewise, the product of $\frac{3}{5}$ and $\frac{5}{3}$ is 1, because

$$\frac{3}{5} \cdot \frac{5}{3} = \frac{15}{15}$$
$$= 1$$

The numbers 2 and $\frac{1}{2}$ are called **reciprocals**, as are the numbers $\frac{3}{5}$ and $\frac{5}{3}$.

> **Definition:** Two numbers are called **reciprocals** if their product is 1.

To find the reciprocal of a number, we write the number as a fraction and *invert* the fraction—that is, the fraction's numerator and denominator trade places.

EXAMPLE 3

Find the reciprocal of **a.** $\frac{4}{7}$, and **b.** 5, and verify the answer.

Solution

a. To find the reciprocal of $\frac{4}{7}$, we invert the fraction $\frac{4}{7}$ to obtain $\frac{7}{4}$. To see that $\frac{4}{7}$ and $\frac{7}{4}$ are reciprocals, we verify that their product is 1:

$$\frac{4}{7} \cdot \frac{7}{4} = \frac{4 \cdot 7}{7 \cdot 4} = 1$$

Progress Check Answers

5. 80

b. To find the reciprocal of 5, we write 5 as the fraction $\frac{5}{1}$ and invert $\frac{5}{1}$ to get $\frac{1}{5}$. To see that $\frac{1}{5}$ is the reciprocal of 5, we verify that the product of 5 and $\frac{1}{5}$ is 1:

$$5 \cdot \frac{1}{5} = \frac{5}{1} \cdot \frac{1}{5} = \frac{5}{5} = 1$$

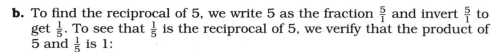

Work Progress Check 6.

Find the reciprocal of $\frac{12}{24}$ and simplify.

To *divide* one fraction by a second, we multiply the first fraction by the reciprocal of the second. Thus, to divide $\frac{7}{8}$ by $\frac{5}{11}$, for example, we multiply $\frac{7}{8}$ by the reciprocal of $\frac{5}{11}$:

$$\frac{7}{8} \div \frac{5}{11} = \frac{7}{8} \cdot \frac{11}{5} \qquad \text{Multiply } \frac{7}{8} \text{ by the reciprocal of } \frac{5}{11}.$$

$$= \frac{7 \cdot 11}{8 \cdot 5}$$

$$= \frac{77}{40}$$

To justify this process, we note that

$$\frac{7}{8} \div \frac{5}{11} \qquad \text{can be written as} \qquad \frac{\frac{7}{8}}{\frac{5}{11}}$$

and then use the fundamental property of fractions:

$$\frac{7}{8} \div \frac{5}{11} = \frac{\frac{7}{8}}{\frac{5}{11}}$$

$$= \frac{\frac{7}{8} \cdot \frac{11}{5}}{\frac{5}{11} \cdot \frac{11}{5}} \qquad \text{Multiply both the numerator and denominator by } \frac{11}{5}.$$

$$= \frac{\frac{7}{8} \cdot \frac{11}{5}}{1} \qquad \text{Do the multiplication in the denominator and simplify.}$$

$$= \frac{7}{8} \cdot \frac{11}{5}$$

Thus, $\frac{7}{8} \div \frac{5}{11}$ can be written as $\frac{7}{8} \cdot \frac{11}{5}$. In general, we have

---

Dividing Fractions: To perform the division $\frac{a}{b} \div \frac{c}{d}$, multiply the first fraction by the reciprocal of the second:

$$\frac{a}{b} \div \frac{c}{d} = \frac{a}{b} \cdot \frac{d}{c} = \frac{a \cdot d}{b \cdot c} \qquad (b \neq 0, c \neq 0, \text{ and } d \neq 0)$$

---

6. 2

Progress Check Answers

EXAMPLE 4

a. $\dfrac{3}{5} \div \dfrac{6}{5} = \dfrac{3}{5} \cdot \dfrac{5}{6}$ Multiply $\dfrac{3}{5}$ by the reciprocal of $\dfrac{6}{5}$.

$\phantom{\dfrac{3}{5} \div \dfrac{6}{5}} = \dfrac{3 \cdot 5}{5 \cdot 6}$ Multiply the fractions.

$\phantom{\dfrac{3}{5} \div \dfrac{6}{5}} = \dfrac{3 \cdot 5}{5 \cdot 2 \cdot 3}$ Factor the denominator.

$\phantom{\dfrac{3}{5} \div \dfrac{6}{5}} = \dfrac{\overset{1}{\cancel{3}} \cdot \overset{1}{\cancel{5}}}{\underset{1}{\cancel{5}} \cdot 2 \cdot \underset{1}{\cancel{3}}}$ Divide out the common factors.

$\phantom{\dfrac{3}{5} \div \dfrac{6}{5}} = \dfrac{1}{2}$

b. $\dfrac{7}{9} \div 5 = \dfrac{7}{9} \div \dfrac{5}{1}$ Write 5 as the fraction $\dfrac{5}{1}$.

$\phantom{\dfrac{7}{9} \div 5} = \dfrac{7}{9} \cdot \dfrac{1}{5}$ Multiply $\dfrac{7}{9}$ by the reciprocal of $\dfrac{5}{1}$.

$\phantom{\dfrac{7}{9} \div 5} = \dfrac{7}{45}$

c. $10 \div \dfrac{15}{7} \div \dfrac{1}{3} = \dfrac{10}{1} \div \dfrac{15}{7} \div \dfrac{1}{3}$ Write 10 as the fraction $\dfrac{10}{1}$.

$\phantom{10 \div \dfrac{15}{7} \div \dfrac{1}{3}} = \dfrac{10}{1} \cdot \dfrac{7}{15} \div \dfrac{1}{3}$ Proceed from left to right. Multiply $\dfrac{10}{1}$ by the reciprocal of $\dfrac{15}{7}$.

$\phantom{10 \div \dfrac{15}{7} \div \dfrac{1}{3}} = \dfrac{10}{1} \cdot \dfrac{7}{15} \cdot \dfrac{3}{1}$ Multiply by the reciprocal of $\dfrac{1}{3}$.

$\phantom{10 \div \dfrac{15}{7} \div \dfrac{1}{3}} = \dfrac{10 \cdot 7 \cdot 3}{1 \cdot 15 \cdot 1}$ Multiply the fractions by multiplying the numerators and multiplying the denominators.

$\phantom{10 \div \dfrac{15}{7} \div \dfrac{1}{3}} = \dfrac{2 \cdot 5 \cdot 7 \cdot 3}{3 \cdot 5}$ Factor.

$\phantom{10 \div \dfrac{15}{7} \div \dfrac{1}{3}} = \dfrac{2 \cdot \overset{1}{\cancel{5}} \cdot 7 \cdot \overset{1}{\cancel{3}}}{\underset{1}{\cancel{3}} \cdot \underset{1}{\cancel{5}}}$ Divide out the common factors.

$\phantom{10 \div \dfrac{15}{7} \div \dfrac{1}{3}} = 14$ Simplify.

Work Progress Check 7.

There is another method of dividing one fraction by another. First, we write the division problem in fractional form. Then we determine the least common multiple of the denominators of the two given fractions. Finally, we use the fundamental property of fractions: Multiply both the numerator and the denominator by that LCM and simplify.

EXAMPLE 5

Perform the division: $\dfrac{13}{34} \div \dfrac{10}{51}$.

Progress Check 7

Divide, and simplify:

a. $\dfrac{5}{3} \div 15$

b. $15 \div \dfrac{5}{3}$

---

7. **a.** $\dfrac{1}{9}$ **b.** 9

Progress Check Answers

Solution

Write the division in fractional form:

$$\frac{13}{34} \div \frac{10}{51} = \frac{\frac{13}{34}}{\frac{10}{51}}$$

Second, find the least common multiple of 34 and 51, the denominators of the two given fractions. Do so by using each prime factor of 34 and 51 the greatest number of times that it appears in either 34 or 51. Proceed as follows:

$$\frac{\frac{13}{34}}{\frac{10}{51}} = \frac{\frac{13}{2 \cdot 17}}{\frac{10}{3 \cdot 17}}$$ Factor the two denominators. Determine that their LCM is $2 \cdot 3 \cdot 17$.

$$= \frac{\frac{13}{2 \cdot 17} \cdot 2 \cdot 3 \cdot 17}{\frac{10}{3 \cdot 17} \cdot 2 \cdot 3 \cdot 17}$$ Multiply numerator and denominator by $2 \cdot 3 \cdot 17$.

$$= \frac{\frac{13 \cdot 2 \cdot 3 \cdot 17}{2 \cdot 17}}{\frac{10 \cdot 2 \cdot 3 \cdot 17}{3 \cdot 17}}$$ Multiply.

$$= \frac{\frac{13 \cdot \cancel{2} \cdot 3 \cdot \cancel{17}}{\cancel{2} \cdot \cancel{17}}}{\frac{10 \cdot 2 \cdot \cancel{3} \cdot \cancel{17}}{\cancel{3} \cdot \cancel{17}}}$$ Divide out common factors.

$$= \frac{13 \cdot 3}{10 \cdot 2}$$

$$= \frac{39}{20}$$

Thus, $\frac{13}{34} \div \frac{10}{51} = \frac{39}{20}$. Work Progress Check 8.

Progress Check 8

Divide and simplify: $\dfrac{\frac{12}{25}}{\frac{9}{10}}$

EXAMPLE 6

Sharon has driven 351 miles, and she estimates that she has completed three-quarters of the total trip. How long is her journey?

Solution

If x represents the total distance that Sharon will drive, then three-fourths of x is 351 miles. Remembering that *of* means *multiply*, we have

$$\frac{3}{4} \cdot x = 351$$

We write this multiplication in the form of the division $x = 351 \div \frac{3}{4}$, and determine x as follows:

8. $\frac{8}{15}$

$$x = 351 \div \frac{3}{4}$$

$$= \frac{351}{1} \div \frac{3}{4} \qquad \text{Write 351 as the fraction } \frac{351}{1}.$$

$$= \frac{351}{1} \cdot \frac{4}{3} \qquad \text{Multiply } \frac{351}{1} \text{ by the reciprocal of } \frac{3}{4}.$$

$$= \frac{351 \cdot 4}{1 \cdot 3} \qquad \text{Multiply the fractions.}$$

$$= \frac{\overset{1}{\cancel{3}} \cdot 117 \cdot 4}{1 \cdot \underset{1}{\cancel{3}}} \qquad \text{Factor 3 from 351 and simplify.}$$

$$= 468 \qquad \text{Simplify.}$$

Sharon's total trip is 468 miles. ■

2.2 Exercises

In Exercises 1–16, perform each multiplication. Simplify answers, if possible.

1. $\frac{1}{2} \cdot \frac{3}{5}$
2. $\frac{3}{4} \cdot \frac{5}{7}$
3. $\frac{5}{9} \cdot \frac{4}{11}$
4. $\frac{7}{10} \cdot \frac{3}{10}$

5. $\frac{4}{3} \cdot \frac{6}{5}$
6. $\frac{7}{8} \cdot \frac{6}{15}$
7. $\frac{5}{12} \cdot \frac{18}{5}$
8. $\frac{5}{4} \cdot \frac{12}{10}$

9. $\frac{21}{30} \cdot \frac{10}{7}$
10. $\frac{15}{16} \cdot \frac{8}{9}$
11. $12 \cdot \frac{5}{6}$
12. $9 \cdot \frac{7}{12}$

13. $\frac{10}{21} \cdot 14$
14. $\frac{7}{63} \cdot 18$
15. $\frac{5}{24} \cdot 16$
16. $21 \cdot \frac{3}{14}$

In Exercises 17–24, find the reciprocal of each number.

17. $\frac{3}{5}$
18. $\frac{13}{7}$
19. $\frac{1}{11}$
20. 7

21. 19
22. $\frac{34}{17}$
23. $\frac{1}{23}$
24. 102

In Exercises 25–40, perform each division. Simplify answers, if possible.

25. $\frac{3}{5} \div \frac{2}{3}$
26. $\frac{4}{5} \div \frac{3}{7}$
27. $\frac{4}{7} \div \frac{3}{11}$
28. $\frac{2}{11} \div \frac{7}{10}$

29. $\frac{3}{4} \div \frac{6}{5}$
30. $\frac{3}{8} \div \frac{15}{28}$
31. $\frac{2}{13} \div \frac{8}{13}$
32. $\frac{4}{7} \div \frac{20}{21}$

33. $\frac{21}{35} \div \frac{3}{14}$
34. $\frac{23}{25} \div \frac{46}{5}$
35. $\frac{42}{30} \div \frac{21}{15}$
36. $\frac{34}{8} \div \frac{17}{4}$

37. $6 \div \frac{3}{14}$
38. $23 \div \frac{46}{5}$
39. $\frac{42}{30} \div 7$
40. $\frac{34}{8} \div 17$

In Exercises 41–60, evaluate each expression. Simplify answers, if possible.

41. $\dfrac{3}{5} \cdot \dfrac{10}{3}$ **42.** $\dfrac{4}{9} \div \dfrac{2}{3}$ **43.** $\dfrac{3}{22} \div \dfrac{6}{33}$ **44.** $\dfrac{2}{13} \cdot \dfrac{39}{10}$

45. $\dfrac{3}{4} \cdot 8$ **46.** $\dfrac{3}{15} \div 15$ **47.** $8 \div \dfrac{4}{5}$ **48.** $15 \div \dfrac{30}{7}$

49. $\dfrac{3}{2} \cdot \dfrac{4}{5} \div \dfrac{1}{3}$ **50.** $\dfrac{3}{5} \div \dfrac{15}{5} \cdot \dfrac{1}{5}$ **51.** $\dfrac{3}{2} \div \dfrac{4}{5} \div \dfrac{1}{3}$ **52.** $\dfrac{3}{12} \div \dfrac{6}{8} \cdot \dfrac{1}{2}$

53. $\dfrac{3}{5} \cdot \dfrac{7}{5} \div \dfrac{21}{25}$ **54.** $\dfrac{3}{5} \div \dfrac{7}{5} \cdot \dfrac{21}{25}$ **55.** $\dfrac{3}{13} \cdot \dfrac{26}{9} \cdot \dfrac{1}{3}$ **56.** $\dfrac{13}{7} \cdot \dfrac{14}{39} \div \dfrac{3}{2}$

57. $\left(\dfrac{3}{4}\right)^2$ **58.** $\left(\dfrac{2}{3}\right)^3$ **59.** $\left(\dfrac{3}{5} \cdot \dfrac{5}{3}\right) \div \dfrac{5}{3}$ **60.** $\left(\dfrac{3}{5} \div \dfrac{5}{3}\right) \cdot \dfrac{5}{3}$

61. The area of a rectangle is the product of its length and width. Find the area of the rectangle in Illustration 1.

62. The volume of a box is the product of its length, width, and depth. Find the volume of the box in Illustration 2.

Illustration 1

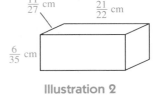

Illustration 2

63. Each vitamin pill contains $\dfrac{2}{7}$ of the daily requirements of vitamin C. What part of the daily requirement will three pills contain?

64. Jim and seven of his friends know they can eat $\dfrac{2}{3}$ of a pizza each. If they order six pizzas, will each person get his share?

65. Each capsule contains $\dfrac{2}{3}$ gram of antibiotic. How many grams of antibiotic are contained in a bottle of 150 capsules?

66. Mount Everest in the Himalayas is the world's highest mountain peak, towering 29,000 feet. Alaska's Mount McKinley is only $\dfrac{7}{10}$ as high. How high is Mount McKinley?

67. If $\dfrac{2}{5}$ of Jim's $3600 tuition bill was paid by a scholarship, how much did Jim pay himself?

68. Out of every 100 students attending private colleges, 64 receive financial aid. Approximately 2,800,000 students attend private schools. How many receive aid?

69. Outfielder Hank Aaron, the all-time leader in home runs and RBIs, got a base hit $\dfrac{61}{200}$ of the times he was at bat. Hank batted 12,364 times in his career. How many were hits?

70. In a shipment of 136 radios, $\dfrac{1}{8}$ were defective. Find the number of defective radios.

71. In a shipment of 473 toasters, $\dfrac{1}{11}$ were defective. Find the number of good toasters.

72. One-seventh of 2625 graduates majored in Computer Science. How many graduated with that major?

73. How many packages containing $\dfrac{5}{8}$ ounce of candy can be made from 45 ounces of candy?

74. A scholarship paid $\dfrac{3}{5}$ of Mia's $5700 tuition bill, and books cost an additional $132. If Mia's parents paid $\dfrac{1}{2}$ of the cost of the books, how much remained for Mia to pay?

75. How many $\frac{3}{8}$-inch-thick slices can be cut from a 24-inch sausage?

76. A bucket is $\frac{3}{5}$ full when it contains 6 gallons. How much water can the bucket hold?

77. A gasoline tank is $\frac{5}{6}$ full when it contains 100 gallons. How much gasoline can the tank hold?

78. A scholarship for $3200 paid $\frac{2}{3}$ of Pedro's tuition. Find the total tuition.

79. To the nearest hundred thousand, Sequoia National Park in California covers 400,000 acres, about $\frac{9}{50}$ of the area of Yellowstone, the largest of our national parks. To the nearest hundred thousand acres, what is the area of Yellowstone?

80. Betty's brother ate $\frac{1}{4}$ of her candy. If 51 pieces of candy remained, how many pieces did her brother eat?

Review Exercises

1. George has 87 shares of stock, but he sells 19. If he then decides he wants a round lot of 100 shares, how many more shares must he buy?

2. There are about 73 apples in one bushel. How many apples are in 17 bushels?

3. To raise money, the Pep Club is holding a raffle. The total cost of the prizes is $1750. How many tickets must be sold at $5 each to break even?

4. Bicycle riding will burn 570 calories per hour. How many hours of bicycle riding will it take to burn 3990 calories?

In Review Exercises 5–8, find the least common multiple of the given numbers.

5. 15, 35   **6.** 50, 105   **7.** 14, 18, 63   **8.** 14, 21, 42

In Review Exercises 9–12, find the greatest common divisor of the given numbers.

9. 15, 35   **10.** 50, 105   **11.** 14, 18, 63   **12.** 14, 21, 42

2.3 Adding and Subtracting Fractions

- Adding Fractions with Like Denominators
- Adding Fractions with Unlike Denominators
- Subtracting Fractions with Like Denominators
- Subtracting Fractions with Unlike Denominators

Adding Fractions with Like Denominators

Addition of whole numbers is used to find the total number of objects when groups of *similar* objects are combined. For example,

3 apples + 2 apples = 5 apples

and

3 pencils + 2 pencils = 5 pencils

In a similar way, it is possible to add fractions. For example,

3 eighths + 2 eighths = 5 eighths

and

3 elevenths + 2 elevenths = 5 elevenths

We would write the addition of these fractions symbolically as

$$\frac{3}{8} + \frac{2}{8} = \frac{5}{8} \quad \text{and} \quad \frac{3}{11} + \frac{2}{11} = \frac{5}{11}$$

In each of these examples, we add the numerators and keep the common denominator. This suggests the rule for adding fractions with the *same denominator*:

Adding Fractions with Like Denominators: To add fractions with the same denominator, add the numerators and keep the common denominator.

$$\frac{a}{d} + \frac{b}{d} = \frac{a+b}{d} \quad (d \neq 0)$$

EXAMPLE 1

Add: **a.** $\frac{7}{13} + \frac{3}{13}$ and **b.** $\frac{7}{15} + \frac{8}{15}$.

Solution

a. Because the fractions $\frac{7}{13}$ and $\frac{3}{13}$ have the same denominator, 13, we add their numerators and keep their common denominator:

$$\frac{7}{13} + \frac{3}{13} = \frac{7+3}{13}$$
$$= \frac{10}{13}$$

b. The fractions $\frac{7}{15}$ and $\frac{8}{15}$ share the common denominator, 15. Add the fractions by adding the numerators and using that denominator.

$$\frac{7}{15} + \frac{8}{15} = \frac{7+8}{15}$$
$$= \frac{15}{15}$$
$$= 1 \qquad \text{Simplify the fraction } \frac{15}{15}.$$

Work Progress Check 9.

Progress Check 9

Add: $\frac{3}{13} + \frac{4}{13} + \frac{5}{13}$.

Adding Fractions with Unlike Denominators

To add fractions with unlike denominators, we use the fundamental property of fractions to write them as equivalent fractions with the *same* denominator.

EXAMPLE 2

Write the fraction $\frac{5}{7}$ as an equivalent fraction with a denominator of 21.

9. $\frac{12}{13}$

Progress Check Answers

Solution

To change the denominator 7 into the denominator 21, we must multiply it by 3. To change the given fraction into an equal fraction with a denominator 21, we use the fundamental property of fractions and multiply both the numerator *and* the denominator by 3:

$$\frac{5}{7} = \frac{5 \cdot 3}{7 \cdot 3} = \frac{15}{21}$$

The fraction $\frac{15}{21}$ has the required denominator, and it is equal to $\frac{5}{7}$. Work Progress Check 10. ∎

To add $\frac{3}{5}$ and $\frac{2}{3}$, we write each fraction as an equivalent fraction having a denominator of 15 and then add the results:

$$\frac{3}{5} + \frac{2}{3} = \frac{3 \cdot 3}{5 \cdot 3} + \frac{2 \cdot 5}{3 \cdot 5}$$

$$= \frac{9}{15} + \frac{10}{15}$$

$$= \frac{9 + 10}{15}$$

$$= \frac{19}{15}$$

Because 15 is the smallest number that can be a common denominator of the fractions, it is called the **least** or **lowest common denominator**, or the **LCD**. The lowest common denominator is the least common multiple of the fractions' denominators.

EXAMPLE 3

Add: $\frac{3}{10} + \frac{5}{14}$.

Solution

We write the fractions as fractions having the same denominator. To find the lowest common denominator (LCD), we find the least common multiple of the denominators 10 and 14. Do so by factoring 10 and 14 and using each prime factor the greatest number of times it appears in either factorization.

$$10 = 2 \cdot 5$$
$$14 = 2 \cdot 7$$
$$\text{LCD} = 2 \cdot 5 \cdot 7 = 70$$

Because 70 is the smallest number that the denominators 10 and 14 divide exactly, write both fractions as fractions with the least common denominator 70.

Progress Check 10

Write $\frac{3}{13}$ as an equivalent fraction with a denominator of 39.

Progress Check Answers

10. $\frac{9}{39}$

$$\frac{3}{10} + \frac{5}{14} = \frac{3 \cdot 7}{10 \cdot 7} + \frac{5 \cdot 5}{14 \cdot 5}$$ Write each fraction as a fraction with a denomimator of 70.

$$= \frac{21}{70} + \frac{25}{70}$$ Add the fractions.

$$= \frac{46}{70}$$

$$= \frac{2 \cdot 23}{2 \cdot 5 \cdot 7}$$ To simplify the fraction, factor both numerator and denominator.

$$= \frac{\overset{1}{\cancel{2}} \cdot 23}{\underset{1}{\cancel{2}} \cdot 5 \cdot 7}$$ Divide out the common factor.

$$= \frac{23}{35}$$

Work Progress Check 11.

Progress Check 11

What is the lowest common denominator needed to add the fractions $\frac{5}{6}$ and $\frac{2}{9}$?

EXAMPLE 4

Add: $\frac{5}{6} + \frac{3}{5} + \frac{2}{15}$.

Solution

We find the lowest common denominator of the three fractions. It is 30, the least common multiple of 6, 5, and 15. We write each fraction as an equivalent fraction with a denominator of 30.

$$\frac{5}{6} + \frac{3}{5} + \frac{2}{15} = \frac{5 \cdot 5}{6 \cdot 5} + \frac{3 \cdot 6}{5 \cdot 6} + \frac{2 \cdot 2}{15 \cdot 2}$$

$$= \frac{25}{30} + \frac{18}{30} + \frac{4}{30}$$

$$= \frac{25 + 18 + 4}{30}$$

$$= \frac{47}{30}$$

Work Progress Check 12.

Progress Check 12

Add: $\frac{5}{6} + \frac{1}{2} + \frac{2}{9} + \frac{5}{18}$. Simplify the answer.

Subtracting Fractions with Like Denominators

To *subtract* two fractions with the *same* denominator, we subtract their numerators and keep their common denominator. For example,

$$\frac{7}{9} - \frac{2}{9} = \frac{7-2}{9} = \frac{5}{9}$$

11. 18
12. $\frac{11}{6}$

Progress Check Answers

> **Subtracting Fractions with Like Denominators:** To subtract fractions with the same denominator, subtract the numerators and keep the common denominator.
> $$\frac{a}{d} - \frac{b}{d} = \frac{a-b}{d} \qquad (d \neq 0)$$

EXAMPLE 5

Subtract: $\frac{9}{17} - \frac{5}{17}$.

Solution

Since the fractions $\frac{9}{17}$ and $\frac{5}{17}$ have the same denominator, 17, we subtract their numerators and keep their common denominator:

$$\frac{9}{17} - \frac{5}{17} = \frac{9-5}{17}$$
$$= \frac{4}{17}$$

Subtracting Fractions with Unlike Denominators

To subtract two fractions with unlike denominators, we write them as equivalent fractions having a common denominator. For example, to subtract $\frac{2}{5}$ from $\frac{3}{4}$, both fractions must be written as fractions with the same denominator. Because 20 is the smallest number that is divisible by both 5 and 4 exactly, the least common denominator is 20. To subtract $\frac{2}{5}$ from $\frac{3}{4}$, we write $\frac{3}{4} - \frac{2}{5}$ and proceed as follows:

$$\frac{3}{4} - \frac{2}{5} = \frac{3 \cdot 5}{4 \cdot 5} - \frac{2 \cdot 4}{5 \cdot 4}$$
$$= \frac{15}{20} - \frac{8}{20}$$
$$= \frac{15-8}{20}$$
$$= \frac{7}{20}$$

EXAMPLE 6

Subtract 5 from $\frac{23}{3}$.

Solution

The subtraction indicated by

$$\frac{23}{3} - 5$$

represents the difference of two fractions, because 5 can be written as $\frac{5}{1}$. Thus,

$$\frac{23}{3} - 5 = \frac{23}{3} - \frac{5}{1}$$

$$= \frac{23}{3} - \frac{5 \cdot 3}{1 \cdot 3} \quad \text{Write } \frac{5}{1} \text{ as a fraction with denominator 3.}$$

$$= \frac{23}{3} - \frac{15}{3}$$

$$= \frac{23 - 15}{3} \quad \text{Subtract the numerators and use the common denominator.}$$

$$= \frac{8}{3}$$

EXAMPLE 7

If $\frac{3}{5} + x = \frac{7}{3}$, find x.

Solution

The addition statement $\frac{3}{5} + x = \frac{7}{3}$ is equivalent to the subtraction statement

$$x = \frac{7}{3} - \frac{3}{5}$$

We perform the subtraction to determine x.

$$x = \frac{7}{3} - \frac{3}{5}$$

$$= \frac{7 \cdot 5}{3 \cdot 5} - \frac{3 \cdot 3}{5 \cdot 3} \quad \text{Change each fraction into an equivalent fraction with lowest common denominator of 15.}$$

$$= \frac{35}{15} - \frac{9}{15}$$

$$= \frac{35 - 9}{15} \quad \text{Subtract the numerators, and use the common denominator.}$$

$$= \frac{26}{15}$$

Thus, $x = \frac{26}{15}$. Work Progress Check 13.

Progress Check 13

If $\frac{4}{7} + x = 7$, find x.

EXAMPLE 8

Use the distributive property to evaluate $4\left(\frac{5}{8} - \frac{1}{4}\right)$.

13. $\frac{45}{7}$

Progress Check Answers

Section 2.3 Adding and Subtracting Fractions

Solution

$$4\left(\frac{5}{8} - \frac{1}{4}\right) = 4 \cdot \frac{5}{8} - 4 \cdot \frac{1}{4}$$ Distribute the multiplication over the addition.

$$= \frac{4}{1} \cdot \frac{5}{8} - \frac{4}{1} \cdot \frac{1}{4}$$

$$= \frac{4 \cdot 5}{1 \cdot 8} - \frac{4 \cdot 1}{1 \cdot 4}$$ Multiply the fractions.

$$= \frac{5}{2} - 1$$ Simplify.

$$= \frac{5}{2} - \frac{2}{2}$$ Write the fractions using the lowest common denominator.

$$= \frac{3}{2}$$ Subtract.

2.3 Exercises

In Exercises 1–8, perform each addition. Simplify answers, if possible.

1. $\frac{3}{5} + \frac{3}{5}$
2. $\frac{4}{7} + \frac{2}{7}$
3. $\frac{4}{7} + \frac{3}{7}$
4. $\frac{2}{11} + \frac{9}{11}$
5. $\frac{23}{17} + \frac{11}{17}$
6. $\frac{23}{15} + \frac{7}{15}$
7. $\frac{9}{25} + \frac{6}{25}$
8. $\frac{3}{16} + \frac{5}{16}$

In Exercises 9–20, write each fraction as an equivalent fraction with the denominator indicated.

9. $\frac{1}{6} = \frac{?}{12}$
10. $\frac{7}{5} = \frac{?}{10}$
11. $\frac{3}{7} = \frac{?}{21}$
12. $\frac{9}{11} = \frac{?}{33}$
13. $\frac{1}{15} = \frac{?}{45}$
14. $\frac{2}{9} = \frac{?}{81}$
15. $\frac{7}{13} = \frac{?}{39}$
16. $\frac{7}{19} = \frac{?}{76}$
17. $5 = \frac{?}{12}$
18. $\frac{1}{24} = \frac{?}{48}$
19. $15 = \frac{?}{3}$
20. $\frac{3}{35} = \frac{?}{105}$

In Exercises 21–40, perform each addition. Simplify answers, if possible.

21. $\frac{3}{5} + \frac{2}{3}$
22. $\frac{4}{3} + \frac{7}{2}$
23. $\frac{9}{4} + \frac{5}{6}$
24. $\frac{2}{15} + \frac{7}{9}$
25. $\frac{7}{10} + \frac{1}{4}$
26. $\frac{7}{25} + \frac{3}{10}$
27. $\frac{5}{14} + \frac{4}{21}$
28. $\frac{2}{33} + \frac{3}{22}$
29. $\frac{3}{11} + \frac{6}{5}$
30. $\frac{7}{5} + \frac{2}{15}$
31. $\frac{5}{6} + \frac{1}{10}$
32. $\frac{3}{8} + \frac{2}{9}$
33. $3 + \frac{3}{4}$
34. $6 + \frac{21}{5}$
35. $\frac{17}{3} + 4$
36. $\frac{13}{9} + 1$
37. $\frac{1}{3} + \frac{1}{2} + \frac{1}{6}$
38. $\frac{3}{5} + \frac{2}{15} + \frac{2}{3}$
39. $\frac{3}{4} + \frac{5}{12} + \frac{1}{3}$
40. $\frac{3}{8} + \frac{1}{6} + \frac{2}{3}$

In Exercises 41–64, perform each subtraction. Simplify answers, if possible.

41. $\dfrac{2}{5} - \dfrac{2}{5}$

42. $\dfrac{5}{7} - \dfrac{2}{7}$

43. $\dfrac{9}{7} - \dfrac{5}{7}$

44. $\dfrac{8}{11} - \dfrac{5}{11}$

45. $\dfrac{16}{21} - \dfrac{7}{21}$

46. $\dfrac{13}{15} - \dfrac{8}{15}$

47. $\dfrac{4}{5} - \dfrac{2}{3}$

48. $\dfrac{5}{7} - \dfrac{1}{2}$

49. $\dfrac{9}{5} - \dfrac{5}{6}$

50. $\dfrac{13}{15} - \dfrac{1}{5}$

51. $\dfrac{7}{10} - \dfrac{1}{4}$

52. $\dfrac{9}{35} - \dfrac{1}{7}$

53. $\dfrac{9}{14} - \dfrac{4}{21}$

54. $\dfrac{2}{15} - \dfrac{1}{12}$

55. $\dfrac{5}{30} - \dfrac{2}{42}$

56. $\dfrac{7}{25} - \dfrac{2}{35}$

57. $\dfrac{5}{12} - \dfrac{1}{18}$

58. $\dfrac{3}{8} - \dfrac{1}{12}$

59. $3 - \dfrac{4}{5}$

60. $6 - \dfrac{7}{9}$

61. $\dfrac{17}{5} - 3$

62. $\dfrac{19}{7} - 2$

63. $35 - \dfrac{35}{2}$

64. $3 - \dfrac{13}{20}$

In Exercises 65–80, evaluate each expression. Simplify answers, if possible.

65. $\dfrac{7}{5} - \dfrac{1}{2}$

66. $\dfrac{9}{7} + \dfrac{2}{3}$

67. $\dfrac{14}{3} - \dfrac{5}{6}$

68. $\dfrac{8}{9} + \dfrac{2}{3}$

69. $\dfrac{13}{15} + \dfrac{4}{3}$

70. $\dfrac{7}{12} - \dfrac{1}{2}$

71. $\dfrac{19}{8} - 2$

72. $\dfrac{12}{7} + 3$

73. $7 + \dfrac{13}{5}$

74. $3 - \dfrac{7}{5}$

75. $\dfrac{9}{8} - \dfrac{1}{2} + \dfrac{1}{4}$

76. $\dfrac{15}{7} - \dfrac{5}{14} + \dfrac{3}{2}$

77. $3 + \dfrac{3}{2} - \dfrac{3}{4}$

78. $\dfrac{5}{4} + \dfrac{5}{8} - \dfrac{1}{2}$

79. $\dfrac{1}{2} + 3 - \dfrac{1}{3}$

80. $5 - \dfrac{3}{5} - \dfrac{3}{2}$

In Exercises 81–88, write each subtraction as an equivalent addition and determine x.

81. $x - \dfrac{2}{5} = \dfrac{3}{5}$

82. $x - \dfrac{2}{7} = \dfrac{5}{7}$

83. $x - \dfrac{3}{10} = \dfrac{2}{5}$

84. $x - \dfrac{5}{8} = \dfrac{3}{4}$

85. $x - \dfrac{5}{9} = \dfrac{5}{6}$

86. $x - \dfrac{2}{10} = \dfrac{4}{15}$

87. $x - \dfrac{2}{7} = 3$

88. $x - 2 = \dfrac{1}{3}$

In Exercises 89–96, write each addition as an equivalent subtraction and determine x.

89. $x + \dfrac{1}{5} = \dfrac{2}{5}$

90. $x + \dfrac{2}{7} = \dfrac{6}{7}$

91. $x + \dfrac{1}{12} = \dfrac{1}{6}$

92. $x + \dfrac{3}{8} = \dfrac{5}{4}$

93. $x + \frac{1}{6} = \frac{2}{9}$ _____

94. $x + \frac{3}{35} = \frac{9}{14}$ _____

95. $x + \frac{5}{7} = 4$ _____

96. $x + 2 = \frac{7}{3}$ _____

97. A rectangle is $\frac{3}{5}$ meter wide and $\frac{2}{3}$ meter long. Find its perimeter.

98. A rectangle is $\frac{2}{9}$ foot wide and $\frac{5}{6}$ foot long. Find its perimeter.

99. The edge of a square is $\frac{17}{5}$ centimeters. Find its perimeter.

100. The three sides of a triangle are $\frac{7}{8}$ inch, $\frac{1}{2}$ inch, and $\frac{3}{4}$ inch. Find its perimeter.

101. A birdfeed mixture contains $\frac{1}{2}$ bushel of sunflower seeds, $\frac{1}{5}$ bushel of safflower seeds, and $\frac{2}{3}$ bushel of millet. How much seed is in the mixture?

102. A fertilizer contains $\frac{1}{3}$ pound of potash, $\frac{1}{4}$ pound of phosphate, $\frac{5}{12}$ pound of nitrate, and $\frac{5}{6}$ pound of inert material. How much does the package of fertilizer weigh?

103. If an industrial accident releases more than one pound of the toxic chemical vinyl chloride into the environment, the Environmental Protection Agency must be notified. In one such accident, three bottles of vinyl chloride were broken, releasing $\frac{3}{5}$ pound, $\frac{1}{3}$ pound, and $\frac{1}{10}$ pound. Must the EPA be notified?

104. The country's most populous state is California, with $\frac{3}{25}$ of the nation's residents. Second in size is New York, with slightly less than $\frac{2}{25}$ of the population, followed by Texas, with almost $\frac{7}{100}$. Approximately what fractional part of the total population lives in these three states?

In Exercises 105–112, perform the indicated operations.

105. $\left(\frac{2}{5} + \frac{1}{5}\right) \cdot \frac{5}{6}$

106. $\frac{2}{5} + \left(\frac{1}{5} \cdot \frac{5}{6}\right)$

107. $\left(\frac{1}{3} - \frac{1}{6}\right) \cdot \frac{12}{8}$

108. $\frac{1}{3} - \left(\frac{1}{6} \cdot \frac{12}{8}\right)$

109. $\left(\frac{5}{2} - \frac{5}{6}\right) \cdot 3$

110. $\frac{5}{2} - \left(\frac{5}{6} \cdot 3\right)$

111. $\frac{13}{2} - \left(3 \cdot \frac{4}{5}\right)$

112. $\frac{1}{2} \cdot \left(2 - \frac{1}{3}\right)$

In Exercises 113–124, evaluate each expression using the distributive property. Write the answer in lowest terms.

113. $7\left(\frac{1}{7} + \frac{6}{7}\right)$

114. $4\left(\frac{5}{8} - \frac{3}{8}\right)$

115. $5\left(5 + \frac{7}{10}\right)$

116. $8\left(3 - \frac{15}{16}\right)$

117. $10\left(\frac{9}{20} - \frac{3}{10}\right)$

118. $16\left(\frac{3}{8} - \frac{3}{16}\right)$

119. $15\left(3 + \frac{3}{5}\right)$

120. $12\left(5 - \frac{5}{6}\right)$

121. $\frac{3}{5}(25 - 10)$

122. $\frac{7}{8}(24 + 16)$

123. $\frac{5}{7}\left(\frac{7}{2} + 7\right)$

124. $\frac{3}{10}\left(10 - \frac{10}{3}\right)$

Review Exercises

1. Is every whole number equal to a fraction?

2. Is every fraction equal to a whole number?

3. Does $\frac{26}{2}$ represent a prime number?

4. Write the whole numbers less than 10.

5. Write the prime numbers less than 10.

6. Write $\frac{13}{21}$ as a fraction with denominator of 63.

In Review Exercises 7–10, indicate whether each statement is true.

7. $\dfrac{6}{9} = \dfrac{10}{15}$ **8.** $\dfrac{6}{8} = \dfrac{27}{36}$ **9.** $\dfrac{3}{5} + \dfrac{4}{10} = \dfrac{7}{15}$ **10.** $\dfrac{2}{5} \cdot \dfrac{5}{2} = 1$

2.4   Mixed Numbers; Comparing Fractional Expressions

- Mixed Numbers
- Arithmetic of Mixed Numbers
- Comparing Fractional Expressions

Mixed Numbers

Fractions such as $\frac{3}{5}$, in which the numerator is less than the denominator, are called **proper fractions**. Fractions such as $\frac{13}{5}$, in which the numerator is greater than the denominator, are called **improper fractions**. An improper fraction such as $\frac{13}{5}$ can be written as the sum of a whole number and a proper fraction:

$$\dfrac{13}{5} = \dfrac{10 + 3}{5} \qquad 13 = 10 + 3.$$

$$= \dfrac{10}{5} + \dfrac{3}{5}$$

$$= 2 + \dfrac{3}{5}$$

The result $2 + \frac{3}{5}$ is often written as $2\frac{3}{5}$, a form called a **mixed number**. We read $2\frac{3}{5}$ as "two and three-fifths." It is important to remember that

$$2\dfrac{3}{5} = 2 + \dfrac{3}{5}$$

and that $2\frac{3}{5}$ does *not* mean $2 \cdot \frac{3}{5}$.

There is another relation between the improper fraction $\frac{13}{5}$ and the mixed number $2\frac{3}{5}$. Recall that the fraction $\frac{13}{5}$ can be interpreted as a division, in which the dividend is 13 and the divisor is 5. We use long division to divide 13 by 5:

$$\text{Divisor} \rightarrow 5\overline{)13} \begin{array}{l} \leftarrow \text{Quotient} \\ \leftarrow \text{Dividend} \end{array}$$
$$\phantom{\text{Divisor} \rightarrow 5)}\underline{10}$$
$$\phantom{\text{Divisor} \rightarrow 5)1}3 \leftarrow \text{Remainder}$$

We have seen that the fraction $\frac{13}{5}$ is equal to $2 + \frac{3}{5}$, or $2\frac{3}{5}$. We now see that the fraction $\frac{13}{5}$ is equal to $quotient + \dfrac{remainder}{divisor}$ because in the previous long division, the quotient is 2, the remainder is 3, and the divisor is 5. That is,

$$\dfrac{13}{5} = quotient + \dfrac{remainder}{divisor} = 2 + \dfrac{3}{5} = 2\dfrac{3}{5}$$

This example illustrates the following fact:

> If one whole number (the *dividend*) is divided by another whole number (the *divisor*) to obtain a *quotient* and a *remainder*, then
> $$\frac{dividend}{divisor} = quotient + \frac{remainder}{divisor}$$

EXAMPLE 1

Write the fraction $\frac{213}{27}$ as a mixed number.

Solution

The fraction represents a division. The *dividend* is 213, and the *divisor* is 27. We begin by performing a long division and writing the answer in the form

$$quotient + \frac{remainder}{divisor}$$

$$\text{Divisor} \to 27 \overline{\smash{\big)}\, 213} \quad \begin{array}{l} \leftarrow \text{Quotient} \\ \leftarrow \text{Dividend} \end{array}$$
$$\phantom{27 \overline{\smash{\big)}\,}} \underline{189}$$
$$\phantom{27 \overline{\smash{\big)}\,2} } 24 \leftarrow \text{Remainder}$$

The *quotient* is 7, and the *remainder* is 24. Thus,

$$\frac{dividend}{divisor} = quotient + \frac{remainder}{divisor}$$

$$\frac{213}{27} = 7 + \frac{24}{27}$$

$$= 7 + \frac{\overset{1}{\cancel{3}} \cdot 8}{\underset{1}{\cancel{3}} \cdot 9} \qquad \text{Factor numerator and denominator and simplify.}$$

$$= 7 + \frac{8}{9}$$

As a mixed number, $\frac{213}{27} = 7\frac{8}{9}$. ∎

EXAMPLE 2

Write the mixed number $3\frac{7}{8}$ as an improper fraction.

Solution

The mixed number $3\frac{7}{8}$ represents the sum $3 + \frac{7}{8}$. We determine that sum by finding a common denominator and adding the fractions:

$$3\frac{7}{8} = 3 + \frac{7}{8}$$

$$= \frac{3}{1} + \frac{7}{8}$$

$$= \frac{3 \cdot 8}{1 \cdot 8} + \frac{7}{8}$$

$$= \frac{3 \cdot 8 + 7}{8}$$

$$= \frac{24 + 7}{8}$$

$$= \frac{31}{8}$$

The mixed number $3\frac{7}{8}$ is equal to the improper fraction $\frac{31}{8}$. ■

Example 2 provides a quick method of converting the mixed number $3\frac{7}{8}$ into an improper fraction. We multiply the denominator of the fractional part of $3\frac{7}{8}$ by the whole-number part, and add the numerator of the fractional part. This result, $8 \cdot 3 + 7$, or 31, is the numerator of the desired improper fraction. The denominator of the improper fraction is 8, which is the denominator of the fractional part of $3\frac{7}{8}$:

$$3\frac{7}{8} = \frac{8 \cdot 3 + 7}{8} = \frac{31}{8}$$

Similarly,

$$5\frac{4}{9} = \frac{9 \cdot 5 + 4}{9} = \frac{49}{9}$$

Work Progress Check 14.

Arithmetic of Mixed Numbers

To add or subtract mixed numbers, we combine their whole-number parts, combine their fractional parts, and convert the answer into mixed-number form.

EXAMPLE 3

A recipe for salad dressing calls for $2\frac{3}{4}$ cups of vegetable oil and $1\frac{1}{3}$ cups of vinegar. How many cups of liquid are required?

Solution

The total amount of liquid is the sum of $2\frac{3}{4}$ cups and $1\frac{1}{3}$ cups. To add these mixed numbers, add their whole-number parts, and add their fractional parts.

Progress Check 14

Change $12\frac{11}{12}$ into an improper fraction.

14. $\frac{155}{144}$

$$2\frac{3}{4} + 1\frac{1}{3} = 2 + \frac{3}{4} + 1 + \frac{1}{3}$$

$$= 3 + \frac{3}{4} + \frac{1}{3} \qquad \text{Add: } 2 + 1 = 3.$$

$$= 3 + \frac{3 \cdot 3}{4 \cdot 3} + \frac{1 \cdot 4}{3 \cdot 4} \qquad \text{The common denominator is 12.}$$

$$= 3 + \frac{9}{12} + \frac{4}{12}$$

$$= 3 + \frac{9+4}{12}$$

$$= 3 + \frac{13}{12}$$

$$= 3 + 1 + \frac{1}{12} \qquad \frac{13}{12} = \frac{12}{12} + \frac{1}{12} = 1 + \frac{1}{12}$$

$$= 4 + \frac{1}{12} \qquad \text{Add the whole numbers.}$$

$$= 4\frac{1}{12}$$

Thus, the recipe calls for $4\frac{1}{12}$ cups of liquid. ■

To multiply or divide mixed numbers, we convert them to improper fractions and multiply or divide. Finally, we change the answer back into a mixed number.

EXAMPLE 4

The perimeter of a square is $17\frac{1}{2}$ feet. Find the length of a side of the square.

Solution

The four sides of a square are of equal length, and the perimeter is the sum of those four lengths. Therefore, the length of each side of the square is one-fourth of the perimeter. That is, the length of each side is one-fourth of $17\frac{1}{2}$ feet, or $(\frac{1}{4})(17\frac{1}{2})$ feet. To do this calculation, we write $17\frac{1}{2}$ as an improper fraction and multiply by one-fourth:

$$\text{One side of the square} = \frac{1}{4} \text{ of the perimeter}$$

$$= \frac{1}{4} \cdot 17\frac{1}{2} \qquad \text{The word } of \text{ means } times.$$

$$= \frac{1}{4} \cdot \frac{35}{2} \qquad \begin{array}{l}\text{Write } 17\frac{1}{2} \text{ as an improper fraction:} \\ 17\frac{1}{2} = \frac{2 \cdot 17 + 1}{2} = \frac{35}{2}\end{array}$$

$$= \frac{1 \cdot 35}{4 \cdot 2} \qquad \text{Multiply.}$$

$$= \frac{35}{8}$$

Finally, we perform a long division to change the improper fraction $\frac{35}{8}$ into the mixed number $4\frac{3}{8}$. The length of each side of the square is $4\frac{3}{8}$ feet. ■

EXAMPLE 5

Each side of the square garden in Figure 2-5 is $7\frac{5}{6}$ meters long. Find the length of fencing required to enclose the garden.

Solution

To find the length of the fence, we find the perimeter of the square. Since the four sides of a square are the same length, the perimeter is equal to 4 times the length of one side. To find the product $4 \cdot 7\frac{5}{6}$, we write the mixed number as an improper fraction and multiply:

$$\begin{aligned} 4 \cdot 7\frac{5}{6} &= 4 \cdot \frac{47}{6} && 7\frac{5}{6} = \frac{6 \cdot 7 + 5}{6} = \frac{42 + 5}{6} = \frac{47}{6} \\ &= \frac{4}{1} \cdot \frac{47}{6} && \text{Write 4 as } \frac{4}{1}. \\ &= \frac{4 \cdot 47}{6} && \text{Multiply.} \\ &= \frac{2 \cdot 2 \cdot 47}{2 \cdot 3} && \text{Factor.} \\ &= \frac{\overset{1}{\cancel{2}} \cdot 2 \cdot 47}{\underset{1}{\cancel{2}} \cdot 3} && \text{Divide out the common factor.} \\ &= \frac{94}{3} && \text{Simplify.} \\ &= 31\tfrac{1}{3} \end{aligned}$$

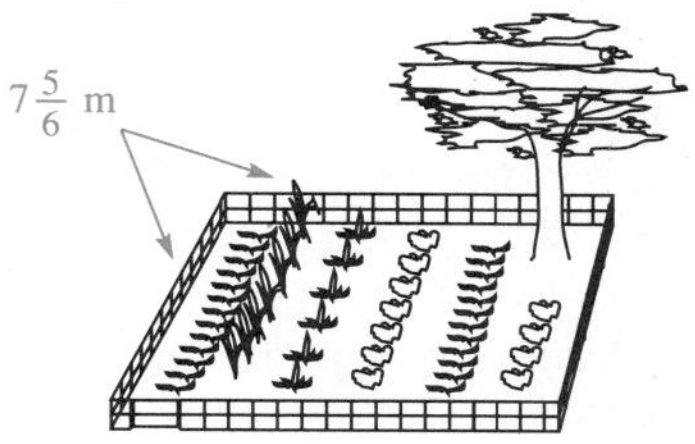

Figure 2-5

The length of the fence is $31\frac{1}{3}$ meters. ■

EXAMPLE 6

A rectangle is $5\frac{1}{3}$ meters long and $3\frac{3}{5}$ meters wide. Find its area.

Solution

The area A of a rectangle is given by the formula $A = lw$, where l is the length of the rectangle and w is the width. Because the given length and width are mixed numbers and we are to find their product, we convert them into improper fractions.

$$\begin{aligned} l &= 5\tfrac{1}{3} & w &= 3\tfrac{3}{5} \\ &= 5 + \frac{1}{3} & &= 3 + \frac{3}{5} \\ &= \frac{15}{3} + \frac{1}{3} & &= \frac{15}{5} + \frac{3}{5} \\ &= \frac{16}{3} & &= \frac{18}{5} \end{aligned}$$

Thus, the length l is $\frac{16}{3}$ meters and the width w is $\frac{18}{5}$ meters. We substitute $\frac{16}{3}$ for l and $\frac{18}{5}$ for w in the formula $A = lw$.

$$A = lw$$
$$= \frac{16}{3} \cdot \frac{18}{5}$$
$$= \frac{16 \cdot 18}{3 \cdot 5}$$
$$= \frac{16 \cdot \overset{1}{\cancel{3}} \cdot 6}{\underset{1}{\cancel{3}} \cdot 5}$$
$$= \frac{96}{5}$$

We perform the indicated division to express $\frac{96}{5}$ as a mixed number. Because the quotient is 19 and the remainder is 1, the fraction $\frac{96}{5}$ is equal to $19 + \frac{1}{5}$, or the mixed number $19\frac{1}{5}$. Thus,

$$A = 19\frac{1}{5}$$

The area of the rectangle is $19\frac{1}{5}$ square meters. Work Progress Check 15.

Progress Check 15

Find the area of the entire rectangle. (*Hint:* The length is the sum of two mixed numbers.)

[Rectangle diagram: $4\frac{1}{4}$ m and $7\frac{1}{2}$ m on top; $3\frac{3}{4}$ m on side]

EXAMPLE 7

How many $2\frac{5}{6}$-foot sections can be cut from a rope that is $38\frac{2}{3}$ feet long?

Solution

It may be that not all of the rope is usable: after cutting as many sections as possible, we could be left with a too-short piece of scrap. To determine the number of sections possible, we divide the length of the rope by the length of a section. If the result is a mixed number, we discard the fractional part, because a fractional part of a $2\frac{5}{6}$-foot section is not usable.

$$\frac{38\frac{2}{3}}{2\frac{5}{6}} = \frac{\frac{116}{3}}{\frac{17}{6}}$$
Write each mixed number as a fraction.

$$= \frac{\frac{116}{3} \cdot 6}{\frac{17}{6} \cdot 6}$$
Use the fundamental property of fractions: Multiply both numerator and denominator by 6, the LCD of $\frac{116}{3}$ and $\frac{17}{6}$.

$$= \frac{\frac{116}{3} \cdot \frac{6}{1}}{\frac{17}{6} \cdot \frac{6}{1}}$$

$$= \frac{116 \cdot 2}{17}$$
Simplify.

$$= \frac{232}{17}$$
Multiply the factors in the numerator.

$$= 13\frac{11}{17}$$
Write $\frac{232}{17}$ as a mixed number.

15. $44\frac{1}{16}$ square meters

Progress Check Answers

Because the answer is a mixed number, and the correct answer must be a whole number, we discard the fractional part. We can cut 13 sections from the rope, and $\frac{11}{17}$ of one section will be left over. ∎

Comparing Fractional Expressions

To determine which of two unequal fractions is greater, we first change the fractions to equivalent fractions with the same denominator. Then, by comparing numerators, we can see which fraction is greater: the greater fraction has the greater numerator.

EXAMPLE 8

Which fraction is larger: $\frac{5}{8}$ or $\frac{17}{28}$?

Solution

Write the fractions as equivalent fractions with the same denominator. To find a common denominator, we find the least common multiple of the denominators 8 and 28. We begin by factoring 8 and 28 and forming the LCD by using each factor the largest number of times it appears in either factorization.

$$8 = 2^3 \qquad 28 = 2^2 \cdot 7$$

Because 2 appears as a factor three times, and 7 appears as a factor once, the least common denominator is $2^3 \cdot 7$, or 56. We change each fraction into an equivalent fraction with a denominator of 56:

$$\frac{5}{8} = \frac{5}{2^3} \qquad\qquad \frac{17}{28} = \frac{17}{2^2 \cdot 7}$$

$$= \frac{5 \cdot 7}{2^3 \cdot 7} \qquad\qquad = \frac{17 \cdot 2}{2^2 \cdot 7 \cdot 2}$$

$$= \frac{35}{56} \qquad\qquad = \frac{34}{56}$$

Finally, we compare the fractions by comparing their numerators. Because 35 is greater than 34, $\frac{35}{56}$ is greater than $\frac{34}{56}$, and therefore,

$$\frac{5}{8} \quad \text{is greater than} \quad \frac{17}{28} \qquad ∎$$

To show that one number a is less than or greater than another number b, we use **inequality symbols**. The symbol $<$ is read "is less than," and the symbol $>$ is read "is greater than."

> **Definition:**
> $a < b$ means that a is less than b
> $a > b$ means that a is greater than b

The result of Example 8 can be written as $\frac{5}{8} > \frac{17}{28}$.

EXAMPLE 9

Which is smaller: $7\frac{13}{15}$ or $\frac{79}{10}$?

Solution

We change the numbers to equivalent improper fractions with the same common denominator. We begin by changing the mixed number $7\frac{13}{15}$ into an improper fraction.

$$\begin{aligned}
7\frac{13}{15} &= 7 + \frac{13}{15} \\
&= \frac{7}{1} + \frac{13}{15} \\
&= \frac{7 \cdot 15}{1 \cdot 15} + \frac{13}{15} \qquad \text{Find a common denominator.} \\
&= \frac{105 + 13}{15} \qquad \text{Add the fractions.} \\
&= \frac{118}{15}
\end{aligned}$$

To compare $\frac{118}{15}$ and $\frac{79}{10}$, we write each as an equivalent fraction with a denominator 30, because 30 is the least common multiple of 15 and 10. (Verify this.) Then we compare numerators:

$$7\frac{13}{15} = \frac{118}{15} \qquad\qquad \frac{79}{10} = \frac{79 \cdot 3}{10 \cdot 3}$$

$$\phantom{7\frac{13}{15}} = \frac{118 \cdot 2}{15 \cdot 2} \qquad\qquad = \frac{237}{30}$$

$$\phantom{7\frac{13}{15}} = \frac{236}{30}$$

The first of these fractions is the smaller. Thus,

$$7\frac{13}{15} < \frac{79}{10}$$

2.4 Exercises

In Exercises 1–8, perform each division. Express answers in quotient $+ \dfrac{\text{remainder}}{\text{divisor}}$ form.

1. $\dfrac{23}{5}$
2. $\dfrac{44}{7}$
3. $\dfrac{45}{11}$
4. $\dfrac{73}{9}$
5. $\dfrac{53}{10}$
6. $\dfrac{95}{17}$
7. $\dfrac{33}{8}$
8. $\dfrac{99}{13}$

In Exercises 9–24, write each improper fraction as a mixed number.

9. $\dfrac{23}{5}$
10. $\dfrac{44}{7}$
11. $\dfrac{45}{11}$
12. $\dfrac{73}{9}$

13. $\dfrac{53}{10}$
14. $\dfrac{95}{17}$
15. $\dfrac{33}{8}$
16. $\dfrac{99}{13}$
17. $\dfrac{57}{9}$
18. $\dfrac{119}{15}$
19. $\dfrac{210}{17}$
20. $\dfrac{157}{19}$
21. $\dfrac{253}{11}$
22. $\dfrac{357}{23}$
23. $\dfrac{234}{117}$
24. $\dfrac{185}{37}$

In Exercises 25–36, write each mixed number as an improper fraction.

25. $1\dfrac{1}{2}$
26. $2\dfrac{3}{4}$
27. $3\dfrac{7}{8}$
28. $3\dfrac{5}{7}$
29. $5\dfrac{5}{11}$
30. $4\dfrac{6}{13}$
31. $5\dfrac{2}{3}$
32. $2\dfrac{9}{10}$
33. $7\dfrac{5}{9}$
34. $12\dfrac{7}{13}$
35. $25\dfrac{21}{31}$
36. $23\dfrac{19}{23}$

In Exercises 37–52, find each sum or difference. Express each answer as a mixed number.

37. $2\dfrac{1}{2} + 3\dfrac{2}{3}$
38. $3\dfrac{3}{4} + 5\dfrac{1}{3}$
39. $5\dfrac{3}{5} + 3\dfrac{1}{2}$
40. $2\dfrac{5}{7} + 5\dfrac{3}{5}$
41. $1\dfrac{1}{2} - 1\dfrac{1}{3}$
42. $4\dfrac{6}{7} - 3\dfrac{2}{5}$
43. $3\dfrac{5}{9} - 2\dfrac{2}{3}$
44. $7\dfrac{1}{4} - 5\dfrac{1}{2}$
45. $5\dfrac{4}{9} + 3\dfrac{1}{3}$
46. $5\dfrac{3}{8} - 3\dfrac{5}{8}$
47. $17\dfrac{11}{16} - 9\dfrac{3}{8}$
48. $17\dfrac{1}{4} + 7\dfrac{3}{8}$
49. $12\dfrac{3}{5} + 29\dfrac{7}{10}$
50. $35\dfrac{13}{32} - 20\dfrac{5}{8}$
51. $30\dfrac{43}{100} - 29\dfrac{43}{50}$
52. $57\dfrac{5}{48} - 31\dfrac{7}{36}$

In Exercises 53–72, find each product or quotient. Write all improper-fraction answers as mixed numbers.

53. $\left(2\dfrac{1}{2}\right)\left(3\dfrac{2}{3}\right)$
54. $\left(3\dfrac{3}{4}\right)\left(5\dfrac{1}{3}\right)$
55. $3\dfrac{1}{2} \cdot 2\dfrac{1}{3}$
56. $5\dfrac{1}{4} \cdot 5\dfrac{1}{2}$
57. $\left(2\dfrac{3}{5}\right)\left(2\dfrac{2}{5}\right)$
58. $\left(7\dfrac{3}{4}\right)\left(2\dfrac{1}{2}\right)$
59. $\dfrac{2\dfrac{1}{2}}{1\dfrac{1}{2}}$
60. $\dfrac{5\dfrac{1}{4}}{2\dfrac{1}{2}}$
61. $3\dfrac{1}{2} \div 2\dfrac{1}{2}$
62. $3\dfrac{3}{4} \div 5\dfrac{1}{3}$
63. $\dfrac{3\dfrac{1}{3}}{1\dfrac{5}{6}}$
64. $\dfrac{12\dfrac{3}{4}}{2\dfrac{3}{8}}$
65. $7\dfrac{3}{5} \cdot 7\dfrac{1}{3}$
66. $9\dfrac{3}{10} \cdot 3\dfrac{1}{2}$
67. $4\dfrac{1}{2} \div 2\dfrac{1}{4}$
68. $5\dfrac{7}{8} \div 5\dfrac{3}{4}$
69. $\dfrac{1\dfrac{7}{11}}{5\dfrac{3}{22}}$
70. $\dfrac{2\dfrac{9}{16}}{1\dfrac{3}{8}}$
71. $4\dfrac{1}{4} \div 2\dfrac{1}{2}$
72. $5\dfrac{3}{4} \div 5\dfrac{7}{8}$

In Exercises 73–76, evaluate each expression by first performing the indicated addition or subtraction, and then the multiplication. Write all improper-fraction answers as mixed numbers.

73. $3\left(2\dfrac{1}{3} + 3\dfrac{1}{3}\right)$ **74.** $5\left(4\dfrac{1}{5} - 1\right)$ **75.** $\left(1\dfrac{1}{5}\right)\left(8\dfrac{1}{2} + 1\dfrac{1}{3}\right)$ **76.** $\left(2\dfrac{1}{17}\right)\left(3\dfrac{1}{4} - 1\dfrac{1}{8}\right)$

In Exercises 77–80, evaluate each expression. Write all improper-fraction answers as mixed numbers.

77. $\dfrac{9\dfrac{5}{14} - 1\dfrac{1}{7}}{2\dfrac{1}{2}}$ **78.** $\dfrac{2\dfrac{1}{2} + 3\dfrac{1}{10}}{2\dfrac{1}{10}}$ **79.** $\dfrac{1\dfrac{1}{7}}{9\dfrac{1}{3} - \dfrac{4}{21}}$ **80.** $\dfrac{1\dfrac{1}{2}}{1\dfrac{1}{8} + 2\dfrac{3}{8}}$

In Exercises 81–88, determine which number of each pair is the smaller.

81. $\dfrac{11}{2}, 5\dfrac{1}{3}$ **82.** $3\dfrac{2}{5}, \dfrac{19}{5}$ **83.** $6\dfrac{5}{6}, 6\dfrac{18}{21}$ **84.** $3\dfrac{3}{8}, 3\dfrac{13}{36}$

In Exercises 85–88, determine which number of each pair is the larger.

85. $\dfrac{28}{3}, \dfrac{77}{8}$ **86.** $\dfrac{23}{11}, \dfrac{27}{13}$ **87.** $\dfrac{52}{9}, 5\dfrac{11}{15}$ **88.** $2\dfrac{5}{8}, 2\dfrac{12}{19}$

89. One pail holds $2\dfrac{1}{2}$ gallons of water, and another holds $3\dfrac{1}{3}$ gallons. How much water can be carried in both pails?

90. A gas tank holds 15 gallons of gasoline. If there are now $8\dfrac{1}{3}$ gallons in the tank, how much more is needed to fill the tank?

91. After three sections, each $5\dfrac{3}{5}$ feet long, are cut from a long rope, $7\dfrac{1}{2}$ feet remain. How long was the original rope?

92. A $21\dfrac{1}{2}$-foot board must be cut in half to get it down the basement stairs. From those two pieces, as many $3\dfrac{1}{2}$-foot sections are cut as possible. What is the total length of the remaining scrap?

93. A $22\dfrac{7}{10}$-meter rope is cut into five equal pieces. Find the length of each piece.

94. A $14\dfrac{2}{3}$-yard cable is cut into six equal pieces. Find the length of each piece.

95. A rectangle is $2\dfrac{3}{5}$ feet wide and $5\dfrac{7}{8}$ feet long. Find the perimeter of the rectangle.

96. A rectangle is $3\dfrac{2}{5}$ feet wide and $7\dfrac{3}{5}$ feet long. Find the area of the rectangle.

97. A carpenter wants to cut two boards, each $19\dfrac{3}{4}$ inches long, from a single board that is 60 inches long. If each sawcut uses $\dfrac{1}{8}$ inch, how long is the remaining board?

98. A $3\dfrac{1}{4}$-inch by $8\dfrac{1}{2}$-inch rectangle is cut from the top of an $8\dfrac{1}{2}$-by-11-inch piece of typing paper. Find the dimensions of the remaining paper.

99. A $2\dfrac{5}{8}$-inch-by-11-inch rectangle is cut from the side of an $8\dfrac{1}{2}$-by-11-inch piece of typing paper. Find the dimensions of the remaining paper.

100. Find the area of the smaller rectangle in Exercise 98.

101. Find the perimeter of the larger rectangle in Exercise 98.

102. Find the perimeter of the smaller rectangle in Exercise 99.

103. Find the area of the larger rectangle in Exercise 99.

104. Find the perimeter of the rectangle in Illustration 1.

105. Find the area of the rectangle in Illustration 1.

106. Find the total of the shaded areas in Illustration 1.

107. Find the total of the unshaded areas in Illustration 1.

108. Jane walks for exercise. She has recorded the number of miles she has walked each day for one week. Use the information in Table 1 to find the total distance Jane walked that week.

109. The base of a triangle is $3\frac{2}{3}$ feet, and the height is $2\frac{1}{2}$ feet. Find the area of the triangle. Recall that the area of a triangle is given by the formula $A = \frac{1}{2}bh$.

110. A tank holds $74\frac{3}{4}$ gallons. Its contents will be hauled away in a $3\frac{1}{4}$-gallon bucket. How many times will the bucket have to be filled?

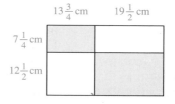

Illustration 1

Table 1

Day	Miles Walked
Monday	$2\frac{1}{2}$
Tuesday	$1\frac{1}{3}$
Wednesday	$2\frac{2}{3}$
Thursday	$3\frac{1}{4}$
Friday	$\frac{1}{2}$
Saturday	$1\frac{3}{4}$
Sunday	2

Review Exercises

In Review Exercises 1–12, perform the indicated operations.

1. $147 + 769$
2. $1567 - 789$
3. $376 \cdot 57$
4. $7303 \div 67$
5. $13^2 \cdot 5$
6. $7^3 \cdot 3$
7. $\dfrac{7}{12} \cdot \dfrac{6}{28}$
8. $\dfrac{3}{11} \div \dfrac{9}{22}$
9. $\dfrac{31}{12} + \dfrac{17}{6}$
10. $\dfrac{6^2}{9}$
11. $\dfrac{34}{11} - 2$
12. $6^2 + 9$

Mathematics in Music

Pythagoras experimented with stringed instruments. He recognized that a full-length string and one $\frac{3}{4}$ as long sounded pleasing together. So also did a string $\frac{2}{3}$ as long. Because $\frac{2}{3} \div \frac{3}{4}$ is $\frac{8}{9}$, Pythagoras must have been excited: $\frac{8}{9}$ is $\frac{2^3}{3^2}$, a fraction filled with masculine 2's and feminine 3's.

The first note of the Pythagorean scale is sounded by a string $\frac{8}{9}$ as long as the original; the second by a string $(\frac{8}{9})^2$, or $\frac{64}{81}$, as long, and so on, to a string that sounds the octave, a string $\frac{1}{2}$ as long. Actually, this process leaves a discrepancy—two gaps of $\frac{243}{256}$. These, Pythagoras called halftones, and placed them between the 3rd and 4th notes of the scale, and between the 7th and the octave. Although the modern scale is tuned somewhat differently, the basic structure has survived for 2500 years.

Chapter Summary

Key Words

cross multiplying (2.1)
denominator (2.1)
equivalent fractions (2.1)
fraction (2.1)
improper fraction (2.4)

inequality symbols (2.4)
least common denominator (2.3)
lowest common denominator (2.3)

lowest terms (2.1)
mixed number (2.4)
numerator (2.1)
proper fraction (2.4)
reciprocal (2.2)

reduce a fraction (2.1)
simplify a fraction (2.1)

Key Ideas

(2.1) **The fundamental property of fractions:**

$$\frac{a \cdot x}{b \cdot x} = \frac{a}{b} \quad (b \neq 0, x \neq 0)$$

To simplify a fraction, first factor the numerator and denominator; then divide out any common factors.

Equivalent fractions:

$$\frac{a}{b} = \frac{c}{d} \text{ means that } ad = bc$$

(2.3) To add (or subtract) two fractions with the same denominators, add (or subtract) their numerators and use their common denominator.

To add or subtract fractions with unlike denominators, write the fractions as fractions with a common denominator, add or subtract their numerators, and use the common denominator.

(2.2) To multiply two fractions, multiply their numerators and multiply their denominators. Be sure to factor the numerator and the denominator, and divide out any common factors.

To divide one fraction by another (Method 1), multiply the first fraction by the reciprocal of the second fraction.

To divide one fraction by another (Method 2), write the division as a fraction. Then multiply numerator and denominator by the LCD of the given fractions.

(2.4) $$\frac{dividend}{divisor} = quotient + \frac{remainder}{divisor}$$

$a < b$ means that a is less than b
$a > b$ means that a is greater than b

To compare two fractional expressions, write them as fractions with the same denominator. The fraction with the smaller numerator is the smaller fraction.

Chapter 2 Review Exercises

[2.1] In Review Exercises 1–4, simplify each fraction.

1. $\frac{45}{27}$

2. $\frac{121}{11}$

3. $\frac{22}{99}$

4. $\frac{14}{42}$

[2.2] In Review Exercises 5–8, perform each indicated operation. Simplify answers, if possible.

5. $\frac{31}{15} \cdot \frac{10}{62}$

6. $\frac{18}{21} \div \frac{6}{7}$

7. $\frac{21}{15} \div \frac{10}{49}$

8. $\frac{48}{30} \cdot \frac{6}{9}$

[2.3] In Review Exercises 9–12, write each fraction as an equivalent fraction with the denominator indicated.

9. $\frac{5}{7}$; 21

10. $\frac{13}{17}$; 34

11. $\frac{7}{31}$; 217

12. $\frac{11}{19}$; 95

In Review Exercises 13–16, perform the indicated operation. Simplify answers, if possible.

13. $\dfrac{1}{3} + \dfrac{1}{7}$ **14.** $\dfrac{2}{3} - \dfrac{3}{5}$ **15.** $\dfrac{9}{10} - \dfrac{7}{15}$ **16.** $\dfrac{8}{35} + \dfrac{2}{21}$

In Review Exercises 17–18, write each subtraction as an equivalent addition and find x.

17. $x - \dfrac{1}{2} = \dfrac{3}{2}$ **18.** $x - \dfrac{3}{5} = \dfrac{7}{5}$

In Review Exercises 19–20, write each addition as an equivalent subtraction and find x.

19. $x + \dfrac{3}{7} = 3$ **20.** $x + \dfrac{2}{3} = 2$

[2.4] In Review Exercises 21–24, write each fraction as a mixed number.

21. $\dfrac{27}{7}$ **22.** $\dfrac{37}{9}$ **23.** $\dfrac{123}{11}$ **24.** $\dfrac{157}{13}$

In Review Exercises 25–28, write each mixed number as an improper fraction.

25. $8\dfrac{3}{7}$ **26.** $7\dfrac{4}{9}$ **27.** $12\dfrac{5}{12}$ **28.** $17\dfrac{7}{10}$

In Review Exercises 29–32, simplify each expression by performing the indicated operations.

29. $\dfrac{1}{2}\left(\dfrac{3}{2} - \dfrac{1}{4}\right)$ **30.** $2\left(1 - \dfrac{1}{2}\right)^2$ **31.** $\dfrac{\left(\dfrac{1}{2} \cdot \dfrac{1}{3}\right) - \dfrac{1}{8}}{\dfrac{1}{2} \cdot \dfrac{1}{12}}$ **32.** $\dfrac{\dfrac{1}{2}\left(1 + \dfrac{1}{3}\right)}{\dfrac{1}{3}\left(1 - \dfrac{1}{2}\right)}$

In Review Exercises 33–36, indicate which expression is larger.

33. $\dfrac{19}{12}, \dfrac{5}{8}$ **34.** $\dfrac{17}{21}, \dfrac{9}{11}$ **35.** $3\dfrac{9}{11}, 3\dfrac{9}{13}$ **36.** $\dfrac{47}{21}, \dfrac{56}{25}$

37. A business-sized envelope weighs $\dfrac{1}{12}$ ounce, and each sheet of an advertisement weighs $\dfrac{1}{9}$ ounce. Can eight sheets be mailed for the one-ounce postage?

38. The perimeter of a square is $6\dfrac{1}{2}$ cm. Find the length of one side.

39. In 1985, $16\dfrac{3}{16}$ million workers were employed by national, state, or local government. This was $\dfrac{7}{8}$ of the government employment in 1990. How many worked for the government in 1990?

40. The 1980 population of Los Angeles County was $7\dfrac{1}{2}$ million, which was $\dfrac{5}{6}$ of the population in 1990. What was the population in 1990?

Chapter 2 Test

1. What fractional part of the pie has been eaten?

2. What fractional part of one dollar is one nickel?

3. Bill has finished 1275 pages of a 1530-page book. What fractional part of the book remains to be read?

4. Express the fraction $\frac{45}{63}$ in lowest terms.

5. Are the fractions $\frac{30}{105}$ and $\frac{16}{56}$ equal? Indicate *yes* or *no*.

6. Multiply: $\frac{21}{28} \cdot \frac{12}{9}$.

7. Multiply: $\frac{18}{45} \cdot 5$.

8. Divide: $\frac{160}{210} \div \frac{4}{7}$.

9. Divide: $\frac{36}{54} \div 6$.

10. Divide: $15 \div \frac{45}{24}$.

11. Write $\frac{11}{21}$ as a fraction with a denominator of 189.

12. Add: $\frac{5}{7} + \frac{9}{7}$.

13. Add: $\frac{4}{15} + \frac{7}{10}$.

14. Subtract: $\frac{13}{18} - \frac{5}{30}$.

15. Subtract: $\dfrac{29}{13} - 2$.

16. Find x if $\dfrac{7}{3} + x = \dfrac{9}{2}$.

17. Write the fraction $\dfrac{72}{13}$ as a mixed number.

18. Write the mixed number $7\dfrac{3}{7}$ as an improper fraction.

19. Simplify: $\dfrac{3 \cdot 5 + 1}{3(5 + 1)} + \dfrac{1}{9}$.

20. Which is larger: $\dfrac{47}{35}$ or $\dfrac{27}{20}$?

21. An electrician has a wire that is $7\dfrac{1}{4}$ feet long. He uses $\dfrac{2}{3}$ of it and discards the rest. How long is the piece that was not used?

22. A developer has 17 acres of land that she will divide into one-half- and one-quarter-acre lots. If there are to be 19 one-half-acre lots, how many one-quarter-acre lots will there be?

23. The base of the Great Pyramid of Cheops is approximately square. Each of its four sides measures $215\dfrac{1}{3}$ meters. Find the perimeter of the base.

24. At the time of the building of the Great Pyramid, the Egyptian unit of measure was the royal cubit, approximately $\dfrac{1}{2}$ meter. Find the perimeter of the base in royal cubits. (See question 23.)

3 Decimals

Mathematics for Fun

The number 142857 is interesting. When 142857 is multiplied by 2, the answer contains the same digits, in the same order in a circular pattern:

$$142857 \times 2 = 285714$$

When multiplied by 3, it happens again:

$$142857 \times 3 = 428571$$

Try multiplying 142857 by 4, by 5, and by 6. Do the same digits still appear in the same circular pattern? Fractions and *decimals*, the topic of this chapter, will help us find a larger number with this same strange property.

Decimals, the subject of this chapter, are similar to fractions; they also are used to represent numbers that are not whole numbers.

3.1 Decimals and Place Value; Rounding

- Decimal Fractions
- Comparing Decimals
- Rounding Decimals

Decimal Fractions

In the **decimal numeration system**, each of the digits $0, 1, \ldots, 9$ has a **place value** determined by the column in which it appears. For example, in **decimal notation** the number 345 represents 3 *hundreds* + 4 *tens* + 5 *ones*. The decimal numeration system extends to **decimal fractions**, in which the digits may also represent *tenths, hundredths, thousandths,* and so on. For example, the number

  35.278

is written in **expanded notation** as

  3 *tens* + 5 *ones* + 2 *tenths* + 7 *hundredths* + 8 *thousandths*

The period to the right of the units position is called the **decimal point**; it separates the whole-number part of the number (35) from the fractional part (.278).

To explain how we would read the number 35.278, we write it using fractions, add the fractions, and write the result as a mixed number:

$$35.278 = 35 + \frac{2}{10} + \frac{7}{100} + \frac{8}{1000}$$

$$= 35 + \frac{2 \cdot 100}{10 \cdot 100} + \frac{7 \cdot 10}{100 \cdot 10} + \frac{8}{1000}$$

$$= 35 + \frac{278}{1000}$$

$$= 35\frac{278}{1000}$$

We read the decimal 35.278 as we would read the mixed number:

Thirty-five and two hundred seventy-eight thousandths

As a general rule, we read the fractional part of a decimal by using the name of the position of the decimal's last digit. Because the last digit of the decimal 15.79, for example, is in the *hundredths* position, we read the decimal 15.79 as *fifteen and seventy-nine hundredths*. Likewise, 0.3456 is read *three thousand four hundred fifty-six ten thousandths*.

EXAMPLE 1

a. 12.6 is read *twelve and six tenths*. As a mixed number, 12.6 would be written $12\frac{6}{10}$, or $12\frac{3}{5}$.

b. 0.004 is read *four thousandths*. As a fraction, it is $\frac{4}{1000}$, or $\frac{1}{125}$.

c. 7.5 is read *seven and five tenths*. As a mixed number, it is $7\frac{5}{10}$, or $7\frac{1}{2}$.

d. 0.25 is read *twenty-five hundredths*. As a fraction, it is $\frac{25}{100}$, or $\frac{1}{4}$.

e. 0.250 is read *two hundred fifty thousandths*. As a fraction, it is $\frac{250}{1000}$, or $\frac{1}{4}$.

Work Progress Check 1.

Progress Check 1

a. Write *fifty-six ten-thousandths* as a decimal.

b. Write 7.003 in words.

Comparing Decimals

We can see the relative sizes of decimal numbers on a **number line** like that in Figure 3-1. The left end of the line is labeled 0, and points equally spaced to the right are labeled 1, 2, 3, ... Decimal fractions mark the points in between.

Figure 3-1

The labels of points on the number line are called the **coordinates** of the points. For example, halfway between the points with coordinates 2 and 3 is the point with coordinate $2\frac{1}{2}$, or 2.5. One-quarter of the way from 1 to 2 is the point with coordinate 1.25. The point three-fourths of the way from 0 to 1 has the coordinate 0.75, because

$$\frac{3}{4} = \frac{75}{100} = 0.75$$

1. a. 0.0056 **b.** Seven and three thousandths

As we move to the right on the number line, the numbers grow larger. Because the point with coordinate 2 lies to the right of the point with coordinate 1.25, 2 is greater than 1.25. Likewise, 1.25 is greater than 1, and 1 is greater than 0.75. Figure 3-2 is an enlarged portion of the number line, showing the relative positions of several points with various decimal coordinates.

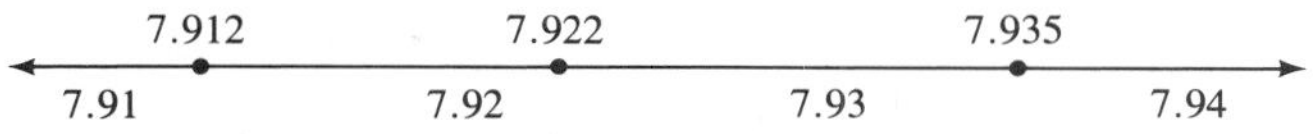

```
         7.912      7.922         7.935
  ←───────•──────────•──────────────•──────────→
    7.91       7.92       7.93         7.94
```

Figure 3-2

Because the point in Figure 3-2 with coordinate 7.92 lies to the right of the point with coordinate 7.91, the number 7.92 is greater than 7.91, and we write

$7.92 > 7.91$

Likewise,

$7.922 > 7.92$

$7.922 > 7.912$

$7.93 \ > 7.92$

$7.91 \ < 7.912$

$7.922 < 7.93$

$7.93 \ < 7.935$

EXAMPLE 2

Arrange the numbers 2.35, 2.345, 1.537, 3.435, and 2.1453 in increasing order.

Solution

We first look at the leftmost digits—those in the ones position. The number with the smallest ones digit is 1.537; it is the smallest of the given numbers. Likewise, because 3 is the largest number in the ones position, 3.435 is the largest of the given numbers. The three remaining numbers,

2.**3**5,   2.**3**45,   and   2.**1**453

have a ones digit of 2. We sort them by looking at the next position to the right, the tenths place. Because 1 is smaller than 3, the smallest of these three numbers is 2.1453.

To decide which of the final two numbers

2.3**5**   and   2.3**4**5

is the smaller, we look at the next position to the right, the hundredths place. There you see the digits 5 and 4. Because 4 is the smaller, the number 2.345 is smaller than 2.35.

Written in increasing order, the five numbers are

1.537,   2.1453,   2.345,   2.35,   and   3.435   ■

Rounding Decimals

The number line provides insight to the process of **rounding** decimals to a position to the right of the decimal point. The number line in Figure 3-3 shows that 5.237 is closer to 5.24 than it is to 5.23. The number 5.237, **rounded to the nearest hundredth**, is 5.24. We also say that 5.237, **rounded to two decimal places**, is 5.24.

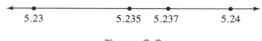

Figure 3-3

> **Rounding Decimals:** To round a number to the nearest tenth, hundredth, thousandth, etc., look at the first digit to the right of that decimal place. If that digit is less than 5, drop it and all the following digits. If it is 5 or larger, add 1 to the digit in the position in which you wish to round, and drop all of the following digits.

EXAMPLE 3

Round the decimal 89.9745 **a.** to the nearest tenth, **b.** to the nearest hundredth, and **c.** to the nearest thousandth.

Solution

a. To round 89.9745 to the nearest tenth (or to one decimal place), we look at the digit in the hundredths position, the 7. Because 7 is greater than 5, we must round up. We do so by adding one to the 9 in the tenths place (which requires a carry into the ones position) and dropping the digits to the right. The result is 90.0. Note that we keep the 0 in the tenths position to show that the number has been rounded to the nearest tenth.

b. To round 89.9745 to the nearest hundredth (or to two decimal places), we look at the digit in the thousandths position, the 4. Because 4 is less than 5, we round down. We do so by dropping the digits to the right of the hundredths position. The result is 89.97.

c. To round 89.9745 to the nearest thousandth (or to three decimal places), we look at the digit in the ten thousandths position. Because it is a 5, we round up. The result is 89.975.

Work Progress Check 2.

Progress Check 2

a. Round 37.95 to the nearest tenth.

b. Round 37.95 to the nearest ten.

2. **a.** 38.0 **b.** 40
Progress Check Answers

3.1 Exercises

In Exercises 1–8, write each number in expanded notation, and indicate how the number would be read.

1. 25.4
2. 3.29
3. 0.16
4. 0.347
5. 5.374
6. 0.009
7. 10.0001
8. 11.111

In Exercises 9–16, write each number in standard decimal notation.

9. three and two hundred fifty-six thousandths

10. three hundred two and fifty-six thousandths

11. three hundred fifty-six thousandths

12. three hundred fifty-six thousand

13. 7 tens + 7 tenths + 7 thousandths

14. 4 hundred + 4 hundredths

15. $3 + \dfrac{5}{10} + \dfrac{7}{100} + \dfrac{1}{1000}$

16. $9 + \dfrac{5}{10} + \dfrac{7}{1000}$

In Exercises 17–24, write each decimal as a mixed number or a proper fraction, and simplify.

17. 2.5   **18.** 4.8   **19.** 3.4   **20.** 5.2

21. 0.15   **22.** 0.04   **23.** 0.125   **24.** 0.625

In Exercises 25–28, arrange the numbers of each group in increasing order.

25. 3.22, 3.2, 2.23, 2.32, 2.33 _____

26. 3.1112, 2.1121, 3.1211, 3.2111, 3.1 _____

27. 0.3201, 1.3102, 1.3012, 0.3210 _____

28. 1.234348, 1.234346, 1.133347 _____

In Exercises 29–36, round each number to the nearest tenth.

29. 3.375   **30.** 3.964   **31.** 0.4392   **32.** 0.3267

33. 0.0135   **34.** 0.004   **35.** 0.255   **36.** 0.667

In Exercises 37–44, round each number to the nearest hundredth.

37. 3.375   **38.** 3.964   **39.** 0.4392   **40.** 0.3267

41. 0.0135   **42.** 0.004   **43.** 0.255   **44.** 0.667

In Exercises 45–52, round each number to the nearest thousandth.

45. 3.3753   **46.** 3.9649   **47.** 0.4392   **48.** 0.3267

49. 0.0195   **50.** 0.0099   **51.** 0.2558   **52.** 0.6673

In Exercises 53–56, round 14.1738 to the nearest . . .

53. one   **54.** ten   **55.** tenth   **56.** hundredth

In Exercises 57–60, round 59.9053 to the nearest . . .

57. ten   **58.** hundred   **59.** tenth   **60.** thousandth

Review Exercises

In Review Exercises 1–6, perform each addition.

1. 435
    + 27

2. 579
    + 98

3. 298
    + 578

4. 678
      17
    + 479

5. 379
    560
    + 700

6. 3794
      98
    + 4527

In Review Exercises 7–12, perform each subtraction.

7. 579
    − 35

8. 780
    − 90

9. 500
    − 488

10. 675
    − 386

11. 1681
    − 779

12. 4787
    − 3699

13. Last month, Maria worked 45 hours during each of 3 weeks and 35 hours during the fourth week. How many hours did Maria work last month?

14. A rectangular garden is twice as long as it is wide. If it is 37 feet wide, find its perimeter.

3.2 Adding and Subtracting Decimals

- Adding Decimals
- Subtracting Decimals
- Mixed Operations

Adding Decimals

We add and subtract decimal numbers in a vertical format, exactly as we add and subtract whole numbers. To add the numbers represented by the decimals 23.726 and 4.59, for example, we write them in a column, being sure to align the decimal points.

$$\begin{array}{r} 23.726 \\ +4.59 \\ \hline . \end{array}$$

Write the numerals in a column, with the decimal points aligned. Align the decimal point of the answer, too.

$$\begin{array}{r} 23.72\boxed{6} \\ +4.59 \\ \hline .\boxed{6} \end{array}$$

Add the digits in the thousandths column.

$$\begin{array}{r} 23.7\overset{1}{\boxed{2}}6 \\ +4.5\boxed{9} \\ \hline .\boxed{1}6 \end{array}$$

Add the digits in the hundredths column. Place the result in the hundredths position of the answer. $2 + 9 = 11$. Write 1 in the hundredths position of the answer and carry 1 to the tenths position.

$$\begin{array}{r} 2\overset{1}{3}.\overset{1}{7}26 \\ +4.59 \\ \hline .316 \end{array}$$

Add the digits in the tenths column, including the carry. $1 + 7 + 5 = 13$. Place the 3 in the tenths position of the answer and carry 1 to the ones column.

94   Chapter 3   Decimals

$$\begin{array}{r}\overset{11}{2\,3.7\,2\,6}\\+4.5\,9\\\hline 8.3\,1\,6\end{array}$$ Add the digits in the ones column. The result is 8. There is no carry.

$$\begin{array}{r}\overset{11}{2\,3.7\,2\,6}\\+4.5\,9\\\hline 2\,8.3\,1\,6\end{array}$$ Add the digits in the tens column.

The sum of 23.726 and 4.59 is 28.316. Work Progress Check 3.

Progress Check 3

Find the sum of 143.5, 216.3, and 640.2.

EXAMPLE 1

A rectangle is 19.6 feet wide and 37.3 feet long. Find its perimeter.

Solution

To find the perimeter of a rectangle, we find the sum of the lengths of its four sides. We perform the calculations in a vertical format.

$$\begin{array}{r}\overset{3\,1}{3\,7.3}\leftarrow \text{Length}\\19.6\leftarrow \text{Width}\\37.3\leftarrow \text{Length}\\+\,19.6\leftarrow \text{Width}\\\hline 113.8\end{array}$$

The perimeter of the rectangle is 113.8 feet.

Subtracting Decimals

To subtract the number 37.8 from the number 53.13, for example, we write them in a column with the decimal points aligned.

$$\begin{array}{r}5\,3\,.\,1\,3\\-\,3\,7\,.\,8\\\hline .\end{array}$$ Write the numerals in a column, with the decimal points aligned. Align the decimal point of the answer, too.

$$\begin{array}{r}5\,3\,.\,1\,\boxed{3}\\-\,3\,7\,.\,8\\\hline .\boxed{3}\end{array}$$ Subtract the hundredths digits: $3 - 0 = 3$. Place the result, 3, in the hundredths position of the answer.

$$\begin{array}{r}5\,\overset{2}{\cancel{3}}\,.\,\boxed{{}_1 1}\,3\\-\,3\,7\,.\,\boxed{8}\\\hline .\,\boxed{3}\,3\end{array}$$ To subtract in the tenths column, borrow ten more tenths from the ones column: $11 - 8 = 3$.

$$\begin{array}{r}\overset{4}{\cancel{5}}\,\overset{12}{\cancel{3}}\,.\,{}_1 1\,3\\-\,3\,\boxed{7}\,.\,8\\\hline \boxed{5}\,.\,3\,3\end{array}$$ To subtract in the ones column, borrow ten more ones from the tens column: $12 - 7 = 5$.

$$\begin{array}{r}\overset{4}{\cancel{5}}\,\overset{12}{\cancel{3}}\,.\,{}_1 1\,3\\-\,3\,7\,.\,8\\\hline \boxed{1}\,5\,.\,3\,3\end{array}$$ Subtract in the tens column.

Thus, $53.13 - 37.8 = 15.33$.

3. 1000.0
Progress Check Answers

EXAMPLE 2

If Jennifer has $137.23 in her savings account, and she withdraws $47.80, what will be her final balance?

Solution

From Jennifer's current balance, we subtract the amount of the withdrawal.

$$\begin{array}{r} 137.23 \\ -47.80 \\ \hline \end{array}$$ Align the decimal points.

$$\begin{array}{r} 137.23 \\ -47.80 \\ \hline .3 \end{array}$$ Subtract in the hundredths column (the pennies column): $3 - 0 = 3$.

$$\begin{array}{r} 13\overset{6}{\cancel{7}}.{}^{1}23 \\ -47.80 \\ \hline .43 \end{array}$$ To subtract in the tenths column (the dimes column), borrow ten dimes from the dollars column: $12 - 8 = 4$.

$$\begin{array}{r} 1\overset{2}{\cancel{3}}\overset{16}{\cancel{7}}.{}^{1}23 \\ -47.80 \\ \hline 9.43 \end{array}$$ To subtract in the ones column (the dollars column), borrow ten dollars from the tens column: $16 - 7 = 9$.

$$\begin{array}{r} 1\overset{2}{\cancel{3}}\overset{16}{\cancel{7}}.{}^{1}23 \\ -47.80 \\ \hline 89.43 \end{array}$$ Subtract in the tens column: 12 tens − 4 tens = 8 tens.

Jennifer's final balance is $89.43. Work Progress Check 4.

Progress Check 4

Subtract: $432.10 - 345.67$.

Mixed Operations

EXAMPLE 3

Last month, Gary won $325 playing the state lottery, and he won another $255 yesterday. He has spent $535.50 on lottery tickets. What is Gary's profit?

Solution

From Gary's total winnings, we subtract the cost of the tickets. To find his total winnings, we find the sum $325 + 255$:

$$\begin{array}{r} \overset{1}{}325 \\ +255 \\ \hline 580 \end{array}$$

4. 86.43

Progress Check Answers

96   Chapter 3   Decimals

Progress Check 5

Mia agreed to a total price of $11,347 for her new car. She received a $500 factory rebate, got $5750 for her old car, and paid $47.75 in various fees in addition to the total price of the car. How much cash did Mia provide?

---

5. $5144.75

Progress Check Answers

Gary won a total of $580. From this total, we subtract $535.50, the cost of the tickets. Because we are no longer dealing with whole dollars, we must write the total winnings as $580.00.

$$\begin{array}{r} 5\,8\,0\,.\,0\,0 \\ -\ 5\,3\,5\,.\,5\,0 \\ \hline 4\,4\,.\,5\,0 \end{array}$$

To subtract in the tenths column, borrow from the *tens* column.

Gary's profit is $44.50. The calculations of this example can be written in horizontal form as

Profit in dollars = 325 + 255 − 535.50 = 44.50

The calculations are performed left to right. Work Progress Check 5.

3.2 Exercises

In Exercises 1–12, perform each addition.

1. 43.7
 + 93.9

2. 45.73
 + 29.57

3. 152.62
 + 58.4

4. 776.03
 + 24.97

5. 27.3 + 14.8

6. 5.39 + 19.12

7. 3.26 + 19.1

8. 1.347 + 9.8

9. 147.2 + 19.28

10. 155.97 + 99.1

11. 153.76 + 18

12. 1347 + 12.88

In Exercises 13–24, perform each subtraction.

13. 51.0
 − 23.9

14. 73.62
 − 58.66

15. 103.7
 −   9.86

16. 500.03
 −   1.047

17. 47.7 − 34.8

18. 15.49 − 13.5

19. 0.36 − 0.003

20. 1.447 − 0.7

21. 153.2 − 20.08

22. 1443.9 − 8.05

23. 17.76 − 14.92

24. 198.4 − 10.66

In Exercises 25–32, perform each operation.

25. 251.3 − 30.08

26. 363.7 + 7.09

27. 143.26 + 254.9

28. 179.04 − 92.9

29. 1277 + 370.3

30. 1391.4 − 48.5

31. 1737.76 − 13.002

32. 144.4 + 33.5666

In Exercises 33–36, find the value of each expression. Operations are performed from left to right.

33. 131.5 + 31.3 − 72.5

34. 47.55 + 24.72 − 47.29

35. 335.37 − 121.7 + 57.9

36. 345.01 − 137.2 − 56.09

In Exercises 37–40, find the perimeter of the rectangle with the given dimensions.

37. length = 37.2 feet, width = 21.7 feet

38. length = 128.9 feet, width = 37.0 feet

39. length = 175.27 meters, width = 137.90 meters

40. length = 49.07 kilometers, width = 38.95 kilometers

41. A rope has been cut into three pieces, measuring 37.5, 88.7, and 107.0 meters. Find the length of the original rope.

42. Deposits of $156.78, $326.27, and $512 have been made to a savings account that contained $1378.55. Find the balance after the deposits.

43. George's checking account contains $267.31. Find his balance if he writes a check for $27.39.

44. If Susan's checking account balance is $352.92 and she writes a check for $376.25, by how much will the account be overdrawn?

45. Diane's savings account contains $357.93. After she deposits $231 and withdraws $472.53, what is her final balance?

46. A bag of candy weighs 37.5 ounces. If 17.8 ounces have been removed, how much remains?

47. A piece of wire exactly 23.9 centimeters long is cut from a piece that is 120.0 centimeters long. Find the length of the remaining wire.

48. Before the holidays, Dave weighed 195.5 pounds, but he soon gained 14.5 pounds. If Dave follows his doctor's advice and loses 25 pounds, what will he weigh?

In Exercises 49–52, refer to the prices in Illustrations 1 and 2.

Illustration 1

Illustration 2

49. Ernie purchased two packages of briefs and three packages of crew-neck T-shirts. What did the purchases total?

50. Carlos bought one package each of boy's briefs and crew-neck T-shirts for his son. What did the purchases total?

51. Andre had two 24-exposure and one 36-exposure rolls of film developed at Walgreens. How much did it cost?

52. MiLee needs to shoot 72 exposures. How much will she save on processing if she shoots two 36-exposure rolls rather than three 24-exposure rolls?

Review Exercises

In Review Exercises 1–6, perform each multiplication.

1. 317 × 37

2. 249 × 85

3. 472 × 12

4. 381 × 46

5. 507 × 250

6. 290 × 208

In Review Exercises 7–12, perform each division.

7. $57\overline{)2166}$ **8.** $43\overline{)4171}$ **9.** $38\overline{)2090}$ **10.** $91\overline{)8281}$

11. $159\overline{)3172}$ **12.** $172\overline{)5291}$

13. Jim used 107.2 kilowatt-hours of electricity in January. In February, he used 93.7 and in March, 110.9 kilowatt-hours. What was his total usage for the three months?

14. If Jim used 113.7 kilowatt-hours of electricity in April, determine the increase above March. (See Review Exercise 13.)

3.3 Multiplying and Dividing Decimals

- Multiplication of Decimals
- Multiplication by a Power of 10
- Division of Decimals
- Dividing by a Power of 10
- Changing Fractions into Decimals

Multiplication of Decimals

We multiply decimals in a vertical format, using a method very similar to that used for multiplying whole numbers. To multiply the decimals 5.25 and 3.7, for example, we proceed as follows:

$\begin{array}{r} 5.25 \\ \times\ \ 3.7 \end{array}$	Write the factors in a column, aligning the right edges of the factors; there is no need to align the decimal points in multiplication.
$\begin{array}{r} \overset{3}{\ \ \ }\\ 5.2\boxed{5} \\ \times\ \ 3.\boxed{7} \end{array}$	Multiply 5 by 7. The product is 35. Place 5 in the right-most column and carry 3.
$\begin{array}{r} \overset{1}{\ }\overset{3}{\ \ }\\ 5.\boxed{2}5 \\ \times\ \ 3.\boxed{7} \\ \hline 7\ 5 \end{array}$	Multiply 2 by 7. The product is 14. To 14, add the carried 3 to get 17. Place the 7 below the line and carry the 1.
$\begin{array}{r} \overset{1}{\ }\overset{3}{\ \ }\\ \boxed{5}.2\ 5 \\ \times\ \ 3.\boxed{7} \\ \hline 3\ 6\ 7\ 5 \end{array}$	Multiply 5 by 7. The product is 35. To this, add the carried 1 to get 36. Write 36.
$\begin{array}{r} \overset{1}{\ \ }\\ 5.2\boxed{5} \\ \times\ \ \boxed{3}.7 \\ \hline 3\ 6\ 7\ 5 \\ 5 \end{array}$	Erase the digits that were carried. Multiply 5 by 3. The product is 15. Because the place value of the 3 is ten times that of the 7, the next row is positioned one place to the left. Place the 5 as indicated and carry 1.
$\begin{array}{r} \overset{1}{\ \ }\\ 5.\boxed{2}\ 5 \\ \times\ \ \boxed{3}.7 \\ \hline 3\ 6\ 7\ 5 \\ 7\ 5 \end{array}$	Multiply 3 by 2. The product is 6. Add the carry, 1. Place 7 on the bottom row as indicated.
$\begin{array}{r} \boxed{5}.2\ 5 \\ \times\ \ \boxed{3}.7 \\ \hline 3\ 6\ 7\ 5 \\ 1\ 5\ 7\ 5 \end{array}$	Multiply 5 by 3 and write the product, 15, in the bottom row, as indicated.
$\begin{array}{r} 5.2\ 5 \\ \times\ \ 3.7 \\ \hline 3\ 6\ 7\ 5 \\ 1\ 5\ 7\ 5\ \ \\ \hline 1\ 9\ 4\ 2\ 5 \end{array}$	Draw another line beneath the two completed rows and add the two rows.

In the final step, we position the decimal point correctly in the answer. By rounding the factors 5.25 and 3.7 to the nearest whole number, we estimate that the product is approximately 5×4, or 20. To make our answer close to 20, we mark off three digits from the right and place the decimal point between the 9 and the 4. The product is 19.425.

Another method can be used to position the decimal point in the answer. We do so by counting the total number of digits that appear to the right of the decimal points in the two factors. We then position the decimal point in the answer so that the number of digits appearing to its right is this total. In this example, the factor 5.23 has two digits to the right of its decimal point, and the other factor, 3.7, has one digit to the right of its decimal point. Thus, there are a total of three digits to the right of the decimal points. On the bottom row, we position the decimal point so that three digits appear to its right.

```
       5.2 5
   ×     3.7
     3 6 7 5
   1 5 7 5
   1 9.4 2 5
       ↑ 3 digits
```

In the factors, there are a total of three digits to the right of the decimal points.

We count off three digits from the right of the result on the bottom line and place the decimal point there.

The product of 5.25 and 3.7 is 19.425.

EXAMPLE 1

Find the product of 2.57 and 1.35.

Solution

```
       2.5 7
   ×   1.3 5
     1 2 8 5    Multiply 257 by 5.
       7 7 1    Multiply 257 by 3.
     2 5 7      Multiply 257 by 1.
     3.4 6 9 5  Add. Mark off four digits to the right of the decimal
                point.
```

Work Progress Check 6. ∎

Progress Check 6

Find the product:
32.05×47.20.

Multiplication by a Power of 10

It is easy to multiply a decimal by a power of 10 (by 10, by 100, by 1000, etc.). We simply move the decimal point. As an example, we multiply 1.234 by 10^2. That is, we find the product 1.234×100.

```
       1.2 3 4
   ×     1 0 0
       0 0 0 0    Multiply 1234 by 0 ones.
     0 0 0 0      Multiply 1234 by 0 tens.
   1 2 3 4        Multiply 1234 by 1 hundred.
   1 2 3.4 0 0    Add. Mark off three digits to the right of the decimal
                  point.
```

The product of 1.234 and 100 is 123.4. To multiply 1.234 by 10^2, or 100, we simply move the decimal point two places to the right:

$$1.234 \times 100 = 1.23.4$$

6. 1512.7600 or 1512.76

Likewise, to multiply 1.234 by 10^3, or 1000, we move the decimal point three places to the right:

$1.234 \times 1000 = 1\underset{\smile}{.234}.$

> **Multiplying a Decimal by a Power of 10:** To multiply a decimal by a whole-number power of 10, move the decimal point to the right.
>
> For example, to multiply by 10^2, move the decimal point two places to the right. To multiply by 10^7, move the decimal point seven places to the right.

Work Progress Check 7.

Progress Check 7

Multiply: $.000234 \times 10^6$.

Division of Decimals

We divide decimals in the same format we used for long division of whole numbers, except that we treat the remainder differently and we must be careful about the positions of the decimal points. One of the first steps in the long division process involves moving some decimal points. To understand why, we consider the division $4.536 \div 1.08$. First we write the division as the fraction

$$\frac{4.536}{1.08}$$

and then as an equivalent fraction with a whole-number denominator. We do this by using the fundamental property of fractions and multiplying the fraction by 1 in the form $\frac{100}{100}$. We multiply both numerator and denominator by 100:

$$\frac{4.536}{1.08} = \frac{4.536 \cdot 100}{1.08 \cdot 100} = \frac{453.6}{108}$$

We note that the decimal points in *both* the divisor and the dividend have been moved two places to the right, and that move has turned the divisor into a whole number.

To divide 4.536 by 1.08, we proceed as follows.

$1.0\,8\,)\overline{4.5\,3\,6}$ Place the divisor and the dividend as indicated. The quotient will appear above the long division symbol.

$1\,0\,8.\,)\overline{4\,5\,3.6}$ Move the decimal point of the divisor two places to the right, to make the divisor a whole number. Move the decimal point of the dividend two places to the right also.

$1\,0\,8.\,)\overline{4\,5\,3\overset{.}{.}6}$ Place the decimal point of the quotient directly above the decimal point in the dividend.

To begin the division process, we ask, *How many times does 108 divide into 4?* The answer is *zero. How many times does 108 divide 45?* Again, the answer is *zero. How many times does 108 divide 453?* Now, the answer is 4. We place a 4 in the quotient and continue as follows:

7. 234

$$\begin{array}{r} 4.\\ 108.\overline{)453.6}\\ 432 \end{array}$$
Estimate that the number of times 108 divides 453 is 4. Place 4 in the position indicated. Multiply 108 by 4, and place the product 432 as indicated.

$$\begin{array}{r} 4.\\ 108.\overline{)453.6}\\ 432\downarrow\\ \overline{216} \end{array}$$
Subtract and bring down the 6.

$$\begin{array}{r} 4.2\\ 108.\overline{)453.6}\\ 432\downarrow\\ \overline{216}\\ 216 \end{array}$$
108 divides into 216 exactly 2 times. Place 2 in the quotient and multiply 108 by 2, placing the product, 216, as indicated.

$$\begin{array}{r} 4.2\\ 108.\overline{)453.6}\\ 432\downarrow\\ \overline{216}\\ \underline{216}\\ 0 \end{array}$$
Subtract. Because the remainder is 0, the division process stops.

When 4.536 is divided by 1.08, the quotient is 4.2. We check this result by verifying that *divisor · quotient = dividend*, or that

$(1.08)(4.2) = 4.536$

Sometimes the long division process does not end, because a subtraction never results in a zero. In such cases, we round the quotient to some specified number of decimal places. To do so, we carry the division one place beyond the specified number of places and then follow the rules for rounding discussed earlier. The next example illustrates this situation.

EXAMPLE 2

On a 230-mile trip, Heidi's car used 7.3 gallons of gas. How many miles per gallon did her car get? Round the answer to the nearest tenth.

Solution

To find how many miles the car gets per gallon, we divide the number of miles traveled by the number of gallons used, using a long division.

$$7.3\overline{)230.0}$$

Because we are to round the answer to the nearest tenth, we carry the quotient one place beyond tenths, to the hundredths position. We move the decimal points of both divisor and dividend one place to the right, and then attach two zeros after the decimal point of the dividend so that we carry out the division far enough to round correctly.

$$73.\overline{)2300.00}$$
Move the decimal points of both divisor and dividend one place to the right. Add two zeros after the decimal point.

To begin the division process, we ask, *How many times does 73 divide into 2? Into 23? Into 230?* The number 230 is the first number that 73 will divide more than zero times. We continue:

$$\begin{array}{r}3.\\73\overline{)2300.00}\\219\\\hline 110\end{array}$$

How many times does 73 divide into 230? About 3. Place 3 as indicated, multiply, subtract, and bring down the 0.

$$\begin{array}{r}31.\\73\overline{)2300.00}\\219\\\hline 110\\73\\\hline 370\end{array}$$

73 divides into 110 only 1 time. Place 1 as indicated, multiply, subtract, and bring down another 0.

$$\begin{array}{r}31.5\\73\overline{)2300.00}\\219\\\hline 110\\73\\\hline 370\\365\\\hline 50\end{array}$$

370 divided by 73 is 5. Place 5 as indicated, multiply, subtract, and bring down another 0.

$$\begin{array}{r}31.50\\73\overline{)2300.00}\\219\\\hline 110\\73\\\hline 370\\365\\\hline 50\end{array}$$

Divide 50 by 73. The answer is 0. Place 0 in the quotient, as indicated.

We have carried the computation one digit beyond what is required, and because that digit, 0, is less than 5, we round down. Heidi's car gets 31.5 miles per gallon. Work Progress Check 8. ∎

Progress Check 8

Divide: $254.32 \div 6.8$.

Dividing by a Power of 10

To multiply a decimal by a power of 10, we move the decimal point to the *right*. To divide by a power of 10, we move the decimal point to the *left*. To divide by 10^1, or 10, we move the decimal point one place to the left; to divide by 10^2, or 100, we move it two places to the left; and so on.

To illustrate why this is true, we use long division to divide 123.4 by 10^1, or 10:

$$\begin{array}{r}12.34\\10\overline{)123.40}\\10\\\hline 23\\20\\\hline 34\\30\\\hline 40\\40\\\hline 0\end{array}$$

Thus, to divide 123.4 by 10^1, we move the decimal point one place to the left:

$$123.4 \div 10 = 12.3.4$$

One place left

8. 37.4

Progress Check Answers

EXAMPLE 3

a. $123.4 \div 100 = 1.23.4 = 1.234$
   Two places left

b. $123.4 \div 1000 = .123.4 = .1234$
   Three places left

> **Dividing a Decimal by a Power of 10:** To divide a decimal by a whole-number power of 10, move the decimal point to the left.
> For example, to divide by 10^2, move the decimal point two places to the left. To divide by 10^7, move the decimal point seven places to the left.

Changing Fractions into Decimals

Any fraction can be written in decimal form. For example, to change $\frac{3}{8}$ to a decimal, we divide 3 by 8 to obtain 0.375:

```
    0.3 7 5
 8)3.0 0 0
   2 4 ↓ ↓
     6 0
     5 6
       4 0
       4 0
          0
```

Because the division ends, the quotient is a **terminating decimal**. If we change a fraction such as $\frac{4}{15}$ to a decimal, we obtain the **repeating decimal** 0.2666..., in which the digit 6 repeats forever:

```
     0.2 6 6 ...
 15)4.0 0 0
    3 0 ↓ ↓
    1 0 0
      9 0
      1 0 0
        9 0
        1 0
```

It is true that the decimal forms of all fractions are either terminating decimals or repeating decimals. For practical purposes, repeating decimals are rounded to an appropriate number of decimal places.

3.3 Exercises

In Exercises 1–26, perform each multiplication.

1. 43.5
    × 2.9

2. 37.9
    × 9.2

3. 12.8
    × 2.59

4. .7795
    × 7.2

5. 2909
    × .002

6. 1.005
    × 25.0

7. 37.9 × 34.7
8. 251.3 · 19.1
9. (14.6)(21.0)
10. 0.1347 × 0.98
11. 577.6 · 93.12
12. 15.79 · 9.3
13. (133.01)(18)
14. (1379)(21.8)
15. (1678.9)(73.50)
16. 175.390 · 93
17. 7135.31 × 12
18. 2389 · 98.6
19. 137.8 × 10
20. (1.5590)(100)
21. 3.5231 · 1000
22. 0.28 × 1000
23. (1.008)(10)
24. 6.5002 · 100
25. 235.1 · 1000
26. 0.367 × 1000

In Exercises 27–46, perform each division.

27. 4.7)16.92
28. 5.8)16.82
29. 9.6)54.72
30. 5.7)5.472
31. 4.34)65.1
32. 0.19)826.5
33. 2.3)41.4
34. 0.18)4.140
35. 16.536 ÷ 3.12
36. 11.088 ÷ 7.92
37. 339.57 ÷ 23.1
38. 3.684 ÷ 122.8
39. 3.88 ÷ 10
40. 15.930 ÷ 100
41. 3.5781 ÷ 1000
42. 0.678 ÷ 1000
43. 0.083 ÷ 10
44. 175.60 ÷ 100
45. 23.791 ÷ 1000
46. 13500.6 ÷ 1000

In Exercises 47–58, perform each operation.

47. 35.02 · 2.9
48. 37.12 ÷ 100
49. 37.12 · 1000
50. 617.76 ÷ 31.2
51. 19.8 · 31.2
52. (7.310)(100)
53. 91.77 ÷ 2.1
54. 91.77 ÷ 100
55. 91.77 ÷ 43.7
56. 18.32 × 100
57. 228.12 ÷ 12
58. (45.1)(0.35)

59. To paint the living room, Luis needs two gallons of satin finish for the walls and one gallon of enamel for the trim. If he also buys the 3-piece paint kit, what will be the total bill? Refer to Illustration 1.

60. Rhonda will need seventeen 4 × 8-foot panels to finish a family room in her basement. She can't decide between Royal Oak and Winter Oak. What would be the difference in the cost? Refer to Illustration 2.

Illustration 1

Plywood 4 x 8 PANELING			
WHITE HEMLOCK	9.89	NATURAL BIRCH	11.99
CASCADE BLUE	9.89	ROYAL OAK	12.59
WESTMINISTER PECAN	10.99	CHEYENNE	15.59
WINTER OAK	10.99	DANUBEE ROSE	15.79

Illustration 2

61. Each 3-pack advertised in Illustration 3 provides 84 exposures. Wendy has $35 to spend on film. If she buys all the same kind, how many exposures of each of the three film speeds can she purchase?

62. How much more money would Wendy need to purchase three packages of 400 speed film? See Exercise 61.

63. When Bill filled the gas tank, his odometer read 67,352. When the odometer read 67,723 he needed 17.5 gallons of gas to refill the tank. To the nearest tenth, how many miles per gallon does Bill's car get?

64. There are approximately 18.7 million Americans working for some form of government, out of a total population of 248 million. To the nearest tenth, how many Americans are there for each government worker?

Illustration 3

Review Exercises

In Review Exercises 1–4, perform each division to find the quotient and remainder. Then write the fraction as a mixed number.

1. $\dfrac{27}{6}$
2. $\dfrac{43}{19}$
3. $\dfrac{270}{35}$
4. $\dfrac{972}{27}$

In Review Exercises 5–8, write each mixed number as an improper fraction.

5. $5\dfrac{5}{9}$
6. $7\dfrac{7}{12}$
7. $9\dfrac{8}{9}$
8. $13\dfrac{1}{13}$

In Review Exercises 9–12, find the prime factorization of each number.

9. 900
10. 1764
11. 2448
12. 12100

13. How many 3.75-inch pieces of wire can be cut from a wire that is 108 inches long?

14. In Review Exercise 13, how long is the unusable scrap?

3.4 Applications of Decimals

- Formulas for Rectangles and Squares
- Formulas for Circles
- Formulas for Triangles

In the applications of mathematics, many quantities are related to each other. For example, distance traveled is related to the time spent traveling as well as to the speed. The **area** of a rectangle is related to its length and its width. The interest earned in a bank account is related to the amount of money on deposit, to the rate of interest, and to the time that the money is left on deposit. A **formula** is a mathematical expression used to express the relationships between quantities.

Formulas for Rectangles and Squares

As we have seen, the area A of a rectangle is related to the rectangle's length l and its width w by the formula $A = l \cdot w$:

Formula for the Area of a Rectangle: If l is the length of a rectangle and w is its width, the area A is

$A = l \cdot w$

A **square** is a rectangle in which the length and the width are equal: All four sides are the same length. If we let s represent the length of any side, then the area of a square is $A = s \cdot s$, or $A = s^2$.

Formula for the Area of a Square: If s is the length of one side of a square, the area A is

$A = s^2$

EXAMPLE 1

To the nearest square meter, find the area of the L-shaped garden shown in Figure 3-4.

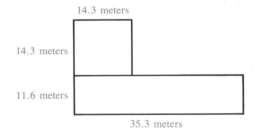

Figure 3-4

Solution

The area of the garden is the sum of two areas—the area of the square and the area of the rectangle. Because each side of the square is 14.3 meters, we substitute 14.3 for s in the formula for the area of a square.

$A = s^2$ The formula for the area of a square.
$A = (14.3)^2$ Substitute 14.3 for s.
$ = 204.49$ To square 14.3, multiply 14.3 by itself.

The area of the square is 204.49 square meters.

The length of the rectangle is 35.3 meters, and its width is 11.6 meters. We substitute 35.3 for l and 11.6 for w in the formula for the area of a rectangle.

$A = l \cdot w$ The formula for the area of a rectangle.
$A = (35.3)(11.6)$ Substitute 35.3 for l and 11.6 for w.
$ = 409.48$

The area of the rectangle is 409.48 square meters.

The area of the L-shaped garden is the sum of these two areas.

Total area = area of square + area of rectangle
Total area = 204.49 + 409.48
= 613.97

The garden covers 613.97 square meters. Rounded to the nearest square meter, the area of the garden is 614 square meters. Work Progress Check 9.

Progress Check 9

Find the area of a rectangle that is 5.3 feet long and 2.7 feet wide.

Formulas for Circles

A **circle** is the geometric shape of round objects such as rings, hula hoops, and bicycle tires.

The **radius of a circle** is the distance from the center of the circle to the circle itself. The radius of the circle in Figure 3-5 is 7 inches.

The **diameter of a circle** is the distance measured directly across the circle, through its center. The diameter of the circle in Figure 3-5 is 14 inches, or two times the radius. Note that the radius is also one-half of the diameter. Thus we have the following fact:

Figure 3-5

If the radius of a circle is r and the diameter of a circle is d, then

$$d = 2r \quad \text{and} \quad r = \frac{d}{2}$$

The **circumference of a circle** is the distance around the circle, and the **area of a circle** is the amount of surface contained within the circle. Formulas for the circumference and the area of a circle involve the number **pi**, denoted by the Greek letter π. The value of π is an unending decimal:

$\pi = 3.14159265\ldots$

For most calculations, we round π to the nearest hundredth. In this text, we will use the value 3.14 for π.

For most calculations, we may assume that

$\pi = 3.14$

The circumference C of a circle with radius r (or diameter d) is given by the following formulas.

---

9. 14.31 square feet

Progress Check Answers

Formulas for the Circumference of a Circle: If r is the radius of a circle, then the circumference C is

$$C = 2\pi r$$

If d is the diameter of a circle, then the circumference C is

$$C = \pi d$$

The **area** A of a circle with radius r is given by another formula.

Formula for the Area of a Circle: If r is the radius of a circle, then the area A is

$$A = \pi r^2$$

EXAMPLE 2

The diameter of a circle is 54 centimeters. Find **a.** the circumference of the circle rounded to the nearest centimeter and **b.** the area of the circle rounded to the nearest square centimeter.

Solution

a. To find the circumference, we substitute 54 for d and 3.14 for π in the formula for the circumference of a circle.

$$C = \pi d$$
$$C = (3.14)(54)$$
$$= 169.56$$

Rounded to the nearest centimeter, the circumference of the circle is approximately 170 centimeters.

b. To find the area of the circle, we first determine its radius. Because the radius of a circle is one-half of its diameter, we have

$$r = \frac{d}{2}$$
$$r = \frac{54}{2} \quad \text{Substitute 54 for } d.$$
$$= 27 \quad \text{Divide.}$$

Because $r = 27$, we substitute 27 for r and 3.14 for π in the formula for the area of a circle.

$$A = \pi r^2$$
$$A = (3.14)(27)^2$$
$$= (3.14)(729) \quad \text{Evaluate the exponential factor first.}$$
$$= 2289.06$$

Rounded to the nearest square centimeter, the area of the circle is 2289 square centimeters.

Work Progress Check 10.

Progress Check 10

The diameter of a circle is 12.6 cm.

a. Rounded to the nearest tenth of a square centimeter, find its area.

b. Rounded to the nearest tenth of a centimeter, find its circumference.

10. a. 124.6 sq cm **b.** 39.6 cm

Progress Check Answers

Formulas for Triangles

A **triangle** is a three-sided figure like that in Figure 3-6. The **base** of that triangle is the 8.2-centimeter length of its bottom side. The **height** of the triangle is the 6.3-centimeter distance measured straight down to the base. The **area of a triangle** is one-half of the product of its base b and its height h.

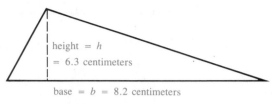

Figure 3-6

The area A of a triangle is given by the formula

Formula for the Area of a Triangle: If b is the base of a triangle and h is its height, then the area A is

$$A = \frac{1}{2}bh$$

EXAMPLE 3

Find the area of the triangle in Figure 3-6.

Solution

The base of the triangle is 8.2 centimeters, and the height is 6.3 centimeters. We substitute 8.2 for b and 6.3 for h in the formula for the area of a triangle and simplify.

$A = \frac{1}{2}bh$

$A = \frac{1}{2}(8.2)(6.3)$ Perform the multiplications from left to right: $\frac{1}{2} \cdot 8.2 = 4.1$.

$= (4.1)(6.3)$

$= 25.83$

The area of the triangle is 25.83 square centimeters. Work Progress Check 11.

We summarize these formulas as follows:

A carpenter cuts ten triangular sheets of plywood. Each triangle has a base of 7.8 feet, and a height of 4.2 feet. Find the total area of the triangles.

11. 163.8 sq ft

Progress Check Answers

> **Formulas from Geometry:**
> The area A of a rectangle with length l and width w: $A = lw$
> The area A of a square with side s: $A = s^2$
> The circumference C of a circle with radius r: $C = 2\pi r$
> The circumference C of a circle with diameter d: $C = \pi d$
> The area A of a circle with radius r: $A = \pi r^2$
> The area A of a triangle with base b and height h: $A = \frac{1}{2}bh$

3.4 Exercises

In Exercises 1–4, find the area of a rectangle with the given dimensions. Round answers to one decimal place.

1. Length = 13.5 meters, width = 7.3 meters

2. Length = 12.7 feet, width = 9.32 feet

3. Length = 23.9 miles, width = 14.3 miles

4. Length = 0.42 inch, width = 0.58 inch

In Exercises 5–12, find the area of a square with the given side. Round answers to one decimal place.

5. 14.1 meters

6. 15.1 centimeters

7. 25.9 inches

8. 0.91 foot

9. 5.97 millimeters

10. 2.83 meters

11. 2.59 inches

12. 1.091 meters

In Exercises 13–20, find the circumference of a circle with the given dimensions. Round answers to the nearest tenth of a unit. (Use $\pi = 3.14$.)

13. Radius 12 inches

14. Diameter 19 meters

15. Diameter 29 feet

16. Radius 192 centimeters

17. Radius 2.38 millimeters

18. Radius 14.09 inches

19. Diameter 19.5 feet

20. Diameter 109.1 meters

In Exercises 21–28, find the area of a circle with the given dimensions. Round answers to the nearest tenth of a square unit. (Use $\pi = 3.14$.)

21. Radius 6 inches

22. Diameter 28 meters

23. Radius 29 feet

24. Radius 192 centimeters

25. Radius 2.38 millimeters

26. Radius 14.09 inches

27. Diameter 29.6 feet

28. Diameter 126.2 meters

In Exercises 29–36, find the area of a triangle with the given dimensions. Round answers to the nearest square unit.

29. Base = 16 meters, height = 7 meters

30. Base = 12 feet, height = 3 feet

31. Base = 2.9 meters, height = 13 meters

32. Base = 1.41 inches, height = 1.08 inches

33. Base = 6.32 meters, height = 7.1 meters

34. Base = 1.2 miles, height = 1 mile

35. Base = 7.21 meters, height = 1.7 meters

36. Base = 2.12 feet, height = 1.3 feet

In Exercises 37–44, find the area of each figure. Round answers to the nearest tenth of a square unit.

37.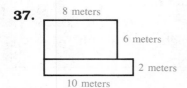

38.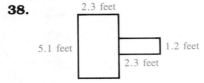

39.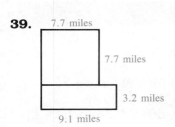

40.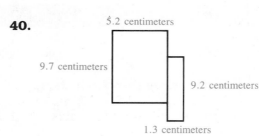

41.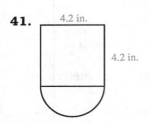

42.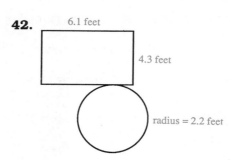

43.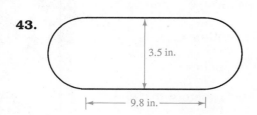

44.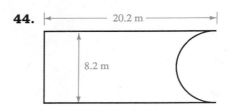

Review Exercises

In Review Exercises 1–12, perform each operation.

1. $23.576 \cdot 10$

2. $254.26 \cdot 1000$

3. $46.397 \cdot 100$

4. $4.2 \cdot 10$

5. $0.351 \cdot 10^3$

6. $26.26 \cdot 10^4$

7. $236.78 \div 10$

8. $578.2 \div 100$

9. $0.378 \div 1000$

10. $0.724 \div 10$

11. $3.567 \div 10^3$

12. $7.164 \div 10^4$

Mathematics for Fun

By dividing 1 by 7, the fraction $\frac{1}{7}$ can be changed into the unending decimal 0.**142857**142857.... Notice two things about this decimal: It repeats in blocks of 6 digits, and 6 is one less than the prime number, 7. The 6 digits of this block are the interesting number 142857.

This is the key to finding other interesting numbers that have the same digit-repeating property: If p is a prime number and the decimal form of the fraction $\frac{1}{p}$ repeats in blocks of one-less-than-p digits, then that block of digits is another interesting number.

The next interesting number contains 18 digits. It is the repeating block produced by the fraction $\frac{1}{19}$. Few calculators can handle this problem, but with pencil and paper, you can. The block begins with a zero; don't ignore it. When this 18-digit number is multiplied by any number from 1 through 18, the same digits appear in a circular pattern.

What happens when you multiply this interesting number by 19?

Chapter Summary

Key Words

area (3.4)
base (3.4)
circle (3.4)
circumference (3.4)
coordinate (3.1)
decimal fraction (3.1)
decimal notation (3.1)
diameter (3.4)
expanded notation (3.1)
formula (3.4)
height (3.4)
number line (3.1)
pi (3.4)
place value (3.1)
radius (3.4)
rounding (3.1)
square (3.4)
triangle (3.4)

Key Ideas

(3.2) Decimal numbers can be added and subtracted in a vertical format, as if they were whole numbers. When you write them in a column, align the decimal points in a vertical column.

(3.4)
$A = l \cdot w$ The area of a rectangle
$A = s^2$ The area of a square
$C = 2\pi r$
   or $C = \pi d$ The circumference of a circle
$A = \pi r^2$ The area of a circle
$A = \frac{1}{2}bh$ The area of a triangle

(3.3) The number of places to the right of the decimal point in a product is the total of the number of decimal places to the right of the decimal points in the numbers being multiplied.

To multiply (or divide) by a whole-number power of 10, we move the decimal point to the right (or left).

To divide a decimal by a decimal, the decimal point must be moved the same number of places in the divisor and the dividend so that the divisor is a whole number.

To write a fraction as a decimal, perform the division indicated by the fraction.

Chapter 3 Review Exercises

[3.1] In Review Exercises 1–4, write each number in expanded notation, and indicate how the number would be read.

1. 43.7   **2.** 302.02   **3.** 31.106   **4.** 0.579

In Review Exercises 5–8, write each number in standard decimal notation.

5. Two hundred fifty-three and fifty-three thousandths _____

6. Two hundred and fifty-three thousandths _____

7. Two hundred fifty-three thousandths _____

8. Five and five hundredths _____

In Review Exercises 9–10, indicate which number is the smaller.

9. 341.003, 341.03 _____

10. 0.030211, 0.030155 _____

In Review Exercises 11–14, round each number to the indicated accuracy.

11. 34.043, to the nearest hundredth _____

12. 39.958, to the nearest hundredth _____

13. 0.57392, to the nearest tenth _____

14. 0.7999, to the nearest thousandth _____

[3.2] In Review Exercises 15–16, perform each addition.

15. 54.2
   + 53.6
   _____

16.   7.13
   + 6.95
   _____

In Review Exercises 17–18, perform each subtraction.

17.   1.254
   − .763
   _____

18. 17.67
   − 9.49
   _____

[3.3] In Review Exercises 19–20, perform each multiplication.

19. 37.605
   ×   6.5
   _____

20. 4948.92
   ×   3.1
   _____

In Review Exercises 21–22, perform each division.

21. 4.56)$\overline{12.312}$

22. 0.58)$\overline{13.746}$

[3.4]

23. Find the perimeter of a rectangle that is 34.7 feet long and 21.9 feet wide.

24. Find the area of a rectangle that is 34.7 feet long and 21.9 feet wide.

25. Find the circumference of a circle with a radius of 13.76 inches. Round the answer to the nearest tenth of an inch. Use $\pi = 3.14$.

26. Find the area of a circle with a radius of 2.03 meters. Round the answer to the nearest hundredth of a square meter. Use $\pi = 3.14$.

27. A triangle has a base of 8.2 inches and a height of 22.6 inches. Find its area.

28. A square has a side of length 24.75 inches. Find the area.

Chapter 3 Test

1. Write 27.05 in expanded notation.

2. Write in decimal notation: *three and twenty-one thousandths*.

3. Write in decimal notation: $3 + \frac{5}{10} + \frac{3}{100}$.

4. Write in decimal notation: $7 + \frac{307}{1000}$.

5. Find the largest of these numbers: 35.0302, 35.0203, 34.0303.

6. Round 35.71099 to the nearest hundredth.

7. Round 35.71099 to the nearest thousandth.

8. Add: 34.57 + 1.256.

9. Subtract: 245.35 − 189.2.

10. Multiply: 23.578 · 0.24.

11. Divide: 123.354 ÷ 5.34.

12. Multiply 99.99 by 1000.

13. Divide 99.99 by 10,000.

14. Multiply 0.25467 by 10^4.

15. Divide 23.567 by 10^3.

16. Find the area of a rectangle with a length of 31.2 inches and a width of 14.7 inches.

17. Find the circumference of a circle with a radius of 192.1 feet. Round to the nearest tenth. _____

18. Find the area of a circle with a radius of 13.7 meters. Use $\pi = 3.14$ and round the answer to the nearest tenth. _____

19. Find the area of a square 39.2 centimeters on a side. _____

20. A triangle has a base of 24.4 meters and a height of 14.2 meters. Find the area. _____

4 Percent

Mathematics in Merchandising

After fifty years, America is still sweet on M&Ms. These rainbow-colored candies have changed only slightly since they were first introduced in 1940. Back then, M&Ms were slightly larger than they are now and were packaged in a paper tube. One of the original colors—violet—has now been replaced by tan. When M&M Peanut chocolate candies were introduced in 1954, they were all brown.

Each day, over a million candy pieces are manufactured, with more browns than any other color. Market research has determined that's what people want: "People tend to prefer a somewhat more subdued appearance for their chocolate assortments," claims the External Relations Department of the M&M–Mars Company. The current mix contains 30% brown, with red and yellow tied at 20% each, and 10% each for orange, green, and tan.

Do the reds taste better? Although carefully controlled taste tests in darkened rooms show that all colors taste the same, most children pick out the red ones first. In a batch of 5 million M&Ms, about how many would be red?

The calculation requires finding 20% of 5 million. In this chapter, we will discuss the topic of *percent*.

A department store advertisement promises *50 percent savings*. Sales tax is *5 percent*. A savings account pays *6 percent interest*. A brand of toothpaste promises *23 percent fewer cavities*. General Motors declares a *10 percent stock dividend*. These examples illustrate the importance of *percent*, which is the topic of this chapter.

4.1 Percents, Fractions, and Decimals

- Changing a Percent to a Fraction
- Changing a Percent to a Decimal
- Changing a Fraction to a Percent
- Changing a Decimal to a Percent

The word *percent* means *per hundred*. A **percent** represents a fraction with a denominator of 100. 20 percent is the fraction $\frac{20}{100}$. A percent can also be written as a decimal. Because $\frac{20}{100} = 0.20$, we can write 20 percent as 0.20. The symbol **%** stands for the word *percent*, so we can write 20% instead of 20 percent. Here are some other examples:

Percent	Fraction	Decimal
1%	$\frac{1}{100}$	0.01
50%	$\frac{50}{100} = \frac{1}{2}$	0.50
75%	$\frac{75}{100} = \frac{3}{4}$	0.75
100%	$\frac{100}{100} = 1$	1.00
150%	$\frac{150}{100} = \frac{3}{2} = 1\frac{1}{2}$	1.50

Changing a Percent to a Fraction

To change a percent to a fraction, we replace the percent sign with $\frac{1}{100}$ and multiply.

EXAMPLE 1

a. $24\% = 24 \cdot \dfrac{1}{100}$ *Percent* means *per hundred*; replace the % sign with $\frac{1}{100}$ and multiply.

$= \dfrac{24}{100}$

$= \dfrac{2 \cdot 2 \cdot 2 \cdot 3}{2 \cdot 2 \cdot 5 \cdot 5}$ Factor both numerator and denominator.

$= \dfrac{\cancel{2} \cdot \cancel{2} \cdot 2 \cdot 3}{\cancel{2} \cdot \cancel{2} \cdot 5 \cdot 5}$ Remove the common factors.

$= \dfrac{6}{25}$

b. $112\% = 112 \cdot \dfrac{1}{100}$ Replace the % sign with $\frac{1}{100}$ and multiply.

$= \dfrac{112}{100}$

$= \dfrac{\cancel{2} \cdot \cancel{2} \cdot 2 \cdot 2 \cdot 7}{\cancel{2} \cdot \cancel{2} \cdot 5 \cdot 5}$ Factor the numerator and denominator and simplify.

$= \dfrac{28}{25}$

As a mixed number, $\dfrac{28}{25} = 1\dfrac{3}{25}$.

c. $7.5\% = 7.5 \cdot \dfrac{1}{100}$ Multiply 7.5 by $\dfrac{1}{100}$.

$= \dfrac{7.5}{100}$

$= 0.075$ To divide 7.5 by 100, move the decimal point two places to the left.

$= \dfrac{75}{1000}$ 0.075 means 75 thousandths.

$= \dfrac{3 \cdot \overset{1}{\cancel{25}}}{40 \cdot \underset{1}{\cancel{25}}}$ Factor the numerator and denominator, and divide out the common factor.

$= \dfrac{3}{40}$

d. $\dfrac{3}{4}\% = \dfrac{3}{4} \cdot \dfrac{1}{100}$ Replace the % sign with $\dfrac{1}{100}$ and multiply.

$= \dfrac{3}{400}$ Multiply the fractions.

e. $2\dfrac{7}{8}\% = 2\dfrac{7}{8} \cdot \dfrac{1}{100}$ Replace the % sign with $\dfrac{1}{100}$ and multiply.

$= \dfrac{23}{8} \cdot \dfrac{1}{100}$ Write $2\dfrac{7}{8}$ as an improper fraction.

$= \dfrac{23}{800}$ Multiply.

Work Progress Check 1.

Changing a Percent to a Decimal

To write a percent as a decimal, we still multiply by $\dfrac{1}{100}$ and drop the % sign. However, multiplying by $\dfrac{1}{100}$ is equivalent to dividing by 100. To divide by 100, we move the decimal point two places to the left and then drop the % symbol. For example, to write 24% as a decimal, we divide 24 by 100 by moving the decimal point in 24 two places to the left and then drop the % symbol:

$24\% = .24. = .24$

Progress Check 1

Determine which is larger, $3\dfrac{4}{5}\%$ or 0.040.

Progress Check 2

Change 0.01% to a decimal.

EXAMPLE 2

a. $35\% = 0.35. = 0.35$ Divide 35 by 100.

b. $237\% = 2.37. = 2.37$

c. $7\% = 0.07. = 0.07$ An extra 0 is needed as a placeholder in the tenths position.

d. $0.423\% = 0.00.423 = 0.00423$

Work Progress Check 2.

1. $3\dfrac{4}{5}\% = \dfrac{19}{500} = \dfrac{38}{1000}$;
0.040 = $\dfrac{40}{1000}$; 0.040 is larger.

2. 0.0001

Section 4.1   Percents, Fractions, and Decimals

Changing a Fraction to a Percent

Because changing a percent to a fraction involves *dividing* by 100 and *dropping* the symbol %, it is reasonable that changing a fraction to a percent involves *multiplying* by 100 and *inserting* the % symbol.

EXAMPLE 3

a. $\dfrac{3}{20} = \dfrac{3}{20} \cdot 100\%$ Multiply by 100%, which is 1.

$\phantom{\dfrac{3}{20}} = \dfrac{3}{20} \cdot \dfrac{100}{1}\%$

$\phantom{\dfrac{3}{20}} = \dfrac{3 \cdot 100}{20}\%$ Multiply the fractions.

$\phantom{\dfrac{3}{20}} = \dfrac{3 \cdot 5 \cdot \cancel{20}^{1}}{\cancel{20}_{1}}\%$ Factor the numerator and denominator and simplify.

$\phantom{\dfrac{3}{20}} = 15\%$

b. $\dfrac{7}{8} = \dfrac{7}{8} \cdot 100\%$ Multiply by 100%, which is 1.

$\phantom{\dfrac{7}{8}} = \dfrac{7}{8} \cdot \dfrac{100}{1}\%$

$\phantom{\dfrac{7}{8}} = \dfrac{7 \cdot 100}{8}\%$ Multiply the fractions.

$\phantom{\dfrac{7}{8}} = \dfrac{7 \cdot \cancel{4}^{1} \cdot 25}{\cancel{4}_{1} \cdot 2}\%$ Factor and remove common factors.

$\phantom{\dfrac{7}{8}} = \dfrac{175}{2}\%$

$\phantom{\dfrac{7}{8}} = 87\dfrac{1}{2}\%$ Write $\dfrac{175}{2}$ as a mixed number.

Work Progress Check 3.

Progress Check 3

Change $\dfrac{43}{50}$ to a percent.

Changing a Decimal to a Percent

We change a decimal into a percent just as we changed a fraction into a percent: We multiply the decimal by 100 and insert the % symbol. To multiply a decimal by 100, we move the decimal point two places to the right.

EXAMPLE 4

a. $0.75 = 0.\underset{\smile}{75}.\%$ Multiply by 100 by moving the decimal point two places to the right, and then insert the % symbol.

$ = 75\%$

b. $7.0 = 7.\underset{\smile}{00}.\%$ Multiply by 100 and insert the % symbol.

$ = 700\%$

3. 86%

Progress Check Answers

We can change a fraction into a percent by using long division to change the fraction into a decimal, and then multiplying by 100% by moving the decimal point and inserting a percent symbol.

EXAMPLE 5

Use long division to change the fraction $\frac{3}{8}$ into a percent.

Solution

To change $\frac{3}{8}$ into a percent, we perform a long division to change $\frac{3}{8}$ to a decimal:

$$\begin{array}{r} 0.375 \\ 8\overline{)3.000} \\ \underline{2\ 4\ \ } \\ 60 \\ \underline{56} \\ 40 \\ \underline{40} \\ 0 \end{array}$$

Thus, $\frac{3}{8}$ is equal to 0.375. To change the decimal to a percent, we multiply 0.375 by 100 (by moving the decimal point two places to the right) and insert a % sign.

$0.375 = 0.37.5\%$ Multiply by 100 and insert the % symbol.

$ = 37.5\%$

$ = 37\frac{1}{2}\%$

Thus, the fraction $\frac{3}{8}$ is equal to $37\frac{1}{2}\%$. Work Progress Check 4. ∎

Progress Check 4

Change $\frac{5}{16}$ into a percent.

4. $\frac{5}{16} = 0.3125 = 31\frac{1}{4}\%$

Progress Check Answers

4.1 Exercises

In Exercises 1–24, change each percent into an equivalent fraction. Simplify the fraction. Write improper fractions as mixed numbers.

1. 25%   **2.** 50%   **3.** 75%   **4.** 100%

5. 20%   **6.** 40%   **7.** 60%   **8.** 80%

9. 5%   **10.** 10%   **11.** 35%   **12.** 85%

13. 170%   **14.** 184%   **15.** 275%   **16.** 330%

17. 0.170%   **18.** 18.4%   **19.** 2.75%   **20.** 3.30%

21. $\frac{1}{2}\%$ **22.** $\frac{7}{8}\%$ **23.** $2\frac{1}{4}\%$ **24.** $7\frac{3}{20}\%$

In Exercises 25–48, change each percent into an equivalent decimal.

25. 32%   **26.** 57%   **27.** 98%   **28.** 86%

29. 49%   **30.** 99%   **31.** 138%   **32.** 250%

33. 4.7%   **34.** 19.5%   **35.** 48.7%   **36.** 99.9%

37. 0.57%   **38.** 0.71%   **39.** 892%   **40.** 976%

41. $\frac{1}{4}$%   **42.** $12\frac{1}{2}$%   **43.** $3\frac{3}{4}$%   **44.** $15\frac{3}{5}$%

45. $\frac{2}{5}$%   **46.** $\frac{3}{8}$%   **47.** $\frac{3}{25}$%   **48.** $\frac{21}{50}$%

In Exercises 49–72, change each fraction into an equivalent percent.

49. $\frac{17}{100}$ **50.** $\frac{37}{100}$ **51.** $\frac{93}{100}$ **52.** $\frac{47}{100}$

53. $\frac{7}{10}$ **54.** $\frac{3}{10}$ **55.** $\frac{9}{20}$ **56.** $\frac{7}{50}$

57. $\frac{3}{5}$ **58.** $\frac{7}{5}$ **59.** $\frac{9}{25}$ **60.** $\frac{3}{25}$

61. $\frac{99}{75}$ **62.** $\frac{333}{75}$ **63.** $\frac{76}{5}$ **64.** $\frac{71}{2}$

65. $\frac{99}{25}$ **66.** $\frac{42}{15}$ **67.** $\frac{67}{5}$ **68.** $\frac{69}{12}$

69. $\frac{13}{40}$ **70.** $\frac{7}{60}$ **71.** $\frac{8}{75}$ **72.** $\frac{13}{45}$

In Exercises 73–96, change each decimal into an equivalent percent.

73. 0.35   **74.** 0.52   **75.** 0.77   **76.** 0.65

77. 0.99   **78.** 0.195   **79.** 0.787   **80.** 0.999

81. 0.835   **82.** 4.7   **83.** 8.2   **84.** 98.6

85. 14.92   **86.** 1.776   **87.** 88.8   **88.** 999.0

89. 83.5   **90.** 47.0   **91.** 0.0082   **92.** 0.0986

93. 100   **94.** 2000   **95.** 0.001   **96.** 0.0057

Review Exercises

In Review Exercises 1–8, perform each operation.

1. $0.35 \cdot 100$
2. $0.55 \cdot 200$
3. $0.75 \cdot 400$
4. $0.80 \cdot 150$
5. $350 \div 0.35$
6. $570 \div 0.57$
7. $960 \div 0.32$
8. $470 \div 0.20$
9. What fractional part of 350 is 245?
10. What fractional part of 650 is 520?
11. What fractional part of 650 is 650?
12. Last year, 183 out of 750 students were enrolled full-time. This year, 67 out of 250 students are full-time students. In which year was the fraction of full-time students greater?

4.2 Problems Involving Percents

- Finding a Percentage
- Finding the Percent One Number Is of Another
- Finding a Number from a Given Percent of It

Percent problems involve answering questions such as

- What is 30% of 1000?
- 300 is 30% of what amount?
- What percent of 1000 is 300?

Percent problems such as these require multiplying or dividing by percents, or dividing two numbers to obtain a percent.

Finding a Percentage

Recall that the word *of* often means *multiply*. Therefore, 30% *of* 1000 is 300, for example, because

30% of $1000 = 30\% \cdot 1000$
$\phantom{30\% \text{ of } 1000} = (0.30)(1000)$ Change 30% to the decimal 0.30.
$\phantom{30\% \text{ of } 1000} = 300$ Perform the multiplication.

In the sentence

30% of 1000 is 300

30% is called the **rate**, the number 1000 is called the **base**, and the product 300 is called a **percentage**. Thus, we have

Definition: The product of a rate r and a base b is called a **percentage**. If p is the percentage, then

$r \cdot b = p$

EXAMPLE 1

a. 75% of 300 = 75% · 300
= (0.75)(300)
= 225

Thus, 75% of the base 300 is the percentage 225.

b. 42% of 200 = 42% · 200
= (0.42)(200)
= 84

Thus, 84 is the percentage obtained when we find 42% of 200.

c. 150% of 240 = 150% · 240
= (1.50)(240)
= 360

Thus, 360 is the percentage obtained when we find 150% of 240. ∎

EXAMPLE 2

Find $33\frac{1}{3}$% of 390.

Solution

Here it is easier to work with a fraction than with a percent. To change $33\frac{1}{3}$% into a fraction, we proceed as follows:

$$33\frac{1}{3}\% = \frac{100}{3}\% \qquad \text{Change } 33\frac{1}{3} \text{ into an improper fraction.}$$

$$= \frac{100}{3} \cdot \frac{1}{100} \qquad \text{Change the percent to a fraction.}$$

$$= \frac{1}{3}$$

Then,

$$33\frac{1}{3}\% \text{ of } 390 = \frac{1}{3} \text{ of } 390$$

$$= \frac{1}{3} \cdot 390 \qquad \text{The word } of \text{ means } times.$$

$$= 130$$

Thus, $33\frac{1}{3}$% of 390 is 130. Work Progress Check 5. ∎

Progress Check 5

Find 30% of 30.

5. 9

Finding the Percent One Number Is of Another

In Chapter 1, we found that multiplication and division are closely related: Recall that the multiplication

$$r \cdot b = p$$

is equivalent to the division

$$r = \frac{p}{b}$$

We use this fact to answer the question *what percent of 1000 is 300?* In this example, the percentage p is 300, the base b is 1000, and we must find a rate r. We first write the formula

$$r \cdot b = p$$

as

$$r \cdot 1000 = 300$$

We change the multiplication into an equivalent division, and we write the fraction as a percent:

$$r = \frac{300}{1000}$$
$$= \frac{30}{100}$$
$$= 0.30$$
$$= 30\%$$

Thus, to find the percent r that a percentage p is of a base b, we write the multiplication $r \cdot b = p$ as an equivalent division

$$r = \frac{p}{b}$$

and convert the fraction to a percent.

EXAMPLE 3

What percent of 340 is 119?

Solution

The base b is 340, the percentage p is 119, and the rate r is given by

$$r = \frac{p}{b}$$
$$r = \frac{119}{340}$$

We convert the fraction $\frac{119}{340}$ into a percent by first using long division to change the fraction to a decimal:

```
        0.35
340)119.00
    102 0
     17 00
     17 00
         0
```

We then convert the decimal 0.35 into a percent by multiplying it by 100%:

$$0.35 = 0.35 \cdot 100\%$$
$$= 35\%$$

To multiply by 100, move the decimal point two places to the right.

Work Progress Check 6.

Progress Check 6

a. What percent is 625 of 125?

b. What percent of 625 is 125?

6. **a.** 500% **b.** 20%
Progress Check Answers

Section 4.2 Problems Involving Percents

Finding a Number from a Given Percent of It

We can find a number if we know a percent of that number. For example, to answer the question *30% of what number is 300?* we must find the base, b, if the rate r is 30% and the percentage p is 300. Because *rate · base = percentage*, we can translate the question into a multiplication statement.

30% of what number is 300?

$30\% \cdot b = 300$ The word *of* means *times*, and *is* means *equals*.

$0.30 \cdot b = 300$ Change 30% to a decimal.

$b = \dfrac{300}{0.30}$ $0.30 \cdot b = 300$ is equivalent to $b = \dfrac{300}{0.30}$.

$= 1000$ Perform the division $.30\overline{)300.00}$.

Thus, 300 is 30% of 1000.

EXAMPLE 4

Find each number if   **a.** 75% of the number is 450   and   **b.** 120% of the number is 1560.

Solution

a. Let x represent the required number. Then

75% of x is 450

$75\% \cdot x = 450$

$0.75x = 450$ Change 75% to a decimal.

$x = \dfrac{450}{0.75}$ Write the multiplication as an equivalent division.

$= 600$ Perform the division.

Thus, 75% of 600 is 450.

b. Let x represent the required number. Then

120% of x is 1560

$120\% \cdot x = 1560$

$1.20x = 1560$ Change 120% to a decimal.

$x = \dfrac{1560}{1.20}$ Write the multiplication as an equivalent division.

$= 1300$ Perform the division.

Thus, 120% of 1300 is 1560.

Work Progress Check 7.

Progress Check 7

45% of a number is 27. Find the number.

Progress Check Answers

7. 60

4.2 Exercises

In Exercises 1–16, find each percentage.

1. 35% of 100
2. 72% of 100
3. 75% of 400
4. 50% of 300
5. 52% of 350
6. 86% of 150
7. 50% of 78.2
8. 15% of 1.8

9. 41% of 250 **10.** 82% of 64 **11.** 150% of 200 **12.** 100% of 370

13. 0.4% of 29 **14.** 0.64% of 175 **15.** 350% of 400 **16.** 1000% of 1000

In Exercises 17–32, find each number.

17. 20% of what number is 40? **18.** 15% of what number is 30?

19. 50% of what number is 90? **20.** 75% of what number is 60?

21. 28% of what number is 42? **22.** 44% of what number is 143?

23. 18% of what number is 72? **24.** 96% of what number is 216?

25. 4.8 is 48% of what number? **26.** 5.2 is 26% of what number?

27. 44 is 8% of what number? **28.** 72 is 12% of what number?

29. 133 is 35% of what number? **30.** 13.3 is 3.5% of what number?

31. 22 is 1.6% of what number? **32.** 36 is 2.4% of what number?

In Exercises 33–48, find each percent.

33. What percent is 45 of 90? **34.** What percent is 19 of 76?

35. What percent is 14 of 70? **36.** What percent is 23 of 92?

37. What percent is 71.5 of 357.5? **38.** What percent is 13.2 of 254?

39. 0.32 is what percent of 4? **40.** 3.6 is what percent of 28.8?

41. 34 is what percent of 17? **42.** 39 is what percent of 13?

43. 18.9 is what percent of 18.9? **44.** 0.38 is what percent of 3.8?

45. 17 is what percent of 51? **46.** 26 is what percent of 13?

47. 3.95 is what percent of 0.0395? **48.** 0.47 is what percent of 4.7?

In Exercises 49–60, find each number.

49. 34% of what number is 9.52? **50.** 19% of what number is 7.79?

51. What percent is 63 of 90? **52.** What percent is 7.2 of 40?

53. 0.64 is what percent of 8? **54.** What percent is 1558 of 3800?

55. What percent is 30.1 of 107.5? **56.** 3.9 is what percent of 52?

57. 32% of 112.5 is what number? **58.** 132% of 112.5 is what number?

59. 266 is 35% of what number? **60.** 26.6 is 0.35% of what number?

61. 30% of Rita's mathematics class are boys. If there are 60 students in the class, how many are boys?

62. 90% of a factory's 2760 workers are members of a union. How many workers are union?

63. Sales tax is 5% of the purchase price. What tax is paid on a $37 lamp?

64. Of all donations to a certain charity, $2\frac{1}{2}$% is used to pay for fund-raising. If the charity collected $237,000 last year, how much was spent on fund-raising?

65. The 5% sales tax on a microwave oven amounts to $13.50. What is the oven's selling price?

66. 20% of Bill's class will receive a grade of A. Bill will give seven A's. How many are in his class?

67. 18% of hospital patients stay for less than 2 days. If 1008 patients last year stayed for less than 2 days, what total number of patients did the hospital treat?

68. The average price of homes in one neighborhood increased 8% since last year, an increase of $7800. What was the average price of a home last year?

69. Out of 1300 magazine subscribers surveyed, 390 approve of the magazine's new graphic design. What percent approve?

70. Sales tax on a $12 compact disk is $0.72. At what rate is sales tax computed?

71. Out of 9200 shoppers, 4140 indicated that they were pleased with the store's service. What percent were *not* pleased?

72. In 1992, three-time Cy Young Award winner Tom Seaver was voted into baseball's Hall of Fame with nominations on 425 of 430 ballots. In the first Hall of Fame election in 1936, Ty Cobb received nominations on 222 of 226 ballots. Who received the greater percent, Seaver or Cobb?

73. Pitcher Rollie Fingers was nominated for the Hall of Fame on 81.2% of the 430 ballots cast in 1992. How many votes did Fingers receive?

74. In the 1992 Hall of Fame election, how many ballots were *not* cast for Rollie Fingers? See Exercise 73.

Review Exercises

In Review Exercises 1–8, perform each operation. Remember the order of operations.

1. $300 + 0.30 \cdot 300$

2. $200 + 0.50 \cdot 200$

3. $350 + 0.75 \cdot 350$

4. $150 + 0.90 \cdot 150$

5. $475 - 0.35 \cdot 475$

6. $550 - 0.57 \cdot 550$

7. $190 - 0.32 \cdot 190$

8. $770 - 0.20 \cdot 770$

4.3 Applications of Percents

- *Finding the Markdown or the Percent of Decrease*
- *Finding the Markup or the Percent of Increase*
- *Investing Money at Simple Interest*

Many financial calculations involving percent are related to finance. Applications involving *markdown, markup*, selling items at a *discount*, and *investing money* require calculations with percents.

The advertisement in Figure 4-1 indicates the sale price and the original price of a shirt. The difference between the sale price and the original price is called the **discount**: The shirt is discounted $7.00. If the discount is expressed as a percent of the selling price, it is called the **rate of discount** or the **markdown**: The shirt's markdown is 25%.

Figure 4-1

EXAMPLE 1

Jennifer buys a $570 stereo system at a 25% discount. What will she pay?

Solution

A 25% discount means that Jennifer will pay less than the regular price. She receives a discount of 25% of the regular price, $570. We calculate her savings:

25% of $570 = (0.25)($570) = $142.50

The sale price of the stereo is the original price less the discount:

Sale price = $570 − 142.50 = $427.50

Jennifer pays $427.50 for the stereo.

Finding the Markdown or the Percent of Decrease

If a television set that normally sells for $125 is on sale for $100, the price has decreased by $25. This $25 decrease is $\frac{25}{125}$ of the *original price*, $125. We convert the fraction $\frac{25}{125}$ to a percent to find the markdown, or the **percent of decrease**:

$$\frac{25}{125} = \frac{25}{125} \cdot \frac{100}{1}\%$$
$$= \frac{25 \cdot 100}{125}\%$$
$$= \frac{\overset{1}{\cancel{25}} \cdot \overset{5}{\cancel{25}} \cdot 4}{\underset{1}{\cancel{25}} \cdot \underset{1}{\cancel{5}}}\%$$
$$= 20\%$$

The price of the television set has decreased 20%.

Definition: If a number a is decreased to a number b, then we find the **percent of decrease**, or the **percent by which b is less than a**, by dividing the amount of decrease, $a - b$, by the *larger number*, a, and writing that quotient as a percent.

EXAMPLE 2

Find the percent by which the second number is smaller than the first:
a. 400, 250   and   **b.** 50, 30

Solution

a. We determine how much 250 is less than 400: It is the difference

$$400 - 250 = 150$$

Next, we determine the fractional part that decrease is of the larger number, 400. It is the fraction

$$\frac{150}{400} \quad \text{or} \quad \frac{3}{8}$$

Finally, we write $\frac{3}{8}$ as a percent:

$$\frac{3}{8} = 0.375 \cdot 100\% = 37.5\%$$

Thus, 250 is 37.5% less than 400.

b. To determine the amount by which 30 is less than 50, we find the fractional part their difference is of the larger number, 50. Then we change the fraction to a percent:

$$\frac{50 - 30}{50} = \frac{20}{50}$$
$$= \frac{2}{5} \cdot 100\%$$
$$= \frac{200}{5}\%$$
$$= 40\%$$

Thus, 30 is 40% less than 50.

EXAMPLE 3

William bought a camera on sale for $384.20. The original price was $452. What was the percent of discount?

Solution

By paying $384.20 for a $452 camera, William saved $452.00 − $384.20, or $67.80. To find the percent of discount, express as a percent the fractional part that $67.80 is of the *original* price, $452.

$$\frac{67.80}{452} \cdot 100\% = 0.15 \cdot 100\% = 15\%$$

William received a 15% discount. Work Progress Check 8.

Progress Check 8

An ad proclaims: *Save 50%!!! Buy two, get one free!* What percent are you really saving?

---

8. $33\frac{1}{3}\%$

Progress Check Answers

Finding the Markup or the Percent of Increase

To make a profit, a merchant must sell an item for more than he paid for it. The increase in price expressed as a percent of the merchant's *cost* is called the **markup based on cost**, or just the **markup**.

EXAMPLE 4

A store manager buys toasters wholesale for $21 and sells them at a 17% markup. What is the retail price?

Solution

The manager must sell the toasters for more than he paid for them. Thus, to the $21 wholesale price, the manager adds 17% of the wholesale price to get the retail price:

$$\text{Retail price} = \text{wholesale price} + 17\% \text{ of the wholesale price}$$
$$= 21 + (0.17)(21)$$
$$= 21 + 3.57$$
$$= 24.57$$

The retail price for a toaster is $24.57.

If the price of a television set increases from $100 to $125, the price has increased by $25. This $25 increase is a certain fractional part of the *original price*, $100. That fractional part is $\frac{25}{100}$, which we convert to a percent to find the **percent of increase**:

$$\frac{25}{100} = 25\%$$

The price of the television set has increased 25%.

> **Definition:** If the number a is increased to a number b, then the **percent of increase**, or the **percent by which b is greater than a**, is found by dividing the amount of increase, $b - a$, by the *smaller number*, a, and writing that quotient as a percent.

EXAMPLE 5

By what percent is the first number larger than the second?   **a.** 400, 250,   and   **b.** 30.8, 17.5

a. We determine the amount by which 400 exceeds 250: It is the difference

$$400 - 250 = 150$$

Next we determine the fractional part that increase is of the *smaller* number, 250. It is the fraction

$$\frac{150}{250} \quad \text{or} \quad \frac{3}{5}$$

Finally, we write $\frac{3}{5}$ as a percent: $\frac{3}{5} = 60\%$. Thus, 400 exceeds 250 by 60%.

b. To determine the amount by which 30.8 exceeds 17.5, we find the fractional part that their difference is of the smaller number, 17.5. We then change the fraction to a percent:

$$\frac{30.8 - 17.5}{17.5} = \frac{13.3}{17.5}$$ Divide the increase by the smaller number.

$$= \frac{13.3}{17.5} \cdot 100\%$$ Change the fraction to a percent.

$$= .76 \cdot 100\%$$ Perform the division: $17.5\overline{)13.3}$.

$$= 76\%$$

Thus, 17.5 has increased by 76% to become 30.8. ■

When the price of a television set increases from $100 to $125, the percent of increase is 25%. When the price *decreases* from $125 to $100, the percent of decrease is 20%. These different results occur because the percent of increase is a percent of the original (smaller) price, $100. The percent of decrease, however, is a percent of the original (larger) price, $125.

Work Progress Check 9.

Progress Check 9

a. By what percent is 75 smaller than 150?

b. By what percent is 150 larger than 75?

Investing Money at Simple Interest

A savings institution pays **interest** for the privilege of using a depositor's money. The amount of interest depends on the amount deposited, called the **principal**, and on the time that the money remains in the account. The account bears interest at a certain **annual rate**, expressed as a percent. One method of calculating interest is called **simple interest**.

> **Formulas for Simple Interest:** If a principal P is placed on deposit for t years in an account paying interest at an annual rate r, the interest I earned is given by the formula
>
> $$I = Prt$$
>
> and the total amount A on deposit is the principal plus the interest, or
>
> $$A = P + Prt$$

EXAMPLE 6

If $350 is deposited in an account paying simple interest at an annual rate of 7.5%, how much will be in the account in 42 months?

Solution

The formula for finding the total amount on deposit requires knowing the time in *years*. We must change 42 months into years. We do so by dividing

9. a. 50%   **b.** 100%

Progress Check Answers

Progress Check 10

Which pays more interest, $750 invested for 18 months at 7.5%, or $800 invested for 1 year at 10%?

---

10. The first investment

42 by 12 to obtain 3.5, or $3\frac{1}{2}$ years. The original principal P is $350, the annual rate of interest r (as a decimal) is 0.075, and the time t is 3.5 years. We substitute these numbers into the formula for finding the total amount on deposit:

$$A = P + Prt$$
$$= 350 + (350)(0.075)(3.5) \quad \text{Substitute 350 for } P, 0.075 \text{ for } r, \text{ and } 3.5 \text{ for } t.$$
$$= 350 + 91.875 \quad \text{Perform the multiplication first.}$$
$$= 441.875$$

After $3\frac{1}{2}$ years, the account will contain $441.88. Work Progress Check 10.

4.3 Exercises

In Exercises 1–12, find the percent of decrease of the second number from the first.

1. 50, 25   **2.** 84, 42   **3.** 63, 42   **4.** 50, 20

5. 200, 116   **6.** 150, 18   **7.** 150, 87   **8.** 925, 777

9. 95, 81.7   **10.** 360, 295.2   **11.** 570, 165.3   **12.** 85.5, 32.49

In Exercises 13–24, find the percent of increase of the second number over the first. That is, find the percent by which the larger number exceeds the smaller.

13. 25, 50   **14.** 42, 84   **15.** 42, 63   **16.** 20, 50

17. 350, 420   **18.** 250, 355   **19.** 150, 213   **20.** 150, 282

21. 1.75, 3.29 **22.** 32.5, 37.7   **23.** 3.5, 65.8   **24.** 195, 382.2

25. All hair care products are on sale for 20% off the regular price. Shampoo regularly sells for $6.00. What is the sale price?

26. Magazine subscriptions are advertised at savings of 40% off the newsstand price. If a magazine would cost $30 per year at the newsstand, what will a subscription cost?

27. Jim receives a bill from his plumber for $756, with a note that reads *5% discount if paid within 10 days*. What should Jim pay within 10 days?

28. At the January sales, sheets and pillowcases were marked 25% off. If sheets are regularly $37 each, and a pair of pillowcases are regularly $24, how much will four sheets and two pairs of pillowcases cost, on sale?

29. Camera Mart purchased cameras wholesale for $370 each. If the store manager wants a profit of 23%, at what price should he sell the cameras?

30. When a fast-food restaurant added tacos to the menu, weekly sales increased from $18,750 to $22,125. What is the percent of increase?

31. Econo Foods marks up its stock by 25%. Find the selling price of one can of beans that costs the store 38.4 cents.

32. A hardware store owner pays $37 for electric drills and sells them at a 42% markup. Find the selling price.

33. The owner of an appliance store paid $590 for a console television and tried to sell it at a 35% markup. When it didn't sell, he put up a sign: *SALE! $33\frac{1}{3}$% OFF!* Find the sale price.

34. Furniture is marked up 75% over wholesale. Find the selling price of a chair that costs $378 wholesale.

35. Refer to Illustration 1. The net weight of a regular-size Power Stick® is 2.5 ounces. Is the labeling accurate?

36. Refer to Illustration 1 and Exercise 35. Find the percent of *decrease* of net weight when a user goes back to a regular-size Power Stick®.

37. Gary deposits $800 in an account that pays simple interest at an annual rate of 8%. How much interest will he earn in 2 years?

38. Ginny deposits $1200 in an account that pays simple interest at an annual rate of 11%. How much interest will she earn in 1 year?

39. Juan deposits $2000 in an account that pays simple interest at an annual rate of 14%. If he leaves the money on deposit for 18 months, how much interest will Juan earn?

40. Jill deposits $790 in an account that pays simple interest at an annual rate of 9%. How much will be in the account after 5 years?

41. George deposits $1000 in an account paying simple interest at an annual rate of 12%. How much will be in the account in 20 years?

42. Betty deposits $1500 in an account paying simple interest at an annual rate of 18%. How much will be in the account in 50 years?

43. The average price of homes in one neighborhood dropped 8% since last year, a decrease of $7800. What was the average price of a home last year?

44. In Exercise 43, what is the average price of a home this year?

45. The advertisement in Illustration 2 claims that a $1059.99 piece of gold jewelry is on sale for $476.99. Is the advertisement accurate?

46. Refer to Illustration 2. Does "50% off plus a 10% bonus" mean "60% off"? Find the selling price of a $1059.99 bracelet on sale for 60% off.

Illustration 1

Illustration 2

Review Exercises

In Review Exercises 1–12, perform each operation.

1. $3.2(5.8 + 2.2)$

2. $5.7(3.7 + 0.3)$

3. $0.31(15.5 - 3.5)$

4. $(3.2)(5.8) + 1.1$

5. $(41.1)(2.5) - 20.5$

6. $(3.7)(2.1) + (1.2)(3)$

7. $(3.2)(2.0) \div 0.8$

8. $(2.2)^2 - 1$

9. $12 - (3.1)^2$

10. $\dfrac{1.3 + 7.7}{1.5}$

11. $\dfrac{77.1 - 19.9}{0.4}$

12. $\dfrac{37.2 + 19.3}{5.0}$

Mathematics in Merchandising

To determine the number of red M&Ms, find 20% of 5 million. Do not reach for your calculator. Do the calculation mentally. Twenty percent is twenty one-hundredths, or one-fifth. One-fifth of *five* million is *one* million. The number of red candies is 1,000,000.

In 1976, concerns over the effects of red dye No. 2 prompted M&M–Mars Company to discontinue the popular red candies. Although red M&Ms never contained that dye, the company feared that the public would think they did. In spite of angry letters from loyal consumers, red M&Ms disappeared until 1987.

Chapter Summary

Key Words

annual rate (4.3)
base (4.2)
discount (4.3)
interest (4.3)
markdown (4.3)
markup (4.3)
percent (4.1)
percent of decrease (4.3)
percent of increase (4.3)
percentage (4.2)
principal (4.3)
rate (4.2)
simple interest (4.3)

Key Ideas

(4.1) To change a percent to a fraction, replace the percent sign by a denominator of 100 and simplify the fraction.

To change a percent to a decimal, drop the percent symbol and move the decimal point two places to the left.

To change a decimal to a percent, move the decimal point two places to the right and insert the % symbol.

To change a fraction to a percent, perform the long division to write the fraction as a decimal. Then move the decimal point two places to the right and insert the % symbol.

(4.2) To find a certain percent of a number, multiply the number by that percent expressed as a decimal.

To find the percent that one number is of another, divide the one number by the other and change that quotient into a percent.

To find a number from a given percent of it, divide the percentage by the percent.

(4.3) If the number a increases (or decreases) to a number b, then the percent of increase (or decrease) is found by dividing the amount of increase (or decrease) by the original number, a, and writing that quotient as a percent.

If a principal P is placed on deposit for t years in an account paying simple interest at an annual rate r, the interest I earned is given by the formula

$$I = Prt$$

The total amount A on deposit is the principal plus the interest, or

$$A = P + Prt$$

Chapter 4 Review Exercises

[4.1] In Review Exercises 1–4, change each percent to a fraction written in lowest terms.

1. 43% 2. 32% 3. 76.5% 4. 5.5%

In Review Exercises 5–8, change each percent to a decimal.

5. 47% 6. 39% 7. 55.9% 8. 155.5%

In Review Exercises 9–12, change each fraction to a percent.

9. $\frac{37}{50}$ 10. $\frac{29}{25}$ 11. $\frac{5}{2}$ 12. $\frac{21}{5}$

In Review Exercises 13–16, change each decimal to a percent.

13. 0.47 14. 0.99 15. 0.099 16. 9.90

[4.2] In Review Exercises 17–20, find each percentage.

17. 27% of 100 18. 4.3% of 1000 19. 0.5% of 250 20. $\frac{1}{2}$% of 500

In Review Exercises 21–28, find each number.

21. 15% of what number is 100?
22. 75% of what number is 90?
23. 20% of what number is 45?
24. 55% of what number is 55?
25. What percent of 25 is 5?
26. What percent is 6 of 60?
27. What percent is 90 of 450?
28. What percent of 35 is 10?

[4.3] In Review Exercises 29–30, find the percent of increase of the larger number over the smaller.

29. 45, 9
30. 36.2, 1.81

In Review Exercises 31–32, find the percent of decrease of the smaller number from the larger.

31. 45, 9
32. 36.2, 1.81

33. If Frank does not pay his electric bill on time, he must pay an extra 3%. He received a bill for $72 and payed it late. How much did he pay?

34. The wholesale price of a sleeping bag is $42.50. If the store has a 40% markup, what is the selling price?

35. Find the discount reflected in the advertisement in Illustration 1.

36. Find the percent discount reflected in the advertisement in Illustration 1.

37. Find the amount of simple interest that an investment of $1300 will earn at an annual rate of 12%, if invested for 4 years.

38. What will be the total value of the account in Review Exercise 37?

Illustration 1

Name _____ Section _____

Chapter 4 Test

1. Express 46% as a fraction and simplify. _____

2. Express 4.2% as a fraction and simplify. _____

3. Express 68% as a decimal. _____

4. Express 5.3% as a decimal. _____

5. Express 0.57 as a percent. _____

6. Express 5.7 as a percent. _____

7. Express $\frac{3}{40}$ as a percent. _____

8. Express $\frac{1}{3}$ as a percent. _____

9. Find 45% of 300. _____

10. Find 28% of 950. _____

11. 60% of a number is 420. Find the number. _____

12. 7.2% of a number is 14.4. Find the number. _____

13. What percent of 250 is 75? _____

14. What percent is 80 of 16? _____

15. What is the percent of increase of 56 over 7? _____

16. What is the percent of increase of 16.1 over 2.3? _____

17. What is the percent of decrease if 350 drops to 273? _____

Chapter 4 Test   137

18. By what percent is 17.3 less than 51.9? _____

19. A $126 tape deck is discounted $33\frac{1}{3}$%. Find the sale price. _____

20. The wholesale cost of a radio is $38. The markup is 23%. Find the selling price. _____

21. $4000 is invested at simple interest for 5 years at 10%. Find the interest earned. _____

22. $5000 is invested at 12% simple interest. Find the total value of the investment after $6\frac{1}{2}$ years. _____

5 Measurement, Estimation, and Reading Graphs

Mathematics in Industry

The Pacific Telesis Group is a family of companies that provides products and services for the telecommunications industry worldwide. One of these companies is PacTel Cellular. Specializing in radio communication and cellular telephone networks, this company has experienced rapid growth. For the annual report to the company's shareholders, the directors need to present many financial facts in an understandable and easily digested form. For example, they must present the fact that revenues have increased from $156 million in 1987 to $256 million in 1988, to $379 million in 1989, to $493 million in 1990, and to $581 million in 1991. In this chapter, we will learn about *graphs*, one method of presenting such numbers in a visual form.

We begin the chapter by discussing the measurement of distances using *units of measurement* such as the *meter, inch, centimeter, mile, kilometer,* and *foot.* These dimensions of objects are often used in calculations. Often, the results of these calculations must be very precise, and at other times rough *estimates* are sufficient. Measured or calculated values are often described in pictures called *graphs*. Measuring, estimating, and reading graphs are the topics of this chapter.

5.1 Measurement

- *Measuring Distances*
- *American Units of Length*
- *Metric Units of Length*

Measuring Distances

We use various measuring devices to measure sizes and distances. The most familiar of these, the ruler, is usually one foot long. Since there are 12 inches in one foot, the ruler is divided into 12 equal distances of one inch. Each inch on a ruler is subdivided further into halves of an inch, quarters of an inch, and eighths of an inch. Several distances are measured on a magnified portion of a ruler in Figure 5-1.

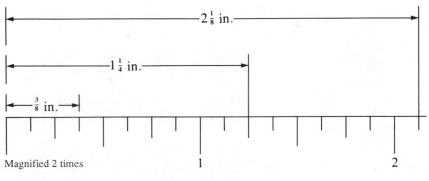

Figure 5-1

EXAMPLE 1

To the nearest $\frac{1}{4}$ inch, measure the nail in Figure 5-2.

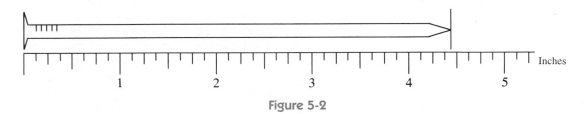

Figure 5-2

Solution

We place one end of the nail even with the left end (the zero end) of the ruler, as in Figure 5-2. The right end of the nail is beyond the $4\frac{1}{4}$-inch mark but does not reach the $4\frac{1}{2}$-inch mark. It is closer to $4\frac{1}{2}$. To the nearest $\frac{1}{4}$ inch, the nail is $4\frac{1}{2}$ inches long. ∎

EXAMPLE 2

To the nearest $\frac{1}{8}$ inch, measure the length of the paper clip in Figure 5-3.

Figure 5-3

Solution

We place one end of the paper clip even with the left end of the ruler. The right end of the paper clip is between the $1\frac{3}{8}$-inch mark and the $1\frac{1}{2}$-inch mark. It is closer to $1\frac{3}{8}$. To the nearest $\frac{1}{8}$ inch, the paper clip is $1\frac{3}{8}$ inches long. Work Progress Check 1. ∎

EXAMPLE 3

To the nearest $\frac{1}{8}$ inch, measure the length of the tear-off portion of the movie ticket in Figure 5-4.

Progress Check 1

To the nearest $\frac{1}{8}$ inch, find the width of one white piano key.

Progress Check Answers

1. $\frac{7}{8}$ in.

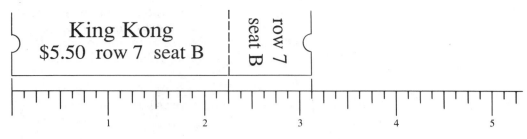

Figure 5-4

Solution

One end of the ticket in Figure 5-4 is aligned with the zero end of the ruler. The length of the entire ticket is $3\frac{1}{8}$ inches. The length of the longer portion of the ticket is $2\frac{1}{4}$ inches. The length of the tear-off portion is the *difference* between these two lengths. To find that length, subtract $2\frac{1}{4}$ from $3\frac{1}{8}$ as follows:

$$3\frac{1}{8} - 2\frac{1}{4} = \frac{25}{8} - \frac{9}{4} \quad \text{Change the mixed numbers to improper fractions.}$$

$$= \frac{25}{8} - \frac{9 \cdot 2}{4 \cdot 2} \quad \text{Write the fractions with a common denominator.}$$

$$= \frac{25}{8} - \frac{18}{8}$$

$$= \frac{25 - 18}{8} \quad \text{Subtract the fractions.}$$

$$= \frac{7}{8} \quad \text{Simplify.}$$

The length of the tear-off portion of the ticket is $\frac{7}{8}$ inch. ∎

Each point on a ruler, like each point on a number line, has a number associated with it: the distance of that point from 0. As Example 3 illustrates, the distance between any two points on a ruler (or on a number line) is the difference of the numbers associated with those points.

We use two systems of units to measure length: **American** units and **metric** units. Almost all countries except the United States use the metric system exclusively. Even in the United States, American units are used for nonscientific purposes; most scientific work uses the metric system. The most common American units of length are the *inch, foot, yard,* and *mile.* Metric units of length include the *millimeter, centimeter, meter,* and *kilometer.*

American Units of Length

Inches, feet, yards, and miles are related in the following ways.

American Units of Length:

12 inches (in.) = 1 foot (ft)

3 feet = 1 yard (yd)

5280 feet = 1 mile (mi)

To convert a length from one unit of measure to another, we multiply the length by a **unit conversion factor** and use properties of fractions to simplify. To find the unit conversion factor to convert between feet and inches, for example, we begin with this basic fact:

12 in. = 1 ft

We divide both of these equal expressions by 1 foot:

$$\frac{12 \text{ in.}}{1 \text{ ft}} = \frac{1 \text{ ft}}{1 \text{ ft}}$$

$$\frac{12 \text{ in.}}{1 \text{ ft}} = 1 \qquad \text{A nonzero number divided by itself is 1, so } \frac{1 \text{ ft}}{1 \text{ ft}} = 1.$$

We use the unit conversion factor $\frac{12 \text{ in.}}{1 \text{ ft}}$ to convert a length in feet to an equivalent length in inches. Because this fraction is equal to 1, multiplying a length by the fraction does not change the length. Only the units of its measure are changed.

EXAMPLE 4

Convert $\frac{3}{4}$ foot to inches.

Solution

We multiply $\frac{3}{4}$ ft by 1 in the form of the unit conversion factor $\frac{12 \text{ in.}}{1 \text{ ft}}$.

$$\frac{3}{4} \text{ ft} = \frac{3}{4} \text{ ft} \cdot 1$$

$$= \frac{3}{4} \text{ ft} \cdot \frac{12 \text{ in.}}{1 \text{ ft}}$$

$$= \frac{3 \cdot 12}{4 \cdot 1} \cdot \frac{\text{ft}}{\text{ft}} \cdot \text{in.}$$

$$= \frac{3 \cdot 12}{4} \cdot \text{in.} \qquad \text{The fraction } \frac{\text{ft}}{\text{ft}} \text{ represents 1.}$$

$$= \frac{3 \cdot 3 \cdot \cancel{4}}{\cancel{4}} \cdot \text{in.} \qquad \text{Factor and simplify.}$$

$$= 9 \text{ in.}$$

Thus, $\frac{3}{4}$ foot is equal to 9 inches. ∎

To find the unit conversion factor to convert inches to feet, we begin with the relation

1 ft = 12 in.

and divide each of these equal expressions by 12 inches:

$$\frac{1 \text{ ft}}{12 \text{ in.}} = \frac{12 \text{ in.}}{12 \text{ in.}}$$

$$\frac{1 \text{ ft}}{12 \text{ in.}} = 1 \qquad \text{A number divided by itself is 1, so } \frac{12 \text{ in.}}{12 \text{ in.}} = 1.$$

Multiplying a length in inches by the unit conversion factor $\frac{1 \text{ ft}}{12 \text{ in.}}$ will convert that length to an equivalent length in feet.

EXAMPLE 5

Convert 54 inches to feet.

Solution

Change 54 inches to feet by multiplying 54 by the unit conversion factor $\frac{1 \text{ ft}}{12 \text{ in.}}$.

$$54 \text{ in.} = 54 \text{ in.} \cdot \frac{1 \text{ ft}}{12 \text{ in.}}$$

$$= \frac{54}{1} \cdot \frac{1}{12} \cdot \frac{\text{in.}}{\text{in.}} \cdot \text{ft}$$

$$= \frac{54}{12} \text{ ft} \qquad \text{The fraction } \frac{\text{in.}}{\text{in.}} \text{ is 1.}$$

$$= 4.5 \text{ ft} \qquad \text{Perform the division.}$$

Thus, 54 inches is 4.5 feet, or $4\frac{1}{2}$ feet. ∎

EXAMPLE 6

Jim is $75\frac{1}{2}$ inches tall. What is Jim's height in feet and inches?

Solution

To change Jim's height into units of feet, change $75\frac{1}{2}$ inches to the improper fraction $\frac{151}{2}$ inches, and multiply by 1 in the form of the unit conversion factor $\frac{1 \text{ ft}}{12 \text{ in.}}$.

$$\frac{151}{2} \text{ in.} = \frac{151}{2} \text{ in.} \cdot 1$$

$$= \frac{151}{2} \text{ in.} \cdot \frac{1 \text{ ft}}{12 \text{ in.}}$$

$$= \frac{151 \cdot 1}{2 \cdot 12} \cdot \frac{\text{in.}}{\text{in.}} \cdot \text{ft}$$

$$= \frac{151}{24} \text{ ft} \qquad \text{The fraction } \frac{\text{in.}}{\text{in.}} \text{ is 1.}$$

$$= 6\frac{7}{24} \text{ ft} \qquad \text{Express } \frac{151}{24} \text{ as a mixed number.}$$

Jim is $6\frac{7}{24}$ feet tall. To express this height in feet and inches, we must change $\frac{7}{24}$ foot into inches. To do so, multiply $\frac{7}{24}$ foot by the unit conversion factor $\frac{12 \text{ in.}}{1 \text{ ft}}$.

$$\frac{7}{24} \text{ ft} = \frac{7}{24} \text{ ft} \cdot \frac{12 \text{ in.}}{1 \text{ ft}}$$

$$= \frac{7 \cdot 12}{24} \text{ in.}$$

$$= \frac{7 \cdot \overset{1}{\cancel{12}}}{\underset{2}{\cancel{24}}} \text{ in.} \qquad \text{Simplify the fraction.}$$

$$= \frac{7}{2} \text{ in.}$$

$$= 3\frac{1}{2} \text{ in.} \qquad \text{Write } \frac{7}{2} \text{ as a mixed number.}$$

Thus, $\frac{7}{24}$ foot is $3\frac{1}{2}$ inches, and we have determined that Jim is 6 feet, $3\frac{1}{2}$ inches tall. Work Progress Check 2.

Progress Check 2

a. Find the unit conversion factor to convert feet to yards.

b. Find the unit conversion factor to convert yards to feet.

Metric Units of Length

The basic metric unit of length is the meter. One meter is about 39 inches, slightly more than one yard. One **millimeter** is one thousandth of one meter. The lead of a mechanical pencil is approximately one millimeter thick. One **centimeter** is one hundredth of a meter. A nickel is about 2 centimeters in diameter. One **kilometer** is one thousand meters. One kilometer is about $\frac{3}{5}$ mile. Less common metric units are the **hectometer**, the **dekameter**, and the **decimeter**.

The metric units of length are related in the following ways.

Metric Units of Length:

$$1 \text{ kilometer (km)} = 1000 \text{ meters (m)} \qquad = 10^3 \text{ m}$$
$$1 \text{ hectometer (hm)} = 100 \text{ meters} \qquad = 10^2 \text{ m}$$
$$1 \text{ dekameter (dam)} = 10 \text{ meters} \qquad = 10^1 \text{ m}$$
$$1 \text{ decimeter (dm)} = \frac{1}{10} \text{ meter}$$
$$1 \text{ centimeter (cm)} = \frac{1}{100} \text{ meter} \qquad = \frac{1}{10^2} \text{ m}$$
$$1 \text{ millimeter (mm)} = \frac{1}{1000} \text{ meter} \qquad = \frac{1}{10^3} \text{ m}$$

The metric system has an advantage over the American system: Unit conversion involves multiplying or dividing by 10, or 100, or 1000, and so on. For example,

$$1 \text{ meter} = 100 \text{ centimeters}$$
$$1 \text{ centimeter} = \frac{1}{100} \text{ meter}$$
$$1 \text{ centimeter} = 10 \text{ millimeters}$$
$$1 \text{ millimeter} = \frac{1}{10} \text{ centimeter}$$

2. **a.** $\frac{1 \text{ yd}}{3 \text{ ft}}$ **b.** $\frac{3 \text{ ft}}{1 \text{ yd}}$

EXAMPLE 7

a. Change 3.2 kilometers to dekameters.   **b.** Change 35.7 centimeters to meters.

Solution

a. First, we find a unit conversion factor to change kilometers to dekameters. Because there are 10 dekameters in one hectometer and 10 hectometers in one kilometer, there are $10 \cdot 10$, or 100, dekameters in one kilometer. Thus, we have

$$1 \text{ km} = 100 \text{ dam}$$
$$\frac{1 \text{ km}}{1 \text{ km}} = \frac{100 \text{ dam}}{1 \text{ km}}$$
$$1 = \frac{100 \text{ dam}}{1 \text{ km}}$$

To convert 3.2 kilometers to dekameters, we multiply by the unit conversion factor $\frac{100 \text{ dam}}{1 \text{ km}}$.

$$3.2 \text{ km} = 3.2 \text{ km} \cdot \frac{100 \text{ dam}}{1 \text{ km}}$$
$$= 3.2 \cdot 100 \text{ km} \cdot \frac{\text{dam}}{\text{km}}$$
$$= 320 \text{ dam}$$

Thus, 3.2 kilometers is equal to 320 dekameters.

b. To change 35.7 centimeters to meters, we find a unit conversion factor. Because there is 1 meter per 100 centimeters, the unit conversion factor is $\frac{1 \text{ m}}{100 \text{ cm}}$.

$$35.7 \text{ cm} = 35.7 \text{ cm} \cdot \frac{1 \text{ m}}{100 \text{ cm}}$$
$$= \frac{35.7}{100} \text{ cm} \cdot \frac{\text{m}}{\text{cm}}$$
$$= \frac{35.7}{100} \text{ m}$$
$$= 0.357 \text{ m}$$

Thus, 35.7 centimeters is 0.357 meter. ∎

Conversions between metric units are easily done by moving a decimal point. Because there are 1000 millimeters in 1 meter, for example, we can change 7.2 meters to millimeters by multiplying 7.2 by 1000. We do so by moving the decimal point three places to the right:

$$7.2 \text{ m} = 7.2 \text{ m} \cdot 1000 \frac{\text{mm}}{\text{m}}$$
$$= (7.2)(1000) \text{ mm}$$
$$= 7.200. \text{ mm}$$
$$= 7200 \text{ mm}$$

Likewise, to change 8.9 millimeters to centimeters, we divide 8.9 by 10, because there are 10 millimeters in one centimeter. We perform the division by moving the decimal point one place to the left:

8.9 mm = .8.9 cm

= 0.89 cm

The prefixes *kilo-*, *hecto-*, and so on, are used in all metric units of measure. For example, a *milli*gram is $\frac{1}{1000}$ gram, and a *kilo*watt is 1000 watts. Table 5-1 gives the movement of the decimal point required to change from one unit of metric measure to another.

For example, to use the chart to change 5.7 hectometers to decimeters, we find the row of the table with the prefix *hecto-* on the left, then find the column with the heading *deci-*. In that row and column, we read **3→**. This indicates that the decimal point should be moved 3 places to the right: 5.7 hectometers = 5700 decimeters.

Work Progress Check 3.

Progress Check 3

Change 34.7 kilograms to milligrams.

3. 34,700,000 mg

Progress Check Answers

Table 5-1

FROM \ TO	kilo-	hecto-	deka-	unit	deci-	centi-	milli-
kilo-		1→	2→	3→	4→	5→	6→
hecto-	1←		1→	2→	3→	4→	5→
deka-	2←	1←		1→	2→	3→	4→
unit	3←	2←	1←		1→	2→	3→
deci-	4←	3←	2←	1←		1→	2→
centi-	5←	4←	3←	2←	1←		1→
milli-	6←	5←	4←	3←	2←	1←	

5.1 Exercises

In Exercises 1–12, use a ruler with a scale in inches to measure each object to the nearest $\frac{1}{8}$ inch.

1. The width of a dollar bill

2. The length of a dollar bill

3. The diameter of a quarter

4. The diameter of a penny

5. The length of a sheet of typing paper

6. The width of a sheet of typing paper

7. The height of an aluminum pop can

8. The height of a stack of 10 dimes

9. The distance between sprocket holes along one edge of a sheet of computer paper

10. The length of a new, unsharpened wooden pencil

11. The length of one octave on a piano keyboard

12. The distance between the Q and P keys on a standard typewriter keyboard, measured center to center

In Exercises 13–18, use a metric ruler to measure each object to the nearest millimeter.

13. The length of a dollar bill

14. The width of a dollar bill

15. The diameter of a nickel

16. The diameter of a quarter

17. The distance between sprocket holes along one edge of a sheet of computer paper

18. The height of a stack of 10 dimes

In Exercises 19–24, use a metric ruler to measure each object to the nearest centimeter.

19. The length of a sheet of typing paper

20. The width of a sheet of typing paper

21. The length of a new, unsharpened wooden pencil

22. The height of an aluminum pop can

23. The length of one octave on a piano keyboard

24. The distance between the Q and P keys on a standard typewriter keyboard, measured center to center

In Exercises 25–40, convert each length from feet to inches.

25. 4 ft **26.** 7 ft **27.** 9 ft **28.** 12 ft

29. 5 ft **30.** 24 ft **31.** 10 ft **32.** 15 ft

33. $3\frac{1}{2}$ ft **34.** $2\frac{2}{3}$ ft **35.** $5\frac{1}{4}$ ft **36.** $6\frac{3}{4}$ ft

37. $7\frac{3}{4}$ ft **38.** $5\frac{2}{3}$ ft **39.** $8\frac{5}{12}$ ft **40.** $16\frac{5}{6}$ ft

In Exercises 41–56, convert each length from inches to feet. Express the answer as a mixed number if necessary.

41. 1 in. **42.** 2 in. **43.** $\frac{1}{2}$ in. **44.** $3\frac{1}{3}$ in.

45. 6 in. **46.** 4 in. **47.** 8 in. **48.** 9 in.

49. 16 in. **50.** 30 in. **51.** 14 in. **52.** 28 in.

53. 56 in. **54.** 44 in. **55.** 76 in. **56.** 94 in.

In Exercises 57–68, convert each length to feet and inches.

57. 106 in. **58.** 350 in. **59.** 131 in. **60.** 123 in.

61. 1210 in. **62.** 1330 in. **63.** 54 in. **64.** 68 in.

65. 123 in. **66.** 141 in. **67.** 89 in. **68.** 115 in.

In Exercises 69–76, convert each measurement from yards to feet. Express the answer as a mixed number if necessary.

69. 5 yd **70.** 7 yd **71.** $2\frac{1}{3}$ yd **72.** $4\frac{2}{3}$ yd

73. $3\frac{1}{2}$ yd **74.** $1\frac{1}{4}$ yd **75.** $2\frac{3}{4}$ yd **76.** $6\frac{1}{8}$ yd

In Exercises 77–84, convert each measurement from yards to feet and inches.

77. $3\frac{1}{2}$ yd **78.** $5\frac{3}{4}$ yd **79.** $1\frac{1}{8}$ yd **80.** $4\frac{3}{8}$ yd

81. $2\frac{1}{3}$ yd **82.** $5\frac{1}{4}$ yd **83.** $5\frac{1}{6}$ yd **84.** $4\frac{5}{6}$ yd

In Exercises 85–96, convert each measurement between feet, yards, and miles.

85. 5280 ft = _____ mi

86. 10,560 ft = _____ mi

87. 2640 ft = _____ mi

88. 1320 ft = _____ mi

89. $\frac{1}{4}$ mi = _____ ft

90. $\frac{3}{8}$ mi = _____ ft

91. 1.75 mi = _____ ft

92. 1.625 mi = _____ ft

93. 1 mi = _____ yd

94. $1\frac{1}{2}$ mi = _____ yd

95. $\frac{2}{3}$ mi = _____ yd

96. $\frac{1}{6}$ mi = _____ yd

In Exercises 97–134, convert each measurement between the given metric units.

97. 3 m = _____ cm

98. 5 m = _____ cm

99. 5.7 m = _____ cm

100. 7.36 km = _____ dam

101. 0.31 dm = _____ cm

102. 0.57 m = _____ cm

103. 73.2 m = _____ dm

104. 55.9 m = _____ cm

105. 76.8 hm = _____ mm

106. 165.7 km = _____ m

107. 4.72 cm = _____ dm

108. 0.593 cm = _____ dam

109. 453.2 cm = _____ m

110. 675.3 cm = _____ m

111. 0.325 dm = _____ m

112. 0.0034 mm = _____ m

113. 3.75 cm = _____ mm

114. 0.074 cm = _____ mm

115. 0.125 m = _____ mm

116. 134 m = _____ hm

117. 3.25 cm = _____ mm

118. 5.04 cm = _____ mm

119. 675 dam = _____ cm

120. 0.00777 cm = _____ dam

121. 638.3 m = _____ hm

122. 6.77 cm = _____ m

123. 6.3 mm = _____ cm

124. 6.77 mm = _____ cm

125. 695 dm = _____ m

126. 6789 cm = _____ dm

127. 5689 m = _____ km

128. 0.0579 km = _____ mm

129. 576.2 mm = _____ dm

130. 65.78 km = _____ dam

131. 6.45 dm = _____ km

132. 6.57 cm = _____ mm

133. 658.23 m = _____ km

134. 0.0068 hm = _____ km

Review Exercises

In Review Exercises 1–8, round each number to the accuracy indicated.

1. 3673.263; nearest hundred
2. 3673.263; nearest ten
3. 3673.263; nearest hundredth
4. 3673.263; nearest tenth
5. 0.100602; nearest thousandth
6. 0.100602; nearest hundredth
7. 0.09999; nearest tenth
8. 0.09999; nearest one

5.2 Estimation

- *Estimating Calculations*
- *Estimating Costs*

Estimating Calculations

In situations where exact answers are not required or cannot be obtained easily, approximations or **estimates** are acceptable and useful. To find an estimate of a calculation instead of a precise value, we perform an easier calculation with numbers that are close to the actual values.

EXAMPLE 1

Find a good whole-number estimate of $(7\frac{1}{8})(8\frac{9}{10}) + 9\frac{7}{8}$.

Solution

The number $7\frac{1}{8}$ is close to 7, the number $8\frac{9}{10}$ is close to 9, and the number $9\frac{7}{8}$ is close to 10. We find a good estimate by performing the easier calculation

$$(7)(9) + 10 = 63 + 10$$
$$= 73$$

A good whole-number estimate of the exact value is 73. ■

EXAMPLE 2

The four walls of a room are $11\frac{1}{2}$ feet long and 8 feet 2 inches high. Estimate the total wall area.

Solution

The area of each wall is the product of its length and its width.

Area = length · width

Since estimation does not require exact calculations, we estimate each wall to be about 10 feet long and about 8 feet high. The area of each wall is approximately 10 · 8 square feet, or 80 square feet. The number 80 is close to 100, and because there are four such walls, an estimate of the total area is about 4 · 100, or 400 square feet. ■

If we had used the actual measured dimensions of the room in Example 2, we would have calculated a total area of about 376 square feet. Our estimate of 400 provides a good answer with much less effort.

Estimating Costs

EXAMPLE 3

A roll of wallpaper costs $23.95 and covers about 33 square feet. Jill's living room measures $14\frac{1}{2}$ feet by 12 feet and has an 8-foot ceiling. Three windows measure 3 feet by 5 feet each, and the room has one doorway 5 feet wide and 7 feet high. Jill can afford to spend $600 for redecorating. Can she afford to paper her living room?

Solution

Exact calculations are not required. For estimation purposes, we assume that the room is 15 feet by 10 feet and has a 10-foot ceiling. The area of each long wall is $15 \cdot 10$ square feet, or 150 square feet. The area of each short wall is $10 \cdot 10$ square feet, or 100 square feet. The total wall area is

$$2 \cdot 150 + 2 \cdot 100 \text{ square feet, or } 500 \text{ square feet}$$

Each roll of wallpaper covers about 25 square feet. We choose 25 because it is close to 33, and divides 500 evenly. Jill will need

$$\frac{500}{25} \text{ rolls, or about 20 rolls}$$

At $25 per roll, 20 rolls will cost Jill $25 \cdot 20$, or $500.

The estimate is high, because we didn't bother to subtract the areas of the doorway and windows, because each roll covers more than 25 square feet, and because the height of the ceiling is actually less than 10 feet. Greater accuracy is wasted effort, because even this generous estimate is within Jill's budget. Work Progress Check 4. ■

Progress Check 4

Tina can't decide whether to put curtains in each of her living room's 8 windows or to install miniblinds. Curtains would cost $47.85, and blinds $107, per window. Estimate the cost of each option.

EXAMPLE 4

The evening news reported that *one million people crowded the sidewalks, watching the parade*. If the spectators crowded one mile of the parade route, is the newspaper report reasonable?

Solution

One mile is about 5000 feet. If we estimate that the sidewalks on each side of the street are 20 feet wide, the total sidewalk area on both sides of the street is about $2 \cdot 20 \cdot 5000$ square feet, or about 200,000 square feet. If each person stood on just one square foot of sidewalk (and that would be *very* crowded!) there would only be 200,000 spectators. The news report is greatly exaggerated; a more reasonable estimate might be 50,000. ■

4. $400; $800

5.2 Exercises

In Exercises 1–22, determine a reasonable estimate of the exact answer. Do not find the exact answer. Answers may vary.

1. Estimate the floor area of a room 29 feet $2\frac{1}{4}$ inches long and 12 feet $3\frac{3}{8}$ inches wide.

2. Estimate the floor area of a room 31 feet $5\frac{3}{4}$ inches long and 20 feet $7\frac{5}{8}$ inches wide.

3. Estimate the wall area of a room 32.7 feet long and 28.5 feet wide, with a $7\frac{1}{2}$-foot ceiling.

4. Estimate the wall area of a room 22 feet long and 18 feet wide, with an 8-foot ceiling.

5. Each of West High School's 105 classrooms contains an average of 37 students. How many students attend West High?

6. Jennifer wants to carpet the dining room, which is 27 feet by 17 feet. How many *square yards* of carpet will she need?

7. Bart wants to tile the kitchen, which is 19.3 feet by 23.2 feet. How many one-square-foot tiles will he need?

8. Marketing surveys show that a typical supermarket customer spends between $50 and $120. One store keeps five cash registers open, and each handles 10 customers per hour. Estimate the amount of money the store receives in one hour.

9. One vitamin pill supplies 28% of the recommended daily requirement of vitamin C and 11% of the daily requirement of vitamin B_2. If Betty takes enough pills to guarantee receiving at least 100% of the vitamin C needed, how much vitamin B_2 will she receive?

10. Maria and her three friends had lunch at a restaurant. They ordered two hamburgers at $4.75 each, one chef salad for $2.95, and one club sandwich for $2.50. Three coffees at 65 cents each and one iced tea at 75 cents finished the meal. With tax and a $3 tip, will $20 cover the expenses?

11. A personal computer costs $4750, and the required color monitor costs $529. Juan needs to buy several of these systems for the office, but expenditures in excess of $30,000 need approval from the Board of Directors. How many systems can he buy without asking the Board?

12. Sue knows that electricity costs about 12 cents per kilowatt-hour and that each month she uses about 689 kilowatt-hours. Last month, her electric bill was $137.72. Is that reasonable?

13. Brent is in the grocery store with only $10. In his cart are two gallons of milk, three frozen TV dinners, and a frozen pizza. Milk is $1.49 per gallon, TV dinners are $2.29 each, and the pizza is $3.49. Does Brent have enough money to pay for the groceries?

14. There are five first-grade classes in Spring Creek Elementary School, with enrollments ranging from 27 to 34 children. Every day, each child gets a carton of milk for snack time. The school orders a week's supply of milk, which is delivered in 48-carton cases. Approximately how many cases are needed each week for the first-grade classes?

15. Stan plans to fence in his $92\frac{1}{2}$-foot-by-117-foot yard. He wants a redwood fence, but that is twice as expensive as galvanized steel. A steel fence can be installed for $7.50 per foot, plus $27.95 for each of two gates. How much would the redwood fence cost?

16. Because the weathered wood of a barn is so rough, one gallon of paint covers only 200 square feet on the first coat. On the second coat, one gallon covers 500 square feet. The side walls of the barn are 47 feet by 21 feet, and the two ends are 28 feet by 32 feet. How many gallons of paint are needed to give the barn two coats?

17. A sport coat regularly sells for $195.95. It is marked 22% off. What is the sale price?

18. A dishwasher regularly sells for $470. It is marked 28% off. What is the sale price?

19. Airline fares will increase 12% next month. Tickets for a flight to Seattle now sell for $236. What will the ticket cost next month?

20. A 6% sales tax is added to the cost of a $14,367 car. What will be the total cost of the car?

21. Forty-three percent of the voters stayed home during the last election. How many of the city's 359,578 registered voters voted?

22. Last year, a charity drive raised $2,989,755. This year, drive organizers hope to increase that by 20%. What is this year's goal?

Review Exercises

In Review Exercises 1–10, perform each operation.

1. What percent of 280 is 70?
2. 31 is what percent of 124?
3. What is 15% of 80?
4. What is $33\frac{1}{3}$% of 963?
5. What number is 15% greater than 80?
6. What number is $33\frac{1}{3}$% less than 963?
7. 4.2 is 75% of what number?
8. 5.1 is 6.8% of what number?
9. A price increases from $150 to $200. What is the percent markup?
10. A price decreases from $200 to $150. What is the percent markdown?

5.3 Reading Graphs and Tables

- Reading Bar Graphs
- Reading Pictographs
- Reading Pie Graphs
- Reading Line Graphs
- Reading Histograms and Frequency Polygons
- Reading Data from Tables

Lists of numbers and other numerical facts are often presented attractively in the form of charts or **graphs**. For example, the table in Figure 5-5(a), the **bar graph** in Figure 5-5(b), and the **pie graph** in Figure 5-5(c) all describe how people rate certain television programs. In the bar graph, the length of each bar represents the percent of responses in each category. In the pie graph, the size of each region represents the percent of response. The two graphs tell the story more quickly and more clearly than the table of numbers.

It is easy to see from either of the graphs that the largest percent of those surveyed rated the programs *excellent*, and that the responses *good* and *fair* were tied for last place. That same information is available in the chart of Figure 5-5(a), but it is not as easy to see at a glance.

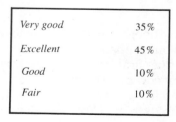

Very good	35%
Excellent	45%
Good	10%
Fair	10%

(a)

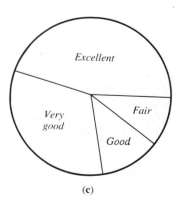

(b)   (c)

Figure 5-5. *Ratings of daytime soap operas*

Reading Bar Graphs

EXAMPLE 1

The bar graph in Figure 5-6 exhibits the total income generated by three sectors of the economy in each of three years. The height of each bar represents income in billions of dollars. Read the graph to answer the following questions.

a. What income was generated by retail sales in 1980?

b. Which sector of the economy consistently generates the most income?

c. By what amount did income from the wholesale sector increase between the years 1970 and 1990?

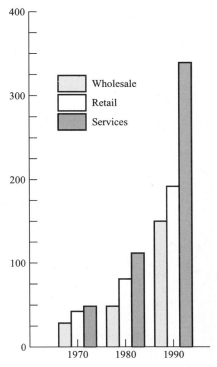

Figure 5-6. *National income by industry, in billions of dollars*

Solution

a. The second group of bars indicates income in 1980. From the key, we determine that the center bar of each group of three bars indicates sales in the retail sector of the economy. The height of that bar, measured against the scale on the left side of the graph, is approximately 80. Numbers on the scale represent billions of dollars, so the retail income generated in 1980 was about $80 billion.

b. In each group of bars, the rightmost bar is the tallest. That bar, according to the key, represents income from the service sector of the economy. Therefore, services consistently generate the most income.

c. According to the key, the leftmost bar in each group represents income from the wholesale sector. Measured against the scale, wholesale generated about $25 billion in 1970, and $150 billion in 1990. The amount of increase in income is the difference of these two quantities:

$$\$150 \text{ billion} - \$25 \text{ billion} = \$125 \text{ billion}$$

Wholesale income increased $125 billion between 1970 and 1990.

EXAMPLE 2

The bar graph in Figure 5-7 indicates the numbers of cars of various types purchased in Dale County for two consecutive years. **a.** Which styles of car have shown a decrease in sales over last year? **b.** Which style of car showed the greatest increase in sales from last year to this year?

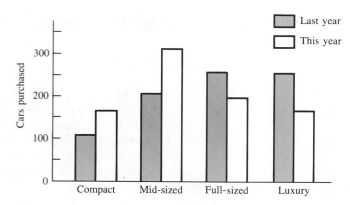

Figure 5-7. *Cars purchased in Dale County*

Solution

a. In Figure 5-7, each pair of touching bars represents sales of a particular model of car. The left bar in each pair gives last year's sales for that model, and the right bar represents this year's sales. Only for full-sized and luxury cars is the left bar (representing last year's sales) taller than the right bar (representing this year's sales). Thus, full-sized and luxury cars have shown a decrease in sales over last year.

b. Sales of compact and mid-sized cars have increased over last year, because for these models the right bar is taller than the left. The difference in the heights of the bars for each model represents the amount of increase. That increase is greater for mid-sized cars. Of all models, mid-sized cars have shown the greatest increase in sales over last year.

Reading Pictographs

A **pictograph** is a variation of the bar graph in which the bars are composed of pictures, and each picture represents a quantity. In Figure 5-8, the key indicates that each picture of a pizza represents 50 pizzas ordered during final exam week. The top row contains three complete and one partial pizza, indicating that the men's residence ordered $3 \cdot 50$, or 150 pizzas, plus about $\frac{1}{4}$ of 50, or 13 pizzas. The men's residence ordered 163 pizzas. The women's residence ordered $4\frac{1}{2} \cdot 50$, or 225 pizzas.

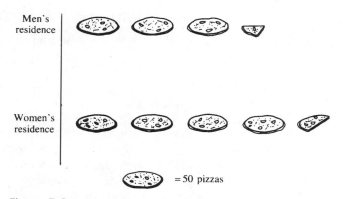

Figure 5-8. *Pizzas ordered during final exam week*

Reading Pie Graphs

EXAMPLE 3

The pie graph in Figure 5-9 shows the contributions of various countries to the world's production of gold. The entire circle represents the world's total production of gold in 1991, and the sizes of the segments of the graph illustrate the parts of that total contributed by the various nations. Thus, Canada produced 7.2% of the world's gold. We can use the graph to answer the following questions:

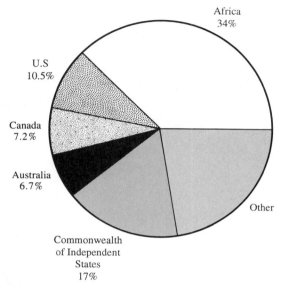

Figure 5-9. World gold production, 1991

a. What percent of the total was the combined production of the United States and Canada?

b. What percent of the total production came from sources other than the nations listed?

c. In 1991, the world's total production of gold was 56.3 million ounces. How many ounces did Australia produce?

Solution

a. According to the graph, the United States produced 10.5% of the total, and Canada produced 7.2%. The combined percent produced by these nations is the sum of 10.5% and 7.2%, or 17.7%. Together, the United States and Canada produced 17.7% of the gold produced in 1991.

b. To find the percent of gold produced by the nations that are not listed, add the contributions of the listed nations and subtract that total from 100%.

$$100\% - (34\% + 10.5\% + 7.2\% + 6.7\% + 17\%) = 100\% - 75.4\%$$
$$= 24.6\%$$

In 1991, nations other than those listed produced 24.6% of the world's total production of gold.

c. From the graph, determine that Australia produced 6.7% of the world's total gold. Because that total was 56.3 million ounces, Australia's share was, in millions of ounces,

$$6.7\% \text{ of } 56.3 = (0.067)(56.3)$$
$$= 3.7721$$

Rounded to the nearest tenth of a million, Australia produced 3.8 million ounces of gold. ∎

Reading Line Graphs

Another type of graph, called a **line graph**, is useful for describing how quantities change with time. From such a graph, we can determine when the quantity is increasing and when it is decreasing.

EXAMPLE 4

The line graph in Figure 5-10 shows how automobile production has varied in the years since 1900. From the graph, we can answer the following questions:

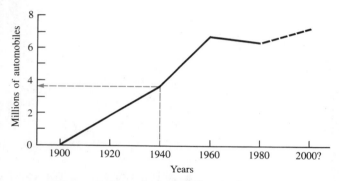

Figure 5-10. *Automobile production*

a. How many automobiles were manufactured in 1940?
b. How many were manufactured in 1950?
c. Over which 20-year span did automobile production increase most rapidly?
d. When did production decrease?
e. Why is a dotted line used for the portion of the graph between 1990 and 2000?

Solution

a. To find the number of autos produced in 1940, we determine the height of the graph at the point directly above the label 1940. We follow the dashed line in Figure 5-10 from the label 1940 straight up to the graph, and then directly over to the scale. There, we read 3.7. Because the scale indicates *millions* of automobiles, approximately 3.7 million, or 3,700,000, autos were produced in 1940.

b. To find the number of autos produced in 1950, we find 1950 on the horizontal line at the bottom of the graph. Although the point is not labeled, it lies halfway between 1940 and 1960. From that point, we move up to the graph and then sideways to the scale on the left, where we read 5.2. Thus, in 1950 approximately 5.2 million autos were manufactured.

c. Between 1940 and 1960, the upward tilt of the graph is greatest. In those years, the production of autos increased most rapidly.

d. The graph rises upward during all years *except* between 1960 and 1990. There, the graph drops, indicating that auto production decreased over the 30 years between 1960 and 1990.

e. Because the year 2000 is still in the future, the production levels indicated by the graph are only a guess, or a *projection*, and the dotted line indicates that the numbers are not certain. The projected manufacturing level in 2000 is 7.2 million automobiles, but that figure may turn out to be wrong. ■

EXAMPLE 5

The graph in Figure 5-11 describes the movements of two trains. The horizontal axis represents time, and the vertical axis represents the distance that the trains have traveled.   **a.** How are the trains moving at time A?   **b.** At what time (A, B, C, D, or E) are both trains stopped?   **c.** At what times have both trains gone the same distance?   **d.** At time E, which train is moving faster?

Figure 5-11

Solution

From the key, we find that the movement of Train 1 is described by the solid line and that of Train 2 by the broken line.

a. At time A on the graph, the broken line is rising. This indicates that the distance traveled by Train 2 is increasing: At time A, Train 2 is moving. At time A, the solid line is horizontal. This indicates that the distance traveled by Train 1 is not changing: At time A, Train 1 is stopped.

b. To find the time at which both trains are stopped, we find the times at which both the solid and the broken lines are horizontal, for at that time distances are not changing. At time B, both trains are stopped.

c. The height of a line at any instant represents the distance a train has traveled at that instant. Both trains will have traveled the *same* distance at any time when the two lines are the same height—that is, at any time when the lines *cross*. This occurs at times C and E.

d. At point E, the solid line is rising more rapidly than the broken line. This indicates that the distance traveled by Train 1 is increasing more rapidly than that traveled by Train 2. At time E, Train 1 is traveling faster.

Work Progress Check 5. ■

Progress Check 5

In Figure 5-11, what is Train 1 doing at time D?

---

5. Train 1, which had been stopped, is beginning to move.

Progress Check Answers

EXAMPLE 6

The line graph in Figure 5-12 shows the annual earnings of the Big Three automobile manufacturers: General Motors, Ford, and Chrysler. From the graph, answer the following questions:

a. How much money did General Motors make in 1982?

b. Which company made the least money in 1987?

c. In what years did the earnings of Ford Motor Corporation surpass those of General Motors?

d. Which company's earnings increased the most during the years 1984 to 1988?

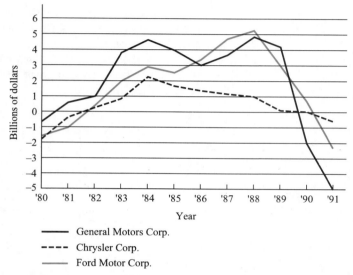

Figure 5-12. *Automobile manufacturers: Corporate earnings*

Solution

a. From the key, we determine that the earnings of General Motors are represented by the solid line. To find the company's earnings in 1982, we move vertically upward from the label '82 until we reach the heavy line. Moving to the left from that point, we reach the number 1 on the vertical scale. Because that scale indicates *billions* of dollars, we have determined that General Motors earned $1 billion in 1982.

b. To determine which company earned the least money in 1987, we move upward from the label '87 until we reach the lowest of the three graphs. That graph is a dashed line, which represents the earnings of Chrysler Corporation. In the year 1987, Chrysler earned the least money of the Big Three.

c. We look at the key to determine that Ford's earnings are represented by the dotted line and that General Motors' are given by the solid line. We then look at the graph to determine the years in which the dotted line is *above* the solid line. That happened in 1986, 1987, 1988, 1990, and 1991. In those years, Ford's earnings exceeded those of General Motors.

d. Find the years 1984 and 1988 on the *year* axis, and determine which of the three graphs has risen the most during that time. Chrysler's earnings steadily decreased during that time, and General Motors' earnings increased slightly. The earnings of Ford Motor Corporation increased the most, from approximately $3 billion in 1984 to $5.3 billion in 1988.

Reading Histograms and Frequency Polygons

A pharmaceutical company is sponsoring a series of reruns of old Western movies. Their marketing department must choose from three advertisements: children talking about Chipmunk Vitamins, a college student catching a quick breakfast and a TurboPill Vitamin, or a gray-haired grandmother taking Silver Sunset Vitamins with a glass of warm milk.

Marketing conducts a survey of the viewing audience, recording the age of each viewer in a sample of 631. They count the number of viewers in the 6-to-15 group, the 16-to-25 group, and so on, and graph the data as the **histogram** in Figure 5-13. The vertical axis, labeled *frequency*, indicates the number of viewers in each age group. For example, the histogram indicates that there are 105 viewers in the 36-to-45 age group.

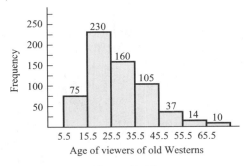

Figure 5-13

A histogram is similar to a bar graph, but several differences are important: the bars of a histogram always touch, data values never fall at the edge of a bar, the widths are the same, and the width of each bar represents a numeric value. The width of each bar in Figure 5-13 represents an age span of 10 years.

Because the greatest number of viewers are in the 16-to-25 age group, the marketing department decides to push TurboPills in commercials appealing to active young adults.

EXAMPLE 7

To determine if three pieces of carry-on luggage should be allowed, an airline weighs the luggage of 2260 passengers and presents the data as the histogram in Figure 5-14.   **a.** How many passengers carry luggage in the 8-to-11 pound range?   **b.** How many are in the 12-to-19 pound range?   **c.** How many passengers carry luggage weighing 13 pounds?

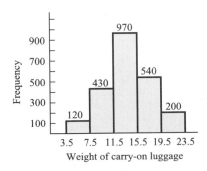

Figure 5-14

Solution

a. Locate the bar corresponding to the 8-to-11 pound range: It is the second bar, with edges at 7.5 and 11.5 pounds. Use the height of the bar (or the number written there) to determine that 430 passengers carry such luggage.

b. The 12-to-19 pound range is covered by two bars. The total number of passengers with luggage in this range is 970 + 540, or 1510.

c. It is likely that some luggage weighs 13 pounds, but the number cannot be found from a histogram. A histogram reports **grouped data**, and information on individual passengers is not available.

A line graph called a **frequency polygon** conveys the same information as a histogram. To construct a frequency polygon from the histogram of Figure 5-13, we first mark and join the center points at the top of each bar as in Figure 5-15**(a)**. On the horizontal axis, we label each point with the middle value of each of the bars. We finally erase the bars to make the frequency polygon in Figure 5-15**(b)**.

Figure 5-15

Reading Data from Tables

Data are often presented in the form of tables in which information is organized in several rows and columns. To read such a table, we must determine which row and column contains the information we need.

Suppose that we want to send an $8\frac{1}{2}$-pound package by priority mail to a friend living in postal zone 4. Postal rates for priority mail appear in Figure 5-16. To determine the cost of mailing the package, we find the *row* of the postage table for priority mail for a package that does not exceed 9 pounds. We find the *column* for zone 4. At the intersection of this row and column, we read the number 7.00. It will cost us $7.00 to mail the package.

U.S. PRIORITY MAIL We Deliver.

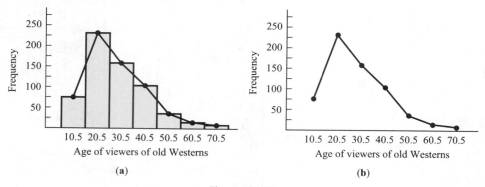

Weight, up to but not exceeding—pound(s)	Local 1, 2, and 3	Zones				
		4	5	6	7	8
1	2.90	2.90	2.90	2.90	2.90	2.90
2	2.90	2.90	2.90	2.90	2.90	2.90
3	4.10	4.10	4.10	4.10	4.10	4.10
4	4.65	4.65	4.65	4.65	4.65	4.65
5	5.45	5.45	5.45	5.45	5.45	5.45
6	5.55	5.75	6.10	6.85	7.65	8.60
7	5.70	6.10	6.70	7.55	8.50	9.65
8	5.90	6.50	7.30	8.30	9.40	10.70
9	6.10	7.00	7.95	9.05	10.25	11.75
10	6.35	7.55	8.55	9.80	11.15	12.80
11	6.75	8.05	9.20	10.55	12.05	13.80
12	7.15	8.55	9.80	11.30	12.90	14.85

Figure 5-16

5.3 Exercises

In Exercises 1–6, refer to the bar graph in Illustration 1.

1. Which source supplied the least energy in 1973?

2. Which energy source remained unchanged between 1973 and 1990?

3. What percent of electrical energy was produced by oil in 1973?

4. Which source provided about 8% of electrical energy in 1990?

5. What was the approximate *percent of increase* in the use of energy from coal?

6. What was the approximate *percent of increase* in the use of nuclear power between 1973 and 1990?

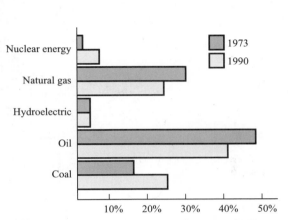

Illustration 1. *Changing sources of electricity*

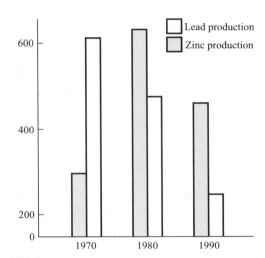

Illustration 2. *World lead and zinc production, in thousand metric tons*

In Exercises 7–12, refer to the bar graph in Illustration 2.

7. The world production of lead in 1970 was approximately equal to the production of zinc in another year. In what other year was that?

8. The world production of zinc in 1990 was approximately equal to the production of lead in another year. In what other year was that?

9. In what year was the production of zinc about one-half that of lead?

10. In what year was the production of zinc about twice that of lead?

11. By how many metric tons did the production of zinc increase between 1970 and 1980?

12. By how many metric tons did the production of lead decrease between 1980 and 1990?

In Exercises 13–16, refer to the bar graph in Illustration 3.

13. In which categories of moving violation have arrests decreased since last month?

14. Last month, which violation occurred most often?

15. This month, which violation occurred least often?

16. Which violation has shown the greatest decrease in number of arrests since last month?

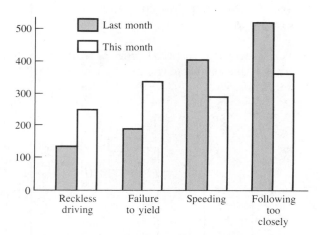

Illustration 3. *Arrests for moving violations*

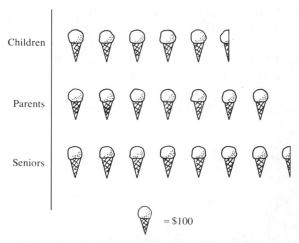

Illustration 4. *Ice cream bought at Barney's Cafe*

In Exercises 17–20, refer to the pictograph in Illustration 4.

17. Which group (children, parents, or seniors) spent the most money on ice cream at Barney's Cafe?

18. How much money did parents spend on ice cream?

19. How much *more* money did seniors spend than parents?

20. How much *more* money did seniors spend than children?

In Exercises 21–26, refer to the pie graph in Illustration 5.

21. Two of the seven languages considered are spoken by groups of about the same size. Which languages are they?

22. Of the languages considered in the graph, which is spoken by the greatest number of people?

23. Do more people speak Russian or English?

24. What percent of the world's population speak Russian or English?

25. What percent of the world's population speak a language other than these seven?

26. What percent of the world's population do not speak either French or German?

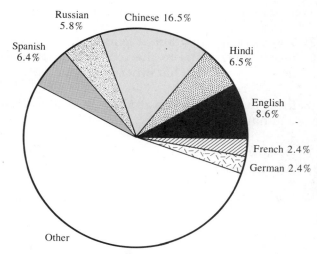

Illustration 5. *World languages, and the percents of the population that speak them*

162   Chapter 5   Measurement, Estimation, and Reading Graphs

In Exercises 27–32, refer to the pie graph in Illustration 6.

27. What percent of total energy sources is nuclear energy?

28. What percent of total energy sources are represented by coal and crude oil combined?

29. By what percent does energy derived from coal exceed that derived from crude oil?

30. By what percent does energy derived from coal exceed that derived from nuclear sources?

31. Solar energy is not listed on the graph. Fill in the blank: Solar energy accounts for less than _____ percent of total energy sources.

32. If in the next ten years, production of nuclear energy tripled (was multiplied by 3), and other sources remained the same, what percent of total energy sources would nuclear energy be?

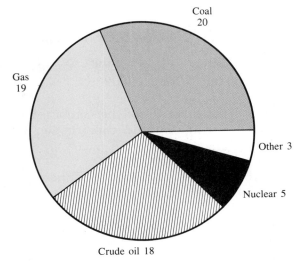

Illustration 6. *Annual energy use (figures indicate quadrillion BTUs)*

In Exercises 33–38, refer to the line graph in Illustration 7.

33. How many nuclear power plants were operational in 1977?

34. How many nuclear power plants were operational in 1987?

35. How many new nuclear power plants came on line between 1977 and 1987?

36. What is the percent of increase of nuclear power plants between 1977 and 1987?

37. In which two-year span was the rate of increase in nuclear plants smallest?

38. In what year was there a sharp increase in the number of nuclear plants?

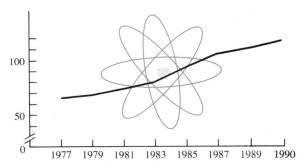

Illustration 7. *Number of operating U.S. nuclear power plants*

In Exercises 39–44, refer to the line graph in Illustration 8.

39. Which runner ran faster at the start of the race?

40. Which runner stopped to rest first?

41. Which runner dropped the baton and had to go back to get it?

42. At what times (A, B, C, or D) is Runner 1 stopped and Runner 2 running?

43. Describe what is happening at time D.

44. Which runner won the race?

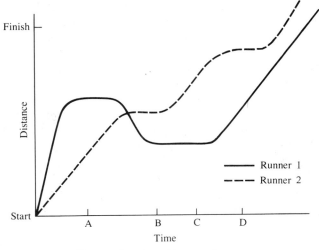

Illustration 8. *Five-mile run*

Section 5.3 Exercises 163

In Exercises 45–50, refer to the line graph in Illustration 9.

45. What were the average weekly earnings in mining for the year 1975?

46. What were the average weekly earnings in construction for the year 1980?

47. What is the estimate of the combined weekly earnings in mining and construction for the year 1990?

48. In approximately what year did miners begin to earn more than construction workers?

49. In the period from 1965 to 1990, which workers received the greatest increase in wages?

50. In what five-year interval did wages in mining increase most rapidly?

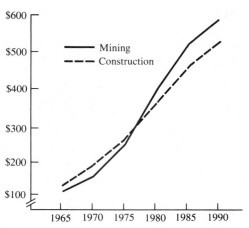

Illustration 9. *Average weekly earnings in mining and construction*

In Exercises 51–54, refer to Illustrations 10 and 11.

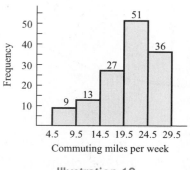

Illustration 10

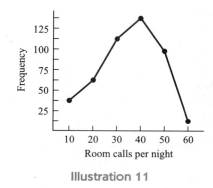

Illustration 11

51. An insurance company collected data on the number of miles its employees drive to and from work. The data are presented in the histogram in Illustration 10. How many employees commute from 15 to 19 miles each week?

52. How many employees commute 14 miles or less each week? (See Illustration 10.)

53. Community Hospital is considering adding more nursing staff during the night shift. They surveyed the medical staff to determine the number of room calls during the night, then constructed the histogram in Illustration 11. On how many nights were there about 30 room calls?

54. About how many room calls could the staff expect on the most busy evening shifts? (See Illustration 11.)

In Exercises 55–58, refer to the postal rate table in Figure 5-16 (page 160).

55. Find the cost of using priority mail to send a package weighing $7\frac{1}{4}$ pounds to zone 3.

56. Find the cost of using priority mail to send a package weighing $2\frac{1}{4}$ pounds to zone 5.

57. Juan wants to send a 6-pound, 1-ounce package to a friend living in zone 2 and a 3-pound, 2-ounce package to a friend in zone 8. If he uses priority mail, how much will it cost?

58. Jenny wants to send a birthday gift and an anniversary gift to her brother who lives in zone 6. One package weighs 2 pounds, 2 ounces, and the other weighs 3 pounds, 4 ounces. If she uses priority mail, how much will she save by sending both gifts as one package instead of two?

In Exercises 59–62, refer to the federal income tax tables in Illustration 12.

1991 Tax Rate Schedules

Schedule X—Use if your filing status is **Single**

If the amount on Form 1040, line 37, is: Over—	But not over—	Enter on Form 1040, line 38	of the amount over—
$0	$20,350	.... 15%	$0
20,350	49,300	$3,052.50 + 28%	20,350
49,300	....	11,158.50 + 31%	49,300

Schedule Y-1—Use if your filing status is **Married filing jointly or Qualifying widow(er)**

If the amount on Form 1040, line 37, is: Over—	But not over—	Enter on Form 1040, line 38	of the amount over—
$0	$34,000	.... 15%	$0
34,000	82,150	$5,100.00 + 28%	34,000
82,150	....	18,582.00 + 31%	82,150

Illustration 12

59. Raul has a taxable income of $57,100, is married, and files jointly. Compute his tax.

60. Herb is single and has a taxable income of $79,250. Compute his tax.

61. Angelina is single and has a taxable income of $53,000. If she gets married, she will gain another deduction, which will reduce her income by $2000, and she could file a joint return. How much would she save in tax by getting married? Assume that her husband is unemployed.

62. Alan (with a taxable income of $53,000) married a woman with a taxable income of $75,000. They filed a joint return. Would they have saved on their taxes if they had both stayed single?

Review Exercises

In Review Exercises 1–4, perform the indicated operations.

1. $4 + 2 \cdot 3^2$
2. $2 + 3^2 \cdot 4$
3. $\dfrac{4(3 + 2) - 5}{12 - 2(3)}$
4. $\dfrac{(4^2 - 3) + 7}{5(2 + 3) - 1}$

5. Write the prime numbers between 10 and 30.
6. Write the first ten composite numbers.
7. Find the prime-factored form of 292.
8. Find the prime-factored form of 1994.
9. Is 1994 divisible by 3?
10. Is 1992 divisible by 3?
11. Find the least common multiple of 35, 42, and 30.
12. Find the greatest common divisor of 35, 42, and 30.

5.4 Mean, Median, and Mode

- The Mean, or the Average
- The Median
- The Mode

As we have seen, one of the purposes of a graph is to communicate many numbers in a form that is easy to understand. There are other ways of describing lists of numbers compactly. We often need to find *one* number that is in some way typical of all of the numbers in a list. We will consider three ways of finding such a typical number: the *mean* or average, the *median*, and the *mode*.

The Mean, or the Average

Evelyn has taken five examinations this semester, with scores of 87, 73, 89, 92, and 84. To find out how she is doing, she calculates the **mean**, or the **average**, of these grades by finding the sum of the grades and then dividing by 5:

$$\text{Evelyn's mean} = \frac{87 + 73 + 89 + 92 + 84}{5}$$

$$= \frac{425}{5}$$

$$= 85$$

The mean score is 85. On some of the exams, Evelyn did better than 85, but on other exams, she did worse. Yet 85 is a typical score, and a good indication of her performance in the class.

> **Definition:** The **mean**, or the **average**, of several values is the sum of those values divided by the number of values.
>
> $$\text{Mean (or average)} = \frac{\text{sum of the values}}{\text{number of values}}$$

EXAMPLE 1

The week's sales in three departments of the Tog Shoppe are given in the chart:

	Men's Department	Women's Department	Children's Department
Monday	$2315	$3135	$1110
Tuesday	2020	2310	890
Wednesday	1100	3206	1020
Thursday	2000	2115	880
Friday	955	1570	1010
Saturday	850	2100	1000

Find the mean of the daily sales in the women's department for this one week.

Solution

Use a calculator to add the sales in the women's department for the week. Then divide the sum of those six values by 6.

$$\text{Mean sales in the women's department} = \frac{3135 + 2310 + 3206 + 2115 + 1570 + 2100}{6}$$

$$= \frac{14{,}436}{6}$$

$$= 2406$$

The mean of the week's daily sales in the women's department is $2406. Work Progress Check 6.

Progress Check 6

In Example 1, find the mean daily sales in all departments for Wednesday.

EXAMPLE 2

In January, Bob drove a total of 4805 miles. On the average, how many miles did he drive each day?

Solution

To find the average number of miles driven each day, we find the sum of the miles driven each day and divide by the number of days. We do not know each individual day's mileage, but we are given the total, 4805. Because there are 31 days in January, we divide 4805 by 31:

6. $1775.33

Progress Check Answers

$$\text{Average number of miles per day} = \frac{\text{total miles driven}}{\text{number of days}}$$

$$= \frac{4805}{31}$$

$$= 155 \qquad \text{Use a calculator.}$$

On the average, Bob drove 155 miles each day.

The Median

There are situations in which the mean is not a typical representative of the values in a list. For example, suppose the weekly earnings of four workers in a small business are $140, $150, $190, and $120. Suppose also that the owner of the company pays himself $5000. The mean salary is given by

$$\text{Mean salary} = \frac{140 + 150 + 190 + 120 + 5000}{5}$$

$$= \frac{5600}{5}$$

$$= 1120$$

When asked about working conditions, the owner smiles and says, "Our employees earn an average of $1120 each week." Clearly, the mean does not fairly represent the workers' salaries.

A better representative of the company's typical salary is the **median**: the salary in the middle when all the numbers are arranged by size.

120   140   **150**   190   5000
↑
the middle salary

The typical worker earns $150 a week, far less than the mean salary.

If there are an even number of values in the list, then there is no middle value. In that case, the median is the mean of the two numbers closest to the middle. For example, there is no middle number in the list 2, 5, **6, 8**, 13, 17. The two numbers closest to the middle are **6** and **8**. The median is the mean of 6 and 8: the median is $\frac{6+8}{2}$, or 7.

> **Definition:** The **median** of several values is the middle value. To find the median,
>
> 1. Arrange the values in increasing order.
> 2. If there are an odd number of values, the median is the value in the middle.
> 3. If there are an even number of values, the median is the average of the two values that are closest to the middle.

EXAMPLE 3

On the first exam, there were three scores of 59, four scores of 77, and scores of 43, 47, 53, 60, 68, 82, and 97. Find the median score.

Solution

Arrange the 14 scores in increasing order:

43 47 53 59 59 59 **60 68** 77 77 77 77 82 97

List the three scores of 59 and the four scores of 77 separately.

Because there is an even number of scores, the median is the mean of the two scores closest to the middle: the 60 and the 68.

The median is $\frac{60 + 68}{2}$, or 64. Work Progress Check 7. ∎

Progress Check 7

Find five numbers that have the same mean and median.

The Mode

Jim is in a hardware store, looking at a display of 20 outdoor thermometers. Twelve of them read 68°, and the remaining 8 have different readings—some higher, some lower. Jim wants to choose an accurate thermometer. Should he choose a thermometer with a reading that is closest to the *mean* of all 20, or to their *median*? Neither. Instead, he chooses one of the 12 reading 68°, figuring that any of those that agree will likely be correct.

Jim has chosen yet another value that is typical of all of the values. By choosing that temperature that appears most often, he has chosen the **mode** of the 20 numbers.

> **Definition:** The **mode** of several values is the single value that occurs most often. If two or more values are tied for that honor, then there is no mode.
> The mode of several values is also called the *modal* value.

EXAMPLE 4

Find the mode of these values: 3, 6, 5, 7, 3, 7, 2, 4, 3, 5, 3, 7, 8, 7, 3, 7, 6, 3, 4.

Solution

To determine the mode of the numbers in the list, make a chart of the distinct numbers that appear, and make tally marks to record the number of times they occur.

2	3	4	5	6	7	8
/	##### /	//	//	//	#### /	/

Because 3 occurs more times than any other number, 3 is the mode. Work Progress Check 8. ∎

Progress Check 8

What is the mode of all the numbers 1 through 100?

---

7. One answer is 1, 2, 3, 4, 5. Mean = Median = 3

8. There is no mode.

5.4 Exercises

In Exercises 1–6, determine the mean of the given numbers.

1. 3, 4, 7, 7, 8, 11, 16
2. 13, 15, 17, 17, 15, 13
3. 5, 9, 12, 35, 37, 45, 60, 77
4. 0, 0, 3, 4, 7, 9, 12
5. 15, 7, 12, 19, 27, 17, 19, 35, 20
6. 45, 67, 42, 35, 86, 52, 91, 102

In Exercises 7–12, determine the median of the given numbers.

7. 2, 5, 9, 9, 9, 17, 29
8. 16, 18, 27, 29, 35, 47
9. 4, 7, 2, 11, 5, 4, 9, 17
10. 0, 0, 3, 4, 0, 0, 3, 4, 5
11. 18, 17, 2, 9, 21, 23, 21, 2
12. 5, 13, 5, 23, 43, 56, 32, 45

In Exercises 13–18, determine the mode (if any) of the given numbers.

13. 3, 5, 7, 3, 5, 4, 6, 7, 2, 3, 1, 4
14. 12, 12, 17, 17, 12, 13, 17, 12
15. 5, 9, 12, 35, 37, 45, 60
16. 0, 3, 0, 2, 7, 0, 6, 0, 3, 4, 2, 0
17. 23.1, 22.7, 23.5, 22.7, 34.2, 22.7
18. $\frac{1}{2}, \frac{1}{3}, \frac{1}{3}, 2, \frac{1}{2}, 2, \frac{1}{5}, \frac{1}{2}, 5, \frac{1}{3}$

19. A survey of soft-drink machines indicates the following prices for a can of pop (in cents): 50, 60, 50, 50, 40, 45, 50, 45, 50, 50, 65, 75, 55, 75, 100, 50, 45, 75. What is the mean price of a can of pop?

20. Several computer stores reported differing prices for toner cartridges for a laser printer (in dollars): 51, 55, 73, 75, 72, 70, 53, 59, and 75. What is the mean price of a toner cartridge?

21. In Exercise 19, what is the median price for a can of pop?

22. In Exercise 20, what is the median price for a toner cartridge?

23. In Exercise 19, what is the modal price for a can of pop?

24. In Exercise 20, what is the mode of the prices for a toner cartridge?

25. Frank's algebra grade is based on the average of four exams, which will count equally. His grades are 75, 80, 90, and 85. What is his average?

26. Temperatures are recorded at hourly intervals, starting at 12:00 midnight. They are: 53, 53, 57, 58, 59, 59, 60, 62, 64, 66, 68, 70, 71, 75, 77, 77, 79, 72, 70, 64, 61, 59, 53, and 51. To the nearest degree, find the average temperature of the period from midnight to 11:00 A.M.

27. See Exercise 25. If Frank's professor decides to count the fourth examination double, what would Frank's average be?

28. In Exercise 26, what is the average temperature for the 24-hour period recorded?

29. There are 37 company cars used by an insurance company's sales force. Last June, those cars logged a total of 98,790 miles. On the average, how many miles did each car travel that month?

30. The Hinrichs family spent $519 on groceries last April. On the average, how much did they spend each day?

31. What was the average number of miles driven daily for each car in Exercise 29?

32. The Hinrichs family (Exercise 30) has five members. What is the average spent for groceries for one family member, for one day?

33. Mike has received the same score on each of five exams. His mean score is 85. What is the median score, and what is the modal score?

34. The class's scores on the last history exam were 57, 59, 61, 63, 63, 63, 70, 87, 89, 95, and 100. Betty got the score of 70 and claims it is "better than average." Which of the three is she better than: mean, median, or mode?

35. Jim received scores of 37, 53, and 78 on three quizzes. His sister, Ginger, received scores of 53, 57, and 58. Who had the better average? Whose grades were more consistent?

36. What is the average of all whole numbers from 1 to 20, inclusive?

Review Exercises

In Review Exercises 1–4, find each number.

1. Find the greatest common divisor of 36 and 81.

2. Find the greatest common divisor of 58 and 102.

3. Find the least common multiple of 45 and 75.

4. Find the least common multiple of 48 and 56.

In Review Exercises 5–10, perform each operation.

5. $\dfrac{3}{4} \cdot \dfrac{2}{9} = $ _____

6. $\dfrac{2}{15} \div \dfrac{4}{5} = $ _____

7. $\dfrac{18}{5} + \dfrac{12}{5} = $ _____

8. $\dfrac{7}{12} - \dfrac{5}{12} = $ _____

9. $\dfrac{8}{5} + \dfrac{3}{10} = $ _____

10. $\dfrac{5}{6} - \dfrac{1}{12} = $ _____

Mathematics in Industry

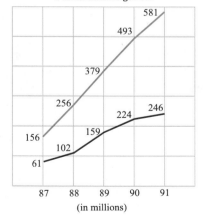

PacTel cellular growth
(in millions)

■ Revenues
■ Operating cash flow (operating income before depreciation and amortization)

Illustration 1

To describe the growth of PacTel Cellular since 1987, Pacific Telesis Group included a line graph in its 1991 Annual Report. That graph is reproduced in Illustration 1 with the permission of Pacific Telesis. The graph communicates on two levels: at a glance, the upward trends of both lines show that the company is prospering. On closer inspection, detailed numeric information of that growth is available for those stockholders who are interested.

1. What was the increase in the company's revenues from 1989 to 1990?
2. What was the percent of increase in the company's revenues between 1989 and 1990?
3. What was the company's operating cash flow in 1991?
4. What one-year interval showed the greatest dollar increase in the company's operating cash flow?

Answers: **1.** $114 million   **2.** 30%   **3.** $246 million   **4.** 1989–1990

Chapter Summary

Key Words

American units (5.1)
average (5.4)
bar graph (5.3)
centimeter (5.1)
estimate (5.2)
foot (5.1)
frequency polygon (5.3)
histogram (5.3)
inch (5.1)
kilometer (5.1)
line graph (5.3)
mean (5.4)
median (5.4)
meter (5.1)
metric units (5.1)
mile (5.1)
millimeter (5.1)
mode (5.4)
pictograph (5.3)
pie graph (5.3)
unit conversion factor (5.1)
units of measurement (5.1)
yard (5.1)

Key Ideas

(5.1) **American Units of Length**

12 inches (in.) = 1 foot (ft)
3 feet = 1 yard (yd)
5280 feet = 1 mile (mi)

Metric Units of Length

1 kilometer (km) = 1000 meters (m) = 10^3 m
1 hectometer (hm) = 100 meters = 10^2 m
1 dekameter (dam) = 10 meters = 10^1 m
1 decimeter (dm) = $\frac{1}{10}$ meter
1 centimeter (cm) = $\frac{1}{100}$ meter = $\frac{1}{10^2}$ m
1 millimeter (mm) = $\frac{1}{1000}$ meter = $\frac{1}{10^3}$ m

(5.2) In situations that do not require exact answers, it is acceptable to estimate answers.

(5.3) Numerical information may be given in the form of a graph.

(5.4) mean (or average) = $\dfrac{\text{sum of the values}}{\text{number of the values}}$

The median of several values in an ordered list is the middle value, or the average of the two values that are closest to the middle.

The mode of several values is the single value that occurs most often.

Chapter 5 Review Exercises

[5.1] In Review Exercises 1–2, measure each object to the nearest $\frac{1}{4}$ inch.

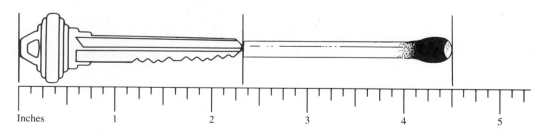

1. Find the length of the key in the illustration.

2. Find the length of the matchstick in the illustration.

In Review Exercises 3–4, measure each object to the nearest millimeter.

3. Find the diameter of a dime.

4. Find the length in millimeters of 1 inch.

In Review Exercises 5–18, convert each length to the units specified.

5. 5.5 ft = _____ in.

6. 2.25 ft = _____ in.

7. 16 in. = _____ ft

8. 27 in. = _____ ft

9. 1.6 mi = _____ ft

10. 0.25 mi = _____ ft

11. 1.6 m = _____ cm

12. 0.25 m = _____ cm

13. 143 cm = _____ m

14. 6.37 cm = _____ m

15. 1670 mm = _____ cm

16. 355 mm = _____ cm

17. 4570 mm = _____ m

18. 3.55 m = _____ mm

In Review Exercises 19–22, estimate the answer.

19. An $85 jacket is marked 45% off. Estimate its cost.

20. Computed at 5%, what will the sales tax be on a $468.29 purchase?

21. Sue and her five friends ordered dinners that ranged from $4.39 to $5.79, and then they had desserts that cost about $1.50. With a $4 tip and 5% tax, will $30 cover it?

22. Electricity costs about 11 cents per kilowatt-hour, and each month Martha uses about 950 kilowatt-hours. Last month's electric bill was $97.72. Is that reasonable?

In Review Exercises 23–26, refer to Illustration 1 to answer each question.

23. How many coupons were redeemed in 1987?

24. Between what two years were the number of redeemed coupons essentially unchanged?

25. Between which two years did the number of redeemed coupons increase the most?

26. What was the percent of the greatest one-year increase?

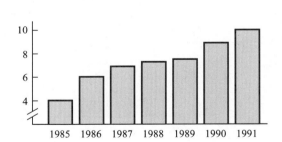

Illustration 1. *Number of supermarket coupons redeemed, in billions*

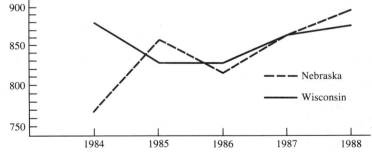

Illustration 2. *Annual egg production, in millions of eggs*

In Review Exercises 27–30, refer to Illustration 2.

27. How many eggs were produced in Wisconsin in 1985?

28. How many eggs were produced in Nebraska in 1987?

29. In what year was the egg production of Wisconsin equal to that of Nebraska?

30. What was the total egg production of Wisconsin and Nebraska in 1988?

In Review Exercises 31–34, refer to Illustration 3.

31. How much money is spent monthly, per prisoner, to pay for the prison staff?

32. How much money is spent monthly, per prisoner, on office costs?

33. What percent of the monthly allotment is spent on one prisoner's food?

34. What percent of the monthly allotment is spent on one prisoner's recreation and training?

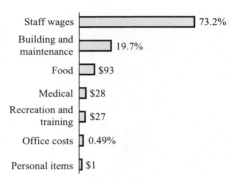

Illustration 3. *Where do prison dollars go? Keeping one prisoner for one month costs $2266.*

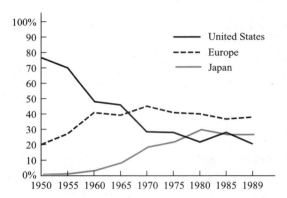

Illustration 4. *Share of world motor vehicle production, 1950–1989*

In Review Exercises 35–38, refer to Illustration 4.

35. In what year did automobile production in Europe first exceed that of the United States?

36. In what year did automobile production in Japan first exceed that of the United States?

37. What percent of world production came from the United States in the year that U.S. production equaled that of Europe?

38. In what year did the U.S. produce 50% of the world's automobiles?

In Review Exercises 39–40, refer to Illustration 5.

39. A survey of the television viewing habits of 320 households produced the histogram in Illustration 5. How many households watch between 6 to 15 hours of TV each week?

40. How many households watch 11 hours or more each week?

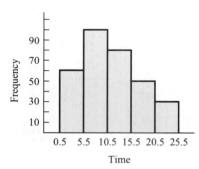

Illustration 5. *Hours per week watching television*

In Review Exercises 41–42, refer to the table in Illustration 6.

Wind speed	35°F	30°F	25°F	20°F	15°F	10°F	5°F	0°F	−5°F	−10°F	−15°F	−20°F	−25°F	−30°F
5 mph	33°	27°	21°	16°	12°	7°	0°	−5°	−10°	−15°	−21°	−26°	−31°	−36°
10 mph	22	16	10	3	−3	−9	−15	−22	−27	−34	−40	−46	−52	−58
15 mph	16	9	−2	−5	−11	−18	−25	−31	−38	−45	−51	−58	−65	−72
20 mph	12	4	−3	−10	−17	−24	−31	−39	−46	−53	−60	−67	−74	−81
25 mph	8	1	−7	−15	−22	−29	−36	−44	−51	−59	−66	−74	−81	−88
30 mph	6	−2	−10	−18	−25	−33	−41	−49	−56	−64	−71	−79	−86	−93
35 mph	4	−4	−12	−20	−27	−35	−43	−52	−58	−67	−74	−82	−89	−97
40 mph	3	−5	−13	−21	−29	−37	−45	−53	−60	−69	−76	−84	−92	−100
45 mph	2	−6	−14	−22	−30	−38	−46	−54	−62	−70	−78	−85	−93	−102

Illustration 6. *Determining the wind-chill temperature*

41. What is the wind-chill temperature on a 10° F day, when a 15 mph wind is blowing?

42. The wind-chill temperature is −25° F, but the outdoor temperature is 15° F. How fast is the wind blowing?

43. Find the mean, median, and mode of the following data:

23, 25, 25, 32, 32, 32, 47, 48

44. Rank the mean, median, and mode in order from largest to smallest for the following data:

100, 1000, 1000, 1500, 2000, 3004

Chapter 5 Test

1. To the nearest $\frac{1}{4}$ inch, find the length of the key in Illustration 1. _____

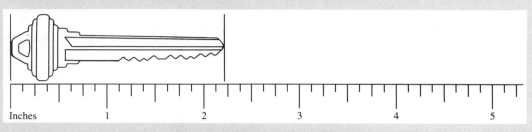

Illustration 1

2. Express 22 inches as a distance in feet. (Your answer will be a mixed number.) _____

3. Express $7\frac{1}{3}$ feet as a distance in inches. _____

4. Express 7 feet 4 inches as a distance in inches. _____

5. Express 7 feet 4 inches as a distance in feet. (Your answer will be a mixed number.) _____

6. How many feet are there in 6 yards? _____

7. How many yards are there in 6 feet? _____

8. How many feet are there in 1 mile? _____

9. How many miles are there in 7920 feet? _____

10. How many yards are there in 1 mile? _____

11. Change 4.7 meters into centimeters. _____

12. Change 4.7 centimeters into meters. _____

13. Change 37.2 millimeters into meters. _____

14. Change 37.2 millimeters into centimeters. _____

15. How many dekameters are in 0.31 kilometers? _____

16. How many decimeters are in 0.31 kilometers? _____

In Test Questions 17–18, refer to Illustration 2.

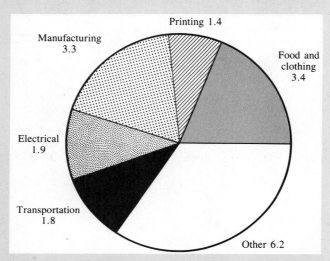

Illustration 2. *Employees in industry (in millions)*

17. Approximately what percent of all employees are in the food and clothing industries? _____

18. Among workers in food and clothing, 2.4 million are in food, and the rest are in clothing. What percent of all workers are in clothing? _____

In Test Questions 19–22, use the information given in Illustration 3.

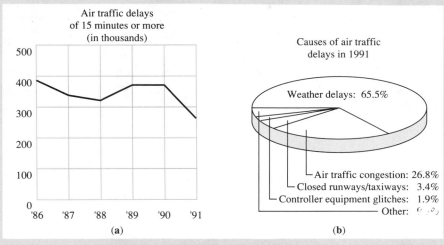

Illustration 3

19. How many air traffic delays occurred in 1991? _____

20. How many air traffic delays in 1991 were due to the weather? _____

21. Which year was worst for air traffic delays? _____

22. The percent for *other* causes for delays in 1991 appears smudged. What should the value be? _____

In Test Questions 23–26, refer to Illustration 4 and choose the best answer from the following statements. Insert the letter of your choice in the blank.

A. Both bicyclists are moving, and bicyclist 1 is faster than bicyclist 2.
B. Both bicyclists are moving, and bicyclist 2 is faster than bicyclist 1.
C. Bicyclist 1 is stopped, and bicyclist 2 is not.
D. Bicyclist 2 is stopped, and bicyclist 1 is not.
E. Both bicyclists are stopped.

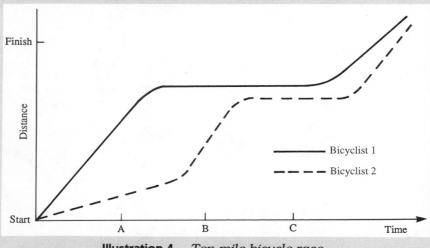

Illustration 4. *Ten-mile bicycle race*

23. Indicate what is happening at time A. _____

24. Indicate what is happening at time B. _____

25. Indicate what is happening at time C. _____

26. Which bicyclist won the race? _____

27. Find the mean, median, and mode of the following ages. _____
8, 10, 13, 15, 15, 20, 22, 22, 22, 30, 40, 59

6 Basic Concepts of Geometry

Mathematics for Fun

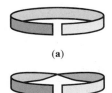

Illustration 1

One of the strangest aspects of geometry is called *topology*. Although this topic is covered in advanced courses, you can experiment with one concept of topology, a strange figure called a *Möbius strip*.

To construct a Möbius strip, begin with a strip of paper (adding-machine tape works well) as in Illustration 1(**a**). Before joining the ends, give one end a half-twist, to form the loop in Illustration 1(**b**).

It is difficult to think of a sheet of paper that does not have two sides, or of a river that does not have two banks. Yet the Möbius strip you have just made has only one side and one edge! Use a pencil to mark a path along the strip as you pull it through your fingers. When you come back to your starting point, you will have marked its one and only side. Look on what appears to be the "other" side—it has been marked, too. Similarly, you can show that the Möbius strip has only one edge. This curiosity has a practical use: a Möbius-strip conveyor belt lasts twice as long, because it wears evenly on "both" sides.

The word *geometry* comes from the Greek words *geo*, meaning earth, and *metron*, meaning measure. Geometry had its origins in early Egypt, where earth measure was made necessary by the annual flooding of the Nile River and the subsequent need for yearly surveying. The Greek mathematician Euclid, who lived about 300 B.C., is considered to be the first person to collect the isolated facts known about geometry and put them into an organized form. His work produced 13 volumes, called the *Elements*. Everything we will discuss in this chapter is work attributed to Euclid.

6.1 Some Undefined Terms and Basic Definitions

- Properties of Points, Lines, and Planes
- Angles

A study of geometry requires logical reasoning. In a logical reasoning system (often called a **deductive reasoning system**), we cannot define all terms. For example, if we try to define the word *set*, we might use the synonym *collection*. But then we must define the word *collection*. To do so, we might use the word *group*. But then we must define *group*. Perhaps we would do so by using synonyms such as *class*, *bunch*, or *pack*. But then

---

Some material in this chapter is adapted from *Elementary Geometry*, 3rd Edition, by R. David Gustafson and Peter D. Frisk. Copyright ©1991 by John Wiley & Sons, Inc. Adapted by permission of John Wiley & Sons, Inc.

we must define these words. The method of finding synonyms to define a term is a never-ending process, and we will eventually be forced to use the word we are trying to define. However, because we all know what is meant by the word *set*, we will simply accept the word without giving it a formal definition.

Other words we will accept as undefined terms are:

point, line, plane, between, length, width, depth, figure, and *surface*

Euclid thought of a **point** as a *figure* that has position but no *length, width,* or *depth*. We often represent points as dots drawn on paper, such as point A in Figure 6-1**(a)**. Points are always named with capital letters. A **line** is a figure with length, but no width or depth. Figure 6-1**(b)** shows a line passing through two points B and C. The line BC is often denoted by the symbol $\overleftrightarrow{BC}$. A line is a set of points that continues forever in opposite directions.

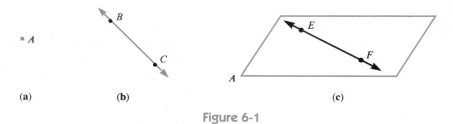

Figure 6-1

A **plane** is a flat *surface* with length and width, but no depth. In Figure 6-1**(c)**, $\overleftrightarrow{EF}$ lies in the plane AB.

Figure 6-2 shows a line passing through points A, B, and C. When several points such as A, B, and C all lie on the same line, we say that the points are **collinear**. On the line, the point B is *between* points A and C. However, point A is not between points B and C, and point C is not between points A and B.

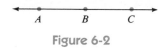

Figure 6-2

We can define other words in a more formal way. For example, a **ray** and a **line segment** are both defined as portions of a line.

> **Definition:** A **ray** is a portion of a line that begins at some point (say, A) and continues forever in one direction. Point A is called the **endpoint** of the ray.

A ray AB with endpoint A and a ray BA with endpoint B are shown in Figures 6-3**(a)** and 6-3**(b)**, respectively. These figures show that each ray has exactly one endpoint. The ray AB with endpoint A is denoted by the symbol $\overrightarrow{AB}$, and the ray BA with endpoint B is denoted by the symbol $\overrightarrow{BA}$.

> **Definition:** Let A and B be two points on a line. The **line segment AB** is the portion of the line that consists of points A and B and all points that are between A and B. The points A and B are called the **endpoints** of the line segment.

The line segment AB in Figure 6-3**(c)** shows that a line segment has two endpoints, point A and point B. The line segment AB is denoted by the symbol $\overline{AB}$.

Figure 6-3

Every line segment has a point midway between its endpoints, called its **midpoint**. When a point (say, M) divides a line segment into two segments of equal length, we say that point M **bisects** the segment. In Figure 6-4,

$m(\overline{AM}) = 4 - 1$
$= 3$

Read $m(\overline{AM})$ as "the measure, or length, of line segment $\overline{AM}$."

and

$m(\overline{MB}) = 7 - 4$
$= 3$

Figure 6-4

Since segments AM and MB are of equal length, point M bisects $\overline{AB}$.

Properties of Points, Lines, and Planes

1. If A and B are two points, they determine exactly one line, the line AB.

2. If A, B, and C are three noncollinear points, they determine a single plane—say, the plane MN. If two or more points lie in the same plane, they are called **coplanar**. Thus, points A, B, and C are coplanar.

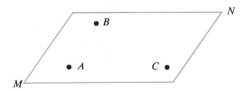

3. If two distinct lines l_1 and l_2 intersect (have points in common), they intersect in *exactly one* point. If two planes intersect, they intersect in a line.

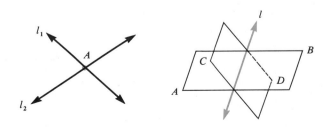

4. If two lines l_1 and l_2 lie in the same plane (say, plane *MN*) and do not intersect, they are called **parallel lines**. In symbols, we say $l_1 \parallel l_2$, where the symbol $\parallel$ is read as "is parallel to." If two planes do not intersect, they are called **parallel planes**. Plane *AB* $\parallel$ plane *CD*.

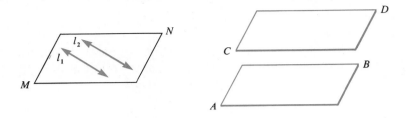

Angles

Another basic geometric figure is the *angle*.

> **Definition:** An **angle** is a figure formed by two rays with a common endpoint. The common endpoint is called the **vertex** of the angle, and the rays are called **sides** of the angle.

There are three different ways to name an angle. One way is to use letters to label three points on the angle—the vertex and a point on each side. We can call the angle in Figure 6-5 $\angle BAC$ (read as "angle *BAC*") or $\angle CAB$ (read as "angle *CAB*"). When we use three letters to name an angle, it is important that the letter name of the angle's vertex be placed between the other two letters.

A second way to name an angle is to use the name of the vertex only. The angle in Figure 6-5 can be called $\angle A$. This simple way of naming an angle can be used when only one angle exists that could have that name.

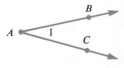

Figure 6-5

A third way to name an angle is to write a small number or a small letter inside the angle. The angle in Figure 6-5 can be called $\angle 1$.

Line segments are commonly measured in inches, feet, yards, meters, centimeters, and so on. Angles are often measured in degrees, denoted with the degree symbol °. The approximate number of **degrees** in an angle can be found by using a protractor such as the one shown in Figure 6-6.

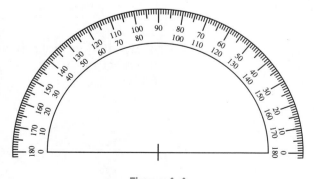

Figure 6-6

If a circle is divided into 360 equal parts and lines are drawn from the center of the circle to two consecutive points of division, the angle formed by those lines measures 1°. In Figure 6-7,

m(∠ABC) = 30°   Read as "The measure of angle *ABC* is 30 degrees."
m(∠ABD) = 60°
m(∠ABE) = 110°
m(∠ABF) = 150°

and

m(∠ABG) = 180°

If we read the protractor from the left, we see that m(∠GBF) = 30°.

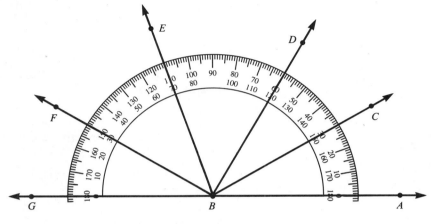

Figure 6-7

When two angles such as ∠ABC and ∠GBF have equal measures, we say that the angles are **congruent**, and we write

∠ABC ≅ GBF   Read as "∠ABC is congruent to ∠GBF."

When a ray whose endpoint is the vertex of an angle divides the angle into two congruent angles, the ray **bisects** the angle. In Figure 6-7, $\overrightarrow{BC}$ is the bisector of ∠ABD.

Angles are classified by their size.

> ### Definition:
> An **acute angle** is an angle whose measure is greater than 0° but less than 90°.
>
> A **right angle** is an angle whose measure is 90°.
>
> An **obtuse angle** is an angle whose measure is greater than 90° but less than 180°.
>
> A **straight angle** is an angle whose measure is 180°.

EXAMPLE 1

Refer to Figure 6-8.

a. ∠AFB, ∠BFC, ∠CFD, and ∠BFD are acute angles because the measure of each angle is greater than 0° and less than 90°.

b. ∠EFA and ∠AFC are right angles because the measure of each angle is 90°.

c. ∠EFB, ∠EFD, and ∠AFD are obtuse angles because the measure of each angle is greater than 90° but less than 180°.

d. ∠EFC is a straight angle because its measure is 180°.

Work Progress Check 1.

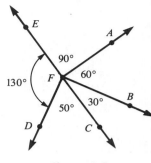

Figure 6-8

If two lines intersect and form "square corners," they are called **perpendicular lines**.

> **Definition:** **Perpendicular lines** are lines that intersect and form right angles.

The lines l_1 and l_2 in Figure 6-9 are perpendicular. The lines l_1 and l_3 are not. To say that two lines such as l_1 and l_2 are perpendicular, we can write

$l_1 \perp l_2$ Read as "Line l_1 is perpendicular to line l_2."

In the figure, the symbol ⌐ is used to show a right angle.

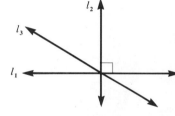

Figure 6-9

Progress Check 1

Refer to the figure and identify each angle as acute, right, obtuse, or straight.

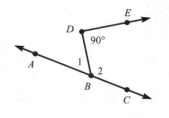

1. ∠1 is acute, ∠2 is obtuse, ∠BDE is a right angle, ∠ABC is a straight angle

Progress Check Answers

6.1 Exercises

1. Try to define the word *good*. What synonyms did you use? How many synonyms can you think of?

2. Try to define the word *defeated*. What synonyms did you use? How many synonyms can you think of?

3. How many endpoints does a ray have?

4. How many endpoints does a line segment have?

5. How many endpoints does a line have?

6. How many lines can contain the same point?

7. How many planes can contain the same line?

8. How many points do two parallel planes have in common?

In Exercises 9–18, refer to Illustration 1.

9. m($\overline{AC}$) = _____

10. m($\overline{BE}$) = _____

11. m($\overline{CE}$) = _____

12. m($\overline{BD}$) = _____

13. m($\overline{CD}$) = _____

14. m($\overline{DE}$) = _____

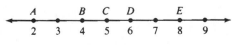

Illustration 1

15. Find the midpoint of $\overline{AD}$.

16. Find the midpoint of $\overline{BE}$.

17. Does point D bisect $\overline{BE}$?

18. Does point B bisect $\overline{AD}$?

In Exercises 19–20, refer to Illustration 2.

19. Explain why $\angle AGF$ cannot be called $\angle G$.

20. Explain why $\overline{AB}$ and $\overline{CD}$ are not perpendicular.

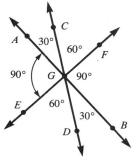

Illustration 2

In Exercises 21–36, refer to Illustration 2 and tell whether each statement is true. If a statement is false, explain why.

21. The ray GF has point G as its endpoint.

22. The line segment AG has no endpoints.

23. The line CD has three endpoints.

24. The line EF is perpendicular to the line AB.

25. Points E, G, and F are collinear.

26. Points A, G, and E are collinear.

27. Point D is the vertex of $\angle DGB$.

28. The vertex of $\angle EGC$ is point G.

29. Point G is between points C and D.

30. Point F is between points C and B.

31. $m(\angle AGC) = m(\angle BGD)$

32. $m(\angle AGF) = m(\angle BGE)$

33. $\angle FGB \cong \angle EGA$

34. $\angle AGC \cong \angle FGC$

35. $\overrightarrow{GE}$ bisects $\angle AGB$.

36. $\overrightarrow{GC}$ bisects $\angle AGF$.

In Exercises 37–44, refer to Illustration 2 and classify each angle as an acute angle, a right angle, an obtuse angle, or a straight angle.

37. $\angle AGC$

38. $\angle EGA$

39. $\angle FGD$

40. $\angle BGA$

41. $\angle BGE$

42. $\angle AGD$

43. $\angle DGC$

44. $\angle DGB$

45. How many line segments can be drawn connecting two points?

46. How many line segments can be drawn connecting exactly two of the points shown in Illustration 3?

47. How many line segments can be drawn connecting exactly two of the points shown in Illustration 4?

48. How many line segments can be drawn connecting exactly two of the points shown in Illustration 5?

Illustration 3   Illustration 4   Illustration 5

Section 6.1 Exercises 187

Review Exercises

1. Convert $3\frac{2}{3}$ to an improper fraction.
2. Calculate 2^4.
3. Add: $\frac{1}{2} + \frac{2}{3} + \frac{3}{4}$
4. Subtract: $\frac{3}{4} - \frac{1}{8} - \frac{1}{3}$
5. Multiply: $\frac{5}{8} \cdot \frac{2}{15} \cdot \frac{6}{5}$
6. Divide: $\frac{12}{17} \div \frac{4}{34}$
7. Calculate: $0.43 + 0.4 + \frac{1}{4}$
8. Calculate: $\frac{3}{5} - 0.2 + 0.75$
9. Multiply: 2.34×0.032
10. Find the product: $\frac{2}{5}(0.4)(0.36)$

6.2 Special Pairs of Angles

- Adjacent Angles
- Vertical Angles

Two types of angles that play important roles in geometry are **complementary** and **supplementary angles**.

> **Definition:** Two angles are called **complementary angles** if the sum of their measures is 90°. If two angles are complementary, either angle is called the **complement** of the other.

> **Definition:** Two angles are called **supplementary angles** if the sum of their measures is 180°. If two angles are supplementary, either angle is called the **supplement** of the other.

EXAMPLE 1

a. Angles measuring 60° and 30° are complementary angles because the sum of their measures is 90°. An angle of 30° is the complement of an angle of 60°, and an angle of 60° is the complement of an angle of 30°.

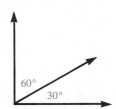

188 Chapter 6 Basic Concepts of Geometry

b. Angles measuring 130° and 50° are supplementary because the sum of their measures is 180°. An angle of 130° is the supplement of an angle of 50°, and an angle of 50° is the supplement of an angle of 130°.

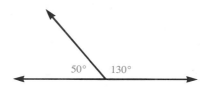

c. Because the sum of two angles measuring 70° and 50° is neither 90° nor 180°, the angles are neither complementary nor supplementary.

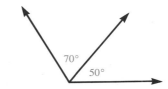

d. Angles measuring 40°, 60°, and 80° are not supplementary angles even though the sum of their measures is 180°. This is because there are *three* angles whose sum is 180°. The definition of supplementary angles requires *two* angles whose sum is 180°.

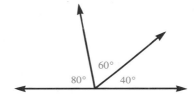

EXAMPLE 2

Find   **a.** the complement   and   **b.** the supplement of an angle measuring 35°.

Solution

a. To find the complement of a 35° angle, subtract 35° from 90° to obtain 55°. Because the sum of the measures of a 35° angle and a 55° angle is 90°, the complement of a 35° angle is a 55° angle.

b. To find the supplement of a 35° angle, subtract 35° from 180° to obtain 145°. Because the sum of the measures of a 35° angle and a 145° angle is 180°, the supplement of a 35° angle is a 145° angle.

Work Progress Check 2.

Progress Check 2

Find the complement and supplement of an angle measuring 50°.

2. 40°, 130°

Progress Check Answers

Adjacent Angles

Angles that are placed side by side are called **adjacent angles**.

> **Definition:** Two angles are called **adjacent angles** if, and only if, they have a common vertex and a common side lying between them.

Section 6.2 Special Pairs of Angles   189

EXAMPLE 3

Consider the angles shown in Figure 6-10.

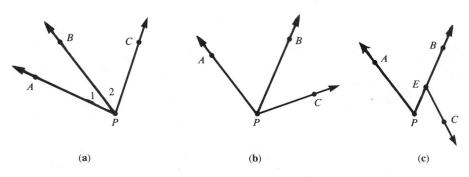

Figure 6-10

a. In Figure 6-10(a), ∠1 and ∠2 are adjacent angles because they have a common vertex at point P and the common side $\overrightarrow{PB}$ lying between them. ∠1 and ∠2 are side-by-side.

b. In Figure 6-10(b), ∠APB and ∠APC are not adjacent angles. While it is true that they have a common vertex, point P, their common side $\overrightarrow{PA}$ is not between the angles. These angles are not side by side. One angle is on top of the other.

c. In Figure 6-10(c), ∠APB and ∠BEC are not adjacent angles because they do not have a common vertex.

Work Progress Check 3.

The definition of adjacent angles contains the words *if, and only if*. This terminology emphasizes the fact that the definition is reversible. This means that the definition is equivalent to the following two *If..., then...* sentences.

- If two angles are adjacent angles, then they have a common vertex and a common side lying between them.
- If two angles have a common vertex and a common side lying between them, then they are adjacent angles.

Progress Check 3

In the figure, which pairs of angles are adjacent?

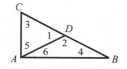

Vertical Angles

Definition: Nonadjacent angles formed by two intersecting lines are called **vertical angles**.

In Figure 6-11, ∠1 and ∠2 are a pair of vertical angles, as are ∠3 and ∠4. It can be proved that vertical angles are always congruent. To illustrate that vertical angles are congruent, we refer to Figure 6-11, in which m(∠2) = 30°. Because the measure of every straight angle is 180°,

$$30° + m(∠4) = 180°$$
$$m(∠4) = 150°$$

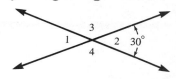

Figure 6-11

3. ∠1 and ∠2, ∠5 and ∠6

Progress Check Answers

and

$$30° + m(\angle 3) = 180°$$
$$m(\angle 3) = 150°$$

Thus, the vertical angles, $\angle 3$ and $\angle 4$, are congruent. Since $m(\angle 1) = 30°$, $\angle 1$ and $\angle 2$ are also congruent.

The previous discussion justifies the following **theorem**.

> **Theorem:** If two angles are vertical angles, they are congruent.

EXAMPLE 4

In Figure 6-12, $\overleftrightarrow{AD}$, $\overleftrightarrow{BE}$, and $\overleftrightarrow{CF}$ are lines. Find **a.** $m(\angle EGD)$, **b.** $m(\angle AGE)$, and **c.** $m(\angle AGD)$.

Solution

a. Because $\angle BGA$ and $\angle EGD$ form a pair of vertical angles, they have equal measures. Thus,

$$m(\angle EGD) = 100°$$

b. Because $\angle DGC$ and $\angle AGF$ are vertical angles,

$$m(\angle AGF) = 50°$$

Thus,

$$m(\angle AGE) = m(\angle AGF) + m(\angle FGE)$$
$$= 50° + 30°$$
$$= 80°$$

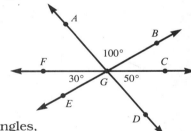

Figure 6-12

c. Because $\angle EGF$ and $\angle BGC$ are vertical angles,

$$m(\angle BGC) = 30°$$

Thus,

$$m(\angle AGD) = m(\angle AGB) + m(\angle BGC) + m(\angle CGD)$$
$$= 100° + 30° + 50°$$
$$= 180°$$

Work Progress Check 4.

Any sentence in the form of an *If...*, *then...* statement is called a **conditional statement**. We have shown that the following conditional statement is true:

*If **two angles are vertical angles**, then **they are congruent**.*

If the **hypothesis** (the information after the word *if*) and the **conclusion** (the information after the word *then*) are interchanged, a new statement is formed that is called the **converse** of the original statement. The converse of the above statement is

*If **two angles are congruent**, then **they are vertical angles**.*

This conditional statement is false. The angles shown in Figure 6-13 are congruent, but they are not vertical angles.

Progress Check 4

a. Find $m(\angle 1)$.

b. Find $m(\angle 2)$.

c. Find $m(\angle 3)$.

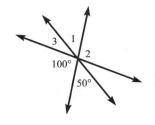

4. **a.** 50° **b.** 100° **c.** 30°

Progress Check Answers

A conditional statement and its converse can both be true. For example,

If two lines are perpendicular, then they intersect and form right angles.

and

If two lines intersect and form right angles, then they are perpendicular.

are both true statements.

Because the converse of a true conditional statement might or might not be true, we cannot assume the converse of a statement without proving that it is true.

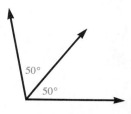

Figure 6-13

6.2 Exercises

1. Find the complement of a 30° angle.

2. Find the supplement of a 30° angle.

3. Find the supplement of a 105° angle.

4. Find the complement of a 75° angle.

5. Explain why an angle measuring 105° cannot have a complement.

6. Explain why an angle measuring 210° cannot have a supplement.

7. Find the supplement of the complement of an 85° angle.

8. Find the complement of the supplement of a 115° angle.

In Exercises 9–16, refer to Illustration 1 and tell whether each statement is true. If a statement is false, explain why.

9. ∠AGB is adjacent to ∠BGC.

10. ∠DGC is adjacent to ∠DGB.

11. ∠FGB is adjacent to ∠FGA.

12. ∠EGC is adjacent to ∠CGB.

13. ∠AGF and ∠DGC are vertical angles.

14. ∠FGE and ∠BGA are vertical angles.

15. ∠AGB and ∠BGC are vertical angles.

16. ∠AGC and ∠DGF are vertical angles.

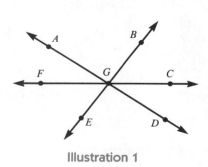

Illustration 1

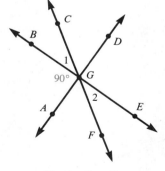

Illustration 2

In Exercises 17–28, refer to Illustration 2 and tell whether each statement is true.

17. m(∠1) = m(∠2)

18. m(∠FGB) = m(∠CGE)

19. ∠AGB ≅ ∠DGE

20. ∠CGD ≅ ∠CGB

192   Chapter 6   Basic Concepts of Geometry

21. ∠AGF ≅ ∠FGE

22. ∠AGB ≅ ∠BGD

23. ∠FGA and ∠AGC are supplementary angles.

24. ∠AGB and ∠BGC are complementary angles.

25. ∠AGF and ∠2 are complementary angles.

26. ∠AGB and ∠EGD are supplementary angles.

27. ∠EGD and ∠DGB are supplementary angles.

28. ∠DGC and ∠AGF are complementary angles.

In Exercises 29–32, refer to Illustration 3.

29. Name all pairs of adjacent angles.

30. Name all pairs of vertical angles, if any.

31. Name all pairs of complementary angles, if any.

32. Name all pairs of supplementary angles, if any.

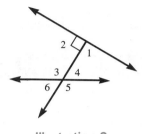

Illustration 3

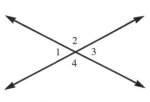

Illustration 4

In Exercises 33–36, refer to Illustration 4, in which m(∠1) = 50°.

33. m(∠4) = _____

34. m(∠3) = _____

35. m(∠1) + m(∠2) + m(∠3) = _____

36. m(∠2) + m(∠4) = _____

In Exercises 37–40, refer to Illustration 5, in which m(∠1) + m(∠3) + m(∠4) = 180°, m(∠3) = m(∠4), and m(∠4) = m(∠5).

37. m(∠1) = _____

38. m(∠2) = _____

39. m(∠3) = _____

40. m(∠6) = _____

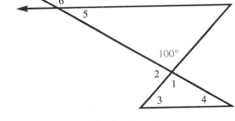

Illustration 5

41. Write the converse of the statement, *If my eyes are open, then I can see.* Is the converse statement true?

42. Write the converse of the statement, *If it is raining, then the sidewalks are wet.* Is the converse statement true?

43. Write the converse of the statement, *If I eat apples, then I will be healthy.* Is the converse statement true?

44. Write the converse of the statement, *If I breathe, then I am alive.* Is the converse statement true?

Section 6.2 Exercises 193

Review Exercises

1. Divide: $1.242 \div 0.23$
2. Divide: $215.539 \div 32.17$
3. Change $\frac{4}{5}$ to a decimal.
4. Change $\frac{2}{3}$ to a decimal. Round to the nearest hundredth.
5. Write without using exponents: $\left(\frac{3}{4}\right)^3$.
6. Write without using exponents: $\left(\frac{2}{5}\right)^2$.

In Review Exercises 7–10, perform the calculations.

7. $3 + 2 \cdot 4$
8. $5 \cdot 3 + 4 \cdot 2$
9. Find 30% of 60.
10. 65 is 25% of what number?

6.3 Parallel Lines and Transversals

- Special Angles
- Theorems About Parallel Lines

Two lines are parallel if they lie in the same plane and do not intersect. If two lines lie in different planes and do not intersect, they are called **skew lines**. For example, a line drawn on the floor of a room will never intersect a line drawn on the ceiling. However, these lines are not parallel unless they are in the same plane.

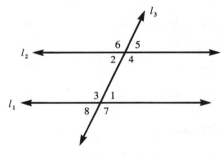

Figure 6-14

Special Angles

If a line l_3, called a **transversal**, intersects lines l_1 and l_2 as in Figure 6-14, several pairs of angles are formed. These angles can be classified as follows:

Definition: If two lines in the same plane are cut by a transversal, the nonadjacent angles on opposite sides of the transversal and on the interior of the two lines are called **alternate interior angles**.

In Figure 6-14, $\angle 1$ and $\angle 2$ are alternate interior angles, as are $\angle 3$ and $\angle 4$.

Definition: If two lines in the same plane are cut by a transversal, the angles on the same side of the transversal and in corresponding positions with respect to the two lines are called **corresponding angles**.

In Figure 6-14, ∠1 and ∠5, ∠3 and ∠6, ∠4 and ∠7, and ∠2 and ∠8 are pairs of corresponding angles.

> **Definition:** If two lines in the same plane are cut by a transversal, nonadjacent angles on opposite sides of the transversal and on the exterior of the two lines are called **alternate exterior angles**.

In Figure 6-14, ∠5 and ∠8, and ∠7 and ∠6 are alternate exterior angles.

Theorems About Parallel Lines

The following **postulate** (a statement that is accepted without proof) provides a basic fact about alternate interior angles.

> **Postulate:** If two parallel lines are cut by a transversal, pairs of alternate interior angles are congruent.

If we use the previous postulate, we can prove theorems involving other angles formed by transversals intersecting parallel lines.

> **Theorem:** If two parallel lines are cut by a transversal, pairs of corresponding angles are congruent.

To prove this theorem, we refer to Figure 6-15, in which transversal l_3 intersects parallel lines l_1 and l_2. We will show that the corresponding angles ∠1 and ∠3 are congruent. A similar argument will establish that the other pairs of corresponding angles are also congruent. Here is the proof:

Because ∠1 and ∠2 are alternate interior angles formed by a transversal intersecting two parallel lines, they are congruent. Thus,

$$m(\angle 1) = m(\angle 2)$$

Because ∠2 and ∠3 are vertical angles, and vertical angles are congruent,

$$m(\angle 3) = m(\angle 2)$$

Because m(∠1) and m(∠3) both equal m(∠2), they are equal to each other:

$$m(\angle 1) = m(\angle 3)$$

Thus, the corresponding angles ∠1 and ∠3 have equal measures and are congruent. □

With a similar argument, we could prove the following theorem.

Theorem: If two parallel lines are cut by a transversal, pairs of alternate exterior angles are congruent.

It is easy to prove another theorem.

Theorem: If two parallel lines are cut by a transversal, pairs of interior angles on the same side of the transversal are supplementary.

To prove this theorem, we refer to Figure 6-16, in which transversal l_3 intersects parallel lines l_1 and l_2. $\angle 3$ and $\angle 2$ form a pair of interior angles on the same side of the transversal. We will show that $\angle 3$ and $\angle 2$ are supplementary. A similar argument will establish that $\angle 1$ and $\angle 4$ are supplementary. Here is the proof:

Because $\angle 1$ and $\angle 2$ are alternate interior angles formed by a transversal intersecting two parallel lines, they are congruent, and

$$m(\angle 1) = m(\angle 2)$$

Because $\angle 1$ and $\angle 3$ form a straight angle,

1. $m(\angle 1) + m(\angle 3) = 180°$

If we substitute $m(\angle 2)$ for $m(\angle 1)$ in Equation 1, we have

$$m(\angle 2) + m(\angle 3) = 180°$$

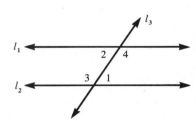

Figure 6-16

Because the sum of the measures of $\angle 2$ and $\angle 3$ is 180°, the angles are supplementary. □

EXAMPLE 1

Refer to Figure 6-17, in which l_3 is a transversal that intersects parallel lines l_1 and l_2 and $m(\angle 1) = 120°$. Find **a.** $m(\angle 4)$, **b.** $m(\angle 5)$, and **c.** $m(\angle 3)$.

Solution

a. Because $\angle 1$ and $\angle 4$ are alternate interior angles formed by a transversal intersecting two parallel lines, their measures are equal. We are given that $m(\angle 1) = 120°$. Thus,

$$m(\angle 4) = 120°$$

b. Because $\angle 1$ and $\angle 5$ are corresponding angles formed by a transversal intersecting two parallel lines, their measures are equal. We are given that $m(\angle 1) = 120°$. Thus,

$$m(\angle 5) = 120°$$

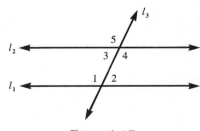

Figure 6-17

c. Because ∠1 and ∠3 are interior angles on the same side of a transversal that intersects two parallel lines, they are supplementary. We are given that m(∠1) = 120°. Thus,

$$120° + m(\angle 3) = 180°$$
$$m(\angle 3) = 60°$$

EXAMPLE 2

Refer to Figure 6-18, in which $l_1 \parallel l_2$ and l_3 is a transversal. Find m(∠1).

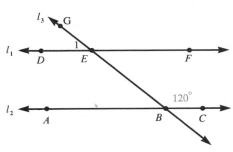

Figure 6-18

Solution

Because ∠CBE and ∠1 are supplementary angles (can you see why?), the sum of their measures is 180°. So m(∠CBE) + m(∠1) = 180°, and

$$120° + m(\angle 1) = 180°$$
$$m(\angle 1) = 60°$$

Thus, m(∠1) = 60°. Work Progress Check 5.

We will prove one more theorem.

Theorem: If a transversal is perpendicular to one of two parallel lines, it is also perpendicular to the other line.

To prove this theorem, we refer to Figure 6-19, in which l_3 is a transversal intersecting parallel lines l_1 and l_2. We shall assume that l_3 is perpendicular to l_1. Because perpendicular lines intersect and form right angles, and the measure of a right angle is 90°,

$$m(\angle 1) = 90°$$

Because ∠1 and ∠2 are alternate interior angles formed by a transversal intersecting two parallel lines, they are congruent, and

2. $m(\angle 1) = m(\angle 2)$

If we substitute 90° for m(∠1) in Equation 2, we see that

$$m(\angle 2) = 90°$$

Because m(∠2) = 90°, ∠2 is a right angle, as are ∠3, ∠4, and ∠5. Thus,

$$l_3 \perp l_2$$

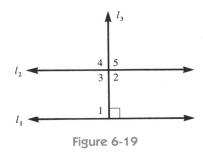

Figure 6-19

Progress Check 5

If $l_1 \parallel l_2$, then

a. Find m(∠1).

b. Find m(∠2).

c. Find m(∠3).

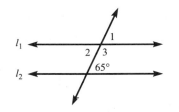

---

5. a. 65° **b.** 65° **c.** 115°

Progress Check Answers

The following theorems can also be shown to be true.

> **Theorem:** If two lines in the same plane are cut by a transversal so that a pair of
> - alternate interior angles
> - corresponding angles
> - alternate exterior angles
>
> are congruent, the lines are parallel.

> **Theorem:** If two lines in the same plane are cut by a transversal so that interior angles on the same side of the transversal are supplementary, the lines are parallel.

> **Theorem:** If two lines in the same plane are perpendicular to the same transversal, they are parallel.

Progress Check 6

Tell whether $l_1 \parallel l_2$ if

a. $m(\angle 1) = m(\angle 2)$

b. $m(\angle 1) = m(\angle 3)$

c. $m(\angle 3) = m(\angle 5)$

d. $m(\angle 3) = m(\angle 4)$

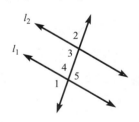

6. **a.** not necessarily **b.** yes
   c. yes **d.** not necessarily

EXAMPLE 3

Refer to Figure 6-20, in which l_1 and l_2 are in the same plane and l_3 intersects l_1 and l_2. Tell whether $l_1 \parallel l_2$ if

a. $m(\angle 1) = m(\angle 3)$

b. $\angle 1$ is supplementary to $\angle 2$

c. $m(\angle 1) = m(\angle 2)$

d. $\angle 3$ is supplementary to $\angle 4$

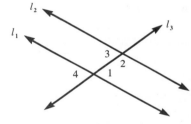

Figure 6-20

Solution

a. Because $\angle 1$ and $\angle 3$ are alternate interior angles with equal measures, the lines are parallel.

b. Because $\angle 1$ and $\angle 2$ are interior angles on the same side of the transversal and they are supplementary, the lines are parallel.

c. $\angle 1$ and $\angle 2$ are interior angles on the same side of a transversal. The lines will be parallel if these two angles are supplementary. The fact that the angles have equal measure does not guarantee that the lines will be parallel. The lines might or might not be parallel.

d. $\angle 3$ and $\angle 4$ are corresponding angles. If these angles had equal measures, the lines would be parallel. The fact that they are supplementary does not guarantee that the lines will be parallel.

Work Progress Check 6.

6.3 Exercises

In Exercises 1–10, refer to Illustration 1 and classify each pair of angles as alternate interior angles, corresponding angles, alternate exterior angles, or interior angles on the same side of the transversal.

1. ∠1 and ∠3   **2.** ∠2 and ∠3   **3.** ∠7 and ∠3   **4.** ∠2 and ∠6

5. ∠1 and ∠4   **6.** ∠2 and ∠4   **7.** ∠8 and ∠4   **8.** ∠8 and ∠6

9. ∠7 and ∠5   **10.** ∠1 and ∠5

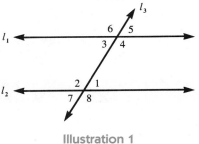

Illustration 1

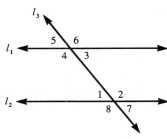

Illustration 2

In Exercises 11–20, use the given information and tell whether lines l_1 and l_2 in Illustration 2 are parallel.

11. m(∠1) = m(∠4)

12. m(∠1) = m(∠3)

13. m(∠1) = m(∠5)

14. m(∠1) = m(∠6)

15. m(∠2) = m(∠4)

16. m(∠2) = m(∠5)

17. ∠1 is supplementary to ∠4.

18. ∠1 is supplementary to ∠6.

19. m(∠6) = m(∠8)

20. m(∠5) = m(∠8)

In Exercises 21–24, refer to Illustration 3, in which lines l_1 and l_2 are parallel.

21. Find m(∠1).   **22.** Find m(∠2).   **23.** Find m(∠3).   **24.** Find m(∠4).

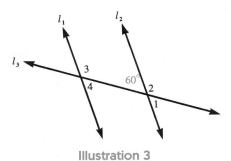

Illustration 3

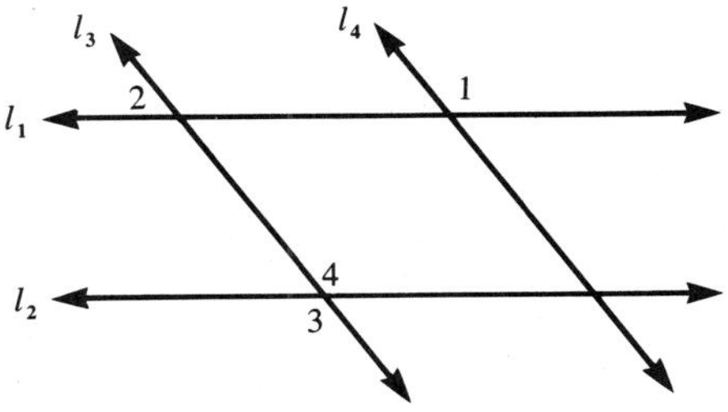

Illustration 4

In Exercises 25–28, refer to Illustration 4, in which $l_1 \parallel l_2$ and $l_3 \parallel l_4$.

25. Find m(∠1).   **26.** Find m(∠2).   **27.** Find m(∠3).   **28.** Find m(∠4).

In Exercises 29–34, refer to Illustration 5.

29. If m(∠1) = m(∠4), can $\overline{ED}$ intersect $\overline{AB}$? Explain.

30. If m(∠1) = m(∠2), can $\overline{ED}$ intersect $\overline{AB}$? Explain.

31. If m(∠1) = m(∠3), can $\overline{ED}$ intersect $\overline{AB}$? Explain.

32. If m(∠1) = m(∠5), can $\overline{ED}$ intersect $\overline{AB}$? Explain.

33. If m(∠6) = m(∠5), can $\overline{ED}$ intersect $\overline{AB}$? Explain.

34. If m(∠2) = m(∠6), can $\overline{ED}$ intersect $\overline{AB}$? Explain.

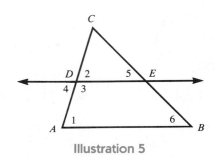

Illustration 5

Review Exercises

In Review Exercises 1–4, tell which of the fractions is the largest.

1. $\frac{1}{2}, \frac{1}{3}, \frac{1}{4}$

2. $\frac{2}{3}, \frac{3}{4}, \frac{7}{12}$

3. $\frac{8}{15}, \frac{13}{30}, \frac{3}{5}$

4. $\frac{11}{9}, \frac{13}{11}, \frac{6}{5}$

In Review Exercises 5–8, insert one of the symbols < or > to make the following comparisons true.

5. 3 _____ 10

6. 6.26 _____ 6.25

7. 9.55 _____ 9.43

8. $\frac{7}{8}$ _____ $\frac{15}{16}$

In Review Exercises 9–10, tell which is larger.

9. 12% of 5500   or   15% of 5200

10. 23% of 1723   or   25% of 1500

6.4 Triangles and Other Polygons

- The Sum of the Angles of a Triangle
- The Sum of the Interior Angles of a Polygon

Three geometric figures appear in Figure 6-21. Figure 6-21**(a)** shows a **closed figure** with all three sides of equal length. (It is called a closed figure because all of its sides connect.) Figure 6-21**(b)** shows a closed figure with four sides. Figure 6-21**(c)** shows a closed figure with five sides. Point M is said to be in the **interior** of the five-sided figure, and point N is said to be in its **exterior**.

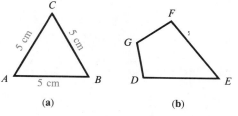

Figure 6-21

200 Chapter 6 Basic Concepts of Geometry

The figures in Figure 6-21 are called **polygons**. The three-sided polygon in Figure 6-21**(a)** is a **triangle**, the four-sided polygon in Figure 6-21**(b)** is a **quadrilateral**, and the five-sided polygon in Figure 6-21**(c)** is a **pentagon**. The points where the sides of a polygon connect are called **vertices** of the polygon.

The vertices of △ABC (read as "triangle ABC") shown in Figure 6-21**(a)** are the points A, B, and C. The vertices of the quadrilateral DEFG are the points D, E, F, and G. The vertices of the pentagon are the points H, I, J, K, and L.

There are special triangles that derive their names from the types of angles that are contained within them.

> **Definition:**
> A **right triangle** is a triangle with one right angle.
> An **acute triangle** is a triangle with three acute angles.
> An **obtuse triangle** is a triangle with one obtuse angle.

A right triangle is shown in Figure 6-22**(a)**, an acute triangle in Figure 6-22**(b)**, and an obtuse triangle in Figure 6-22**(c)**. The longest side of a right triangle is called the **hypotenuse**.

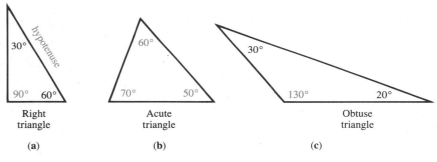

Figure 6-22

Other triangles are named according to the relationships of their sides.

> **Definition:** An **equilateral triangle** is a triangle with all sides of equal length.

All angles in an equilateral triangle have the same measure. Thus, an equilateral triangle is also **equiangular**. An equiangular triangle is also equilateral. Thus, an equilateral triangle has three congruent sides and three congruent angles.

> **Definition:** An **isosceles triangle** is a triangle that has at least two sides of equal length. The third side is called its **base**.

Definition: A **scalene triangle** is a triangle in which no two sides are of equal length.

An equilateral triangle, an isosceles triangle, and a scalene triangle are shown in Figure 6-23.

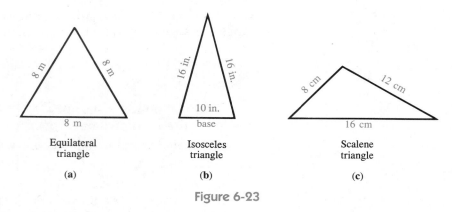

Figure 6-23

EXAMPLE 1

Find the perimeter of the equilateral triangle shown in Figure 6-23(a).

Solution

Recall that the perimeter of a triangle is the distance around it. Thus, the perimeter of the triangle shown in the figure is

$P = 8 + 8 + 8 = 24$

The perimeter is 24 meters.

EXAMPLE 2

The perimeter of the isosceles triangle in Figure 6-24 with a side 12 meters long is 50 meters. Find the length of its base.

Solution

Since one of the sides of equal length is 12 meters long and the perimeter is 50 meters long, we have

$12 + 12 + m(\text{base}) = 50$
$24 + m(\text{base}) = 50$
$m(\text{base}) = 26$

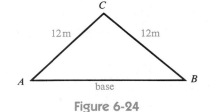

Figure 6-24

Thus, the length of the base of the triangle is 26 meters. Work Progress Check 7.

Progress Check 7

The perimeter of an isosceles triangle is 60 meters. If one of its congruent sides is 12 meters long, how long is its base?

7. 36 meters

The Sum of the Angles of a Triangle

We can use our knowledge of parallel lines to prove a basic theorem about triangles.

> **Theorem:** The sum of the measures of the angles of any triangle is 180°.

To prove this theorem, we refer to the triangle in Figure 6-25. Through point C, we draw the line that is parallel to the line AB. This forms two pairs of congruent alternate interior angles.

$$m(\angle 1) = m(\angle 2) \quad \text{and} \quad m(\angle 3) = m(\angle 4)$$

Because $\angle 2 + \angle 5 + \angle 4$ is a straight angle,

3. $m(\angle 2) + m(\angle 5) + m(\angle 4) = 180°$

Substituting $m(\angle 1)$ for $m(\angle 2)$ and $m(\angle 3)$ for $m(\angle 4)$ in Equation 3, we have

$$m(\angle 1) + m(\angle 5) + m(\angle 3) = 180°$$

Thus, the sum of the angles of $\triangle ABC$ is 180°.

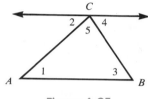

Figure 6-25

EXAMPLE 3

In Figure 6-26, find $m(\angle B)$.

Solution

The sum of the measures of $\angle A$, $\angle B$, and $\angle C$ must be 180°:

4. $m(\angle A) + m(\angle B) + m(\angle C) = 180°$

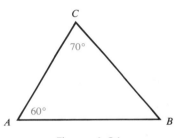

Figure 6-26

Because $m(\angle A) = 60°$ and $m(\angle C) = 70°$, we substitute these values into Equation 4 and solve for $m(\angle B)$.

$$60° + m(\angle B) + 70° = 180°$$
$$m(\angle B) + 130° = 180°$$
$$m(\angle B) = 50°$$

Thus, $m(\angle B) = 50°$.

EXAMPLE 4

In Figure 6-27, find the measure of
a. $\angle 1$, **b.** $\angle 2$, and **c.** $\angle 3$.

Solution

a. In $\triangle DBC$, the sum of $m(\angle 1)$, $m(\angle B)$, and $m(\angle BCD)$ must be 180°. Thus,

$$m(\angle 1) + 70° + 50° = 180°$$
$$m(\angle 1) + 120° = 180°$$
$$m(\angle 1) = 60°$$

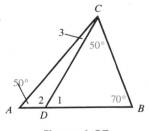

Figure 6-27

b. Because ∠1 and ∠2 form the straight angle ∠ADB, we have m(∠1) + m(∠2) = 180°. Thus,

$$60° + m(\angle 2) = 180° \quad \text{Substitute 60° for m(∠1).}$$
$$m(\angle 2) = 120°$$

c. In △ADC, the sum of m(∠A), m(∠2), and m(∠3) must be 180°. Thus,

$$m(\angle A) + m(\angle 2) + m(\angle 3) = 180°$$
$$50° + 120° + m(\angle 3) = 180° \quad \text{Substitute 50° for m(∠A)}$$
$$170° + m(\angle 3) = 180° \quad \text{and 120° for m(∠2).}$$
$$m(\angle 3) = 10°$$

Work Progress Check 8.

Progress Check 8

In the triangle, find

a. m(∠1).

b. m(∠2).

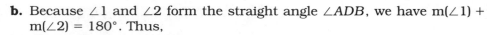

The Sum of the Interior Angles of a Polygon

We can find the sum of the angles of any polygon. For example, to find the sum of the angles of the quadrilateral in Figure 6-28, we pick any vertex (say, A) and draw the diagonal $\overline{AC}$ to form triangles △ABC and △ADC. In △ABC, we have

$$m(\angle 1) + m(\angle B) + m(\angle 2) = 180°$$

and in △ADC, we have

$$m(\angle 4) + m(\angle D) + m(\angle 3) = 180°$$

If we add the angles in these two triangles together, we obtain

$$m(\angle 1) + m(\angle 4) + m(\angle B) + m(\angle D) + m(\angle 2) + m(\angle 3) = 360°$$

But because m(∠1) + m(∠4) = m(∠BAD) and m(∠2) + m(∠3) = m(∠DCB), we have

$$m(\angle BAD) + m(\angle B) + m(\angle D) + m(\angle DCB) = 360°$$

Thus, the sum of the measures of the angles of the quadrilateral is 360°.

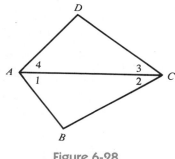

Figure 6-28

A generalization of this discussion leads to the following theorem:

Theorem: The sum of the measures of the interior angles of any polygon is given by the formula

$$S = (n - 2)180°$$

where n is the number of sides of the polygon.

8. a. 50° **b.** 10°

EXAMPLE 5

Find the sum of the measures of the interior angles of a **hexagon** (a six-sided polygon).

Solution

Because a hexagon has six sides, substitute 6 for n in the formula and simplify.

$$S = (n - 2)180°$$
$$S = (6 - 2)180°$$
$$S = 4(180°)$$
$$S = 720°$$

The sum of the measures of the interior angles of a hexagon is 720°. Work Progress Check 9.

Progress Check 9

Find the sum of the measures of the interior angles of a 24-sided polygon.

9. 3960°

Progress Check Answers

6.4 Exercises

In Exercises 1–8, classify each triangle as a right triangle, an acute triangle, or an obtuse triangle.

1.
2.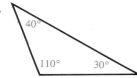
3.
4.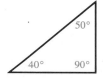

5. △ABC with m(∠A) = 30°, m(∠B) = 70°, and m(∠C) = 80°

6. △DEF with m(∠D) = 40°, m(∠E) = 50°, and m(∠F) = 90°

7. △MNP with m(∠M) = 20°, m(∠N) = 120°, and m(∠P) = 40°

8. △RST with m(∠R) = 40°, m(∠S) = 70°, and m(∠T) = 70°

In Exercises 9–16, classify each triangle as an equilateral triangle, an isosceles triangle, or a scalene triangle.

9.
10.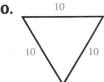
11.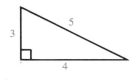
12.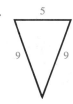

13. △ABC with m($\overline{AB}$) = 6 inches, m($\overline{BC}$) = 8 inches, and m($\overline{CA}$) = 10 inches

14. △DEF with m($\overline{DE}$) = 32 centimeters, m($\overline{EF}$) = 32 centimeters, and m($\overline{FD}$) = 32 centimeters

15. △DEF with m($\overline{DE}$) = 12 meters, m($\overline{EF}$) = 10 meters, and m($\overline{FD}$) = 12 meters

16. △ABC with m($\overline{AB}$) = 100 centimeters, m($\overline{BC}$) = 1.5 meters, and m($\overline{CA}$) = 1 meter

17. Find the perimeter of an isosceles triangle with a base of length 21 centimeters and sides of length 32 centimeters.

18. The perimeter of an isosceles triangle is 80 meters. If the length of one of the congruent sides is 22 meters, how long is the base?

19. The perimeter of an equilateral triangle is 85 feet. Find the length of each side.

20. An isosceles triangle with sides measuring 49.3 centimeters has a perimeter of 121.7 centimeters. Find the length of the base.

In Exercises 21–28, the measures of two angles of △ABC are given. Find the measure of the third angle.

21. m(∠A) = 30°, m(∠B) = 60°.
m(∠C) = _____

22. m(∠A) = 45°, m(∠C) = 105°.
m(∠B) = _____

23. m(∠B) = 100°, m(∠A) = 35°.
m(∠C) = _____

24. m(∠B) = 33°, m(∠C) = 77°.
m(∠A) = _____

25. m(∠C) = 97°, m(∠B) = 32°.
m(∠A) = _____

26. m(∠C) = 103°, m(∠A) = 65°.
m(∠B) = _____

27. m(∠A) = 25.5°, m(∠B) = 63.8°.
m(∠C) = _____

28. m(∠B) = 67.25°, m(∠C) = 72.5°.
m(∠A) = _____

In Exercises 29–32, refer to Illustration 1.

29. m(∠1) = _____

30. m(∠2) = _____

31. m(∠A) = _____

32. If m(∠ACB) = 80°, m(∠B) = _____

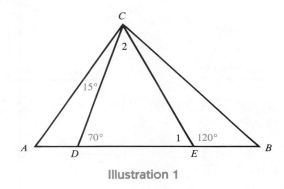

Illustration 1

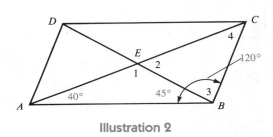

Illustration 2

In Exercises 33–36, refer to Illustration 2.

33. m(∠1) = _____

34. m(∠2) = _____

35. If m(∠ABC) = 120°, m(∠3) = _____

36. m(∠4) = _____

37. Find the sum of the measures of the interior angles of a pentagon (a five-sided polygon).

38. Find the sum of the measures of the interior angles of an octagon (an eight-sided polygon).

39. Find the sum of the measures of the interior angles of a decagon (a ten-sided polygon).

40. Find the sum of the measures of the interior angles of a dodecagon (a twelve-sided polygon).

Review Exercises

1. Find 20% of 110.

2. Find 15% of 50.

3. Find 20% of $\frac{6}{11}$.

4. Find 30% of $\frac{3}{5}$.

5. What percent of 200 is 80?

6. What percent of 500 is 100?

7. 20% of what number is 500?

8. 33% of what number is 21?

9. Perform the calculation: $0.85 \div 2(0.25)$.

10. Perform the calculations: $3.25 + 12 \div 0.4 \cdot 2$.

6.5 Congruent Triangles

- SSS ≅ SSS
- ASA ≅ ASA, AAS ≅ AAS
- SAS ≅ SAS

If two triangles have the same area and the same shape, they are said to be **congruent**. In Figure 6-29, $\triangle ABC$ and $\triangle DEF$ are congruent, and we can write

$$\triangle ABC \cong \triangle DEF \quad \text{Read as "}\triangle ABC \text{ is congruent to } \triangle DEF.\text{"}$$

Because each pair of **corresponding parts** (matching parts such as $\angle C$ and $\angle F$) in $\triangle DEF$ and $\triangle ABC$ have the same measure, they are congruent. In congruent triangles ABC and DEF, we have

$$\angle A \cong \angle D$$
$$\angle B \cong \angle E$$
$$\angle C \cong \angle F$$
$$\overline{BC} \cong \overline{EF}$$
$$\overline{AC} \cong \overline{DF}$$

and

$$\overline{AB} \cong \overline{DE}$$

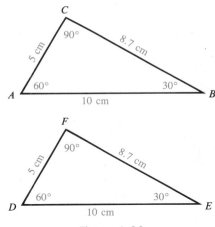

Figure 6-29

In $\triangle ABC$ of Figure 6-30, side CB is opposite $\angle A$, and $\angle A$ is opposite side CB. Likewise, side AC and $\angle B$ are opposite each other, and side AB and $\angle C$ are opposite each other. In congruent triangles, corresponding angles are always opposite corresponding sides, and corresponding sides are always opposite corresponding angles.

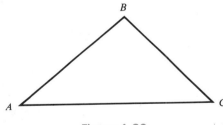

Figure 6-30

Congruent triangles can be defined as follows:

> **Definition:** Two triangles are called **congruent triangles** if, and only if, all of their corresponding parts are congruent.

SSS ≅ SSS

To show that two triangles are congruent, we could show that each of the six pairs of corresponding parts are congruent and then use the previous definition. However, this would be more work than is necessary. If we choose the corresponding parts in a special way, we can establish that two triangles are congruent by using only three pairs of corresponding parts. The following three postulates indicate which pairs of corresponding parts are necessary to prove that triangles are congruent.

> **Postulate:** If three sides of one triangle are congruent, respectively, to three sides of a second triangle, the triangles are congruent. **(SSS ≅ SSS)**

The triangles in Figure 6-31 are congruent because three sides of one triangle are congruent to three sides of the other triangle **(SSS ≅ SSS)**.

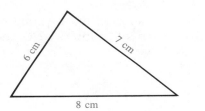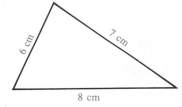

Figure 6-31

ASA ≅ ASA, AAS ≅ AAS

> **Postulate:** If two angles and a side of one triangle are congruent, respectively, to two angles and a side of a second triangle, the triangles are congruent. **(ASA ≅ ASA, AAS ≅ AAS)**

The triangles in Figure 6-32 are congruent because two angles and the side between them of one triangle are congruent to two angles and the side between them of the second triangle **(ASA ≅ ASA)**.

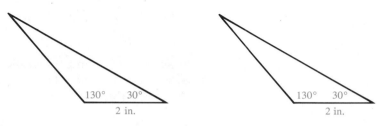

Figure 6-32

In the previous example, the side did not have to be between the angles. The triangles in Figure 6-33 are also congruent (**AAS ≅ AAS**).

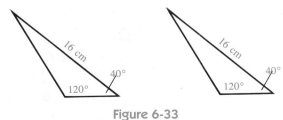

Figure 6-33

SAS ≅ SAS

> **Postulate:** If two sides and the angle between them in one triangle are congruent, respectively, to two sides and the angle between them in a second triangle, the triangles are congruent. (**SAS ≅ SAS**)

The triangles in Figure 6-34 are congruent because two sides and the angle between them in one triangle are congruent to two sides and the angle between them in the other triangle (**SAS ≅ SAS**). *In this case, the angle must be between the two sides.*

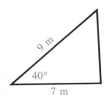

Figure 6-34

EXAMPLE 1

Prove that the triangles shown in Figure 6-35 are congruent.

Solution

Because m($\overline{BC}$) = 10 cm and m($\overline{DC}$) = 10 cm,

$\overline{BC} \cong \overline{DC}$

Because m($\overline{AC}$) = 5 cm and m($\overline{EC}$) = 5 cm,

$\overline{AC} \cong \overline{EC}$

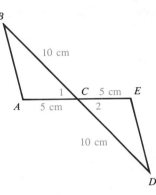

Figure 6-35

Because ∠1 and ∠2 are vertical angles, their measures are equal, and

∠1 ≅ ∠2

Because two sides and the angle between them of △ABC are congruent to two sides and the angle between them of △EDC, the triangles are congruent **(SAS ≅ SAS)**:

△ABC ≅ △EDC

EXAMPLE 2

Prove that the triangles shown in Figure 6-36 are congruent.

Solution

Because m(∠ABD) = 40° and m(∠CDB) = 40°, we have

∠ABD ≅ ∠CDB

Because m(∠ADB) = 50° and m(∠CBD) = 50°, we have

∠ADB ≅ ∠CBD

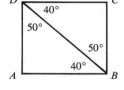

Figure 6-36

Because △ABD and △CDB share side BD and m($\overline{BD}$) = m($\overline{DB}$),

$\overline{BD} \cong \overline{DB}$

Because two angles and a side of △ABD are congruent to two angles and a side of △CDB, the triangles are congruent **(ASA ≅ ASA)**. Work Progress Check 10.

We often use congruent triangles to prove parts of two triangles congruent. For example, in Figure 6-37, we can show that ∠1 ≅ ∠2 as follows. We see from the figure that m(∠A) = m(∠B) = 60° and that m(∠ADC) = m(∠BDC) = 90°. Thus,

∠A ≅ ∠B   and   ∠ADC ≅ ∠BDC

Because $\overline{CD}$ is identical to $\overline{CD}$, we know that m($\overline{CD}$) = m($\overline{CD}$) and that

$\overline{CD} \cong \overline{CD}$

We have shown that two angles and a side of △ADC are congruent to two angles and a side of △BDC **(AAS ≅ AAS)**. Thus, the triangles are congruent. Because the triangles are congruent, their corresponding parts are congruent. Thus,

∠1 ≅ ∠2

In the isosceles triangle shown in Figure 6-38, the sides of equal length are shown with slash marks: m($\overline{AC}$) = m($\overline{BC}$). The two angles, ∠A and ∠B, that are opposite the sides of equal length are called **base angles**. Angle ACB, which is opposite the base, is called the **vertex** angle.

By using congruent triangles, we can prove that the base angles of this isosceles triangle are congruent. To do so, we refer to the figure, in which we are given that m($\overline{AC}$) = m($\overline{BC}$). Thus,

$\overline{AC} \cong \overline{BC}$

Progress Check 10

Tell why each pair of triangles are congruent.

10. SAS ≅ SAS, ASA ≅ ASA, SSS ≅ SSS

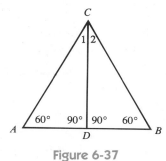

Figure 6-37

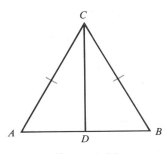

Figure 6-38

We now locate the midpoint of $\overline{AB}$, which we will call point D. Because a midpoint divides a segment into two segments of equal length, we have

$$\overline{AD} \cong \overline{BD}$$

Because two points determine a line segment, we can draw $\overline{CD}$, and because m($\overline{CD}$) = m($\overline{CD}$), we have

$$\overline{CD} \cong \overline{CD}$$

We have now shown that three sides of $\triangle ADC$ are congruent to three sides of $\triangle BDC$ (**SSS $\cong$ SSS**). Thus, the triangles are congruent. Because they are congruent, their corresponding parts are congruent, and

$$\angle A \cong \angle B$$

This proves the following theorem.

Theorem: The base angles of an isosceles triangle are congruent.

EXAMPLE 3

The measure of one of the base angles of an isosceles triangle is 70°. How large is the vertex angle?

Solution

We refer to Figure 6-39, in which $\triangle ABC$ is an isosceles triangle with base angles $\angle A$ and $\angle B$. Since one of the base angles, say $\angle A$, measures 70°, so does the other. Thus, m($\angle B$) = 70°. Since the sum of the measures of the three angles of a triangle must be 180°, we have

$$m(\angle A) + m(\angle B) + m(\angle C) = 180°$$
$$70° + 70° + m(\angle C) = 180°$$
$$140° + m(\angle C) = 180°$$
$$m(\angle C) = 40°$$

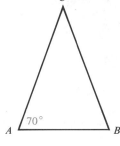

Figure 6-39

The measure of the vertex angle is 40°. Work Progress Check 11.

Progress Check 11

The measure of one of the base angles of an isosceles triangle is 65°. How large is the vertex angle?

11. 50°

Progress Check Answers

If two angles in a triangle are congruent, it can be shown that the sides opposite those angles are also congruent. Thus, if a triangle has two angles with equal measure, the triangle is isosceles.

Theorem: If two angles in a triangle are congruent, the sides opposite those angles are congruent.

EXAMPLE 4

Use the information in Figure 6-40 to show that $\triangle ABC$ is an isosceles triangle.

Solution

Because the sum of the measures of the angles in $\triangle DCB$ is equal to $180°$, we have

$$20° + 50° + m(\angle DCB) = 180°$$
$$m(\angle DCB) = 110°$$

Because $\angle DCB$ and $\angle BCA$ form a straight angle,

$$m(\angle DCB) + m(\angle BCA) = 180°$$
$$110° + m(\angle BCA) = 180° \quad \text{Substitute } 110° \text{ for } m(\angle DCB).$$
$$m(\angle BCA) = 70°$$

Because $m(\angle BCA) = 70°$ and $m(\angle A) = 70°$, two angles in $\triangle ABC$ are congruent. By the previous theorem, the sides opposite those angles are congruent, and

$$m(\overline{AB}) = m(\overline{CB})$$

Since $\triangle ABC$ has two sides of equal measure, it is isosceles. Work Progress Check 12. ■

Figure 6-40

Progress Check 12

Is $\triangle ABD$ isosceles?

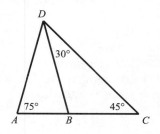

12. yes

Progress Check Answers

6.5 Exercises

In Exercises 1–6, refer to Illustration 1. Pairs of congruent sides of congruent triangles are shown with slash marks. Complete each statement involving a correspondence between parts of the two triangles.

1. Side AC corresponds to side _____.
2. Side DE corresponds to side _____.
3. Side BC corresponds to side _____.
4. $\angle A$ corresponds to _____.
5. $\angle E$ corresponds to _____.
6. $\angle F$ corresponds to _____.

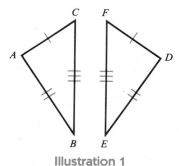

Illustration 1

In Exercises 7–12, the given information refers to Illustration 2. In each exercise, tell whether the triangles are congruent. If the triangles are congruent, explain why.

7. $\angle A \cong \angle D$
   $\angle B \cong \angle E$
   $\overline{AB} \cong \overline{DE}$

8. $\angle A \cong \angle D$
   $\angle B \cong \angle E$
   $\angle C \cong \angle F$

9. $\angle A \cong \angle D$
   $\overline{BC} \cong \overline{EF}$
   $\overline{AC} \cong \overline{DF}$

10. $\overline{AC} \cong \overline{DF}$
    $\overline{AB} \cong \overline{DE}$
    $\overline{BC} \cong \overline{EF}$

11. $\overline{AB} \cong \overline{DE}$
    $\angle B \cong \angle E$
    $\overline{BC} \cong \overline{EF}$

12. $\overline{BC} \cong \overline{EF}$
    $\angle C \cong \angle F$
    $\angle A \cong \angle D$

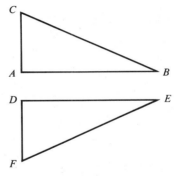

Illustration 2

In Exercises 13–16, use the information shown in Illustration 3, in which $\angle A \cong \angle D$.

13. Explain why $\triangle ABC \cong \triangle DEF$.

14. m($\overline{EF}$) = _____

15. m($\angle C$) = _____

16. m($\angle B$) = _____

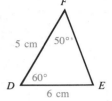

Illustration 3

In Exercises 17–20, use the information shown in Illustration 4.

17. Explain why $\triangle ABC \cong \triangle DEC$.

18. m($\overline{DE}$) = _____

19. m($\angle CDE$) = _____

20. m($\angle B$) = _____

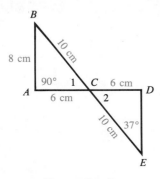

Illustration 4

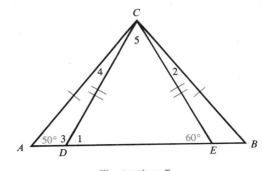

Illustration 5

In Exercises 21–26, use the information in Illustration 5, in which m($\overline{AC}$) = m($\overline{BC}$) and m($\overline{DC}$) = m($\overline{EC}$).

21. m($\angle 1$) = _____

22. m($\angle B$) = _____

23. m($\angle 3$) = _____

24. m($\angle 4$) = _____

25. m($\angle 5$) = _____

26. m($\angle ACB$) = _____

In Exercises 27–32, use the information in Illustration 6, in which m($\overline{AC}$) = m($\overline{BC}$) and m($\overline{AD}$) = m($\overline{BD}$).

27. m(∠1) = _____

28. m(∠CAB) = _____

29. m(∠CBA) = _____

30. m(∠2) = _____

31. m(∠D) = _____

32. m(∠C) = _____

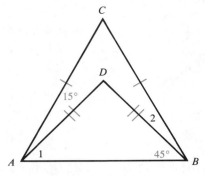

Illustration 6

In Exercises 33–38, use the information shown in Illustration 7, in which m($\overline{DC}$) = m($\overline{EC}$).

33. m(∠1) = _____

34. m(∠2) = _____

35. m(∠A) = _____

36. m(∠3) = _____

37. m(∠B) = _____

38. m($\overline{BC}$) = _____

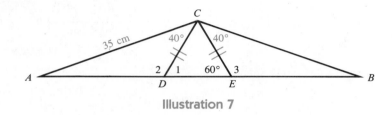

Illustration 7

Review Exercises

In Review Exercises 1–10, perform the calculations.

1. $2 + \frac{3}{5} - \frac{2}{3}$

2. $3 - \frac{7}{4} + \frac{2}{3}$

3. $0.45 + \frac{5}{6} - \frac{2}{5}$

4. $0.65 - 0.07 - 0.12$

5. $0.56(2.34 - 0.43)$

6. $2.5(0.23)^2$

7. $3\frac{3}{5} + 2\frac{3}{4} - 1\frac{1}{2}$

8. $3\frac{1}{2} - 1\frac{2}{3} + 4$

9. $7\frac{3}{2} \div \frac{3}{4}$

10. $6\frac{1}{2} \div 5\frac{2}{3}$

6.6 Parallelograms

- Rectangles
- Rhombuses
- Squares

Until now we have dealt principally with triangles and their properties. We now broaden the discussion by considering other geometric figures.

> **Definition:** A **parallelogram** is a quadrilateral whose opposite sides are parallel.

If $\overline{AB} \parallel \overline{CD}$ and $\overline{AD} \parallel \overline{CB}$, the quadrilateral in Figure 6-41 is a parallelogram.

Figure 6-41

The following theorem lists several more properties of parallelograms.

> **Theorem:**
> 1. A **diagonal** of a parallelogram divides the parallelogram into two congruent triangles.
> 2. Opposite sides of a parallelogram have the same measure.
> 3. Opposite angles in a parallelogram have the same measure.

To prove the previous theorem, we refer to Figure 6-42, in which $\overline{AC}$ is a diagonal of $\square ABCD$ (read as "parallelogram $ABCD$").

Because the opposite sides of a parallelogram are parallel,

$$\overline{AB} \parallel \overline{CD} \quad \text{and} \quad \overline{AD} \parallel \overline{CB}$$

Because $\angle 1$ and $\angle 2$ are alternate interior angles formed by a transversal intersecting parallel lines $\overleftrightarrow{AB}$ and $\overleftrightarrow{CD}$, their measures are equal, and

$$\angle 1 \cong \angle 2$$

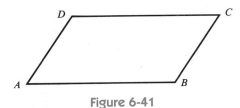

Figure 6-42

Because $\angle 4$ and $\angle 3$ are alternate interior angles formed by a transversal intersecting parallel lines $\overleftrightarrow{AD}$ and $\overleftrightarrow{CB}$, their measures are equal, and

$$\angle 4 \cong \angle 3$$

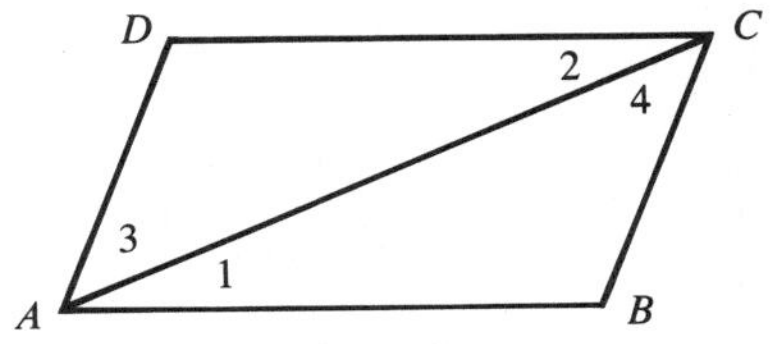

Because $m(\overline{AC}) = m(\overline{CA})$,

$$\overline{AC} \cong \overline{CA}$$

Because two angles and a side of $\triangle ABC$ are congruent to two angles and a side of $\triangle CDA$, the triangles are congruent:

$$\triangle ABC \cong \triangle CDA$$

Thus, the diagonal divides the parallelogram into two congruent triangles, and the first part of the theorem is proved.

To prove Parts 2 and 3 of the theorem, we simply make use of the fact that corresponding parts of congruent triangles are congruent. □

EXAMPLE 1

Refer to ▱ABCD in Figure 6-43 and use the given information to find **a.** m(∠1), **b.** m(∠2), and **c.** m(∠C).

Solution

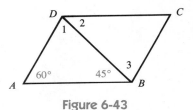

Figure 6-43

a. Because the sum of the measures of the angles in any triangle is 180°,

$$60° + 45° + m(\angle 1) = 180°$$
$$105° + m(\angle 1) = 180°$$
$$m(\angle 1) = 75°$$

b. Quadrilateral ABCD is a parallelogram with $\overline{AB} \parallel \overline{CD}$. Because ∠ABD and ∠2 are alternate interior angles formed by a transversal intersecting two parallel lines, their measures are equal. Thus,

$$m(\angle 2) = 45°$$

c. Quadrilateral ABCD is a parallelogram with $\overline{AD} \parallel \overline{CB}$. Because ∠1 and ∠3 are alternate interior angles formed by a transversal intersecting two parallel lines, their measures are equal. From Part a, we know that m(∠1) = 75°. Thus,

$$m(\angle 3) = 75°$$

From Part b, we know that m(∠2) = 45°. Because the sum of the angles of a triangle must equal 180°, we have

$$m(\angle 2) + m(\angle 3) + m(\angle C) = 180°$$
$$45° + 75° + m(\angle C) = 180°$$ Substitute 45° for m(∠2) and 75° for m(∠3).
$$120° + m(\angle C) = 180°$$
$$m(\angle C) = 60°$$

The fact that m(∠C) = m(∠A) = 60° illustrates that opposite angles in a parallelogram have the same measure. Work Progress Check 13. ■

Progress Check 13

In ▱ABCD, find

a. m($\overline{AD}$)

b. m(∠1)

c. m(∠A)

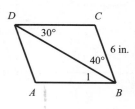

Rectangles

There are three special types of parallelograms. One of them is the **rectangle**.

Definition: A **rectangle** is a parallelogram with a right angle.

13. **a.** 6 in. **b.** 30° **c.** 110°

Progress Check Answers

It can be shown that every rectangle has four right angles. A rectangle is shown in Figure 6-44.

Because a rectangle is a parallelogram, it has all of the properties of a parallelogram, plus some additional ones.

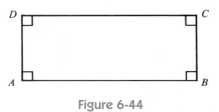

Figure 6-44

Properties of a Rectangle:

- All angles in a rectangle are right angles.
- Opposite sides of a rectangle are parallel.
- Opposite sides of a rectangle have equal measure.
- The diagonals of a rectangle are of equal length.
- If the diagonals of a parallelogram are of equal length, the parallelogram is a rectangle.

EXAMPLE 2

A carpenter intends to build a small rectangular shed with an 8-by-12-foot base. How can she make sure the rectangular foundation is square?

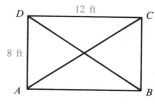

Figure 6-45

Solution

Refer to Figure 6-45. The carpenter should use a tape measure to find m($\overline{AC}$) and m($\overline{BD}$). If these diagonals are of equal length, the figure will be a rectangle and have four right angles. Work Progress Check 14. ∎

Rhombuses

A second special parallelogram is the **rhombus**.

Definition: A **rhombus** is a parallelogram with all sides of equal length.

A rhombus is shown in Figure 6-46. Because a rhombus is a parallelogram, it has all of the properties of a parallelogram plus some additional ones.

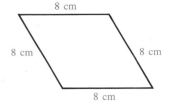

Figure 6-46

Progress Check 14

In rectangle $ABCD$,

a. If m($\overline{AC}$) = 20 centimeters, find m($\overline{BD}$).

b. Find m($\angle 1$).

c. Find m($\angle 2$).

d. Find m($\angle 3$).

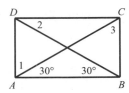

14. **a.** 20 cm **b.** 60° **c.** 30°
    d. 60°

Progress Check Answers

Section 6.6 Parallelograms 217

> **Properties of a Rhombus:**
> - All sides of a rhombus are of equal length.
> - Opposite angles of a rhombus are of equal measure.
> - Opposite sides of a rhombus are parallel.
> - The diagonals of a rhombus are perpendicular.
> - If the diagonals of a parallelogram are perpendicular, the parallelogram is a rhombus.
> - The diagonals of a rhombus bisect its angles.

EXAMPLE 3

In Figure 6-47, □ABCD is a rhombus. Find
a. m(∠1), **b.** m(∠2), **c.** m($\overline{BC}$), and **d.** m(∠3).

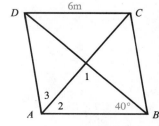

Figure 6-47

Solution

a. The diagonals of a rhombus are perpendicular. Thus,

m(∠1) = 90°

b. Because the sum of the measures of the angles of a triangle must equal 180° and m(∠1) = 90°,

m(∠2) + 40° + 90° = 180°
m(∠2) + 130° = 180°
m(∠2) = 50°

c. Because all sides of a rhombus are of equal length,

m($\overline{BC}$) = 6 meters

d. Because the diagonals of a rhombus bisect its angles, $\overline{AC}$ divides ∠BAD into two angles of equal measure. Thus, m(∠2) = m(∠3). Because m(∠2) = 50° (from Part b), it follows that

m(∠3) = 50°

Work Progress Check 15.

Progress Check 15

In rhombus ABCD, find

a. m($\overline{DC}$)

b. m(∠1)

c. m(∠2)

d. m(∠3)

e. m(∠4)

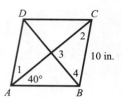

Squares

The third special parallelogram is the **square**, which can be defined in two ways:

> **Definition:**
> A **square** is a rectangle with adjacent sides of equal length.
> A **square** is a rhombus with a right angle.

Because a square is both a rectangle and a rhombus, it has the properties of both the rectangle and the rhombus.

15. **a.** 10 in. **b.** 40° **c.** 40°
    d. 90° **e.** 50°

Properties of a Square:

- Opposite sides of a square are parallel.
- All sides of a square are of equal length.
- All angles in a square are right angles.
- The diagonals of a square are of equal length.
- The diagonals of a square are perpendicular.
- If the diagonals of a rhombus are of equal length, the rhombus is a square.
- If the diagonals of a rectangle are perpendicular, the rectangle is a square.
- The diagonals of a square bisect its angles.

6.6 Exercises

In Exercises 1–8, identify each parallelogram as a rectangle, a rhombus, or a square. Some parallelograms will fit into more than one category.

1.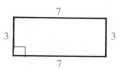
2.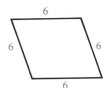
3.
4.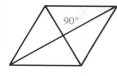
5.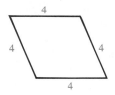
6.
7.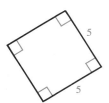
8.

In Exercises 9–12, refer to Illustration 1, in which ▱ABCD is a rectangle.

9. m(∠1) = _____
10. m(∠3) = _____
11. m(∠2) = _____
12. If m($\overline{AC}$) = 8 centimeters, m($\overline{DB}$) = _____

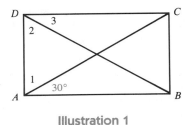

Illustration 1

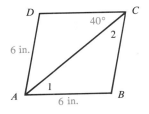

Illustration 2

In Exercises 13–16, refer to Illustration 2, in which ▱ABCD is a rhombus.

13. m(∠1) = _____
14. m(∠2) = _____
15. m(∠B) = _____
16. m($\overline{DC}$) = _____

In Exercises 17–22, refer to Illustration 3, in which ⏥ABCD is a rhombus.

17. m(∠1) = _____ 18. m(∠2) = _____

19. m(∠3) = _____ 20. m(∠4) = _____

21. m(∠5) = _____ 22. m(∠DCB) = _____

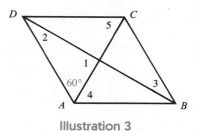

Illustration 3

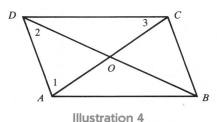

Illustration 4

In Exercises 23–34, refer to Illustration 4, in which quadrilateral ABCD is a parallelogram.

23. If m($\overline{AB}$) = 10 centimeters, m($\overline{CD}$) = _____ 24. If m($\overline{CB}$) = 6 inches, m($\overline{AD}$) = _____

25. If m(∠ABC) = 70°, m(∠CDA) = _____ 26. If m(∠DAB) = 110°, m(∠BCD) = _____

27. If m(∠BAD) = 90° and m($\overline{DB}$) = 10 inches, m($\overline{AC}$) = _____ 28. If m($\overline{AD}$) = m($\overline{DC}$) and m(∠1) = 50°, m(∠2) = _____

29. If m($\overline{AC}$) = m($\overline{BD}$) and m(∠1) = 40°, m(∠3) = _____ 30. If m(∠DAB) = 90° and m($\overline{DB}$) = 7 centimeters, m($\overline{AC}$) = _____

31. If $\overline{DB} \perp \overline{AC}$ and m($\overline{CB}$) = 5 centimeters, m($\overline{DC}$) = _____ 32. If $\overline{DB} \perp \overline{AC}$ and m(∠1) = 30°, m(∠ADC) = _____

33. If $\overline{DB} \perp \overline{AC}$ and m(∠1) = 45°, is ⏥ABCD a rectangle? Explain. 34. If m($\overline{AC}$) = m($\overline{BD}$) and $\overline{AC} \perp \overline{BD}$, is ⏥ABCD a square? Explain.

Review Exercises

In Review Exercises 1–4, tell which number is the greater.

1. $\frac{11}{13}, \frac{13}{15}$ 2. $\frac{21}{23}, \frac{25}{28}$ 3. 0.4356, 0.4346   4. $\left(\frac{3}{7}\right)^2, \left(\frac{2}{5}\right)^2$

In Review Exercises 5–8, estimate the value of each quantity.

5. $\frac{2(0.994)(1.978)}{3.9978}$ 6. $\frac{(3.9978)^2 + 1.9978}{2.9992}$ 7. $\frac{1.968 + 2(5.0012)}{6.001}$ 8. $\frac{3(4.014) - 2(1.13)}{4.982}$

In Review Exercises 9–10, round each number to the nearest hundredth.

9. 23.3456   10. 465.649223

6.7 Circles

- Central Angles
- Inscribed Angles
- Tangent Lines

We can think of a circle as a tire or a ring. More formally, we make this definition:

> **Definition:** A **circle** is the set of all points in a plane that lie a fixed distance from a point called its **center**.

The line segment drawn from the center of a circle to a point on the circle is called a **radius**. The plural of *radius* is *radii*. From the definition of a circle, it follows that all radii of the same circle are congruent.

> **Definition:** A **chord** of a circle is a line segment connecting two points on the circle.

> **Definition:** A **diameter** of a circle is a chord that passes through the center of the circle.

It can be shown that a diameter of a circle is twice as long as a radius. Thus, if d is the length of a diameter of a circle and r is the length of a radius,

$$d = 2r$$

Each of the previous definitions is illustrated in Figure 6-48, in which O is the center of the circle.

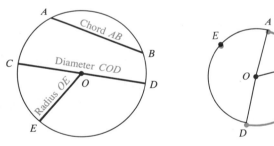

Figure 6-48   Figure 6-49

A continuous portion of a circle is called an **arc**. For example, in Figure 6-49, the portion of the circle from point A to point B is $\widehat{AB}$, read as "arc AB." $\widehat{CD}$ is the portion of the circle from point C to point D.

An arc that is exactly half of a circle is called a **semicircle**.

> **Definition:** A **semicircle** is an arc of a circle whose endpoints lie at the extremities of a diameter.

If point O is the center of the circle in Figure 6-49, then $\overline{AD}$ is a diameter and $\overparen{AED}$ is a semicircle. The middle letter E is used to distinguish semicircle $\overparen{AED}$ from semicircle $\overparen{ABCD}$.

An arc that is shorter than a semicircle is called a **minor arc**. An arc that is longer than a semicircle is called a **major arc**. In Figure 6-49,

$\overparen{AE}$ is a minor arc   and $\overparen{ABCDE}$ is a major arc.

Central Angles

Definition: A **central angle of a circle** is an angle whose vertex is the center of the circle and whose sides are radii of the circle.

In Figure 6-50, $\angle 1$, $\angle 2$, $\angle 3$, and $\angle 4$ are central angles. In the figure, $\angle 1$ is said to *intercept* $\overparen{BC}$, and $\overparen{BC}$ is said to be *intercepted* by $\angle 1$. A similar statement is true for the other angles and their intercepted arcs.

Arcs are often measured in degrees.

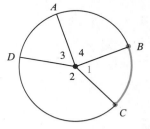

Figure 6-50

Definition: The **measure of an arc** is the number of degrees in the central angle that intercepts the arc.

In Figure 6-51, $\angle AOB$, $\angle BOC$, $\angle COD$, and $\angle DOA$ are central angles with measures of 70°, 110°, 70°, and 110°, respectively. Thus,

$m(\overparen{AB}) = 70°$, $m(\overparen{BC}) = 110°$,
$m(\overparen{CD}) = 70°$, $m(\overparen{DA}) = 110°$

From the definition of the measure of an arc, it follows that central angles with equal measures have arcs with equal measures.

Because the sum of the measures of the four central angles shown in Figure 6-51 is 360°, it follows that the measure of the arc of the entire circle is also 360°.

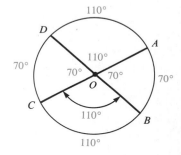

Figure 6-51

Inscribed Angles

Other important angles associated with circles are **inscribed angles**.

Definition: An **inscribed angle** is an angle whose vertex is on a circle and whose sides are chords of the circle.

In Figure 6-52, ∠1 and ∠2 are inscribed angles.

The measure of an inscribed angle is also related to the measure of the arc it intercepts.

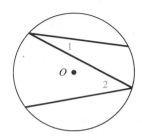

Figure 6-52

> **Theorem:** The measure of an inscribed angle in a circle is half the number of degrees of its intercepted arc.

EXAMPLE 1

In Figure 6-53, ∠AOB is a central angle with a measure of 100°. Find the measure of ∠C.

Solution

We are given that ∠AOB is a central angle with a measure of 100°. Because an arc has the same measure as its central angle, m($\widehat{AB}$) = 100°. Because ∠C is an inscribed angle that intercepts $\widehat{AB}$, its measure must be one-half of 100°. Thus,

$$m(\angle C) = 50°$$

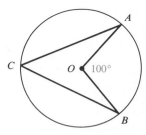

Figure 6-53

Work Progress Check 16.

Progress Check 16

In circle O, find

a. m($\widehat{AB}$)

b. m(∠1)

Tangent Lines

Lines that are often associated with circles are **tangent lines**.

> **Definition:** A **tangent** is a line that intersects a circle in exactly one point.

In Figure 6-54, $\overleftrightarrow{AC}$ is a tangent to circle O. The point B where the tangent intersects the circle is called a **point of tangency**.

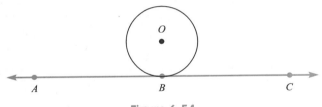

Figure 6-54

16. a. 80° **b.** 80°

Progress Check Answers

There are several theorems that involve tangent lines.

> **Theorems:**
> - A radius drawn to a point of tangency is perpendicular to the tangent.
> - Two tangents drawn to a circle from a point outside the circle are of equal length.
> - The measure of an angle formed by a tangent and a chord is half the measure of its intercepted arc.

Each of the previous theorems is illustrated in Figure 6-55.

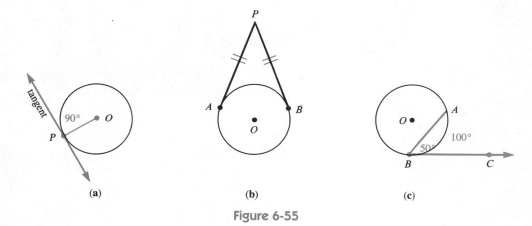

Figure 6-55

EXAMPLE 2

Refer to Figure 6-56, in which O is the center of the circle and $\overline{AB}$ is tangent to circle O at point B. Find **a.** m($\angle 1$), **b.** m($\angle 2$), and **c.** m($\angle A$).

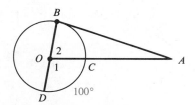

Figure 6-56

Solution

a. Because $\angle 1$ is a central angle that intercepts an arc measuring $100°$, and a central angle has the same number of degrees as its intercepted arc,

m($\angle 1$) = 100°

Progress Check 17

$\overline{AB}$ and $\overline{AD}$ are tangent to circle O. Find

a. $m(\overline{AD})$

b. $m(\angle ABC)$

c. $m(\angle 1)$

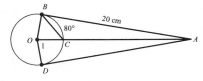

17. a. 20 cm b. 40° c. 80°
Progress Check Answers

b. Because the sum of ∠1 and ∠2 is a straight angle, ∠1 and ∠2 are supplementary:

1. $m(\angle 1) + m(\angle 2) = 180°$

From Part **a** we know that $m(\angle 1) = 100°$. Substituting 100° for $m(\angle 1)$ in Equation 1 and solving for $m(\angle 2)$ gives

$$100° + m(\angle 2) = 180°$$
$$m(\angle 2) = 80°$$

c. Because the radius $\overline{OB}$ is drawn to the tangent $\overline{AB}$ at the point of tangency (point B), radius $\overline{OB}$ is perpendicular to the tangent, and

$$m(\angle OBA) = 90°$$

Because the sum of the angles of a triangle must equal 180°,

$$m(\angle 2) + m(\angle OBA) + m(\angle A) = 180°$$

Substituting 80° for $m(\angle 2)$ and 90° for $m(\angle OBA)$ and solving for $m(\angle A)$ gives

$$80° + 90° + m(\angle A) = 180°$$
$$170° + m(\angle A) = 180°$$
$$m(\angle A) = 10°$$

Work Progress Check 17.

6.7 Exercises

In Exercises 1–10, refer to Illustration 1, in which $\overline{AE}$ and $\overline{BE}$ are tangents to circle O.

1. Name each radius.

2. Name each diameter.

3. Name each chord.

4. Name each central angle.

5. Name each inscribed angle.

6. Name each tangent line.

7. Name each minor arc.

8. Name each semicircle.

9. Name each major arc.

10. Name all angles formed by tangent AE and chord AD.

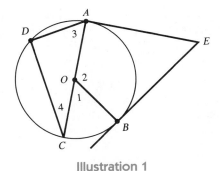

Illustration 1

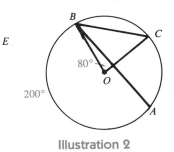

Illustration 2

In Exercises 11–14, refer to Illustration 2, in which O is the center of the circle.

11. $m(\widehat{BC}) = $ _____

12. $m(\widehat{AC}) = $ _____

13. $m(\angle C) = $ _____

14. $m(\angle ABC) = $ _____

Section 6.7 Exercises 225

In Exercises 15–20, refer to Illustration 3, in which O is the center of the circle.

15. m(∠BOC) = _____ **16.** m(∠BAC) = _____

17. If m(∠DAC) = 40°, m($\widehat{DC}$) = _____ **18.** If m(∠DAB) = 65°, m($\widehat{DC}$) = _____

19. If m(∠DAB) = 65°, m($\widehat{AB}$) = _____ **20.** If m(∠DAB) = 65°, m(∠D) = _____

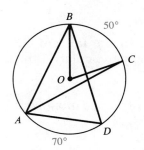

Illustration 3

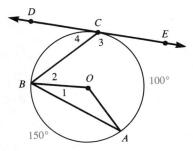

Illustration 4

In Exercises 21–26, refer to Illustration 4, in which O is the center of the circle and $\overline{DCE}$ is a tangent.

21. m(∠O) = _____ **22.** m(∠CBA) = _____

23. m(∠1) = _____ **24.** m(∠2) = _____

25. m(∠3) = _____ **26.** m(∠4) = _____

In Exercises 27–32, refer to Illustration 5, in which $\overline{BD}$ and $\overline{CD}$ are tangent to circle O.

27. m(∠1) = _____ **28.** m($\overline{CD}$) = _____

29. m(∠2) = _____ **30.** m(∠D) = _____

31. m(∠3) = _____ **32.** m(∠EBD) = _____

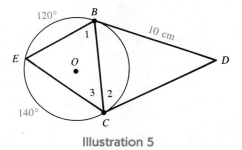

Illustration 5

Illustration 6

In Exercises 33–36, refer to Illustration 6.

33. m(∠1) = _____ **34.** m(∠2) = _____

35. m(∠3) = _____ **36.** m($\widehat{AD}$) = _____

226   Chapter 6   Basic Concepts of Geometry

In Exercises 37–40, refer to Illustration 7.

37. If m($\widehat{BD}$) = 60° and m($\widehat{AE}$) = 100°, m($\angle C$) = _____

38. If m($\angle C$) = 30° and m($\widehat{BD}$) = 80°, m($\widehat{AE}$) = _____

39. If m($\angle C$) = 42° and m($\widehat{AE}$) = 120°, m($\widehat{BD}$) = _____

40. If m($\widehat{BD}$) = 15° and m($\widehat{AE}$) = 3[m($\widehat{BD}$)], m($\angle C$) = _____

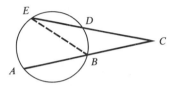

Illustration 7

Review Exercises

1. Carlos bought six pencils at $0.60 each and a notebook for $1.25. He gave the clerk a $5 bill. How much change did he receive?

2. Mary bought three pairs of socks at $3.29 each and a pair of shoes for $39.95. Can she buy these clothes with three $20 bills?

3. George bought three packages of golf balls for $1.99 each, a package of tees for $0.49, and a golf glove for $6.95. How much did he spend?

4. Lisa bought four compact disks at $9.99 each, three tapes for $6.95 each, and a carrying case for compact disks for $10.25. How much did she spend?

5. Juan has $100. Can he afford to buy a jacket costing $39.95 and two pairs of jeans costing $29.95 each?

6. Lisa had $496 in her checking account before she wrote checks for $127, $312, and $50. Thinking she might have overdrawn her account, she rushed to the bank with a deposit of $100. Was the trip necessary?

Mathematics for Fun

The Möbius strip has other curious properties. Try to cut a Möbius strip in half, lengthwise. When cut in half, most objects produce two pieces. What happens to the Möbius strip? What happens if you cut it in half again? What happens to a Möbius strip if it is cut lengthwise along a line that is one-third of the way from its edge? What happens if you give the original strip of paper a full twist before joining the ends? How about $1\frac{1}{2}$ twists? Experiment. That is one way that new mathematics is discovered.

Chapter Summary

Key Words

acute angle (6.1)
acute triangle (6.4)
adjacent angles (6.2)
alternate exterior angles (6.3)
alternate interior angles (6.3)
angle (6.1)
arc (6.7)
base angles of an isosceles triangle (6.5)
central angle (6.7)
chord (6.7)
circle (6.7)
closed figure (6.4)

collinear points (6.1)
complementary angles (6.2)
conclusion (6.2)
conditional statements (6.2)
congruent angles (6.1)
congruent triangles (6.5)
converse (6.2)
coplanar points (6.1)
corresponding angles (6.3)
degrees (6.1)
diagonal (6.6)
diameter (6.6)

endpoint (6.1)
equiangular triangle (6.4)
equilateral triangle (6.4)
exterior (6.4)
hexagon (6.3)
hypotenuse (6.4)
hypothesis (6.2)
inscribed angle (6.7)
interior (6.4)
isosceles triangle (6.4)
line (6.1)
line segment (6.1)
major arc (6.7)
measure of an arc (6.7)

midpoint (6.1)
minor arc (6.7)
obtuse angle (6.1)
obtuse triangle (6.4)
parallel lines (6.1)
parallel planes (6.1)
parallelogram (6.6)
pentagon (6.4)
perpendicular lines (6.1)
plane (6.1)
point (6.1)
point of tangency (6.7)
polygons (6.4)
postulate (6.3)
quadrilateral (6.4)

radius (6.7)
ray (6.1)
rectangle (6.6)
rhombus (6.6)
right angle (6.1)
right triangle (6.4)

scalene triangle (6.4)
semicircle (6.7)
skew lines (6.3)
square (6.6)
straight angle (6.1)

supplementary
  angles (6.2)
tangent to a circle (6.7)
theorem (6.2)
transversal (6.3)

triangle (6.4)
vertex angle of an
  isosceles triangle (6.5)
vertex of an angle (6.1)
vertical angles (6.2)

Key Ideas

(6.1) Two lines intersect in a point.

Two planes intersect in a line.

Perpendicular lines intersect and form four right angles.

(6.2) Two angles are complementary if the sum of their measures is 90°.

Two angles are supplementary if the sum of their measures is 180°.

Vertical angles are congruent.

(6.3) If two parallel lines are cut by a transversal, all

- alternate interior angles are congruent.
- alternate exterior angles are congruent.
- corresponding angles are congruent.
- interior angles on the same side of the transversal are supplementary.

Two lines are parallel if they are cut by a transversal and the

- alternate interior angles are congruent.
- alternate exterior angles are congruent.
- corresponding angles are congruent.
- interior angles on the same side of the transversal are supplementary.

(6.4) The sum of the measures of the angles in any triangle is 180°.

The sum of the measures of the angles in any polygon is

$$S = (n - 2)180°$$

(6.5) Two triangles are congruent if, and only if,

- three sides of one triangle are congruent, respectively, to three sides of the other triangle.
- two angles and a side of one triangle are congruent, respectively, to two angles and a side of the other triangle.
- two sides and the angle between them are congruent, respectively, to two sides and the angle between them in the other triangle.

The base angles of an isosceles triangle are congruent.

If two angles of a triangle are congruent, the triangle is isosceles.

(6.6) In a parallelogram,

- opposite sides are parallel.
- the diagonal divides the parallelogram into two congruent triangles.
- opposite sides are of equal length.
- opposite angles have the same measure.

In a rectangle,

- opposite sides are parallel.
- all angles are right angles.
- opposite sides are of equal length.
- the diagonals are of equal length.

In a rhombus,

- opposite sides are parallel.
- all sides are of equal length.
- the diagonals are perpendicular.
- the diagonals bisect its angles.

A square has all of the properties of a rectangle and a rhombus.

(6.7) The measure of an arc is the number of degrees in the central angle that intercepts the arc.

The measure of an inscribed angle is half the number of degrees of its intercepted arc.

A radius drawn to a point of tangency is perpendicular to the tangent.

Tangents drawn to a circle from a point outside the circle are of equal length.

The measure of an angle formed by a tangent and a chord is half the measure of its intercepted arc.

Chapter 6 Review Exercises

[6.1–6.2] In Review Exercises 1–4, refer to Illustration 1.

1. m($\overline{AC}$) = _____

2. m($\overline{BC}$) = _____

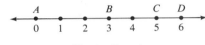

Illustration 1

3. Find the midpoint of $\overline{AD}$.

4. Does point B bisect $\overline{AD}$?

In Review Exercises 5–10, refer to Illustration 2.

5. Name each right angle.

6. Name each acute angle.

7. Name each obtuse angle.

8. Name each straight angle.

9. Name each pair of adjacent angles.

10. Explain why there are no vertical angles in Illustration 2.

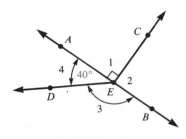

Illustration 2

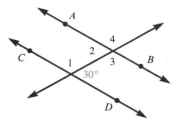

Illustration 3

11. Find the complement of an angle that measures 35°.

12. Find the supplement of an angle that measures 35°.

[6.3] In Review Exercises 13–16, refer to Illustration 3, in which $\overleftrightarrow{AB}$ ∥ $\overleftrightarrow{CD}$.

13. m(∠1) = _____

14. m(∠2) = _____

15. m(∠4) = _____

16. m(∠3) = _____

In Review Exercises 17–20, refer to Illustration 4, in which $l_1 \parallel l_2$.

17. m(∠1) = _____   **18.** m(∠2) = _____

19. m(∠3) = _____   **20.** m(∠4) = _____

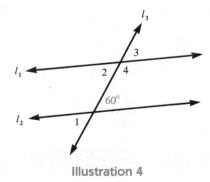

Illustration 4

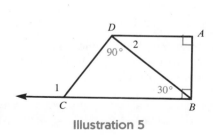

Illustration 5

[6.4] In Review Exercises 21–24, refer to Illustration 5, in which $\overline{AD} \parallel \overline{BC}$.

21. m(∠2) = _____   **22.** m(∠ABD) = _____

23. m(∠DCB) = _____   **24.** m(∠1) = _____

In Review Exercises 25–28, refer to Illustration 6, in which m($\overline{AB}$) = m($\overline{CB}$).

25. m(∠A) = _____   **26.** m(∠1) = _____

27. m(∠2) = _____   **28.** m(∠3) = _____

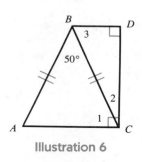

Illustration 6

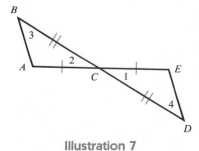

Illustration 7

29. Find the sum of the measures of the angles in a hexagon (a six-sided figure).

30. Find the sum of the measures of the angles in an octagon (an eight-sided figure).

[6.5] In Review Exercises 31–34, refer to Illustration 7, in which m($\overline{BC}$) = m($\overline{DC}$) and m($\overline{AC}$) = m($\overline{EC}$).

31. Explain why $\triangle ABC \cong \triangle EDC$.

32. If m(∠E) = 70°, m(∠A) = _____.

33. If m($\overline{DE}$) = 6 centimeters, m($\overline{AB}$) = _____.

34. If m($\overline{BC}$) = 3 inches, m($\overline{BD}$) = _____.

In Review Exercises 35–38, refer to Illustration 8, in which m($\overline{AC}$) = m($\overline{BC}$) and m($\overline{AD}$) = m($\overline{BD}$).

35. m(∠1) = _____ **36.** m(∠2) = _____

37. m(∠3) = _____ **38.** m(∠4) = _____

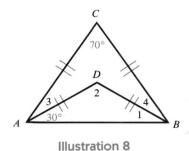

Illustration 8

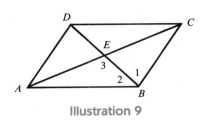

Illustration 9

[6.6] In Review Exercises 39–42, refer to ▱ABCD in Illustration 9.

39. If m(∠DAB) = 75°, m(∠BCD) = _____. **40.** If m(∠DAB) = 75° and m(∠1) = 70°, m(∠2) = _____.

41. If m($\overline{AB}$) = m($\overline{AD}$), m(∠3) = _____. **42.** If m($\overline{AC}$) = m($\overline{BD}$), m(∠ADC) = _____.

[6.7] In Review Exercises 43–52, refer to Illustration 10, in which $\overline{AB}$ ∥ $\overline{CD}$, chords AOC and DOB are diameters, DE is tangent to circle O at point D, and EC is tangent to the circle at point C.

43. m(∠1) = _____ **44.** m(∠A) = _____

45. m($\widehat{BC}$) = _____ **46.** m(∠ACD) = _____

47. m($\widehat{AD}$) = _____

48. m($\widehat{DC}$) = _____

49. m(∠3) = _____

50. m(∠2) = _____

51. If m($\overline{DE}$) = 18 centimeters, m($\overline{EC}$) = _____

52. m(∠4) = _____

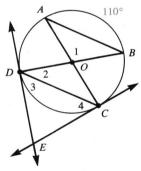

Illustration 10

Name _____ Section _____

Chapter 6 Test

1. In Illustration 1, m($\overline{AB}$) = ?

   Illustration 1

2. Find the complement of an angle measuring 67°.

3. Find the supplement of an angle measuring 117°.

In Problems 4–10, refer to Illustration 2, in which $l_1 \parallel l_2$ and m($\overline{AB}$) = m($\overline{CB}$).

4. m(∠1) = ?

5. m(∠2) = ?

6. m(∠3) = ?

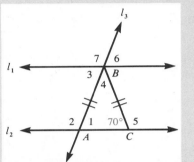

7. m(∠4) = ?

8. m(∠5) = ?

9. m(∠6) = ?

Illustration 2

10. m(∠7) = ?

11. If the measures of two angles in a triangle are 65° and 85°, find the measure of the third angle.

12. If the measures of two angles in a triangle are 40° and 100°, is the triangle an equilateral, isosceles, or scalene triangle?

13. Find the sum of the measures of the angles in a parallelogram.

14. Find the sum of the measures of the angles in a decagon (a ten-sided polygon).

In Problems 15–16, refer to Illustration 3.

15. If m($\overline{AB}$) = 8 centimeters, m($\overline{CB}$) = ?

16. If m($\overline{AC}$) = 30 centimeters, m($\overline{AD}$) = ?

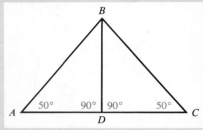

Illustration 3

17. Each acute angle of a parallelogram measures 35°. Find the measure of one obtuse angle.

18. One side of a rhombus measures 17 cm. Find the perimeter of the rhombus.

In Problems 19–22, refer to Illustration 4, in which $\overline{AC}$ and $\overline{BC}$ are tangents to circle O.

19. m($\widehat{AB}$) = ?

20. m($\angle 1$) = ?

21. m($\angle 2$) = ?

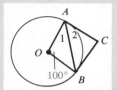

Illustration 4

22. m($\angle C$) = ?

In Problems 23–30, refer to Illustration 5, in which $\overline{CD}$ and $\overline{AD}$ are tangents to circle O.

23. m($\angle BAC$) = ?

24. m($\angle 2$) = ?

25. m($\angle 3$) = ?

26. m($\angle 4$) = ?

27. m($\widehat{AC}$) = ?

28. m($\widehat{AB}$) = ?

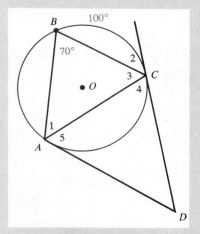

Illustration 5

29. m($\angle 5$) = ?

30. m($\angle D$) = ?

7 Proportion, Similar Triangles, Area, and Volume

Mathematics in Merchandising

Packaging can be deceptive, and what appears to be a good bargain may not be. For example, Pizza Palace offers a choice of two medium pizzas for $12 or one large pizza for $10. Are two 11-inch medium pizzas a better deal than one 14-inch large pizza?

One of the topics of this chapter provides a mathematical method for choosing the best pizza deal.

To prepare fuel for a lawn mower, gasoline must be mixed with oil in the ratio of 50 to 1. The ratio of men to women at a certain college is 3 to 2. In 14-karat jewelry, gold is mixed with other metals in the ratio of 14 to 10. To begin this chapter, we will learn how ratios and a related concept, proportions, can be used to solve a variety of problems.

7.1 Ratio and Proportion

- Ratios
- Proportions
- Solving Proportions

Ratios

Definition: A **ratio** is the comparison of two numbers by their indicated quotient.

A ratio is a fraction. Some examples of ratios are

$$\frac{7}{8}, \quad \frac{21}{24}, \quad \text{and} \quad \frac{117}{223}$$

The fraction $\frac{7}{8}$ can be read as "the ratio of 7 to 8," the fraction $\frac{21}{24}$ can be read as "the ratio of 21 to 24," and the fraction $\frac{117}{223}$ can be read as "the ratio of 117 to 223." Because the fractions $\frac{7}{8}$ and $\frac{21}{24}$ represent equal numbers, they are **equal ratios**.

The material in this chapter is adapted from *Elementary Geometry*, 3rd edition, by R. David Gustafson and Peter D. Frisk. Copyright © 1991 by John Wiley & Sons, Inc. Adapted by permission of John Wiley & Sons, Inc.

EXAMPLE 1

Express each phrase as a ratio in lowest terms:

a. The ratio of 15 to 12
b. The ratio of 3 inches to 7 inches
c. The ratio of 2 feet to 1 yard
d. The ratio of 6 ounces to 1 pound

Solution

a. The ratio of 15 to 12 can be written as the fraction $\frac{15}{12}$. Expressed in lowest terms, it is $\frac{5}{4}$.

b. The ratio of 3 inches to 7 inches can be written as the fraction $\frac{3 \text{ inches}}{7 \text{ inches}}$, or just $\frac{3}{7}$.

c. The ratio of 2 feet to 1 yard should not be written as the fraction $\frac{2}{1}$. To express the ratio as a pure number, we must use the same units. Because there are 3 feet in 1 yard, the proper ratio is $\frac{2 \text{ feet}}{3 \text{ feet}}$, or just $\frac{2}{3}$.

d. To express the ratio as a pure number, we must use the same units. Since there are 16 ounces in 1 pound, the proper ratio is $\frac{6 \text{ ounces}}{16 \text{ ounces}}$, which simplifies to $\frac{3}{8}$.

Work Progress Check 1.

Progress Check 1

Express each phrase as a ratio in lowest terms:

a. The ratio of 8 centimeters to 12 centimeters

b. The ratio of 8 ounces to 2 pounds

Proportions

Definition: A **proportion** is a statement that two ratios are equal.

Some examples of proportions are

$$\frac{1}{2} = \frac{3}{6}, \quad \frac{3}{7} = \frac{9}{21}, \quad \text{and} \quad \frac{8}{1} = \frac{40}{5}$$

The proportion $\frac{1}{2} = \frac{3}{6}$ can be read as "1 is to 2 as 3 is to 6," the proportion $\frac{3}{7} = \frac{9}{21}$ can be read as "3 is to 7 as 9 is to 21," and the proportion $\frac{8}{1} = \frac{40}{5}$ can be read as "8 is to 1 as 40 is to 5."

In the proportion $\frac{1}{2} = \frac{3}{6}$, the numbers 1 and 6 are called the **extremes** of the proportion, and the numbers 2 and 3 are called the **means**. The product of the extremes and the product of the means in this proportion are equal:

$$1 \cdot 6 = 6 \quad \text{and} \quad 3 \cdot 2 = 6$$

This example illustrates a general rule:

Theorem: In any proportion, the product of the extremes is equal to the product of the means.

1. **a.** $\frac{2}{3}$ **b.** $\frac{1}{4}$

Progress Check Answers

EXAMPLE 2

Show that the product of the extremes and the product of the means are equal in the following proportions: **a.** $\frac{3}{7} = \frac{9}{21}$ and **b.** $\frac{8}{1} = \frac{40}{5}$

Solution

a. The product of the extremes is $3 \cdot 21 = 63$. The product of the means is $9 \cdot 7 = 63$. Thus, they are equal.

b. The product of the extremes is $8 \cdot 5 = 40$. The product of the means is $40 \cdot 1 = 40$. Thus, they are equal. ∎

When one pair of numbers (such as 2 and 3) and another pair (such as 8 and 12) form a proportion, we say that they are **proportional**. To show that 2, 3 and 8, 12 are proportional, we determine whether

$$\frac{2}{3} = \frac{8}{12}$$

is a proportion. To do so, we check to see if the product of the extremes is equal to the product of the means:

$$2 \cdot 12 = 24 \quad 8 \cdot 3 = 24$$

Since they are equal, the numbers are proportional.

Solving Proportions

Suppose we know three numbers in the proportion

$$\frac{?}{5} = \frac{24}{20}$$

and we wish to find the unknown number. To do so, we can let the missing number be represented by a variable (say, x), multiply the extremes and multiply the means, set them equal, and find x:

$$\frac{x}{5} = \frac{24}{20}$$
$$20 \cdot x = 24 \cdot 5 \quad \text{The product of the extremes is equal to the product of the means.}$$
$$20 \cdot x = 120$$

Because this multiplication is equivalent to the division $x = \frac{120}{20}$, we have

$$x = 6$$

The unknown number is 6.

EXAMPLE 3

Solve the proportion $\frac{12}{18} = \frac{3}{x}$ for x.

Solution

$$\frac{12}{18} = \frac{3}{x}$$

$12 \cdot x = 3 \cdot 18$ The product of the extremes equals the product of the means.

$12 \cdot x = 54$ Simplify.

$x = \frac{54}{12}$ Replace the multiplication with a division.

$x = \frac{9}{2}$ Simplify.

Thus, x represents the fraction $\frac{9}{2}$. Work Progress Check 2.

Progress Check 2

Solve the proportion $\frac{15}{x} = \frac{20}{32}$.

EXAMPLE 4

If 5 tomatoes cost $1.15, how much will 16 tomatoes cost?

Solution

Let c represent the cost of 16 tomatoes. The ratio of the numbers of tomatoes is the same as the ratio of their costs. We express this relationship as a proportion and solve for c.

$$\frac{5}{16} = \frac{1.15}{c}$$

$5 \cdot c = 1.15(16)$ The product of the extremes is equal to the product of the means.

$5 \cdot c = 18.4$ Do the multiplication.

$c = \frac{18.4}{5}$ Replace the multiplication with a division.

$c = 3.68$ Simplify.

Sixteen tomatoes will cost $3.68.

Progress Check Answers

2. $x = 24$

7.1 Exercises

In Exercises 1–20, express each phrase as a ratio in lowest terms.

1. 5 to 7
2. 3 to 5
3. 17 to 34
4. 19 to 38
5. 22 to 33
6. 14 to 21
7. 4 ounces to 12 ounces
8. 3 inches to 15 inches
9. 12 minutes to 1 hour
10. 8 ounces to 1 pound
11. 3 days to 1 week
12. 2 quarts to 1 gallon
13. 4 inches to 2 yards
14. 1 mile to 5280 feet
15. 3 pints to 2 quarts
16. 4 dimes to 8 pennies
17. 6 nickels to 1 quarter
18. 3 people to 12 people
19. 3 meters to 12 centimeters
20. 3 dollars to 3 quarters

In Exercises 21–28, tell whether each statement is a proportion.

21. $\dfrac{9}{7} = \dfrac{81}{70}$ **22.** $\dfrac{5}{2} = \dfrac{20}{8}$ **23.** $\dfrac{7}{3} = \dfrac{14}{6}$ **24.** $\dfrac{13}{19} = \dfrac{65}{95}$

25. $\dfrac{9}{19} = \dfrac{38}{80}$ **26.** $\dfrac{40}{29} = \dfrac{29}{22}$ **27.** $\dfrac{10.4}{3.6} = \dfrac{41.6}{14.4}$ **28.** $\dfrac{13.23}{3.45} = \dfrac{39.96}{11.35}$

In Exercises 29–36, solve for the variable in each proportion.

29. $\dfrac{2}{3} = \dfrac{x}{6}$ **30.** $\dfrac{3}{6} = \dfrac{x}{8}$ **31.** $\dfrac{5}{10} = \dfrac{3}{c}$ **32.** $\dfrac{7}{14} = \dfrac{2}{x}$

33. $\dfrac{6}{x} = \dfrac{8}{4}$ **34.** $\dfrac{4}{x} = \dfrac{2}{8}$ **35.** $\dfrac{x}{3} = \dfrac{9}{3}$ **36.** $\dfrac{x}{2} = \dfrac{18}{6}$

In Exercises 37–48, set up and solve the required proportion.

37. Three pints of yogurt cost $1. How much will 51 pints cost?

38. Sport shirts are on sale at two for $25. How much will five shirts cost?

39. Garden seeds are on sale at three packets for 50 cents. How much will 39 packets cost?

40. A recipe for spaghetti sauce requires four 16-ounce bottles of ketchup to make two gallons of sauce. How many bottles of ketchup are needed to make 10 gallons of sauce?

41. A car gets 42 miles per gallon of gas. How much gas is needed to drive 315 miles?

42. A truck gets 12 miles per gallon of gas. How far can the truck go on 17 gallons of gas?

43. Bill earns $412 for a 40-hour week. Last week he missed 10 hours of work. How much did he get paid?

44. An HO-scale model railroad engine is 9 inches long. The HO scale is 87 feet to 1 foot. How long is a real engine?

45. An N-scale model railroad caboose is 3.5 inches long. The N scale is 169 feet to 1 foot. How long is a real caboose?

46. Standard dollhouse scale is 1 inch to 1 foot. Heidi's dollhouse is 32 inches wide. How wide would it be if it were a real house?

47. A school board has determined that there should be 3 teachers for every 50 students. How many teachers are needed for an enrollment of 2700 students?

48. In a scale drawing, a 280-foot antenna tower is drawn 7 inches high. The building next to it is drawn 2 inches high. How tall is the actual building?

49. The instructions on a can of oil intended to be added to lawn mower gasoline read:

Recommended	Gasoline	Oil
50 to 1	6 gal	16 oz

Are these instructions correct? (*Hint:* There are 128 ounces in 1 gallon.)

Review Exercises

1. Change $\frac{9}{10}$ to a percent.
2. Change $\frac{7}{8}$ to a percent.
3. Change $33\frac{1}{3}\%$ to a fraction.
4. Change 75% to a fraction.
5. Find 30% of 1600.
6. Find $\frac{1}{2}\%$ of 520.
7. Maria bought a dress for 25% off the original price of $98. How much did the dress cost?
8. Bill purchased a shirt on sale for $17.50. What was the original cost of the shirt if it was on sale at 30% off?
9. Ricardo purchased a pair of shoes for 30% off the original price of $59, and a pair of boots for 40% off the original price of $79. What did he have to pay?
10. Anita purchased a purse for 25% off the original price of $39.95 and a pair of gloves for 50% off the original price of $6.95. What total price did she have to pay?

7.2 Similar Triangles

Two triangles with the same shape are **similar triangles**. Since congruent triangles have the same shape, they are similar triangles. However, similar triangles are not necessarily congruent. For two similar triangles to be congruent, they must have the same area. Two similar triangles that are not congruent are shown in Figure 7-1. In symbols, we write

$\triangle ABC \sim \triangle DEF$ Read as "Triangle *ABC* is similar to triangle *DEF*."

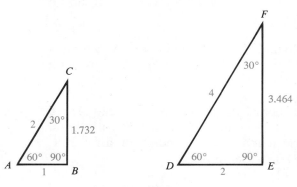

Figure 7-1

Definition: Two triangles are **similar triangles** if, and only if,
- three angles of one triangle are congruent to three angles of the second triangle, and
- the measures of all corresponding sides are in proportion.

In the similar triangles in Figure 7-1, the following relationships are true:

$$\angle A \cong \angle D \qquad \angle B \cong \angle E \qquad \angle C \cong \angle F$$

$$\frac{m(\overline{AB})}{m(\overline{DE})} = \frac{m(\overline{BC})}{m(\overline{EF})} = \frac{m(\overline{CA})}{m(\overline{FD})}$$

We will use the following postulate to show that two triangles are similar.

> **Postulate:** Two triangles are similar if, and only if, two angles of one triangle are congruent to two angles of the other triangle.

EXAMPLE 1

In Figure 7-2, $\overline{AB} \parallel \overline{DE}$. Show that $\triangle ABC \sim \triangle DEC$.

Solution

$\angle A$ and $\angle D$ form a pair of alternate interior angles, and since $\overline{AB} \parallel \overline{DE}$,

$$\angle A \cong \angle D$$

$\angle 1$ and $\angle 2$ form a pair of vertical angles, and since vertical angles are always congruent,

$$\angle 1 \cong \angle 2$$

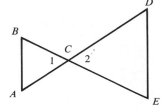

Figure 7-2

Because two angles of $\triangle ABC$ are congruent to two angles of $\triangle DEC$, the triangles are similar. ∎

EXAMPLE 2

Refer to Figure 7-3, in which $\triangle ABC$ and $\triangle DEF$ have $\angle A \cong \angle D$ and $\angle C \cong \angle F$. Find **a.** $m(\overline{DE})$ and **b.** $m(\overline{EF})$.

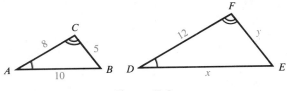

Figure 7-3

Solution

Let $m(\overline{DE}) = x$ and $m(\overline{EF}) = y$. Because $\angle A \cong \angle D$ and $\angle C \cong \angle F$, two angles of $\triangle ABC$ are congruent to two angles of $\triangle DEF$, and the triangles are similar. Thus, the lengths of their corresponding sides are in proportion:

a. $\dfrac{m(\overline{AC})}{m(\overline{DF})} = \dfrac{m(\overline{AB})}{m(\overline{DE})}$

and

b. $\dfrac{m(\overline{AC})}{m(\overline{DF})} = \dfrac{m(\overline{BC})}{m(\overline{EF})}$

Substituting 8 for m($\overline{AC}$), 12 for m($\overline{DF}$), 10 for m($\overline{AB}$), and x for m($\overline{DE}$) in Equation 1 and finding x gives

$\dfrac{8}{12} = \dfrac{10}{x}$

$8x = 120$ In a proportion, the product of the extremes is equal to the product of the means.

$x = \dfrac{120}{8}$ Change the multiplication to a division.

$x = 15$ Simplify.

Substituting 8 for m($\overline{AC}$), 12 for m($\overline{DF}$), 5 for m($\overline{BC}$), and y for m($\overline{EF}$) in Equation 2 and finding y gives

$\dfrac{8}{12} = \dfrac{5}{y}$

$8y = 60$ In a proportion, the product of the extremes is equal to the product of the means.

$x = \dfrac{60}{8}$ Change the multiplication to a division.

$x = \dfrac{15}{2}$ Simplify.

Thus, m($\overline{DE}$) = 15 units, and m($\overline{EF}$) = $7\tfrac{1}{2}$ units.

EXAMPLE 3

In Figure 7-4, $\overline{DE} \parallel \overline{AB}$. Find x.

Solution

Since $\overline{DE} \parallel \overline{AB}$, the corresponding angles $\angle A$ and $\angle 1$ and the corresponding angles $\angle B$ and $\angle 2$ are congruent. Thus, two angles of $\triangle ABC$ are congruent to two angles of $\triangle DEC$, and the triangles are similar. Because the measures of corresponding sides of similar triangles are in proportion,

$\dfrac{m(\overline{DC})}{m(\overline{AC})} = \dfrac{m(\overline{EC})}{m(\overline{BC})}$

So

$\dfrac{8}{12} = \dfrac{x}{16}$

$128 = 12 \cdot x$

$\dfrac{128}{12} = x$

$\dfrac{32}{3} = x$

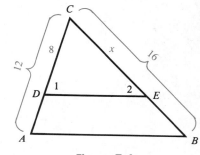

Figure 7-4

Thus, $x = 10\tfrac{2}{3}$ units. Work Progress Check 3.

Progress Check 3

In the figure, $AB \parallel DE$. Find m($\overline{AC}$).

3. 19.2

EXAMPLE 4

Two Boy Scouts wish to know the width of the river represented in Figure 7-5. How can they find the width without crossing to the other side?

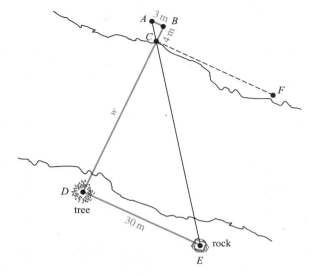

Figure 7-5

Solution

The boys can measure sides AB and BC directly. They can estimate the measure of side DE indirectly by measuring $\overline{CF}$ (F is on the bank straight across from E, and C is on the bank straight across from D). The boys find that

$\text{m}(\overline{AB}) = 3$ meters
$\text{m}(\overline{CB}) = 4$ meters

and

$\text{m}(\overline{CF}) = \text{m}(\overline{DE}) = 30$ meters

Because $\angle ACB$ and $\angle ECD$ are vertical angles, they are congruent. Because $\angle B$ and $\angle D$ are right angles, they are congruent. Since two angles of $\triangle ABC$ are congruent to two angles of $\triangle EDC$,

$\triangle ABC \sim \triangle EDC$

Thus, the boys can form the following proportion and solve for w:

$\dfrac{3}{30} = \dfrac{4}{w}$
$120 = 3w$
$40 = w$

The river is 40 meters wide.

7.2 Exercises

In Exercises 1–6, tell whether each statement is true. If a statement is false, explain why.

1. All equilateral triangles are similar.
2. All right triangles are similar.
3. All similar triangles are right triangles.
4. All similar triangles are isosceles triangles.
5. Some scalene triangles are similar.
6. Congruent triangles are similar.

In Exercises 7–10, the measures of certain parts of △ABC and △DEF are given. Draw the triangles and explain why they are similar.

7. m(∠A) = m(∠D) and m(∠B) = m(∠E)
8. m(∠A) = m(∠D), m(∠A) = m(∠B), and m(∠A) = m(∠E)
9. m(∠A) = m(∠D), m(∠A) = m(∠B), and m(∠D) = m(∠E)
10. m(∠A) = 60°, m(∠D) = 60°, m(∠B) = 50°, and m(∠F) = 70°

In Exercises 11–14, refer to the similar triangles in Illustration 1, in which ∠A ≅ ∠D and ∠B ≅ ∠E.

11. If m($\overline{AC}$) = 7 cm, m($\overline{DF}$) = 4 cm, and m($\overline{AB}$) = 15 cm, find m($\overline{DE}$).
12. If m($\overline{CB}$) = 12 cm, m($\overline{AB}$) = 15 cm, and m($\overline{DE}$) = 7 cm, find m($\overline{FE}$).
13. If m($\overline{AC}$) = 21 cm, m($\overline{BC}$) = 25 cm, and m($\overline{EF}$) = 8 cm, find m($\overline{DF}$).
14. If m($\overline{DE}$) = 6 cm, m($\overline{AC}$) = 12 cm, and m($\overline{DF}$) = 4 cm, find m($\overline{AB}$).

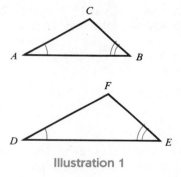

Illustration 1

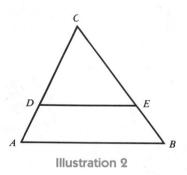

Illustration 2

In Exercises 15–20, refer to Illustration 2, in which $\overline{DE}$ ∥ $\overline{AB}$.

15. If m($\overline{DC}$) = 9, m($\overline{AC}$) = 12, and m($\overline{CE}$) = 6, find m($\overline{BC}$).
16. If m($\overline{DC}$) = 7, m($\overline{CE}$) = 6, and m($\overline{CB}$) = 12, find m($\overline{AC}$).
17. If m($\overline{AD}$) = 5, m($\overline{DC}$) = 12, and m($\overline{CB}$) = 204, find m($\overline{CE}$).
18. If m($\overline{AC}$) = 15, m($\overline{CE}$) = 16, and m($\overline{EB}$) = 4, find m($\overline{DC}$).
19. If m($\overline{BC}$) = 65, m($\overline{EB}$) = 10, and m($\overline{AC}$) = 52, find m($\overline{DC}$).
20. If m($\overline{AC}$) = 42, m($\overline{AD}$) = 6, and m($\overline{BC}$) = 56, find m($\overline{EB}$).

In Exercises 21–26, refer to Illustration 3, in which m(∠ACB) = 90° and m(∠CDB) = 90°.

21. If m(∠A) = 40°, find m(∠1).

22. If m(∠2) = 60°, find m(∠B).

23. Show that ∠A ≅ ∠2 and ∠B ≅ ∠1.

24. Show that △ADC ~ △ACB.

25. Show that △BDC ~ △BCA.

26. Show that △ADC ~ △CDB.

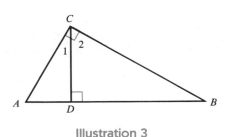

Illustration 3

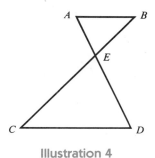

Illustration 4

In Exercises 27–28, refer to Illustration 4.

27. If m($\overline{AE}$) = 4, m($\overline{DE}$) = 6, m($\overline{BE}$) = 3, and $\overline{AB} \parallel \overline{CD}$, find m($\overline{CE}$).

28. If m($\overline{AE}$) = 4, m($\overline{DE}$) = 6, m($\overline{BC}$) = 8, and $\overline{AB} \parallel \overline{CD}$, find m($\overline{BE}$).

29. A building casts a shadow of 50 feet at the same time a person casts a shadow of 8 feet. If the person is 6 feet tall, how tall is the building?

30. A person $5\frac{1}{2}$ feet tall casts a 3-foot shadow when a flagpole casts a 60-foot shadow. How tall is the flagpole?

31. A straight road going up a constant grade rises 3 feet in the first 50 feet of roadway. If a car travels 1000 feet from the bottom of the hill to the top, how high is the hill?

32. A ski hill has a run $\frac{1}{2}$ mile long. If the hill drops 5 feet as the skier slides 9 feet, how high is the hill? (*Hint:* 5280 feet = 1 mile.)

33. Sally is 5 feet tall. To measure the height of a smokestack, she asks her 6-foot boyfriend to help her. They stand as in Illustration 5, with her line of sight to the top of the stack being tangent to his head. If the boyfriend stands 126 feet from the smokestack and the girl is 3 feet from her friend, how tall is the smokestack?

34. Find the width of the river shown in Illustration 6.

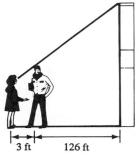

Illustration 5

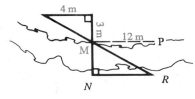

Illustration 6

35. An airplane ascends 100 feet as it flies 1000 feet. How much altitude will it gain if it flies 1 mile?

36. In a landing approach, an airplane descends 45 feet as it flies 500 feet. How far does the plane fly as it descends 66 feet?

37. While walking in the woods in northern Wisconsin, you come upon a huge white pine tree that has been estimated to be over 400 years old. A small pine tree 3 feet tall casts a shadow of 2 feet when the large pine casts a shadow of 116 feet. Find the height of the tree.

38. To estimate the height of the Space Needle in Seattle, Juan notes that the Space Needle casts a shadow of approximately 173 feet when a 7-foot pole casts a shadow of 2 feet. Approximately how tall is the Space Needle?

Review Exercises

In Review Exercises 1–4, identify the base and the exponent in each expression.

1. 7^6
2. 32^4
3. $\left(\frac{2}{3}\right)^2$
4. $\left(\frac{4}{3}\right)^3$

In Review Exercises 5–10, write each expression without using exponents.

5. $3(4)^3$
6. $(3 \cdot 2)^4$
7. $3^3 \cdot 4^2$
8. $3^3 + 4^2$
9. $2 + 4 \cdot 3^2$
10. $23 - (5 - 1)^2 + 3^2$

7.3 Areas of Rectangles and Parallelograms

- Area of a Rectangle
- Area of a Parallelogram

We now discuss finding areas of rectangles and parallelograms. Later, we will find areas of other polygons. Technically, a polygon does not *have* an area; it *encloses* an area. However, it is common to speak of the area of a polygon.

The **area of a polygon** is the measure of the amount of surface enclosed by the polygon. Knowing how to find areas is helpful when solving problems such as estimating the cost of carpeting or the cost of painting a house.

Area of a Rectangle

We can use a scale drawing to find the area of a rectangle. For example, each square in the rectangle *ABCD* shown in Figure 7-6 represents an area of one square foot—a surface enclosed by a square measuring one foot on each side. It is easy to count the number of squares bounded by rectangle *ABCD*. Because there are 7 rows with 10 squares in each row, there are 70 squares. We say that the area of the rectangle is 70 square feet. The result is often written as 70 ft².

Because many of the squares in △*ABC* shown in Figure 7-7 are incomplete, we cannot find its exact area by counting squares. However, we can approximate the area by adding the area enclosed by complete squares to an estimate of the area enclosed by incomplete squares.

Fortunately, there are better ways to find areas than counting squares. Geometry will enable us to develop formulas that will give exact areas of many polygons.

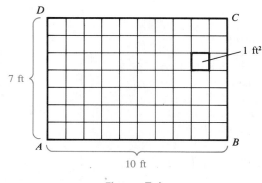

Figure 7-6

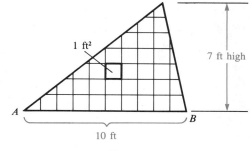

Figure 7-7

In the rectangle shown in Figure 7-6, $\overline{AB}$ is called the **length** and $\overline{AD}$ is called the **width**. Because the rectangle has 7 rows of 10 squares each, it contains 70 squares and has an area of 70 ft². Thus, the area of the rectangle is the product of its length and width.

> **Postulate:** The **area of a rectangle** is given by the formula
> $$A = lw$$
> where l is the length and w is the width of the rectangle.

EXAMPLE 1

Find the area of the rectangle $ABCD$ shown in Figure 7-8.

Solution

From the figure, the length is 12 units, and the width is 8 units. We find the area of the rectangle by using the formula and substituting 12 for l and 8 for w.

$A = lw$
$A = 12 \cdot 8$
$A = 96$

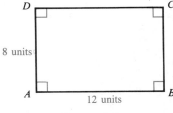

Figure 7-8

The area of the rectangle is 96 square units. Work Progress Check 4.

Progress Check 4

Find the area of a rectangle with dimensions of 6 inches by 2 feet.

EXAMPLE 2

The living room/dining room area of a house has the floor plan shown in Figure 7-9. If carpet costs $29 per square yard, including pad and installation, how much will it cost to carpet the room? Assume there is no waste.

4. 144 in.²

Section 7.3 Areas of Rectangles and Parallelograms 247

Solution

To find the cost, we must find the area. Since $\overline{CF}$ divides the room into two rectangles, the area of the living room and the dining room are found by multiplying their lengths by their widths. To find the area of the living room, we proceed as follows:

Area of living room $= l \cdot w$
$= 7 \cdot 4$
$= 28$

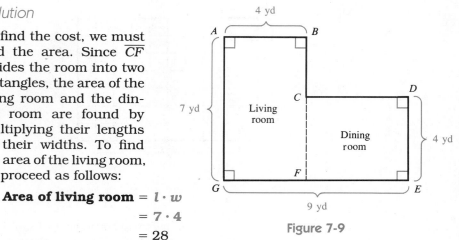

Figure 7-9

The area of the living room is 28 yd².

To find the area of the dining room, we find its length by subtracting 4 yards from 9 yards to obtain 5 yards. Its width is 4 yards. Because the area of the dining room is the product of its length and width,

Area of dining room $= l \cdot w$
$= 5 \cdot 4$
$= 20$

The area of the dining room is 20 yd².

The total area to be carpeted is the sum of these two areas, which is 28 yd² + 20 yd², or 48 yd². At $29 per square yard, the cost to carpet the room will be 48 · $29, or $1392. ∎

Area of a Parallelogram

To develop a formula for the area of a parallelogram, we need the following definition.

> **Definition:** An **altitude of a parallelogram** is a line segment drawn from a vertex of the parallelogram perpendicular to a non-adjacent side or its extension.

Altitudes $\overline{DQ}$, $\overline{CP}$, and $\overline{DR}$ appear in Figure 7-10. Altitude $\overline{DQ}$ is drawn to side $\overline{AB}$, and altitudes $\overline{CP}$ and $\overline{DR}$ are drawn to extensions of sides $\overline{AB}$ and $\overline{BC}$, respectively. An altitude is often called a **height**, and the side to which it is drawn is often called a **base**. We will use the word *height* to indicate the length of an altitude.

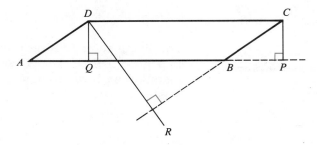

Figure 7-10

To prove the formula used for finding the area of a parallelogram, we refer to □ABCD, shown in Figure 7-11. We draw a perpendicular line segment from point A to the extension of line DC and a perpendicular from point B to line DC. This forms rectangle ABFE with length b and width h. It also forms right triangles △AED and △BFC, with ∠E and ∠BFC as right angles. Thus,

$$\angle E \cong \angle BFC$$

Because opposite sides of a parallelogram are of equal length,

$$\overline{AD} \cong \overline{BC}$$

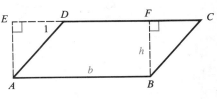

Figure 7-11

Because $\overleftrightarrow{EC}$ is a transversal intersecting the parallel lines $\overleftrightarrow{AD}$ and $\overleftrightarrow{BC}$, ∠1 and ∠C are congruent corresponding angles:

$$\angle 1 \cong \angle C$$

Because two angles and a side of △AED are congruent to two angles and a side of △BFC, the triangles are congruent. Because these congruent triangles have equal areas, it follows that

$$A(\square ABCD) = A(\text{rectangle } ABFE)$$

Read $A(\square ABCD)$ as "the area of parallelogram ABCD."

Because A(rectangle ABFE) is the product of its length b and its width h,

$$A(\text{rectangle } ABFE) = bh$$

Thus,

$$A(\square ABCD) = bh$$

where b is the length of the base of the parallelogram and h is its height. This proves the following theorem.

Theorem: The **area of a parallelogram** is given by the formula

$$A = bh$$

where b is the length of the base and h is the height of the parallelogram.

Progress Check 5

Find the area of □ABCD.

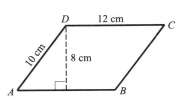

5. 96 cm²

Progress Check Answers

EXAMPLE 3

Find the area of the parallelogram shown in Figure 7-12.

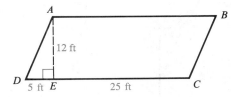

Figure 7-12

Solution

The length of the base of the parallelogram is 5 feet + 25 feet, or 30 feet. The height is 12 feet. We substitute 30 for b and 12 for h in the formula for the area of a parallelogram and simplify:

$$A = bh$$
$$= 30 \cdot 12$$
$$= 360$$

The area of the parallelogram is 360 ft². Work Progress Check 5. ∎

Section 7.3 Areas of Rectangles and Parallelograms

7.3 Exercises

In Exercises 1–6, find the area of each parallelogram.

1.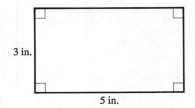

2.

3.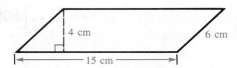

4.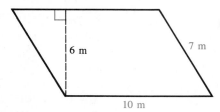

5.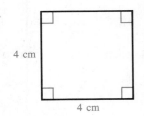

6.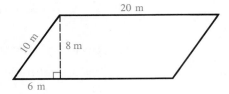

In Exercises 7–10, find the area of each figure.

7.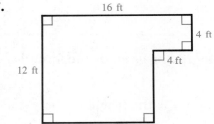

8.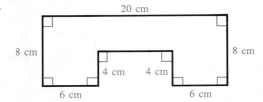

9.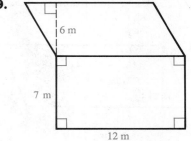

10.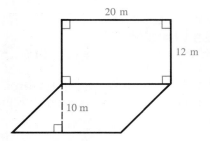

In Exercises 11–14, find the area of each shaded region.

11.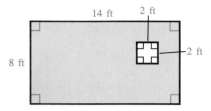

12.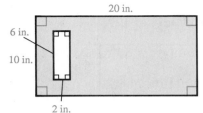

13.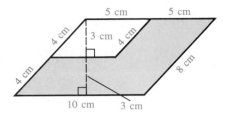

14.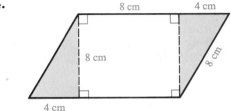

15. Find the area of the parallelogram shown in Illustration 1.

16. Find the area of the parallelogram shown in Illustration 2.

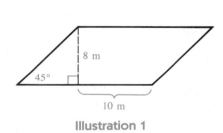

Illustration 1

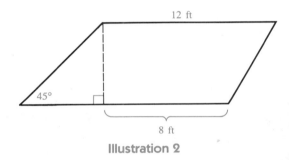

Illustration 2

17. How many square feet are there in one square yard?

18. How many square inches are there in one square foot?

19. How many square centimeters are there in one square meter?

20. How many square meters are there in one square kilometer?

21. A room is a rectangle that is 24 feet long and 15 feet wide. At $30 per square yard, how much will it cost to carpet the room? (Assume no waste.)

22. Luz's living room is a rectangle measuring 30 by 18 feet. At $32 per square yard, how much will it cost to carpet the room? (Assume no waste.)

23. The basement room in Sam's house is a rectangle measuring 14 by 20 feet. Vinyl floor tiles that are 1 ft^2 cost $1.29 each. How much will the tile cost if Sam decides to tile the floor? (Ignore any waste.)

24. The north wall of a barn is a rectangle 23 feet high and 72 feet long. There are five windows in the wall, each 4 by 6 feet. If a gallon of paint will cover 300 ft^2, how many gallons must the painter buy?

25. A rectangle has a length of 4 feet 2 inches and a width of 2 feet 4 inches. Find its area in square feet.

26. A rectangle has a length of 8.3 meters and a width of 7.1 meters. Find its area in square centimeters.

Review Exercises

1. What number is 25% of 42?
2. 32 is what % of 60?
3. 20% of what number is 50?
4. 86% of what number is 142?
5. 26 is 2.5% of what number?
6. What number is 17.3% of 42?
7. 55 is what % of 125?
8. 12.3 is what % of 96.8?
9. 75% of what number is 12?
10. Find 36.8% of 562.

7.4 Areas of Triangles and Trapezoids

- Area of a Triangle
- Area of a Trapezoid

Area of a Triangle

To find the area of a triangle, we refer to $\triangle ABC$ shown in Figure 7-13. We draw $\overline{CD}$ parallel to $\overline{AB}$ and draw $\overline{AD}$ parallel to $\overline{BC}$. These two lines intersect at point D. Since the opposite sides of quadrilateral $ABCD$ are parallel, quadrilateral $ABCD$ is a parallelogram with base $\overline{AB}$ and height h. Thus,

$$A(\square ABCD) = bh$$

However, because diagonal $\overline{AC}$ divides $\square ABCD$ into two congruent triangles,

$$A(\triangle ABC) = \frac{1}{2}[A(\square ABCD)]$$
$$= \frac{1}{2}bh$$

This proves the following theorem:

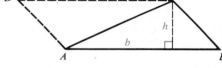

Figure 7-13

Theorem: The **area of a triangle** is given by the formula

$$A = \frac{1}{2}bh$$

where b is the length of the base and h is the height of the triangle.

EXAMPLE 1

Find the area of a triangle with a base 8 inches long and a height of 5 inches.

Solution

Refer to Figure 7-14. The area of a triangle is found by substituting 8 for b and 5 for h in the formula for the area of a triangle

$$A = \frac{1}{2}bh$$

$$A = \frac{1}{2}(8)(5)$$

$$= 20$$

The area of the triangle is 20 in^2.

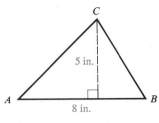

Figure 7-14

EXAMPLE 2

Find the area of a triangle with a height of 13 feet and a base of 8 feet.

Solution

Refer to Figure 7-15. We substitute 8 for b and 13 for h in the formula for the area of a triangle and simplify.

$$A = \frac{1}{2}bh$$

$$A = \frac{1}{2}(8)(13)$$

$$= 4(13)$$

$$= 52$$

The area of the triangle is 52 ft^2. Work Progress Check 6.

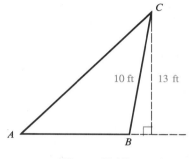

Figure 7-15

Progress Check 6

Find the area of $\triangle ABC$.

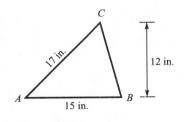

Area of a Trapezoid

Definition: A **trapezoid** is a quadrilateral with exactly two sides parallel.
The parallel sides are called **bases**, and the nonparallel sides are called **legs**.

To develop a formula for finding the area of a trapezoid, we refer to the trapezoid in Figure 7-16, with bases $\overline{AB}$ and $\overline{DC}$ and height h. The area of trapezoid $ABCD$ is the sum of the areas of $\triangle ABD$ and $\triangle CDB$:

$$A(\text{trapezoid } ABCD) = A(\triangle ABD) + A(\triangle CDB)$$

$$= \frac{1}{2}bh + \frac{1}{2}b'h \quad \text{Read } b' \text{ as "}b\text{ prime."}$$

Later, we will show that this formula can be written as

$$A = \frac{1}{2}h(b + b')$$

This suggests the following theorem:

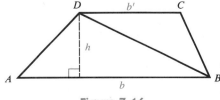

Figure 7-16

6. 90 in.^{2}

Progress Check Answers

> **Theorem:** The **area of a trapezoid** is given by the formula
>
> $$A = \frac{1}{2}h(b + b')$$
>
> where h is the height of the trapezoid and b and b' represent the length of each base.

EXAMPLE 3

A trapezoid has bases that measure 6 and 10 inches. If the height of the trapezoid is 1 foot, what is its area?

Solution

We refer to Figure 7-17. In this example, $b = 10$ and $b' = 6$. It is incorrect to say that $h = 1$, because the height of 1 foot must be expressed in inches to be consistent with the units of the bases. Thus, we substitute 10 for b, 6 for b', and 12 for h in the formula for finding the area of a trapezoid, and simplify:

$$A = \frac{1}{2}h(b + b')$$

$$A = \frac{1}{2}(12)(10 + 6)$$

$$= 6(16)$$

$$= 96$$

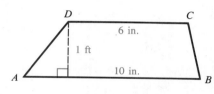

Figure 7-17

The area of the trapezoid is 96 in². Work Progress Check 7.

Progress Check 7

Find the area of trapezoid $ABCD$.

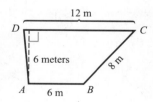

EXAMPLE 4

The wall at one end of an attic room has the shape of a trapezoid. The wall is 8 feet high at one end and 3 feet high at the other. If the room is 10 feet wide, find the area of the wall.

Solution

Refer to Figure 7-18. The wall forms the shape of a trapezoid with bases of length 8 and 3 feet and a height of 10 feet. Substituting these values into the formula for the area of a trapezoid gives

$$A = \frac{1}{2}h(b + b')$$

$$A = \frac{1}{2}(10)(8 + 3)$$

$$= 5(11)$$

$$= 55$$

The area of the wall is 55 ft².

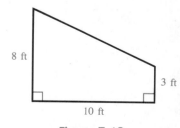

Figure 7-18

7. 54 m²

7.4 Exercises

In Exercises 1–8, find the area of each figure.

1.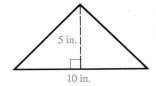

2.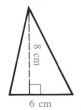

3.

4.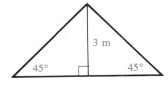

5.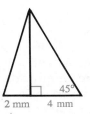

6.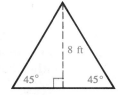

7.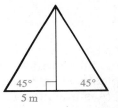

8.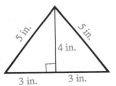

In Exercises 9–12, find the area of each trapezoid.

9.

10.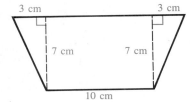

11.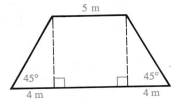

12.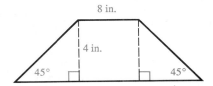

In Exercises 13–16, find the area of each figure.

13.

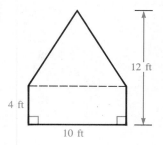

14.

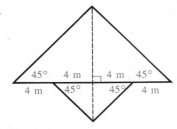

15.

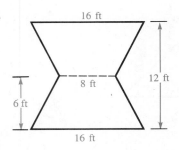

16.

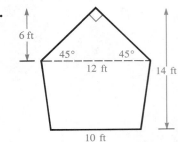

17. If nylon sail fabric costs $12 per square yard, how much would the fabric cost to make a sail that is in the shape of a triangle with a base of 12 feet and a height of 24 feet?

18. The gable end of a warehouse is an isosceles triangle with a height of 4 yards and a base of 23 yards. It will require one coat of primer and one coat of finish to paint the triangle. Primer costs $17 per gallon, and the finish coat costs $23 per gallon. If one gallon covers 300 square feet, how much will it cost to paint the gable?

19. A homeowner can have a landscape company sod his front lawn for $1.17 per square foot. The front lawn is in the shape of a trapezoid with bases of 20 yards and 15 yards and a height of 9 yards. What will it cost to sod the lawn?

20. A swimming pool is in the shape of a trapezoid with dimensions as shown in Illustration 1. How many square meters of plastic sheeting are required to cover the pool? (Assume no waste.)

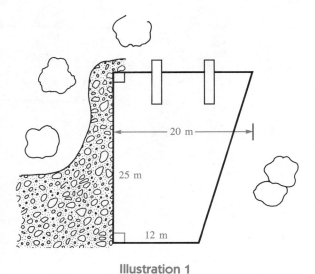

Illustration 1

256   Chapter 7   Proportion, Similar Triangles, Area, and Volume

Review Exercises

In Review Exercises 1–4, estimate the answer to each problem.

1. $\dfrac{0.95 \cdot 3.89}{2.997}$
2. 21% of 42
3. 32% of 60
4. $\dfrac{4.966 + 5.001}{2.994}$

In Review Exercises 5–8, perform each calculation.

5. $2.3 \cdot 34.5$
6. $14.98 \div 3.5$
7. $343.54 \div 8.9$
8. $0.236 \cdot 6.4$

7.5 Areas of Circles

- Circumference of a Circle
- Area of a Circle

The formula for the area of a circle involves the number **pi**. This number, denoted by the Greek letter π, is equal to $3.141592653589\ldots$.

Circumference of a Circle

Since early history, mathematicians have known that the ratio of the distance around a circle, called its **circumference**, divided by the length of its diameter is approximately 3. Today, we know that this ratio is π:

1. $\pi = \dfrac{C}{D}$

where C is the circumference of a circle and D is the length of its diameter. If we write Equation 1 as a multiplication, we have the following theorem.

> **Theorem:** The **circumference, C, of a circle** is given by the formula
>
> $C = \pi D$
>
> where D is the diameter of the circle.

Because a diameter of a circle is twice as long as a radius, we can substitute $2r$ for D in the formula $C = \pi D$ to obtain another formula for the circumference:

$C = 2\pi r$

where C is the circumference of the circle and r is the radius.

Area of a Circle

To develop a formula for finding the area of a circle, we refer to Figure 7-19. If we divide the circle in Figure 7-19**(a)** into an even number of pie-shaped pieces and then regroup these pieces as in Figure 7-19**(b)**, we have a figure that looks like a parallelogram. The figure has a base that is half as long as the circumference of the circle, and its height is approximately the same length as a radius of the circle.

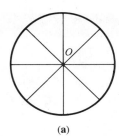
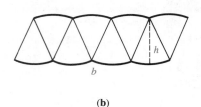

(a)   (b)

Figure 7-19

If we divide the circle into more and more pie-shaped pieces, the figure will look more and more like a parallelogram. As the number of pie-shaped pieces becomes extremely large, the figure becomes a parallelogram, and we can find its area by using the formula for finding the area of a parallelogram.

$A = bh$

Furthermore, as the number of pie-shaped pieces becomes extremely large, b becomes very close to $\frac{1}{2}C$, and h becomes very close to r. Thus,

$$A = bh = \frac{1}{2}Cr = \frac{1}{2}(2\pi r)r = \pi r^2$$

Because the areas of the pie-shaped pieces do not change when we rearrange them, we have the following theorem.

> **Theorem:** The **area of a circle** is given by the formula
> $A = \pi r^2$
> where r is the radius of the circle.

EXAMPLE 1

Find the area of the circle shown in Figure 7-20.

Solution

Because the length of a diameter is given to be 10 centimeters and the length of a diameter is twice the length of a radius, the length of a radius is 5 centimeters. To find the area of the circle, we substitute 5 for r in the formula for the area of a circle.

$A = \pi r^2$

$A = \pi(5^2)$

$= 25\pi$

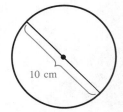

Figure 7-20

The area of the circle is 25π cm². Work Progress Check 8. ■

EXAMPLE 2

Orange paint is available in gallons that cost $19 each. Each gallon will cover 375 ft². How much will the paint cost to cover a helicopter pad 60 feet in diameter?

Progress Check 8

Find the area of a circle with a diameter of 12 feet. Use $\pi \approx 3.14$ and give the answer as a decimal. Read the symbol $\approx$ as "is approximately equal to."

8. 113.04 ft²

Solution

First we calculate the area of the helicopter pad.

$A = \pi r^2$
$A = \pi(\mathbf{30^2})$ Substitute one-half of 60 for r.
$ = 900\pi$
$ \approx 2826$ Substitute 3.14 for π and simplify.

The area of the helicopter pad is approximately 2826 ft². Because each gallon of paint covers 375 ft², the number of gallons needed is found by dividing 2826 by 375.

$$\text{Number of gallons needed} \approx \frac{2826}{375}$$
$$\approx 7.54$$

The painters will need 7.54 gallons, but they must purchase 8 gallons. Thus, the cost of the paint will be 8($19), or $152. ∎

7.5 Exercises

1. Find the circumference of a circle with a diameter of 8 inches.

2. Find the circumference of a circle with a radius of 10 centimeters.

3. Find the diameter of a circle with a circumference of 36π meters.

4. Find the radius of a circle with a circumference of 50π meters.

In Exercises 5–16, find the shaded area of each figure.

5.

6.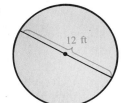

7.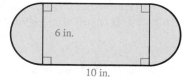

8.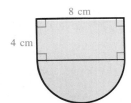

9.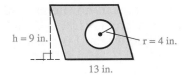

10.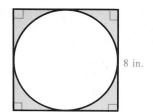

11.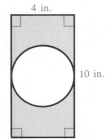

12.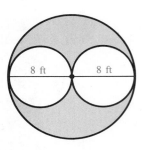

13.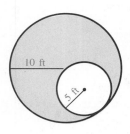

14.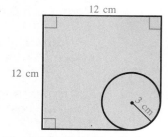

15.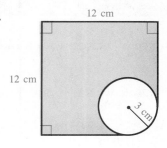

16.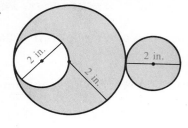

In Exercises 17–24, give each answer to the nearest hundredth ($\pi \approx 3.14$).

17. Round Lake has a circular shoreline 2 miles in diameter. Find the area of the lake.

18. Sam is planning to hike around the lake in Exercise 17. How far will he walk?

19. Joan wants to jog 10 miles on a circular track $\frac{1}{4}$ mile in diameter. How many times must Joan circle the track?

20. The rotunda at a state capitol is a circular area 100 feet in diameter. The legislature wishes to appropriate money to have the rotunda tiled. The lowest bid is $83 per square yard, including installation. How much must the legislature spend?

21. If the radius of a circle is doubled, by what factor is the area increased?

22. If the radius of a circle is tripled, by what factor is the area increased?

23. A steel band is drawn tightly about the earth's equator. The band is then loosened by increasing its length by 10 feet, and the resulting slack is distributed evenly along the band's entire length. How far above the earth's surface is the band? Use $\pi \approx 3.14$. (*Hint:* You do not need to know the circumference of the earth.)

24. Two circles are called **concentric circles** if they have the same center. Find the area of the band between two concentric circles if their diameters are 10 centimeters and 6 centimeters. Use $\pi \approx 3.14$.

Review Exercises

In Review Exercises 1–10, perform the calculations. Write all improper fractions as mixed numbers.

1. $\dfrac{3}{4} + \dfrac{2}{3}$
2. $\dfrac{7}{8} - \dfrac{2}{3}$
3. $3\dfrac{3}{4} + 2\dfrac{1}{3}$
4. $7\dfrac{5}{8} - 2\dfrac{5}{6}$
5. $9\dfrac{7}{9} + 7\dfrac{3}{5}$
6. $6\dfrac{9}{10} - 2\dfrac{1}{4}$
7. $0.24 + 5.2 + 75$
8. $0.876 + 12.2 + 5$
9. $7\dfrac{1}{2} \div 5\dfrac{2}{5}$
10. $5\dfrac{3}{4} \cdot 2\dfrac{5}{6}$

7.6 Surface Area and Volume

- Right Prisms
- Spheres
- Right-Circular Cylinders
- Right-Circular Cones
- Right Pyramids

So far, we have considered **plane figures**—two-dimensional figures that lie in a plane. Other figures, called **three-dimensional figures**, occupy space.

Right Prisms

Two three-dimensional figures, called **right prisms**, are shown in Figure 7-21. The **base** of the prism in Figure 7-21(a) is a triangle, and the base of the prism in Figure 7-21(b) is a rectangle. If the base of a prism is a rectangle, the prism is often called a **rectangular solid**. The rectangular solid in Figure 7-21(b) represents an empty box with length l, width w, and height h. A rectangular solid whose length, width, and height are all the same is called a **cube**.

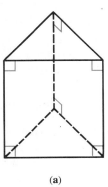
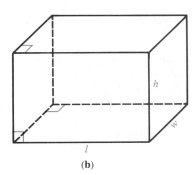

(a)   (b)

Figure 7-21

The **surface area of a rectangular solid** is the sum of the areas of the six rectangular **faces** that are its sides. The **volume of a rectangular solid** is the amount of space it encloses. We assume the following theorem without proof.

Theorem: The **surface area, S**, and the **volume, V, of a rectangular solid** are given by the formulas

$$S = 2lw + 2wh + 2lh$$
$$V = lwh$$

where l is its length, w is its width, and h is its height.

EXAMPLE 1

The oil storage tank shown in Figure 7-22 is a rectangular solid with dimensions of 17 by 10 by 8 feet. Find both the surface area and the volume of the tank.

Solution

To find the surface area, we substitute 17 for l, 10 for w, and 8 for h in the formula for surface area and simplify.

$$S = 2lw + 2wh + 2lh$$
$$S = 2(17)(10) + 2(10)(8) + 2(17)(8)$$
$$= 772$$

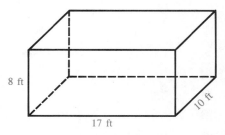

Figure 7-22

The surface area is 772 ft².

To find the volume, we substitute 17 for l, 10 for w, and 8 for h in the formula for volume and simplify.

$$V = lwh$$
$$V = (17)(10)(8)$$
$$= 1360$$

The volume is 1360 ft³. Work Progress Check 9.

To find the surface area of any right prism, we must find the sum of the areas of its faces. To find its volume, we can use the following formula.

Theorem: The **volume of a right prism** is given by the formula

$$V = Bh$$

where B is the area of its base and h is its height.

EXAMPLE 2

Find **a.** the surface area and **b.** the volume of the triangular prism shown in Figure 7-23.

Progress Check 9

Find the surface area and the volume of a rectangular solid with dimensions of 8, 12, and 20 meters.

9. 992 m², 1920 m³

Progress Check Answers

Solution

a. The area of each triangular base is

$$\frac{1}{2}(6)(8) \text{ cm}^2 \quad \text{or} \quad 24 \text{ cm}^2$$

and the areas of the rectangular faces are

$$6(12) \text{ cm}^2, \quad 8(12) \text{ cm}^2,$$
$$\text{and} \quad 10(12) \text{ cm}^2$$

Thus, the surface area, S, including the triangular ends, is

$$S = 2(24) + 72 + 96 + 120$$
$$= 336$$

The surface area of the prism is 336 cm².

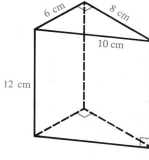

Figure 7-23

b. The volume of the prism is the area of its base multiplied by its height. Since the area of the triangular base is 24 cm² and the height is 12 centimeters, we have

$$A = Bh = 24(12) = 288$$

The volume of the prism is 288 cm³.

Work Progress Check 10.

Progress Check 10

Find the surface area and the volume of the right-triangular prism.

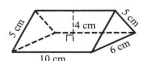

Spheres

A **sphere** is a hollow, round, ball-like *surface*. The points on the sphere shown in Figure 7-24 all lie at a fixed distance r from a point called its **center**. A segment drawn from the center of the sphere to a point on the sphere is called a **radius**.

We accept the formulas for surface area and volume of a sphere without proof.

Figure 7-24

Theorem: The **surface area, S**, and the **volume, V, of a sphere** are given by the formulas

$$S = 4\pi r^2$$
$$V = \frac{4}{3}\pi r^3$$

where r is the radius of the sphere.

EXAMPLE 3

A beach ball has a diameter of 16 inches. How many square inches of plastic material were needed to make the ball? (Disregard any waste.)

10. 184 cm², 120 cm³

Progress Check Answers

Section 7.6 Surface Area and Volume

Solution

Because the radius of the ball is half its diameter, the radius is 8 inches. So we substitute 8 for r in the formula for the surface area of a sphere and simplify.

$$S = 4\pi r^2$$
$$S = 4\pi(8)^2$$
$$= 256\pi$$
$$\approx 804 \quad \text{Substitute 3.14 for } \pi.$$

Approximately 804 in.² of material was required to make the beach ball. ∎

EXAMPLE 4

How many cubic feet of water are needed to fill a spherical tank with a radius of 15 feet?

Solution

Because the tank has a radius of 15 feet, we substitute 15 for r in the formula for the volume of a sphere and simplify.

$$V = \frac{4}{3}\pi r^3$$
$$V = \frac{4}{3}\pi(15)^3$$
$$= \frac{4}{3}\pi(3375)$$
$$= 4500\pi$$
$$\approx 14{,}130 \quad \text{Substitute 3.14 for } \pi.$$

Approximately 14,130 ft³ of water are needed to fill the tank. Work Progress Check 11. ∎

Progress Check 11

Find the surface area and the volume of a sphere with a radius of 2 inches.

Right-Circular Cylinders

A **right-circular cylinder** is a hollow figure like a soup can. The **radius of a cylinder** is the radius of its circular cross section. Its height is the distance between its ends. The radius of the cylinder in Figure 7-25 is r, and its height is h.

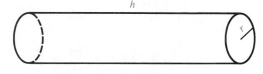

Figure 7-25

The surface area of a right-circular cylinder is the sum of the areas of its curved side and its two ends. Its volume is the amount of space enclosed by the cylinder.

11. 16π in.², $\frac{32}{3}\pi$ in.³

Progress Check Answers

> **Theorem:** The **surface area, S,** and the **volume, V, of a right-circular cylinder** are given by the formulas
>
> $$S = 2\pi rh + 2\pi r^2 \quad \text{(area of sides + area of ends)}$$
> $$V = \pi r^2 h$$
>
> where r is the radius of the cylinder and h is its height.

EXAMPLE 5

A farmer's silo is a right-circular cylinder 50 feet tall, topped with a **hemisphere** (a half-sphere). The radius of the cylinder is 10 feet. Find **a.** the surface area not including the floor and **b.** the volume of the silo.

Solution

a. Refer to the silo shown in Figure 7-26. The total surface area is the sum of two areas. The first is the surface area of the curved side of the cylinder, and the second is the surface area of the dome.

$$\begin{aligned}\textbf{Area of the cylinder's side} &= 2\pi rh \\ &= 2\pi(10)(50) \\ &= 1000\pi\end{aligned}$$

Do not include the ends.

$$\begin{aligned}\textbf{Area of the dome} &= \frac{1}{2}(\text{area of the sphere}) \\ &= \frac{1}{2}(4\pi r^2) \\ &= 2\pi r^2 \\ &= 2\pi 10^2 \\ &= 200\pi\end{aligned}$$

The total surface area of the silo is 1000π ft² + 200π ft², or 1200π ft², which is approximately 3768 ft². ($\pi \approx 3.14$.)

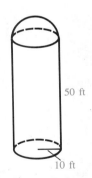

Figure 7-26

b. To find the volume of the silo, find the sum of the volumes of the cylinder and the dome.

Volume of the cylinder + volume of the dome

$$\begin{aligned}&= \pi r^2 h + \frac{1}{2}\left(\frac{4}{3}\pi r^3\right) \\ &= \pi(10^2)50 + \frac{1}{2}\left(\frac{4}{3}\pi 10^3\right) \\ &= 5000\pi + \frac{2000}{3}\pi \\ &= \frac{17{,}000}{3}\pi \\ &\approx 17{,}793\end{aligned}$$

The volume of the silo is approximately 17,793 ft³.

Work Progress Check 12.

Progress Check 12

Find the surface area and the volume of the right-circular cylinder.

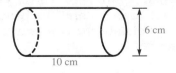

12. 78π cm², 90π cm³

Right-Circular Cones

A **right-circular cone** is shown in Figure 7-27. The height of the cone is h, and the radius of the cone is r—the radius of its circular base. The distance s measured along the side of the cone is called the **slant height**.

The formulas for area and volume of a cone are as follows:

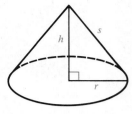

Figure 7-27

> **Theorem:** The **surface area, S**, and the **volume, V, of a right-circular cone** are given by the formulas
>
> $$S = \pi rs + \pi r^2$$
> $$V = \frac{1}{3}\pi r^2 h$$
>
> where r is the length of the radius of its base, h is its height, and s is the length of its slant height.

EXAMPLE 6

Find the surface area of the cone shown in Figure 7-28.

Solution

We substitute 6 for s and $\frac{1}{2}(8)$, or 4, for r in the formula for the surface area of a cone and simplify.

$$S = \pi rs + \pi r^2$$
$$S = \pi(4)(6) + \pi 4^2$$
$$= 24\pi + 16\pi$$
$$= 40\pi$$
$$\approx 125.6 \qquad \text{Use } \pi \approx 3.14.$$

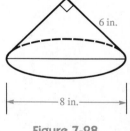

Figure 7-28

The area of the cone is approximately 125.6 square inches. Work Progress Check 13.

Right Pyramids

Two **right pyramids** with a height of h are shown in Figure 7-29. The base of the pyramid in Figure 7-29**(a)** is an equilateral triangle, and the base of the pyramid in Figure 7-29**(b)** is a square. The distance, s, measured along a face of the pyramid is called the **slant height**.

Progress Check 13

Find the volume of the right-circular cone.

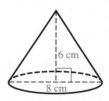

13. 32π cm^3

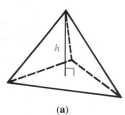
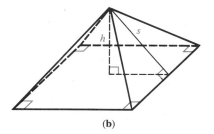

(a) (b)

Figure 7-29

Progress Check 14

Find the volume of a pyramid with a square base with each side 6 meters long and a height of 9 meters.

To find the surface area of a pyramid, we must find the sum of the areas of its faces. We state the formula for the volume of a pyramid without proof.

Theorem: The **volume of a pyramid** is given by the formula

$$V = \frac{1}{3}Bh$$

where B is the area of its base and h is its height.

Work Progress Check 14.

Progress Check Answers

14. 108 m^3

7.6 Exercises

In Exercises 1–12, find the surface area and volume of each figure.

1. A rectangular solid with dimensions of 3 by 4 by 5 centimeters.

2. A rectangular solid with dimensions of 5 by 8 by 10 meters.

3. A right prism whose base is a right triangle with legs of length 3 and 4 meters, whose hypotenuse is 5 meters long, and whose height is 8 meters.

4. A right prism whose base is a right triangle with legs that are 5 and 12 feet long, whose hypotenuse is 13 feet long, and whose height is 10 feet.

5. A sphere with a radius of 9 inches.

6. A sphere with a diameter of 10 feet.

7. A right-circular cylinder with a base of radius 6 meters and a height of 12 meters.

8. A right-circular cylinder with a base of diameter 18 meters and a height of 4 meters.

9. A right-circular cone with a base of diameter 10 centimeters, a height of 12 centimeters, and a slant height of 13 centimeters.

10. A right-circular cone with a base of radius 4 inches, a height of 3 inches, and a slant height of 5 inches.

11. A pyramid with a square base 10 meters on each side, a height of 12 meters, and a slant height of 13 meters.

12. A pyramid with a square base 6 inches on each side, a height of 4 inches, and a slant height of 5 inches.

Section 7.6   Exercises   267

In Exercises 13–16, find the volume of each figure.

13.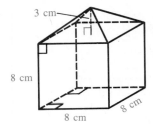

14.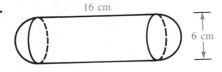

15.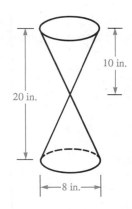

16.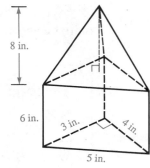

17. A sugar cube measures $\frac{1}{2}$ inch on each edge. How much space does one cube occupy?

18. A classroom is 40 feet long, 30 feet wide, and 9 feet high. Find the number of cubic feet contained in the room.

19. A cylindrical oil tank has a diameter of 6 feet and a length of 7 feet. Find the volume of the tank. Use $\pi \approx 3.14$.

20. A restaurant serves pudding in a conical dish that has a diameter of 3 inches. If the dish is 4 inches deep, how many cubic inches of pudding are in each dish? Use $\pi \approx 3.14$.

21. How much material is needed to manufacture a balloon 10 feet in diameter? (Assume no waste.) Use $\pi \approx 3.14$.

22. How many square yards of wood does Sonia need to build a cubical box that measures 6 feet on each side?

23. The lifting power of a spherical balloon depends on its volume. How many cubic feet of gas will a balloon hold if it is 40 feet in diameter? Use $\pi \approx 3.14$.

24. A box of cereal measures 3 by 8 by 10 inches. The manufacturer plans to market a smaller box that measures $2\frac{1}{2}$ by 7 by 8 inches. By how much will the volume be reduced?

25. Explain why the volume of a cube can be found by using the formula

$$V = s^3$$

where s is the length of one edge.

26. How many cubic inches are in one cubic foot?

27. How many cubic feet are in one cubic yard?

28. How many cubic inches are in one cubic yard?

Review Exercises

In Review Exercises 1–4, perform the indicated operations.

1. $4(6 + 4) - 2^2$ **2.** $5(5 - 2)^2 + 3$ **3.** $5 + 2(6 + 2^3)$ **4.** $3(6 + 3^4) - 2^4$

In Review Exercises 5–8, which of the listed quantities is the largest?

5. $0.34, \dfrac{3}{10}, \dfrac{4}{11}$ **6.** $4\dfrac{3}{5}, 4.61, 4\dfrac{13}{20}$

7. 5% of 42, 6% of 36   **8.** 6.5% of 50, 7.5% of 40

7.7 Inductive and Deductive Reasoning

- *Inductive Reasoning*
- *Deductive Reasoning*

The method used in a laboratory is to conduct an experiment and observe the outcome. After several repetitions and similar outcomes, the scientist will generalize the results into what seems to be a true statement:

- If I heat water to 212°, it will boil.
- If I throw the dish, it will break.
- If I touch the fire, I will be burned.

Inductive Reasoning

When drawing conclusions from specific observations, we are using **inductive reasoning**. The next examples show how inductive reasoning can be used in mathematical thinking.

EXAMPLE 1

Find the next number in the sequence 5, 8, 11, 14, ...

Solution

To discover a pattern in the sequence, we find the difference between each pair of successive numbers

$$8 - 5 = 3$$
$$11 - 8 = 3$$
$$14 - 11 = 3$$

Since the difference between each pair of numbers is 3, we see that there is an increasing pattern in the sequence: Each successive number is 3 greater than the previous one.

$$5$$
$$5 + 3 = 8$$
$$8 + 3 = 11$$
$$11 + 3 = 14$$

Thus, the next number in the pattern is 14 + 3, or 17.

EXAMPLE 2

Find the next number in the sequence 1, 4, 2, 5, 3, 6, 4, ...

Solution

Here, there is an **alternating pattern**: We add 3 to the first number to get the second. Then we subtract 2 to get the third. To get successive terms in the sequence, we alternately add 3 to one number and then subtract 2 from the next number.

$$1$$
$$1 + 3 = 4$$
$$4 - 2 = 2$$
$$2 + 3 = 5$$
$$5 - 2 = 3$$
$$3 + 3 = 6$$
$$6 - 2 = 4$$

Thus, the next number is 4 + 3, or 7. ∎

EXAMPLE 3

Find the next number in the sequence 1, 3, 6, 5, 7, 10, 9, 11, 14, 13, ...

Solution

Here, there is a **circular pattern**: We add 2 to the first number to get the second, add 3 to the second to get the third, and then subtract 1 from the third to get the fourth. To get subsequent numbers, this pattern continues.

$$1 + 2 = 3$$
$$3 + 3 = 6$$
$$6 - 1 = 5$$
$$5 + 2 = 7$$
$$7 + 3 = 10$$
$$10 - 1 = 9$$
$$9 + 2 = 11$$
$$11 + 3 = 14$$
$$14 - 1 = 13$$

Thus, the next number is 13 + 2, or 15. ∎

EXAMPLE 4

Find the next geometric shape in the sequence

 , , , ...

Solution

The first figure is a triangle with three sides and one dot, the second figure is a square with four sides and two dots, and the third figure is a pentagon with five sides and three dots. Thus, we would expect the next figure to be a hexagon with six sides and four dots. ∎

Deductive Reasoning

As opposed to inductive reasoning, **deductive reasoning** moves from the general case to the specific. For example, if we know that the sum of the angles in any triangle is 180°, we know that the sum of the angles of △ABC is 180°. Whenever we apply a general principle to a particular instance, we are using deductive reasoning. As in geometry, a deductive reasoning system is built on four items:

1. **Undefined terms**—terms that we accept without giving them formal meaning
2. **Defined terms**—terms that we define in a formal way
3. **Axioms** or **postulates**—statements that we accept without proof
4. **Theorems**—statements that we can prove with formal reasoning

Many problems can be solved by deductive reasoning. For example, suppose that we plan to enroll in an early morning algebra class and that we know Professors Perry, Miller, and Tveten are scheduled to teach algebra next semester. After some investigating, we find out that Professor Perry teaches only in the afternoon, and Professor Tveten teaches only in the evenings. Without knowing anything about Professor Miller, we can conclude that he will be our teacher, since he is the only remaining possibility.

The following example shows how to use deductive reasoning to solve problems.

EXAMPLE 5

Four professors are scheduled to teach one course each, calculus, statistics, algebra, or trigonometry. They have the following preferences:

1. Professors A and B don't want to teach calculus.
2. Professor C wants to teach statistics.
3. Professor B wants to teach algebra.

Who will teach trigonometry?

Solution

We organize the facts in a chart or a diagram such as the following

A	A	A	A
B	B	B	B
C	C	C	C
D	D	D	D
Calculus	*Algebra*	*Statistics*	*Trigonometry*

Since Professors A and B don't want to teach calculus, we can cross them off the calculus list. Since Professor C wants to teach statistics, we can cross him off of every other list. This leaves Professor D as the only person to teach calculus, so we can cross her off of every other list. Since Professor B wants to teach algebra, we can cross him off of every other list. Thus, the only remaining person left to teach trigonometry is Professor A.

A	A	A	A
B	B	B	B
C	C	C	C
D	D	D	D
Calculus	Algebra	Statistics	Trigonometry

7.7 Exercises

In Exercises 1–12, find the number that comes next in each sequence.

1. 1, 5, 9, 13, . . .
2. 15, 12, 9, 6, . . .
3. 3, 5, 8, 12, . . .
4. 5, 9, 14, 20, . . .
5. 7, 9, 6, 8, 5, 7, 4, . . .
6. 2, 5, 3, 6, 4, 7, 5, . . .
7. 9, 5, 7, 3, 5, 1, . . .
8. 13, 16, 14, 17, 15, 18, . . .
9. 2, 3, 5, 6, 8, 9, . . .
10. 8, 11, 9, 12, 10, 13, . . .
11. 6, 8, 9, 7, 9, 10, 8, 10, 11, . . .
12. 10, 8, 7, 11, 9, 8, 12, 10, 9, . . .

In Exercises 13–14, find the figure that comes next in each sequence.

13. 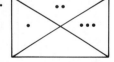 , , 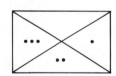 , 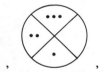 , . . .

14. , , , , 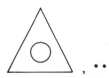 , . . .

In Exercises 15–16, find the missing figure in each sequence.

15. , , , ,

16. , , , ,

In Exercises 17–21, what conclusion(s) can be drawn from each set of information?

17. Four people named John, Luis, Maria, and Paula have occupations as teacher, butcher, baker, and candlestick maker.
  1. John and Paula are married.
  2. The teacher plans to marry the baker in December.
  3. Luis is the baker.

  Who is the teacher?

18. In a zoo, a zebra, a tiger, a lion, and a monkey are to be placed in four cages numbered from 1 to 4, from left to right. The following decisions have been made:
  1. The lion and the tiger should not be side-by-side.
  2. The monkey should be in one of the end cages.
  3. The tiger is to be in cage 4.

  In which cage is the zebra?

19. A Ford, a Buick, a Dodge, and a Mercedes are parked side-by-side.
  1. The Ford is between the Mercedes and the Dodge.
  2. The Mercedes is not next to the Buick.
  3. The Buick is parked on the left end.

  Which car is parked on the right end?

20. Four divers at the Olympics finished first, second, third, and fourth.
  1. Diver A beat diver B.
  2. Diver C placed between divers B and D.
  3. Diver B beat diver D.

  In which order did they finish?

21. A green, a blue, a red, and a yellow flag are hanging on a flagpole.
  1. The blue flag is between the green and yellow flags.
  2. The red flag is next to the yellow flag.
  3. The green flag is above the red flag.

  What is the order of the flags, from top to bottom?

Exercise 22 is much more difficult. It is presented for its entertainment value.

22. Tom, Dick, and Harry each have two occupations: bootlegger, musician, painter, chauffeur, barber, and gardener. From the following facts, find the occupations of each man.
  1. The painter bought a quart of spirits from the bootlegger.
  2. The chauffeur offended the musician by laughing at his mustache.
  3. The chauffeur dated the painter's sister.
  4. Both the musician and the gardener used to go hunting with Tom.
  5. Harry beat both Dick and the painter at Monopoly.
  6. Dick owes the gardener $100.

Mathematics in Merchandising

Refer to the problem on the first page of this chapter. Because both the medium and the large pizzas are the same thickness, we will calculate their areas, and then their **unit cost**, or their cost per square inch. The smaller unit cost will be the better deal. Recall that the area A of a circle of radius r is given by the formula $A = \pi r^2$.

Unit cost, M, of two 11-inch pizzas

$$M = \frac{\text{cost of two pizzas}}{\text{area of two pizzas}}$$

$$= \frac{12}{2\pi r^2}$$

$$= \frac{6}{\pi(5.5)^2} \quad \text{Simplify and substitute one-half of 11 for } r.$$

$$\approx 0.063 \quad \text{Use a calculator.}$$

Unit cost, L, of one 14-inch pizza

$$L = \frac{\text{cost of one pizza}}{\text{area of one pizza}}$$

$$= \frac{10}{\pi r^2}$$

$$= \frac{10}{\pi(7)^2} \quad \text{Substitute one-half of 14 for } r.$$

$$\approx 0.065 \quad \text{Use a calculator.}$$

Because the unit cost of two medium pizzas is slightly less than that of one large, two mediums are a better deal—but not by much!

Chapter Summary

Key Words

altitude (7.3)
area (7.3)
base (7.3)
center (7.6)
circumference (7.5)
cone (7.6)
cube (7.6)
cylinder (7.6)
deductive reasoning (7.7)

extremes of a
   proportion (7.1)
height (7.3)
hemisphere (7.3)
inductive reasoning (7.7)
leg (7.4)
length (7.3)
means of a
   proportion (7.1)

pi (7.5)
plane figures (7.6)
prism (7.6)
proportion (7.1)
proportional (7.1)
pyramid (7.6)
radius (7.6)
rectangular solid (7.6)
ratio (7.1)

similar triangles (7.2)
slant height (7.6)
sphere (7.6)
surface area (7.6)
three-dimensional
   figure (7.6)
trapezoid (7.4)
volume (7.6)
width (7.3)

Key Ideas

(7.1) In a proportion, the product of the means is equal to the product of the extremes.

(7.2) Two triangles are similar if, and only if, two angles of one triangle are congruent to two angles of the other triangle.

(7.3) **Area of a rectangle:** $A = lw$

Area of a parallelogram: $A = bh$

(7.4) **Area of a triangle:** $A = \frac{1}{2}bh$

Area of a trapezoid: $A = \frac{1}{2}h(b + b')$

(7.5) **Circumference of a circle:**

$$C = \pi D \quad \text{or} \quad C = 2\pi r$$

Area of a circle: $A = \pi r^2$

(7.6)

Figure	Surface Area	Volume
Rectangular solid	$S = 2lw + 2wh + 2lh$	$V = lwh$
Prism	Sum of areas of faces	$V = Bh$*
Sphere	$S = 4\pi r^2$	$V = \frac{4}{3}\pi r^3$
Cylinder	$S = 2\pi rh + 2\pi r^2$	$V = \pi r^2 h$
Cone	$S = \pi rs + \pi r^2$	$V = \frac{1}{3}\pi r^2 h$
Pyramid	Sum of areas of faces	$V = \frac{1}{3}Bh$

*B represents the area of the base.

Chapter 7 Review Exercises

[7.1] In Review Exercises 1–2, express each phrase as a ratio in lowest terms.

1. 5 liters to 12 liters

2. 7 ounces to 2 pounds

In Review Exercises 3–4, tell whether each expression is a proportion.

3. $\dfrac{5}{6} = \dfrac{30}{38}$

4. $\dfrac{30}{48} = \dfrac{22.4}{27.2}$

In Review Exercises 5–8, solve each proportion.

5. $\dfrac{x}{4} = \dfrac{6}{8}$

6. $\dfrac{5}{x} = \dfrac{12}{17}$

7. $\dfrac{15}{8} = \dfrac{y}{4}$

8. $\dfrac{9}{5} = \dfrac{18}{y}$

[7.2] In Review Exercises 9–10, refer to Illustration 1.

9. Prove that $\triangle ABC \sim \triangle DEC$.

10. m($\overline{DE}$) =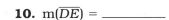

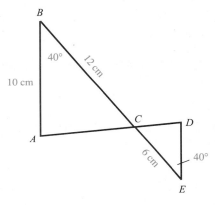

Illustration 1

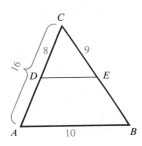

Illustration 2

In Review Exercises 11–14, refer to Illustration 2, in which $\overline{DE} \parallel \overline{AB}$.

11. m($\overline{BC}$) = _____

12. m($\overline{EB}$) = _____

13. m($\overline{DE}$) = _____

14. Find the perimeter of $\triangle ABC$.

In Review Exercises 15–18, refer to Illustration 3.

15. m($\angle 2$) = _____

16. m($\angle CDB$) = _____

17. m($\angle 1$) = _____

18. m($\angle B$) = _____

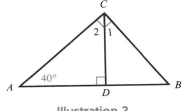

Illustration 3

19. A building casts a shadow of 75 feet at the same time a 6-foot person casts a shadow of 5 feet. How tall is the building?

20. If a 22-foot ladder safely reaches a window 16 feet above the ground, how high will a 30-foot ladder reach? Assume the same angle with the horizontal.

21. A 15-foot ladder rests against the side of a building. The base of the ladder is 9 feet from the wall. If a painter climbs 10 feet up the ladder, how far is he from the wall?

22. The top of a 20-foot slide is 12 feet above the ground. If a girl slides 5 feet down the slide and stops, how far is she above the ground?

[7.3, 7.4, 7.5] In Review Exercises 23–34, solve each problem.

23. Find the area of a square with a side 7 inches long.

24. Find the area of a square with a side 24.3 centimeters long.

25. Find the area of a rectangle with dimensions 12 by 15 feet.

26. Find the area of a rectangle 12 meters long and 13 meters wide.

27. Find the area of a parallelogram with a base of 20 centimeters and a height of 12 centimeters.

28. Find the area of a triangle with a base of 10 inches and a height of 16 inches.

29. Find the area of an isosceles triangle with a base of 25 inches and a height of 12.2 inches.

30. Find the area of a trapezoid with bases measuring 10 and 14 meters and a height measuring 6 meters.

31. Find the circumference of a circle that is 36 feet in diameter. Use $\pi \approx 3.14$ and give the answer to the nearest hundredth.

32. Find the radius of a circle whose circumference is 36π feet.

33. Find the area of a circle with a radius of 7.5 centimeters.

34. Find the diameter of a circle whose area is 25π cm^2.

[7.6] In Review Exercises 35–44, solve each problem.

35. Find the surface area of a cube with an edge of 15 centimeters.

36. Find the volume of a rectangular solid with dimensions of 5 by 7 by 9 meters.

37. Find the volume of a prism that is 12 inches tall and whose base is an equilateral triangle with a height of 6.06 inches and a perimeter of 21 inches.

38. Find the surface area of a sphere with a diameter of 10 meters.

39. Find the volume of a sphere with a diameter of 6 meters.

40. Find the area of a cylinder with a radius of 3 centimeters and a height of 12 centimeters.

41. Find the volume of a cylinder with a diameter of 4 inches and a height of 1 foot.

42. Find the surface area of a cone with a base that is 4 feet in diameter and a slant height of 8 feet.

43. Find the volume of a 6-inch-tall cone with a base 2 inches in diameter.

44. Find the volume of a pyramid that is 100 feet high and whose base is a rectangle with a length of 50 feet and a width of 40 feet.

[7.7] In Review Exercises 45–50, answer each question.

45. Find the next number in the sequence: 12, 8, 11, 7, 10, ...

46. Find the next number in the sequence: 5, 9, 17, 33, ...

47. Find the missing geometric figure in the sequence:

 , , , , , ?

48. Find the missing geometric figure in the sequence:

 , , ,

49. Four animals—a cow, a horse, a pig, and a sheep—are kept in a barn, each in a separate stall.
  1. The cow is in the first stall.
  2. The pig is between the horse and the sheep.
  3. The sheep cannot be next to the cow.

  What animal is in the last stall?

50. George, Sandra, Juan, and Mary all teach in the same college. One teaches mathematics, one teaches English, one teaches history, and one teaches music.
  1. George and Sandra eat lunch with the math teacher.
  2. Juan and Sandra carpool with the English teacher.
  3. George is married to the math teacher.
  4. Mary works in the same building as the history teacher.

  Who is the math teacher?

Name _____ Section _____

Chapter 7 Test

1. Write the expression *7 liters to 21 liters* as a ratio in lowest terms. _____

2. Write the expression *3 centimeters to 1 meter* as a ratio in lowest terms. _____

3. Is the expression $\dfrac{9}{13} = \dfrac{36}{52}$ a proportion? _____

4. Is the expression $\dfrac{2\frac{1}{2}}{1\frac{1}{2}} = \dfrac{10}{6}$ a proportion? _____

5. Solve the proportion $\dfrac{24}{y} = \dfrac{4.8}{3.6}$. _____

6. Solve the proportion $\dfrac{8}{24} = \dfrac{x}{96}$. _____

In Problems 7–8, refer to Illustration 1, in which m(∠A) = m(∠D) and m(∠C) = m(∠F).

7. Find x. _____

8. Find y. _____

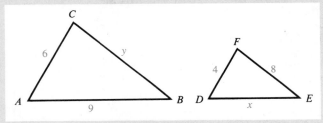

Illustration 1

In Problems 9–11, refer to Illustration 2, in which $\overline{DE} \parallel \overline{AB}$.

9. Are △ABC and △EDC similar? _____

10. Find m($\overline{AB}$). _____

11. Find m($\overline{EC}$). _____

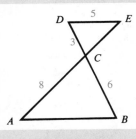

Illustration 2

In Problems 12–14, refer to Illustration 3, in which $\overline{DE} \parallel \overline{AB}$.

12. If m($\overline{DC}$) = 5, m($\overline{AC}$) = 7, and m($\overline{EC}$) = 7, find m($\overline{BC}$). _____

13. If m($\overline{DC}$) = 5, m($\overline{AD}$) = 5, and m($\overline{DE}$) = 6, find m($\overline{AB}$). _____

14. If m($\overline{AC}$) = 12, m($\overline{DC}$) = 4, and m($\overline{CB}$) = 9, find m($\overline{EB}$). _____

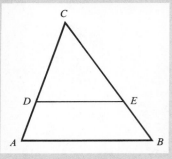

Illustration 3

15. A telephone pole casts a shadow of 36 feet at the same time a yardstick casts a shadow of 8 feet. How tall is the pole? _____

16. A skier drops $\frac{1}{10}$ mile in altitude while skiing $\frac{1}{4}$ mile. How far must the person ski to drop 1 mile in altitude? _____

17. Find the area of a rectangle with dimensions of 8 by 12 feet. _____

18. Find the area of a right triangle with legs measuring 8 inches and 12 inches. _____

19. Find the area of a trapezoid with bases 12 and 15 feet long and a height of 6 feet. _____

20. Find the area of a circle with a diameter of 6 feet. _____

21. Find the volume of a rectangular solid with dimensions 4 by 5 by 6 meters. _____

22. Find the volume of a sphere that is 8 meters in diameter.

23. Find the surface area of a cone that has a base 6 meters in diameter and a slant height of 4 meters.

24. Find the volume of a 10-foot-tall pyramid with a rectangular base 5 feet long and 4 feet wide.

25. Find the next number in the sequence: 10, 7, 9, 6, 8, ...

26. Find the next number in the sequence: 2, 5, 9, 14, ...

27. Find the missing geometric figure in the sequence

□ , △ , ○ , △ , □ , ? , ○

28. Jim, Carlos, Sue, and Lisa are officers of the mathematics club at Pan American University.

   1. Jim and Carlos sing in the choir with the president of the club.
   2. Sue and Lisa are in the same history class as the vice president.
   3. Jim keeps the books, and Lisa keeps the minutes.

Who is the president?

8 Real Numbers and Their Properties

Mathematics for Fun

Three friends chipped in $10 each to share the cost of a gift. After they paid $30, the store owner realized that the gift was on sale for $25. Being honest, the owner said to the clerk, "Here is $5. Find the customers, and give them their money." The dishonest clerk thought, "I can't split $5 among three people. I'll give them one dollar each, and keep two for myself." Thus, the friends paid $27, and the clerk kept $2. However, $27 + $2 is just $29. What happened to the other dollar?

One of the properties of the real numbers will help us explain the missing dollar.

Algebra is an extension of arithmetic. Its origins are found in a papyrus written before 1000 B.C. by an Egyptian priest named Ahmes.

In A.D. 830, one of the greatest mathematicians of Arabian history, al-Khowarazmi, wrote a book called *Ihm al-jabr wa'l muqabalah*, soon shortened to *al-Jabr*. We now know the subject as *algebra*. The French mathematician François Viète (1540–1603) simplified algebra by developing the symbolic notation that we use today.

We begin the study of algebra by discussing the properties of numbers.

8.1 Sets of Numbers and Their Graphs

- Sets of Numbers
- Graphing Sets of Numbers
- Equality and Inequality Symbols

Sets of Numbers

The most basic **set** of numbers is the set of **natural numbers**.

> **Definition:** The **natural numbers** are the numbers
>
> $1, 2, 3, 4, 5, 6, 7, 8, 9, 10, \ldots$

The three dots, called an **ellipsis**, in the definition above indicate that the list continues forever. There is no largest natural number.

The natural numbers that can be divided by 2 are the **even natural numbers**:

2, 4, 6, 8, 10, 12, 14, 16, 18, 20, ...

The natural numbers that cannot be divided by 2 are the **odd natural numbers**:

1, 3, 5, 7, 9, 11, 13, 15, 17, 19, ...

Natural numbers with exactly two divisors are the **prime numbers**.

> **Definition:** A **prime number** is a natural number that is larger than 1 and is exactly divisible only by 1 and by itself.

The prime numbers are the numbers

2, 3, 5, 7, 11, 13, 17, 19, 23, 29, 31, ...

The only even prime is 2. All other primes are odd natural numbers.

Natural numbers with more than two divisors are the **composite numbers.**

> **Definition:** A **composite number** is a natural number greater than 1 that is not a prime number.

The composite numbers are the numbers

4, 6, 8, 9, 10, 12, 14, 15, 16, 18, 20, 21, ...

The natural number 1 is neither a prime number nor a composite number.

The natural numbers together with 0 form the set of **whole numbers**:

> **Definition:** The **whole numbers** are the numbers
>
> 0, 1, 2, 3, 4, 5, 6, 7, 8, 9, 10, ...

Braces are often used to enclose the **elements** (or members) of a set. Each pair of braces is read as "the set of."

The set of natural numbers is $\{1, 2, 3, 4, 5, 6, \ldots\}$.

The set of prime numbers is $\{2, 3, 5, 7, 11, 13, \ldots\}$.

The set of composite numbers is $\{4, 6, 8, 9, 10, 12, \ldots\}$.

The set of whole numbers is $\{0, 1, 2, 3, 4, 5, \ldots\}$.

The number 3 is an element of the set of natural numbers, an element of the set of prime numbers, and an element of the set of whole numbers. However, 3 is *not* an element of the set of composite numbers.

EXAMPLE 1

Classify **a.** 7, **b.** 6, and **c.** 0 as a natural number, even natural number, odd natural number, prime number, composite number, and whole number.

Solution

a. The number 7 is a natural number, an odd natural number, a prime number, and a whole number.

b. The number 6 is a natural number, an even natural number, a composite number, and a whole number.

c. The number 0 is a whole number.

Work Progress Check 1.

Progress Check 1

Consider the set $\{0, 1, 2, 5, 6, 9, 11\}$. List the numbers that are

a. even natural numbers

b. prime numbers

c. composite numbers

Graphing Sets of Numbers

Sets of numbers can be pictured on the **number line**. To construct the number line shown in Figure 8-1, we pick a point on the line, label it 0, and call it the **origin**. We then pick a unit length, mark off points to the right of the origin, and label these points with the natural numbers. The point labeled 6, for example, is 6 units to the right of the origin.

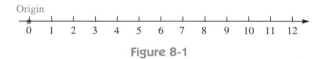

Figure 8-1

Figure 8-2 shows the **graph** of the even natural numbers from 2 to 8. The dot corresponding to the number 4, for example, is called the **graph of 4**. The number 4 is called the **coordinate** of its corresponding point.

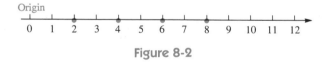

Figure 8-2

EXAMPLE 2

Graph the set of prime numbers between 1 and 20.

Solution

The prime numbers between 1 and 20 are 2, 3, 5, 7, 11, 13, 17, and 19. See Figure 8-3. Work Progress Check 2.

Figure 8-3

Progress Check 2

Graph the composite numbers from 8 to 13 on the number line.

1. a. 2, 6 **b.** 2, 5, 11 **c.** 6, 9

2.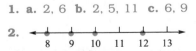

Equality and Inequality Symbols

The **equal sign**, written as =, shows that two expressions represent the same number. Because 4 + 5 and 9 represent the same number,

$4 + 5 = 9$ Read as "The sum of 4 and 5 is 9."

Likewise,

$5 - 3 = 2$ Read as "The difference between 5 and 3 is 2."

$4 \cdot 5 = 20$ Read as "The product of 4 and 5 is 20."

and

$30 \div 6 = 5$ Read as "The quotient obtained when 30 is divided by 6 is 5."

We use **inequality symbols** to show that expressions are *not* equal.

Symbol	Read as
$\neq$	"is not equal to"
$<$	"is less than"
$>$	"is greater than"
$\leq$	"is less than or equal to"
$\geq$	"is greater than or equal to"

EXAMPLE 3

a. $6 \neq 9$ is read as "6 is not equal to 9."

b. $8 < 10$ is read as "8 is less than 10."

c. $12 > 1$ is read as "12 is greater than 1."

d. $5 \leq 5$ is read as "5 is less than or equal to 5." (Because 5 is equal to 5, this is a true statement.)

e. $9 \geq 7$ is read as "9 is greater than or equal to 7." (Because 9 is greater than 7, this is a true statement.)

Statements of inequality can be written so that the inequality symbol points in the opposite direction. For example,

$5 < 7$

is read as "5 is less than 7," and

$7 > 5$

is read as "7 is greater than 5." Both show that 5 is a smaller number than 7. Likewise,

$12 \geq 3$ Read as "12 is greater than or equal to 3."

and

$3 \leq 12$ Read as "3 is less than or equal to 12."

are equivalent statements.

If one point is to the *right* of a second point on the number line, its coordinate is the *greater*. The point with coordinate 5 in Figure 8-3 lies to the right of the point with coordinate 2. Thus, $5 > 2$.

If one point is to the *left* of another, its coordinate is the *smaller*. The point with coordinate 11 is to the left of the point with coordinate 19. Thus, $11 < 19$.

8.1 Exercises

In Exercises 1–8, consider the set {0, 1, 2, 3, 4, 9, 18}. List the numbers that are in each designated set.

1. natural number
2. even natural number
3. odd natural number
4. prime number
5. whole number
6. composite number
7. even and prime
8. odd and composite

In Exercises 9–16, consider the set {0, 1, 2, 5, 10, 15}. List the numbers that are in each designated set.

9. prime number
10. whole number
11. even natural number
12. natural number
13. even and composite
14. odd natural number
15. composite number
16. odd and prime

In Exercises 17–26, graph each set of numbers on the number line.

17. The natural numbers between 2 and 8
18. The prime numbers from 10 to 20
19. The composite numbers from 20 to 26
20. The whole numbers less than 6
21. The even natural numbers greater than 10 but less than 20
22. The even natural numbers that are also prime numbers
23. The numbers that are whole numbers but not natural numbers
24. The prime numbers between 20 and 30
25. The natural numbers between 11 and 25 that are exactly divisible by 6
26. The odd natural numbers between 14 and 24 that are exactly divisible by 3

In Exercises 27–40, place one of the symbols =, <, or > in the box to make a true statement.

27. 5 ☐ 3 + 2
28. 9 ☐ 7
29. 25 ☐ 32
30. 2 + 3 ☐ 17
31. 5 + 7 ☐ 10
32. 3 + 3 ☐ 9 − 3
33. 3 + 2 + 5 ☐ 5 + 2 + 3
34. 8 − 5 ☐ 5 − 2
35. 3 + 9 ☐ 20 − 8
36. 19 − 3 ☐ 8 + 6
37. 4 · 2 ☐ 2 · 4
38. 7 · 9 ☐ 9 · 6
39. 8 ÷ 2 ☐ 4 + 2
40. 0 ÷ 7 ☐ 1

In Exercises 41–46, write each statement as a mathematical expression.

41. Seven is greater than three.

42. Five is less than thirty-two.

43. Seventeen is less than or equal to seventeen

44. Twenty-five is not equal to twenty-three.

45. The result of adding three and four is equal to seven.

46. Thirty-seven is greater than or equal to the result of multiplying three and four.

In Exercises 47–56, rewrite each inequality statement as an equivalent inequality in which the inequality symbol points in the opposite direction.

47. $3 \leq 7$

48. $5 > 2$

49. $6 > 0$

50. $34 \leq 40$

51. $3 + 8 > 8$

52. $8 - 3 < 8$

53. $6 - 2 < 10 - 4$

54. $8 \cdot 2 \geq 8 \cdot 1$

55. $8 \div 2 \geq 9 \div 3$

56. $12 \div 4 < 24 \div 6$

In Exercises 57–62, graph each pair of numbers on a number line. Indicate which number in the pair is the greater and which number lies to the right of the other number on the number line.

57. 3, 6

58. 4, 7

59. 11, 6

60. 12, 10

61. 0, 2

62. 4, 10

63. Explain why there is no greatest natural number.

64. Explain why 2 is the only even prime number.

65. Explain why no natural numbers are both even natural numbers and also odd natural numbers.

66. Find the only natural number that is neither a prime number nor a composite number.

Review Exercises

In Review Exercises 1–8, perform the indicated operations.

1. Add: 132
   45
   73

2. Add: 261
   79
   31

3. Subtract: 321
   173

4. Subtract: 532
   437

5. Multiply: 437
   38

6. Multiply: 529
   42

7. Divide: $37 \overline{)3885}$

8. Divide: $53 \overline{)11607}$

In Review Exercises 9–10, perform the indicated operations.

9. $235 + 517 - 26$

10. $135 + 22 - 156$

11. Sally bought four paintings for $350, $900, $820, and $1000. Later, she sold them for $750, $850, $1250, and $2700, respectively. What was her total profit?

12. An airplane is cruising at 29,000 feet. The pilot receives permission to descend 7000 feet and later to climb 11,000 feet. At what altitude is the plane now cruising?

8.2 Variables and Algebraic Terms

- Algebraic Expressions
- Forming Algebraic Expressions
- Evaluating Algebraic Expressions
- Algebraic Terms

Algebraic Expressions

Variables and numbers can be combined to produce **algebraic expressions**. For example, if x and y are variables, the algebraic expression $x + y$ represents the **sum** of x and y, and the algebraic expression $x - y$ represents their **difference**.

There are many ways to read the sum $x + y$. Some of them are

- The sum of x and y
- x increased by y
- x plus y
- y more than x
- y added to x

There are many ways to read the difference $x - y$. Some of them are

- x less y
- y less than x
- x decreased by y
- x minus y
- The result obtained when y is subtracted from x
- The result of subtracting y from x

Forming Algebraic Expressions

EXAMPLE 1

Let x represent a number. Write an expression that represents **a.** the number that is 5 more than x and **b.** the number 12 decreased by x.

Solution

a. The number "5 more than x" is the number found by adding 5 to x. It is represented by $x + 5$.

b. The number "12 decreased by x" is the number found by subtracting x from 12. It is represented by $12 - x$. ∎

Because the times sign "×" looks like x, it is seldom used in algebra. Instead, a dot, parentheses, or no symbol at all is used to denote multiplication. Each of the following expressions indicates the **product** of x and y.

$$x \cdot y \quad (x)(y) \quad x(y) \quad (x)y \quad xy$$

There are many ways to indicate the product xy in words:

- x multiplied by y
- the product of x and y
- x times y

EXAMPLE 2

Let x represent a number. Find a number that is **a.** twice as large as x, **b.** 5 more than 3 times x, and **c.** 4 less than $\frac{1}{2}$ of x.

Solution

a. The number "twice as large as x" is found by multiplying x by 2. It is represented by $2x$.

b. The number "5 more than 3 times x" is found by adding 5 to the product of 3 and x. It is represented by $3x + 5$.

c. The number "4 less than $\frac{1}{2}$ of x" is found by subtracting 4 from the product of $\frac{1}{2}$ and x. It is represented by $\frac{1}{2}x - 4$. ∎

EXAMPLE 3

Jim has x dimes, y nickels, and 3 quarters. **a.** How many coins does Jim have? **b.** What is the value of the coins?

Solution

a. Because there are x dimes, y nickels, and 3 quarters, there are $x + y + 3$ coins.

b. The value of x dimes is $10x$ cents, the value of y nickels is $5y$ cents, and the value of 3 quarters is $3 \cdot 25$, or 75 cents. The total value of the coins is $(10x + 5y + 75)$ cents.

Work Progress Check 3. ∎

Progress Check 3

If x is a number, write an expression that represents 3 less than twice x.

If x and y represent two numbers and $y \neq 0$, the **quotient** obtained when x is divided by y is denoted by each of the following expressions:

$$x \div y, \quad x/y, \quad \text{and} \quad \frac{x}{y}$$

EXAMPLE 4

Let x and y represent two numbers. Write an algebraic expression that represents the sum obtained when 3 times the first number is added to the quotient obtained when the second number is divided by 6.

Solution

Three times the first number x is denoted as $3x$. The quotient obtained when the second number y is divided by 6 is the fraction $\frac{y}{6}$. Their sum is expressed as $3x + \frac{y}{6}$. ∎

EXAMPLE 5

A 5-foot section is cut from the end of a rope that is l feet long. The remaining rope is then divided into three equal pieces. How long will each equal piece be?

Solution

After a 5-foot section has been cut from one end of l feet of rope, the rope that remains is $(l - 5)$ feet long. When that remaining rope has been cut into three equal pieces, each piece will be $\dfrac{l - 5}{3}$ feet long. ∎

3. $2x - 3$

Evaluating Algebraic Expressions

We can evaluate an algebraic expression if we know the numbers that its variables represent. For example, if x represents the number 12 and y represents the number 5, we can evaluate the expression $x + y + 2$:

$x + y + 2 = 12 + 5 + 2$ Substitute 12 for x and 5 for y.
$ = 19$

EXAMPLE 6

If x represents 8 and y represents 10, evaluate the expressions **a.** $x + y$, **b.** $y - x$, **c.** $3xy$, and **d.** $\dfrac{5x}{y - 5}$.

Solution

We substitute 8 for x and 10 for y in each expression and simplify.

a. $x + y = 8 + 10$
$ = 18$

b. $y - x = 10 - 8$
$ = 2$

c. $3xy = (3)(8)(10)$
$ = (24)(10)$ Do the multiplications from left to right.
$ = 240$

d. $\dfrac{5x}{y - 5} = \dfrac{5 \cdot 8}{10 - 5}$
$\phantom{\dfrac{5x}{y - 5}} = \dfrac{40}{5}$ Simplify the numerator and denominator separately.
$\phantom{\dfrac{5x}{y - 5}} = 8$ Simplify the fraction.

After numbers are substituted for the variables in a product, a dot or parentheses are often needed to indicate the multiplication. This is to ensure that (3)(8)(10), for example, is not mistaken for 3810 and that $5 \cdot 8$ is not mistaken for 58. Work Progress Check 4.

Progress Check 4

If $x = 10$ and $y = 8$, evaluate $\dfrac{4x}{x - y}$.

Algebraic Terms

Numbers without variables, such as 7, 21, and 23, are called **constants**. Expressions with numbers and/or variables such as 37, xyz, or $32t$ are called **algebraic terms**.

Because $3x + 5y$ denotes the sum of two algebraic terms, the expression $3x + 5y$ contains two terms. Likewise, $xy - 7$ contains two terms. The expression $3 + x + 2y$ contains three terms. Its first term is 3, its second term is x, and its third term is $2y$.

Numbers that are part of an indicated product are called **factors**. For example, the product $7x$ has two factors, 7 and x. Either factor is called the **coefficient** of the other. When we speak of the coefficient in a term such as $7x$, however, we generally mean the **numerical coefficient**, which is 7. For example, the numerical coefficient of the term $12xyz$ is 12. The coefficient of such terms as x, ab, or rst is understood to be 1. Thus,

$x = 1x$, $ab = 1ab$,   and $rst = 1rst$

4. 20

Progress Check Answers

EXAMPLE 7

a. The expression $5x + y$ has two terms. The numerical coefficient of its first term is 5. The numerical coefficient of its second term is 1.

b. The expression $17wxyz$ has one term, which contains the five factors 17, w, x, y, and z. Its numerical coefficient is 17.

c. The expression 37 has the one term 37. Its numerical coefficient is 37.

8.2 Exercises

In Exercises 1–18, let x, y, and z represent three numbers. Write an algebraic expression to denote each quantity.

1. The sum of x and y

2. The product of x and y

3. The product of x and twice y

4. The sum of twice x and twice y

5. The difference obtained when x is subtracted from y

6. The difference obtained when twice x is subtracted from y

7. The quotient obtained when y is divided by x

8. The quotient obtained when the sum of x and y is divided by z

9. The sum obtained when the quotient of x divided by y is added to z

10. y decreased by x

11. z less the product of x and y

12. z less than the product of x and y

13. The product of 3, x, and y

14. The quotient obtained when the product of 3 and z is divided by the product of 4 and x

15. The quotient obtained when the sum of x and y is divided by the sum of y and z

16. The quotient obtained when the product of x and y is divided by the sum of x and z

17. The sum of the product xy and the quotient obtained when y is divided by z

18. The number obtained when x decreased by 4 is divided by the product of 3 and the variable y

19. George is enrolled in college for c hours of credit, and Pam is enrolled for four more hours than George. Write an expression that represents the number of hours Pam is taking.

20. Sam's car has 25,000 more miles on its odometer than does his father's new car. If his father's car has traveled m miles, what expression represents the mileage on Sam's car?

21. Write an expression that represents the value of t pencils, each worth 22 cents.

22. Write an expression that represents the value of a pounds of candy worth 95 cents per pound.

23. A rope x feet long is cut into five equal pieces. How long is each piece?

24. A rope 18 feet long is cut into x equal pieces. How long is each piece?

25. Kris has d dollars, and her brother Steven has five dollars more than three times that amount. Write an expression that represents the amount Steven has.

26. Wendy is now x years old. Her sister Bonnie is two years less than twice Wendy's age. Write an expression that represents Bonnie's age.

In Exercises 27–38, write each algebraic expression as an appropriate English phrase.

27. $x + 3$

28. $y - 2$

29. $\dfrac{x}{y}$

30. xz

31. $2xy$

32. $\dfrac{2}{x} + y$

33. $\dfrac{2}{x+y}$

34. $\dfrac{3x}{y+z}$

35. $\dfrac{3+x}{y}$

36. $3 + \dfrac{x}{y}$

37. $\dfrac{xy}{x+y}$

38. $\dfrac{x+y+z}{xyz}$

In Exercises 39–50, let x represent 8, y represent 4, and z represent 2. Evaluate each algebraic expression.

39. $x + z$

40. xyz

41. $y - z$

42. $\dfrac{y}{z}$

43. $3yz$

44. $7xy$

45. $\dfrac{3xy}{z}$

46. $\dfrac{10z}{4}$

47. $\dfrac{x+y+z}{7z}$

48. $\dfrac{y+x+x}{x+z}$

49. $\dfrac{8y}{y-z}$

50. $\dfrac{3z}{x-y}$

In Exercises 51–60, state the number of terms in each algebraic expression and also state the numerical coefficient of the first term.

51. $6d$

52. $4c + 3d$

53. $xy - 4t + 35$

54. xy

55. $3ab + bc - cd - ef$

56. $2xyz + cde - 14$

57. $4xyz + 7xy - z$

58. $5uvw - 4uv + 8uw$

59. $3x + 4y + 2z + 2$

60. $7abc - 9ab + 2bc + a - 1$

In Exercises 61–64, consider the algebraic expression $29xyz + 23xy + 19x$.

61. What are the factors of the third term?

62. What are the factors of the second term?

63. What are the factors of the first term?

64. What factor is common to all three terms?

In Exercises 65–68, consider the algebraic expression $3xyz + 2 \cdot 3xy + 2 \cdot 3 \cdot 3xz$.

65. What are the factors of the first term?

66. What are the factors of the second term?

67. What are the factors of the third term?

68. What factors are common to all three terms?

In Exercises 69–72, consider the algebraic expression $5xy + xt + 8xyt$.

69. Determine the numerical coefficients of each term.

70. What factor is common to all three terms?

71. What factors are common to the first and third terms?

72. What factors are common to the second and third terms?

In Exercises 73–76, consider the algebraic expression $3xy + y + 25xyz$.

73. Determine the numerical coefficient of each term and find their product.

74. Determine the numerical coefficient of each term and find their sum.

75. What factors are common to the first and the third terms?

76. What factor is common to all three terms?

Review Exercises

In Review Exercises 1–4, perform the indicated operation and classify the result into the categories of even natural number, odd natural number, and prime number.

1. $7 - 6$
2. $32 + 0$
3. $\dfrac{31}{7} + \dfrac{18}{7}$
4. $\dfrac{14}{5} \cdot \dfrac{25}{7}$

5. Simplify the fraction $\dfrac{15}{75}$.

6. Simplify the fraction $\dfrac{14}{105}$.

7. Write the number 150 in prime-factored form.

8. Write the number 165 in prime-factored form.

In Review Exercises 9–12, perform the indicated operation and simplify if possible.

9. $\dfrac{52}{7} \cdot \dfrac{14}{13}$
10. $\dfrac{25}{12} \div 15$
11. $\dfrac{13}{6} - \dfrac{7}{5}$
12. $\dfrac{25}{15} + 3$

8.3 Exponents and Order of Operations

- Exponents
- Order of Operations

The number 5 in the product 5(4) indicates that 4 is to be used five times in a sum:

$$5(4) = 4 + 4 + 4 + 4 + 4 = 20$$

The algebraic term $5x$ indicates that x is to be used five times in a sum:

$$5x = x + x + x + x + x$$

Exponents

To show how many times a number is to be used as a *factor* in a product, we use an **exponent**. For example, the 5 in x^5 indicates that x is to be used as a factor five times:

$$x^5 = \overbrace{x \cdot x \cdot x \cdot x \cdot x}^{5 \text{ factors of } x}$$

In the expression x^5, 5 is the **exponent**, x is the **base**, and the entire term is called an **exponential expression** or a **power** of x.

$$x^5 \leftarrow \text{exponent}$$
$$\text{base} \nearrow$$

In terms such as x or y, the exponent is understood to be 1:

$$x = x^1 \quad \text{and} \quad y = y^1$$

In general, we have the following definition:

Definition: If n is a natural number, then

$$x^n = \overbrace{x \cdot x \cdot x \cdot \cdots \cdot x}^{n \text{ factors of } x}$$

294 Chapter 8 Real Numbers and Their Properties

EXAMPLE 1

a. $4^2 = 4 \cdot 4 = 16$ Read 4^2 as "4 squared" or as "4 to the second power."

b. $5^3 = 5 \cdot 5 \cdot 5 = 125$ Read 5^3 as "5 cubed" or as "5 to the third power."

c. $6^4 = 6 \cdot 6 \cdot 6 \cdot 6 = 1296$ Read 6^4 as "6 to the fourth power."

d. $\left(\dfrac{2}{3}\right)^5 = \dfrac{2}{3} \cdot \dfrac{2}{3} \cdot \dfrac{2}{3} \cdot \dfrac{2}{3} \cdot \dfrac{2}{3} = \dfrac{32}{243}$ Read $\left(\dfrac{2}{3}\right)^5$ as "$\dfrac{2}{3}$ to the fifth power."

In the next example, the base of each exponential expression is a variable.

EXAMPLE 2

a. $y^6 = y \cdot y \cdot y \cdot y \cdot y \cdot y$ Read y^6 as "y to the sixth power."

b. $x^3 = x \cdot x \cdot x$ Read x^3 as "x cubed" or as "x to the third power."

c. $z^2 = z \cdot z$ Read z^2 as "z squared" or as "z to the second power."

d. $a^1 = a$ Read a^1 as "a to the first power."

Work Progress Check 5.

Progress Check 5

Write each expression without using exponents.

a. 3^5

b. t^3

Order of Operations

As we have seen, the order in which we do arithmetic is extremely important. To guarantee that calculations will have a single correct result, we must use the familiar set of *priority rules*:

> **Order of Mathematical Operations:** If an expression does not contain grouping symbols,
>
> 1. Find the values of any exponential expressions.
> 2. Do all multiplications and divisions as they are encountered while working from left to right.
> 3. Do all additions and subtractions as they are encountered while working from left to right.
>
> If an expression contains grouping symbols, use the rules above to perform the calculations within each pair of grouping symbols, working from the innermost pair to the outermost pair.
>
> The bar of a fraction is a grouping symbol. Thus, in a fraction, simplify the numerator and the denominator separately. Then simplify the fraction, whenever possible.

Because of these priority rules, $4x^3$ is not equal to $(4x)^3$:

$4x^3 = 4xxx$ but $(4x)^3 = (4x)(4x)(4x) = 4 \cdot 4 \cdot 4xxx = 64x^3$

5. a. 243   **b.** $t \cdot t \cdot t$

EXAMPLE 3

If $x = 2$, find the value of   **a.** $4x^3$ and   **b.** $(4x)^3$.

Solution

a. $4x^3 = 4 \cdot 2^3$ Substitute 2 for x.
         $= 4 \cdot 8$ Evaluate the exponential expression.
         $= 32$ Perform the multiplication.

b. $(4x)^3 = (4 \cdot 2)^3$ Substitute 2 for x.
         $= 8^3$ Perform the multiplication within the parentheses.
         $= 512$ Evaluate the exponential expression.

EXAMPLE 4

Simplify $5^3 + 2(8 - 3 \cdot 2)$.

Solution

$5^3 + 2(8 - 3 \cdot 2) = 5^3 + 2(8 - 6)$ Do the multiplication within the parentheses.
                        $= 5^3 + 2(2)$ Do the subtraction within the parentheses.
                        $= 125 + 2(2)$ Find the value of the exponential expression.
                        $= 125 + 4$ Do the multiplication.
                        $= 129$ Do the addition.

EXAMPLE 5

Simplify $\dfrac{3(3 + 2) + 5}{17 - 3(4)}$.

Solution

$\dfrac{3(3 + 2) + 5}{17 - 3(4)} = \dfrac{3(5) + 5}{17 - 3(4)}$ Simplify the numerator and denominator separately. Begin by doing the addition within the parentheses.

                $= \dfrac{15 + 5}{17 - 12}$ Do the multiplications.

                $= \dfrac{20}{5}$ Do the addition and the subtraction.

                $= 4$ Do the division.

Work Progress Check 6.

Progress Check 6

Simplify $\dfrac{4 + 3(3 + 4)}{16 - 2(3)}$.

6. $\dfrac{5}{2}$

Progress Check Answers

EXAMPLE 6

If $x = 3$ and $y = 4$, evaluate   **a.** $3y + x^2$,   **b.** $3(y + x^2)$,   and
c. $3(y + x)^2$.

Solution

a. $3y + x^2 = 3(4) + 3^2$ Substitute 3 for x and 4 for y.
$= 3(4) + 9$ Evaluate the exponential expression.
$= 12 + 9$ Perform the multiplication.
$= 21$ Perform the addition.

b. $3(y + x^2) = 3(4 + 3^2)$ Substitute 3 for x and 4 for y.
$= 3(4 + 9)$ Evaluate the exponential expression.
$= 3(13)$ Parentheses indicate that the addition is performed next.
$= 39$ Perform the multiplication.

c. $3(y + x)^2 = 3(4 + 3)^2$ Substitute 3 for x and 4 for y.
$= 3(7)^2$ Parentheses indicate that the addition is performed next.
$= 3(49)$ Evaluate the exponential expression.
$= 147$ Perform the multiplication.

EXAMPLE 7

If $x = 4$ and $y = 3$, evaluate $\dfrac{3x^2 - 2y}{2(x + y)}$.

Solution

$$\dfrac{3x^2 - 2y}{2(x + y)} = \dfrac{3(4^2) - 2(3)}{2(4 + 3)}$$

$= \dfrac{3(16) - 2(3)}{2(7)}$ Find the value of 4^2 in the numerator and do the addition in the denominator.

$= \dfrac{48 - 6}{14}$ Do the multiplications.

$= \dfrac{42}{14}$ Do the subtraction.

$= 3$ Do the division.

Work Progress Check 7.

Progress Check 7

If $x = 3$ and $y = 4$, evaluate $\dfrac{2x + y^2}{2(y - x)}$.

Progress Check Answers

7. 11

8.3 Exercises

In Exercises 1–6, find the value of each expression.

1. 4^3
2. 5^2
3. 6^2
4. 7^3
5. 10^4
6. 2^6

In Exercises 7–14, write each expression as the product of several factors.

7. x^2
8. y^3
9. $3z^4$
10. $5t^2$
11. $(5t)^2$
12. $(3z)^4$
13. $5(2x)^3$
14. $7(3t)^2$

In Exercises 15–22, find the value of each expression if $x = 3$ and $y = 2$.

15. $4x^2$ **16.** $4y^3$ **17.** $(5y)^3$ **18.** $(2y)^4$

19. $2x^y$ **20.** $3y^x$ **21.** $(3y)^x$ **22.** $(2x)^y$

In Exercises 23–64, simplify each expression by performing the indicated operations.

23. $3 \cdot 5 - 4$ **24.** $4 \cdot 6 + 5$ **25.** $3(5 - 4)$ **26.** $4(6 + 5)$

27. $3 + 5^2$ **28.** $4^3 - 2^2$ **29.** $(3 + 5)^2$ **30.** $(5 - 2)^3$

31. $2 + 3 \cdot 5 - 4$ **32.** $12 + 2 \cdot 3 + 2$ **33.** $64 \div (3 + 1)$ **34.** $16 \div (5 + 3)$

35. $3(2 + 4)5$ **36.** $(7 + 9) \div 2 \cdot 4 - 15$

37. $(14 + 7) \div 3 \cdot 4$ **38.** $(12 + 8) \div (2 \cdot 5) + 2 \cdot 5$

39. $24 \div 4 \cdot 3 + 3$ **40.** $36 \div 9 \cdot 4 - 2$

41. $49 \div 7 \cdot 7 + 7$ **42.** $100 \div 10 \cdot 10 + 10$

43. $100 \div 10 \cdot 10 \div 100$ **44.** $100 \div (10 \cdot 10) \div 100$

45. $(100 \div 10) \cdot (10 \div 100)$ **46.** $100 \div [10 \cdot (10 \div 100)]$

47. $[(14 + 7) \div 3]4$ **48.** $4[3 + 2(5 - 2)] - 1$

49. $3^2 + 2(1 + 4) - 2$ **50.** $4 \cdot 3 + 2(5 - 2) - 2^3$

51. $5^2 - (7 - 3)^2$ **52.** $3^3 + (3 - 1)^3$

53. $(2 \cdot 3 - 4)^3$ **54.** $(3 \cdot 5 - 2 \cdot 6)^2$

55. $(3^2 - 2^3)7$ **56.** $(3^3 - 2^4)^2$

57. $6[2 + 3(8 - 3 \cdot 2) + 2]$ **58.** $5[10 - (3 \cdot 2 - 3)] - (5 + 2)$

59. $\dfrac{(3 + 5)^2 + 2}{2(8 - 5)}$ **60.** $\dfrac{25 - (2 \cdot 3 - 1)}{2 \cdot 9 - 8}$

61. $\dfrac{(5 - 3)^2 + 2}{4^2 - (8 + 2)}$ **62.** $\dfrac{(4^2 - 2) + 7}{5(2 + 4) - 3^2}$

63. $\dfrac{2[4 + 2(3 - 1)]}{3[3(2 \cdot 3 - 4)]}$ **64.** $\dfrac{6[3 \cdot 7 - 5(3 \cdot 4 - 11)]}{2[4(3 + 2) - 3^2 + 5]}$

In Exercises 65–90, evaluate each expression, given that $x = 3$, $y = 2$, and $z = 4$.

65. $2x - y$ **66.** $2z + y$ **67.** $10 - 2x$ **68.** $15 - 3z$

69. $5z \div 2 + y$ **70.** $5x \div 3 + y$ **71.** $4x - 2z$ **72.** $5y - 3x$

73. $x + yz$ **74.** $3z + x - 2y$ **75.** $3(2x + y)$ **76.** $4(x + 3y)$

77. $(3 + x)y$ **78.** $(4 + z)y$ **79.** $(z + 1)(x + y)$ **80.** $3(z + 1) \div x$

81. $(x + y) \div (z + 1)$
82. $(2x + 2y) \div (3z - 2)$
83. $xyz + z^2 - 4x$
84. $zx + y^2 - 2z$

85. $3x^2 + 2y^2$
86. $3x^2 + (2y)^2$
87. $\dfrac{2x + y^2}{y + 2z}$
88. $\dfrac{2z^2 - y}{2x - y^2}$

89. $\dfrac{2x^3 - (xy - 2)}{2(3y + 5z) - 27}$
90. $\dfrac{x^2[14 - y(x + 2)] - 6}{5[xy - z(5y - 9)]}$

In Exercises 91–94, insert parentheses in $3 \cdot 8 + 5 \cdot 3$ to make its value equal to the given number.

91. 39
92. 117
93. 87
94. 69

In Exercises 95–98, insert parentheses in $4 + 3 \cdot 5 - 3$ to make its value equal to the given number.

95. 14
96. 10
97. 32
98. 16

Review Exercises

In Review Exercises 1–2, place one of the symbols $\leq$ or $>$ in the box to make the statement true.

1. $3 \;\square\; \dfrac{6}{9}$
2. $11 - 3 \;\square\; 1 + 7$

In Review Exercises 3–4, place one of the symbols $\geq$ or $<$ in the box to make the statement true.

3. $\dfrac{7}{3} \;\square\; \dfrac{21}{9}$
4. $\dfrac{20 - 5}{3} \;\square\; 1 + 3 + 5$

In Review Exercises 5–8, let x represent 6, y represent 8, and z represent 0. Evaluate each expression.

5. $x + y - z$
6. xyz
7. $3y + z$
8. $\dfrac{3 + x}{y + 1}$

9. Find the sum of the coefficients of the first and last terms of the expression $5x^3y - 127xy + 12$.

10. A rectangle is 3 feet longer than it is wide. If its width is w feet, what expression represents its length?

11. The focal length, f, of a double-convex thin lens is given by the formula

$$f = \dfrac{rs}{(r + s)(n - 1)}$$

If $r = 8$, $s = 12$, and $n = 1.6$, find f.

12. The total resistance, R, of two resistors in parallel is given by the formula

$$R = \dfrac{rs}{r + s}$$

If $r = 170$ and $s = 255$, find R.

8.4 Real Numbers and Their Absolute Values

- Integers
- Rational Numbers
- Real Numbers
- Graphing Sets of Real Numbers
- Absolute Value

When constructing number lines, we can also mark off points at equal distances to the left of the origin, as in Figure 8-4.

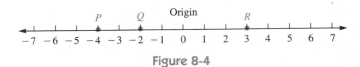

Figure 8-4

The coordinates of points to the left of the origin are called **negative numbers**. Because point P is 4 units to the *left* of the origin, the coordinate of P is -4 (read as "negative 4"). Because point Q is 2 units to the *left* of the origin, the coordinate of Q is -2.

Because point R is 3 units to the *right* of the origin, the coordinate of R is 3. The coordinates of points to the right of the origin are called **positive numbers**. The number 3, for example, is *positive* 3 and can be written as $+3$. Likewise,

$$4 = +4 \quad \text{and} \quad 7 = +7$$

Integers

The coordinates of points shown in Figure 8-4 are called **integers**.

> **Definition:** The set of **integers** is the set
> $$\{\ldots, -5, -4, -3, -2, -1, 0, 1, 2, 3, 4, 5, \ldots\}$$

If an integer can be divided by 2, it is an even integer. Otherwise, it is odd.

$\ldots, -6, -4, -2, 0, 2, 4, 6, \ldots$ are the even integers
$\ldots, -7, -5, -3, -1, 1, 3, 5, 7, \ldots$ are the odd integers

Because the set of natural numbers is included within the set of integers, we say that the set of natural numbers is a **subset** of the set of integers. The set of whole numbers is also a subset of the set of integers.

Because the coordinates of points on the number line become greater as we move from left to right, the negative numbers are less than 0 and the positive numbers are greater than 0. The number 0 is neither positive nor negative. Numbers that are either positive or negative are called **signed numbers**. Each of the following is a true statement about signed numbers:

$-5 < 3$ Read as "-5 is less than 3."
$-4 < -2$ Read as "-4 is less than -2."
$1 \geq -1$ Read as "1 is greater than or equal to -1."

Rational Numbers

Fractions such as $\frac{3}{2}$, $\frac{17}{12}$, and $\frac{-43}{8}$ are not integers. They are examples of numbers that are called **rational numbers**.

Definition: A **rational number** is any number that can be written in the form $\frac{a}{b}$, where a and b are integers and $b \neq 0$.

The decimal 0.5 represents a rational number because it can be written as the quotient of two integers:

$$0.5 = \frac{5}{10} = \frac{1}{2}$$

We must remember that division by 0 is *never allowed*. The symbols $\frac{1}{0}$ and $\frac{0}{0}$ are meaningless.

Because every integer can be written as a fraction with a denominator of 1, the set of integers is a subset of the rational numbers. For example, the integers 6, −4, and 0 are rational numbers because each can be written as a fraction with a denominator of 1:

$$6 = \frac{6}{1}, \quad -4 = \frac{-4}{1}, \quad \text{and} \quad 0 = \frac{0}{1}$$

Since numbers such as $\sqrt{2}$ and π cannot be written as the quotient of two integers, they are not rational numbers. They are examples of **irrational numbers**, a set we will study later in this book.

Real Numbers

Definition: A **real number** is any number that is either a rational number or an irrational number.

EXAMPLE 1

Classify the numbers **a.** 5, **b.** −12, and **c.** 0.25 as a positive number, negative number, integer, rational number, and real number.

Solution

a. The number 5 is a positive number, an integer, a rational number, and a real number.

b. The number −12 is a negative number and an integer. Because it can be written as $\frac{-12}{1}$, it is a rational number. It is also a real number.

c. The number 0.25 can be written as the fraction $\frac{1}{4}$. Thus, 0.25 is a positive number, a rational number, and a real number.

Work Progress Check 8.

Many points on the number line do not have integer coordinates. The point midway between 0 and 1, for example, has the coordinate $\frac{1}{2}$. The point with coordinate $-\frac{3}{2}$ lies midway between −2 and −1 (see Figure 8-5).

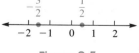

Figure 8-5

Progress Check 8

Consider the set $\{-2, -1, 0, 0.5, 1, 2, \frac{7}{3}, \sqrt{3}\}$.
List the numbers that are

a. Integers

b. Rational numbers

c. Real numbers

8. **a.** −2, −1, 0, 1, 2 **b.** −2, −1, 0, 0.5, 1, 2, $\frac{7}{3}$ **c.** −2, −1, 0, 0.5, 1, 2, $\frac{7}{3}$, $\sqrt{3}$

Progress Check Answers

Graphing Sets of Real Numbers

Graphs of sets of real numbers are intervals on the number line instead of isolated points. For example, the graph of the set of real numbers between −2 and 4 is shown in Figure 8-6.

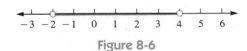

Figure 8-6

The open circles at −2 and 4 show that these points are not included in the graph. However, all numbers between −2 and 4 are included in the graph.

EXAMPLE 2

Graph the real numbers less than −3 or greater than 1.

Solution

The graph of the real numbers less than −3 includes all points that are to the left of −3. The graph of the real numbers greater than 1 includes all points that are to the right of 1. The graph is shown in Figure 8-7.

Figure 8-7

EXAMPLE 3

Graph the set of real numbers from −5 to −1.

Solution

The set of real numbers from −5 to −1 includes both −5 and −1 and all the numbers in between. We use solid circles at −5 and at −1 to show that these points are included in the graph. The graph is shown in Figure 8-8.

Figure 8-8

Work Progress Check 9.

If two real numbers are the same distance from 0 on a number line but on opposite sides of 0, they are called **negatives** or **opposites** of each other. For example,

The negative of 5, written as −(5), is −5

The negative of −5, written as −(−5), is 5

Likewise,

The negative of 9, written as −(9), is −9

The negative of −9, written as −(−9), is 9

The results of −(−5) = 5 and −(−9) = 9 suggest that the negative of the negative of a number is the number itself.

Progress Check 9

a. Graph the real numbers between −3 and 6.

b. Graph the real numbers less than −3 or greater than 6.

9. a.
   b.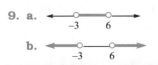

Progress Check Answers

> **The Double Negative Rule:** If x represents a number, then
> $$-(-x) = x$$

EXAMPLE 4

If $r = 7$ and $t = -3$, evaluate   **a.** $-r$,   **b.** $-t$,   **c.** $-(-r)$,   and   **d.** $-(-t)$.

Solution

a. $-r = -(7)$
$ = -7$

b. $-t = -(-3)$
$ = 3$

c. $-(-r) = -(-7)$
$ = 7$

d. $-(-t) = -[-(-3)]$
$ = -(3)$
$ = -3$ ∎

Absolute Value

The distance on the number line between a number and the origin is called the **absolute value** of the number. For example, the distance on the number line between the point with coordinate 5 and the origin is 5 units (see Figure 8-9). Thus, the absolute value of 5, denoted $|5|$, is 5:

$|5| = 5$ Read as "The absolute value of 5 is 5."

The distance on the number line between the point with coordinate -5 and the origin is 5. Again, see Figure 8-9. Thus,

$|-5| = 5$ Read as "The absolute value of -5 is 5."

Because $|-5|$ and $|5|$ are both equal to 5, $|-5| = |5|$. In general, we have

$|x| = |-x|$ The absolute values of a number and its negative are equal.

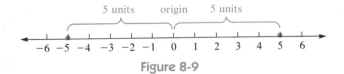

Figure 8-9

The absolute value of a number can be defined another way:

> **Definition:** If $x \geq 0$, then $|x| = x$.
> If $x < 0$, then $|x| = -x$.

This definition shows that if x is positive or 0, then x is its own absolute value. However, if x is negative, then $-x$ (which is positive) is the absolute value. Thus, $|x|$ is always positive or 0:

$|x| \geq 0$

EXAMPLE 5

Evaluate **a.** |6|, **b.** |−6|, **c.** −|7|, **d.** −|−7|, and **e.** |0|.

Solution

a. |6| = 6

b. |−6| = 6

c. −|7| = −(7) = −7

d. −|−7| = −(7) = −7

e. |0| = 0

The absolute value symbol is also a grouping symbol. For example, to evaluate |9 − 7|, we first do the subtraction within the absolute value symbol and then find the absolute value of the difference.

$$|9 − 7| = |2| = 2$$

EXAMPLE 6

Evaluate $\left|\frac{5}{2} + \frac{1}{3}\right|$.

Solution

$\left|\frac{5}{2} + \frac{1}{3}\right| = \left|\frac{5 \cdot 3}{2 \cdot 3} + \frac{1 \cdot 2}{3 \cdot 2}\right|$ Write the fractions as equivalent fractions with the common denominator 6.

$= \left|\frac{15}{6} + \frac{2}{6}\right|$

$= \left|\frac{15 + 2}{6}\right|$ Add the fractions.

$= \left|\frac{17}{6}\right|$

$= \frac{17}{6}$ Find the absolute value of $\frac{17}{6}$.

EXAMPLE 7

If $x = 9$ and $y = 12$, evaluate $3|−x| + 5|y − x|$.

Solution

$3|−x| + 5|y − x| = 3|−9| + 5|12 − 9|$ Substitute values for the variables.

$= 3|−9| + 5|3|$ Evaluate the expressions within the absolute value symbol.

$= 3 \cdot 9 + 5 \cdot 3$ Determine the absolute values.

$= 27 + 15$ Do the multiplications.

$= 42$ Do the addition.

Work Progress Check 10.

Progress Check 10

If $x = 6$ and $y = −8$, evaluate $\frac{2x + |y|}{x + 2|y|}$.

10. $\frac{10}{11}$

8.4 Exercises

In Exercises 1–12, classify each number as a positive number, negative number, integer, rational number, and real number.

1. 4
2. -3
3. -8
4. $\dfrac{1}{2}$
5. $\dfrac{2}{3}$
6. 21
7. 0.75
8. 0.125
9. $-\dfrac{12}{4}$
10. $\dfrac{0}{5}$
11. $-\dfrac{7}{8}$
12. $-\dfrac{17}{5}$

In Exercises 13–30, graph each set of numbers on a number line.

13. The numbers $\dfrac{1}{3}$, $\dfrac{5}{2}$, and $-\dfrac{4}{3}$
14. The numbers $\dfrac{2}{3}$, $\dfrac{5}{3}$, and $-\dfrac{7}{3}$
15. The numbers $-\dfrac{9}{2}$, $-\dfrac{5}{2}$, and $\dfrac{3}{2}$
16. The numbers $\dfrac{1}{4}$, $-\dfrac{1}{4}$, and $\dfrac{5}{4}$
17. All real numbers greater than 2 and less than 5
18. All real numbers less than -2 or greater than -1
19. All real numbers greater than 2
20. All real numbers less than 2
21. All real numbers less than -5
22. All real numbers greater than -5
23. All real numbers from -7 to -2
24. All real numbers from -5 to 2
25. All real numbers between -2 and 4
26. All real numbers between -7 and -1
27. All real numbers less than -7 or greater than -2
28. All real numbers greater than 5 or less than -2
29. All real numbers between -4 and -2 or greater than 0
30. All real numbers between -1 and 1 or less than -6

In Exercises 31–50, evaluate each expression.

31. $|8|$
32. $|9|$
33. $|-8|$
34. $|-9|$
35. $|0|$
36. $|-2|^2$
37. $|3|^2$
38. $|-5|^3$
39. $-|10|$
40. $-|-10|$
41. $-(|-2| + |-3|)$
42. $|-3| - |-2|$
43. $\left|\dfrac{5}{3} + \dfrac{3}{5}\right|$
44. $\left|\dfrac{7}{2} - \dfrac{3}{5}\right|$
45. $3|15 - 8| + 2|13 - 9|$
46. $5|3 - 1|^2 + 2$
47. $2|-3|^3 - |-4|^2$
48. $8|5 - 3| - |-4|^2$
49. $|5|^2 - |-1|^3 - |-2|^2$
50. $|-5^2| - |1^3| - |-3|^2$

51. $-x$ **52.** $-y$ **53.** $-(-y)$ **54.** $-(-x)$

55. $|x|$ **56.** $|y|$ **57.** $-|z|$ **58.** $|-z|$

59. $-|-y|$ **60.** $-|-x|$ **61.** $|x|+|y|$ **62.** $|-x|\cdot|y|$

63. $|x-z|^2$ **64.** $|x+3z|^3$ **65.** $|2x+3|^2$ **66.** $|x^2-x|^3$

67. $\dfrac{|-y+z|}{|2x-1|}$ **68.** $\dfrac{|x+z|}{|x-z|}$ **69.** $\dfrac{|3z+6|}{|5x-7|}$ **70.** $\dfrac{8-|x+2z|}{3|-x|}$

71. $\dfrac{|x^4|+|y|^2}{|y|^2+|xz|-|4|}$ **72.** $\dfrac{(|x|+|y|+|z|)^3}{(|x|+|y|)^2}$ **73.** $\dfrac{|x^2|+2|y|^2}{|z|^7+|x^2z|+|-x|}$ **74.** $\dfrac{(|-x|\cdot|y|-|x|^2)^3}{(|x|+|y|)^2-1}$

75. Explain why the set of even natural numbers is a subset of the rational numbers.

76. Explain why the set of prime numbers is not a subset of the odd natural numbers.

77. Explain why the absolute value of a number is equal to the absolute value of the negative of that number.

78. The absolute value of a number is 2. What might the number be?

Review Exercises

In Review Exercises 1–4, simplify each expression.

1. $3+7^2$ **2.** $(3+7)^2$ **3.** $4\cdot 3^2$ **4.** $(4\cdot 3)^2$

In Review Exercises 5–8, let $x=3$, $y=5$, and $z=1$. Evaluate each expression.

5. $x(y-z)$ **6.** $xy-z$ **7.** x^z+y **8.** $(x^2-y)^x$

In Review Exercises 9–10, graph each set of numbers on the number line.

9. The numbers 0, -3, and $\dfrac{5}{3}$

10. The real numbers greater than -3 and less than or equal to $\dfrac{3}{2}$

11. The volume, V, of a right-circular cylinder is given by

$$V=\pi r^2 h$$

If $r=7$ and $h=1$, find V. Use $\pi\approx 3.14$.

12. How many terms are in the expression $5x^2y+7xy^2+9x+2$?

8.5 Adding and Subtracting Real Numbers

- Adding Numbers with Like Signs
- Adding Numbers with Unlike Signs
- Subtracting Real Numbers

Adding Numbers with Like Signs

Suppose we wish to add +2 and +3. We can represent +2 with an arrow of length 2, pointing to the right. We can represent +3 with an arrow of length 3, also pointing to the right. To find the sum $(+2) + (+3)$, we place the two arrows end to end, as in Figure 8-10. The endpoint of the second arrow is the point with coordinate +5. Thus,

$$(+2) + (+3) = +5$$

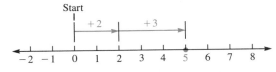

Figure 8-10

The addition problem

$$(-2) + (-3)$$

can be represented by two arrows on the number line, as shown in Figure 8-11. The number -2 can be represented with an arrow that begins at the origin, is 2 units long, and points to the *left*. The number -3 can be represented with an arrow that is 3 units long, continuing from point -2, and also pointing to the *left*. The endpoint of the final arrow is the point -5. Thus,

$$(-2) + (-3) = -5$$

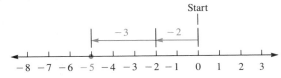

Figure 8-11

These examples provide the basis for the following rule.

> **Adding Real Numbers with Like Signs:** If two real numbers a and b have the same sign, their sum is found by adding their absolute values and using their common sign.

Progress Check 11

Add:

a. $+4 + (+5)$

b. $-4 + (-5)$

EXAMPLE 1

a. $(+4) + (+6) = +(4 + 6)$
$= +10$

b. $(-4) + (-6) = -(4 + 6)$
$= -10$

c. $+5 + (+10) = +(5 + 10)$
$= +15$

d. $-\dfrac{1}{2} + \left(-\dfrac{3}{2}\right) = -\left(\dfrac{1}{2} + \dfrac{3}{2}\right)$
$= -\dfrac{4}{2}$
$= -2$

Work Progress Check 11.

11. **a.** 9 **b.** -9

Adding Numbers with Unlike Signs

Two real numbers with *unlike* signs can be represented by arrows on a number line that point in *opposite* directions. For example,

$$(-6) + (+2)$$

can be represented on a number line as shown in Figure 8-12. Thus,

$$(-6) + (+2) = -4$$

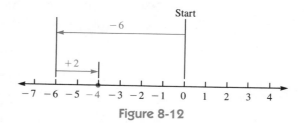

Figure 8-12

The addition problem

$$(+7) + (-4)$$

can be represented on a number line as in Figure 8-13. Thus,

$$(+7) + (-4) = +3$$

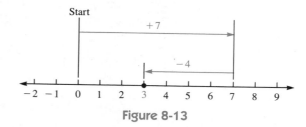

Figure 8-13

These examples provide the basis for the following rule.

Adding Real Numbers with Unlike Signs: If two real numbers a and b have unlike signs, their sum is found by subtracting their absolute values (the smaller from the larger) and using the sign of the number with the greater absolute value.

Progress Check 12

Add:

a. $+5 + (-8)$

b. $-16 + (+11)$

EXAMPLE 2

a. $(+6) + (-5) = +(6 - 5)$
$ = +1$

b. $(-2) + (+3) = +(3 - 2)$
$ = +1$

c. $+6 + (-9) = -(9 - 6)$
$ = -3$

d. $-10 + (+4) = -(10 - 4)$
$ = -6$

Work Progress Check 12.

12. **a.** -3 **b.** -5

EXAMPLE 3

a. $[(+3) + (-7)] + (-4) = [-4] + (-4)$ Do the work within the brackets first.
$= -8$

b. $-3 + [(-2) + (-8)] = -3 + [-10]$ Do the work within the brackets first.
$= -13$

Progress Check 13

Work Progress Check 13.

Simplify: $-5 + [-2 + 8]$.

EXAMPLE 4

If $x = -4$, $y = 5$, and $z = -13$, evaluate   **a.** $x + y$ and   **b.** $2y + z$.

Solution

We substitute -4 for x, 5 for y, and -13 for z and then simplify.

a. $x + y = (-4) + (5)$
$= +(5 - 4)$
$= +1$

b. $2y + z = 2 \cdot 5 + (-13)$
$= 10 + (-13)$
$= -(13 - 10)$
$= -3$

We can add real numbers using a vertical format.

EXAMPLE 5

Add:

a. $+5$
$\underline{+2}$
$+7$

b. $+5$
$\underline{-2}$
$+3$

c. -5
$\underline{+2}$
-3

d. -5
$\underline{-2}$
-7

Words and phrases such as *found, gain, credit, up, increase, forward, rises, in the future,* and *to the right* show a positive direction. Words and phrases such as *lost, loss, debit, down, backward, falls, in the past,* and *to the left* show a negative direction.

EXAMPLE 6

Laura opens a checking account by depositing $350. The bank debits her account $9 for check printing, and Laura writes a check for $22. After these transactions, what is the balance in Laura's account?

Solution

The deposit can be represented by $+350$. A debit of $9 can be represented by -9, and the check written for $22 can be represented by -22. The balance in Laura's account after these three transactions is the sum of 350, -9, and -22.

$+350 + (-9) + (-22) = 341 + (-22)$ Work from left to right.
$= 319$

Laura's balance is $319.

13. 1

Section 8.5 Adding and Subtracting Real Numbers

Subtracting Real Numbers

The expression

$$7 - 4 = 3$$

can be thought of as taking four objects away from seven objects, leaving three objects. Another approach treats the subtraction problem $7 - 4$ as the equivalent addition problem $7 + (-4)$. In either case, the answer is 3.

$$7 - 4 = 3 \quad \text{and} \quad 7 + (-4) = 3$$

We use this idea to define the *difference* when y is subtracted from x.

Finding the Difference of Real Numbers: If x and y are two real numbers,

$$x - y = x + (-y)$$

This rule points out that to *subtract* a second number from a first number, we add the negative (or opposite) of the second number to the first.

EXAMPLE 7

Evaluate **a.** $12 - 4$, **b.** $-13 - 5$, and **c.** $-14 - (-6)$.

Solution

a. $12 - 4 = 12 + (-4)$
$ = 8$

b. $-13 - 5 = -13 + (-5)$
$ = -18$

c. $-14 - (-6) = -14 + [-(-6)]$
$ = -14 + 6 \quad$ Use the double negative rule.
$ = -8$

Work Progress Check 14.

EXAMPLE 8

If $x = -5$ and $y = -3$, evaluate **a.** $\dfrac{y - x}{7 + x}$ and **b.** $\dfrac{6 + x}{y - x} - \dfrac{y - 4}{7 + x}$.

Solution

We substitute -5 for x and -3 for y into each expression and simplify.

a. $\dfrac{y - x}{7 + x} = \dfrac{-3 - (-5)}{7 + (-5)}$

$\phantom{\dfrac{y - x}{7 + x}} = \dfrac{-3 + 5}{2}$

$\phantom{\dfrac{y - x}{7 + x}} = \dfrac{2}{2}$

$\phantom{\dfrac{y - x}{7 + x}} = 1$

Progress Check 14

Subtract:

a. $-11 - (+6)$

b. $-12 - (-4)$

14. **a.** -17 **b.** -8

b. $\dfrac{6+x}{y-x} - \dfrac{y-4}{7+x} = \dfrac{6+(-5)}{-3-(-5)} - \dfrac{-3-4}{7+(-5)}$

$= \dfrac{1}{-3+5} - \dfrac{-3+(-4)}{2}$

$= \dfrac{1}{2} - \dfrac{-7}{2}$

$= \dfrac{1-(-7)}{2}$

$= \dfrac{1+[-(-7)]}{2}$

$= \dfrac{1+7}{2}$

$= \dfrac{8}{2}$

$= 4$ ■

To subtract real numbers vertically, we add the opposite of the number to be subtracted by changing the sign of the lower number (called the **subtrahend**) and adding.

EXAMPLE 9

a. The subtraction $\begin{array}{r} 5 \\ -\underline{-4} \end{array}$ becomes the addition $\begin{array}{r} 5 \\ +\underline{+4} \\ 9 \end{array}$

b. The subtraction $\begin{array}{r} -8 \\ -\underline{+3} \end{array}$ becomes the addition $\begin{array}{r} -8 \\ +\underline{-3} \\ -11 \end{array}$ ■

EXAMPLE 10

Simplify **a.** $3 - [4 + (-6)]$ and **b.** $[-5 + (-3)] - [-2 - (+5)]$.

Solution

a. $3 - [4 + (-6)] = 3 - (-2)$ Do the work within the brackets first.

$= 3 + [-(-2)]$ Use the rule for subtracting two numbers.

$= 3 + 2$ Use the double negative rule.

$= 5$ Add.

b. $[-5 + (-3)] - [-2 - (+5)]$

$= [-5 + (-3)] - [-2 + (-5)]$ Use the rule for subtracting two numbers.

$= -8 - (-7)$ Do the work within the brackets.

$= -8 + [-(-7)]$ Use the rule for subtracting two numbers.

$= -8 + 7$ Use the double negative rule.

$= -1$ Add.

Work Progress Check 15. ■

Progress Check 15

Simplify: $[-4 + (-5)] + [3 + (-5)]$.

15. -11

EXAMPLE 11

At noon, the temperature was 7 degrees above zero. At midnight, the temperature was 4 degrees below zero. Find the difference between these two temperatures.

Solution

A temperature of 7 degrees above zero can be represented as $+7$. A temperature of 4 degrees below zero can be represented as -4. To find the difference between these temperatures, we set up a subtraction and simplify.

$$7 - (-4) = 7 + [-(-4)] \quad \text{Use the rule for subtracting two numbers.}$$
$$= 7 + 4 \quad \text{Use the double negative rule.}$$
$$= 11$$

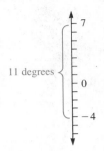

Figure 8-14

The difference between the temperatures is 11 degrees, as shown in Figure 8-14. ∎

8.5 Exercises

In Exercises 1–20, find each sum.

1. $4 + 8$
2. $(-4) + (-2)$
3. $(-3) + (-7)$
4. $(+4) + 11$
5. $6 + (-4)$
6. $5 + (-3)$
7. $9 + (-11)$
8. $10 + (-13)$
9. $(-5) + (-7)$
10. $(-6) + (-4)$
11. $(-4) + (-9)$
12. $(-12) + (-5)$
13. $\dfrac{1}{5} + \left(+\dfrac{1}{7}\right)$
14. $\left(+\dfrac{2}{3}\right) + \left(-\dfrac{1}{4}\right)$
15. $\left(-\dfrac{3}{4}\right) + \left(+\dfrac{2}{3}\right)$
16. $\dfrac{3}{5} + \left(-\dfrac{2}{3}\right)$
17. $\begin{array}{r} 5 \\ +\underline{-4} \end{array}$
18. $\begin{array}{r} -20 \\ +\underline{-17} \end{array}$
19. $\begin{array}{r} -1.3 \\ +\underline{3.5} \end{array}$
20. $\begin{array}{r} 1.3 \\ +\underline{-2.5} \end{array}$

In Exercises 21–34, evaluate each expression.

21. $5 + [4 + (-2)]$
22. $-6 + [(-3) + 8]$
23. $-2 + (-4 + 5)$
24. $5 + [-4 + (-6)]$
25. $[-4 + (-3)] + [2 + (-2)]$
26. $[3 + (-1)] + [-2 + (-3)]$
27. $-4 + [-3 + 2] + (-3)$
28. $5 + [2 + (-5)] + (-2)$
29. $-|-9 + (-3)| + (-6)$
30. $-|8 + (-4)| + 7$
31. $\left|\dfrac{3}{5} + \left(-\dfrac{4}{5}\right)\right|$
32. $\left|\dfrac{1}{6} + \left(-\dfrac{5}{6}\right)\right|$
33. $-5.2 + |-2.5 + (-4)|$
34. $6.8 + |8.6 + (-1.1)|$

In Exercises 35–50, let $x = 2$, $y = -3$, $z = -4$, and $u = 5$. Evaluate each expression.

35. $x + y$ **36.** $x + z$ **37.** $x + z + u$ **38.** $y + z + u$

39. $(x + u) + 3$ **40.** $(y + 5) + x$ **41.** $x + (-1 + z)$ **42.** $-7 + (z + x)$

43. $(x + z) + (u + z)$ **44.** $(z + u) + (x + y)$

45. $x + [5 + (y + u)]$ **46.** $y + \{u + [z + (-6)]\} + y$

47. $|2x + y|$ **48.** $3|x + y + z|$

49. $|x + z| + |x + y + z|$ **50.** $|z + z| + |y + y|$

In Exercises 51–70, find each difference.

51. $8 - 4$ **52.** $-8 - 4$ **53.** $8 - (-4)$ **54.** $-9 - (-5)$

55. $-12 - 5$ **56.** $11 - (+4)$ **57.** $0 - (-5)$ **58.** $0 - 75$

59. $\dfrac{5}{3} - \dfrac{7}{6}$ **60.** $-\dfrac{5}{9} - \dfrac{5}{3}$ **61.** $-5 - \left(-\dfrac{3}{5}\right)$ **62.** $\dfrac{7}{8} - (-3)$

63. $-3\dfrac{1}{2} - 5\dfrac{1}{4}$ **64.** $2\dfrac{1}{2} - \left(-3\dfrac{1}{2}\right)$ **65.** $-6.7 - (-2.5)$ **66.** $25.3 - 17.5$

67. $\begin{array}{r} 8 \\ -\underline{4} \end{array}$ **68.** $\begin{array}{r} 8 \\ -\underline{-3} \end{array}$ **69.** $\begin{array}{r} -10 \\ -\underline{-3} \end{array}$ **70.** $\begin{array}{r} -13 \\ -\underline{5} \end{array}$

In Exercises 71–84, evaluate each quantity.

71. $+3 - [(-4) - 3]$ **72.** $-5 - [4 - (-2)]$

73. $(5 - 3) + (3 - 5)$ **74.** $(3 - 5) - [5 - (-3)]$

75. $5 - [4 + (-2) - 5]$ **76.** $3 - [-(-2) + 5]$

77. $[5 - (-34)] - [-2 + (-23)]$ **78.** $-5 + \{-3 - [-2 - (+4)]\}$

79. $\left(\dfrac{5}{2} - 3\right) - \left(\dfrac{3}{2} - 5\right)$ **80.** $\left(\dfrac{7}{3} - \dfrac{5}{6}\right) - \left[\dfrac{5}{6} - \left(-\dfrac{7}{3}\right)\right]$

81. $(5.2 - 2.5) - (5.25 - 5)$ **82.** $\left(3\dfrac{1}{2} - 2\dfrac{1}{2}\right) - \left[5\dfrac{1}{3} - \left(-5\dfrac{2}{3}\right)\right]$

83. $-|-9 - (-7)| - (-3)$ **84.** $-|8 - (-4)| - 7$

In Exercises 85–100, let $x = -4$, $y = 5$, and $z = -6$. Evaluate each quantity.

85. $x + y$ **86.** $y - z$ **87.** $x - y - z$ **88.** $y + z - x$

89. $x - (y - z)$ **90.** $y + (z - x)$

91. $3 - \{[x + (-3)] - z\}$ **92.** $[(-2) + z - (x - y)] + 10$

93. $|x - y + z|$ **94.** $|y - z - x|$ **95.** $|x|^2 - |y - z|^2$ **96.** $|z - x|^2 + |y|^2$

97. $\dfrac{y - x}{3 - z}$ **98.** $\dfrac{y - z}{3y + x}$ **99.** $\dfrac{x - y}{y} - \dfrac{z}{y}$ **100.** $\dfrac{y}{x - z} - \dfrac{x}{8 + z}$

In Exercises 101–108, let $a = 2$, $b = -3$, and $c = -4$. Evaluate each quantity.

101. $a + b - c$ **102.** $a - b + c$ **103.** $b - (c + a)$ **104.** $c + (a - b)$

105. $\dfrac{a + b}{b - c}$ **106.** $\dfrac{c - a}{-(a + b)}$ **107.** $\dfrac{|b + c|}{a - c}$ **108.** $\dfrac{a - b - c}{|a + b|}$

In Exercises 109–130, solve each problem by finding the sum or difference of two or more signed numbers.

109. Wendy has $25. If she pays Bruce the $17 she owes him, how much will she have left?

110. Scott weighed 150 pounds. After losing 4 pounds, what does Scott weigh?

111. The temperature rose 7 degrees in 1 hour. It then dropped 3 degrees in the next hour. What signed number represents the net change in temperature?

112. Amalia lost 12 pounds and then gained back 5 pounds. What signed number represents her change in weight?

113. One night, the temperature fell from zero to 14 below. By 5:00 P.M. the next day, the temperature had risen 20 degrees. What was the temperature at 5:00 P.M.?

114. In 1897, Joseph Thompson discovered the electron. Fifty-four years later, the first fission reactor was built. Nineteen years before the reactor, James Chadwick discovered the neutron. In what year was the neutron discovered?

115. Akli deposited $212 in a new checking account, wrote a check for $173.30, and deposited another $312.50. Find the balance in Akli's account.

116. An army retreated 2300 meters. After regrouping, it moved forward 1750 meters. The next day, it gained another 1875 meters. What was the army's net gain?

117. A football player gained and lost the following number of yards on six consecutive plays: +5, +7, −5, +1, −2, and −6. Find the net outcome.

118. On January 1, Sally had $437 in the bank. During the month, she had deposits of $25, $37, and $45, and she had withdrawals of $17, $83, and $22. How much was in her account at the end of the month?

119. At the opening bell on Monday, the Dow Jones Industrial Average was 2153. At the close of light trading, the Dow was down 12 points. On Tuesday, news of a half-point drop in interest rates sent prices up. With a volume of 181 million shares, Tuesday's Dow is up 21 points. What was the Dow average after the market closed on Tuesday?

120. On Monday morning, the Dow Jones Industrial Average opened at 2917. For the week, the Dow rose 29 points on Monday and 12 points on Wednesday, but it fell 53 points on Tuesday and 27 points on both Thursday and Friday. Where did the Dow close on Friday?

121. Andy owned 500 shares of Transitronics Corporation before the company declared a two-for-one stock split. After the split, Andy sold 300 shares. How many shares does Andy now own?

122. Dion gained 12 yards on the first play of a football game. However, on the next play, he lost 15 yards. What was Dion's net gain or loss?

123. Tuesday's high and low prices for Transitronics stock were $37\frac{1}{8}$ and $31\frac{5}{8}$. Find the range of prices for the stock.

124. Find the difference between a temperature of 32 degrees above zero and a temperature of 27 degrees above zero.

125. Find the difference between a temperature of 3 degrees below zero and a temperature of 21 degrees below zero.

126. Maria earned $2532 in a part-time business. However, 25% of the earnings went for taxes. What were Maria's net earnings?

127. The Greek mathematician Euclid was alive in 300 B.C. The English mathematician Sir Isaac Newton was alive in A.D. 1700. How many years apart did they live?

128. Juan owed his mother $75 and his brother $32. However, his sister owed Juan $47. Use a signed number to express Juan's financial position.

129. Susan owed her mother $125. Her mother agreed to cancel $70 of this debt. Use a signed number to express Susan's current financial position.

130. Mike owed Paul $350. However, Paul canceled $43 of the debt when Mike tutored him in algebra. Another $75 of the debt was canceled when Mike gave Paul his golf clubs. How much does Mike still owe Paul?

Review Exercises

1. Simplify the fraction $\frac{24}{27}$.

2. Add: $\frac{3}{5} + \frac{1}{3}$.

3. Simplify: $3(7 - 2)$.

4. Simplify: $3(7) - 2$.

In Review Exercises 5–6, let $x = -3$ and $y = 5$. Perform the indicated operations.

5. $y|x| - 2$

6. $|y|(|x| - 2)$

In Review Exercises 7–10, write an algebraic expression that represents the given quantity.

7. The sum of x and twice the absolute value of y

8. Twice the sum of x and the absolute value of y

9. The absolute value of the sum of x and twice y

10. Twice the sum of the absolute values of x and y

11. Jim has one penny, seven nickels, x dimes, and y quarters. What is the total value of his coins?

12. On her first quiz, Latasha earned a grade of k points. On each successive quiz, her grade improved by 2 points. What was her score on the seventh quiz?

8.6 Multiplying and Dividing Real Numbers

- Multiplying Real Numbers
- Dividing Real Numbers

Multiplying Real Numbers

To find rules for multiplying real numbers, we recall that the expression $5 \cdot 4$ means that 4 is to be used as a term in a sum five times:

$$5(4) = +4 + 4 + 4 + 4 + 4 = 20$$

Likewise, the expression 5(−4) means that −4 is to be used as a term in a sum five times:

$$5(-4) = (-4) + (-4) + (-4) + (-4) + (-4) = -20$$

Since multiplying by a positive number means repeated addition, we can assume that multiplying by a negative number means repeated subtraction. The expression (−5)4, for example, means that 4 is to be used as a term in a repeated subtraction five times:

$$(-5)4 = -(4) - (4) - (4) - (4) - (4)$$
$$= (-4) + (-4) + (-4) + (-4) + (-4)$$
$$= -20$$

Likewise, the expression (−5)(−4) means that −4 is to be used as a term in a repeated subtraction five times:

$$(-5)(-4) = -(-4) - (-4) - (-4) - (-4) - (-4)$$
$$= -(-4) + [-(-4)] + [-(-4)] + [-(-4)] + [-(-4)]$$
$$= 4 + 4 + 4 + 4 + 4$$
$$= 20$$

The expression 0(−2) means that −2 is to be used zero times in a repeated addition. Thus,

$$0(-2) = 0$$

This suggests that the product of any number and 0 is 0.

Finally, since (−3)(1) means $-1 + (-1) + (-1)$, we have $(-3)(1) = -3$. This suggests that the product of any number and 1 is the number itself.

Rules for Multiplying Real Numbers:

1. The product of two real numbers with like signs is positive. It is the product of their absolute values.
2. The product of two real numbers with unlike signs is the negative of the product of their absolute values.
3. A number multiplied by 0 is 0: $a \cdot 0 = 0 \cdot a = 0$.
4. A number multiplied by 1 is the number: $a \cdot 1 = 1 \cdot a = a$.

EXAMPLE 1

Find each product: **a.** 4(−7), **b.** (−5)(−4), **c.** (−7)(6), **d.** 8(6), **e.** (−3)(5)(−4),   and   **f.** (−4)(−2)(−3).

Solution

a. $4(-7) = -28$

b. $(-5)(-4) = +20$

c. $(-7)(6) = -42$

d. $8(6) = +48$

e. $(-3)(5)(-4) = (-15)(-4)$

f. $(-4)(-2)(-3) = (8)(-3)$
$\qquad\qquad\qquad = 60$
$\qquad\qquad\qquad\qquad\qquad = -24$

EXAMPLE 2

Find each product: **a.** $\left(-\dfrac{2}{3}\right)\left(-\dfrac{6}{5}\right)$ and   **b.** $\left(\dfrac{3}{10}\right)\left(-\dfrac{5}{9}\right)$.

Solution

a. Because both fractions are negative, the product is positive.

$$\left(-\frac{2}{3}\right)\left(-\frac{6}{5}\right) = +\frac{2}{3} \cdot \frac{6}{5}$$
$$= +\frac{2 \cdot 6}{3 \cdot 5}$$
$$= +\frac{12}{15}$$
$$= +\frac{4}{5}$$

b. Because the fractions are of opposite signs, the product is negative.

$$\left(\frac{3}{10}\right)\left(-\frac{5}{9}\right) = -\frac{3}{10} \cdot \frac{5}{9}$$
$$= -\frac{3 \cdot 5}{10 \cdot 9}$$
$$= -\frac{15}{90}$$
$$= -\frac{1}{6}$$

Progress Check 16

Work Progress Check 16.

Find each product:

a. $3(-5)(-4)$

b. $-\frac{1}{2}\left(-\frac{1}{3}\right)(-12)$

EXAMPLE 3

If $x = -3$, $y = 2$, and $z = 4$, evaluate **a.** $y + xz$ and **b.** $x(y - z)$.

Solution

We substitute -3 for x, 2 for y, and 4 for z in each expression and simplify.

a. $y + xz = 2 + (-3)(4)$
$= 2 + (-12)$
$= -10$

b. $x(y - z) = -3[2 - 4]$
$= -3[2 + (-4)]$
$= -3[-2]$
$= 6$

EXAMPLE 4

If $x = -2$ and $y = 3$, evaluate **a.** $x^2 - y^2$ and **b.** $-x^2$.

Solution

a. $x^2 - y^2 = (-2)^2 - 3^2$ Substitute -2 for x and 3 for y and simplify.
$= 4 - 9$ Simplify the exponential expressions first.
$= -5$ Do the subtraction.

b. $-x^2 = -(-2)^2$ Substitute -2 for x and simplify.
$= -4$ Simplify $(-2)^2$.

16. a. 60 **b.** -2

Progress Check Answers

Section 8.6 Multiplying and Dividing Real Numbers

EXAMPLE 5

If the temperature is dropping 4 degrees each hour, how much warmer was it 3 hours ago?

Solution

A temperature drop of 4 degrees per hour can be represented by -4 degrees per hour. "Three hours ago" can be represented by -3. The temperature 3 hours ago can be represented by the product $(-3)(-4)$. Because

$$(-3)(-4) = +12$$

the temperature was 12 degrees warmer 3 hours ago. ∎

Dividing Real Numbers

We have seen that 8 divided by 4 is 2, because there are two 4's in 8. That is,

$$\frac{8}{4} = 2 \quad \text{because} \quad 2 \cdot 4 = 8$$

In general, the rule

$$\frac{a}{b} = c \quad \text{if and only if} \quad c \cdot b = a$$

is true for any real number a and any nonzero real number b. For example,

$$\frac{+10}{+2} = +5 \quad \text{because } (+5)(+2) = +10$$

$$\frac{-10}{-2} = +5 \quad \text{because } (+5)(-2) = -10$$

$$\frac{+10}{-2} = -5 \quad \text{because } (-5)(-2) = +10$$

$$\frac{-10}{+2} = -5 \quad \text{because } (-5)(+2) = -10$$

These examples suggest that the rules for dividing real numbers are similar to the rules for multiplying real numbers.

Rules for Dividing Real Numbers:

1. The quotient of two real numbers with like signs is the positive quotient of their absolute values.
2. The quotient of two real numbers with unlike signs is the negative of the quotient of their absolute values.
3. Division by zero is undefined.
4. If $a \neq 0$, then $\frac{0}{a} = 0$.
5. $\frac{a}{1} = a$
6. If $a \neq 0$, then $\frac{a}{a} = 1$.

A negative fraction can have a − sign either in front of the fraction or in the numerator of the fraction. For example, because $-\frac{10}{5}$ and $\frac{-10}{5}$ are both equal to −2,

$$-\frac{10}{5} = \frac{-10}{5}$$

The negative fraction $-\frac{10}{5}$ is a rational number, because it can be written as the quotient of two integers.

EXAMPLE 6

Find each quotient: **a.** $\frac{36}{18}$, **b.** $\frac{-44}{11}$, **c.** $\frac{27}{-9}$, and **d.** $\frac{-64}{-8}$.

Solution

a. $\frac{36}{18} = 2$ — The quotient of a positive number and a positive number is the positive quotient of their absolute values.

b. $\frac{-44}{11} = -\frac{44}{11} = -4$ — The quotient of a negative number and a positive number is the negative of the quotient of their absolute values.

c. $\frac{27}{-9} = -\frac{27}{9} = -3$ — The quotient of a positive number and a negative number is the negative of the quotient of their absolute values.

d. $\frac{-64}{-8} = \frac{64}{8} = 8$ — The quotient of a negative number and a negative number is the positive quotient of their absolute values.

Progress Check 17 Work Progress Check 17.

Progress Check 17

Find each quotient:

a. $\frac{-30}{15}$

b. $\frac{-45}{-15}$

EXAMPLE 7

If $x = -64$, $y = 16$, and $z = -4$, evaluate **a.** $\frac{yz}{-x}$, **b.** $\frac{z^3 y}{x}$, and **c.** $\frac{x+y}{-z^2}$.

Solution

We substitute −64 for x, 16 for y, and −4 for z in each expression and simplify.

a. $\frac{yz}{-x} = \frac{16(-4)}{-(-64)}$

$= \frac{-64}{+64}$

$= -1$

b. $\frac{z^3 y}{x} = \frac{(-4)^3(16)}{-64}$

$= \frac{(-64)(16)}{(-64)}$

$= 16$

c. $\frac{x+y}{-z^2} = \frac{-64+16}{-(-4)^2}$

$= \frac{-48}{-16}$

$= 3$

17. **a.** −2 **b.** 3
Progress Check Answers

EXAMPLE 8

If $x = -50$, $y = 10$, and $z = -5$, evaluate **a.** $\dfrac{xyz}{x - 5z}$ and **b.** $\dfrac{3xy + 2yz}{2(x + y)}$.

Solution

We substitute -50 for x, 10 for y, and -5 for z in each expression and simplify.

a. $\dfrac{xyz}{x - 5z} = \dfrac{(-50)(10)(-5)}{-50 - 5(-5)}$

$= \dfrac{(-500)(-5)}{-50 + 25}$

$= \dfrac{2500}{-25}$

$= -100$

b. $\dfrac{3xy + 2yz}{2(x + y)} = \dfrac{3(-50)(10) + 2(10)(-5)}{2(-50 + 10)}$

$= \dfrac{-150(10) + (20)(-5)}{2(-40)}$

$= \dfrac{-1500 - 100}{-80}$

$= \dfrac{-1600}{-80}$

$= 20$

Progress Check 18

Work Progress Check 18.

a. If $x = 2$, $y = 6$, and $z = -3$, evaluate $-\dfrac{xy}{z}$.

b. If $x = 2$, $y = 6$, and $z = -3$, evaluate $\dfrac{5yz^2}{x - z}$.

EXAMPLE 9

Twelve gamblers lost \$336. If they lost equal amounts, how much did each gambler lose?

Solution

A loss of \$336 can be represented by -336. Because there are 12 gamblers, the amount lost by each gambler can be represented by the quotient $\dfrac{-336}{12}$.

$$\dfrac{-336}{12} = -28$$

Each gambler lost \$28.

18. **a.** 4 **b.** 54

8.6 Exercises

In Exercises 1–24, find each product.

1. $(+1)(+3)$
2. $(-2)(-5)$
3. $(-3)(-6)$
4. $(4)(-6)$
5. $(+8)(+4)$
6. $(+8)(-4)$
7. $(-8)(4)$
8. $(-8)(-4)$
9. $(+9)(-6)$
10. $(-9)(6)$
11. $\left(\dfrac{1}{2}\right)(-32)$
12. $\left(-\dfrac{3}{4}\right)(12)$

13. $\left(-\dfrac{3}{4}\right)\left(-\dfrac{8}{3}\right)$ **14.** $\left(-\dfrac{2}{5}\right)\left(\dfrac{15}{2}\right)$ **15.** $(-3)\left(-\dfrac{1}{3}\right)$ **16.** $(5)\left(-\dfrac{2}{5}\right)$

17. $(3)(-4)(-6)$ **18.** $(-1)(-3)(-6)$ **19.** $(-2)(3)(4)$ **20.** $(5)(0)(-3)$

21. $(2)(-5)(-6)(-7)$ **22.** $(-3)(-5)(-5)(-2)$

23. $(-2)(-2)(-2)(-3)(-4)$ **24.** $(-5)(4)(3)(-2)(-1)$

In Exercises 25–48, let $x = -1$, $y = 2$, and $z = -3$. Evaluate each expression.

25. y^2 **26.** x^2 **27.** $-z^2$ **28.** $-xz$

29. xy **30.** yz **31.** $y + xz$ **32.** $z - xy$

33. $(x + y)z$ **34.** $y(x - z)$ **35.** $(x - z)(x + z)$ **36.** $(y + z)(x - z)$

37. $xy + yz$ **38.** $zx - zy$ **39.** xyz **40.** $x^2 y$

41. $y^2 z^2$ **42.** $z^3 y$ **43.** $y(x - y)^2$ **44.** $z(y - x)^2$

45. $x^2(y - z)$ **46.** $y^2(x - z)$ **47.** $(-x)(-y) + z^2$ **48.** $(-x)(-z) - y^2$

In Exercises 49–60, simplify each expression.

49. $\dfrac{8}{-2}$ **50.** $\dfrac{-6}{3}$ **51.** $\dfrac{-10}{-5}$ **52.** $\dfrac{20}{4}$

53. $\dfrac{-16}{4}$ **54.** $\dfrac{-25}{-5}$ **55.** $\dfrac{32}{-16}$ **56.** $\dfrac{18}{-3}$

57. $\dfrac{8 - 12}{-2}$ **58.** $\dfrac{16 - 2}{2 - 9}$ **59.** $\dfrac{20 - 25}{7 - 12}$ **60.** $\dfrac{2(15)^2 - 2}{-2^3 + 1}$

In Exercises 61–72, evaluate each expression if $x = -2$, $y = 3$, $z = 4$, $t = 5$, and $w = -18$.

61. $\dfrac{yz}{x}$ **62.** $\dfrac{zt}{x}$ **63.** $\dfrac{tw}{y}$ **64.** $\dfrac{w}{xy}$

65. $\dfrac{z + w}{x}$ **66.** $\dfrac{xyx}{y - 1}$ **67.** $\dfrac{xtz}{y + 1}$ **68.** $\dfrac{x + y + z}{t}$

69. $\dfrac{wz - xy}{x + y}$ **70.** $\dfrac{x^2 y^3}{yz}$ **71.** $\dfrac{yw + xy}{xt}$ **72.** $\dfrac{tw}{xz - w}$

In Exercises 73–84, evaluate each expression if $x = 4$, $y = -6$, and $z = -3$. Use a calculator.

73. $\dfrac{2x^2 + 2y}{x + y}$ **74.** $\dfrac{y^2 + z^2}{y + z}$ **75.** $\dfrac{2x^2 - 2z^2}{x + z}$ **76.** $\dfrac{8x^3 - 8y^2}{x - z}$

77. $\dfrac{y^3 + 4z^3}{(x + y)^2}$ **78.** $\dfrac{x^2 - 2xz + z^2}{x - y + z}$ **79.** $\dfrac{xy^2 z + x^2 y}{2y - 2z}$ **80.** $\dfrac{(x^2 - 2y)z^2}{-xz}$

81. $\dfrac{xyz - y^2 z}{y(x + z)^4}$ **82.** $\dfrac{-3x^2 - 2z^2 + x^2}{(x + y + z)^2}$

83. $\dfrac{2(x - y)(y - z)(x - z)}{2x - 3y + y}$ **84.** $\dfrac{x^3 y - (yz)^2 - 10y - 2}{x^2 + y^2 + z^3}$

In Exercises 85–92, evaluate each expression if $x = \frac{1}{2}$, $y = -\frac{2}{3}$, and $z = -\frac{3}{4}$.

85. $x + y$ **86.** $y + z$ **87.** $x + y + z$ **88.** $y + x - z$

89. $(x + y)(x - y)$ **90.** $(x - z)(x + z)$ **91.** $(x + y + z)(xyz)$ **92.** $xyz(x - y - z)$

In Exercises 93–102, solve each problem by finding a product or quotient of two signed numbers.

93. If the temperature increases 2 degrees each hour, how much warmer will it be in 3 hours?

94. If the temperature decreases 3 degrees each hour, how much colder will it be in 2 hours?

95. In Las Vegas, Robert lost $30 per hour playing the slot machines. Use a signed number to express Robert's financial condition after he gambled for 15 hours.

96. Rafael worked all day mowing lawns and was paid $8 per hour. If he had $94 at the end of an 8-hour day, how much did he have before he started working?

97. If a drain is emptying a pool at the rate of 12 gallons per minute, how much more water was in the pool 1 hour ago?

98. The flow of water from a pipe is filling a pool at the rate of 23 gallons per minute. How much less water was in the pool 2 hours ago?

99. At a carnival, three boys lost a total of $21. Use a signed number to express how much each lost, assuming that they lost equal amounts.

100. Suppose that the temperature is falling 3 degrees each hour. If the temperature has fallen 18 degrees, what signed number expresses how many hours the temperature has been falling?

101. Suppose the temperature is dropping at the rate of 4 degrees each hour. Use a signed number to show how many hours ago the temperature was 20 degrees warmer.

102. A man lost 37.5 pounds. If he lost 2.5 pounds each week, how long has he been dieting?

Review Exercises

In Review Exercises 1–10, place one of the symbols =, <, or > in the box to make the statement true.

1. $5 + (-3) \;\square\; -5 - (-7)$

2. $-3 - 3 \;\square\; (-2)(-3)$

3. $2(7 - 9) \;\square\; \frac{-12}{4}$

4. $-(-7) \;\square\; -5 + [3 - (-2)]$

5. $|-2[9 - (-1)]| \;\square\; -(-5) + (-2)$

6. $\frac{-3 + 5}{8(5 - 3)} \;\square\; \frac{-1}{2} + \frac{3 - (-2)}{2}$

7. $(-3)^2 \;\square\; (-2)^3$

8. $(5 - 7)^4 \;\square\; (7 - 4)^4$

9. $\left(-\frac{24}{8}\right)^2 \;\square\; |-9|$

10. $3^2 + 2^2 \;\square\; (3 + 2)^2$

11. On the number line, graph the real numbers between -3 and $\frac{3}{2}$.

12. On the number line, graph the real numbers from x to $2x$ if x is the negative of -3.

8.7 Properties of Real Numbers

- The Closure Properties
- The Commutative Properties
- The Associative Properties
- The Distributive Property
- The Identity Elements
- Additive and Multiplicative Inverses

The Closure Properties

The sum, difference, product, or quotient (except for division by zero) of any two real numbers is another real number. The properties are called the **closure properties**.

> **The Closure Properties:** If a and b are real numbers, then
> $a + b$ is a real number
> $a - b$ is a real number
> ab is a real number
> $\frac{a}{b}$ is a real number, provided that $b \neq 0$

EXAMPLE 1

Assume that $x = 8$ and $y = -4$. Find the real number answer to show that **a.** $x + y$, **b.** $x - y$, **c.** xy, and **d.** $\frac{x}{y}$ all represent real numbers.

Solution

We substitute 8 for x and -4 for y in each expression and simplify.

a. $x + y = 8 + (-4)$
$ = 4$

b. $x - y = 8 - (-4)$
$ = 8 + 4$
$ = 12$

c. $xy = 8(-4)$
$ = -32$

d. $\frac{x}{y} = \frac{8}{-4}$
$\phantom{\frac{x}{y}} = -2$

The Commutative Properties

We have seen that the **commutative properties** enable us to add or multiply two numbers in either order.

> **The Commutative Properties:** If a and b are two real numbers, then
> $a + b = b + a$ commutative property of addition
> $ab = ba$ commutative property of multiplication

EXAMPLE 2

Assume that $x = -3$ and $y = 7$. Show that **a.** $x + y = y + x$ and **b.** $xy = yx$.

Solution

a. We show that the sum $x + y$ is the same as the sum $y + x$ by substituting -3 for x and 7 for y in each expression and simplifying.

$$x + y = -3 + 7 = 4 \quad \text{and} \quad y + x = 7 + (-3) = 4$$

b. We show that the product xy is the same number as the product yx by substituting -3 for x and 7 for y in each expression and simplifying.

$$xy = -3(7) = -21 \quad \text{and} \quad yx = 7(-3) = -21$$

The Associative Properties

The **associative properties** are used to find sums and products when more than two numbers are involved.

> **The Associative Properties:** If a, b, and c are real numbers, then
>
> $(a + b) + c = a + (b + c)$ associative property of addition
> $(ab)c = a(bc)$ associative property of multiplication

The associative property of addition permits us to group, or *associate*, the numbers in a sum in any way that we wish. For example,

$$(3 + 4) + 5 = 7 + 5 = 12 \quad \text{and} \quad 3 + (4 + 5) = 3 + 9 = 12$$

The answer is 12 regardless of how we group the three numbers.

The associative property of multiplication permits us to group, or associate, the numbers in a product in any way that we wish. For example,

$$(3 \cdot 4) \cdot 7 = 12 \cdot 7 = 84 \quad \text{and} \quad 3 \cdot (4 \cdot 7) = 3 \cdot 28 = 84$$

The answer is 84 regardless of how we group the three numbers.

The Distributive Property

The **distributive property** shows how to multiply the sum of two numbers by a third number. Because of this property, we can often add first and then multiply, or multiply first and then add. For example, $2(3 + 7)$ can be calculated in two different ways. One way is to perform the indicated addition and then the multiplication:

$$2(3 + 7) = 2(10) = 20$$

The other way is to distribute the multiplication by 2 over the addition by first multiplying each number within the parentheses by 2 and then adding:

$$2(3 + 7) = 2 \cdot 3 + 2 \cdot 7$$
$$= 6 + 14$$
$$= 20$$

Either way, the result is 20.

In general, we have

> **The Distributive Property:** If a, b, and c are real numbers, then
> $$a(b + c) = ab + ac$$

Since multiplication is commutative, the distributive property can be written as

$$(b + c)a = ba + ca$$

EXAMPLE 3

Evaluate each expression in two ways. **a.** $3(5 + 9)$ and **b.** $-2(-7 + 3)$.

Solution

a. $3(5 + 9) = 3(14) = 42$
$3(5 + 9) = 3 \cdot 5 + 3 \cdot 9 = 15 + 27 = 42$

b. $-2(-7 + 3) = -2(-4) = 8$
$-2(-7 + 3) = -2(-7) + (-2)3 = 14 + (-6) = 8$ ∎

EXAMPLE 4

Use the distributive property to write $3(x + 2)$ without using parentheses.

Solution

$3(x + 2) = 3x + 3 \cdot 2$
$ = 3x + 6 \qquad$ Simplify. ∎

The distributive property can be extended to three or more terms. For example, if a, b, c, and d are real numbers, then

$$a(b + c + d) = ab + ac + ad$$

The Identity Elements

The numbers 0 and 1 play special roles in algebra. The number 0 is the only number that can be added to another number (say, a) to give an answer of that same number a:

$$0 + a = a + 0 = a$$

The number 1 is the only number that can be multiplied by another number (say, a) to give an answer of that same number a:

$$1 \cdot a = a \cdot 1 = a$$

Because adding 0 to a number or multiplying a number by 1 leaves that number identically the same, the numbers 0 and 1 are called **identity elements**.

> **The Identity Elements:** The number 0 is called the **identity element for addition**. The number 1 is called the **identity element for multiplication**.

Additive and Multiplicative Inverses

If the sum of two numbers is 0, the numbers are called **negatives**, **opposites**, or **additive inverses** of each other. For example, because $3 + (-3) = 0$, the numbers 3 and -3 are negatives, opposites, or additive inverses of each other. In general, because

$$a + (-a) = 0$$

the numbers represented by a and $-a$ are negatives, opposites, or additive inverses.

If the product of two numbers is 1, the numbers are called **reciprocals**, or **multiplicative inverses**, of each other. For example, because $7(\frac{1}{7}) = 1$, the numbers 7 and $\frac{1}{7}$ are reciprocals. Because $(-0.25)(-4) = 1$, the numbers -0.25 and -4 are reciprocals. In general, because

$$a\left(\frac{1}{a}\right) = 1 \quad \text{provided } a \neq 0$$

the numbers represented by a and $\frac{1}{a}$ are reciprocals or multiplicative inverses.

> **The Additive and Multiplicative Inverse Properties:** Because
> $$a + (-a) = 0$$
> the numbers represented by a and $-a$ are **negatives**, **opposites**, or **additive inverses** of each other. Because
> $$a\left(\frac{1}{a}\right) = 1 \quad \text{provided } a \neq 0$$
> the numbers represented by a and $\frac{1}{a}$ are **reciprocals** or **multiplicative inverses** of each other.

EXAMPLE 5

The property in the right column justifies the statement in the left column.

$3 + 4$ is a real number	The closure property of addition
$\frac{8}{3}$ is a real number	The closure property of division
$3 + 4 = 4 + 3$	The commutative property of addition
$-3 + (2 + 7) = (-3 + 2) + 7$	The associative property of addition
$(5)(-4) = (-4)(5)$	The commutative property of multiplication
$(ab)c = a(bc)$	The associative property of multiplication

$3(a + 2) = 3a + 3 \cdot 2$ The distributive property
$3 + 0 = 3$ The additive identity property
$3(1) = 3$ The multiplicative identity property
$2 + (-2) = 0$ The additive inverse property
$\left(\frac{2}{3}\right)\left(\frac{3}{2}\right) = 1$ The multiplicative inverse property

The properties of the real numbers are summarized as follows:

For all real numbers a, b, and c,

	Addition	Multiplication
Closure properties	$a + b$ is a real number	$a \cdot b$ is a real number
Commutative properties	$a + b = b + a$	$a \cdot b = b \cdot a$
Associative properties	$(a + b) + c = a + (b + c)$	$(ab)c = a(bc)$
Identity properties	$a + 0 = a$	$a \cdot 1 = a$
Inverse properties	$a + (-a) = 0$	$a \cdot \left(\frac{1}{a}\right) = 1$ $a \neq 0$
Distributive property		$a(b + c) = ab + ac$

8.7 Exercises

In Exercises 1–8, assume that $x = 12$ and $y = -2$. Show that each expression represents a real number by finding the real-number answer.

1. $x + y$
2. $y - x$
3. xy
4. $\dfrac{x}{y}$

5. x^2
6. y^2
7. $\dfrac{x}{y^2}$
8. $\dfrac{2x}{3y}$

In Exercises 9–14, assume that $x = 5$ and $y = 7$. Show that both given expressions have the same value.

9. $x + y$; $y + x$
10. xy; yx
11. $3x + 2y$; $2y + 3x$
12. $3xy$; $3yx$
13. $x(x + y)$; $(x + y)x$
14. $xy + y^2$; $y^2 + xy$

In Exercises 15–20, assume that $x = 2$, $y = -3$, and $z = 1$. Show that both given expressions have the same value.

15. $(x + y) + z$; $x + (y + z)$
16. $(xy)z$; $x(yz)$
17. $(xz)y$; $x(yz)$
18. $(x + y) + z$; $y + (x + z)$
19. $x^2(yz^2)$; $(x^2y)z^2$
20. $x(y^2z^3)$; $(xy^2)z^3$

In Exercises 21–30, use the distributive property to write each expression without parentheses.

21. $3(x + y)$ **22.** $4(a + b)$ **23.** $x(x + 3)$ **24.** $y(y + z)$

25. $-x(a + b)$ **26.** $a(x + y)$ **27.** $4(x^2 + x)$ **28.** $-2(a^2 + 3)$

29. $-5(t + 2)$ **30.** $-2a(x + a)$

In Exercises 31–40, give the additive and the multiplicative inverse of each number if possible.

31. 2 **32.** 3 **33.** $\frac{1}{3}$ **34.** $-\frac{1}{2}$

35. 0 **36.** -2 **37.** $-\frac{5}{2}$ **38.** 0.5

39. -0.2 **40.** $\frac{4}{3}$

In Exercises 41–52, state which property of real numbers justifies each statement.

41. $3 + x = x + 3$ **42.** $(3 + x) + y = 3 + (x + y)$

43. $xy = yx$ **44.** $(3)(2) = (2)(3)$

45. $-2(x + 3) = -2x + (-2)(3)$ **46.** $x(y + z) = (y + z)x$

47. $(x + y) + z = z + (x + y)$ **48.** $3(x + y) = 3x + 3y$

49. $5 \cdot 1 = 5$ **50.** $x + 0 = x$

51. $3 + (-3) = 0$ **52.** $9 \cdot \frac{1}{9} = 1$

In Exercises 53–62, use the given property to rewrite the expression in a different form.

53. $3(x + 2)$; distributive property **54.** $x + y$; commutative property of addition

55. y^2x; commutative property of multiplication **56.** $x + (y + z)$; associative property of addition

57. $(x + y)z$; commutative property of addition **58.** $x(y + z)$; distributive property

59. $(xy)z$; associative property of multiplication **60.** $1x$; multiplicative identity property

61. $0 + x$; additive identity property **62.** $5 \cdot \frac{1}{5}$; multiplicative inverse property

Review Exercises

In Review Exercises 1–4, write each English phrase as a mathematical expression.

1. The sum of x and the square of y.

2. Three more than the square of x.

3. The square of three more than x.

4. The sum of x and y is greater than or equal to the cube of z.

In Review Exercises 5–8, fill each blank with the appropriate response.

5. If $x \geq 0$, then $|x| = $ _____.

6. If $x < $ _____, then $|x| = -x$.

7. $x - y = x + ($ _____ $)$.

8. The product of two negative numbers is a _____ number.

In Review Exercises 9–12, let $x = 10$ and $y = -5$. Evaluate each expression and classify the result as a positive number, negative number, integer, and rational number.

9. $x - y$

10. $y^2 - x$

11. $\dfrac{x}{y + 1}$

12. $|x + y|^2$

Mathematics for Fun

The operation of addition is associative, but associativity does not hold if subtraction is involved: $a - (b + c)$ is not equal to $(a - b) + c$.

In the problem on the first page of this chapter, the calculation of the real cost of the gift should be $30 - 5 = 25$, or $30 - (3 + 2) = 25$. The puzzle uses the associative property incorrectly: $30 - (3 + 2)$ is not equal to $(30 - 3) + 2 = 27 + 2 = 29$. There is no missing dollar.

Chapter Summary

Key Words

absolute value (8.4)
additive inverse (8.7)
algebraic expressions (8.2)
algebraic terms (8.2)
associative properties (8.7)
base (8.3)
closure properties (8.7)
coefficient (8.2)
commutative properties (8.7)
composite numbers (8.1)
constants (8.2)
coordinate (8.1)
difference (8.2)
distributive property (8.7)
element (of a set) (8.1)
equal sign (8.1)
even natural number (8.1)
exponential expression (8.3)
exponents (8.3)
factors (8.2)
graph (8.1)
identity elements (8.7)
integers (8.4)
irrational numbers (8.4)
multiplicative inverse (8.7)
natural numbers (8.1)
negative numbers (8.4)
number line (8.1, 8.4)
numerical coefficient (8.2)
odd natural number (8.1)
opposites (8.4)
origin (8.1)
positive numbers (8.4)
power of x (8.3)
prime numbers (8.1)
product (8.2)
quotient (8.2)
rational numbers (8.4)
real numbers (8.4)
reciprocal (8.7)
set (8.1)
signed numbers (8.4)
subset (8.4)
sum (8.2)
variables (8.2)
whole numbers (8.1)

Key Ideas

(8.1) Sets of numbers can be graphed on the number line.

(8.2) If a term has no written numerical coefficient, the numerical coefficient is 1.

(8.3) If n is a natural number, then

$$x^n = \overbrace{x \cdot x \cdot x \cdot x \cdots \cdot x}^{n \text{ factors of } x}.$$

(8.4) If x represents a number, then $-(-x) = x$.

If $x \geq 0$, then $|x| = x$.

If $x < 0$, then $|x| = -x$.

(8.5) If two real numbers x and y have the same sign, their sum is found by adding their absolute values and using their common sign.

If two real numbers x and y have unlike signs, their sum is found by subtracting their absolute values (the smaller from the larger) and using the sign of the number with the greatest absolute value.

If x and y are two real numbers, then $x - y = x + (-y)$.

(8.6) The product of two real numbers with like signs is the positive product of their absolute values.

The product of two real numbers with unlike signs is the negative of the product of their absolute values.

The quotient of two real numbers with like signs is the positive of the quotient of their absolute values.

The quotient of two real numbers with unlike signs is the negative of the quotient of their absolute values.

Division by zero is undefined.

(8.7) **The closure properties:**

$x + y$ is a real number.

$x - y$ is a real number.

xy is a real number.

$\dfrac{x}{y}$ is a real number   provided $y \neq 0$.

The commutative properties:

$x + y = y + x \qquad xy = yx$

The associative properties:

$(x + y) + z = x + (y + z) \qquad (xy)z = x(yz)$

The distributive property:

$x(y + z) = xy + xz$

The identity elements:

0 is the identity for addition.

1 is the identity for multiplication.

The additive and multiplicative inverse properties:

$x + (-x) = 0 \qquad x\left(\dfrac{1}{x}\right) = 1 \qquad$ provided $x \neq 0$

Chapter 8 Review Exercises

[8.1] In Review Exercises 1–4, consider the set of numbers $\{0, 1, 2, 3, 4, 5\}$.

1. Which numbers are natural numbers?

2. Which numbers are prime numbers?

3. Which numbers are odd natural numbers?

4. Which numbers are composite numbers?

5. Draw a number line and graph the composite numbers from 10 to 20.

6. Draw a number line and graph the whole numbers between 15 and 25.

In Review Exercises 7–8, insert one of the symbols =, <, or > to make the statement true.

7. $8 + 3 \square 13 - 3$

8. $\dfrac{15}{4} \square \dfrac{34}{7}$

[8.2] In Review Exercises 9–10, follow the given directions.

9. Let x represent a certain number. Write an algebraic expression denoting a number that is three more than twice x.

10. Let x and y represent two numbers. Write an algebraic expression to denote the difference obtained when twice x is subtracted from the product of 3 and y.

In Review Exercises 11–12, let x represent 6 and y represent 8. Evaluate each expression.

11. $\dfrac{x + y}{x - 4}$

12. $\dfrac{xy - 12}{4 + y}$

13. How many terms does the expression $3x + 4y + 9$ have?

14. What is the numerical coefficient of the term $7xy$?

15. What is the numerical coefficient of the term xy?

16. Find the sum of the numerical coefficients in the expression $2x^3 + 4x^2 + 3x$.

[8.3] In Review Exercises 17–20, assume that x = 2 and y = 3. Find the value of each expression.

17. y^4

18. x^y

19. $x^2 + xy^2$

20. $\dfrac{x^2 + y}{x^3 - 1}$

[8.4] In Review Exercises 21–24, place one of the symbols =, <, or > in the box to make a true statement.

21. $-5 \square 1$

22. $-7 \square -5$

23. $-1 \square -3$

24. $3 \square -3$

25. Division by a certain number is undefined. What is the number?

26. Draw a number line and graph all real numbers less than -2 or greater than 2.

27. Draw a number line and graph all real numbers between -4 and 3.

28. Evaluate $|6|$.

29. Evaluate $-|5|$.

30. Evaluate $|-8|$.

[8.5–8.6] In Review Exercises 31–46, let x = 2, y = −3, and z = −1. Evaluate each expression.

31. $y + z$

32. $x + y$

33. $x + (y + z)$

34. $x - y$

35. $x - (y - z)$

36. $(x - y) - z$

37. xy

38. yz

39. $x(x + z)$

40. xyz

41. $y^2 z + x$

42. $yz^3 + (xy)^2$

43. $\dfrac{xy}{z}$

44. $\dfrac{|xy|}{3z}$

45. $\dfrac{3y^2 - x^2 + 1}{y |z|}$

46. $\dfrac{2y^2 - xyz}{x^2 |yz|}$

[8.7] In Review Exercises 47–56, tell which property of real numbers justifies each statement. Assume that all variables represent real numbers.

47. $x + y$ is a real number

48. $3 \cdot (4 \cdot 5) = (4 \cdot 5) \cdot 3$

49. $3 + (4 + 5) = (3 + 4) + 5$

50. $3(x + 2) = 3 \cdot x + 3 \cdot 2$

51. $a + x = x + a$

52. $3(4 \cdot 5) = (3 \cdot 4) \cdot 5$

53. $3 + (x + 1) = (x + 1) + 3$

54. $x \cdot 1 = x$

55. $17 + (-17) = 0$

56. $x + 0 = x$

Name _____ Section _____

Chapter 8 Test

1. List the prime numbers between 30 and 50.

2. Graph the composite numbers less than 10 on a number line.

3. Graph the real numbers from 5 to 15 on a number line.

4. Let x and y represent two real numbers. Write an algebraic expression to denote the quotient obtained when the product of the two numbers is divided by their sum.

5. Let x and y represent two real numbers. Write an algebraic expression to denote the difference obtained when the sum of x and y is subtracted from the product of 5 and y.

In Problems 6–7, which of the symbols =, <, or > placed in the box will make the statement true?

6. $3(4-2) \square -2(2-5)$

7. $1 + 4 \cdot 3 \square -2(-7)$

8. What is the numerical coefficient of the term $3xy^2$?

9. Evaluate $|-16|$.

10. Evaluate $-|23|$.

11. Evaluate $-|7| + |-7|$.

In Problems 12–17, let $x = -2$, $y = 3$, and $z = 4$. Evaluate each expression.

12. $x + y + z$

13. $\dfrac{z + 4y}{2x}$

14. $|x^y - z|$

15. $x^3 + y^2 + z$ _____

16. $|x - y| - |z|$ _____

17. $|x| - 3|y| - 4|z|$ _____

18. What is the identity element for addition? _____

19. What is the multiplicative inverse of $\frac{1}{5}$? _____

In Problems 20–23, state which property of the real numbers justifies each statement.

20. $(xy)z = z(xy)$ _____

21. $3(x + y) = 3x + 3y$ _____

22. $2 + x = x + 2$ _____

23. $7 \cdot \frac{1}{7} = 1$ _____

24. Jack lives 12 miles from work and 7 miles from the grocery store. If he made x round trips to work and y round trips to the store, how many miles did Jack drive? _____

9 Equations and Inequalities

Mathematics in Retirement

Most corporations provide their employees with a retirement plan. Both the employee and the company contribute money to the plan, and the funds are invested and provide income at retirement. Federal tax code also provides opportunity for self-employed workers to fund a retirement plan. One such plan, called a **SEP** (Simplified Employee Pension), allows an annual contribution that does not exceed 15% of the income available after deductible expenses.

That would seem to be an easy calculation—subtract the expenses from gross income, and take 15% of what's left. Tax code, however, is never so simple. The contribution itself is considered an expense: from gross income, subtract expenses (including the amount of the contribution), and use what's left to calculate the amount of the contribution. It would seem that to determine the contribution, you must first know the contribution!

How is it possible to calculate the maximum annual SEP contribution? Equations, the topic of this chapter, can help with even this taxing problem.

To answer such questions as How many? How far? How fast? How heavy? we make use of **equations**. In this chapter, we discuss this important idea.

9.1 The Addition and Subtraction Properties of Equality

- Checking Solutions
- Solving Equations
- Forming Equations

An **equation** is a statement showing that two expressions are equal. Each of the following statements is an example of an equation.

$$x + 5 = 21 \qquad 3(3x + 4) = 3 + x \qquad 3x^2 - 4x + 5 = 0$$

In the equation $x + 5 = 21$, the expression $x + 5$ is called the **left-hand side**, and 21 is called the **right-hand side**. The letter x is called the **variable** (or the **unknown**) of the equation.

An equation can be true or false. For example, $16 + 5 = 21$ is a true equation, whereas $10 + 5 = 21$ is a **false equation**. An equation such as $2x - 5 = 11$ might be true or false, depending on the value of x. If $x = 8$, the equation is true, because

$$2(8) - 5 = 16 - 5 = 11$$

However, this equation is false for all other values of x.

Any number that makes an equation true when substituted for its variable is said to **satisfy** the equation. Such numbers are called **solutions** or **roots**. Because 8 is the only number that satisfies the equation $2x - 5 = 11$, it is the only solution of the equation.

Checking Solutions

EXAMPLE 1

Verify that 16 is a solution of the equation $x + 5 = 21$.

Solution

We substitute 16 for the variable x in the equation and verify that both sides are equal.

$$x + 5 = 21$$
$$16 + 5 \stackrel{?}{=} 21 \qquad \text{Replace } x \text{ with 16.}$$
$$21 = 21 \qquad \text{Simplify.}$$

Since both sides equal 21, 16 is a solution.

EXAMPLE 2

Is 6 a solution of the equation $3x - 5 = 2x$?

Solution

Substitute 6 for the variable x in the equation and simplify.

$$3x - 5 = 2x$$
$$3 \cdot 6 - 5 \stackrel{?}{=} 2 \cdot 6 \qquad \text{Replace } x \text{ with 6.}$$
$$18 - 5 \stackrel{?}{=} 12 \qquad \text{Simplify.}$$
$$13 = 12 \qquad \text{Simplify.}$$

Because the left- and right-hand sides are *not* equal, 6 is *not* a solution. Work Progress Check 1.

Progress Check 1

Is 8 a solution of $2x + 4 = 3x - 4$?

Solving Equations

To *solve an equation* means to find its solutions. To do so, we use certain properties of equality to isolate the variable on one side of the equation. We begin by considering the addition and subtraction properties of equality. When either of these properties is applied, the resulting equation will have the same solutions as the original equation.

The Addition Property of Equality: Let a, b, and c be real numbers:

If $a = b$, then $a + c = b + c$.

---

1. yes

The **addition property of equality** can be stated in words: *If the same quantity is added to equal quantities, the results will be equal quantities.*

EXAMPLE 3

Solve the equation $x - 5 = 2$.

Solution

To solve for x, we must isolate x on one side of the equation. We do this by using the addition property of equality and adding 5 to both sides.

$$x - 5 = 2$$
$$x - 5 + 5 = 2 + 5 \quad \text{Add 5 to both sides of the equation.}$$
$$x = 7 \quad \text{Simplify.}$$

The solution is 7. We check this solution by substituting 7 for x in the original equation and simplifying.

$$x - 5 = 2$$
$$7 - 5 \stackrel{?}{=} 2 \quad \text{Replace } x \text{ with 7.}$$
$$2 = 2 \quad \text{Simplify.}$$

The solution checks. ∎

The Subtraction Property of Equality: Let a, b, and c be real numbers:

If $a = b$, then $a - c = b - c$.

The **subtraction property of equality** can be stated in words: *If the same quantity is subtracted from equal quantities, the results will be equal quantities.*

EXAMPLE 4

Solve the equation $x + 4 = 9$.

Solution

To isolate x on one side of the equation, we undo the indicated addition of 4 by subtracting 4 from both sides of the equation.

$$x + 4 = 9$$
$$x + 4 - 4 = 9 - 4 \quad \text{Subtract 4 from both sides.}$$
$$x = 5 \quad \text{Simplify.}$$

The solution is 5. We check by substituting 5 for x in the original equation and simplifying.

$$x + 4 = 9$$
$$5 + 4 \stackrel{?}{=} 9 \quad \text{Replace } x \text{ with 5.}$$
$$9 = 9 \quad \text{Simplify.}$$

Progress Check 2

Solve and check each equation.

a. $x - 8 = 12$

b. $x + 7 = 14$

The solution checks. Note that instead of subtracting 4 from both sides, we could just as well have added -4 to both sides. Work Progress Check 2.

Forming Equations

To use algebra to solve *applied* problems, we must understand the problem, form an equation, solve the equation, and check it.

EXAMPLE 5

A stockbroker lost 17 clients who went bankrupt last year. He now has 73 clients. How many did he have before?

Solution

Let c represent the original number of clients. Translate the words of the problem into an equation as follows, and solve the equation.

The original number of clients	minus	17	is	the present number of clients.
c	$-$	17	$=$	73

$$c - 17 = 73$$
$$c - 17 + 17 = 73 + 17 \quad \text{Add 17 to both sides.}$$
$$c = 90 \quad \text{Simplify.}$$

Check: Originally, the stockbroker had 90 clients. After losing 17, he has $90 - 17$, or 73 clients left. The solution checks.

EXAMPLE 6

A 1700-year-old manuscript is 425 years older than the clay jar in which it was found. How old is the jar?

Solution

Let j represent the age of the clay jar. Translate the words of the problem into an equation and solve it.

The age of the manuscript	is	425	more than	the age of the jar.
1700	$=$	425	$+$	j

$$1700 = 425 + j$$
$$1700 - 425 = 425 + j - 425 \quad \text{Subtract 425 from both sides.}$$
$$1275 = j \quad \text{Simplify.}$$

Because 1275 and j represent the same number, the previous result can be written as

$$j = 1275$$

Check: The clay jar is 1275 years old. The manuscript, at age 1700, is 425 years older than the 1275-year-old jar. The solution checks.

2. a. 20 b. 7

EXAMPLE 7

Mike has $192. However, he wants to buy a ten-speed bike that is on sale for $209. How much more money does he need?

Solution

Let x represent the extra money that Mike needs to purchase the bike. Translate the words of the problem into an equation as follows, and solve it.

The amount Mike has	plus	the amount Mike needs	is	the bike's total cost.
192	+	x	=	209

$$192 + x = 209$$
$$192 + x - 192 = 209 - 192 \quad \text{Subtract 192 from both sides.}$$
$$x = 17 \quad \text{Simplify.}$$

Check: Mike needs $17 more. Then he will have $192 + $17, or $209, to buy the bike. The solution checks. ∎

9.1 Exercises

In Exercises 1–8, indicate whether each statement is an equation.

1. $x = 2$
2. $y - 3$
3. $7x < 8$
4. $7 + x = 2$
5. $x + y = 0$
6. $3 - 3y > 2$
7. $1 + 1 = 3$
8. $5 = a + 2$

In Exercises 9–26, indicate whether the indicated number is a solution of the given equation.

9. $x + 2 = 3$; 1
10. $x - 2 = 4$; 6
11. $a - 7 = 0$; 7
12. $x + 4 = 4$; 0
13. $8 - y = y$; -4
14. $10 - c = c$; 5
15. $2x + 32 = 0$; 16
16. $3x - 9 = 0$; -3
17. $z + 7 = z$; -7
18. $n - 9 = n$; 9
19. $2x = x$; 0
20. $3x = 2x$; 0
21. $3k + 5 = 5k - 1$; 3
22. $2s - 1 = s + 7$; 6
23. $\frac{x}{2} + 3 = 5$; 4
24. $x - \frac{x}{7} = 12$; 21
25. $\frac{x-5}{6} + x = x + 1$; 11
26. $\frac{5+x}{10} - x = \frac{1}{2}$; 0

In Exercises 27–46, use the addition property of equality to solve each equation. Check all solutions.

27. $x - 7 = 3$
28. $y - 3 = 7$
29. $a - 2 = -5$
30. $z - 3 = -9$
31. $2 = -5 + b$
32. $3 = -7 + t$
33. $x - 4 = 0$
34. $c - 3 = 0$
35. $y - 7 = 6$
36. $a - 2 = 4$
37. $17 = -15 + x$
38. $16 = -9 + b$

39. $312 = x - 428$ **40.** $x - 307 = -113$ **41.** $x - \dfrac{5}{2} = -\dfrac{5}{2}$ **42.** $y - \dfrac{3}{5} = 0$

43. $z - \dfrac{2}{3} = \dfrac{1}{2}$ **44.** $r - \dfrac{1}{5} = \dfrac{1}{3}$ **45.** $h - \dfrac{1}{3} = 2$ **46.** $w - 3 = -\dfrac{1}{3}$

In Exercises 47–66, use the subtraction property of equality to solve each equation. Check all solutions.

47. $x + 9 = 3$ **48.** $x + 3 = 9$ **49.** $y + 7 = 12$ **50.** $c + 11 = 22$

51. $t + 19 = 28$ **52.** $s + 34 = -45$ **53.** $23 + x = -13$ **54.** $34 + y = 34$

55. $3 = 4 + c$ **56.** $41 = 23 + x$ **57.** $-19 = r + 43$ **58.** $-92 = r + 37$

59. $\dfrac{5}{3} = \dfrac{5}{3} + x$ **60.** $\dfrac{29}{12} = \dfrac{29}{12} + x$ **61.** $d + \dfrac{2}{3} = \dfrac{3}{2}$ **62.** $s + \dfrac{2}{3} = \dfrac{1}{5}$

63. $w + \dfrac{1}{3} = 5$ **64.** $2 = x + \dfrac{3}{5}$ **65.** $r + 2 = -\dfrac{1}{2}$ **66.** $-\dfrac{1}{3} = s + \dfrac{2}{3}$

In Exercises 67–90, solve each equation. Check all solutions.

67. $3 + x = -7$ **68.** $4 + b = -8$ **69.** $y - 5 = 7$ **70.** $z - 9 = 3$

71. $-4 + a = 12$ **72.** $5 + x = 13$ **73.** $13 + x = 34$ **74.** $-23 + x = 19$

75. $-37 + z = 37$ **76.** $-43 + a = -43$ **77.** $-57 = b - 29$ **78.** $-93 = 67 + y$

79. $493 = -313 + x$ **80.** $347 = -4132 + y$ **81.** $137 = x - 323$ **82.** $-773 = -577 + y$

83. $\dfrac{13}{17} = x + \dfrac{13}{17}$ **84.** $\dfrac{23}{71} = \dfrac{23}{71} + x$ **85.** $\dfrac{5}{7} = x - \dfrac{3}{5}$ **86.** $\dfrac{21}{3} = x + \dfrac{7}{4}$

87. $x - \dfrac{13}{2} = \dfrac{21}{4}$ **88.** $\dfrac{2}{13} + x = 1$ **89.** $-\dfrac{5}{17} + y = -3$ **90.** $x + \dfrac{11}{17} = \dfrac{19}{17}$

In Exercises 91–106, translate the words into an equation and solve it. Check all solutions.

91. What number added to 6 gives 11?

92. What number decreased by 4 gives 8?

93. What number increased by 9 gives 27?

94. If a certain number is decreased by 41, the result is 31. Find the number.

95. Three of Jennifer's party invitations were lost in the mail, but 59 were delivered. How many invitations did Jennifer send?

96. Hank graduated from college eight years ago. He is now 31 years old. How old was Hank at graduation?

97. The price of a condominium is $57,500 less than the cost of a house. The house costs $102,700. Find the price of the condominium.

98. Heidi ate 10 pancakes, 6 more than her sister Sarah. How many pancakes did Sarah eat?

99. Maria ate 4 hot dogs less than her brother Kurt. Kurt ate 7 hot dogs. How many hotdogs did Maria eat?

100. Kim and Mary shared the costs of driving to Pittsburgh. Kim paid $12 more than Mary, who chipped in $68. How much did Kim pay?

101. A man needs $345 for a new set of golf clubs. How much more money does he need if he now has $317?

102. A student needs 42 cents more to buy a $4.50 movie ticket. How much money does he have?

103. Heather paid $48.50 for a sweater. Holly bought the same sweater on sale for $9 less. How much did Holly pay for the sweater?

104. Last week, Molly made $543.21 more than Max. How much money did Molly make if Max made $123.45?

105. A woman paid $29 less to have her car fixed at a muffler shop than she would have at a gas station. She would have paid $219 at the gas station. How much did she pay to have her car fixed?

106. A man had to wait 20 minutes for a bus today. Three days ago, he had to wait 15 minutes longer than he did today because four buses passed by without stopping. How long did he wait three days ago?

Review Exercises

In Review Exercises 1–8, perform the indicated operations and classify the result as a whole number, a natural number, an even natural number, or an odd natural number.

1. $3[2 - (-3)]$
2. $3 - 3 \cdot (-2)$
3. $5(-3) + 3^3$
4. $(2 - 4)^4$

5. $\dfrac{2^3 - 14}{3^2 - 3}$
6. $\dfrac{5^2 - 2 \cdot 9}{5 \cdot 7 - 6^2}$
7. $\dfrac{1}{2^2} + \dfrac{3}{6 - 2}$
8. $\dfrac{3 + 5}{3} - \dfrac{5}{7 - 4}$

In Review Exercises 9–12, let $x = 3$, $y = -5$, and $z = 1$. Evaluate each expression.

9. $x(y - z)$
10. $xy - z$
11. $y^x + z$
12. $(x^2 + y)^x$

9.2 The Division and Multiplication Properties of Equality

■ Solving Equations
■ Forming Equations

To solve many equations, we must divide or multiply both sides of the equation by the same nonzero number. When this is done, the resulting equation will have the same solutions as the original equation.

> **The Division Property of Equality:** Let a, b, and c be real numbers, with $c \neq 0$:
>
> If $a = b$, then $\dfrac{a}{c} = \dfrac{b}{c}$.

The **division property of equality** can be stated in words: *If equal quantities are divided by the same nonzero quantity, the results will be equal quantities.*

Solving Equations

EXAMPLE 1

Solve the equation $3x = 6$.

Solution

To isolate x on one side of the equation, we undo the multiplication by 3 by dividing both sides of the equation by 3.

$$3x = 6$$
$$\frac{3x}{3} = \frac{6}{3} \quad \text{Divide both sides by 3.}$$
$$x = 2 \quad \text{Simplify.}$$

The solution of the equation is 2. Check it as follows:

$$3x = 6$$
$$3 \cdot 2 \stackrel{?}{=} 6 \quad \text{Replace } x \text{ with 2.}$$
$$6 = 6 \quad \text{Simplify.}$$

Because the final equation is a true statement, the solution checks. ∎

EXAMPLE 2

Solve the equation $-5x = 15$.

Solution

To isolate x on one side of the equation, we undo the multiplication by -5 by dividing both sides of the equation by -5.

$$-5x = 15$$
$$\frac{-5x}{-5} = \frac{15}{-5} \quad \text{Divide both sides by } -5.$$
$$x = -3 \quad \text{Simplify.}$$

The solution of the equation is -3. Check it as follows:

$$-5x = 15$$
$$-5(-3) \stackrel{?}{=} 15 \quad \text{Replace } x \text{ with } -3.$$
$$15 = 15 \quad \text{Simplify.}$$

Work Progress Check 3. ∎

We can also multiply both sides of an equation by the same nonzero number, and the resulting equation will have the same solutions as the original equation.

Progress Check 3

Solve $-7x = 28$.

3. -4

Progress Check Answers

> **The Multiplication Property of Equality:** Let a, b, and c be real numbers, with $c \neq 0$.
>
> If $a = b$, then $ca = cb$.

The **multiplication property of equality** can be stated in words: *If equal quantities are multiplied by the same nonzero quantity, the results will be equal quantities.*

EXAMPLE 3

Solve the equation $\dfrac{x}{5} = 7$.

Solution

To isolate x on one side, we undo the division by 5 by multiplying both sides of the equation by 5.

$$\dfrac{x}{5} = 7$$

$$5 \cdot \dfrac{x}{5} = 5 \cdot 7 \qquad \text{Multiply both sides by 5.}$$

$$x = 35 \qquad \text{Simplify.}$$

Check:

$$\dfrac{x}{5} = 7$$

$$\dfrac{35}{5} \stackrel{?}{=} 7 \qquad \text{Replace } x \text{ with 35.}$$

$$7 = 7 \qquad \text{Simplify.}$$

Work Progress Check 4.

Progress Check 4

Solve $\dfrac{x}{4} = 2$.

EXAMPLE 4

Solve the equation $-12x + 5 = 1$.

Solution

To solve this equation, we must use both the addition and division properties.

$$-12x + 5 = 1$$

$$-12x + 5 + (-5) = 1 + (-5) \qquad \text{Add } -5 \text{ to both sides.}$$

$$-12x = -4 \qquad \text{Simplify.}$$

$$\dfrac{-12x}{-12} = \dfrac{-4}{-12} \qquad \text{Divide both sides by } -12.$$

$$x = \dfrac{1}{3} \qquad \text{Simplify.}$$

Check:

$$-12x + 5 = 1$$

$$-12\left(\dfrac{1}{3}\right) + 5 \stackrel{?}{=} 1 \qquad \text{Replace } x \text{ with } \dfrac{1}{3}.$$

$$-4 + 5 \stackrel{?}{=} 1 \qquad \text{Simplify.}$$

$$1 = 1 \qquad \text{Simplify.}$$

4. 8

Progress Check Answers

EXAMPLE 5

Solve the equation $5(x - 3) = -10$.

Solution

$$5(x - 3) = -10$$
$$5x - 5 \cdot 3 = -10 \qquad \text{Use the distributive property to remove parentheses.}$$
$$5x - 15 = -10 \qquad \text{Simplify.}$$
$$5x - 15 + 15 = 10 + 15 \qquad \text{Add 15 to both sides.}$$
$$5x = 5 \qquad \text{Simplify.}$$
$$\frac{5x}{5} = \frac{5}{5} \qquad \text{Divide both sides by 5.}$$
$$x = 1 \qquad \text{Simplify.}$$

Check:

$$5(x - 3) = -10$$
$$5(1 - 3) \stackrel{?}{=} -10 \qquad \text{Replace } x \text{ with 1.}$$
$$5(-2) \stackrel{?}{=} -10 \qquad \text{Simplify.}$$
$$-10 = -10 \qquad \text{Simplify.}$$

EXAMPLE 6

Solve the equation $\dfrac{3(x - 2)}{4} = -6$.

Solution

$$\frac{3(x - 2)}{4} = -6$$
$$4 \cdot \frac{3(x - 2)}{4} = 4 \cdot (-6) \qquad \text{Multiply both sides by 4.}$$
$$3(x - 2) = -24 \qquad \text{Simplify.}$$
$$3x - 6 = -24 \qquad \text{Use the distributive property to remove parentheses.}$$
$$3x - 6 + 6 = -24 + 6 \qquad \text{Add 6 to both sides.}$$
$$3x = -18 \qquad \text{Simplify.}$$
$$\frac{3x}{3} = \frac{-18}{3} \qquad \text{Divide both sides by 3.}$$
$$x = -6 \qquad \text{Simplify.}$$

Check:

$$\frac{3(x - 2)}{4} = -6$$
$$\frac{3(-6 - 2)}{4} \stackrel{?}{=} -6 \qquad \text{Replace } x \text{ with } -6.$$
$$\frac{3(-8)}{4} \stackrel{?}{=} -6 \qquad \text{Simplify.}$$
$$\frac{-24}{4} \stackrel{?}{=} -6 \qquad \text{Simplify.}$$
$$-6 = -6 \qquad \text{Simplify.}$$

EXAMPLE 7

Solve the equation $-\frac{2}{3}(4x - 6) = 2$.

Solution

$$-\frac{2}{3}(4x - 6) = 2$$

$$3\left(-\frac{2}{3}\right)(4x - 6) = 3(2) \quad \text{Multiply both sides by 3.}$$

$$-2(4x - 6) = 6 \quad \text{Simplify.}$$

$$-8x + 12 = 6 \quad \text{Remove parentheses.}$$

$$-8x + 12 + (-12) = 6 + (-12) \quad \text{Add } -12 \text{ to both sides.}$$

$$-8x = -6 \quad \text{Simplify.}$$

$$\frac{-8x}{-8} = \frac{-6}{-8} \quad \text{Divide both sides by } -8.$$

$$x = \frac{3}{4} \quad \text{Simplify.}$$

Check:

$$-\frac{2}{3}(4x - 6) = 2$$

$$-\frac{2}{3}\left[4\left(\frac{3}{4}\right) - 6\right] \stackrel{?}{=} 2 \quad \text{Replace } x \text{ with } \frac{3}{4}.$$

$$-\frac{2}{3}(3 - 6) \stackrel{?}{=} 2$$

$$-\frac{2}{3}(-3) \stackrel{?}{=} 2$$

$$2 = 2$$

Work Progress Check 5.

Progress Check 5

Solve $\frac{3}{4}(5x - 3) = 9$.

Forming Equations

EXAMPLE 8

If one-third of a number is decreased by 5, the result is 2. Find the number.

Solution

Let x represent the unknown number. Translate the statement into an equation, solve the equation, and check the solution.

One-third of a certain number	decreased by	5	is	2.
$\frac{1}{3}x$	$-$	5	$=$	2

5. 3

$$\frac{1}{3}x - 5 = 2 \quad \text{The equation to solve.}$$

$$\frac{1}{3}x - 5 + 5 = 2 + 5 \quad \text{Add 5 to both sides.}$$

$$\frac{1}{3}x = 7 \quad \text{Simplify.}$$

$$3\left(\frac{1}{3}x\right) = 3 \cdot 7 \quad \text{Multiply both sides by 3.}$$

$$x = 21 \quad \text{Simplify.}$$

The number is 21.

Check: One-third of 21 is 7. If 7 is decreased by 5, the result is 2. The answer checks. ∎

EXAMPLE 9

Bill's sales commissions this year were $2500 less than three times his last year's commissions. This year, Bill earned $38,000. How much did he earn last year?

Solution

Let c represent Bill's earnings last year. Then translate the words of the problem into an equation. His earnings this year (in dollars) can be expressed in two ways: as 2500 less than $3c$, and as 38,000.

Three	times	last year's earnings	less	$2500	is	$38,000.
3	×	c	−	2500	=	38,000

$$3c - 2500 = 38,000 \quad \text{The equation to solve.}$$

$$3c - 2500 + 2500 = 38,000 + 2500 \quad \text{Add 2500 to both sides.}$$

$$3c = 40,500 \quad \text{Simplify.}$$

$$\frac{3c}{3} = \frac{40,500}{3} \quad \text{Divide both sides by 3.}$$

$$c = 13,500 \quad \text{Simplify.}$$

Bill earned $13,500 in commissions last year.

Check: If Bill's commissions of $38,000 had been $2500 greater, he would have commissions of $38,000 + $2500, or $40,500. Three times last year's commissions is 3($13,500), or $40,500. The solution checks. ∎

EXAMPLE 10

The manager of a store hires a woman to distribute advertising circulars door to door. She will earn $5 a day plus a nickel for every advertisement distributed. After one day's work, she receives $42.50. How many circulars did she distribute?

Solution

Let a represent the number of circulars that the woman distributed. Then translate the words of the problem into an equation. The total payment can be expressed in two ways: as \$5 more than the nickel-apiece cost of distributing the circulars, and as \$42.50.

a ads at \$0.05 each	plus	\$5.00	is	\$42.50.
$0.05a$	+	5.00	=	42.50

$$0.05a + 5.00 = 42.50 \quad \text{The equation to solve.}$$
$$0.05a + 5.00 + (-5.00) = 42.50 + (-5.00) \quad \text{Add } -5.00 \text{ to both sides.}$$
$$0.05a = 37.50 \quad \text{Simplify.}$$
$$\frac{0.05a}{0.05} = \frac{37.50}{0.05} \quad \text{Divide both sides by 0.05.}$$
$$a = 750 \quad \text{Simplify.}$$

The woman distributed 750 advertisements. Check this result. ∎

9.2 Exercises

In Exercises 1–64, solve each equation and check each solution.

1. $3x = 3$
2. $5x = 5$
3. $6x = 15$
4. $4x = 18$

5. $7y = -21$
6. $9y = -27$
7. $-4y = -20$
8. $-8y = -2^5$

9. $-32z = 64$
10. $-17z = -51$
11. $9z + 2 = 11$
12. $3z + 5 = 11$

13. $3x - 2^3 = 1$
14. $7x - 19 = 2$
15. $11x + 17 = -5$
16. $13x - 29 = -3$

17. $43t + 72 = 158$
18. $96t + 23 = -265$
19. $-47 - 21s = 58$
20. $-151 + 13s = -229$

21. $5(x - 3) = 0$
22. $3(x + 7) = 6$
23. $-9(3 + y) = -9$
24. $4(y + 3) = 40$

25. $2(x + 7) = 10$
26. $3(x - 9) = 15$
27. $-4(2 + a) = -8$
28. $-7(2 + b) = 14$

29. $\frac{1}{3}b + 5 = 2$
30. $\frac{1}{5}a - 3 = -4$
31. $\frac{s}{11} + 9 = 6$
32. $\frac{s}{12} + 2 = 4$

33. $\frac{b + 5}{3} = 11$
34. $\frac{2 + a}{11} = 3$
35. $\frac{r + 7}{3} = 4$
36. $\frac{t - 2}{7} = -3$

37. $\frac{3(u - 2)}{5} = 3$
38. $\frac{5(v - 7)}{3} = -5$
39. $\frac{7(x - 4)}{4} = -21$
40. $\frac{9(3 + y)}{5} = -27$

41. $\frac{3x - 12}{2} = 9$
42. $\frac{5x + 10}{7} = 0$
43. $\frac{1}{2}(a + 8) = 1$
44. $\frac{1}{3}(2x - 3) = \frac{-17}{3}$

45. $\frac{2}{3}(5x - 3) = 38$
46. $\frac{3}{2}(3x + 2) = 3$
47. $\frac{3}{4}(7k - 5) = 12$
48. $\frac{5}{3}(6 + 5b) = 10$

49. $\frac{1}{3}(4n - 3) = 5$
50. $\frac{7}{11}(4x + 1) = -7$
51. $\frac{3}{8}(5x + 2) = -3$
52. $\frac{2}{7}(3m - 7) = -2$

53. $\frac{1}{4}\left(\frac{q}{2}+1\right)=1$ **54.** $\frac{1}{2}\left(\frac{a}{3}-4\right)=-3$ **55.** $\frac{2r-10}{6}+7=17$ **56.** $\frac{5a+10}{10}+7=0$

57. $\frac{3z+2}{17}=0$ **58.** $\frac{10t-4}{2}=1$ **59.** $\frac{17k-28}{21}+\frac{4}{3}=0$ **60.** $\frac{5a-2}{3}=\frac{1}{6}$

61. $-\frac{x}{3}-\frac{1}{2}=-\frac{5}{2}$ **62.** $2\left(\frac{17-a}{3}\right)=4$ **63.** $\frac{1}{3}\left(\frac{9-w}{5}\right)=\frac{2}{5}$ **64.** $\frac{3t-5}{5}+\frac{1}{2}=-\frac{19}{2}$

In Exercises 65–90, translate the words into an equation and solve it. Check all solutions.

65. Six less than 3 times a certain number is 9. Find the number.

66. Four times a certain number is increased by 7. The result is 43. Find the number.

67. If 2 is added to twice a certain number, the result is 12. Find the number.

68. If 5 is subtracted from a certain number, the result is 11. Find the number.

69. Four more than 3 times a certain number is 25. Find the number.

70. If a certain number is decreased by 7, and that result is doubled, the number 6 is obtained. Find the original number.

71. When 4 more than a certain number is divided by 3, the result is 4. Find the number.

72. If 12 is decreased by a certain number, and that result is multiplied by 5, the result is 15. Find the number.

73. If 5 is added to a certain number, and that result is divided by 6, the value obtained is 2. Find the number.

74. If a certain number is tripled, increased by 7, and the result is then divided by 2, the number 5 is obtained. Find the original number.

75. If a certain number is tripled, then decreased by 7, and the result is then divided by 2, the number 1 is obtained. Find the original number.

76. Seventeen more than one-third of a certain number is 18. Find the number.

77. The number 9 decreased by one-half of a certain number is 11. Find the number.

78. One-half of the sum of 3 and one-half of a certain number is 4. Find the number.

79. Two-thirds of the movie audience left the theater in disgust. If 78 angry patrons walked out, how many were there originally?

80. Mandy was disturbed to learn that only four-sevenths of her class passed the first exam. If 16 students did pass the exam, how large is the class?

81. Susan's sales in February were $1000 less than three times her sales in January. If her February sales were $5000, how much did she sell in January?

82. Martina is 22, which is twice as old as her brother will be next year. How old is her brother now?

83. When Maggie is 51, she will be 3 times as old as she was last year. How old is she now?

84. If a girl had twice as much money as she has now, she would have enough money to buy a $3.75 movie ticket and an 85-cent box of popcorn. How much money does she have now?

85. In 5 years, Roger will be 3 times as old as Irvin is now. In 5 years, Irvin will be 27. How old is Roger now?

86. Jane won't move to that bigger apartment because its $400 monthly rent is $100 less than twice what she is now paying. What is her current monthly rent?

87. A mechanic charged $20 an hour to repair the water pump on a car, plus $95 for parts. The total bill was $155. How many hours did the repair take?

88. Vacation boarding for a dog at the kennel is $16 plus $12 a day. The stay cost her owners $100. How many days were her owners gone?

89. Wanda's monthly water bill is $3, plus 20 cents for every 100 gallons of water used. Last month, Wanda paid $5.40 for water. How much water did she use?

90. A music group charges $1500 for each performance, plus one-fifth of the total ticket sales. After the concert, the promoter gave the group's agent a check for $2980. How much money did the ticket sales raise?

Review Exercises

In Review Exercises 1–6, find each quantity.

1. Find the perimeter of a rectangle with sides measuring 8.5 cm and 16.5 cm.

2. Find the area of a rectangle with sides measuring 2.3 in. and 3.7 in.

3. Find the area of a triangle with a base of 9 cm and a height of 6 cm.

4. Find the area of a trapezoid with a height of 8.5 in. and bases measuring 6.7 in. and 12.2 in.

5. Find the volume of a rectangular solid with dimensions of 8.2 cm by 7.6 cm by 10.2 cm.

6. Find the volume of a sphere with a radius of 21 ft. (Use $\pi \approx \frac{22}{7}$.)

In Review Exercises 7–12, indicate the property of real numbers or the property of equality that justifies each statement.

7. $3 + 31$ is a real number.

8. $3(x + y) = 3x + 3y$

9. $a + (3 + b) = (3 + b) + a$

10. $a + (3 + b) = (a + 3) + b$

11. $3xy + (-3xy) = 0$

12. If $a + b = c$, then $a + b - x = c - x$.

9.3 Simplifying Expressions to Solve Equations

- Combining Like Terms
- Solving More Equations
- Identities and Impossible Equations

A **term** is either a number or the product of numbers and variables. Some examples of terms are

$7x$, $-3xy$, y^2, and 8

The **numerical coefficient** of the term $7x$ is 7, the numerical coefficient of the term $-3xy$ is -3, and the numerical coefficient of the term y^2 is the understood factor of 1. The number 8 is the numerical coefficient of the term 8.

Definition: **Like terms**, or **similar terms**, are terms with exactly the same variables and exponents.

The terms $3x$ and $5x$ are **like terms**, as are $9x^2$ and $-3x^2$. The terms $4xy$ and $3x^2$ are **unlike terms**, because they have different variables. The terms $4x$ and $5x^2$ are unlike terms, because the variables have different exponents.

Combining Like Terms

The distributive property can be used to combine terms of algebraic expressions that contain sums or differences of like terms. For example, the terms of the algebraic expression $3x + 5x$ can be combined as follows:

$$3x + 5x = (3 + 5)x = 8x$$

Likewise,

$$9xy^2 - 11xy^2 = (9 - 11)xy^2 = -2xy^2$$

These examples suggest the following rule.

> **Combining Like Terms:** To combine like terms, add their numerical coefficients and keep the same variables and exponents.

If the terms of an expression are not like terms, they cannot be combined. Because the terms of the expression $9xy^2 - 11x^2y$, for example, have variables with different exponents, they are unlike terms and cannot be combined.

EXAMPLE 1

Simplify $3(x + 2) + 2(x - 8)$.

Solution

$3(x + 2) + 2(x - 8)$
$= 3x + 3 \cdot 2 + 2x - 2 \cdot 8$ Remove parentheses.
$= 3x + 6 + 2x - 16$ Simplify.
$= 3x + 2x + 6 - 16$ Use the associative and commutative properties of addition to rearrange terms.
$= 5x - 10$ Combine terms.

EXAMPLE 2

Simplify $3(x - 3) - 5(x + 4)$.

Solution

$3(x - 3) - 5(x + 4)$
$= 3(x - 3) + (-5)(x + 4)$ $a - b = a + (-b)$.
$= 3x - 3 \cdot 3 + (-5)x + (-5)4$ Remove parentheses.
$= 3x - 9 + (-5x) + (-20)$ Multiply.
$= -2x - 29$ Combine terms.

Work Progress Check 6.

Progress Check 6

Simplify:

a. $2(x + 2) + 3(x - 4)$

b. $5(x - 4) - 2(x - 1)$

---

6. a. $5x - 8$ **b.** $3x - 18$

Progress Check Answers

Solving More Equations

We must combine like terms to solve the equations in the next examples.

EXAMPLE 3

Solve the equation $3(x + 2) - 5x = 0$.

Solution

$$\begin{align} 3(x + 2) - 5x &= 0 & \\ 3x + 3 \cdot 2 - 5x &= 0 & \text{Remove parentheses.} \\ 3x - 5x + 6 &= 0 & \text{Rearrange terms and simplify.} \\ -2x + 6 &= 0 & \text{Combine terms.} \\ -2x + 6 + (-6) &= 0 + (-6) & \text{Add } -6 \text{ to both sides.} \\ -2x &= -6 & \text{Combine terms.} \\ \frac{-2x}{-2} &= \frac{-6}{-2} & \text{Divide both sides by } -2. \\ x &= 3 & \text{Simplify.} \end{align}$$

Check:

$$\begin{align} 3(x + 2) - 5x &= 0 \\ 3(3 + 2) - 5 \cdot 3 &\stackrel{?}{=} 0 & \text{Replace } x \text{ with 3.} \\ 3 \cdot 5 - 5 \cdot 3 &\stackrel{?}{=} 0 \\ 15 - 15 &\stackrel{?}{=} 0 \\ 0 &= 0 \end{align}$$

EXAMPLE 4

Solve the equation $3(x - 5) = 4(x + 9)$.

Solution

$$\begin{align} 3(x - 5) &= 4(x + 9) & \\ 3x - 15 &= 4x + 36 & \text{Remove parentheses.} \\ 3x - 15 + (-3x) &= 4x + 36 + (-3x) & \text{Add } -3x \text{ to both sides.} \\ 3x - 3x - 15 &= 4x - 3x + 36 & \text{Rearrange terms.} \\ -15 &= x + 36 & \text{Combine terms.} \\ -15 + (-36) &= x + 36 + (-36) & \text{Add } -36 \text{ to both sides.} \\ -51 &= x & \text{Combine terms.} \end{align}$$

or

$$x = -51$$

Check:

$$\begin{align} 3(x - 5) &= 4(x + 9) \\ 3(-51 - 5) &\stackrel{?}{=} 4(-51 + 9) & \text{Replace } x \text{ with } -51. \\ 3(-56) &\stackrel{?}{=} 4(-42) \\ -168 &= -168 \end{align}$$

EXAMPLE 5

Solve the equation $\dfrac{3x + 11}{5} = x + 3$.

Solution

First, we clear the equation of fractions by multiplying both sides of the equation by 5. When multiplying the right-hand side of the equation by 5, be sure to multiply the *entire* right-hand side by 5.

$\dfrac{3x + 11}{5} = x + 3$	
$5 \cdot \dfrac{3x + 11}{5} = 5(x + 3)$	Multiply both sides by 5.
$3x + 11 = 5(x + 3)$	Simplify the left-hand side.
$3x + 11 = 5x + 15$	Remove parentheses.
$3x + 11 + (-11) = 5x + 15 + (-11)$	Add -11 to both sides.
$3x = 5x + 4$	Combine terms.
$3x + (-5x) = 5x + 4 + (-5x)$	Add $-5x$ to both sides.
$-2x = 4$	Combine terms.
$\dfrac{-2x}{-2} = \dfrac{4}{-2}$	Divide both sides by -2.
$x = -2$	Simplify.

Check:

$$\dfrac{3x + 11}{5} = x + 3$$

$$\dfrac{3(-2) + 11}{5} \stackrel{?}{=} (-2) + 3 \quad \text{Replace } x \text{ with } -2.$$

$$\dfrac{-6 + 11}{5} \stackrel{?}{=} 1 \quad \text{Simplify.}$$

$$\dfrac{5}{5} \stackrel{?}{=} 1$$

$$1 = 1$$

EXAMPLE 6

Solve the equation $0.2x + 0.4(50 - x) = 19$.

Solution

Because $0.2 = \dfrac{2}{10}$ and $0.4 = \dfrac{4}{10}$, this equation contains fractions. To clear the equation of these fractions, multiply both sides by 10, the LCD. Then solve the resulting equation.

$0.2x + 0.4(50 - x) = 19$	
$10[0.2x + 0.4(50 - x)] = 10(19)$	Multiply both sides by 10.
$10[0.2x] + 10[0.4(50 - x)] = 10(19)$	Use the distributive property on the left-hand side.
$2x + 4(50 - x) = 190$	Multiply each term by 10.
$2x + 200 - 4x = 190$	Remove parentheses.

$$-2x + 200 = 190 \qquad \text{Combine terms.}$$
$$-2x = -10 \qquad \text{Add } -200 \text{ to both sides.}$$
$$x = 5 \qquad \text{Divide both sides by } -2.$$

 **Progress Check 7**

Verify that the solution checks. Work Progress Check 7.

Solve $\dfrac{-5(x-2)}{3} = x + 6$.

EXAMPLE 7

Solve the equation $x(x - 5) = x^2 + 15$.

Solution

$$x(x - 5) = x^2 + 15$$
$$x^2 - 5x = x^2 + 15 \qquad \text{Remove parentheses.}$$
$$x^2 - 5x + (-x^2) = x^2 + 15 + (-x^2) \qquad \text{Add } -x^2 \text{ to both sides.}$$
$$-5x = 15 \qquad \text{Combine terms.}$$
$$\frac{-5x}{-5} = \frac{15}{-5} \qquad \text{Divide both sides by } -5.$$
$$x = -3 \qquad \text{Simplify.}$$

Verify that the solution checks.

Identities and Impossible Equations

An equation that is true for all values of its variable is called an **identity**. The equation

$$x + x = 2x$$

is an identity because it is true for all values of x.

Because no number can equal a number that is 1 larger than itself, the equation $x = x + 1$ is true for no values of x. Such equations are called **impossible equations** or **contradictions**.

EXAMPLE 8

Solve the equation $3(x + 8) + 5x = 2(12 + 4x)$.

Solution

$$3(x + 8) + 5x = 2(12 + 4x)$$
$$3x + 24 + 5x = 24 + 8x \qquad \text{Remove parentheses.}$$
$$8x + 24 = 24 + 8x \qquad \text{Combine terms.}$$
$$8x + 24 + (-8x) = 24 + 8x + (-8x) \qquad \text{Add } -8x \text{ to both sides.}$$
$$24 = 24 \qquad \text{Combine terms.}$$

The result $24 = 24$ is a true equation, and it is true for every number x. Thus, every value of x is a solution of the original equation. This equation is an identity.

7. -1

Progress Check Answers

EXAMPLE 9

Solve the equation $3(x + 7) - x = 2(x + 10)$.

Solution

$$3(x + 7) - x = 2(x + 10)$$
$$3x + 21 - x = 2x + 20 \quad \text{Remove parentheses.}$$
$$2x + 21 = 2x + 20 \quad \text{Combine terms.}$$
$$2x + 21 + (-2x) = 2x + 20 + (-2x) \quad \text{Add } -2x \text{ to both sides.}$$
$$21 = 20 \quad \text{Combine terms.}$$

The result $21 = 20$ is a false equation; no value of x can make 21 equal to 20. Thus, the original equation has no solution and is an impossible equation. Work Progress Check 8.

Progress Check 8

Classify $7(x - 1) + 1 = 7x - 6$ as an identity or an impossible equation.

8. identity

9.3 Exercises

In Exercises 1–30, simplify each expression, if possible.

1. $3x + 17x$
2. $12y - 15y$
3. $8x^2 - 5x^2$
4. $17x^2 + 3x^2$
5. $9x + 3y$
6. $5x + 5y$
7. $3(x + 2) + 4x$
8. $9(y - 3) + 2y$
9. $5(z - 3) + 2z$
10. $4(y + 9) - 6y$
11. $12(x + 11) - 11$
12. $-3(3 + z) + 2z$
13. $-23(x^2 - 2) + x^2$
14. $5(17 + x) - 27$
15. $3x^3 - 2(x^3 + 5)$
16. $-7z^2 - 2z^3 + z$
17. $8x(x + 3) - 3x^2$
18. $2x + x(x + 3)$
19. $8(y + 7) - 2(y - 3)$
20. $9(z + 2) + 5(3 - z)$
21. $2x + 4(y - x) + 3y$
22. $3y - 6(y + z) + y$
23. $9(x + y) - 9(x - y)$
24. $2(x + z) + 3(x - z)$
25. $(x + 2) + (x - y)$
26. $3z + 2(y - z) + y$
27. $2\left(4x + \dfrac{9}{2}\right) - 3\left(x + \dfrac{2}{3}\right)$
28. $7\left(3x - \dfrac{2}{7}\right) - 5\left(2x - \dfrac{3}{5}\right) + x$
29. $2(7x + 2^2) - 3(3x - 3^2)$
30. $2^2\left(2x - \dfrac{3}{2}\right) + 3^3\left(\dfrac{x}{3} - \dfrac{2}{3}\right)$

In Exercises 31–88, solve each equation, if possible. Check all solutions.

31. $3x + 2 = 2x$
32. $5x + 7 = 4x$
33. $5x + 3 = 4x$
34. $4x + 3 = 5x$
35. $9y - 3 = 6y$
36. $8y + 4 = 4y$
37. $8y - 7 = y$
38. $9y - 8 = y$
39. $3(a + 2) = 4a$
40. $4(a - 5) = 3a$
41. $5(b + 7) = 6b$
42. $8(b + 2) = 9b$
43. $-9x + 3 = 8x + 20$
44. $-11x + 3 = 4x - 27$
45. $-7a + 5 = 11a - 22$
46. $-6a + 7 = 9a - 8$
47. $4x + 8 + 3(x - 2) = -12$
48. $7x + 2 + 4(x - 3) = 12$

354 Chapter 9 Equations and Inequalities

49. $2 + 3(x - 5) = 4(x - 1)$

50. $2 - (4x + 7) = 3 + 2(x + 2)$

51. $10x + 3(2 - x) = 5(x + 2) - 4$

52. $11x + 6(3 - x) = 3$

53. $3(a + 2) = 2(a - 7)$

54. $9(t - 1) = 6(t + 2) - t$

55. $9(x + 11) + 5(13 - x) = 0$

56. $3(x + 15) + 4(11 - x) = 0$

57. $5(x + 2) = 3x - 2$

58. $3(x + 7) = 2(2x + 4) + x - 1$

59. $2 + 3(x - 5) = 4(x - 1)$

60. $2 - (4x + 7) = 3 + 2(x + 2)$

61. $13(b - 9) - 2 = 2(3b - 4) + b$

62. $21(b - 1) + 3 = 4(5b - 6)$

63. $\dfrac{3(t - 7)}{2} = t - 6$

64. $\dfrac{2(t + 9)}{3} = t - 8$

65. $\dfrac{5(2 - s)}{3} = s + 6$

66. $\dfrac{8(5 - s)}{5} = -2s$

67. $\dfrac{4(2x - 10)}{3} = 2(x - 4)$

68. $\dfrac{11(x - 12)}{2} = 9 - 2x$

69. $\dfrac{5(2x - 23) - (6 - x)}{9} = -5\left(x + \dfrac{1}{5}\right)$

70. $\dfrac{2(4x + 40) - (x - 4)}{5} = 7(x - 2)$

71. $3.1(x - 2) = 1.3x + 2.8$

72. $0.6x - 0.8 = 0.8(2x - 1) - 0.7$

73. $2.7(y + 1) = 0.3(3y + 33)$

74. $1.5(5 - y) = 3y + 12$

75. $19.1x - 4(x + 0.3) = -46.5$

76. $18.6x + 7.2 = 1.5(48 - 2x)$

77. $14.3(x + 2) + 13.7(x - 3) = 15.5$

78. $1.25(x - 1) = 0.5(3x - 1) - 1$

79. $x(2x - 3) = 2x^2 + 15$

80. $2x(3x + 4) = 6x^2 + 32$

81. $a(a + 2) = a(a - 4) + 16$

82. $b(b - 1) + 18 = b(b + 5)$

83. $\dfrac{x(2x - 8)}{2} = x(x + 2)$

84. $\dfrac{3x(2x + 1)}{2} = 3x^2 - 6$

85. $2y^2 - 9 = y(y + 3) + y^2$

86. $y(3y - 4) - y^2 = 2y(y + 3) + 20$

87. $\dfrac{x(4x + 3) + 2(x^2 + 9)}{2} = 3x(x + 2)$

88. $x(x + 2) + 24 = \dfrac{x(x + 2) + x(2x - 8)}{3}$

In Exercises 89–100, solve each equation. If it is an identity or an impossible equation, so indicate.

89. $8x + 3(2 - x) = 5(x + 2) - 4$

90. $5(x + 2) = 5x - 2$

91. $s(s + 2) = s^2 + 2s + 1$

92. $21(b - 1) + 3 = 3(7b - 6)$

93. $\dfrac{2(t - 1)}{6} - 2 = \dfrac{t + 2}{6}$

94. $\dfrac{2(2r - 1)}{6} + 5 = \dfrac{3(r + 7)}{6}$

95. $2(3z + 4) = 2(3z - 2) + 13$

96. $x + 7 = \dfrac{2x + 6}{2} + 4$

97. $2(y - 3) - \dfrac{y}{2} = \dfrac{3}{2}(y - 4)$

98. $\dfrac{20 - a}{2} = \dfrac{3}{2}(a + 4)$

99. $\dfrac{3x + 14}{2} = x - 2 + \dfrac{x + 18}{2}$

100. $\dfrac{5(x + 3)}{3} - x = \dfrac{2(x + 8)}{3}$

Review Exercises

In Review Exercises 1–6, let $x = -3$, $y = -5$, and $z = 0$. Evaluate each expression.

1. $x^2 z(y^3 - z)$

2. $z - x^2 y$

3. $y^3 - x$

4. $x^3(y - z^{23})$

5. $\dfrac{x - y^2}{2y - 1 + x}$

6. $\dfrac{2y + 1}{x} - x$

In Review Exercises 7–10, solve each equation. Check all solutions.

7. $3x + 5 = 11$

8. $7 - 8y = -9$

9. $\dfrac{3}{7}(w - 2) - 2 = 1$

10. $\dfrac{2}{13}(21 - z) = 4$

In Review Exercises 11–12, translate the words into an equation and solve it. Then check your solution.

11. What number added to 37 gives 20?

12. Mark needs $27 more to buy a $140 bicycle. How much money does Mark have?

9.4 Literal Equations

- Solving Formulas
- Using Formulas

Equations with several variables are called **literal equations**. Often these equations are **formulas** such as $A = lw$, the formula for finding the area of a rectangle. Suppose we wish to use this formula to find the lengths of several rectangles whose areas and widths are known. It would be tedious to substitute values for A and w into the formula, and then repeatedly solve the formula for l. It would be better to solve the formula $A = lw$ for the variable l first, and then substitute values for A and w and compute l directly.

Solving Formulas

To *solve a formula for a variable* means to isolate that variable on one side of the equation, with all other quantities on the opposite side.

EXAMPLE 1

Solve $A = lw$ for the variable l.

Solution

To isolate l on the left-hand side, we undo the multiplication by w. To do this, we divide both sides of the equation by w and simplify.

$$A = lw$$

$$\frac{A}{w} = \frac{lw}{w} \qquad \text{Divide both sides by } w.$$

$$\frac{A}{w} = l \qquad \text{Simplify.}$$

or

$$l = \frac{A}{w}$$

EXAMPLE 2

The formula $A = \frac{1}{2}bh$ gives the area of a triangle with base b and height h. Solve this formula for b.

Solution

$$A = \frac{1}{2}bh$$

$$2 \cdot A = 2 \cdot \frac{1}{2}bh \qquad \text{Multiply both sides by 2.}$$

$$2A = bh \qquad \text{Simplify.}$$

$$\frac{2A}{h} = \frac{bh}{h} \qquad \text{Divide both sides by } h.$$

$$\frac{2A}{h} = b \qquad \text{Simplify.}$$

or

$$b = \frac{2A}{h}$$

If the area A and the height h of a triangle are known, the base b is given by the formula $b = \frac{2A}{h}$.

EXAMPLE 3

The formula $C = \frac{5}{9}(F - 32)$ is used to convert Fahrenheit temperature readings into their Celsius equivalents. Solve this formula for F.

Solution

$$C = \frac{5}{9}(F - 32)$$

$$\frac{9}{5}C = \frac{9}{5} \cdot \frac{5}{9}(F - 32) \qquad \text{Multiply both sides by } \frac{9}{5}.$$

$$\frac{9}{5}C = 1(F - 32) \qquad \text{Simplify.}$$

$$\frac{9}{5}C = F - 32 \qquad \text{Remove parentheses.}$$

$$\frac{9}{5}C + \mathbf{32} = F - 32 + \mathbf{32} \qquad \text{Add 32 to both sides.}$$

$$\frac{9}{5}C + 32 = F \qquad \text{Combine terms.}$$

$$F = \frac{9}{5}C + 32$$

The formula $F = \frac{9}{5}C + 32$ is used to convert degrees Celsius to degrees Fahrenheit. ∎

EXAMPLE 4

The area A of the trapezoid shown in Figure 9-1 is given by the formula

$$A = \frac{1}{2}(B + b)h$$

where B and b are its bases and h is its height. Solve the formula for b.

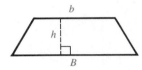

Figure 9-1

Solution

Method 1: $\quad A = \frac{1}{2}(B + b)h$

$$\mathbf{2} \cdot A = \mathbf{2} \cdot \frac{1}{2}(B + b)h \qquad \text{Multiply both sides by 2.}$$

$$2A = Bh + bh \qquad \text{Simplify and remove parentheses.}$$

$$2A + (\mathbf{-Bh}) = Bh + bh + (\mathbf{-Bh}) \qquad \text{Add } -Bh \text{ to both sides.}$$

$$2A - Bh = bh \qquad \text{Simplify.}$$

$$\frac{2A - Bh}{h} = \frac{bh}{h} \qquad \text{Divide both sides by } h.$$

$$\frac{2A - Bh}{h} = b \qquad \text{Simplify.}$$

Method 2: $\quad A = \frac{1}{2}(B + b)h$

$$\mathbf{2} \cdot A = \mathbf{2} \cdot \frac{1}{2}(B + b)h \qquad \text{Multiply both sides by 2.}$$

$$2A = (B + b)h \qquad \text{Simplify.}$$

$$\frac{2A}{h} = \frac{(B + b)h}{h} \qquad \text{Divide both sides by } h.$$

$$\frac{2A}{h} = B + b \qquad \text{Simplify.}$$

$$\frac{2A}{h} + (\mathbf{-B}) = B + b + (\mathbf{-B}) \qquad \text{Add } -B \text{ to both sides.}$$

$$\frac{2A}{h} - B = b \qquad \text{Combine terms.}$$

Although they look different, the results of Methods 1 and 2 are equivalent. Work Progress Check 9. ∎

Progress Check 9

Solve $F = \frac{9}{5}C + 32$ for C.

9. $C = \frac{5}{9}(F - 32)$

Progress Check Answers

Using Formulas

EXAMPLE 5

Solve the formula $P = 2l + 2w$ for l and find l if $P = 56$ and $w = 11$.

Solution

We solve the formula $P = 2l + 2w$ for l and then substitute the given values of P and w to determine the value of l.

$$P = 2l + 2w$$
$$P + (-2w) = 2l + 2w + (-2w) \quad \text{Add } -2w \text{ to both sides.}$$
$$P - 2w = 2l \quad \text{Combine terms.}$$
$$\frac{P - 2w}{2} = \frac{2l}{2} \quad \text{Divide both sides by 2.}$$
$$\frac{P - 2w}{2} = l \quad \text{Simplify.}$$

Then we substitute 56 for P and 11 for w, and simplify.

$$l = \frac{P - 2w}{2}$$
$$l = \frac{56 - 2(11)}{2}$$
$$= \frac{56 - 22}{2}$$
$$= \frac{34}{2}$$
$$= 17$$

Thus, $l = 17$.

EXAMPLE 6

The volume V of the right-circular cone shown in Figure 9-2 is given by the formula

$$V = \frac{1}{3}Bh$$

where B is the area of its circular base and h is its height. Solve the formula for h and find the height of a right-circular cone with a volume of 64 cubic centimeters and a base area of 16 square centimeters.

Solution

$$V = \frac{1}{3}Bh \quad \text{The formula to solve for } h.$$
$$3V = 3 \cdot \frac{1}{3}Bh \quad \text{Multiply both sides by 3.}$$
$$3V = Bh \quad \text{Simplify.}$$

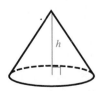

Figure 9-2

Progress Check 10

Find the area of the base of a right-circular cone with a volume of 48 cubic centimeters and a height of 6 centimeters.

$\dfrac{3V}{B} = \dfrac{Bh}{B}$ Divide both sides by B.

$\dfrac{3V}{B} = h$ Simplify.

Now we substitute 64 for V and 16 for B and simplify.

$h = \dfrac{3V}{B}$

$h = \dfrac{3(64)}{16}$

$= 3(4)$

$= 12$

10. 24 square centimeters

The height of the right-circular cone is 12 centimeters. Work Progress Check 10.

9.4 Exercises

In Exercises 1–24, solve each formula for the variable indicated.

1. $E = IR$; for I
2. $i = prt$; for r
3. $V = lwh$; for w
4. $K = A + 32$; for A

5. $P = a + b + c$; for b
6. $P = 4s$; for s
7. $P = 2l + 2w$; for w
8. $d = rt$; for t

9. $A = P + Prt$; for t
10. $A = \dfrac{1}{2}(B + b)h$; for h
11. $C = 2\pi r$; for r
12. $I = \dfrac{E}{R}$; for R

13. $K = \dfrac{wv^2}{2g}$; for w
14. $V = \pi r^2 h$; for h
15. $P = I^2 R$; for R
16. $V = \dfrac{1}{3}\pi r^2 h$; for h

17. $K = \dfrac{wv^2}{2g}$; for g
18. $P = \dfrac{RT}{mV}$; for V
19. $F = \dfrac{GMm}{d^2}$; for M
20. $C = 1 - \dfrac{A}{a}$; for A

21. $F = \dfrac{GMm}{d^2}$; for d^2
22. $y = mx + b$; for x
23. $G = 2(r - 1)b$; for r
24. $F = f(1 - M)$; for M

In Exercises 25–32, solve each formula for the variable indicated and then substitute numbers to find that variable's value.

25. $d = rt$ Find t if $d = 135$ and $r = 45$.

26. $d = rt$ Find r if $d = 275$ and $t = 5$.

27. $i = prt$ Find t if $i = 12$, $p = 100$, and $r = 0.06$.

28. $i = prt$ Find r if $i = 120$, $p = 500$, and $t = 6$.

29. $P = a + b + c$ Find c if $P = 37$, $a = 15$, and $b = 19$.

30. $y = mx + b$ Find x if $y = 30$, $m = 3$, and $b = 0$.

31. $K = \dfrac{1}{2}h(a + b)$ Find h if $K = 48$, $a = 7$, and $b = 5$.

32. $\dfrac{x}{2} + y = z^2$ Find x if $y = 3$ and $z = 3$.

33. The formula $E = IR$, called **Ohm's law**, is used in electronics. Solve for I and then calculate the current I if the voltage E is 48 volts and the resistance R is 12 ohms. Current has units of **amperes**.

34. The volume V of a cone is given by the formula $V = \frac{1}{3}\pi r^2 h$. Solve the formula for h and then calculate the height h if V is 36π cubic inches and the radius r is 6 inches.

35. The circumference C of a circle is given by $C = 2\pi r$, where r is the radius of the circle. Solve the formula for r and then calculate the radius of a circle with a circumference of 17 feet. Use $\pi = 3.14$ and give your answer to the nearest hundredth of a foot.

36. At a simple interest rate r, an amount of money P grows to an amount A in t years according to the formula $A = P(1 + rt)$. Solve the formula for P. After $t = 3$ years, a girl has an amount $A = \$4300$ on deposit. What amount P did she start with? Assume an interest rate of 9% ($r = 0.09$).

37. The power P lost when an electric current I passes through a resistance R is given by the formula $P = I^2 R$. Solve for R. If P is 2700 watts and I is 14 amperes, calculate R to the nearest hundredth of an ohm.

38. The perimeter P of a rectangle with length l and width w is given by the formula $P = 2l + 2w$. Solve this formula for w. If the perimeter of a certain rectangle is 58 meters and its length is 17 meters, find its width.

39. The force of gravitation, F, between two objects with masses m and M and separated by distance d is given by the formula

$$F = \frac{GmM}{d^2}$$

where G is a constant. Solve for m.

40. In thermodynamics, the Gibbs free-energy function is given by

$$G = U - TS + pV$$

Solve this equation for the pressure, p.

41. The approximate length L of a belt joining two pulleys of radii r and R feet with centers D feet apart is given by the formula $L = 2D + 3.25(r + R)$. See Illustration 1. Solve the formula for D. If a 25-foot belt joins pulleys of radius 1 foot and 3 feet, how far apart are the centers of the pulleys?

42. The measure a of an interior angle of a regular polygon with n sides is given by the formula $a = 180°(1 - \frac{2}{n})$. See Illustration 2. Solve the formula for n. How many sides does a regular polygon have if an interior angle is 108°? (*Hint:* Distribute first.)

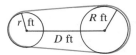

Illustration 1

Illustration 2

Review Exercises

In Review Exercises 1–8, simplify each expression, if possible.

1. $3x - 5x$
2. $2x - 5y + 3x$
3. $2x^2y + 5x^2y^2$
4. $8x^2 + 3(x^2 + 2)$
5. $2(y^3 - x^2) + 5(y^3 + 3) - 10$
6. $7(x - 5) - 7(5 - x)$
7. $\frac{3}{5}(x + 5) - \frac{8}{5}(10 + x)$
8. $\frac{2}{11}(22x - y^2) + \frac{9}{11}y^2$

In Review Exercises 9–12, solve each equation. Check all solutions.

9. $3(x - 5) = 2(x + 3)$
10. $7(3 - 2x) - 3x = 4$
11. $\frac{3x - 5}{13} = x - 5$
12. $3x - 2(5 - x) = 15x$

9.5 Applications of Equations

- Number Problems
- Problems with Too Much Information

In this section, we discuss the solutions to several types of word problems. We will follow these steps as we solve each problem.

> 1. Read the problem and analyze the facts. What information is given? What are you asked to find? A sketch, chart, or diagram will help you visualize the facts of the problem.
> 2. Pick a variable to represent the quantity to be found and write a sentence stating what that variable represents. Express all other important quantities mentioned in the problem as expressions involving this single variable.
> 3. Organize the data and find a way to express a quantity in two different ways.
> 4. Write an equation showing that the two quantities found in Step 3 are equal.
> 5. Solve the equation.
> 6. State the solution or solutions.
> 7. Check the answer in the words of the problem. Have all the questions been answered?

Number Problems

EXAMPLE 1

Six more than twice a certain number is equal to 1 less than 3 times that number. Find the number.

Analysis

We pick a variable such as x to represent the unknown number. Then we can express the other information of the problem in terms of x.

Solution

Let x represent the certain number. Then

$2x + 6$ represents *6 more than twice the certain number*, and

$3x - 1$ represents *1 less than 3 times the number.*

We know that these two quantities are equal.

Six more than twice a number	is equal to	1 less than 3 times that number
$2x + 6$	=	$3x - 1$

$2x + 6 = 3x - 1$ The equation to solve.
$6 = x - 1$ Add $-2x$ to both sides.
$7 = x$ Add 1 to both sides.

The number is 7. Check this solution in the words of the original problem. First, calculate *6 more than twice the number*. The result is 2 · 7 + 6, or 20. Then calculate *1 less than 3 times the number*. This result is 3 · 7 − 1, or 20. Because the results agree, the solution checks. ∎

EXAMPLE 2

Jim wants to cut a 17-foot pipe into three sections. The longest section is to be three times as long as the shortest, and the middle-sized section is to be 2 feet longer than the shortest. How long should each section be?

Analysis

The information in this problem is given in terms of the length of the shortest of the three sections of the pipe. Therefore, let a variable represent the length of that shortest section. Then express the other facts in terms of that variable.

Solution

Let x represent the length of the shortest section. Then

$3x$ represents the length of the longest section, and

$x + 2$ represents the length of the middle-sized section.

Sketch the situation described in the problem, as shown in Figure 9-3.

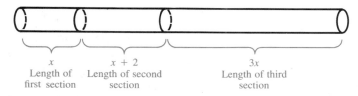

Figure 9-3

The sum of the lengths of these three sections must equal the total length of the pipe.

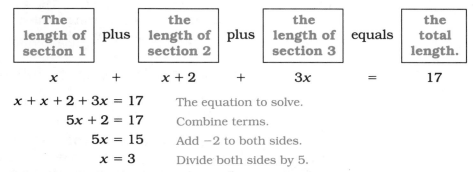

$$x + x + 2 + 3x = 17 \quad \text{The equation to solve.}$$
$$5x + 2 = 17 \quad \text{Combine terms.}$$
$$5x = 15 \quad \text{Add } -2 \text{ to both sides.}$$
$$x = 3 \quad \text{Divide both sides by 5.}$$

The length of the pipe's shortest section is 3 feet. Because the middle-sized section is 2 feet longer than the shortest, it is 5 feet long. Because the longest section is three times as long as the shortest, it is 9 feet long.

Check: Because 3 feet, 5 feet, and 9 feet total 17 feet, the solution checks. ∎

EXAMPLE 3

A truck made five trips to the city landfill, and a larger truck made eight trips. When fully loaded, the larger truck carries 2 tons more garbage than the smaller truck. These two trucks hauled a total of 55 tons. How much garbage can the smaller truck carry in one load?

Analysis

Pick a variable such as x to represent the number of tons that the smaller truck can carry when it is fully loaded. Because the larger truck can carry 2 tons more than the smaller truck, it can carry $(x + 2)$ tons when fully loaded. Since the smaller truck can carry x tons in each load, it can carry $5x$ tons in five loads. Since the larger truck can carry $(x + 2)$ tons in each load, it can carry $8(x + 2)$ tons in eight loads. The total number of tons of garbage hauled can be expressed two ways: as the sum of $5x$ and $8(x + 2)$ and as the number 55.

Solution

Let x represent the capacity of the smaller truck in tons. Then $x + 2$ represents the capacity of the larger truck in tons.

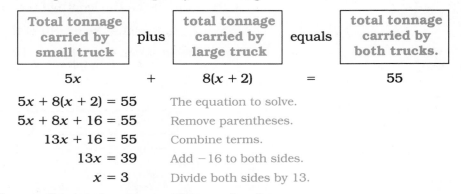

Total tonnage carried by small truck	plus	total tonnage carried by large truck	equals	total tonnage carried by both trucks.
$5x$	$+$	$8(x + 2)$	$=$	55

$$5x + 8(x + 2) = 55 \quad \text{The equation to solve.}$$
$$5x + 8x + 16 = 55 \quad \text{Remove parentheses.}$$
$$13x + 16 = 55 \quad \text{Combine terms.}$$
$$13x = 39 \quad \text{Add } -16 \text{ to both sides.}$$
$$x = 3 \quad \text{Divide both sides by 13.}$$

The smaller truck can carry 3 tons of garbage.

Check: Each load on the small truck is 3 tons. Because the larger truck carries 2 tons more than that, it carries 5 tons. The small truck made five trips carrying 3 tons on each trip, and the larger truck made eight trips carrying 5 tons on each trip. The small truck carried a total of $5 \cdot 3$, or 15 tons, while the larger truck hauled $8 \cdot 5$, or 40 tons. In total they carried $15 + 40$, or 55 tons. The solution checks. ∎

Problems with Too Much Information

EXAMPLE 4

In a fourth-grade class of 32 students, only 29 will be promoted. There are 12 more girls in the class than there are boys, and only 1 of the girls will have to repeat fourth grade. How many boys are in the class?

Analysis

We are asked to find how many boys are currently in the fourth-grade class. It is not important to know how many of them will pass or how many will not. The important information is the total number in the class

(32) and the fact that there are 12 more girls than boys. Let b represent the number of boys in the class. Since there are 12 more girls than boys, the expression $b + 12$ represents the number of girls. Thus the total number of students can be expressed as $b + (b + 12)$ and as 32.

Solution

Let b represent the number of boys in the class. Then $b + 12$ represents the number of girls.

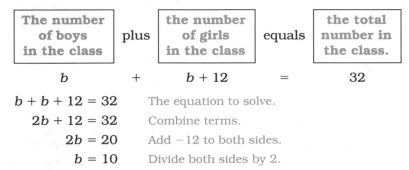

$$b + b + 12 = 32 \quad \text{The equation to solve.}$$
$$2b + 12 = 32 \quad \text{Combine terms.}$$
$$2b = 20 \quad \text{Add } -12 \text{ to both sides.}$$
$$b = 10 \quad \text{Divide both sides by 2.}$$

There are 10 boys in the class.

Check: There are 10 boys in the class and 22 girls (12 more than 10). The total number is $10 + 22$, or 32. The solution checks. ∎

9.5 Exercises

In Exercises 1–30, pick a variable to represent the unknown quantity, set up an equation involving the variable, solve the equation, and check it.

1. If a number is doubled, the result is 3 less than triple the number. Find the number.

2. If a number is multiplied by 4, the result is 7 greater than 3 times the number. Find the number.

3. If 6 is added to a number, the result is equal to 4 times the number. Find the number.

4. If 9 is subtracted from 3 times a number, the result is 1 less than the original number. Find the number.

5. Ten less than 7 times a number is 12 more than 5 times the number. Find the number.

6. Thirteen minus 2 times a number is the same as 4 more than the number. Find the number.

7. One-fifth of a number added to the number itself gives 42. Find the number.

8. One-fourth of a number added to twice the number equals 27. Find the number.

9. Six more than 3 times a certain number is 5 times the number. Find the number.

10. The number 5 decreased by a certain number becomes 13 greater than the number. Find the number.

11. What number is equal to its own double?

12. What number is equal to one-half of itself?

13. A 12-foot board has been cut into two sections, one twice as long as the other. How long is each section?

14. A 20-foot pipe has been cut into two sections, one 3 times as long as the other. How long is each section?

15. A 45-foot length of wire has been cut into three pieces. The longest of the three is 5 times as long as the shortest. The middle-sized section is 3 times as long as the shortest. How long is each piece of wire?

16. A 30-foot steel beam must be cut into two pieces so that the longer piece is 2 feet more than 3 times as long as the shorter piece. How long will each piece be?

17. A 35-foot beam, 1 foot wide and 2 inches thick, is cut into three sections. One section is 14 feet long. Of the remaining two sections, one is twice as long as the other. How long is each of the three sections?

18. Two tanks hold a total of 45 gallons of water. One tank holds 6 gallons more than twice the amount in the other. How much water does each tank hold?

19. The price of a chair is reduced by one-fourth of its retail price, making it one-half of the cost of a matching sofa. The reduced price is $237. What was the chair's retail price?

20. If you buy one bottle of vitamins, you can get a second bottle for half price. Two bottles cost $2.25. What is the usual price for a single bottle of vitamins?

21. Thirty-six children received awards at a science fair. There were three first-place awards, and there were twice as many third-place awards as second-place awards. How many of each award were given?

22. A glazed donut has 50 fewer calories than 5 times the number of calories in a slice of whole wheat bread. The donut and the bread together contain 280 calories. How many calories are in the slice of bread?

23. A slice of pie with a scoop of ice cream contains 850 calories. The number of calories in the pie alone is 100 greater than double the number of calories in the ice cream alone. How many calories are in the ice cream?

24. Juanita buys a biology text and its lab manual for $69.50. The text costs $2 more than four times the cost of the manual. Find the cost of the text.

25. A hat and a coat together cost $100. The coat costs $80 more than the hat. How much does the hat cost?

26. A novel can be purchased in a hardcover edition for $15.95 or in paperback for $4.95. The publisher printed 11 times as many paperbacks as hardcover books, a total of 114,000 copies. How many hardcover books were printed?

27. Concrete contains 3 times as much gravel as cement. How much cement is in 500 pounds of dry concrete mix?

28. Sam gave 3 times as much to a charity as Bill did, and Bob donated $4 more than Bill. If their contributions totaled $109, how much did each contribute?

29. Water is made up of hydrogen and oxygen. The mass of the oxygen in water is 8 times the mass of the hydrogen. How much hydrogen is in 2700 grams of water?

30. A man can buy several 20-cent stamps, using all of the money in his pocket. If he buys 50-cent stamps instead, he can buy six fewer stamps, but he would then receive 30 cents change. How many 20-cent stamps can he buy?

Review Exercises

In Review Exercises 1–6, give an example that illustrates the given property of real numbers.

1. The commutative property of addition
2. The distributive property
3. The associative property of multiplication
4. The closure property of addition
5. The double negative rule
6. The multiplicative identity property

In Review Exercises 7–12, find each number.

7. The absolute value of the negative of 12

8. The negative of the absolute value of 12

9. The negative of the negative of 12

10. The negative of the reciprocal of 12

11. The sum of 12 and its additive inverse

12. The product of 12 and its multiplicative inverse

9.6 More Applications of Equations

- Integer Problems
- Geometric Problems
- Coin Problems
- Break-Even Analysis

In this section, we consider more types of word problems.

Integer Problems

EXAMPLE 1

The sum of two consecutive even integers is 22. Find the integers.

Analysis

Recall that consecutive even integers differ by 2; numbers such as 2, 4, 6, 8, and 10 are consecutive even integers. Hence, if x is an even integer, then the expressions x and $x + 2$ represent two consecutive even integers. The sum of these two integers can be written in two different ways: as $x + (x + 2)$ and as 22.

Solution

Let x represent the first even integer. Then
$x + 2$ represents the next consecutive even integer.

We can form the equation

The first even integer	plus	the second even integer	equals	their sum.
x	$+$	$x + 2$	$=$	22

$$x + x + 2 = 22 \quad \text{The equation to solve.}$$
$$2x + 2 = 22 \quad \text{Combine terms.}$$
$$2x = 20 \quad \text{Add } -2 \text{ to both sides.}$$
$$x = 10 \quad \text{Divide both sides by 2.}$$

The first of the two consecutive even integers is 10. The next even integer is 12. Hence, the two consecutive even integers are 10 and 12.

Check: The numbers 10 and 12 are indeed consecutive even integers, and their sum is 22. The answers check. ■

Geometric Problems

EXAMPLE 2

The length of a rectangle is 4 meters longer than twice its width. If the perimeter of the rectangle is 26 meters, find its dimensions.

Analysis

Recall that the **perimeter** of a rectangle is given by the formula $P = 2l + 2w$, where P represents the perimeter, l represents the length, and w represents the width of the rectangle. Since the length of the rectangle shown in Figure 9-4 is 4 meters longer than twice its width, the length of the rectangle can be represented by the expression $2w + 4$. Thus, its perimeter is $P = 2(2w + 4) + 2w$. The perimeter is also 26.

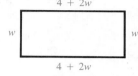

Figure 9-4

Solution

Let w represent the width of the rectangle. Then $2w + 4$ represents the length of the rectangle.

We can form the equation

Two lengths	plus	two widths	equals	the perimeter.
$2(2w + 4)$	+	$2w$	=	26

$$
\begin{aligned}
2(2w + 4) + 2w &= 26 &&\text{The equation to solve.} \\
4w + 8 + 2w &= 26 &&\text{Remove parentheses.} \\
6w + 8 &= 26 &&\text{Combine terms.} \\
6w &= 18 &&\text{Add } -8 \text{ to both sides.} \\
w &= 3 &&\text{Divide both sides by 6.}
\end{aligned}
$$

The width of the rectangle is 3 meters, and the length, $2w + 4$, is 10 meters.

Check: If a rectangle has a width of 3 meters and a length of 10 meters, then the length is 4 meters longer than twice the width ($2 \cdot 3 + 4 = 10$). Furthermore, the perimeter is $2(10) + 2(3)$, or 26 meters. The solution checks. ∎

EXAMPLE 3

The vertex angle of an isosceles triangle is 56°. Find the measure of each base angle.

Analysis

An **isosceles triangle** has two equal sides, which meet to form the **vertex angle**. The angles opposite those sides, called **base angles**, are also equal (see Figure 9-5). If we let x represent the measure of one base angle, then the measure of the other base angle is also x. In any triangle, the sum of the three angles is 180°.

Solution

Let x represent the measure of one base angle. Then x also represents the measure of the other base angle.

We can form the equation

One base angle	plus	the other base angle	plus	the vertex angle	equals	180°.
x	+	x	+	56	=	180

$x + x + 56 = 180$ The equation to solve.
$2x + 56 = 180$ Combine terms.
$2x = 124$ Add -56 to both sides.
$x = 62$ Divide both sides by 2.

The measure of each base angle is 62°.

Check: The measure of each base angle is 62°, and the vertex angle measures 56°. These three angles total 180°. The solution checks.

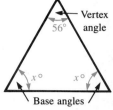

Figure 9-5

EXAMPLE 4

Find the measure of each angle marked in Figure 9-6.

Analysis

The marked angles form a pair of vertical angles, and we have seen that vertical angles are congruent. Thus, the measures of the two angles are equal.

Solution

Set $3x + 10$ equal to $6x - 2$ and solve for x.

$3x + 10 = 6x - 2$
$10 = 3x - 2$
$12 = 3x$
$4 = x$

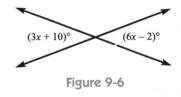

Figure 9-6

Because $x = 4$, one angle has a measure of $[3(4) + 10]° = 22°$. The other has the same measure, since $[6(4) - 2]° = 22°$. Each angle measures 22°.

Coin Problems

EXAMPLE 5

Jill has some pennies, nickels, dimes, and quarters. She has twice as many nickels as dimes, and four times as many quarters as nickels. She has 36 coins in all, and three of them are pennies. How many nickels, dimes, and quarters does Jill have?

Analysis

Suppose d represents the number of dimes. Because Jill has twice as many nickels as dimes, she has $2d$ nickels. Because she has four times as many quarters as nickels, she has $4(2d)$, or $8d$, quarters. Jill also has 3 pennies. The total number of pennies, nickels, dimes, and quarters is 36. The number of coins can be expressed in two ways:

as $(3 + 2d + d + 8d)$ coins and as 36 coins

Solution

Let d represent the number of dimes. Then

$2d$ represents the number of nickels and
$8d$ represents the number of quarters.

We can form the equation

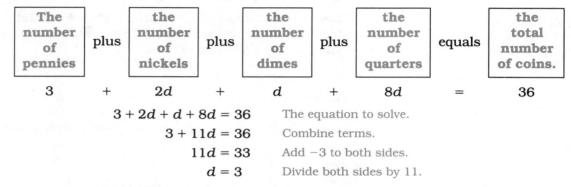

$$\begin{aligned} 3 + 2d + d + 8d &= 36 & &\text{The equation to solve.} \\ 3 + 11d &= 36 & &\text{Combine terms.} \\ 11d &= 33 & &\text{Add } -3 \text{ to both sides.} \\ d &= 3 & &\text{Divide both sides by 11.} \end{aligned}$$

Jill has three dimes. Because she has twice as many nickels as dimes, she has six nickels. Because she has four times as many quarters as nickels, she has 24 quarters.

Check: Because 3 pennies, 6 nickels, 3 dimes, and 24 quarters total 36 coins, the solution checks. ∎

EXAMPLE 6

George has \$2 in nickels, dimes, and quarters. He has 5 times as many quarters as he has nickels and two more dimes than he has quarters. How many of each type of coin does he have?

Analysis

It is important to distinguish between the *number* of coins and the *value* of those coins. If n is the number of nickels, then those nickels are worth $5n$ cents. Because there are $5n$ quarters, their value is $25(5n)$ cents. Finally, there are $(5n + 2)$ dimes (there are two more dimes than quarters, and there are $5n$ quarters). The value of these dimes is $10(5n + 2)$ cents. The total value can be expressed in two ways:

as $5(n) + 25(5n) + 10(5n + 2)$ cents and as 200 cents

Solution

Let n represent the number of nickels. Then

$5n$ represents the number of quarters, and
$5n + 2$ represents the number of dimes.

The value of the nickels (in cents)	plus	the value of the quarters (in cents)	plus	the value of the dimes (in cents)	equals	the total value (in cents).
5(n)	+	25(5n)	+	10(5n + 2)	=	200

$5n + 25(5n) + 10(5n + 2) = 200$ The equation to solve.
$5n + 125n + 50n + 20 = 200$ Remove parentheses.
$180n + 20 = 200$ Combine terms.
$180n = 180$ Add -20 to both sides.
$n = 1$ Divide both sides by 18.

George has one nickel. Because he has 5 times as many quarters, he has five quarters. Because he has two more dimes than quarters, he has seven dimes.

Check: Because the values of one nickel, five quarters, and seven dimes add up to $2, the solution checks. ■

Break-Even Analysis

In any type of manufacturing business, there are two types of costs—**fixed costs** and **unit costs**. Fixed costs do not depend on the amount of product manufactured. Fixed costs would include the cost of plant rental, insurance, and machinery. Unit costs do depend on the amount of product manufactured. Unit costs would include the cost of raw materials and labor.

Break-even analysis is used to find a production level where revenue will just offset the cost of production. When production exceeds the break-even point, the company will make a profit.

EXAMPLE 7

An electronics company has fixed costs of $6405 a week and a unit cost of $75 for each compact disc player manufactured. The company can sell all the CD players it can make at a wholesale price of $90. Find the company's break-even point.

Analysis

Suppose the company manufactures x CD players each week. The cost of manufacturing these players is the sum of the fixed costs and the unit costs. We are given that the fixed costs are $6405 each week. The unit cost is the product of x, the number of CD players manufactured each week, and $75, the cost of manufacturing a single player. Thus, the weekly cost is

$\text{Cost} = 75x + 6405$

Since the company can sell all the machines it can make, the weekly revenue is the product of x, the number of players manufactured (and sold) each week, and $90, the wholesale price of each CD player. Thus, the weekly revenue is

$\text{Revenue} = 90x$

The break-even point is the value of x for which the weekly revenue is equal to the weekly cost.

Solution

Let x represent the number of CD players manufactured each week. Then

$90x$ represents the weekly revenue, and

$75x + 6405$ represents the weekly cost.

Because the break-even point occurs when revenue equals cost, set up and solve the following equation:

$$\boxed{\text{The weekly revenue}} \text{ equals } \boxed{\text{the weekly cost.}}$$

$$90x = 75x + 6405$$
$$15x = 6405 \quad \text{Add } -75x \text{ to both sides.}$$
$$x = 427 \quad \text{Divide both sides by 15.}$$

If the company manufactures 427 CD players, its revenue will equal its costs, and the company will break even. Verify that this is true. ∎

9.6 Exercises

In Exercises 1–16, pick a variable to represent the unknown quantity, set up an equation involving the variable, solve the equation, and check it.

Integer Problems

1. The sum of two consecutive even integers is 54. Find the integers.

2. The sum of two consecutive odd integers is 88. Find the integers.

3. The sum of three consecutive integers is 120. Find the three integers.

4. The sum of three consecutive even integers is 72. Find the three even integers.

5. The sum of an integer and twice the next integer is 23. Find the smaller integer.

6. If 4 times the smallest of three consecutive integers is added to the largest, the result is 112. Find the three integers.

7. The larger of two integers is 10 greater than the smaller. The larger is 3 less than twice the smaller. Find the smaller integer.

8. The smaller of two integers is one-half of the larger, and 11 greater than one-third of the larger. Find the smaller integer.

Geometric Problems

9. The perimeter of a triangle is 57 feet. If all three sides are equal, find the length of each side.

10. The length of a rectangle is 7 centimeters longer than the width. The perimeter is 90 centimeters. Find the dimensions of the rectangle.

11. The width of a rectangle is 11 meters less than the length. The perimeter is 94 meters. Find the dimensions of the rectangle.

12. One of the two equal sides of an isosceles triangle is 4 feet less than the third side. The perimeter is 25 feet. Find the lengths of the sides.

13. The length of a rectangle is 5 inches longer than twice the width. The perimeter is 112 inches. Find the dimensions of the rectangle.

14. One of the two equal angles of an isosceles triangle is 4 times the third angle (the vertex angle). What is the measure of the vertex angle?

15. The three angles of an equilateral triangle are all equal. What is the measure of each?

16. The perimeter of a certain square is twice the perimeter of a certain equilateral (equal-sided) triangle. If one side of the square is 6 inches, what is the length of a side of the triangle?

In Exercises 17–20, refer to Illustration 1, in which $\overleftrightarrow{AC} \parallel \overleftrightarrow{FG}$.

17. Find x.

18. Find m($\angle ABD$).

19. Find m($\angle ABE$).

20. Find m($\angle BDF$).

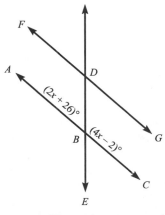

Illustration 1

In Exercises 21–32, pick a variable to represent the unknown quantity, set up an equation involving the variable, solve the equation, and check it.

Coin Problems

21. Liz has some nickels, dimes, and quarters—nine coins in all. She has twice as many dimes as nickels, and one less quarter than she has dimes. How many nickels does she have?

22. A man has twice as many pennies as he has dimes, and four less quarters than pennies. He has 41 coins in all. How many coins of each type does he have?

23. A man has $3.15 in nickels, dimes, and quarters. He has as many nickels as dimes, and three more quarters than dimes. How many of each coin does he have?

24. Carlos has $6.75 in pennies, dimes, and quarters. He has twice as many dimes as pennies, and just six quarters. How many pennies and how many dimes does he have?

25. Margaret has twice as many dimes as Fern has quarters. When they pool their resources, they have $4.05. How many quarters does Fern have?

26. Jeanne has three times as many dimes as quarters, and half as many nickels as dimes. The coins are worth $3.75. How many quarters does Jeanne have?

27. If all of Kim's nickels were dimes, she would be 35 cents richer. How many nickels does she have?

28. If all of Bob's dimes were quarters, he would be $1.05 richer. How many dimes does he have?

Break-Even Analysis

29. A shoe company has fixed costs of $9600 per month and a unit cost of $20 per pair of shoes. The company can sell all the shoes it can make at a wholesale price of $30 per pair. Find the break-even point.

30. A belt company has fixed costs of $5400 per month and a unit cost of $12 per belt. The company can sell every belt it can make at a wholesale price of $15. Find the break-even point.

31. A machine shop has two machines that can mill a certain brass plate. One machine has a setup cost of $500 and a cost per plate of $2, while the other machine has a setup cost of $800 and a cost per plate of $1. How many plates should be manufactured if the cost is to be the same using either machine?

32. A rug manufacturer has two looms for weaving Oriental-style rugs. One loom has a setup cost of $750 and can produce a rug for $115. The other loom has a setup cost of $950 and can produce a rug for $95. How many rugs can be manufactured if it doesn't matter which loom is used?

In Exercises 33–36, a paint manufacturer can choose between two processes for manufacturing house paint, with monthly costs as shown in Table 1. The paint can be sold for $21 per gallon.

33. Find the break-even point for process A.

34. Find the break-even point for process B.

35. If sales will be 8800 gallons per month, which process should the company choose?

36. If sales will be 9000 gallons per month, which process should the company choose?

Table 1

Process	Fixed Costs	Unit Cost (per gallon)
A	$75,000	$11
B	$128,000	$5

Review Exercises

In Review Exercises 1–6, refer to the formulas in Section 7.6.

1. Find the volume of a pyramid that has a height of 6 centimeters and a square base, 10 centimeters on each side.

2. Find the volume of a cone with a height of 6 centimeters and a circular base with radius 6 centimeters.

3. Find the volume of a sphere whose diameter is 14 meters.

4. Find the volume of a cylinder with a height of 8 centimeters and a circular base with diameter of 20 centimeters.

5. Find the number of cubic feet in one cubic yard.

6. Find the number of cubic inches in one cubic foot.

In Review Exercises 7–12, simplify each expression.

7. $3(x + 2) + 4(x - 3)$

8. $4(x - 2) - 3(x + 1)$

9. $\frac{1}{2}(x + 1) - \frac{1}{2}(x + 4)$

10. $\frac{3}{2}\left(x + \frac{2}{3}\right) + \frac{1}{2}(x + 8)$

11. $\frac{2}{3}(x^2 + x) - \frac{1}{3}(2x^2 - x)$

12. $\frac{3}{4}(x^2 + 4) + \frac{1}{4}(x^2 - 12)$

9.7 Even More Applications of Equations

- Investment Problems
- Uniform Motion Problems
- Liquid Mixture Problems
- Dry Mixture Problems

In this section, we consider more types of word problems.

Investment Problems

EXAMPLE 1

Kristy invested part of $12,000 at 9% annual interest, and the rest at 11%. Her annual income from these investments was $1230. How much did she invest at each rate?

Analysis

The interest i earned by an amount p invested at an annual rate r for t years is given by the formula $i = prt$. In this example, $t = 1$ year. Hence, if x dollars were invested at 9%, the interest earned would be $0.09x$. If x dollars were invested at 9%, the rest of the money $(12,000 - x)$ would be invested at 11%. The interest earned on that money would be $0.11(12,000 - x)$. The total interest earned in dollars can be expressed in two ways: as 1230 and as the sum $0.09x + 0.11(12,000 - x)$.

Solution

Let x represent the amount of money invested at 9%. Then $12,000 - x$ represents the amount of money invested at 11%.

Form an equation as follows:

The interest earned at 9%	plus	the interest earned at 11%	equals	the total interest.
$0.09x$	$+$	$0.11(12,000 - x)$	$=$	1230

$$0.09x + 0.11(12,000 - x) = 1230 \quad \text{The equation to solve.}$$
$$9x + 11(12,000 - x) = 123,000 \quad \text{Multiply both sides by 100 to clear the equation of decimals.}$$
$$9x + 132,000 - 11x = 123,000 \quad \text{Remove parentheses.}$$
$$-2x + 132,000 = 123,000 \quad \text{Combine terms.}$$
$$-2x = -9000 \quad \text{Add } -132,000 \text{ to both sides.}$$
$$x = 4500 \quad \text{Divide both sides by } -2.$$

Kristy invested $4500 at 9% and $12,000 - $4500, or $7500, at 11%.

Check: The first investment yielded 9% of $4500, or $405. The second investment yielded 11% of $7500, or $825. Because the total return was $405 + $825, or $1230, the answers check. ■

Uniform Motion Problems

EXAMPLE 2

Chicago, Illinois, and Green Bay, Wisconsin, are about 200 miles apart. A car leaves Chicago traveling toward Green Bay at 55 miles per hour. At the same time, a truck leaves Green Bay bound for Chicago at 45 miles per hour. How long will it take them to meet?

Analysis

Uniform motion problems are based on the formula $d = rt$, where d is the distance traveled, r is the rate, and t is the time. Organize the information of this problem in chart form, as in Figure 9-7.

	r	$\cdot$	t	$=$	d
Car	55		t		$55t$
Truck	45		t		$45t$

Figure 9-7

We know that the two vehicles travel for the same amount of time—say, t hours. The faster car travels $55t$ miles, and the slower truck travels $45t$ miles. The total distance can be expressed in two ways: as the sum $55t + 45t$ and as 200 miles.

Solution

Let t represent the time that each vehicle travels until they meet. Then

$55t$ represents the distance traveled by car

$45t$ represents the distance traveled by the truck

We can form the equation

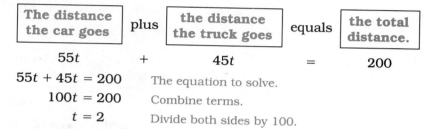

$$55t + 45t = 200$$

$$55t + 45t = 200 \quad \text{The equation to solve.}$$
$$100t = 200 \quad \text{Combine terms.}$$
$$t = 2 \quad \text{Divide both sides by 100.}$$

The vehicles meet after 2 hours.

Check: During those 2 hours, the car travels $55 \cdot 2$, or 110 miles, while the truck travels $45 \cdot 2$, or 90 miles. The total distance traveled is $110 + 90$, or 200 miles. This is the total distance between Chicago and Green Bay. The answer checks. ∎

Liquid Mixture Problems

EXAMPLE 3

A chemist has a solution that is 50% sulfuric acid and another that is 20% sulfuric acid. How much of each should she use to make 12 liters of a 30% solution?

Analysis

See Figure 9-8. If x represents the numbers of liters of the 50% solution required for the mixture, then the rest of the mixture $((12 - x)$ liters) must be the 20% solution. Only 50% of the x liters, and only 20% of the $(12 - x)$ liters, is pure sulfuric acid. The total of these amounts is also the amount of acid in the final mixture, which is 30% of 12 liters.

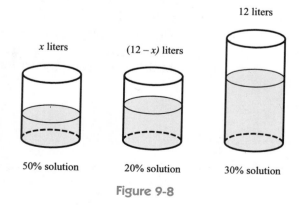

Figure 9-8

Solution

Let x represent the required number of liters of the 50% solution. Then $12 - x$ represents the required number of liters of the 20% solution.

We can form the equation

The acid in the 50% solution	plus	the acid in the 20% solution	equals	the acid in the mixture.
50% of x	+	20% of $(12 - x)$	=	30% of 12

$0.50x + 0.20(12 - x) = 0.30(12)$ The equation to solve.
$5x + 2(12 - x) = 3(12)$ Multiply both sides by 10 to clear the equation of decimals.
$5x + 24 - 2x = 36$ Remove parentheses.
$3x + 24 = 36$ Combine terms.
$3x = 12$ Add -24 to both sides.
$x = 4$ Divide both sides by 3.

The chemist must mix 4 liters of the 50% solution and 8 liters $((12 - 4)$ liters) of the 20% solution. Check these results. ∎

Dry Mixture Problems

EXAMPLE 4

Fancy cashews priced at $6 per pound are not selling because they are too expensive. Filberts are selling at $3 per pound. How many pounds of filberts should be combined with 50 pounds of cashews to obtain a mixture that can be sold at $4 per pound?

Analysis

The money received by selling the filberts and cashews separately should equal the money received by selling the mixture. Suppose x pounds of filberts are used in the mixture. At $3 per pound, they are worth $3x$. At $6 per pound, the 50 pounds of cashews are worth $6 \cdot 50$, or $300. The mixture will weigh $(50 + x)$ pounds, and at $4 per pound, it is worth $4 \cdot (50 + x)$. The value of the ingredients, $\$(3x + 300)$, is equal to the value of the mixture, $\$4(50 + x)$.

Solution

Let x represent the number of pounds of filberts in the mixture.

We can form the equation

The value of the filberts	plus	the value of the cashews	equals	the value of the mixture.
$3x$	+	$6 \cdot 50$	=	$4(50 + x)$

$3x + 6 \cdot 50 = 4(50 + x)$ The equation to solve.
$3x + 300 = 200 + 4x$ Remove parentheses and simplify.
$100 = x$ Add $-3x$ and -200 to both sides.

The storekeeper should use 100 pounds of filberts in the mixture.

Check:

The value of 100 pounds of filberts at $3 per pound is	$300
The value of 50 pounds of cashews at $6 per pound is	$300
The value of the mixture is	$600

The value of 150 pounds of mixture at $4 per pound is $600. The answer checks.

9.7 Exercises

In Exercises 1–28, pick a variable to represent the unknown quantity, set up an equation involving the variable, solve the equation, and check it.

Investment Problems

1. Steve's $24,000 is invested in two accounts, one earning 6% annual interest and the other earning 7%. After 1 year, his combined interest is $1635. How much was invested at each rate?

2. Martha's $18,750 is invested in two accounts, one earning 12% interest and the other earning 10%. After 1 year, her combined interest income is $2117. How much has she invested at each rate?

3. Carol invested equal amounts in each of two investments. One investment pays 8% and the other pays 11%. Her combined interest income for 1 year is $712.50. How much did she invest at each rate?

4. Craig invested equal amounts in each of three accounts, paying 7%, 8%, and 10.5%. His combined yearly interest income is $1249.50. How much did he invest in each account?

5. Marilu invested $7050 in two accounts, one paying 13% annual interest and the other paying 10%. The amount invested at the lower rate was half that invested at the higher rate. How much did she invest at each rate?

6. Equal amounts are invested in accounts paying 11% and 13%. The difference in the annual interests is $150. How much is invested in each account?

7. Twice as much money is invested at 12% annual interest than at 10%. The difference in the annual incomes is $350. How much is invested at each rate?

8. The amount of annual interest earned by $8000 invested at a certain rate is $200 less than $12,000 would earn at a 1% lower rate. At what rate is the $8000 invested? (*Hint:* 1% = 0.01.)

Uniform Motion Problems

9. Two cities, A and B, are 315 miles apart. A car leaves A bound for B at 50 miles per hour. At the same time, another car leaves B and heads toward A at 55 miles per hour. In how many hours will the two cars meet?

10. Granville and Preston are 535 miles apart. A car leaves Preston bound for Granville at 47 miles per hour. At the same time, another car leaves Granville and heads toward Preston at 60 miles per hour. How long will it take them to meet?

11. Two cars leave Peoria at the same time, one heading east at 60 miles per hour and the other west at 50 miles per hour. How long will it take them to be 715 miles apart?

12. Two boats steam out of port, one heading north at 35 knots (nautical miles per hour), the other south at 47 knots. If they leave port at the same time, how long will it take them to be 738 nautical miles apart?

13. Two cars start together and head north, one at 42 miles per hour and the other at 53 miles per hour. In how many hours will the cars be 82.5 miles apart?

14. Two trains are 330 miles apart, and their speeds differ by 20 miles per hour. They travel toward each other and meet in 3 hours. Find the speed of each train.

15. Two planes are 6000 miles apart, and their speeds differ by 200 miles per hour. They travel toward each other and meet in 5 hours. Find the speed of the slower plane.

16. An automobile averaged 40 miles per hour for part of a trip and 50 miles per hour for the remainder. If the 5-hour trip covered 210 miles, for how long did the car average 40 miles per hour?

Mixture Problems

17. How many gallons of fuel costing $1.15 per gallon must be mixed with 20 gallons of a fuel costing $0.85 per gallon to obtain a mixture costing $1 per gallon?

18. Paint costing $19 per gallon is to be mixed with 5 gallons of a $3 per gallon thinner to make a paint that can be sold for $14 per gallon. How much paint will be produced?

19. How many gallons of a 30% salt solution must be mixed with 50 gallons of a 70% solution to obtain a 50% solution?

20. How many gallons of milk containing 4% butterfat should be mixed with 10 gallons of milk containing 1% butterfat to obtain a mixture containing 2% butterfat?

21. A nurse wants to add water to 30 ounces of a 10% alcohol solution to dilute it to an 8% solution. How much water must the nurse add?

22. A chemist wants to mix 2 liters of a 5% silver iodide solution with a 10% solution to get a 7% solution. How many liters of 10% solution must the chemist add?

23. Lemon drops worth $1.90 per pound are to be mixed with jelly beans that cost $1.20 per pound to make 100 pounds of a mixture worth $1.48 per pound. How many pounds of each candy should be used?

24. One grade of tea, worth $3.20 per pound, is to be mixed with another grade worth $2 per pound to make 20 pounds that will sell for $2.72 per pound. How much of each grade of tea must be used?

25. A pound of hard candy is worth $0.30 less than a pound of soft candy. Equal amounts of each are used to make 40 pounds of a mixture that sells for $1.05 per pound. How much is a pound of soft candy worth?

26. Twenty pounds of a candy that sells for $1.70 per pound is mixed with candy that sells for $2 per pound. If the mixture is to sell for $1.80 per pound, how much of the more expensive candy should be used?

27. A store sells regular coffee for $4 a pound and a gourmet coffee for $7 a pound. To get rid of 40 pounds of the gourmet coffee, the shopkeeper makes a gourmet blend that he will put on sale for $5 a pound. How many pounds of regular coffee should be used?

28. A garden store sells Kentucky bluegrass seed for $6 per pound and ryegrass seed for $3 per pound. How much rye must be mixed with 100 pounds of bluegrass to obtain a blend that will sell for $5 per pound?

Review Exercises

In Review Exercises 1–8, graph each set on a number line.

1. The integers between -2 and 4

2. The even integers between 3 and 10

3. All real numbers between -2 and 4

4. All real numbers less than -2 or greater than or equal to 4

5. All negative real numbers greater than or equal to -6

6. All positive numbers less than 3

7. All real numbers less than $\frac{3}{2}$

8. All real numbers greater than $-\frac{5}{3}$

9. The amount, A, on deposit in a bank account bearing simple interest is given by the formula

$$A = P + Prt$$

Determine A when $P = \$1200$, $r = 0.08$, and $t = 3$.

10. The distance, s, that a certain object falls in t seconds is given by the formula

$$s = 350 - 16t^2 + vt$$

Determine s when $t = 4$ and $v = -3$.

11. David needs $3.50 more and Heidi needs $1.50 more to split equally the $13 cost of a pizza. How much money does David have?

12. Find a number that is 1 less than three times itself.

9.8 Inequalities

- Solving Inequalities
- Double or Compound Inequalities

Recall the meaning of the following symbols.

> $<$ means "is less than"
> $>$ means "is greater than"
> $\leq$ means "is less than or equal to"
> $\geq$ means "is greater than or equal to"

An **inequality** is an expression that shows that two quantities are not necessarily equal. A **solution of an inequality** is any number that makes the inequality a true statement. The number 2 is a solution of the inequality

$$x \leq 3$$

because $2 \leq 3$. The inequality $x \leq 3$ has many more solutions, because *any* real number that is less than or equal to 3 will satisfy the inequality.

We can use a graph on the number line to show the solutions of the inequality $x \leq 3$. The colored arrow in Figure 9-9 indicates all those points with coordinates that satisfy the inequality $x \leq 3$. The solid circle at the point with coordinate 3 indicates that the number 3 is a solution of the inequality $x \leq 3$.

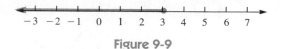

Figure 9-9

The graph of the inequality $x > 1$ appears in Figure 9-10.

Figure 9-10

The colored arrow indicates all those points whose coordinates satisfy the inequality $x > 1$. The open circle at the point with coordinate 1 indicates that 1 is not a solution of the inequality $x > 1$.

Solving Inequalities

To solve inequalities, we often need to use the addition, subtraction, **multiplication**, and **division properties of inequalities**. When we use any of these properties, the resulting inequality will have the same solutions as the original inequality.

The Addition Property of Inequality: Let a, b, and c be real numbers:

If $a < b$, then $a + c < b + c$.

Similar statements can be made for $>$, $\leq$, and $\geq$.

The **addition property of inequality** can be stated in words: *If any quantity is added to both sides of an inequality, the resulting inequality has the same direction as the original inequality.*

The Subtraction Property of Inequality: Let a, b, and c be real numbers:

If $a < b$, then $a - c < b - c$.

Similar statements can be made for $>$, $\leq$, and $\geq$.

The **subtraction property of inequality** can be stated in words: *If any quantity is subtracted from both sides of an inequality, the resulting inequality has the same direction as the original inequality.*

Note that to *subtract* a number a from both sides of an inequality, we could instead *add* the *negative* of a to both sides.

EXAMPLE 1

Solve the inequality $2x + 5 > x - 4$ and graph its solution on a number line.

Solution

To isolate the x on the left-hand side of the $>$ sign, we proceed as if solving equations.

$$2x + 5 > x - 4$$
$$2x + 5 - 5 > x - 4 - 5 \quad \text{Subtract 5 from both sides.}$$
$$2x > x - 9 \quad \text{Combine terms.}$$
$$2x - x > x - 9 - x \quad \text{Subtract } x \text{ from both sides.}$$
$$x > -9 \quad \text{Combine terms.}$$

The graph of this solution (see Figure 9-11) includes all points to the right of −9 but does not include −9 itself, since ">" means "strictly greater than." For that reason, we use an open circle at −9.

Figure 9-11

Progress Check 11

Solve $2x - 1 < x + 2$ and graph the solution on a number line.

Work Progress Check 11.

If both sides of the true inequality $2 < 5$ are multiplied by a *positive* number, such as 3, another true inequality results.

$2 < 5$
$3 \cdot 2 < 3 \cdot 5$ Multiply both sides by 3.
$6 < 15$ Simplify to get a true inequality.

However, if both sides of the inequality $2 < 5$ are multiplied by a negative number, such as −3, the direction of the inequality symbol must be reversed to produce another true inequality.

$2 < 5$
$-3 \cdot 2 > -3 \cdot 5$ Multiply both sides by the *negative* number −3 and reverse the direction of the inequality.
$-6 > -15$ Simplify to get a true inequality.

Thus, if both sides of an inequality are multiplied by a *positive* number, the solutions of the resulting inequality remain the same. However, if both sides of an inequality are multiplied by a *negative* number, the direction of the resulting inequality must be reversed. This is stated formally as follows.

The Multiplication Property of Inequality: Let a, b, and c be real numbers:

If $a < b$ and $c > 0$,   then $ac < bc$.
If $a < b$ and $c < 0$,   then $ac > bc$.

There is a similar property for division.

The Division Property of Inequality: Let a, b, and c be real numbers:

If $a < b$ and $c > 0$,   then $\dfrac{a}{c} < \dfrac{b}{c}$.

If $a < b$ and $c < 0$,   then $\dfrac{a}{c} > \dfrac{b}{c}$.

To *divide* both sides of an inequality by a nonzero number c, we could instead *multiply* both sides by $\frac{1}{c}$.

The multiplication and division properties of inequality are also true for $\leq$, $>$, and $\geq$.

11.

EXAMPLE 2

Solve the inequality $3x + 7 \leq -5$ and graph the solution.

Solution

$$3x + 7 \leq -5$$
$$3x + 7 + (-7) \leq -5 + (-7) \quad \text{Add } -7 \text{ to both sides.}$$
$$3x \leq -12 \quad \text{Combine terms.}$$
$$\frac{3x}{3} \leq \frac{-12}{3} \quad \text{Divide both sides by 3.}$$
$$x \leq -4$$

The solution of $3x + 7 \leq -5$ consists of all real numbers less than, and also including, -4. The solid circle at -4 in the graph in Figure 9-12 shows that -4 is one of the solutions.

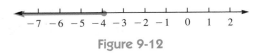

Figure 9-12

EXAMPLE 3

Solve the inequality $5 - 3x \leq 14$ and graph the solution.

Solution

$$5 - 3x \leq 14$$
$$5 - 3x + (-5) \leq 14 + (-5) \quad \text{Add } -5 \text{ to both sides.}$$
$$-3x \leq 9 \quad \text{Combine terms.}$$
$$\frac{-3x}{-3} \geq \frac{9}{-3} \quad \text{Divide both sides by } -3 \text{ and reverse the direction of the } \leq \text{ symbol.}$$
$$x \geq -3$$

In the last step, both sides of the inequality were divided by -3. Because -3 is negative, the direction of the inequality was *reversed*. The graph of the solution appears in Figure 9-13. The solid circle at -3 shows that -3 is one of the solutions.

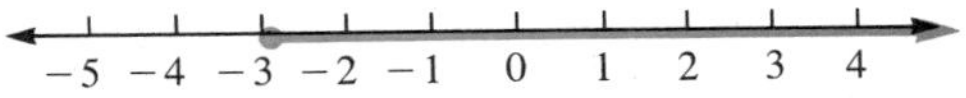

Figure 9-13

EXAMPLE 4

Solve the inequality $x(x - 8) < x^2 + 3(x + 11)$ and graph the solution.

Solution

$$x(x-8) < x^2 + 3(x+11)$$
$$x^2 - 8x < x^2 + 3x + 33 \quad \text{Remove parentheses.}$$
$$-8x < 3x + 33 \quad \text{Add } -x^2 \text{ to both sides.}$$
$$-11x < 33 \quad \text{Add } -3x \text{ to both sides.}$$
$$x > -3 \quad \text{Divide both sides by } -11 \text{ and reverse the direction of the inequality sign.}$$

The graph of the solution appears in Figure 9-14. The open circle at -3 shows that -3 is not a solution.

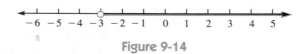

Figure 9-14

Progress Check 12

Work Progress Check 12.

Solve $8 - 4x \geq 40$ and graph the solution on the number line.

Double or Compound Inequalities

Two inequalities can often be combined into a **double inequality** or **compound inequality** to show that numbers lie *between* two fixed values. The inequality $2 < x < 5$, for example, indicates that x is greater than 2 and that x is *also* less than 5. The solution of the double inequality $2 < x < 5$ consists of all numbers that lie *between* 2 and 5. The graph of this set, called an **interval**, appears in Figure 9-15.

Figure 9-15

EXAMPLE 5

Solve the inequality $-4 < 2(x-1) \leq 4$ and graph the solution.

Solution

$$-4 < 2(x-1) \leq 4$$
$$-4 < 2x - 2 \leq 4 \quad \text{Remove parentheses.}$$
$$-2 < 2x \leq 6 \quad \text{Add 2 to all three parts.}$$
$$-1 < x \leq 3 \quad \text{Divide all three parts by 2.}$$

The graph of the solution appears in Figure 9-16.

Progress Check 13

Solve $-6 \leq 3(x+1) < 12$ and graph the solution on a number line.

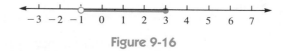

Figure 9-16

Work Progress Check 13.

EXAMPLE 6

A labor union requires that its members work no more than 45 hours per week. How many minutes may a member work in a week?

Progress Check Answers

Solution

Let x represent the number of minutes a union member may work each week. Because there are 60 minutes in 1 hour, there are $60 \cdot 45$, or 2700, minutes in a 45-hour workweek. Thus,

$$x \leq 2700$$

Of course, no one can work less than 0 minutes per week, so the solution is given by the double inequality

$$0 \leq x \leq 2700$$

A union member may work from 0 to 2700 minutes each week. The graph of this solution appears in Figure 9-17.

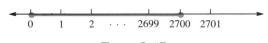

Figure 9-17

EXAMPLE 7

The perimeter of a certain equilateral triangle is no less than 15 feet. How long could a side be?

Solution

Recall that each side of an equilateral triangle is the same length and that the perimeter of a triangle is the sum of the lengths of its three sides. Let x represent the length of one side of the triangle. Then $x + x + x$ represents the perimeter. Because the perimeter is to be no less than 15 feet (the same as "greater than or equal to 15 feet"), we have

$$x + x + x \geq 15$$
$$3x \geq 15 \quad \text{Combine terms.}$$
$$x \geq 5 \quad \text{Divide both sides by 3.}$$

Each side of the triangle must be at least 5 feet long.

9.8 Exercises

In Exercises 1–40, solve each inequality and graph the solution.

1. $x + 2 > 5$
2. $x + 5 \geq 2$
3. $-x - 3 \leq 7$
4. $-x - 9 > 3$
5. $3 + x < 2$
6. $5 + x \geq 3$
7. $2x - 3 \leq 5$
8. $-3x - 5 < 4$
9. $-3x - 7 > -1$
10. $-5x + 7 \leq 12$
11. $-4x + 1 > 17$
12. $7x - 9 > 5$
13. $2x + 9 \leq x + 8$
14. $3x + 7 \leq 4x - 2$
15. $9x + 13 \geq 8x$
16. $7x - 16 < 6x$
17. $8x + 4 > 6x - 2$
18. $7x + 6 \geq 4x$
19. $5x + 7 < 2x + 1$
20. $7x + 2 > 4x - 1$
21. $7 - x \leq 3x - 1$
22. $2 - 3x \geq 6 + x$
23. $9 - 2x > 24 - 7x$
24. $13 - 17x < 34 - 10x$
25. $3(x - 8) < 5x + 6$
26. $9(x - 11) > 13 + 7x$
27. $8(5 - x) \leq 10(8 - x)$
28. $17(3 - x) \geq 3 - 13x$
29. $x(5x - 5) > 5x^2 + 15$
30. $x(x - 7) < x^2 + 3x + 20$
31. $x(x + 8) \leq x^2 + 24$
32. $x(5 + 3x) \leq 15 + 3x^2$

33. $89x^2 - 178 > 89x(x - 1)$

34. $31x^2 + 124 > 3.1x(10x + 20)$

35. $\frac{5}{2}(7x - 15) + x \geq \frac{13}{2}x - \frac{3}{2}$

36. $\frac{5}{3}(x + 1) \leq -x + \frac{2}{3}$

37. $\frac{3x - 3}{2} < 2x + 2$

38. $\frac{x + 7}{3} \geq x - 3$

39. $\frac{2(x + 5)}{3} \leq 3x - 6$

40. $\frac{3(x - 1)}{4} > x + 1$

In Exercises 41–60, solve each inequality and graph the solution.

41. $2 < x - 5 < 5$
42. $3 < x - 2 < 7$
43. $-5 < x + 4 \leq 7$
44. $-9 \leq x + 8 < 1$
45. $0 \leq x + 10 \leq 10$
46. $-8 < x - 8 < 8$
47. $4 < -2x < 10$
48. $-4 \leq -4x < 12$
49. $-3 \leq \frac{x}{2} \leq 5$
50. $-12 \leq \frac{x}{3} < 0$
51. $3 \leq 2x - 1 < 5$
52. $4 < 3x - 5 \leq 7$
53. $0 < 10 - 5x \leq 15$
54. $1 \leq -7x + 8 \leq 15$
55. $-6 < 3(x + 2) < 9$
56. $-18 \leq 9(x - 5) < 27$
57. $x^2 + 3 \leq x(x - 3) \leq x^2 + 9$
58. $x^2 < x(x + 4) \leq x^2 + 8$
59. $3 - x < 5 < 7 - x$
60. $x + 1 < 2x + 3 < x + 5$

In Exercises 61–76, express each solution as an inequality.

61. The perimeter of an equilateral triangle is at most 57 feet. What could be the length of a side? (*Hint:* All three sides of an equilateral triangle are equal.)

62. The perimeter of a square is no less than 68 centimeters. How long can a side be?

63. The land elevations in Nevada range from the 13,143-foot height of Boundary Peak to the Colorado River at 470 feet. To the nearest tenth, find the range of these elevations in miles. (*Hint:* 1 mile is 5280 feet.)

64. A teacher requires that students do homework at least 2 hours a day. How many minutes should a student work each week?

65. A pilot plans to fly at an altitude of between 17,500 and 21,700 feet. To the nearest tenth, what will be the range of altitudes in miles? (*Hint:* There are 5280 feet in 1 mile.)

66. A doctor advises Jake to exercise at least 15 minutes but less than 30 minutes per day. How many hours of exercise will Jake get each week?

67. To hold the temperature of a room between 19° and 22° Celsius, what Fahrenheit temperatures must be maintained? (*Hint:* Fahrenheit temperature F and Celsius temperature C are related by the formula $C = \frac{5}{9}(F - 32)$.)

68. To melt iron, the temperature of a furnace must be at least 1540°C but no more than 1650°C. What range of Fahrenheit temperatures must be maintained?

69. The radii of phonograph records must lie between 5.9 and 6.1 inches. What variation in circumference can occur? (*Hint:* The circumference of a circle is given by the formula $C = 2\pi r$, where r is the radius. Let $\pi = 3.14$.)

70. The heights of children in three fourth-grade classes range from 42 to 57 inches. What is the range of their heights in centimeters? (*Hint:* There are 2.54 centimeters in 1 inch.)

71. The normal weight for a 6-foot, 2-inch man is between 150 and 190 pounds. To the nearest hundredth, what would such a person weigh in kilograms? (*Hint:* There are 2.2 pounds in one kilogram.)

72. Jim's car averages between 19 and 23 miles per gallon. At $1.20 per gallon, what might Jim expect to pay for fuel on a 350-mile trip?

73. Chen weighs 40 pounds more than his mother and 10 pounds less than his father. Altogether, they weigh no more than 420 pounds and no less than 390 pounds. What might Chen weigh?

74. The time required to assemble a television set at the factory is 2 hours. A stereo receiver requires only 1 hour. The labor force at the factory can supply at least 640 and at most 810 hours of assembly time per week. When the factory is producing 3 times as many television sets as stereos, how many stereos could be manufactured in 1 week?

75. A rectangle's length is 3 feet less than twice its width, and its perimeter is between 24 and 48 feet. What might be its width?

76. A rectangle's width is 8 feet less than 3 times its length, and its perimeter is between 8 and 16 feet. What might be its length?

Review Exercises

In Review Exercises 1–6, simplify each expression, if possible.

1. $3x^2 - 2(y^2 - x^2)$
2. $5(xy + 2) - 3xy - 8$
3. $x(2x - y) - 3(x^2 - xy)$
4. $7x(x - 2) + x^2(2 + y)$

5. $\frac{1}{3}(x + 6) - \frac{4}{3}(x - 9)$
6. $\frac{4}{5}x(y + 1) - \frac{9}{5}y(x - 1)$

In Review Exercises 7–10, let $x = -\frac{2}{3}$ and $z = \frac{1}{3}$. Evaluate each expression.

7. $x^2 - z^2$
8. $(x + z)^2$
9. $\dfrac{x - z}{x + z}$
10. $\dfrac{x}{z} - \dfrac{z}{x}$

11. A man spent $16,500 for a car and a boat. The car cost $500 more than three times the cost of the boat. Find the cost of the car.

12. On the last exam, one-half of the class received a grade of C, one-third a grade of B, and one-twelfth a grade of D. No one failed, and three students received an A. How many students are in the class?

Mathematics in Retirement

To determine the maximum allowable annual contribution to a SEP, we first subtract fixed expenses from gross income. This is the net income, before the deduction for the retirement contribution; call this income N, and let C be the maximum allowable contribution. This contribution is 15% of the net income that remains after subtracting the contribution. Thus, C is 15% of (N − C), or

$$C = 0.15(N - C)$$

Solve this literal equation for C.

$C = 0.15N - 0.15C$	Remove parentheses.
$C + 0.15C = 0.15N$	Add 0.15C to both sides.
$1.15C = 0.15N$	Combine terms.
$C = \dfrac{0.15}{1.15}N$	Divide both sides by 1.15.
$C = .1304N$	Simplify.

Thus, the maximum allowable annual contribution to a SEP plan is slightly more than 13% of net taxable income.

The information in this example is basically true, but tax law covers many special cases and includes many exceptions. It is always wise to consult a tax professional for advice.

Chapter Summary

Key Words

addition property of equality (9.1)
addition property of inequality (9.8)
base angles of an isosceles triangle (9.6)
break-even analysis (9.6)
compound inequality (9.8)
contradictions (9.3)
division property of equality (9.2)
division property of inequality (9.8)
double inequality (9.8)
equation (9.1)
fixed cost (9.6)
formula (9.4)
identity (9.3)
impossible equations (9.3)
inequality (9.8)
interval (9.8)
isosceles triangle (9.6)
like terms (9.3)
literal equation (9.4)
multiplication property of equality (9.2)
multiplication property of inequality (9.8)
numerical coefficient (9.3)
perimeter (9.6)
root of an equation (9.1)
similar terms (9.3)
solution of an equation (9.1)
solution of an inequality (9.8)
subtraction property of equality (9.1)
subtraction property of inequality (9.8)
unknown (9.1)
unlike terms (9.3)
variable (9.1)
vertex angle of an isosceles triangle (9.6)

Key Ideas

(9.1) Any real number can be added to (or subtracted from) both sides of an equation to form another equation with the same solutions as the original equation.

(9.2) Both sides of an equation can be multiplied (or divided) by any *nonzero* real number to form another equation with the same solutions as the original equation.

(9.3) Like terms can be combined by adding their numerical coefficients and using the same variables and exponents.

(9.4) A literal equation, or formula, can often be solved for any of its variables.

(9.5–9.7) Equations are useful in solving many applied problems.

(9.8) Inequalities are solved by techniques similar to those used to solve equations, with this important exception: *If both sides of an inequality are multiplied or divided by a negative number, the direction of the inequality must be reversed.*

The solution of an inequality can be graphed on the number line.

Chapter 9 Review Exercises

[9.1] In Review Exercises 1–8, indicate whether the indicated number is a solution of the given equation.

1. $3x + 7 = 1$; -2
2. $5 - 2x = 3$; -1
3. $2(x + 3) = x$; -3
4. $5(3 - x) = 2 - 4x$; 13
5. $3\left(\dfrac{x}{2} - 1\right) = 9$; 8
6. $5\left(\dfrac{x}{3} + 3\right) = 40$; 15
7. $3(x + 5) = 2(x - 3)$; -21
8. $2(x - 7) = x + 14$; 0

[9.1–9.2] In Review Exercises 9–32, solve each equation. Check all solutions.

9. $x + 7 = 3$
10. $5 - x = 2$
11. $2x - 5 = 13$
12. $3x + 4 = -8$
13. $5y + 6 = 21$
14. $5y - 9 = 1$
15. $12z + 4 = -8$
16. $17z + 3 = 20$

17. $13 - 13t = 0$ **18.** $10 + 7t = -4$ **19.** $23a - 43 = 3$ **20.** $84 - 21a = -63$

21. $3x + 7 = 3 + x$ **22.** $7 - 9x = 15 - x$ **23.** $\dfrac{b+3}{4} = 2$ **24.** $\dfrac{b-7}{2} = -2$

25. $\dfrac{x-8}{5} = 1$ **26.** $\dfrac{x+10}{2} = -1$ **27.** $\dfrac{2(y-1)}{4} = 2$ **28.** $\dfrac{3(y+4)}{11} = 3$

29. $\dfrac{1}{2}(x+7) = 11$ **30.** $\dfrac{1}{3}(x-9) = 7$ **31.** $\dfrac{3}{4}\left(\dfrac{a}{3} + 3\right) = 6$ **32.** $\dfrac{3}{4}\left(\dfrac{b}{7} - 2\right) = 0$

[9.3] In Review Exercises 33–44, simplify each expression, if possible.

33. $5x + 9x$ **34.** $7a + 12a$ **35.** $18b - 13b$ **36.** $21x - 23x$

37. $5x + 7y$ **38.** $19x - 19$ **39.** $y^2 + 3(y^2 - 2)$ **40.** $2x^2 - 2(x^2 - 2)$

41. $7(x + 2) + 2(x - 7)$ **42.** $2(3 - x) + x - 6y$

43. $2^3(x + 9) - 8(x + 3^2)$ **44.** $2^3 + 3k + 3(k - 2)$

In Review Exercises 45–56, solve each equation. Check all solutions.

45. $5x + 7 = 4x$ **46.** $7x - 9 = 8x$

47. $8a + 2 = 2a + 8$ **48.** $12x + 3 = 5x - 11$

49. $2x - 19 = 2 - x$ **50.** $5b - 19 = 2b + 20$

51. $3x + 20 = 5 - 2x$ **52.** $9x + 100 = 7x + 18$

53. $10(t - 3) = 3(t + 11)$ **54.** $2(5x - 7) = 2(x - 35)$

55. $x(x + 6) = x^2 + 6$ **56.** $3y(y + 2) = 2(y^2 + y) + y^2$

[9.4] In Review Exercises 57–68, solve each equation for the variable indicated.

57. $E = IR$; for R **58.** $i = prt$; for t **59.** $P = I^2 R$; for R **60.** $d = rt$; for r

61. $V = lwh$; for h **62.** $y = mx + b$; for m **63.** $V = \pi r^2 h$; for h **64.** $A = 2\pi rh$; for r

65. $F = \dfrac{GMm}{d^2}$; for G **66.** $P = \dfrac{RT}{mV}$; for m **67.** $T = n(V - 3)$; for V **68.** $T = n(V - 3)$; for n

[9.5–9.7] In Review Exercises 69–84, translate each problem into an equation, solve it, and check the solution.

69. What number added to 19 gives 11?

70. If twice a number is decreased by 7, the result is 9. Find the original number.

71. If Mark had 10 cents more than twice the amount he has, he would have enough to pay $9.80 for a pizza. How much money does Mark have?

72. Sue's electric rate is $17.50 per month, plus 18 cents for every kilowatt-hour of energy used. Her bill in July was $43.96. How many kilowatt-hours did she use that month?

73. The installation of rain gutters on Bert's house will cost $35, plus $1.50 per foot. He expects to pay $162.50 to replace the gutters. How many feet of gutter does he need?

74. The sum of two consecutive odd integers is 44. Find the two integers.

75. If it were 10 degrees warmer today, it would be 5 degrees colder than twice today's temperature. Find today's temperature.

76. A 45-foot rope is to be cut into three sections. One section is to be 15 feet long. Of the remaining sections, one is to be 2 feet less than 3 times the length of the other. Find the length of the shortest section.

77. The perimeter of a rectangle is 84 inches. If the length is 3 inches more than twice the width, how wide is the rectangle?

78. Costs for machining an automotive part run $668.50 per week, plus variable costs of $1.25 per unit. The company can sell as many as it can make for $3 each. Find the break-even point.

79. A company can manufacture baseball caps on either of two machines, with costs as shown in the following table. At their projected sales level, the company finds that costs on the two machines are equal. What are the expected sales?

Machine	Setup Cost	Unit Cost (per cap)
A	$85	$3
B	$105	$2.50

80. Juanita invests $27,000 for 1 year. She invests part of it in a junk bond fund paying 12% interest and the remaining amount in a cash management fund paying 14%. The total interest on her two investments is $3460. How much does she invest at each rate?

81. A bicycle path is 5 miles long. Jim starts at one end, walking at the rate of 3 miles per hour. At the same time, his friend Jerry bicycles from the other end, traveling at 12 miles per hour. In how many minutes will they meet?

82. A store manager mixes candy worth 90 cents per pound with gumdrops worth $1.50 per pound to make 20 pounds of a mixture worth $1.20 per pound. How many pounds of each kind of candy does the manager use?

83. After a 6% raise, Theresa now makes $18,550. What was her salary before the raise?

84. After sales tax, a $35 purchase will cost $37.45. What is the sales tax rate?

[9.8] In Review Exercises 85–94, solve each inequality and graph its solution.

85. $3x + 2 < 5$

86. $-5x - 8 > 7$

87. $5x - 3 \geq 2x + 9$

88. $7x + 1 \leq 8x - 5$

89. $5(3 - x) \leq 3(x - 3)$

90. $3(5 - x) \geq 2x$

91. $8 < x + 2 < 13$

92. $0 \leq 2 - 2x < 6$

93. $x^2 < x(x + 1) \leq x^2 + 9$

94. $x^2 + 4 \geq x^2 + x \geq x^2 + 3$

Chapter 9 Test

In Problems 1–4, state whether the indicated number is a solution of the given equation.

1. $5x + 3 = -2$; -1

2. $3(x + 2) = 2x$; -6

3. $-3(2 - x) = 0$; -2

4. $x(x + 1) = x^2 + 1$; 1

In Problems 5–10, solve each equation.

5. $8x + 2 = -14$

6. $3 = 5 - 2x$

7. $23 - 5(x + 10) = -12$

8. $x(x + 5) = x^2 + 3x - 8$

9. $\dfrac{3(x - 6)}{2} = 6x$

10. $\dfrac{5}{3}(x - 7) = 15(x + 1)$

In Problems 11–16, simplify each expression.

11. $x + 5(x - 3)$

12. $3x - 5(2 - x)$

13. $x(x - 3) + 2x^2 - 3x$

14. $2^2(x^2 - 3^2) - x(x + 36)$

15. $-3x(x + 3) + 3x(x - 3)$

16. $-4x(2x - 5) - 7x(4x + 1)$

In Problems 17–22, solve each equation for the variable indicated.

17. $d = rt$; for t

18. $P = 2l + 2w$; for l

19. $A = 2\pi rh$; for h

20. $A = P + Prt$; for r

21. $P = \dfrac{RT}{v}$; for v

22. $A = \dfrac{1}{3}\pi r^2 h$; for h

23. A chocolate chip cookie has two-thirds of the calories of a glass of milk. Jim has a glass of milk and 2 cookies and consumes 385 calories. How many calories are in a glass of milk?

24. The sum of two consecutive odd integers is 36. Find the integers.

25. Karen invested part of $13,750 at 7% annual interest and the rest at 6%. After 1 year, she received $910 in interest. How much did she invest at the lower rate?

26. A car leaves Rockford at the rate of 65 miles per hour, bound for Madison. At the same time, a truck leaves Madison at the rate of 55 miles per hour, bound for Rockford. If the cities are 72 miles apart, how long will it take for the car and the truck to meet?

27. How many liters of water must be added to 30 liters of a 10% brine solution to dilute it to an 8% solution?

In Problems 28–30, solve each inequality and graph its solution.

28. $8x - 20 \geq 4$

29. $x^2 - x(x + 7) > 14$

30. $-4 \leq 2(x + 1) < 10$

10 Polynomials

Mathematics in Medicine

The red cells in blood pick up oxygen in the lungs and carry it to all parts of the body. Each red cell is a tiny disc with an approximate radius of 0.00015 inch. Because the amount of oxygen carried depends on the surface area of the cells, and the cells are so tiny, a very great number is needed—25 trillion in an average adult.

What is the total surface area of all the red blood cells in the body? The calculations involve numbers very large and very small. A topic of this chapter, **scientific notation**, will make the calculations easier.

We now develop rules for finding products, powers, and quotients of exponential expressions, and then discuss how to add, subtract, multiply, and divide polynomials.

10.1 Natural-Number Exponents

- The Product Rule for Exponents
- The Power Rules for Exponents
- The Quotient Rule for Exponents

We have used exponents to show repeated multiplication. For example,

$2^5 = 2 \cdot 2 \cdot 2 \cdot 2 \cdot 2 = 32 \qquad (-7)^3 = (-7)(-7)(-7) = -343$

$x^4 = x \cdot x \cdot x \cdot x \qquad\qquad -y^5 = -y \cdot y \cdot y \cdot y \cdot y$

These examples suggest a definition for x^n, where n is a natural number.

> **Definition:** If n is a natural number, then
> $$x^n = \overbrace{x \cdot x \cdot x \cdot \cdots \cdot x}^{n \text{ factors of } x}$$

In the exponential expression x^n, x is called the **base**, and n is called the **exponent**. The entire expression is called a **power of x**.

$x^n \leftarrow$ exponent

base

A natural-number exponent tells how many times the base is to be used as a factor in a product. An exponent of 1 indicates that its base is to be used as a factor one time, an exponent of 2 indicates that its base is to be used as a factor two times, and so on.

$$3^1 = 3, \quad (-y)^1 = -y, \quad (-4z)^2 = (-4z)(-4z), \quad \text{and} \quad (t^2)^3 = t^2 \cdot t^2 \cdot t^2$$

EXAMPLE 1

Show that -2^4 and $(-2)^4$ are different numbers.

Solution

We write each expression without exponents to see that we obtain different results.

$$-2^4 = -(2^4) \qquad \text{Do the exponentiation first.}$$
$$= -(2 \cdot 2 \cdot 2 \cdot 2)$$
$$= -16$$
$$(-2)^4 = (-2)(-2)(-2)(-2) = 16$$

Since $-16 \neq 16$, it follows that $-2^4 \neq (-2)^4$. ∎

EXAMPLE 2

Write each expression without using exponents: **a.** r^3, **b.** $(-2s)^4$, and **c.** $(\frac{1}{3}ab)^5$.

Solution

a. $r^3 = r \cdot r \cdot r$

b. $(-2s)^4 = (-2s)(-2s)(-2s)(-2s)$

c. $\left(\dfrac{1}{3}ab\right)^5 = \left(\dfrac{1}{3}ab\right)\left(\dfrac{1}{3}ab\right)\left(\dfrac{1}{3}ab\right)\left(\dfrac{1}{3}ab\right)\left(\dfrac{1}{3}ab\right)$

Progress Check 1

Write each expression without using exponents.

a. $3(xy)^2$

b. $(3xy)^2$

Work Progress Check 1. ∎

The Product Rule for Exponents

To multiply x^3 by x^2, we note that the expression x^2 means that x is to be used as a factor two times and the expression x^3 means that x is to be used as a factor three times.

$$x^2 x^3 = \underbrace{x \cdot x}_{\text{2 factors of } x} \cdot \underbrace{x \cdot x \cdot x}_{\text{3 factors of } x}$$
$$= \underbrace{x \cdot x \cdot x \cdot x \cdot x}_{\text{5 factors of } x}$$
$$= x^5$$

In general, we have

$$x^m \cdot x^n = \underbrace{x \cdot x \cdot x \cdots x}_{m \text{ factors of } x} \cdot \underbrace{x \cdot x \cdot x \cdot x \cdots x}_{n \text{ factors of } x}$$
$$= \underbrace{x \cdot x \cdot x \cdot x \cdot x \cdot x \cdots x \cdot x \cdot x}_{m + n \text{ factors of } x}$$
$$= x^{m+n}$$

1. a. $3(xy)(xy)$ **b.** $(3xy)(3xy)$

Progress Check Answers

These examples justify the rule for multiplying exponential expressions: *To multiply two exponential expressions with the same base, we keep the base and add the exponents.*

The Product Rule for Exponents: If m and n are natural numbers, then

$$x^m x^n = x^{m+n}$$

EXAMPLE 3

a. $x^3 x^4 = x^{3+4}$
$ = x^7$

b. $y^2 y^4 = y^{2+4}$
$ = y^6$

c. $3 \cdot 3^3 = 3^1 3^3$
$ = 3^{1+3}$
$ = 3^4$

d. $x^2 x^3 x^6 = (x^2 x^3) x^6$
$ = (x^{2+3}) x^6$
$ = x^5 x^6$
$ = x^{5+6}$
$ = x^{11}$

e. $(2y^3)(3y^2) = 2(3) y^3 y^2$
$ = 6 y^{3+2}$
$ = 6 y^5$

f. $(4x)(-3x^2) = 4(-3) x x^2$
$ = -12 x^{1+2}$
$ = -12 x^3$

The product rule for exponents applies only to exponential expressions with the same base. An expression such as $x^2 y^3$ cannot be simplified, because x^2 and y^3 have different bases.

Work Progress Check 2.

Progress Check 2

Write each expression with a single exponent.

a. $y^2 y^3 y^4$

b. $(2x^2)(-3x^4)$

The Power Rules for Exponents

To find another rule for exponents, we consider the expression $(x^3)^4$. The expression $(x^3)^4$ can be written as $x^3 \cdot x^3 \cdot x^3 \cdot x^3$. Because each of the four factors of x^3 contains three factors of x, there are $4 \cdot 3$, or 12, factors of x. Thus, the product is x^{12}.

$(x^3)^4 = x^3 \cdot x^3 \cdot x^3 \cdot x^3$

$ = \overbrace{\underbrace{x \cdot x \cdot x}_{x^3} \cdot \underbrace{x \cdot x \cdot x}_{x^3} \cdot \underbrace{x \cdot x \cdot x}_{x^3} \cdot \underbrace{x \cdot x \cdot x}_{x^3}}^{12 \text{ factors of } x}$

$ = x^{12}$

In general, we have

$(x^m)^n = \overbrace{x^m \cdot x^m \cdot x^m \cdot \cdots \cdot x^m}^{n \text{ factors of } x^m}$

$ = \overbrace{x \cdot x \cdot x \cdot x \cdot x \cdot x \cdot x \cdot \cdots \cdot x}^{mn \text{ factors of } x}$

$ = x^{mn}$

2. a. y^9 **b.** $-6x^6$

Progress Check Answers

These examples justify the rule for raising an exponential expression to a power: *To raise an exponential expression to a power, we keep the same base and multiply the exponents.*

> **The First Power Rule for Exponents:** If m and n are natural numbers, then
> $$(x^m)^n = x^{mn}$$

EXAMPLE 4

a. $(2^3)^7 = 2^{3 \cdot 7}$
$= 2^{21}$

b. $(y^5)^2 = y^{5 \cdot 2}$
$= y^{10}$

c. $(z^7)^7 = z^{7 \cdot 7}$
$= z^{49}$

d. $(u^x)^y = u^{x \cdot y}$
$= u^{xy}$

Work Progress Check 3.

Write each expression with a single exponent.

a. $(x^3)^2$

b. $(x^7)^3$

In the next example, both the product and power rules of exponents are applied.

EXAMPLE 5

a. $(x^2 x^5)^2 = (x^7)^2$
$= x^{14}$

b. $(yy^6 y^2)^3 = (y^9)^3$
$= y^{27}$

c. $(z^2)^4 (z^3)^3 = z^8 z^9$
$= z^{17}$

d. $(x^3)^2 (x^5 x^2)^3 = x^6 (x^7)^3$
$= x^6 x^{21}$
$= x^{27}$

Work Progress Check 4.

Write $(x^3)^2(xx^2)$ as an expression with a single exponent.

To find two more power rules for exponents, we consider $(2x)^3$ and $(\frac{2}{x})^3$.

$(2x)^3 = (2x)(2x)(2x)$
$= (2 \cdot 2 \cdot 2)(x \cdot x \cdot x)$
$= 2^3 x^3$
$= 8x^3$

$\left(\dfrac{2}{x}\right)^3 = \left(\dfrac{2}{x}\right)\left(\dfrac{2}{x}\right)\left(\dfrac{2}{x}\right) \quad (x \neq 0)$
$= \dfrac{2 \cdot 2 \cdot 2}{x \cdot x \cdot x}$
$= \dfrac{2^3}{x^3}$
$= \dfrac{8}{x^3}$

These examples suggest that *to raise a product to a power, we raise each factor of the product to the power,* and *to raise a fraction to a power, we raise both the numerator and denominator to the power.*

3. **a.** x^6 **b.** x^{21}

4. x^9

More Power Rules for Exponents: If n is a natural number, then

$$(xy)^n = x^n y^n \quad \text{and if } y \neq 0, \text{ then} \quad \left(\frac{x}{y}\right)^n = \frac{x^n}{y^n}$$

EXAMPLE 6

a. $(ab)^4 = a^4 b^4$

b. $(3c)^3 = 3^3 c^3 = 27c^3$

c. $(x^2 y^3)^5 = (x^2)^5 (y^3)^5 = x^{10} y^{15}$

d. $(-2x^3 y)^2 = (-2)^2 (x^3)^2 y^2 = 4x^6 y^2$

e. $\left(\dfrac{4}{k}\right)^3 = \dfrac{4^3}{k^3} = \dfrac{64}{k^3}$

f. $\left(\dfrac{3x^2}{2y^3}\right)^5 = \dfrac{3^5 (x^2)^5}{2^5 (y^3)^5} = \dfrac{243 x^{10}}{32 y^{15}}$

Work Progress Check 5. ■

Progress Check 5

Write each expression without using parentheses.

a. $(xy^2)^3$

b. $(-2a^2 b^3)^3$

The Quotient Rule for Exponents

To find a rule for dividing exponential expressions we simplify the fraction $\dfrac{4^5}{4^2}$.

$$\frac{4^5}{4^2} = \frac{4 \cdot 4 \cdot 4 \cdot 4 \cdot 4}{4 \cdot 4}$$

Note that the exponent in the numerator is greater than the exponent in the denominator.

$$= \frac{\overset{1}{\cancel{4}} \cdot \overset{1}{\cancel{4}} \cdot 4 \cdot 4 \cdot 4}{\underset{1}{\cancel{4}} \cdot \underset{1}{\cancel{4}}}$$

$$= 4^3$$

The result, 4^3, has a base of 4 and an exponent of $5 - 2$, or 3. This suggests that *to divide exponential expressions with the same base, we keep the base and subtract the exponents.*

The Quotient Rule for Exponents: If m and n are natural numbers, $m > n$ and $x \neq 0$, then

$$\frac{x^m}{x^n} = x^{m-n}$$

EXAMPLE 7

If there are no divisions by 0, then

a. $\dfrac{x^4}{x^3} = x^{4-3} = x^1 = x$

b. $\dfrac{8y^2 y^6}{4y^3} = \dfrac{8y^8}{4y^3} = \dfrac{8}{4} y^{8-3} = 2y^5$

---

5. **a.** $x^3 y^6$ **b.** $-8a^6 b^9$

Progress Check Answers

c. $\dfrac{a^3 a^5 a^7}{a^4 a} = \dfrac{a^{15}}{a^5} = a^{15-5} = a^{10}$

d. $\dfrac{(a^3 b^4)^2}{ab^5} = \dfrac{a^6 b^8}{ab^5} = a^{6-1}b^{8-5} = a^5 b^3$

Progress Check 6

Simplify $\dfrac{x^4 x^3}{xx^2}$.

Work Progress Check 6.

We summarize the rules for positive exponents as follows:

Properties of Exponents: If n is a natural number, then

$$x^n = \overbrace{x \cdot x \cdot x \cdot \cdots \cdot x}^{n \text{ factors of } x}$$

If m and n are natural numbers and there are no divisions by 0,

$$x^m x^n = x^{m+n}, \qquad (x^m)^n = x^{mn}, \qquad (xy)^n = x^n y^n,$$

$$\left(\dfrac{x}{y}\right)^n = \dfrac{x^n}{y^n}, \qquad \dfrac{x^m}{y^n} = x^{m-n} \quad (\text{provided } m > n)$$

6. x^4

10.1 Exercises

In Exercises 1–12, identify the base and the exponent in each expression.

1. 4^3
2. $(-5)^2$
3. x^5
4. y^8
5. $(2y)^3$
6. $(-3x)^2$
7. $-x^4$
8. $(-x)^4$
9. x
10. (xy)
11. $2x^3$
12. $-3y^6$

In Exercises 13–20, write each expression without using exponents.

13. 5^3
14. -4^5
15. x^7
16. $3x^3$
17. $-4x^5$
18. $(-2y)^4$
19. $(3t)^5$
20. $a^3 b^2$

In Exercises 21–28, write each expression using exponents.

21. $2 \cdot 2 \cdot 2$
22. $5 \cdot 5$
23. $x \cdot x \cdot x \cdot x$
24. $y \cdot y \cdot y \cdot y \cdot y \cdot y$
25. $(2x)(2x)(2x)$
26. $(-4y)(-4y)$
27. $-4t \cdot t \cdot t \cdot t$
28. $5 \cdot u \cdot u$

In Exercises 29–36, evaluate each expression.

29. 5^4
30. $(-3)^3$
31. $2^2 + 3^2$
32. $2^3 - 2^2$
33. $5^4 - 4^3$
34. $2(4^3 + 3^2)$
35. $-5(3^4 + 4^3)$
36. $-5^2(4^3 - 2^6)$

In Exercises 37–54, write each expression as an expression involving only one exponent.

37. $x^4 x^3$
38. $y^5 y^2$
39. $x^5 x^2$
40. yy^3

41. tt^2
42. $w^3 w^4$
43. $a^3 a^4 a^5$
44. $b^2 b^3 b^5$

45. $y^3(y^2 y^4)$
46. $(y^4 y)y^6$
47. $4x^2(3x^5)$
48. $-2y(y^3)$

49. $(-y^2)(4y^3)$
50. $(-4x^3)(-5x)$
51. $6x^3(2x^2)(x^4)$
52. $-2x(x^2)(3x)$

53. $(-2a^3)(3a^2)(-a)$
54. $(-a^3)(-a^2)(-a)$

In Exercises 55–70, write each expression as an expression involving only one exponent.

55. $(3^2)^4$
56. $(4^3)^3$
57. $(y^5)^3$
58. $(b^3)^6$

59. $(a^3)^7$
60. $(b^2)^3$
61. $(x^2 x^3)^5$
62. $(y^3 y^4)^4$

63. $(3zz^2 z^3)^5$
64. $(4t^3 t^6 t^2)^2$
65. $(x^5)^2 (x^7)^3$
66. $(y^3 y)^2 (y^2)^2$

67. $(r^3 r^2)^4 (r^3 r^5)^2$
68. $(s^2)^3 (s^3)^2 (s^4)^4$
69. $(s^3)^3 (s^2)^2 (s^5)^4$
70. $(yy^3)^3 (y^2 y^3)^4 (y^3 y^3)^2$

In Exercises 71–90, write each expression without using parentheses.

71. $(xy)^3$
72. $(uv^2)^4$
73. $(r^3 s^2)^2$
74. $(a^3 b^2)^3$

75. $(4ab^2)^2$
76. $(3x^2 y)^3$
77. $(-2r^2 s^3 t)^3$
78. $(-3x^2 y^4 z)^2$

79. $\left(\dfrac{a}{b}\right)^3$
80. $\left(\dfrac{r^2}{s}\right)^4$
81. $\left(\dfrac{x^2}{y^3}\right)^5$
82. $\left(\dfrac{u^4}{v^2}\right)^6$

83. $\left(\dfrac{-2a}{b}\right)^5$
84. $\left(\dfrac{2t}{3}\right)^4$
85. $\dfrac{(a^3 a^5)^3}{(a^2)^4}$
86. $\dfrac{(x^2 x^7)^3}{(x^3 x^2)^2}$

87. $\left(\dfrac{y^3 y}{2yy^2}\right)^2$
88. $\left(\dfrac{3t^3 t^4 t^5}{4t^2 t^6}\right)^3$
89. $\left(\dfrac{-2r^2 r^3}{3r^4 r}\right)^3$
90. $\left(\dfrac{-6y^4 y^2}{5y^3 y^5}\right)^2$

In Exercises 91–102, simplify each expression.

91. $\dfrac{x^5}{x^3}$
92. $\dfrac{a^6}{a^3}$
93. $\dfrac{y^3 y^4}{yy^2}$
94. $\dfrac{b^4 b^5}{b^2 b^3}$

95. $\dfrac{12a^2 a^3 a^4}{4(a^4)^2}$
96. $\dfrac{16(aa^2)^3}{2a^2 a^3}$
97. $\dfrac{(ab^2)^3}{(ab)^2}$
98. $\dfrac{(m^3 n^4)^3}{(mn^2)^3}$

99. $\dfrac{20(r^4 s^3)^4}{6(rs^3)^3}$
100. $\dfrac{15(x^2 y^5)^5}{21(x^3 y)^2}$
101. $\dfrac{17(x^4 y)^8}{34(x^5 y^2)^4}$
102. $\dfrac{35(r^3 s^2 t)^2}{49r^6 s^4 t^2}$

Review Exercises

1. Graph the real numbers from -4 to $\frac{5}{2}$.

2. Graph the real numbers less than -3 or greater than 5.

In Review Exercises 3–6, perform the indicated operation.

3. $\frac{5}{6} + \frac{2}{9}$

4. $\frac{1}{2} - \frac{1}{5}$

5. $\frac{2}{7} \cdot \frac{21}{6}$

6. $\frac{1}{5} \div \frac{3}{15}$

In Review Exercises 7–10, translate each algebraic expression into an English phrase.

7. $3(x + y)$

8. $(x + y)(x - y)$

9. $|x - y|$

10. $\frac{x + y}{xy}$

In Review Exercises 11–12, translate each English phrase into an algebraic expression.

11. Three greater than the absolute value of twice x

12. The sum of the numbers y and z, decreased by the sum of their squares

10.2 More on Exponents

- The Zero Exponent
- Negative Exponents

We now develop the quotient rule when the exponents in the numerator and denominator are equal, and also when the denominator exponent is greater than the numerator exponent.

The Zero Exponent

If we apply the quotient rule to the fraction $\frac{5^3}{5^3}$, where the exponents in the numerator and denominator are equal, we obtain

$$\frac{5^3}{5^3} = 5^{3-3} = 5^0$$

However, because any nonzero number divided by itself is equal to 1, we have

$$\frac{5^3}{5^3} = 1$$

To make the results of 5^0 and 1 consistent, we shall define 5^0 to be equal to 1. In general, we have the following definition.

Definition: If x is any nonzero real number, then
$$x^0 = 1$$

400 Chapter 10 Polynomials

EXAMPLE 1

a. $\left(\dfrac{1}{13}\right)^0 = 1$

b. $(-0.115)^0 = 1$

c. $\dfrac{4^2}{4^2} = 4^{2-2}$
$= 4^0$
$= 1$

d. $\dfrac{x^5}{x^5} = x^{5-5} \quad (x \neq 0)$
$= x^0$
$= 1$

e. $3x^0 = 3(1) = 3$

f. $(3x)^0 = 1$

g. $\dfrac{6^n}{6^n} = 6^{n-n}$
$= 6^0$
$= 1$

h. $\dfrac{y^m}{y^m} = y^{m-m} \quad (y \neq 0)$
$= y^0$
$= 1$

Note in Parts e and f that $3x^0 \neq (3x)^0$.

Negative Exponents

If we apply the quotient rule to the fraction $\dfrac{6^2}{6^5}$, where the exponent in the numerator is less than the exponent in the denominator, we obtain

$$\dfrac{6^2}{6^5} = 6^{2-5} = 6^{-3}$$

However, we know that

$$\dfrac{6^2}{6^5} = \dfrac{\cancel{6} \cdot \cancel{6}}{\cancel{6} \cdot \cancel{6} \cdot 6 \cdot 6 \cdot 6} = \dfrac{1}{6^3}$$

To make the results of 6^{-3} and $\dfrac{1}{6^3}$ consistent, we define 6^{-3} to be equal to $\dfrac{1}{6^3}$. In general, we have

Definition: If x is any nonzero number and n is a natural number, then

$$x^{-n} = \dfrac{1}{x^n}$$

Progress Check 7

Write each expression without 0 or negative exponents.

a. $(3xy^0)^2$

b. $(x^2x^3)^{-2}$

EXAMPLE 2

Express each quantity without using negative exponents or parentheses.

a. $3^{-5} = \dfrac{1}{3^5}$

b. $x^{-4} = \dfrac{1}{x^4} \quad (x \neq 0)$

c. $(2x)^{-2} = \dfrac{1}{(2x)^2} \quad (x \neq 0)$
$= \dfrac{1}{4x^2}$

d. $(x^3x^2)^{-3} = (x^5)^{-3} \quad (x \neq 0)$
$= \dfrac{1}{(x^5)^3}$
$= \dfrac{1}{x^{15}}$

Work Progress Check 7.

7. a. $9x^2$ **b.** $\dfrac{1}{x^{10}}$

Progress Check Answers

Because of the definitions of negative and zero exponents, the product, power, and quotient rules are also true for all integral exponents.

Properties of Exponents: If m and n are integers and there are no divisions by 0, then

$$x^m x^n = x^{m+n}, \quad (x^m)^n = x^{mn}, \quad (xy)^n = x^n y^n, \quad \left(\frac{x}{y}\right)^n = \frac{x^n}{y^n},$$

$$x^0 = 1 \ (x \neq 0), \quad x^{-n} = \frac{1}{x^n}, \quad \frac{x^m}{x^n} = x^{m-n}$$

EXAMPLE 3

Simplify each quantity and express the result without using negative exponents.

a. $(x^{-3})^2 = x^{-6} \quad (x \neq 0)$

$\phantom{(x^{-3})^2} = \dfrac{1}{x^6}$

b. $\dfrac{x^3}{x^7} = x^{3-7} \quad (x \neq 0)$

$\phantom{\dfrac{x^3}{x^7}} = x^{-4}$

$\phantom{\dfrac{x^3}{x^7}} = \dfrac{1}{x^4}$

c. $\dfrac{y^{-4}y^{-3}}{y^{-20}} = \dfrac{y^{-7}}{y^{-20}} \quad (y \neq 0)$

$\phantom{\dfrac{y^{-4}y^{-3}}{y^{-20}}} = y^{-7-(-20)}$

$\phantom{\dfrac{y^{-4}y^{-3}}{y^{-20}}} = y^{-7+20}$

$\phantom{\dfrac{y^{-4}y^{-3}}{y^{-20}}} = y^{13}$

d. $\dfrac{12a^3 b^4}{4a^5 b^2} = 3a^{3-5}b^{4-2} \quad (a \neq 0, b \neq 0)$

$\phantom{\dfrac{12a^3 b^4}{4a^5 b^2}} = 3a^{-2}b^2$

$\phantom{\dfrac{12a^3 b^4}{4a^5 b^2}} = \dfrac{3b^2}{a^2}$

Work Progress Check 8.

The properties of exponents discussed thus far are also true for exponential expressions with variables in an exponent.

EXAMPLE 4

a. $x^m x^{3m} = x^{m+3m}$

$\phantom{x^m x^{3m}} = x^{4m}$

b. $\dfrac{y^{2m}}{y^{4m}} = y^{2m-4m} \quad (y \neq 0)$

$\phantom{\dfrac{y^{2m}}{y^{4m}}} = y^{-2m}$

$\phantom{\dfrac{y^{2m}}{y^{4m}}} = \dfrac{1}{y^{2m}}$

c. $a^{2m-1} a^{2m} = a^{2m-1+2m}$

$\phantom{a^{2m-1} a^{2m}} = a^{4m-1}$

d. $(b^{m+1})^{2m} = b^{(m+1)2m}$

$\phantom{(b^{m+1})^{2m}} = b^{2m^2 + 2m}$

Progress Check 8

Simplify each expression.

a. $\dfrac{x^{-2} x^3}{x^{-5}}$

b. $\dfrac{18x^2 y^5}{6x^5 y^2}$

8. **a.** x^6 **b.** $\dfrac{3y^3}{x^3}$

Progress Check Answers

10.2 Exercises

In Exercises 1–64, simplify each expression. Write each answer without using parentheses or negative exponents.

1. $2^5 \cdot 2^{-2}$
2. $10^2 \cdot 10^{-4} \cdot 10^5$
3. $4^{-3} \cdot 4^{-2} \cdot 4^5$
4. $3^{-4} \cdot 3^5 \cdot 3^{-3}$
5. $\dfrac{3^5 \cdot 3^{-2}}{3^3}$
6. $\dfrac{6^2 \cdot 6^{-3}}{6^{-2}}$
7. $\dfrac{2^5 \cdot 2^7}{2^6 \cdot 2^{-3}}$
8. $\dfrac{5^{-2} \cdot 5^{-4}}{5^{-6}}$
9. $2x^0$
10. $(2x)^0$
11. $(-x)^0$
12. $-y^0$
13. $\left(\dfrac{a^2 b^3}{ab^4}\right)^0$
14. $\dfrac{2}{3}\left(\dfrac{xyz}{x^2 y}\right)^0$
15. $\dfrac{x^0 - 5x^0}{2x^0}$
16. $\dfrac{4a^0 + 2a^0}{3a^0}$
17. x^{-2}
18. y^{-3}
19. b^{-5}
20. c^{-4}
21. $(2y)^{-4}$
22. $(-3x)^{-1}$
23. $(ab^2)^{-3}$
24. $(m^2 n^3)^{-2}$
25. $\dfrac{y^4}{y^5}$
26. $\dfrac{t^7}{t^{10}}$
27. $\dfrac{(r^2)^3}{(r^3)^4}$
28. $\dfrac{(b^3)^4}{(b^5)^4}$
29. $\dfrac{y^4 y^3}{y^4 y^{-2}}$
30. $\dfrac{x^{12} x^{-7}}{x^3 x^4}$
31. $\dfrac{a^4 a^{-2}}{a^2 a^0}$
32. $\dfrac{b^0 b^3}{b^{-3} b^4}$
33. $(ab^2)^{-2}$
34. $(c^2 d^3)^{-2}$
35. $(x^2 y)^{-3}$
36. $(-xy^2)^{-4}$
37. $(x^{-4} x^3)^3$
38. $(y^{-2} y)^3$
39. $(y^3 y^{-2})^{-2}$
40. $(x^{-3} x^{-2})^2$
41. $(a^{-2} b^{-3})^{-4}$
42. $(y^{-3} z^5)^{-6}$
43. $(-2x^3 y^{-2})^{-5}$
44. $(-3u^{-2} v^3)^{-3}$
45. $\left(\dfrac{a^3}{a^{-4}}\right)^2$
46. $\left(\dfrac{a^4}{a^{-3}}\right)^3$
47. $\left(\dfrac{b^5}{b^{-2}}\right)^{-2}$
48. $\left(\dfrac{b^{-2}}{b^3}\right)^{-3}$
49. $\left(\dfrac{4x^2}{3x^{-5}}\right)^4$
50. $\left(\dfrac{-3r^4 r^{-3}}{r^{-3} r^7}\right)^3$
51. $\left(\dfrac{12y^3 z^{-2}}{3y^{-4} z^3}\right)^2$
52. $\left(\dfrac{6xy^3}{3x^{-1} y}\right)^3$
53. $\left(\dfrac{2x^3 y^{-2}}{4xy^2}\right)^7$
54. $\left(\dfrac{9u^2 v^3}{18u^{-3} v}\right)^4$
55. $\left(\dfrac{14u^{-2} v^3}{21u^{-3} v}\right)^4$
56. $\left(\dfrac{-27u^{-5} v^{-3} w}{18u^3 v^{-2}}\right)^4$
57. $\left(\dfrac{6a^2 b^3}{2ab^2}\right)^{-2}$
58. $\left(\dfrac{3r^2 s^{-2} t}{15r^{-3} s^3}\right)^{-3}$
59. $\left(\dfrac{12a^2 b^3 c^{-4}}{18a^{-1} b^2 c}\right)^{-3}$
60. $\left(\dfrac{14x^{-2} y^2 z^{-2}}{21x^3 y^{-1}}\right)^{-2}$
61. $\dfrac{(2x^{-2} y)^{-3}}{(4x^2 y^{-1})^3}$
62. $\dfrac{(ab^{-2} c)^2}{(a^{-2} b)^{-3}}$
63. $\dfrac{(17x^5 y^{-5} z)^{-3}}{(17x^{-5} y^3 z^2)^{-4}}$
64. $\dfrac{16(x^{-2} yz)^{-2}}{(2x^{-3} z^0)^4}$

In Exercises 65–80, write each expression with a single exponent.

65. $x^{2m} x^m$
66. $y^{3m} y^{2m}$
67. $u^{2m} v^{3n} u^{3m} v^{-3n}$
68. $r^{2m} s^{-3} r^{3m} s^3$
69. $y^{3m+2} y^{-m}$
70. $x^{m+1} x^m$
71. $\dfrac{y^{3m}}{y^{2m}}$
72. $\dfrac{z^{4m}}{z^{2m}}$
73. $\dfrac{x^{3n}}{x^{6n}}$
74. $\dfrac{x^m}{x^{5m}}$
75. $(x^{m+1})^2$
76. $(y^{2m})^{m+1}$
77. $(x^{3-2n})^{-4}$
78. $(y^{1-n})^{-3}$
79. $(y^{2-n})^{-4}$
80. $(x^{3-4n})^{-2}$

Review Exercises

1. If $a = -2$ and $b = 3$, evaluate $\dfrac{3a^2 + 4b + 8}{a + 2b^2}$.

2. Evaluate $|-3 + 5 \cdot 2|$.

In Review Exercises 3–4, solve each equation.

3. $3(x - 4) = 12$

4. $\dfrac{2}{3}(4x - 3) = 6$

5. If 5 times a certain number is increased by 6, the result is 51. Find the number.

6. Solve the equation $\dfrac{5(2 - x)}{6} = \dfrac{x + 6}{2}$.

7. Solve the equation $P = L + \dfrac{s}{f}i$ for s.

8. A man wants to cut a 33-foot rope into three pieces. He wants the shortest piece to be 8 feet shorter than the middle-sized piece and 13 feet shorter than the longest piece. After the rope is cut, how long is each piece?

In Review Exercises 9–10, solve each inequality and graph the solution set.

9. $-4(r + 1) > 2(r - 3)$

10. $2 < -2(x + 3) < 10$

11. A woman buys two cars for a total of $41,400. One of them costs $2400 more than twice the other. Find the cost of each.

12. The sum of two consecutive even numbers is 54. Find the numbers.

10.3 Scientific Notation

- Changing Numbers to Scientific Notation
- Changing Numbers to Standard Notation
- Using Scientific Notation in Computations

Scientists deal with extremely large and extremely small numbers. For example, the distance from the earth to the sun is approximately 150,000,000 kilometers, and ultraviolet light emitted from a mercury arc has a wavelength of approximately 0.000025 centimeter. To work with such numbers, scientists often use a compact form of notation called scientific notation.

> **Definition:** A number is written in **scientific notation** if it is written as the product of a number between 1 (including 1) and 10 and a power of 10.

Changing Numbers to Scientific Notation

EXAMPLE 1

Change 150,000,000 to scientific notation.

Solution

We must write 150,000,000 as a product of a number between 1 and 10 and a power of 10. Note that 1.5 lies between 1 and 10. To obtain 150,000,000 from 1.5, the decimal point must be moved 8 places to the right. Because multiplying a number by 10 moves the decimal point 1 place to the right, we can do this by multiplying 1.5 by 10 eight times. Thus, 150,000,000 written in scientific notation is

1.5×10^8

EXAMPLE 2

Change 0.000025 to scientific notation.

Solution

We must express 0.000025 as a product of a number between 1 and 10 and a power of 10. Note that 2.5 lies between 1 and 10. To obtain 0.000025 from 2.5, the decimal point must be moved 5 places to the left. We can do this by dividing 2.5 by 10^5, which is equivalent to multiplying 2.5 by $\frac{1}{10^5}$, or by 10^{-5}. Thus, 0.000025 written in scientific notation is

2.5×10^{-5}

EXAMPLE 3

Write   **a.** 235,000   and   **b.** 0.00000235 in scientific notation.

Solution

a. $235{,}000 = 2.35 \times 10^5$, because $2.35 \times 10^5 = 235{,}000$ and 2.35 is between 1 and 10.

b. $0.00000235 = 2.35 \times 10^{-6}$, because $2.35 \times 10^{-6} = 0.00000235$ and 2.35 is between 1 and 10.

Work Progress Check 9.

Progress Check 9

Write each number in scientific notation.

a. 93,000,000

b. 0.00037

Changing Numbers to Standard Notation

Any number written in scientific notation can be changed to **standard notation**. For example, to write 9.3×10^7 in standard notation, we multiply 9.3 by 10^7.

$$9.3 \times 10^7 = 9.3 \times 10{,}000{,}000$$
$$= 93{,}000{,}000$$

EXAMPLE 4

Write   **a.** 3.4×10^5 and   **b.** 2.1×10^{-4} in standard notation.

9. a. 9.3×10^7 **b.** 3.7×10^{-4}

Progress Check Answers

Solution

a. $3.4 \times 10^5 = 3.4 \times 100{,}000$
$= 340{,}000$

b. $2.1 \times 10^{-4} = 2.1 \times \dfrac{1}{10^4}$
$= 2.1 \times \dfrac{1}{10{,}000}$
$= 0.00021$

Progress Check 10

Write each number in standard notation.

a. 2.7×10^5

b. 4.2×10^{-3}

Work Progress Check 10.

Each of the following numbers is written in both scientific and standard notation. In each case, the exponent gives the number of places that the decimal point moves, and the sign of the exponent indicates the direction that it moves.

$5.32 \times 10^5 = 532000.$	5 places to the right.
$2.37 \times 10^6 = 2370000.$	6 places to the right.
$8.95 \times 10^{-4} = 0.000895$	4 places to the left.
$8.375 \times 10^{-3} = 0.008375$	3 places to the left.
$9.77 \times 10^0 = 9.77$	No movement of the decimal point.

EXAMPLE 5

Write 432.0×10^5 in scientific notation.

Solution

The number 432.0×10^5 is not written in scientific notation, because 432.0 is not a number between 1 and 10. To write the number in scientific notation, proceed as follows:

$432.0 \times 10^5 = 4.32 \times 10^2 \times 10^5$ Write 432.0 in scientific notation.
$= 4.32 \times 10^7$ Simplify.

Using Scientific Notation in Computations

We can use scientific notation to simplify fractions such as

$$\dfrac{(3200)(25{,}000)}{0.00040}$$

that contain large and/or small numbers. First we write each number in scientific notation and then do the arithmetic on the numbers and the exponential expressions separately.

$$\dfrac{(3200)(25{,}000)}{0.00040} = \dfrac{(3.2 \times 10^3)(2.5 \times 10^4)}{4.0 \times 10^{-4}}$$
$$= \dfrac{(3.2)(2.5)}{4.0} \times \dfrac{10^3 \, 10^4}{10^{-4}}$$
$$= \dfrac{8.0}{4.0} \times 10^{3+4-(-4)}$$
$$= 2.0 \times 10^{11}$$
$$= 200{,}000{,}000{,}000$$

Because the final answer contains so many digits, this computation cannot be done with a nonscientific calculator.

10. a. 270,000 **b.** 0.0042

EXAMPLE 6

In a vacuum, light travels 1 meter in approximately 0.000000003 second. How long does it take for light to travel 500 kilometers?

Solution

Because 1 kilometer is equal to 1000 meters, the length of time for light to travel 500 kilometers (500 · 1000 meters) is given by

$$(0.000000003)(500)(1000) = (3 \times 10^{-9})(5 \times 10^2)(1 \times 10^3)$$
$$= 3(5) \times 10^{-9+2+3}$$
$$= 15 \times 10^{-4}$$
$$= 1.5 \times 10^1 \times 10^{-4}$$
$$= 1.5 \times 10^{-3}$$
$$= 0.0015$$

Light travels 500 kilometers in approximately 0.0015 second.

10.3 Exercises

In Exercises 1–12, write each number in scientific notation.

1. 23,000
2. 4750
3. 1,700,000
4. 290,000
5. 0.062
6. 0.00073
7. 0.0000051
8. 0.04
9. 42.5×10^2
10. 0.3×10^3
11. 0.25×10^{-2}
12. 25.2×10^{-3}

In Exercises 13–24, write each number in standard notation.

13. 2.3×10^2
14. 3.75×10^4
15. 8.12×10^5
16. 1.2×10^3
17. 1.15×10^{-3}
18. 4.9×10^{-2}
19. 9.76×10^{-4}
20. 7.63×10^{-5}
21. 25×10^6
22. 0.07×10^3
23. 0.51×10^{-3}
24. 617×10^{-2}

25. The distance from the Earth to the nearest star outside our solar system is approximately 25,200,000,000,000 miles. Express this number in scientific notation.

26. The speed of sound in air is 33,100 centimeters per second. Express this number in scientific notation.

27. The distance from Mars to the Sun is approximately 1.14×10^8 miles. Express this number in standard notation.

28. The distance from Venus to the Sun is approximately 6.7×10^7 miles. Express this number in standard notation.

29. One meter is approximately 0.00622 mile. Use scientific notation to express this number.

30. One angstrom is 1×10^{-7} millimeter. Express this number in standard notation.

In Exercises 31–36, use scientific notation to simplify each expression. Give all answers in standard notation.

31. $(3.4 \times 10^2)(2.1 \times 10^3)$

32. $(4.1 \times 10^{-3})(3.4 \times 10^4)$

33. $\dfrac{9.3 \times 10^2}{3.1 \times 10^{-2}}$

34. $\dfrac{7.2 \times 10^6}{1.2 \times 10^8}$

35. $\dfrac{96{,}000}{(12{,}000)(0.00004)}$

36. $\dfrac{(0.48)(14{,}400{,}000)}{96{,}000{,}000}$

37. The distance from Mercury to the Sun is approximately 3.6×10^7 miles. Use scientific notation to express this distance in feet. (*Hint:* 5280 feet = 1 mile.)

38. The mass of one proton is approximately 1.7×10^{-24} gram. Use scientific notation to express the mass of 1 million protons.

39. The speed of sound in air is approximately 3.3×10^4 centimeters per second. Use scientific notation to express this speed in kilometers per second. (*Hint:* 100 centimeters = 1 meter and 1000 meters = 1 kilometer.)

40. One light-year is approximately 5.87×10^{12} miles. Use scientific notation to express this distance in feet. (*Hint:* 5280 feet = 1 mile.)

Review Exercises

1. List the prime numbers between 10 and 30.

2. If $y = -3$, find the value of $-5y^2$.

3. Evaluate $(3 + 4 \cdot 3) \div 5$.

4. Evaluate $\dfrac{3a^2 - 2b}{2a + 2b}$ if $a = 4$ and $b = 3$.

In Review Exercises 5–6, tell which property of real numbers justifies each statement.

5. $5 + z = z + 5$

6. $7(u + 3) = 7u + 7 \cdot 3$

In Review Exercises 7–8, solve each equation.

7. $3(x - 4) - 6 = 0$

8. $8(3x - 5) - 4(2x + 3) = 12$

In Review Exercises 9–10, solve each inequality and graph the solution set.

9. $5(x + 4) \geq 3(x + 2)$

10. $3 < 3x + 4 \leq 10$

11. The perimeter of a square is not less than 44 feet and not greater than 60 feet. What are the possible lengths of the square's edge?

12. The average acceleration of a moving object is given by the formula

$$a = \dfrac{V - v}{t}$$

Solve the formula for v.

408   Chapter 10   Polynomials

10.4 Polynomials

- Degree of a Polynomial
- Polynomial Notation

Expressions such as

$$3x, \quad 4y^2, \quad -8x^2y^3, \quad \text{and} \quad 25$$

that contain constant and/or variable factors are called **algebraic terms**. The numerical coefficients of the first three of these terms are 3, 4, and -8, respectively. Because $25 = 25x^0$, the number 25 is considered to be the numerical coefficient of the term 25.

If we add or subtract terms to form many-termed expressions, we form mathematical expressions called **polynomials**.

> **Definition:** A **polynomial** is an algebraic expression that is the sum of one or more terms containing whole number exponents on its variables.

The expressions

$$8xy^2t, \quad 3x + 2, \quad 4y^2 - 2y + 3, \quad \text{and} \quad 3a - 4b - 4c + 8d$$

are examples of polynomials. The expression $2x^3 - 3y^{-2}$, however, is not a polynomial, because the second term contains a negative exponent on a variable base.

A polynomial with exactly one term is called a **monomial**. A polynomial with exactly two terms is called a **binomial**. A polynomial with exactly three terms is called a **trinomial**. Here are some examples of each:

Monomials	Binomials	Trinomials
$5x^2y$	$3u^3 - 4u^2$	$-5t^2 + 4t + 3$
$-6x$	$18a^2b + 4ab$	$27x^3 - 6x - 2$
29	$-29z^{17} - 1$	$-32r^6 + 7y^3 - z$

Work Progress Check 11.

Progress Check 11

Classify each polynomial as a monomial, binomial, or trinomial.

a. $2x^2 + 3x$

b. $3x^2 - 2x - 7$

c. $7x^2y$

Degree of a Polynomial

The monomial $7x^6$ is called a **monomial of sixth degree** or a **monomial of degree 6**, because the variable x occurs as a factor six times. The monomial $3x^3y^4$ is a monomial of the seventh degree, because the variables x and y occur as factors a total of seven times. Other examples are:

$-2x^3$ is a monomial of degree 3
$47x^2y^3$ is a monomial of degree 5
$18x^4y^2z^8$ is a monomial of degree 14
8 is a monomial of degree 0, because $8 = 8x^0$

These examples illustrate the following definition.

> **Definition:** If a is a nonzero constant, the **degree of the monomial** ax^n is n. The degree of a monomial containing several variables is the sum of the exponents of those variables.

11. a. binomial b. trinomial
    c. monomial

Section 10.4 Polynomials 409

Because each term of a polynomial is a monomial, we define the degree of a polynomial by considering the degrees of each of its terms.

> **Definition:** The **degree of a polynomial** is the same as the degree of its term with the largest degree.

For example,

$x^2 + 2x$ is a binomial of degree 2, because the degree of its first term is 2 and the degree of its other term is less than 2.

$3x^3y^2 + 4x^4y^4 - 3x^3$ is a trinomial of degree 8, because the degree of its second term is 8 and the degree of each of its other terms is less than 8.

$25x^4y^3z^7 - 15xy^8z^{10} - 32x^8y^8z^3 + 4$ is a polynomial of degree 19, because its second and third terms are of degree 19. Its other terms have degrees less than 19.

Work Progress Check 12.

Progress Check 12

Give the degree of each polynomial.

a. $-7x^3y^4 + x^4y^2 - 8x^2y^4z^5$

b. $25x^2y^4 - 32xy^6 + 12x^3y^3$

c. 1000

Polynomial Notation

Polynomials that contain a single variable are often denoted by symbols such as

$P(x)$ Read as "P of x."

and

$Q(t)$ Read as "Q of t."

where the letter within the parentheses represents the variable of the polynomial. The symbol $P(x)$ does not show the product of P and x. Instead, it represents a polynomial in the variable x. $P(x)$ and $Q(t)$ could represent the polynomials

$$P(x) = 3x + 4$$
$$Q(t) = -3t^2 + 4t - 5$$

The symbol $P(x)$ is convenient to use because it provides a way to indicate the value of a polynomial in x at different values of x. If $P(x) = 3x + 4$, for example, then $P(1)$ represents the value of the polynomial $P(x) = 3x + 4$ when $x = 1$.

$$P(x) = 3x + 4$$
$$P(1) = 3(1) + 4 = 7$$

Likewise, if $Q(t) = -3t^2 + 4t - 5$, then $Q(-2)$ represents the value of the polynomial $Q(t) = -3t^2 + 4t - 5$ when $t = -2$.

$$Q(t) = -3t^2 + 4t - 5$$
$$Q(-2) = -3(-2)^2 + 4(-2) - 5$$
$$= -3(4) - 8 - 5$$
$$= -12 - 8 - 5$$
$$= -25$$

12. a. 11 **b.** 7 **c.** 0

Progress Check Answers

EXAMPLE 1

Consider the polynomial $P(z)$, where $P(z) = 3z^2 + 2$. Find **a.** $P(0)$, **b.** $P(2)$, **c.** $P(-3)$, and **d.** $P(s)$.

Solution

a. $P(z) = 3z^2 + 2$
$P(0) = 3(0)^2 + 2$
$= 2$

b. $P(z) = 3z^2 + 2$
$P(2) = 3(2)^2 + 2$
$= 3(4) + 2$
$= 12 + 2$
$= 14$

c. $P(z) = 3z^2 + 2$
$P(-3) = 3(-3)^2 + 2$
$= 3(9) + 2$
$= 27 + 2$
$= 29$

d. $P(z) = 3z^2 + 2$
$P(s) = 3s^2 + 2$

EXAMPLE 2

Consider the polynomial $Q(y)$, where $Q(y) = 2y^2 - y - 7$. Find **a.** $Q(0)$, **b.** $Q(2)$, **c.** $Q(-1)$, and **d.** $Q(-x)$.

Solution

a. $Q(y) = 2y^2 - y - 7$
$Q(0) = 2(0)^2 - (0) - 7$
$= -7$

b. $Q(y) = 2y^2 - y - 7$
$Q(2) = 2(2)^2 - (2) - 7$
$= 8 - 2 - 7$
$= -1$

c. $Q(y) = 2y^2 - y - 7$
$Q(-1) = 2(-1)^2 - (-1) - 7$
$= 2 + 1 - 7$
$= -4$

d. $Q(y) = 2y^2 - y - 7$
$Q(-x) = 2(-x)^2 - (-x) - 7$
$= 2x^2 + x - 7$

EXAMPLE 3

Consider the polynomial $P(x)$, where $P(x) = x^3 + 1$. Find **a.** $P(2t)$, **b.** $P(-3y)$, and **c.** $P(s^4)$.

Solution

a. $P(x) = x^3 + 1$
$P(2t) = (2t)^3 + 1$
$= 8t^3 + 1$

b. $P(x) = x^3 + 1$
$P(-3y) = (-3y)^3 + 1$
$= -27y^3 + 1$

c. $P(x) = x^3 + 1$
$P(s^4) = (s^4)^3 + 1$ Substitute s^4 for x.
$= s^{12} + 1$

Work Progress Check 13.

Progress Check 13

Consider the polynomial
$P(x) = -2x^2 - x + 4$. Find

a. $P(0)$

b. $P(1)$

c. $P(-2t)$

13. a. 4 **b.** 1 **c.** $-8t^2 + 2t + 4$

10.4 Exercises

In Exercises 1–12, classify each polynomial as a monomial, binomial, or trinomial, if possible.

1. $3x + 7$
2. $3y - 5$
3. $3y^2 + 4y + 3$
4. $3xy$
5. $3z^2$
6. $3x^4 - 2x^3 + 3x - 1$
7. $5t - 32$
8. $9x^2y^3z^4$
9. $s^2 - 23s + 31$
10. $12x^3 - 12x^2 + 36x - 3$
11. $3x^5 - 2x^4 - 3x^3 + 17$
12. x^3

In Exercises 13–24, give the degree of each polynomial.

13. $3x^4$
14. $3x^5 - 4x^2$
15. $-2x^2 + 3x^3$
16. $-5x^5 + 3x^2 - 3x$
17. $3x^2y^3 + 5x^3y^5$
18. $-2x^2y^3 + 4x^3y^2z$
19. $-5r^2s^2t - 3r^3st^2 + 3$
20. $4r^2s^3t^3 - 5r^2s^8$
21. $x^{12} + 3x^2y^3z^4 - 4x^7$
22. 17^2x
23. 38
24. -25

In Exercises 25–32, let $P(x) = 5x - 3$. Find each value.

25. $P(2)$
26. $P(0)$
27. $P(-1)$
28. $P(-2)$
29. $P(w)$
30. $P(t)$
31. $P(-y)$
32. $P(2t)$

In Exercises 33–40, let $Q(z) = -z^2 - 4$. Find each value.

33. $Q(0)$
34. $Q(1)$
35. $Q(-1)$
36. $Q(-2)$
37. $Q(r)$
38. $Q(-u)$
39. $Q(3s)$
40. $Q(-2x)$

In Exercises 41–48, let $R(y) = y^2 - 2y + 3$. Find each value.

41. $R(0)$
42. $R(3)$
43. $R(-2)$
44. $R(-1)$
45. $R(-b)$
46. $R(t)$
47. $R\left(-\frac{1}{4}w\right)$
48. $R\left(\frac{1}{2}u\right)$

In Exercises 49–64, let $P(x) = 5x - 2$. Find each value.

49. $P\left(\frac{1}{5}\right)$
50. $P\left(\frac{1}{10}\right)$
51. $P(u^2)$
52. $P(-v^4)$
53. $P(-4z^6)$
54. $P(10x^7)$
55. $P(x^2y^2)$
56. $P(x^3y^3)$
57. $P(x + h)$
58. $P(x - h)$
59. $P(x) + P(h)$
60. $P(x) - P(h)$
61. $P(2y + z)$
62. $P(-3r + 2s)$
63. $P(2y) + P(z)$
64. $P(-3r) + P(2s)$

Review Exercises

In Review Exercises 1–2, solve each equation.

1. $5(u - 5) + 9 = 2(u + 4)$
2. $8(3a - 5) - 12 = 4(2a + 3)$

In Review Exercises 3–4, solve each inequality and graph the solution set.

3. $-4(3y + 2) \leq 20$
4. $-5 < 3t + 4 \leq 13$

5. The monthly cost of electricity in a certain city is $7 plus 11 cents per kilowatt-hour used. How many kilowatt-hours are used in a month when the bill is $68.60?

6. A rectangle with a perimeter of 28 inches is 6 inches longer than it is wide. Find its dimensions.

In Review Exercises 7–10, write each expression without using parentheses or negative exponents.

7. $(x^2 x^4)^3$
8. $(a^2)^3 (a^3)^2$
9. $\left(\dfrac{y^2 y^5}{y^4}\right)^3$
10. $\left(\dfrac{2t^3}{t}\right)^{-4}$

11. The estimated mass of the planet Jupiter is 1.9×10^{27} kilograms. Express this number in standard notation.

12. Solve the equation $y - 3 = m(x - 2)$ for x.

10.5 Adding and Subtracting Polynomials

- Adding Monomials
- Subtracting Monomials
- Adding Polynomials
- Subtracting Polynomials

Adding Monomials

Like terms are terms with the same variables and the same exponents. Because of the distributive property, we can combine like terms by adding their coefficients and using the same variables and exponents. For example,

$$2y + 5y = (2 + 5)y = 7y$$
$$-3x^2 + 7x^2 = (-3 + 7)x^2 = 4x^2$$

Likewise,

$$4x^3y^2 + 9x^3y^2 = 13x^3y^2$$
$$4r^2s^3t^4 + 7r^2s^3t^4 = 11r^2s^3t^4$$

Thus, to add like monomials together, we simply combine like terms.

EXAMPLE 1

a. $5xy^3 + 7xy^3 = 12xy^3$

b. $-7x^2y^2 + 6x^2y^2 + 3x^2y^2$
$= -x^2y^2 + 3x^2y^2$
$= 2x^2y^2$

c. $(2x^2)^2 + (3x)^4 = 4x^4 + 81x^4$
$= 85x^4$

d. $2(x + y) + 3(x + y) = 5(x + y)$
$= 5x + 5y$

Subtracting Monomials

Recall from Section 8.5 that

$$x - y = x + (-y)$$

Thus, to subtract one monomial from another, we can add the negative of the monomial that is to be subtracted.

EXAMPLE 2

a. $8x^2 - 3x^2 = 8x^2 + (-3x^2)$
$ = 5x^2$

b. $6x^3y^2 - 9x^3y^2 = 6x^3y^2 + (-9x^3y^2)$
$ = -3x^3y^2$

c. $-3r^2st^3 - 5r^2st^3 = -3r^2st^3 + (-5r^2st^3)$
$ = -8r^2st^3$

Work Progress Check 14.

Progress Check 14

a. Add: $3x^2y^3 + 4x^2y^3 + x^2y^3$.

b. Subtract: $8a^3b^2 - 6a^3b^2$.

Adding Polynomials

Because of the distributive property, we can remove parentheses enclosing terms when the sign preceding the parentheses is a + sign. We simply drop the parentheses.

$+(3x^2 + 3x - 2) = +1(3x^2 + 3x - 2)$
$ = 1(3x^2) + 1(3x) + 1(-2)$
$ = 3x^2 + 3x + (-2)$
$ = 3x^2 + 3x - 2$

Polynomials are added by removing parentheses, if necessary, and then combining any like terms that are contained within the polynomials.

EXAMPLE 3

$(3x^2 - 3x + 2) + (2x^2 + 7x - 4) = 3x^2 - 3x + 2 + 2x^2 + 7x - 4$
$ = 3x^2 + 2x^2 - 3x + 7x + 2 + (-4)$
$ = 5x^2 + 4x - 2$

To make the addition easier, problems such as Example 3 are often written with the terms aligned vertically.

$3x^2 - 3x + 2$
$\underline{2x^2 + 7x - 4}$
$5x^2 + 4x - 2$

EXAMPLE 4

Add:

$4x^2y + 8x^2y^2 - 3x^2y^3$
$\underline{3x^2y - 8x^2y^2 + 8x^2y^3}$
$7x^2y + 5x^2y^3$

14. **a.** $8x^2y^3$ **b.** $2a^3b^2$

Subtracting Polynomials

We can also remove parentheses enclosing several terms when the sign preceding the parentheses is a − sign. We simply drop the minus sign and the parentheses and *change the sign of every term within the parentheses.*

$$-(3x^2 + 3x - 2) = -1(3x^2 + 3x - 2)$$
$$= -1(3x^2) + (-1)(3x) + (-1)(-2)$$
$$= -3x^2 + (-3x) + 2$$
$$= -3x^2 - 3x + 2$$

This suggests that the way to subtract polynomials is to remove parentheses and combine like terms.

EXAMPLE 5

a. $(3x - 4) - (5x + 7) = 3x - 4 - 5x - 7$
$$= -2x - 11$$

b. $(3x^2 - 4x - 6) - (2x^2 - 6x + 12) = 3x^2 - 4x - 6 - 2x^2 + 6x - 12$
$$= x^2 + 2x - 18$$

c. $(-4rt^3 + 2r^2t^2) - (-3rt^3 + 2r^2t^2) = -4rt^3 + 2r^2t^2 + 3rt^3 - 2r^2t^2$
$$= -rt^3$$

To subtract polynomials in vertical form, we add the negative of the **subtrahend** (the bottom polynomial) to the **minuend** (the top polynomial).

EXAMPLE 6

Subtract $3x^2y - 2xy^2$ from $2x^2y + 4xy^2$.

Solution

We write the problem in vertical form, change the signs of the terms of the subtrahend, and add:

$$\begin{array}{r} 2x^2y + 4xy^2 \\ -\underline{3x^2y - 2xy^2} \end{array} \quad \rightarrow \quad \begin{array}{r} 2x^2y + 4xy^2 \\ +\underline{-3x^2y + 2xy^2} \\ -x^2y + 6xy^2 \end{array}$$

EXAMPLE 7

Subtract $6xy^2 + 4x^2y^2 - x^3y^2$ from $-2xy^2 - 3x^3y^2$.

Solution

$$\begin{array}{r} -2xy^2 \qquad\qquad -3x^3y^2 \\ -\underline{6xy^2 + 4x^2y^2 - x^3y^2} \end{array} \rightarrow \begin{array}{r} -2xy^2 \qquad\qquad -3x^3y^2 \\ +\underline{-6xy^2 - 4x^2y^2 + x^3y^2} \\ -8xy^2 - 4x^2y^2 - 2x^3y^2 \end{array}$$

Work Progress Check 15.

Progress Check 15

a. Add:
$(3x^2 + 2x - 7) + (2x^2 - x + 9)$.

b. Subtract:
$(-2x^2 - x + 2) - (3x^2 + x - 2)$.

---

15. **a.** $5x^2 + x + 2$
    b. $-5x^2 - 2x + 4$

Progress Check Answers

We can remove parentheses enclosing several terms when a monomial precedes the parentheses. We use the distributive property and multiply every term within the parentheses by that monomial. For example, to add $3(2x + 5)$ and $2(4x - 3)$, we proceed as follows:

$$3(2x + 5) + 2(4x - 3) = 6x + 15 + 8x - 6$$
$$= 6x + 8x + 15 - 6$$
$$= 14x + 9$$

Progress Check 16

Simplify: $2(x^3 + x^2 + 1) - 3(x^2 + 1) + 4(x^3 + 2)$.

16. $6x^3 - x^2 + 7$

Progress Check Answers

EXAMPLE 8

a. $3(x^2 + 4x) + 2(x^2 - 4) = 3x^2 + 12x + 2x^2 - 8$
$$= 5x^2 + 12x - 8$$

b. $8(y^2 - 2y + 3) - 4(2y^2 + y - 3) = 8y^2 - 16y + 24 - 8y^2 - 4y + 12$
$$= -20y + 36$$

c. $-4x(xy^2 - xy + 3) - x(xy^2 - 2) + 3(x^2y^2 + 2x^2y)$
$$= -4x^2y^2 + 4x^2y - 12x - x^2y^2 + 2x + 3x^2y^2 + 6x^2y$$
$$= -2x^2y^2 + 10x^2y - 10x$$

Work Progress Check 16.

10.5 Exercises

In Exercises 1–12, tell whether the terms are like or unlike terms. If they are like terms, add them.

1. $3y, 4y$
2. $3x^2, 5x^2$
3. $3x, 3y$
4. $3x^2, 6x$
5. $3x^3, 4x^3, 6x^3$
6. $-2y^4, -6y^4, 10y^4$
7. $-5x^3y^2, 13x^3y^2$
8. $23, 12x$
9. $-23t^6, 32t^6, 56t^6$
10. $32x^5y^3, -21x^5y^3, -11x^5y^3$
11. $-x^2y, xy, 3xy^2$
12. $4x^3y^2z, -6x^3y^2z, 2z^3y^2z$

In Exercises 13–30, simplify each expression if possible.

13. $4y + 5y$
14. $-2x + 3x$
15. $-8t^2 - 4t^2$
16. $15x^2 + 10x^2$
17. $32u^3 - 16u^3$
18. $25xy^2 - 7xy^2$
19. $18x^5y^2 - 11x^5y^2$
20. $17x^6y - 22x^6y$
21. $3rst + 4rst + 7rst$
22. $-2ab + 7ab - 3ab$
23. $-4a^2bc + 5a^2bc - 7a^2bc$
24. $(xy)^2 + 4x^2y^2 - 2x^2y^2$
25. $(3x)^2 - 4x^2 + 10x^2$
26. $(2x)^4 - (3x^2)^2$
27. $5x^2y^2 + 2(xy)^2 - (3x^2)y^2$
28. $-3x^3y^6 + 2(xy^2)^3 - (3x)^3y^6$
29. $(-3x^2y)^4 + (4x^4y^2)^2 - 2x^8y^4$
30. $5x^5y^{10} - (2xy^2)^5 + (3x)^5y^{10}$

416   Chapter 10   Polynomials

In Exercises 31–62, perform the indicated operations and simplify.

31. $(3x + 7) + (4x - 3)$

32. $(2y - 3) + (4y + 7)$

33. $(4a + 3) - (2a - 4)$

34. $(5b - 7) - (3b + 5)$

35. $(2x + 3y) + (5x - 10y)$

36. $(5x - 8y) - (2x + 5y)$

37. $(-8x - 3y) - (11x + y)$

38. $(-4a + b) + (5a - b)$

39. $(3x^2 - 3x - 2) + (3x^2 + 4x - 3)$

40. $(3a^2 - 2a + 4) - (a^2 - 3a + 7)$

41. $(2b^2 + 3b - 5) - (2b^2 - 4b - 9)$

42. $(4c^2 + 3c - 2) + (3c^2 + 4c + 2)$

43. $(2x^2 - 3x + 1) - (4x^2 - 3x + 2) + (2x^2 + 3x + 2)$

44. $(-3z^2 - 4z + 7) + (2z^2 + 2z - 1) - (2z^2 - 3z + 7)$

45. $2(x + 3) + 3(x + 3)$

46. $5(x + y) + 7(x + y)$

47. $-8(x - y) + 11(x - y)$

48. $-4(a - b) - 5(a - b)$

49. $2(x^2 - 5x - 4) - 3(x^2 - 5x - 4) + 6(x^2 - 5x - 4)$

50. $7(x^2 + 3x + 1) + 9(x^2 + 3x + 1) - 5(x^2 + 3x + 1)$

51. Add: $3x^2 + 4x + 5$
$\phantom{\text{Add: }}2x^2 - 3x + 6$

52. Add: $2x^3 + 2x^2 - 3x + 5$
$\phantom{\text{Add: }}3x^3 - 4x^2 - x - 7$

53. Add: $2x^3 - 3x^2 + 4x - 7$
$\phantom{\text{Add: }}-9x^3 - 4x^2 - 5x + 6$

54. Add: $-3x^3 + 4x^2 - 4x + 9$
$\phantom{\text{Add: }}2x^3 \phantom{{}+ 4x^2} + 9x - 3$

55. Add: $-3x^2y + 4xy + 25y^2$
$\phantom{\text{Add: }}5x^2y - 3xy - 12y^2$

56. Add: $-6x^3z - 4x^2z^2 + 7z^3$
$\phantom{\text{Add: }}-7x^3z + 9x^2z^2 - 21z^3$

57. Subtract: $3x^2 + 4x - 5$
$\phantom{\text{Subtract: }}-2x^2 - 2x + 3$

58. Subtract: $3y^2 - 4y + 7$
$\phantom{\text{Subtract: }}6y^2 - 6y - 13$

59. Subtract: $4x^3 + 4x^2 - 3x + 10$
$\phantom{\text{Subtract: }}5x^3 - 2x^2 - 4x - 4$

60. Subtract: $3x^3 + 4x^2 + 7x + 12$
$\phantom{\text{Subtract: }}-4x^3 + 6x^2 + 9x - 3$

61. Subtract: $-2x^2y^2 - 4xy + 12y^2$
$\phantom{\text{Subtract: }}10x^2y^2 + 9xy - 24y^2$

62. Subtract: $25x^3 - 45x^2z + 31xz^2$
$\phantom{\text{Subtract: }}12x^3 + 27x^2z - 17xz^2$

63. Find the sum when $x^2 + x - 3$ is added to the sum of $2x^2 - 3x + 4$ and $3x^2 - 2$.

64. Find the sum when $3y^2 - 5y + 7$ is added to the sum of $-3y^2 - 7y + 4$ and $5y^2 + 5y - 7$.

65. Find the difference when $t^3 - 2t^2 + 2$ is subtracted from the sum of $3t^3 + t^2$ and $-t^3 + 6t - 3$.

66. Find the difference when $-3z^3 - 4z + 7$ is subtracted from the sum of $2z^2 + 3z - 7$ and $-4z^3 - 2z - 3$.

67. Find the sum when $3x^2 + 4x - 7$ is added to the sum of $-2x^2 - 7x + 1$ and $-4x^2 + 8x - 1$.

68. Find the difference when $32x^2 - 17x + 45$ is subtracted from the sum of $23x^2 - 12x - 7$ and $-11x^2 + 12x + 7$.

In Exercises 69–78, simplify each expression.

69. $2(x + 3) + 4(x - 2)$

70. $3(y - 4) - 5(y + 3)$

71. $-2(x^2 + 7x - 1) - 3(x^2 - 2x + 7)$

72. $-5(y^2 - 2y - 2) + 6(2y^2 + 2y + 4)$

73. $2(2y^2 - 2y + 2) - 4(3y^2 - 4y - 1) + 4y(y^2 - y - 1)$

74. $-4(z^2 - 5z) - 5(4z^2 - 1) + 6(2z - 3)$

75. $2a(ab^2 - b) - 3b(a + 2ab) + b(b - a + a^2b)$

76. $3y(xy + y) - 2y^2(x - 4 + y) + 2(y^3 + y^2)$

77. $-4xy^2(x + y + z) - 2x(xy^2 - 4y^2z) - 2y(8xy^2 - 1)$

78. $-3uv(u - v^2 + w) + 4w(uv + w) - 3w(w + uv)$

In Exercises 79–80, let $P(x) = 3x - 5$. Find each value.

79. $P(x + h) + P(x)$

80. $P(x + h) - P(x)$

Review Exercises

1. On the number line, graph the real numbers between -5 and 5.

2. Evaluate $-|3 - 5|$.

In Review Exercises 3–6, let $a = 3$, $b = -2$, $c = -1$, and $d = 2$. Evaluate each expression.

3. $ab + cd$

4. $ad + bc$

5. $a(b + c)$

6. $d(b + a)$

In Review Exercises 7–8, solve each equation.

7. $3(a + 2) - (2 - a) = a - 5$

8. $\dfrac{3}{2}x = 7(x + 11)$

9. A rectangle with a perimeter of 72 feet is 6 feet longer than it is wide. Find its area.

10. Solve the inequality $-4(2x - 9) \geq 12$ and graph the solution set.

11. The **kinetic energy** of a moving object is given by the formula

$$K = \dfrac{mv^2}{2}$$

Solve the formula for m.

12. A rectangular garden is 3 feet longer than twice its width. Its perimeter is 48 feet. Find its area.

10.6 Multiplying Polynomials

- Multiplying Monomials
- Multiplying Polynomials by Monomials
- Multiplying Polynomials
- Multiplying Binomials
- Multiplying Binomials to Solve Equations

Multiplying Monomials

To multiply two monomials such as $4x^2$ and $-2x^3$, we use the commutative and associative properties of multiplication to group the numerical factors and the variable factors together and multiply.

$$4x^2(-2x^3) = 4(-2)x^2x^3$$
$$= -8x^5$$

EXAMPLE 1

a. $3x^5(2x^5) = 3(2)x^5x^5$
$= 6x^{10}$

b. $-2a^2b^3(5ab^2) = -2(5)a^2ab^3b^2$
$= -10a^3b^5$

c. $-4y^5z^2(2y^3z^3)(3yz) = -4(2)(3)y^5y^3yz^2z^3z$
$= -24y^9z^6$

The previous examples suggest the following rule.

> **Multiplying Monomials:** To multiply two monomials, first multiply the numerical factors and then multiply the variable factors.

Multiplying Polynomials by Monomials

To find the product of a monomial and a polynomial, we use the distributive property. To multiply $x + 4$ by $3x$, for example, we proceed as follows:

$3x(x + 4) = 3x \cdot x + 3x \cdot 4$
$= 3x^2 + 12x$

EXAMPLE 2

a. $2a^2(3a^2 - 4a) = 2a^2 \cdot 3a^2 - 2a^2 \cdot 4a$
$= 6a^4 - 8a^3$

b. $-2xz^2(2x - 3z + 2x^2z^2) = -2xz^2(2x) - (-2xz^2)(3z) + (-2xz^2)(2x^2z^2)$
$= -4x^2z^2 + 6xz^3 - 4x^3z^4$

The results of Example 2 suggest the following rule:

> **Multiplying Polynomials by Monomials:** To multiply a polynomial with more than one term by a monomial, remove parentheses and simplify.

Progress Check 17

Multiply:

a. $(-3x^2y^3)(4x^3y^4)$

b. $-2a^2b(3ab^2 - 4a^3b + ab^3)$

Work Progress Check 17.

Multiplying Polynomials

We must use the distributive property more than once to multiply a polynomial by a binomial. For example, to multiply $3x^2 + 3x - 5$ by $2x + 3$, we proceed as follows:

$(2x + 3)(3x^2 + 3x - 5) = (2x + 3)3x^2 + (2x + 3)3x - (2x + 3)5$
$= 3x^2(2x + 3) + 3x(2x + 3) - 5(2x + 3)$
$= 6x^3 + 9x^2 + 6x^2 + 9x - 10x - 15$
$= 6x^3 + 15x^2 - x - 15$

17. **a.** $-12x^5y^7$
   b. $-6a^3b^3 + 8a^5b^2 - 2a^3b^4$

Progress Check Answers

EXAMPLE 3

a. $(2x - 4)(3x + 5) = (2x - 4)3x + (2x - 4)5$
$= 2x \cdot 3x - 4 \cdot 3x + 2x \cdot 5 - 4 \cdot 5$
$= 6x^2 - 12x + 10x - 20$
$= 6x^2 - 2x - 20$

b. $(3x - 2y)(2x + 3y) = (3x - 2y)2x + (3x - 2y)3y$
$= 3x \cdot 2x - 2y \cdot 2x + 3x \cdot 3y - 2y \cdot 3y$
$= 6x^2 - 4xy + 9xy - 6y^2$
$= 6x^2 + 5xy - 6y^2$

c. $(3y + 1)(3y^2 + 2y + 2) = (3y + 1)3y^2 + (3y + 1)2y + (3y + 1)2$
$= 3y \cdot 3y^2 + 1 \cdot 3y^2 + 3y \cdot 2y + 1 \cdot 2y$
$\quad + 3y \cdot 2 + 1 \cdot 2$
$= 9y^3 + 3y^2 + 6y^2 + 2y + 6y + 2$
$= 9y^3 + 9y^2 + 8y + 2$

The results of Example 3 suggest the following rule:

Multiplying Polynomials: To multiply one polynomial by another, multiply every term of one polynomial by every term of the other polynomial and combine like terms.

It is often convenient to organize the work vertically.

EXAMPLE 4

a. Multiply:

$$\begin{array}{r} 2x - 4 \\ 3x + 2 \\ \hline \end{array}$$

$3x(2x - 4) \rightarrow \quad 6x^2 - 12x$
$2(2x - 4) \rightarrow \quad \quad\quad + 4x - 8$
$\quad\quad\quad\quad\quad\quad\quad \overline{6x^2 - 8x - 8}$

b. Multiply:

$$\begin{array}{r} 3a^2 - 4a + 7 \\ 2a + 5 \\ \hline \end{array}$$

$2a(3a^2 - 4a + 7) \rightarrow \quad 6a^3 - 8a^2 + 14a$
$5(3a^2 - 4a + 7) \rightarrow \quad\quad\quad\; + 15a^2 - 20a + 35$
$\quad\quad\quad\quad\quad\quad\quad\quad\quad \overline{6a^3 + 7a^2 - 6a + 35}$

c. Multiply:

$$\begin{array}{r} 3y^2 - 5y + 4 \\ - 4y^2 - 3 \\ \hline \end{array}$$

$-4y^2(3y^2 - 5y + 4) \rightarrow \quad -12y^4 + 20y^3 - 16y^2$
$-3(3y^2 - 5y + 4) \rightarrow \quad\quad\quad\quad\quad\quad\quad - 9y^2 + 15y - 12$
$\quad\quad\quad\quad\quad\quad\quad\quad\quad\quad \overline{-12y^4 + 20y^3 - 25y^2 + 15y - 12}$

Work Progress Check 18.

Progress Check 18

Multiply:
$(3x - 2y)(x^2 + 3xy - 2y^2)$.

Progress Check Answers

18. $3x^3 + 7x^2y - 12xy^2 + 4y^3$

Multiplying Binomials

We can use a short-cut method, called the **FOIL** method, to multiply one binomial by another. FOIL is an acronym for First terms, Outer terms, Inner terms, and Last terms. To use the FOIL method to multiply $2a - 4$ by $3a + 5$, we

1. multiply the First terms $2a$ and $3a$ to obtain $6a^2$,
2. multiply the Outer terms $2a$ and 5 to obtain $10a$,
3. multiply the Inner terms -4 and $3a$ to obtain $-12a$, and
4. multiply the Last terms -4 and 5 to obtain -20.

Then we simplify the resulting polynomial, if possible.

$$(2a - 4)(3a + 5) = 2a(3a) + 2a(5) + (-4)(3a) + (-4)(5)$$
$$= 6a^2 + 10a - 12a - 20 \quad \text{Simplify.}$$
$$= 6a^2 - 2a - 20 \quad \text{Combine terms.}$$

EXAMPLE 5

Use the FOIL method to find each product.

a. $(3x + 4)(2x - 3) = 3x(2x) + 3x(-3) + 4(2x) + 4(-3)$
$= 6x^2 - 9x + 8x - 12$
$= 6x^2 - x - 12$

b. $(2y - 7)(5y - 4) = 2y(5y) + 2y(-4) + (-7)(5y) + (-7)(-4)$
$= 10y^2 - 8y - 35y + 28$
$= 10y^2 - 43y + 28$

c. $(2r - 3s)(2r + t) = 2r(2r) + 2r(t) - 3s(2r) - 3s(t)$
$= 4r^2 + 2rt - 6rs - 3st$

EXAMPLE 6

Simplify each expression.

Progress Check 19

Multiply: $4(3x + 2)(2x - 3)$.

a. $3(2x - 3)(x + 1) = 3(2x^2 + 2x - 3x - 3)$ Use FOIL to multiply the binomials.
$= 3(2x^2 - x - 3)$ Combine terms.
$= 6x^2 - 3x - 9$ Use the distributive property to remove parentheses.

b. $(x + 1)(x - 2) - 3x(x + 3) = x^2 - 2x + x - 2 - 3x^2 - 9x$
$= -2x^2 - 10x - 2$ Combine like terms.

19. $24x^2 - 20x - 24$

Work Progress Check 19.

The products discussed in the next example are called **special products**.

EXAMPLE 7

Use the FOIL method to find each special product.

a. $(x + y)^2 = (x + y)(x + y)$

$ = x^2 + xy + xy + y^2$

$ = x^2 + 2xy + y^2$

The square of the sum of two quantities has three terms: the square of the first quantity, plus twice the product of the quantities, and the square of the second quantity.

b. $(x - y)^2 = (x - y)(x - y)$

$ = x^2 - xy - xy + y^2$

$ = x^2 - 2xy + y^2$

The square of the difference of two quantities has three terms: the square of the first quantity, minus twice the product of the quantities, and the square of the second quantity.

c. $(x + y)(x - y) = x^2 - xy + xy - y^2$

$ = x^2 - y^2$

The product of a sum and a difference of two quantities is a binomial. It is the product of the first quantities minus the product of the second quantities.

Binomials that have the same terms but different signs between them are often called **conjugate binomials**. For example, the *conjugate* of $2x + 12$ is $2x - 12$, and the conjugate of $ab - c$ is $ab + c$. ∎

The Special Products:

$(x + y)^2 = x^2 + 2xy + y^2$

$(x - y)^2 = x^2 - 2xy + y^2$

$(x + y)(x - y) = x^2 - y^2$

Multiplying Binomials to Solve Equations

To solve an equation such as $(x + 2)(x + 3) = x(x + 7)$, we use the **FOIL** method on the left-hand side and the distributive property on the right-hand side to remove parentheses, and proceed as follows:

$$(x + 2)(x + 3) = x(x + 7)$$
$$x^2 + 3x + 2x + 6 = x^2 + 7x$$
$$5x + 6 = 7x \quad \text{Add } -x^2 \text{ to both sides and combine terms.}$$
$$6 = 2x \quad \text{Add } -5x \text{ to both sides.}$$
$$3 = x \quad \text{Divide both sides by 2.}$$

Check: $(x + 2)(x + 3) = x(x + 7)$
$(3 + 2)(3 + 3) \stackrel{?}{=} 3(3 + 7)$ Replace x with 3.
$5(6) \stackrel{?}{=} 3(10)$ Perform the additions within parentheses.
$30 = 30$

EXAMPLE 8

Solve the equation $(x + 5)(x + 4) = (x + 9)(x + 10)$.

Solution

$$(x + 5)(x + 4) = (x + 9)(x + 10)$$
$$x^2 + 4x + 5x + 20 = x^2 + 10x + 9x + 90 \quad \text{Use the FOIL method to remove parentheses.}$$
$$9x + 20 = 19x + 90 \quad \text{Add } -x^2 \text{ to both sides and combine terms.}$$
$$20 = 10x + 90 \quad \text{Add } -9x \text{ to both sides.}$$
$$-70 = 10x \quad \text{Add } -90 \text{ to both sides.}$$
$$-7 = x \quad \text{Divide both sides by 10.}$$

Check: $(x + 5)(x + 4) = (x + 9)(x + 10)$
$(-7 + 5)(-7 + 4) \stackrel{?}{=} (-7 + 9)(-7 + 10)$ Replace x with -7.
$(-2)(-3) \stackrel{?}{=} (2)(3)$ Perform the additions within parentheses.
$6 = 6$

Work Progress Check 20.

Progress Check 20

Solve:
$(x + 2)(x + 3) = (x - 1)(x + 4)$.

20. -5

10.6 Exercises

In Exercises 1–12, find each product.

1. $(3x^2)(4x^3)$ **2.** $(-2a^3)(3a^2)$ **3.** $(3b^2)(-2b)(4b^3)$ **4.** $(3y)(2y^2)(-y^4)$

5. $(2x^2y^3)(3x^3y^2)$ **6.** $(-x^3y^6z)(x^2y^2z^7)$ **7.** $(x^2y^5)(x^2z^5)(-3y^2z^3)$ **8.** $(-r^4st^2)(2r^2st)(rst)$

9. $(x^2y^3)^5$ **10.** $(a^3b^2c)^4$ **11.** $(a^3b^2c)(abc^3)^2$ **12.** $(xyz^3)(xy^2z^2)^3$

In Exercises 13–30, find each product.

13. $3(x + 4)$ **14.** $-3(a - 2)$ **15.** $-4(t + 7)$ **16.** $6(s^2 - 3)$

17. $3x(x - 2)$ **18.** $4y(y + 5)$ **19.** $-2x^2(3x^2 - x)$ **20.** $4b^3(2b^2 - 2b)$

21. $3xy(x + y)$ **22.** $-4x^2(3x^2 - x)$ **23.** $2x^2(3x^2 + 4x - 7)$ **24.** $3y^3(2y^2 - 7y - 8)$

25. $-3x^3y^3(x^2y^3 - x^3y^2 + xy^2)$

26. $7rst(r^2 + s^2 - t^2)$

27. $-abc^3(a^3 + b^3 - c^3)$

28. $4x^5y^3z^2(3x^2 - 4y^2z - 4)$

29. $(3xy)(-2x^2y^3)(x + y)$

30. $(-2a^2b)(-3a^3b^2)(3a - 2b)$

In Exercises 31–54, use the FOIL method to find each product.

31. $(a + 4)(a + 5)$ **32.** $(y - 3)(y + 5)$ **33.** $(3x - 2)(x + 4)$ **34.** $(t + 4)(2t - 3)$

35. $(2a + 4)(3a - 5)$ **36.** $(2b - 1)(3b + 4)$ **37.** $(3x - 5)(2x + 1)$ **38.** $(2y - 5)(3y + 7)$

39. $(x + 3)(2x - 3)$ **40.** $(2x + 3)(2x - 5)$ **41.** $(2t + 3s)(3t - s)$ **42.** $(3a - 2b)(4a + b)$

43. $(x + y)(x - 2y)$ **44.** $(a - b)(2a - b)$ **45.** $(-2r - 3s)(2r + 7s)$ **46.** $(-4a + 3)(-2a - 3)$

47. $(-2a - 4b)(3a + 2b)$ **48.** $(3u + 2v)(-2u + 5v)$ **49.** $(4t - u)(-3t + u)$ **50.** $(-3t + 2s)(2t - 3s)$

51. $(x + y)(x + z)$ **52.** $(a - b)(x + y)$ **53.** $(u + v)(u + 2t)$ **54.** $(x - 5y)(a + 2y)$

In Exercises 55–60, find each product.

55. $\begin{array}{r} 4x + 3 \\ \underline{x + 2} \end{array}$ **56.** $\begin{array}{r} 5r + 6 \\ \underline{2r - 1} \end{array}$ **57.** $\begin{array}{r} 4x - 2y \\ \underline{3x + 5y} \end{array}$ **58.** $\begin{array}{r} 5r + 6s \\ \underline{2r - s} \end{array}$

59. $\begin{array}{r} x^2 + x + 1 \\ \underline{x - 1} \end{array}$ **60.** $\begin{array}{r} 4x^2 - 2x + 1 \\ \underline{2x + 1} \end{array}$

In Exercises 61–80, find each special product.

61. $(x + 4)(x + 4)$ **62.** $(a + 3)(a + 3)$ **63.** $(t - 3)(t - 3)$ **64.** $(z - 5)(z - 5)$

65. $(r + 4)(r - 4)$ **66.** $(b + 2)(b - 2)$ **67.** $(x + 5)^2$ **68.** $(y - 6)^2$

69. $(2s + 1)(2s + 1)$ **70.** $(3t - 2)(3t - 2)$ **71.** $(4x + 5)(4x - 5)$ **72.** $(5z + 1)(5z - 1)$

73. $(3r + 4s)(3r + 4s)$ **74.** $(2u - 3v)(2u - 3v)$ **75.** $(x - 2y)^2$ **76.** $(3a + 2b)^2$

77. $(2a - 3b)^2$ **78.** $(2x + 3y)^2$ **79.** $(4x + 5y)(4x - 5y)$ **80.** $(6p + 5q)(6p - 5q)$

In Exercises 81–90, find each product.

81. $2(x - 4)(x + 1)$

82. $-3(2x + 3y)(3x - 4y)$

83. $3a(a + b)(a - b)$

84. $-2r(r + s)(r + s)$

85. $-3y^2z(y + 2z)(2y - z)$

86. $4t^3u^2(3t - 2u)(t + 2u)$

87. $(4t + 3)(t^2 + 2t + 3)$

88. $(3x + y)(2x^2 - 3xy + y^2)$

89. $(-3x + y)(x^2 - 8xy + 16y^2)$

90. $(3x - y)(x^2 + 3xy - y^2)$

In Exercises 91–100, simplify each expression.

91. $2t(t + 2) + 3t(t - 5)$

92. $3y(y + 2) + (y + 1)(y - 1)$

93. $3xy(x + y) - 2x(xy - x)$

94. $(a + b)(a - b) - (a + b)(a + b)$

95. $(x + y)(x - y) + x(x + y)$

96. $(2x - 1)(2x + 1) + x(2x + 1)$

97. $(x + 2)^2 - (x - 2)^2$

98. $(x - 3)^2 - (x + 3)^2$

99. $(2s - 3)(s + 2) + (3s + 1)(s - 3)$

100. $(3x + 4)(2x - 2) - (2x + 1)(x + 3)$

In Exercises 101–110, solve each equation.

101. $(s - 4)(s + 1) = s^2 + 5$

102. $(y - 5)(y - 2) = y^2 - 4$

103. $z(z + 2) = (z + 4)(z - 4)$

104. $(z + 3)(z - 3) = z(z - 3)$

105. $(x + 4)(x - 4) = (x - 2)(x + 6)$

106. $(y - 1)(y + 6) = (y - 3)(y - 2) + 8$

107. $(a - 3)^2 = (a + 3)^2$

108. $(b + 2)^2 = (b - 1)^2$

109. $4 + (2y - 3)^2 = (2y - 1)(2y + 3)$

110. $7s^2 + (s - 3)(2s + 1) = (3s - 1)^2$

111. If 3 less than a certain number is multiplied by 4 more than the number, the product is 6 less than the square of the number. Find the number.

112. The difference between the squares of two consecutive positive integers is 11. Find the integers.

113. The difference between the squares of two consecutive odd positive integers is 32. Find the integers.

114. The sum of the squares of three consecutive odd integers is 112 less than 3 times the square of the largest integer. Find the integers.

115. In major league baseball, the distance between bases is 30 feet greater than it is in softball. The bases in major league baseball mark the corners of a square that has an area 4500 square feet greater than for softball. Find the distance between the bases in baseball.

116. Two square sheets of cardboard differ in area by 44 square inches. An edge of the larger square is 2 inches greater than an edge of the smaller square. Find the length of an edge of the smaller square.

117. The radius of one circle is 1 inch greater than the radius of another circle, and their areas differ by 4π square inches. Find the radius of the smaller circle.

118. The radius of one circle is 3 meters greater than the radius of another circle, and their areas differ by 15π square meters. Find the radius of the larger circle.

Review Exercises

In Review Exercises 1–4, list the number in the set $\{-\frac{1}{2}, 2, 4, \frac{7}{8}\}$ that satisfies the given condition.

1. Negative number
2. Prime number
3. Composite number
4. Integer

In Review Exercises 5–8, tell which property of real numbers justifies each statement.

5. $3(x + 5) = 3x + 3 \cdot 5$

6. $(x + 3) + y = x + (3 + y)$

7. $3(ab) = (ab)3$

8. $a + 0 = a$

9. Solve the equation $\frac{5}{3}(5y + 6) - 10 = 0$.

10. Solve the equation $F = \frac{GMm}{d^2}$ for m.

11. The **parsec**, a unit of distance used in astronomy, is 3×10^{16} meters. The distance to Betelgeuse, a star in the constellation Orion, is 160 parsecs. Use scientific notation to express this distance in meters.

12. What number is one greater than one-half of itself?

10.7 Dividing Polynomials by Monomials

- Dividing Monomials
- Dividing Polynomials by Monomials

Dividing by a number is equivalent to multiplying by its reciprocal. For example, dividing the number 8 by 2 gives the same answer as multiplying 8 by $\frac{1}{2}$.

$$\frac{8}{2} = 4 \quad \text{and} \quad \frac{1}{2} \cdot 8 = 4$$

In general, the following is true.

> If $b \neq 0$, then $\frac{a}{b} = \frac{1}{b} \cdot a$.

Recall that to simplify a fraction, we write both the numerator and denominator as the product of several factors and then divide out all common factors:

$$\frac{4}{6} = \frac{2 \cdot 2}{2 \cdot 3} = \frac{\cancel{2} \cdot 2}{\cancel{2} \cdot 3} = \frac{2}{3}$$

$$\frac{20}{25} = \frac{4 \cdot 5}{5 \cdot 5} = \frac{4 \cdot \cancel{5}}{\cancel{5} \cdot 5} = \frac{4}{5}$$

Dividing Monomials

To divide monomials, we can use the previous method for simplifying arithmetic fractions or use the rules of exponents.

EXAMPLE 1

Simplify a. $\dfrac{x^2 y}{xy^2}$ and b. $\dfrac{-8a^3 b^2}{4ab^3}$.

Solution

Method for arithmetic fractions	Using the rules of exponents
a. $\dfrac{x^2y}{xy^2} = \dfrac{xxy}{xyy}$	a. $\dfrac{x^2y}{xy^2} = x^{2-1}y^{1-2}$
$\quad = \dfrac{\cancel{x}x\cancel{y}}{\cancel{x}y\cancel{y}}$	$\quad = x^1 y^{-1}$
$\quad = \dfrac{x}{y}$	$\quad = \dfrac{x}{y}$
b. $\dfrac{-8a^3b^2}{4ab^3} = \dfrac{-2 \cdot 4aaabb}{4abbb}$	b. $\dfrac{-8a^3b^2}{4ab^3} = \dfrac{-2^3 a^3 b^2}{2^2 ab^3}$
$\quad = \dfrac{-2 \cdot \cancel{4}a a \cancel{a} \cancel{b} \cancel{b}}{\cancel{4} \cancel{a} b b \cancel{b}}$	$\quad = -2^{3-2} a^{3-1} b^{2-3}$
	$\quad = -2^1 a^2 b^{-1}$
$\quad = \dfrac{-2a^2}{b}$	$\quad = \dfrac{-2a^2}{b}$

Work Progress Check 21.

Progress Check 21

Simplify: $\dfrac{8a^5b^7}{12a^3b^{10}}$.

Dividing Polynomials by Monomials

To divide a polynomial by a monomial, we rewrite the division as a product, remove parentheses, and simplify each resulting fraction.

EXAMPLE 2

Simplify $\dfrac{9x + 6y}{3xy}$.

Solution

$\dfrac{9x + 6y}{3xy} = \dfrac{1}{3xy}(9x + 6y)$

$\qquad = \dfrac{9x}{3xy} + \dfrac{6y}{3xy}$ Remove parentheses.

$\qquad = \dfrac{3}{y} + \dfrac{2}{x}$ Simplify each fraction.

EXAMPLE 3

Simplify $\dfrac{6x^2y^2 + 4x^2y - 2xy}{2xy}$.

Solution

$\dfrac{6x^2y^2 + 4x^2y - 2xy}{2xy} = \dfrac{1}{2xy}(6x^2y^2 + 4x^2y - 2xy)$

$\qquad = \dfrac{6x^2y^2}{2xy} + \dfrac{4x^2y}{2xy} - \dfrac{2xy}{2xy}$ Remove parentheses.

$\qquad = 3xy + 2x - 1$ Simplify each fraction.

21. $\dfrac{2a^2}{3b^3}$

Progress Check Answers

EXAMPLE 4

Simplify $\dfrac{12a^3b^2 - 4a^2b + a}{6a^2b^2}$.

Solution

$$\dfrac{12a^3b^2 - 4a^2b + a}{6a^2b^2}$$

$$= \dfrac{1}{6a^2b^2}(12a^3b^2 - 4a^2b + a)$$

$$= \dfrac{12a^3b^2}{6a^2b^2} - \dfrac{4a^2b}{6a^2b^2} + \dfrac{a}{6a^2b^2} \qquad \text{Remove parentheses.}$$

$$= 2a - \dfrac{2}{3b} + \dfrac{1}{6ab^2} \qquad \text{Simplify each fraction.}$$

Work Progress Check 22.

Progress Check 22

Simplify:
$$\dfrac{12x^4yz^2 - 4x^3y^3z^2 + 16x^4y^3z}{8x^3y^3z}.$$

22. $\dfrac{3xz}{2y^2} - \dfrac{z}{2} + 2x$

Progress Check Answers

10.7 Exercises

In Exercises 1–12, simplify each fraction.

1. $\dfrac{5}{15}$
2. $\dfrac{64}{128}$
3. $\dfrac{-125}{75}$
4. $\dfrac{-98}{21}$

5. $\dfrac{120}{160}$
6. $\dfrac{70}{420}$
7. $\dfrac{-3612}{-3612}$
8. $\dfrac{-288}{-112}$

9. $\dfrac{-90}{360}$
10. $\dfrac{8432}{-8432}$
11. $\dfrac{5880}{2660}$
12. $\dfrac{-762}{366}$

In Exercises 13–40, perform each division by simplifying each fraction. Write all answers without using negative or zero exponents.

13. $\dfrac{xy}{yz}$
14. $\dfrac{a^2b}{ab^2}$
15. $\dfrac{r^3s^2}{rs^3}$
16. $\dfrac{y^4z^3}{y^2z^2}$

17. $\dfrac{8x^3y^2}{4xy^3}$
18. $\dfrac{-3y^3z}{6yz^2}$
19. $\dfrac{12u^5v}{-4u^2v^3}$
20. $\dfrac{16rst^2}{-8rst^3}$

21. $\dfrac{-16r^3y^2}{-4r^2y^4}$
22. $\dfrac{35xyz^2}{-7x^2yz}$
23. $\dfrac{-65rs^2t}{15r^2s^3t}$
24. $\dfrac{112u^3z^6}{-42u^3z^6}$

25. $\dfrac{x^2x^3}{xy^6}$
26. $\dfrac{(xy)^2}{x^2y^3}$
27. $\dfrac{(a^3b^4)^3}{ab^4}$
28. $\dfrac{(a^2b^3)^3}{a^6b^6}$

29. $\dfrac{15(r^2s^3)^2}{-5(rs^5)^3}$
30. $\dfrac{-5(a^2b)^3}{10(ab^2)^3}$
31. $\dfrac{-32(x^3y)^3}{128(x^2y^2)^3}$
32. $\dfrac{68(a^6b^7)^2}{-96(abc^2)^3}$

33. $\dfrac{(5a^2b)^3}{(2a^2b^2)^3}$
34. $\dfrac{-(4x^3y^3)^2}{(x^2y^4)^8}$
35. $\dfrac{-(3x^3y^4)^3}{-(9x^4y^5)^2}$
36. $\dfrac{(2r^3s^2t)^2}{-(4r^2s^2t^2)^2}$

37. $\dfrac{(a^2a^3)^4}{(a^4)^3}$
38. $\dfrac{(b^3b^4)^5}{(bb^2)^2}$
39. $\dfrac{(z^3z^{-4})^3}{(z^{-3})^2}$
40. $\dfrac{(t^{-3}t^5)}{(t^2)^{-3}}$

In Exercises 41–54, perform each indicated division.

41. $\dfrac{6x + 9y}{3xy}$

42. $\dfrac{8x + 12y}{4xy}$

43. $\dfrac{5x - 10y}{25xy}$

44. $\dfrac{2x - 32}{16x}$

45. $\dfrac{3x^2 + 6y^3}{3x^2y^2}$

46. $\dfrac{4a^2 - 9b^2}{12ab}$

47. $\dfrac{15a^3b^2 - 10a^2b^3}{5a^2b^2}$

48. $\dfrac{9a^4b^3 - 16a^3b^4}{12a^2b}$

49. $\dfrac{4x - 2y + 8z}{4xy}$

50. $\dfrac{5a^2 + 10b^2 - 15ab}{5ab}$

51. $\dfrac{12x^3y^2 - 8x^2y - 4x}{4xy}$

52. $\dfrac{12a^2b^2 - 8a^2b - 4ab}{4ab}$

53. $\dfrac{-25x^2y + 30xy^2 - 5xy}{-5xy}$

54. $\dfrac{-30a^2b^2 - 15a^2b - 10ab^2}{-10ab}$

In Exercises 55–64, simplify each numerator and perform the indicated division.

55. $\dfrac{5x(4x - 2y)}{2y}$

56. $\dfrac{9y^2(x^2 - 3xy)}{3x^2}$

57. $\dfrac{(-2x)^3 + (3x^2)^2}{6x^2}$

58. $\dfrac{(-3x^2y)^3 + (3xy^2)^3}{27x^3y^4}$

59. $\dfrac{4x^2y^2 - 2(x^2y^2 + xy)}{2xy}$

60. $\dfrac{-5a^3b - 5a(ab^2 - a^2b)}{10a^2b^2}$

61. $\dfrac{(3x - y)(2x - 3y)}{6xy}$

62. $\dfrac{(2m - n)(3m - 2n)}{-3m^2n^2}$

63. $\dfrac{(a + b)^2 - (a - b)^2}{2ab}$

64. $\dfrac{(x - y)^2 + (x + y)^2}{2x^2y^2}$

Review Exercises

In Review Exercises 1–2, let $P(x) = 3x^2 + x$.

1. Find $P(4)$.

2. Find $P(-2)$.

3. Write the number 0.000265 in scientific notation.

4. Write the number 5.67×10^3 in standard notation.

In Review Exercises 5–8, simplify each expression.

5. $(3x^2)^0$

6. $(a^2b^3a^4b^5)^3$

7. $\dfrac{8x^4y^3}{4x^5y}$

8. $\left(\dfrac{9r^2st^3}{3r^4st^2}\right)^2$

9. Solve the equation $x(x + 2) = (x + 3)^2 - 1$.

10. Solve $y = mx + b$ for m.

11. The product of two consecutive integers is 19 greater than the square of the smaller integer. Find the integers.

12. The total surface area, A, of a box with dimensions l, w, and d (see Illustration 1) is given by the formula

$$A = 2lw + 2wd + 2ld$$

If $A = 202$ square inches, $l = 9$ inches, and $w = 5$ inches, find d.

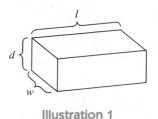

Illustration 1

10.8 Dividing Polynomials by Polynomials

- Dividing Polynomials
- Writing Exponents in Descending Order
- Missing Terms

If we use pencil and paper to divide 156 by 12, the work would look like this:

$$\begin{array}{r} 13 \\ 12{\overline{\smash{\big)}\,156}} \\ \underline{12} \\ 36 \\ \underline{36} \\ 0 \end{array}$$

To find a process for dividing a polynomial by a polynomial with more than one term, we will use a slightly different format for the division. We begin by noting that the **dividend** (the number 156) can be written as

$$156 = 100 + 50 + 6$$

and that the **divisor** (the number 12) can be written as

$$12 = 10 + 2$$

The division can then be accomplished as follows:

Step 1

$$10 + 2{\overline{\smash{\big)}\,100 + 50 + 6}}^{\,10}$$

How many times does 10 divide 100? $100 \div 10 = 10$. Place the 10 above the division symbol.

Step 2

$$\begin{array}{r} 10 \\ 10 + 2{\overline{\smash{\big)}\,100 + 50 + 6}} \\ 100 + 20 \end{array}$$

Multiply each term in the divisor by 10. Place the product under $100 + 50$ as indicated and draw a line.

Step 3

$$\begin{array}{r} 10 \\ 10 + 2{\overline{\smash{\big)}\,100 + 50 + 6}} \\ \underline{100 + 20} \\ 30 + 6 \end{array}$$

Subtract $100 + 20$ from $100 + 50$ by adding the negative of $100 + 20$ to $100 + 50$. Bring down the next term.

Step 4

$$\begin{array}{r} 10 + 3 \\ 10+2\overline{)100+50+6} \\ 100+20 \\ \hline 30+6 \end{array}$$

How many times does 10 divide 30? $30 \div 10 = +3$. Place the +3 above the division symbol.

Step 5

$$\begin{array}{r} 10 + 3 \\ 10+2\overline{)100+50+6} \\ 100+20 \\ \hline 30+6 \\ 30+6 \end{array}$$

Multiply each term in the divisor by 3. Place the product under the $30 + 6$ as indicated and draw a line.

Step 6

$$\begin{array}{r} 10 + 3 \\ 10+2\overline{)100+50+6} \\ 100+20 \\ \hline 30+6 \\ \underline{30+6} \\ 0 \end{array}$$

Subtract $30 + 6$ from $30 + 6$ by adding the negative of $30 + 6$.

Because the **remainder** is 0, the **quotient** is $10 + 3$, or 13.

Step 7

We check the answer by verifying that the product of the divisor and the quotient is equal to the dividend:

$12 \cdot 13 = 156$

The answer checks.

Dividing Polynomials

We will now see how to use this method to divide one polynomial by another. The first example closely parallels the previous discussion.

EXAMPLE 1

Divide $x^2 + 5x + 6$ by $x + 2$.

Solution

Step 1

$$\begin{array}{r} x \\ x+2\overline{)x^2+5x+6} \end{array}$$

How many times does x divide x^2? $x^2 \div x = x$. Place the x above the division symbol.

Step 2

$$\begin{array}{r} x \\ x+2\overline{)x^2+5x+6} \\ x^2+2x \end{array}$$

Multiply each term in the divisor by x. Place the product under $x^2 + 5x$ as indicated and draw a line.

Step 3

$$x + 2 \overline{\smash{)}\begin{aligned}x^2 + 5x + 6\\ \underline{x^2 + 2x}\\ 3x + 6\end{aligned}}$$

Subtract $x^2 + 2x$ from $x^2 + 5x$ by adding the negative of $x^2 + 2x$ to $x^2 + 5x$. Bring down the next term.

Step 4

$$x + 2 \overline{\smash{)}\begin{aligned}x + 3\\ x^2 + 5x + 6\\ \underline{x^2 + 2x}\\ 3x + 6\end{aligned}}$$

How many times does x divide $3x$? $3x \div x = +3$. Place the $+3$ above the division symbol.

Step 5

$$x + 2 \overline{\smash{)}\begin{aligned}x + 3\\ x^2 + 5x + 6\\ \underline{x^2 + 2x}\\ 3x + 6\\ 3x + 6\end{aligned}}$$

Multiply each term in the divisor by 3. Place the product under the $3x + 6$ as indicated and draw a line.

Step 6

$$x + 2 \overline{\smash{)}\begin{aligned}x + 3\\ x^2 + 5x + 6\\ \underline{x^2 + 2x}\\ 3x + 6\\ \underline{3x + 6}\\ 0\end{aligned}}$$

Subtract $3x + 6$ from $3x + 6$ by adding the negative of $3x + 6$.

The quotient is $x + 3$.

Step 7

We check the answer by verifying that the product of $x + 2$ and $x + 3$ is $x^2 + 5x + 6$.

$$(x + 2)(x + 3) = x^2 + 3x + 2x + 6$$
$$= x^2 + 5x + 6$$

EXAMPLE 2

Divide: $\dfrac{6x^2 - 7x - 2}{2x - 1}$.

Solution

Step 1

$$2x - 1 \overline{\smash{)}\begin{aligned}3x\\ 6x^2 - 7x - 2\end{aligned}}$$

How many times does $2x$ divide $6x^2$? $6x^2 \div 2x = 3x$. Place the $3x$ above the division symbol.

Step 2

$$2x - 1 \overline{\smash{)}\begin{aligned}3x\\ 6x^2 - 7x - 2\\ \underline{6x^2 - 3x}\end{aligned}}$$

Multiply each term in the divisor by $3x$. Place the product under $6x^2 - 7x$ as indicated and draw a line.

Step 3

$$2x - 1 \overline{\smash{)}\begin{array}{l} \phantom{6x^2-7x-{}}3x \\ 6x^2 - 7x - 2 \\ \underline{6x^2 - 3x} \\ \phantom{6x^2 -{}}-4x - 2 \end{array}}$$

Subtract $6x^2 - 3x$ from $6x^2 - 7x$ by adding the negative of $6x^2 - 3x$ to $6x^2 - 7x$. Bring down the next term.

Step 4

$$2x - 1 \overline{\smash{)}\begin{array}{l} 3x - 2 \\ 6x^2 - 7x - 2 \\ \underline{6x^2 - 3x} \\ \phantom{6x^2 -{}}-4x - 2 \end{array}}$$

How many times does $2x$ divide $-4x$? $-4x \div 2x = -2$. Place the -2 above the division symbol.

Step 5

$$2x - 1 \overline{\smash{)}\begin{array}{l} 3x - 2 \\ 6x^2 - 7x - 2 \\ \underline{6x^2 - 3x} \\ \phantom{6x^2 -{}}-4x - 2 \\ \phantom{6x^2 -{}}\underline{-4x + 2} \end{array}}$$

Multiply each term in the divisor by -2. Place the product under the $-4x - 2$ as indicated and draw a line.

Step 6

$$2x - 1 \overline{\smash{)}\begin{array}{l} 3x - 2 \\ 6x^2 - 7x - 2 \\ \underline{6x^2 - 3x} \\ \phantom{6x^2 -{}}-4x - 2 \\ \phantom{6x^2 -{}}\underline{-4x + 2} \\ \phantom{6x^2 - 7x -{}}-4 \end{array}}$$

Subtract $-4x + 2$ from $-4x - 2$ by adding the negative of $-4x + 2$.

In this example, the quotient is $3x - 2$, and the remainder is -4. It is common to write the answer in the following way:

$$3x - 2 + \frac{-4}{2x - 1}$$

where the fraction $\frac{-4}{2x - 1}$ is formed by dividing the remainder by the divisor.

Step 7

To check the answer, we multiply $3x - 2 + \frac{-4}{2x - 1}$ by $2x - 1$. The product should be the dividend.

$$(2x - 1)\left(3x - 2 + \frac{-4}{2x - 1}\right) = (2x - 1)(3x - 2) + (2x - 1)\left(\frac{-4}{2x - 1}\right)$$
$$= (2x - 1)(3x - 2) - 4$$
$$= 6x^2 - 4x - 3x + 2 - 4$$
$$= 6x^2 - 7x - 2$$

Because the result is the dividend, the answer checks. Work Progress Check 23. ∎

Progress Check 23

Divide:

23. $2x + 4 + \dfrac{-1}{3x + 1}$

Progress Check Answers

Writing Exponents in Descending Order

EXAMPLE 3

Divide $4x^2 + 2x^3 + 12 - 2x$ by $x + 3$.

Solution

The division process works most efficiently if the terms in the divisor and the dividend are written with their exponents in descending order. This means that the term involving the highest power of x appears first, the term involving the second highest power of x appears second, and so on.

$$\begin{array}{r} 2x^2 - 2x + 4 \\ x+3 \overline{\smash{)}\, 2x^3 + 4x^2 - 2x + 12} \\ \underline{2x^3 + 6x^2} \\ -2x^2 - 2x \\ \underline{-2x^2 - 6x} \\ +4x + 12 \\ \underline{+4x + 12} \end{array}$$

Use the commutative property of addition to write the terms of the dividend in descending order.

Check:

$$(x + 3)(2x^2 - 2x + 4) = 2x^3 - 2x^2 + 4x + 6x^2 - 6x + 12$$
$$= 2x^3 + 4x^2 - 2x + 12 \quad \blacksquare$$

Missing Terms

EXAMPLE 4

Divide: $\dfrac{x^2 - 4}{x + 2}$.

Solution

Note that the binomial $x^2 - 4$ does not have a term involving x. To perform this division, we must either include the term $0x$ or leave a space for it.

$$\begin{array}{r} x - 2 \\ x+2 \overline{\smash{)}\, x^2 + 0x - 4} \\ \underline{x^2 + 2x} \\ -2x - 4 \\ \underline{-2x - 4} \end{array}$$

Check:

$$(x + 2)(x - 2) = x^2 - 2x + 2x - 4$$
$$= x^2 - 4 \quad \blacksquare$$

EXAMPLE 5

Divide $x^3 + y^3$ by $x + y$.

Solution

Write $x^3 + y^3$, leaving spaces for any missing terms, and proceed as follows:

$$\begin{array}{r}x^2 - xy + y^2\\ x+y\overline{)x^3+y^3}\\ \underline{x^3 + x^2y}\\ -x^2y\\ \underline{-x^2y - xy^2}\\ +xy^2 + y^3\\ \underline{xy^2 + y^3}\end{array}$$

Check:

$(x + y)(x^2 - xy + y^2) = x^3 - x^2y + xy^2 + x^2y - xy^2 + y^3$
$= x^3 + y^3$

Work Progress Check 24.

Progress Check 24

Divide: $a - b\overline{)2a^2 - b^2}$.

24. $2a + 2b + \dfrac{b^2}{a-b}$

10.8 Exercises

1. Divide $x^2 + 4x + 4$ by $x + 2$.

2. Divide $x^2 - 5x + 6$ by $x - 2$.

3. Divide $y^2 + 13y + 12$ by $y + 1$.

4. Divide $z^2 - 7z + 12$ by $z - 3$.

5. Divide $a^2 + 2ab + b^2$ by $a + b$.

6. Divide $a^2 - 2ab + b^2$ by $a - b$.

In Exercises 7–12, perform each division.

7. $\dfrac{6a^2 + 5a - 6}{2a + 3}$

8. $\dfrac{8a^2 + 2a - 3}{2a - 1}$

9. $\dfrac{3b^2 + 11b + 6}{3b + 2}$

10. $\dfrac{3b^2 - 5b + 2}{3b - 2}$

11. $\dfrac{2x^2 - 7xy + 3y^2}{2x - y}$

12. $\dfrac{3x^2 + 5xy - 2y^2}{x + 2y}$

In Exercises 13–24, rearrange the terms so that the powers of x are in descending order, and divide.

13. $5x + 3\overline{)11x + 10x^2 + 3}$

14. $2x - 7\overline{)-x - 21 + 2x^2}$

15. $4 + 2x\overline{)-10x - 28 + 2x^2}$

16. $1 + 3x\overline{)9x^2 + 1 + 6x}$

17. $2x - y\overline{)xy - 2y^2 + 6x^2}$

18. $2y + x\overline{)3xy + 2x^2 - 2y^2}$

19. $x + 3y\overline{)2x^2 - 3y^2 + 5xy}$

20. $2x - 3y\overline{)2x^2 - 3y^2 - xy}$

21. $3x - 2y\overline{)-10y^2 + 13xy + 3x^2}$

22. $2x + 3y\overline{)-12y^2 + 10x^2 + 7xy}$

23. $4x + y\overline{)-19xy + 4x^2 - 5y^2}$

24. $x - 4y\overline{)5x^2 - 4y^2 - 19xy}$

In Exercises 25–30, perform each division.

25. $2x + 3\overline{)2x^3 + 7x^2 + 4x - 3}$

26. $2x - 1\overline{)2x^3 - 3x^2 + 5x - 2}$

27. $3x + 2\overline{)6x^3 + 10x^2 + 7x + 2}$

28. $4x + 3\overline{)4x^3 - 5x^2 - 2x + 3}$

29. $2x + y\overline{)2x^3 + 3x^2y + 3xy^2 + y^3}$

30. $3x - 2y\overline{)6x^3 - x^2y + 4xy^2 - 4y^3}$

In Exercises 31–40, perform each division. If there is a remainder, leave the answer in quotient + $\frac{\text{remainder}}{\text{divisor}}$ form.

31. $\dfrac{2x^2 + 5x + 2}{2x + 3}$

32. $\dfrac{3x^2 - 8x + 3}{3x - 2}$

33. $\dfrac{4x^2 + 6x - 1}{2x + 1}$

34. $\dfrac{6x^2 - 11x + 2}{3x - 1}$

35. $\dfrac{x^3 + 3x^2 + 3x + 1}{x + 1}$

36. $\dfrac{x^3 + 6x^2 + 12x + 8}{x + 2}$

37. $\dfrac{2x^3 + 7x^2 + 4x + 3}{2x + 3}$

38. $\dfrac{6x^3 + x^2 + 2x + 1}{3x - 1}$

39. $\dfrac{2x^3 + 4x^2 - 2x + 3}{x - 2}$

40. $\dfrac{3y^3 - 4y^2 + 2y + 3}{y + 3}$

In Exercises 41–50, perform each division.

41. $\dfrac{x^2 - 1}{x - 1}$

42. $\dfrac{x^2 - 9}{x + 3}$

43. $\dfrac{4x^2 - 9}{2x + 3}$

44. $\dfrac{25x^2 - 16}{5x - 4}$

45. $\dfrac{x^3 - 1}{x + 1}$

46. $\dfrac{x^3 - 8}{x - 2}$

47. $\dfrac{16a^2 - 9b^2}{4a + 3b}$

48. $\dfrac{25a^2 - 4b^2}{5a - 2b}$

49. $3x - 4 \overline{) 15x^3 - 29x^2 + 16}$

50. $2y + 3 \overline{) 23y^2 + 6y^3 + 27y}$

Review Exercises

1. List the composite numbers from 20 to 30.

2. Graph the prime numbers between 20 and 30 on a number line.

In Review Exercises 3–4, let $a = -2$ and $b = -3$. Evaluate each expression.

3. $4a - 2b$

4. $\dfrac{3a^2 + 2b^2}{3(a - b)}$

In Review Exercises 5–8, write each expression as an equivalent expression without absolute value symbols.

5. $|25|$

6. $|-32 - 5|$

7. $-|3 - 5|$

8. $-|16(-3)|$

In Review Exercises 9–10, simplify each expression.

9. $3(2x^2 - 4x + 5) + 2(x^2 + 3x - 7)$

10. $-2(y^3 + 2y^2 - y) - 3(3y^3 + y)$

11. The area of the ring between two circles of radius r and R (see Illustration 1) is given by

$$A = \dfrac{22}{7}(R + r)(R - r)$$

If $r = 3$ inches and $R = 17$ inches, find A.

12. Laura bought a color TV set for $502.90, which included a 7% sales tax. What was the selling price of the TV before the tax was added?

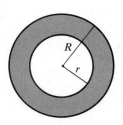

Illustration 1

436　Chapter 10　Polynomials

Mathematics in Medicine

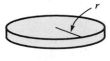

Illustration 1

Recall that the area of a circle is given by the formula $A = \pi r^2$. The bulk of the surface area of the red blood cell in Illustration 1 is contained on its top and bottom. That area is $2\pi r^2$, twice the area of one circle. If there are N discs, their total surface area, T, will be N times the surface area of a single disc: $T = 2N\pi r^2$.

To find the total surface area of the oxygen-carrying red cells, first express the given quantities in scientific notation.

$$\text{radius} = r = 0.00015 \text{ in.} = 1.5 \times 10^{-4} \text{ in.}$$
$$\text{quantity} = N = 25 \text{ trillion}$$
$$= 2.5 \times 10^{13}$$

Then substitute these values into the formula for total surface area.

$$T = 2N\pi r^2$$
$$T = 2(2.5 \times 10^{13})(3.14)(1.4 \times 10^{-4})^2$$
$$= 2(2.5)(3.14)(1.4)^2 \times 10^{13} \times 10^{-8}$$
$$= 31 \times 10^5$$
$$= 3.1 \times 10^6$$
$$= 3{,}100{,}000$$

The total surface area of the red blood cells is over 3 million square inches, or 21,500 square feet, almost one-half the area of a football field!

Chapter Summary

Key Words

algebraic terms (10.4)
base (10.1)
binomial (10.4)
conjugate
  binomial (10.6)
degree of a
  monomial (10.4)

degree of a
  polynomial (10.4)
dividend (10.8)
divisor (10.8)
exponent (10.1)
FOIL (10.6)
like terms (10.5)

minuend (10.5)
monomial (10.4)
polynomial (10.4)
power (10.1)
quotient (10.8)
remainder (10.8)
scientific notation (10.3)

special products (10.6)
standard notation (10.3)
subtrahend (10.5)
trinomial (10.4)

Key Ideas

(10.1)–(10.2) **Properties of exponents.** If n is a natural number, then

$$x^n = \overbrace{x \cdot x \cdot x \cdot x \cdot \cdots \cdot x}^{n \text{ factors of } x}$$

If m and n are integers and there are no divisions by 0,

$$x^m x^n = x^{m+n}, \quad (x^m)^n = x^{mn}, \quad (xy)^n = x^n y^n$$

$$\left(\frac{x}{y}\right)^n = \frac{x^n}{y^n}, \quad \frac{x^m}{x^n} = x^{m-n}, \quad x^0 = 1, \quad x^{-n} = \frac{1}{x^n}$$

(10.4) If $P(x)$ is a polynomial in x, then $P(r)$ is the value of the polynomial when $x = r$.

(10.3) A number is written in scientific notation if it is written as the product of a number between 1 (including 1) and 10 and an integer power of 10.

(10.5) When adding or subtracting polynomials, combine like terms by adding or subtracting the numerical coefficients and using the same variables and the same exponents.

(10.6) To multiply two monomials, first multiply the numerical factors and then multiply the variable factors.

To multiply a polynomial with more than one term by a monomial, multiply each term of the polynomial by the monomial and simplify.

To multiply one polynomial by another, multiply each term of one polynomial by each term of the other polynomial and simplify.

To multiply two binomials, use the FOIL method.

Special products:

$(x + y)^2 = x^2 + 2xy + y^2$
$(x - y)^2 = x^2 - 2xy + y^2$
$(x + y)(x - y) = x^2 - y^2$

(10.7) To simplify a fraction, divide out all factors common to the numerator and the denominator of the fraction.

To divide a polynomial by a monomial, rewrite the division as a product, use the distributive law to remove parentheses, and simplify each resulting fraction.

(10.8) Use long division to divide one polynomial by another.

Chapter 10 Review Exercises

[10.1–10.2] In Review Exercises 1–8, evaluate each expression.

1. 5^3
2. 3^5
3. $(-8)^2$
4. -8^2
5. $3^2 + 2^2$
6. $(3 + 2)^2$
7. $3(3^3 + 3^3)$
8. $1^{17} + 17^1$

In Review Exercises 9–24, perform the indicated operation and simplify.

9. $x^3 x^2$
10. $x(x^2 y)$
11. $y^7 y^3$
12. $x^0 y^5$
13. $2b^3 b^4 b^5$
14. $(-z^2)(z^3 y^2)$
15. $(4^4 s)s^2$
16. $-3y(y^5)$
17. $(x^2 x^3)^3$
18. $(2x^2 y)^2$
19. $(3x^0)^2$
20. $(3x^2 y^2)^0$
21. $\dfrac{x^7}{x^3}$
22. $\left(\dfrac{x^2 y}{xy^2}\right)^2$
23. $\dfrac{8(y^2 x)^2}{2^3 (yx^2)^2}$
24. $\dfrac{(5x^0 y^2 z^3)^3}{25(yz)^5}$

In Review Exercises 25–32, write each expression without using negative exponents or parentheses.

25. $x^{-2} x^3$
26. $y^4 y^{-3}$
27. $\dfrac{x^3}{x^{-7}}$
28. $(x^{-3} x^{-4})^{-2}$
29. $\dfrac{x^3}{x^7}$
30. $\left(\dfrac{x^2}{x}\right)^{-5}$
31. $\left(\dfrac{3s}{6s^2}\right)^3$
32. $\left(\dfrac{15z^4}{5z^3}\right)^{-2}$

[10.3] In Review Exercises 33–40, write each number in scientific notation.

33. 728
34. 9370
35. 0.0136
36. 0.00942
37. 7.61
38. 795×10^3
39. 0.012×10^{-2}
40. 600×10^2

In Review Exercises 41–46, write each number in standard notation.

41. 7.26×10^5 **42.** 3.91×10^{-4} **43.** 2.68×10^0 **44.** 5.76×10^1

45. 731×10^{-2} **46.** 0.498×10^3

[10.4] In Review Exercises 47–50, give the degree of each polynomial.

47. $13x^7$ **48.** $5^3x + x^2$ **49.** $-3x^5 + x - 1$ **50.** $9x + 21x^3$

In Review Exercises 51–54, let $P(x) = 3x + 2$. Find each value.

51. $P(3)$ **52.** $P(0)$ **53.** $P(-2)$ **54.** $P(2t)$

In Review Exercises 55–58, let $P(x) = 5x^4 - x$. Find each value.

55. $P(3)$ **56.** $P(0)$ **57.** $P(-2)$ **58.** $P(2t)$

[10.5] In Review Exercises 59–68, simplify each expression, if possible.

59. $3x + 5x - x$

60. $2x + 3y$

61. $(xy)^2 + 3x^2y^2$

62. $-2x^2yz + 3yx^2z$

63. $3x^2y^0 + 2x^2$

64. $2(x + 7) + 3(x + 7)$

65. $(3x^2 + 2x) + (5x^2 - 8x)$

66. $(7a^2 + 2a - 5) - (3a^2 - 2a + 1)$

67. $3(9x^2 + 3x + 7) + 2(2x^2 - 8x + 3) - 2(11x^2 - 5x + 9)$

68. $4(4x^3 + 2x^2 - 3x - 8) - 5(2x^3 - 3x + 8)$

[10.6] In Review Exercises 69–88, find each product.

69. $(2x^2y^3)(5xy^2)$ **70.** $(xyz^3)(x^3z)^2$ **71.** $5(x + 3)$ **72.** $3(2x + 4)$

73. $x^2(3x^2 - 5)$ **74.** $2y^2(y^2 + 5y)$ **75.** $-x^2y(y^2 - xy)$ **76.** $-3xy(xy - x)$

77. $(x + 3)(x + 2)$ **78.** $(2x + 1)(x - 1)$ **79.** $(3a - 3)(2a + 2)$ **80.** $6(a - 1)(a + 1)$

81. $(a - b)(2a + b)$ **82.** $(3x - y)(2x + y)$ **83.** $(-3a - b)(3a - b)$ **84.** $(x + 5)(x - 5)$

85. $(y - 2)(y + 2)$ **86.** $(x + 4)^2$ **87.** $(y - 3)^2$ **88.** $y(y + 1)^2$

In Review Exercises 89–94, solve each equation.

89. $x^2 + 3 = x(x + 3)$

90. $x^2 + x = (x + 1)(x + 2)$

91. $(x + 2)(x - 5) = (x - 4)(x - 1)$

92. $(x - 1)(x - 2) = (x - 3)(x + 1)$

93. $x^2 + x(x + 2) = x(2x + 1) + 1$

94. $(x + 5)(3x + 1) = x^2 + (2x - 1)(x - 5)$

[10.7–10.8] In Review Exercises 95–104, perform each division.

95. $\dfrac{3x + 6y}{2xy}$

96. $\dfrac{14xy - 21x}{7xy}$

97. $\dfrac{15a^2bc + 20ab^2c - 25abc^2}{-5abc}$

98. $\dfrac{(x+y)^2 + (x-y)^2}{-2xy}$

99. $x + 2 \overline{\smash{)}x^2 + 3x + 5}$

100. $x - 1 \overline{\smash{)}x^2 - 6x + 5}$

101. $x + 3 \overline{\smash{)}2x^2 + 7x + 3}$

102. $3x - 1 \overline{\smash{)}3x^2 + 14x - 5}$

103. $2x - 1 \overline{\smash{)}6x^3 + x^2 - 1}$

104. $3x + 1 \overline{\smash{)}-13x - 4 + 9x^3}$

Name _____   Section _____

Chapter 10 Test

1. Use exponents to rewrite $2xxxyyyy$. _____

2. Evaluate $3^2 + 5^3$. _____

In Problems 3–6, write each expression as an expression containing only one exponent.

3. $y^2(yy^3)$ _____

4. $(-3b^2)(2b^3)(-b^2)$ _____

5. $(2x^3)^5$ _____

6. $(2rr^2r^3)^3$ _____

In Problems 7–10, simplify each expression. Write answers without using parentheses or negative exponents.

7. $3x^0$ _____

8. $2y^{-5}$ _____

9. $\dfrac{y^2}{yy^{-2}}$ _____

10. $\left(\dfrac{a^2b^{-1}}{4a^3b^{-2}}\right)^{-3}$ _____

11. Express 28,000 in scientific notation. _____

12. Express 0.0025 in scientific notation. _____

13. Express 7.4×10^3 in standard notation. _____

14. Express 9.3×10^{-5} in standard notation. _____

15. Identify $3x^2 + 2$ as a monomial, binomial, or trinomial. _____

16. Find the degree of the polynomial $3x^2y^3z^4 + 2x^3y^2z - 5x^2y^3z^5$. _____

17. If $P(x) = x^2 + x - 2$, find $P(-2)$.

18. Simplify $(xy)^2 + 5x^2y^2 - 3x^2y^2$.

19. Remove parentheses and simplify: $-6(x - y) + 2(x + y) - 3(x + 2y)$.

20. Remove parentheses and simplify: $-2(x^2 + 3x - 1) - 3(x^2 - x + 2) + 5(x^2 + 2)$.

21. Add: $\quad 3x^3 + 4x^2 - x - 7$
$\phantom{\text{Add: }}\underline{2x^3 - 2x^2 + 3x + 2}$

22. Subtract: $\quad 2x^2 - 7x + 3$
$\phantom{\text{Subtract: }}\underline{3x^2 - 2x - 1}$

In Problems 23–27, find each product.

23. $(-2x^3)(2x^2y)$

24. $3y^2(y^2 - 2y + 3)$

25. $(2x - 5)(3x + 4)$

26. $(3x - y)(3x + y)$

27. $(2x - 3)(x^2 - 2x + 4)$

28. Solve the equation $(a + 2)^2 = (a - 3)^2$.

29. Simplify the fraction $\dfrac{8x^2y^3z^4}{16x^3y^2z^4}$.

30. Perform the division $\dfrac{6a^2 - 12b^2}{24ab}$.

31. Divide: $2x + 3 \overline{)2x^2 - x - 6}$.

32. Divide: $x - y \overline{)x^3 - y^3}$.

11 Factoring Polynomials

Mathematics in Geometry

The square in Illustration 1 shows how one area can be broken apart to form another figure with the same total area. The area of the large square is $(a + b)^2$ and the total area of the four sections is $a^2 + 2ab + b^2$. This is no surprise, for you learned in Chapter 10 that $(a + b)^2 = a^2 + 2ab + b^2$.

The area of the shaded region in Illustration 2 is the area of the larger square, minus the area of the smaller square. The area of the shaded region is also $4ab$. This, you will see, illustrates geometrically one technique of *factoring*, the topic of this chapter.

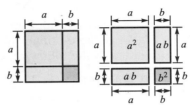

Illustration 1   Illustration 2

We now reverse the operation of multiplication and show how to find the factors of a given product. This process is called **factoring**.

11.1 Factoring Out the Greatest Common Factor

- Writing Numbers in Prime-Factored Form
- The Greatest Common Factor
- Factoring Monomials from Polynomials

If a natural number a divides a natural number b, then a is called a **factor** of b. The natural-number factors of 8, for example, are 1, 2, 4, and 8, because each of these numbers divides 8.

Recall that a natural number greater than 1 whose only factors are 1 and the number itself is called a **prime number**. Since the only natural-number factors of 17 are 1 and 17, 17 is a prime number.

Writing Numbers in Prime-Factored Form

To factor a natural number means to write it as the product of other natural numbers.

> **Definition:** A natural number is said to be in **prime-factored form** if it is written as the product of factors that are prime numbers.

The right-hand sides of the equations

$$42 = 2 \cdot 3 \cdot 7$$
$$60 = 2^2 \cdot 3 \cdot 5$$
$$90 = 2 \cdot 3^2 \cdot 5$$

show the prime-factored forms or **prime factorizations** of 42, 60, and 90. A theorem called the *fundamental theorem of arithmetic* points out that there is *exactly one* prime factorization for every natural number.

The Greatest Common Factor

The largest natural number that divides each of several natural numbers is called the **greatest common factor** or the **greatest common divisor** of these numbers. For example, 6 is the greatest common factor of 42, 60, and 90, because

$$\frac{42}{6} = 7, \qquad \frac{60}{6} = 10, \qquad \text{and} \qquad \frac{90}{6} = 15$$

and no natural number greater than 6 divides 42, 60, and 90.

Algebraic monomials also have greatest common factors. The right-hand sides of the following equations

$$6a^2b^3 = 2 \cdot 3 \cdot a \cdot a \cdot b \cdot b \cdot b$$
$$4a^3b^2 = 2 \cdot 2 \cdot a \cdot a \cdot a \cdot b \cdot b$$
$$18a^2b = 2 \cdot 3 \cdot 3 \cdot a \cdot a \cdot b$$

show the unique prime factorizations of $6a^2b^3$, $4a^3b^2$, and $18a^2b$. Because all three of these monomials have one factor of 2, two factors of a, and one factor of b in common, their greatest common factor is

$$2 \cdot a \cdot a \cdot b \qquad \text{or} \qquad 2a^2b$$

To find the greatest common factor of several monomials, we follow these steps:

> **Finding the Greatest Common Factor:**
> 1. Find the prime factorization of each monomial.
> 2. Use each common factor the least number of times it appears in any one monomial.
> 3. Find the product of the factors found in Step 2 to obtain the greatest common factor.

Recall that the distributive property provides the way to multiply a polynomial by a monomial. For example,

$$3x^2(2x - 3y) = 3x^2 \cdot 2x - 3x^2 \cdot 3y$$
$$= 6x^3 - 9x^2y$$

Factoring Monomials from Polynomials

Given a polynomial such as $6x^3 - 9x^2y$, we can factor each monomial and use the distributive property in reverse to factor the polynomial.

$$6x^3 - 9x^2y = 3x^2 \cdot 2x - 3x^2 \cdot 3y$$
$$= 3x^2(2x - 3y)$$

Because $3x^2$ is the greatest common factor of the terms $6x^3$ and $-9x^2y$, this process is called **factoring out the greatest common factor**.

EXAMPLE 1

Factor $6x + 9$.

Solution

To find the greatest common factor, find the prime factorization of $6x$ and 9.

$$6x = 3 \cdot 2 \cdot x$$
$$9 = 3 \cdot 3$$

The greatest common factor in $6x$ and 9 is 3, and we can factor it out.

$$6x + 9 = 3 \cdot 2x + 3 \cdot 3$$
$$= 3(2x + 3) \qquad \text{Use the distributive property.}$$

Check this result by verifying that $3(2x + 3) = 6x + 9$. Work Progress Check 1. ■

Progress Check 1

Factor $4x + 16$.

EXAMPLE 2

Factor $12y^2 + 20y$.

Solution

To find the greatest common factor, find the prime factorization of $12y^2$ and $20y$.

$$12y^2 = 2 \cdot 2 \cdot 3 \cdot y \cdot y$$
$$20y = 2 \cdot 2 \cdot 5 \cdot y$$

The greatest common factor in $12y^2$ and $20y$ is $2 \cdot 2 \cdot y$, or $4y$, and we can factor it out.

$$12y^2 + 20y = 4y \cdot 3y + 4y \cdot 5$$
$$= 4y(3y + 5) \qquad \text{Use the distributive property.}$$

Check this result by verifying that $4y(3y + 5) = 12y^2 + 20y$. ■

EXAMPLE 3

Factor $35a^3b^2 - 14a^2b^3$.

---

1. $4(x + 4)$

Progress Check Answers

Solution

To find the greatest common factor, find the prime factorization of $35a^3b^2$ and $-14a^2b^3$.

$$35a^3b^2 = 5 \cdot 7 \cdot a \cdot a \cdot a \cdot b \cdot b$$
$$-14a^2b^3 = -2 \cdot 7 \cdot a \cdot a \cdot b \cdot b \cdot b$$

The greatest common factor in $35a^3b^2$ and $-14a^2b^3$ is $7 \cdot a \cdot a \cdot b \cdot b$, or $7a^2b^2$, and we can factor it out.

$$35a^3b^2 - 14a^2b^3 = 7a^2b^2 \cdot 5a - 7a^2b^2 \cdot 2b$$
$$= 7a^2b^2(5a - 2b) \quad \text{Use the distributive property.}$$

Check this result by verifying that $7a^2b^2(5a - 2b) = 35a^3b^2 - 14a^2b^3$.

EXAMPLE 4

Factor $a^2b^2 - ab$.

Solution

Factor out the greatest common factor, which is ab.

$$a^2b^2 - ab = \mathbf{ab} \cdot ab - \mathbf{ab} \cdot 1$$
$$= \mathbf{ab}(ab - 1) \quad \text{Use the distributive property.}$$

It is important to understand where the 1 comes from. The last term of the binomial $a^2b^2 - ab$ has an implied coefficient of 1. When the ab is factored out, this coefficient of 1 must be written.

Check the result by verifying that $ab(ab - 1) = a^2b^2 - ab$. Work Progress Check 2.

EXAMPLE 5

Factor $12x^3y^2z + 6x^2yz - 3xz$.

Solution

Factor out the greatest common factor, which is $3xz$.

$$12x^3y^2z + 6x^2yz - 3xz = \mathbf{3xz} \cdot 4x^2y^2 + \mathbf{3xz} \cdot 2xy - \mathbf{3xz} \cdot 1$$
$$= \mathbf{3xz}(4x^2y^2 + 2xy - 1)$$

Check the result by verifying that $3xz(4x^2y^2 + 2xy - 1) = 12x^3y^2z + 6x^2yz - 3xz$.

EXAMPLE 6

Factor -1 out of the trinomial $-a^3 + 2a^2 - 4$.

Solution

$$-a^3 + 2a^2 - 4$$
$$= (-1)a^3 + (-1)(-2a^2) + (-1)4 \quad \text{Note that } (-1)(-2a^2) = +2a^2.$$
$$= -1(a^3 - 2a^2 + 4)$$
$$= -(a^3 - 2a^2 + 4)$$

Check the result by verifying that $-(a^3 - 2a^2 + 4) = -a^3 + 2a^2 - 4$.

Progress Check 2

Factor $x^3y^3 + x^2y^2$.

2. $x^2y^2(xy + 1)$

Progress Check Answers

EXAMPLE 7

Factor out the negative of the greatest common factor in $-18a^2b + 6ab^2 - 12a^2b^2$.

Solution

The greatest common factor in the trinomial is $6ab$. Since we want to factor out the negative of this greatest common factor, factor out $-6ab$:

$$-18a^2b + 6ab^2 - 12a^2b^2 = (-6ab)3a - (-6ab)b + (-6ab)2ab$$
$$= -6ab(3a - b + 2ab)$$

Check the result by verifying that $-6ab(3a - b + 2ab) = -18a^2b + 6ab^2 - 12a^2b^2$. Work Progress Check 3.

Progress Check 3

Factor $-4x^2y^3 + 8x^3y^2 - 12xy$.

3. $-4xy(xy^2 - 2x^2y + 3)$
Progress Check Answers

11.1 Exercises

In Exercises 1–12, find the prime factorization of each number.

1. 12   **2.** 24   **3.** 15   **4.** 20

5. 40   **6.** 62   **7.** 98   **8.** 112

9. 225   **10.** 144   **11.** 288   **12.** 968

In Exercises 13–20, finish each factorization.

13. $4x + 20$, $4(\quad)$

14. $3y - 15$, $3(\quad)$

15. $ab - ac$, $\underline{\quad}(b - c)$

16. $rs + rt$, $\underline{\quad}(s + t)$

17. $2x^2 + 4x^2y$, $2x^2(\quad)$

18. $3a^2b^2 - 12ab$, $3ab(\quad)$

19. $7a^2b^3 + 14ab^2$, $7ab^2(\quad)$

20. $10p^2q - 15p^4q^2$, $5p^2q(\quad)$

In Exercises 21–48, factor out the greatest common factor.

21. $3x + 6$

22. $2y - 10$

23. $xy - xz$

24. $uv + ut$

25. $t^2 + 2t$

26. $b^3 - 3b$

27. $2r^4 - 4r^2$

28. $a^3 + 3a^2$

29. $a^3b^3z^3 - a^2b^3z^2$

30. $r^3s^6t^9 + r^2s^2t^2$

31. $24x^2y^3z^4 + 8xy^2z^3$

32. $3x^2y^3 - 9x^4y^3z$

33. $12uvw^3 - 18uv^2w^2$

34. $14xyz - 16x^2y^2z$

35. $3x + 3y - 6z$

36. $2x - 4y + 8z$

37. $ab + ac - ad$

38. $rs - rt + ru$

39. $4y^2 + 8y - 2xy$

40. $3x^2 - 6xy + 9xy^2$

41. $12r^2 - 3rs + 9r^2s^2$

42. $6a^2 - 12a^3b + 36ab$

43. $abx - ab^2x + abx^2$

44. $a^2b^2x^2 + a^3b^2x^2 - a^3b^3x^3$

45. $4x^2y^2z^2 - 6xy^2z^2 + 12xyz^2$

46. $32xyz + 48x^2yz + 36xy^2z$

47. $70a^3b^2c^2 + 49a^2b^3c^3 - 21a^2b^2c^2$

48. $8a^2b^2 - 24ab^2c + 9b^2c^2$

In Exercises 49–60, factor out -1 from each polynomial.

49. $-a - b$ **50.** $-x - 2y$ **51.** $-2x + 4$ **52.** $-3x + 9$

53. $-2a + 3b$ **54.** $-2x + 5y$ **55.** $-3m - 4n + 1$ **56.** $-3r + 2s - 3$

57. $-3xy + 2z + 5w$ **58.** $-4ab + 3c - 5d$

59. $-3abc - 6abd + 9abe$ **60.** $-6xyz + 12x^2yz - 15x^2y^2z^2$

In Exercises 61–70, factor each polynomial by factoring out the greatest common factor, including -1.

61. $-3x^2y - 6xy^2$ **62.** $-4a^2b^2 + 6ab^2$

63. $-4a^2b^3 + 12a^3b^2$ **64.** $-25x^4y^3z^2 + 30x^2y^3z^4$

65. $-4a^2b^2c^2 + 14a^2b^2c - 10ab^2c^2$ **66.** $-10x^4y^3z^2 + 8x^3y^2z - 20x^2y$

67. $-14a^6b^6 + 49a^2b^3 - 21ab$ **68.** $-35r^9s^9t^9 + 25r^6s^6t^6 + 75r^3s^3t^3$

69. $-5a^2b^3c + 15a^3b^4c^2 - 25a^4b^3c$ **70.** $-7x^5y^4z^3 + 49x^5y^5z^4 - 21x^6y^4z^3$

Review Exercises

In Review Exercises 1–6, multiply the binomials.

1. $(3x + 2)(2y + 1)$ **2.** $(2r - 3)(r + t)$ **3.** $(2a + 1)(b - 1)$ **4.** $(m - 4n)(m + 2)$

5. $(3p - q)(2r - q)$ **6.** $(2u + v)(3m - v)$

In Review Exercises 7–12, remove parentheses and simplify.

7. $3(x + y) + a(x + y)$ **8.** $x(y + 1) + 5(y + 1)$

9. $(x + 3)(x + 1) - y(x + 1)$ **10.** $x(x^2 + 2) - y(x^2 + 2)$

11. $(3x - y)(x^2 - 2) + 3(x^2 - 2)$ **12.** $(x - 5y)(a + 2) - (x - 5y)b$

11.2 Factoring by Grouping

- Factoring Polynomials with Four Terms
- Factoring Polynomials with More Than Four Terms

Sometimes we can factor out a polynomial as the greatest common factor. For example, we can factor $a + b$ out of $(a + b)x + (a + b)y$. To do so, we note that $a + b$ is a common factor of both $(a + b)x$ and $(a + b)y$, and it can be factored out.

$$(a + b)x + (a + b)y = (a + b) \cdot x + (a + b) \cdot y$$
$$= (a + b)(x + y) \quad \text{Use the distributive property.}$$

We can check the result by verifying that $(a + b)(x + y) = (a + b)x + (a + b)y$.

EXAMPLE 1

Factor $a + 3$ out of the expression $(a + 3) + (a + 3)^2$.

Solution

Recognize that $a + 3$ is equal to $(a + 3)1$ and that $(a + 3)^2$ is equal to $(a + 3)(a + 3)$. Then factor out $a + 3$ and simplify.

$$(a + 3) + (a + 3)^2 = (a + 3)1 + (a + 3)(a + 3)$$
$$= (a + 3)[1 + (a + 3)]$$
$$= (a + 3)(a + 4)$$

EXAMPLE 2

Factor $6a^2b^2(x + 2y) - 9ab(x + 2y)$.

Solution

$6a^2b^2(x + 2y) - 9ab(x + 2y)$
$= (x + 2y)(6a^2b^2 - 9ab)$ Factor out $(x + 2y)$.
$= (x + 2y)3ab(2ab - 3)$ Factor out $3ab$ from $6a^2b^2 - 9ab$.
$= 3ab(x + 2y)(2ab - 3)$ Use the commutative property of multiplication.

Work Progress Check 4.

Progress Check 4

Factor $12x^3y^2(2x - y) + 18x^2y^4(2x - y)$.

Factoring Polynomials with Four Terms

Suppose we wish to factor

$$ax + ay + cx + cy$$

Although no factor is common to all four terms, there is a common factor of a in $ax + ay$ and a common factor of c in $cx + cy$. We can factor out the a and the c to obtain

$$ax + ay + cx + cy = a(x + y) + c(x + y)$$
$$= (x + y)(a + c)$$

This result can be checked by multiplication.

$$(x + y)(a + c) = ax + cx + ay + cy$$
$$= ax + ay + cx + cy$$

Thus, $ax + ay + cx + cy$ factors as $(x + y)(a + c)$. This type of factoring is called **factoring by grouping**.

EXAMPLE 3

Factor $2c + 2d - cd - d^2$.

Solution

$2c + 2d - cd - d^2$
$= 2(c + d) - d(c + d)$ Factor out 2 from $2c + 2d$ and factor out $-d$ from $-cd - d^2$.
$= (c + d)(2 - d)$ Factor out $c + d$.

Progress Check Answers

4. $6x^2y^2(2x - y)(2x + 3y^2)$

Check by multiplication:
$$(c + d)(2 - d) = 2c - cd + 2d - d^2$$
$$= 2c + 2d - cd - d^2$$

EXAMPLE 4

Factor $x^2y - ax - xy + a$.

Solution

$x^2y - ax - xy + a$
$= x(xy - a) - 1(xy - a)$ Factor out x from $x^2y - ax$ and factor out -1 from $-xy + a$.
$= (xy - a)(x - 1)$ Factor out $xy - a$.

Check by multiplication. Work Progress Check 5.

Progress Check 5

Factor $ac + bc + 4a + 4b$.

Factoring Polynomials with More Than Four Terms

The method of factoring by grouping often works for polynomials with more than four terms.

EXAMPLE 5

Progress Check 6

Factor $2ax + 2ay + 2az + bx + by + bz$.

Factor $5am - 5bm + 5cm + 3an - 3bn + 3cn$.

Solution

Factor $6m$ from the first three terms and $3n$ from the last three terms to obtain

$6am - 6bm + 6cm + 3an - 3bn + 3cn$
$= 6m(a - b + c) + 3n(a - b + c)$

Then factor out the common factor of $(a - b + c)$.

5. $(a + b)(c + 4)$

6. $(x + y + z)(2a + b)$

$6am - 6bm + 6cm + 3an - 3bn + 3cn = (a - b + c)(6m + 3n)$

Progress Check Answers

Check by multiplication. Work Progress Check 6.

11.2 Exercises

In Exercises 1–12, finish each factorization.

1. $(a + b)3 + (a + b)x$, $(a + b)($)
2. $(p - q)m + (p - q)n$, $(p - q)($)
3. $5(x - y) - b(x - y)$, $(x - y)($)
4. $x(x + 1) - 5(x + 1)$, $(x + 1)($)
5. $3(2a - b) - x(2a - b)$, $(2a - b)($)
6. $y(2a + b) + x(2a + b)$, $(2a + b)($)
7. $z(y + 3) + (y + 3)^2$, $(y + 3)($)
8. $r(t + 4) - (t + 4)^2$, $(t + 4)($)
9. $2(x^2 + y) + a(x^2 + y)$, $(x^2 + y)($)
10. $6x(c - d) + 6y(c - d)$, $6(c - d)($)
11. $6x^2(p + q) - 9y^2(p + q)$, $3(p + q)($)
12. $6x^2y^2(p - q) - 3xy(p - q)$, $3xy(p - q)($)

In Exercises 13–30, factor each expression.

13. $(x + y)2 + (x + y)b$

14. $(a - b)c + (a - b)d$

15. $3(x + y) - a(x + y)$

16. $x(y + 1) - 5(y + 1)$

17. $3(r - 2s) - x(r - 2s)$

18. $x(a + 2b) + y(a + 2b)$

19. $(x - 3)^2 + (x - 3)$

20. $(2t + 4) - (2t + 4)^2$

21. $2x(a^2 + b) + 2y(a^2 + b)$

22. $3x(c - 3d) + 6y(c - 3d)$

23. $3x^2(r + 3s) - 6y^2(r + 3s)$

24. $9a^2b^2(3x - 2y) - 6ab(3x - 2y)$

25. $3x(a + b + c) - 2y(a + b + c)$

26. $2m(a - 2b + 3c) - 21xy(a - 2b + 3c)$

27. $14x^2y(r + 2s - t) - 21xy(r + 2s - t)$

28. $15xy^3(2x - y + 3z) + 25xy^2(2x - y + 3z)$

29. $(x + 3)(x + 1) - y(x + 1)$

30. $x(x^2 + 2) - y(x^2 + 2)$

In Exercises 31–48, factor each expression.

31. $2x + 2y + ax + ay$

32. $bx + bz + 5x + 5z$

33. $7r + 7s - kr - ks$

34. $9p - 9q + mp - mq$

35. $xr + xs + yr + ys$

36. $pm - pn + qm - qn$

37. $2ax + 2bx + 3a + 3b$

38. $3xy + 3xz - 5y - 5z$

39. $2ab + 2ac + 3b + 3c$

40. $3ac + a + 3bc + b$

41. $2x^2 + 2xy - 3x - 3y$

42. $3ab + 9a - 2b - 6$

43. $3tv - 9tw + uv - 3uw$

44. $ce - 2cf + 3de - 6df$

45. $9mp + 3mq - 3np - nq$

46. $ax + bx - a - b$

47. $mp - np - m + n$

48. $6x^2u - 3x^2v + 2yu - yv$

In Exercises 49–56, factor each expression. Factor out all common factors first if they exist.

49. $ax^3 + bx^3 + 2ax^2y + 2bx^2y$

50. $x^3y^2 - 2x^2y^2 + 3xy^2 - 6y^2$

51. $4a^2b + 12a^2 - 8ab - 24a$

52. $-4abc - 4ac^2 + 2bc + 2c^2$

53. $x^3 + 2x^2 + x + 2$

54. $y^3 - 3y^2 - 5y + 15$

55. $x^3y - x^2y - xy^2 + y^2$

56. $2x^3z - 4x^2z + 32xz - 64z$

In Exercises 57–64, factor each expression completely.

57. $x^2 + xy + x + 2x + 2y + 2$

58. $ax + ay + az + bx + by + bz$

59. $am + bm + cm - an - bn - cn$

60. $x^2 + xz - x - xy - yz + y$

61. $ad - bd - cd + 3a - 3b - 3c$

62. $ab + ac - ad - b - c + d$

63. $ax^2 - ay + bx^2 - by + cx^2 - cy$

64. $a^2x - bx - a^2y + by + a^2z - bz$

In Exercises 65–74, factor each expression completely. You may have to rearrange some terms first.

65. $2r - bs - 2s + br$

66. $5x + ry + rx + 5y$

67. $ax + by + bx + ay$

68. $mr + ns + ms + nr$

69. $ac + bd - ad - bc$

70. $sx - ry + rx - sy$

71. $ar^2 - brs + ars - br^2$

72. $a^2bc + a^2c + abc + ac$

73. $ab + 3 + a + 3b$

74. $xy + 7 + y + 7x$

Review Exercises

In Review Exercises 1–4, simplify each expression. Write all results without using negative exponents.

1. $u^3 u^2 u^4$

2. $\dfrac{y^6}{y^8}$

3. $\dfrac{a^3 b^4}{a^2 b^5}$

4. $(3x^5)^0$

5. Write 0.00045 in scientific notation.

6. Write 6.28×10^4 in standard notation.

In Review Exercises 7–10, multiply the binomials.

7. $(a + b)(a - b)$

8. $(2r + s)(2r - s)$

9. $(3x + 2y)(3x - 2y)$

10. $(4x^2 + 3)(4x^2 - 3)$

In Review Exercises 11–12, write each statement as an algebraic expression.

11. The quotient obtained when the sum of the numbers x and y is divided by their product.

12. The product of the numbers r and s, decreased by twice their sum.

11.3 Factoring the Difference of Two Squares

- The Difference of Two Squares
- The Sum of Two Squares

Whenever we multiply a binomial of the form $x + y$ by a binomial of the form $x - y$, we obtain another binomial:

$$(x + y)(x - y) = x^2 - y^2$$

The Difference of Two Squares

The binomial $x^2 - y^2$ is called the **difference of two squares** because x^2 is the square of x and y^2 is the square of y.

The difference of the squares of two quantities such as x and y always factors into the sum of those two quantities multiplied by the difference of those two quantities.

> **Factoring the Difference of Two Squares:**
>
> $x^2 - y^2 = (x + y)(x - y)$

To factor $x^2 - 9$, we note that $x^2 - 9$ can be written in the form $x^2 - 3^2$ and that $x^2 - 3^2$ is the difference of the squares of x and 3. Thus, it factors into the product of (x plus 3) and (x minus 3).

$$x^2 - 9 = x^2 - 3^2$$
$$= (x + 3)(x - 3)$$

We can check this result by verifying that $(x + 3)(x - 3) = x^2 - 9$.

EXAMPLE 1

Factor $4y^4 - 25z^2$.

Solution

Because $4y^4 - 25z^2$ can be written in the form $(2y^2)^2 - (5z)^2$, it represents the difference of the squares of $2y^2$ and $5z$. Thus, it factors into the sum of these two quantities times the difference of these two quantities.

$$4y^4 - 25z^2 = (2y^2)^2 - (5z)^2$$
$$= (2y^2 + 5z)(2y^2 - 5z)$$

Check by multiplication. Work Progress Check 7. ∎

Progress Check 7

Factor $16x^2 - 25y^4$.

We can often factor out a greatest common factor before factoring the difference of two squares. For example, to factor $8x^2 - 32$, we begin by factoring out the greatest common factor of 8 and then factor the resulting difference of two squares.

$$8x^2 - 32 = 8(x^2 - 4) \quad \text{Factor out 8.}$$
$$= 8(x^2 - 2^2) \quad \text{Write 4 as } 2^2.$$
$$= 8(x + 2)(x - 2) \quad \text{Factor the difference of two squares.}$$

We can verify this result by multiplication:

$$8(x + 2)(x - 2) = 8(x^2 - 4) = 8x^2 - 32$$

EXAMPLE 2

Factor $2a^2x^3y - 8b^2xy$.

---

7. $(4x + 5y^2)(4x - 5y^2)$

Progress Check Answers

Solution

$$2a^2x^3y - 8b^2xy = 2xy(a^2x^2 - 4b^2) \quad \text{Factor out } 2xy.$$
$$= 2xy[(ax)^2 - (2b)^2]$$
$$= 2xy(ax + 2b)(ax - 2b) \quad \text{Factor the difference of two squares.}$$

Use multiplication to verify this result. Work Progress Check 8.

Sometimes we must factor a difference of two squares more than once to factor a polynomial completely. For example, $625a^4 - 81b^4$ can be written in the form $(25a^2)^2 - (9b^2)^2$. Thus, it factors as

$$625a^4 - 81b^4 = (25a^2)^2 - (9b^2)^2$$
$$= (25a^2 + 9b^2)(25a^2 - 9b^2)$$

The factor $25a^2 - 9b^2$, however, can be written in the form $(5a)^2 - (3b)^2$ and can be factored as $(5a + 3b)(5a - 3b)$. Thus, the complete factorization is

$$625a^4 - 81b^4 = (25a^2 + 9b^2)(5a + 3b)(5a - 3b)$$

The Sum of Two Squares

The binomial $25a^2 + 9b^2$ is called the **sum of two squares** because it can be written in the form $(5a)^2 + (3b)^2$. Such binomials cannot be factored if we are limited to integral coefficients. Polynomials that do not factor over the integers are called **prime polynomials**.

EXAMPLE 3

Factor $2x^4y - 32y$.

Solution

$$2x^4y - 32y = 2y \cdot x^4 - 2y \cdot 16$$
$$= 2y(x^4 - 16) \quad \text{Factor out } 2y.$$
$$= 2y(x^2 + 4)(x^2 - 4) \quad \text{Factor } x^4 - 16.$$
$$= 2y(x^2 + 4)(x + 2)(x - 2) \quad \text{Factor } x^2 - 4. \text{ Note that } x^2 + 4 \text{ does not factor.}$$

Work Progress Check 9.

EXAMPLE 4

Factor $2x^3 - 8x + 2yx^2 - 8y$.

Solution

$$2x^3 - 8x + 2yx^2 - 8y$$
$$= 2(x^3 - 4x + yx^2 - 4y) \quad \text{Factor out 2.}$$
$$= 2[x(x^2 - 4) + y(x^2 - 4)] \quad \text{Factor out } x \text{ from } x^3 - 4x \text{ and factor out } y \text{ from } yx^2 - 4y.$$
$$= 2[(x^2 - 4)(x + y)] \quad \text{Factor out } x^2 - 4.$$
$$= 2(x + 2)(x - 2)(x + y) \quad \text{Factor } x^2 - 4.$$

Check by multiplication.

Progress Check 8

Factor $16x^2y^2 - 36b^2$.

Progress Check 9

Factor $2x^4 - 162$.

8. $4(2xy + 3b)(2xy - 3b)$

9. $2(x^2 + 9)(x + 3)(x - 3)$

11.3 Exercises

In Exercises 1–12, finish each factorization.

1. $x^2 - 9$, $(x + 3)($ $)$
2. $a^2 - 64$, $(a + 8)($ $)$
3. $t^2 - 1$, $(t + 1)($ $)$
4. $p^2 - 100$, $(p + 10)($ $)$
5. $r^2 - 4$, $($ $)(r - 2)$
6. $s^2 - 121$, $($ $)(s - 11)$
7. $q^2 - 144$, $(q + 12)($ $)$
8. $t^2 - 169$, $(t + 13)($ $)$
9. $x^2 - 400$, $($ $)(x - 20)$
10. $x^2 - 196$, $($ $)(x - 14)$
11. $a^2 - 625$, $($ $)(a - 25)$
12. $b^2 - 256$, $($ $)(b - 16)$

In Exercises 13–30, factor each expression, if possible. If a polynomial is prime, so indicate.

13. $x^2 - 16$
14. $x^2 - 25$
15. $y^2 - 49$
16. $y^2 - 81$
17. $4y^2 - 49$
18. $9z^2 - 4$
19. $9x^2 - y^2$
20. $4x^2 - z^2$
21. $25t^2 - 36u^2$
22. $49u^2 - 64v^2$
23. $16a^2 - 25b^2$
24. $36a^2 - 121b^2$
25. $a^2 + b^2$
26. $121a^2 - 144b^2$
27. $a^4 + 4b^2$
28. $9y^2 + 16z^2$
29. $49y^2 - 225z^4$
30. $196x^4 - 169y^2$

In Exercises 31–46, factor each expression completely. Factor out any common monomial factors first.

31. $8x^2 - 32y^2$
32. $2a^2 - 200b^2$
33. $2a^2 - 8y^2$
34. $32x^2 - 8y^2$
35. $3r^2 - 12s^2$
36. $45u^2 - 20v^2$
37. $x^3 - xy^2$
38. $a^2b - b^3$
39. $4a^2x - 9b^2x$
40. $4b^2y - 16c^2y$
41. $3m^3 - 3mn^2$
42. $2p^2q - 2q^3$
43. $4x^4 - x^2y^2$
44. $9xy^2 - 4xy^4$
45. $2a^3b - 242ab^3$
46. $50c^4d^2 - 8c^2d^4$

In Exercises 47–58, factor each expression completely.

47. $x^4 - 81$
48. $y^4 - 625$
49. $a^4 - 16$
50. $b^4 - 256$
51. $a^4 - b^4$
52. $m^4 - 16n^4$
53. $81r^4 - 256s^4$
54. $x^8 - y^4$
55. $a^4 - b^8$
56. $16y^8 - 81z^4$
57. $x^8 - y^8$
58. $x^8y^8 - 1$

In Exercises 59–78, factor each expression completely.

59. $2x^4 - 2y^4$
60. $a^5 - ab^4$
61. $a^4b - b^5$
62. $m^5 - 16mn^4$
63. $48m^4n - 243n^5$
64. $2x^4y - 512y^5$
65. $3a^5y + 6ay^5$
66. $2p^{10}q - 32p^2q^5$
67. $3a^{10} - 3a^2b^4$
68. $2x^9y + 2xy^9$
69. $2x^8y^2 - 32y^6$
70. $3a^8 - 243a^4b^8$

71. $a^6b^2 - a^2b^6c^4$
72. $a^2b^3c^4 - a^2b^3d^4$
73. $a^2b^7 - 625a^2b^3$
74. $16x^3y^4z - 81x^3y^4z^5$
75. $243r^5s - 48rs^5$
76. $1024m^5n - 324mn^5$
77. $16(x - y)^2 - 9$
78. $9(x + 1)^2 - y^2$

In Exercises 79–88, factor each expression completely.

79. $a^3 - 9a + 3a^2 - 27$
80. $b^3 - 25b - 2b^2 + 50$
81. $y^3 - 16y - 3y^2 + 48$
82. $a^3 - 49a + 2a^2 - 98$
83. $3x^3 - 12x + 3x^2 - 12$
84. $2x^3 - 18x - 6x^2 + 54$
85. $3m^3 - 3mn^2 + 3am^2 - 3an^2$
86. $ax^3 - axy^2 - bx^3 + bxy^2$
87. $2m^3n^2 - 32mn^2 + 8m^2 - 128$
88. $2x^3y + 4x^2y - 98xy - 196y$

Review Exercises

In Review Exercises 1–10, multiply the binomials.

1. $(x + 6)(x + 6)$
2. $(y - 7)(y - 7)$
3. $(a - 3)(a - 3)$
4. $(r + 8)(r + 8)$
5. $(x + 4y)(x + 5y)$
6. $(r - 2s)(r - 3s)$
7. $(m + 3n)(m - 2n)$
8. $(a - 3b)(a + 4b)$
9. $(u - 3v)(u - 5v)$
10. $(x + 4y)(x - 6y)$

11. In the study of the flow of fluids, Bernoulli's Law is given by

$$\frac{p}{w} + \frac{v^2}{2g} + h = k$$

Solve the equation for p.

12. Solve Bernoulli's Law for h. (See Review Exercise 11.)

11.4 Factoring Trinomials with Lead Coefficients of 1

- Factoring Perfect Square Trinomials
- Factoring Nonperfect Square Trinomials
- Prime Polynomials

The product of two binomials is often a trinomial. For example,

$$(x + 3)(x + 3) = x^2 + 6x + 9$$
$$(x - 4y)(x - 4y) = x^2 - 8xy + 16y^2$$

and

$$(3x - 4)(2x + 3) = 6x^2 + x - 12$$

Because the product of two binomials can be a trinomial, we should not be surprised that many trinomials factor into the product of two binomials.

Factoring Perfect Square Trinomials

Many trinomials can be factored by using the following two special product formulas, first discussed in Section 10.6.

$$(x + y)(x + y) = x^2 + 2xy + y^2$$
$$(x - y)(x - y) = x^2 - 2xy + y^2$$

The trinomials $x^2 + 2xy + y^2$ and $x^2 - 2xy + y^2$ are called **perfect square trinomials**, because each one can be written as the square of a binomial.

1. $x^2 + 2xy + y^2 = (x + y)(x + y) = (x + y)^2$
2. $x^2 - 2xy + y^2 = (x - y)(x - y) = (x - y)^2$

To factor a perfect square trinomial such as $x^2 + 8x + 16$, we note that the trinomial can be written in the form

$$x^2 + 2(x)(4) + 4^2$$

and that if $y = 4$, this form matches the left-hand side of Equation 1. Thus,

$$x^2 + 8x + 16 = x^2 + 2(x)(4) + 4^2$$
$$= (x + 4)(x + 4)$$

This result can be verified by multiplication by using the **FOIL** method.

$$(x + 4)(x + 4) = x^2 + 4x + 4x + 16$$
$$= x^2 + 8x + 16 \qquad \text{Combine like terms.}$$

Likewise, the perfect square trinomial $a^2 - 4ab + 4b^2$ can be written in the form

$$a^2 - 2(a)(2b) + (2b)^2$$

If $x = a$ and $y = 2b$, this form matches the left-hand side of Equation 2. Thus,

$$a^2 - 4ab + 4b^2 = a^2 - 2(a)(2b) + (2b)^2$$
$$= (a - 2b)(a - 2b)$$

This result can also be verified by multiplication.

Factoring Nonperfect Square Trinomials

Because the trinomial $x^2 + 5x + 6$ is not a perfect square trinomial, it cannot be factored by using a special product formula. However, it can be factored into the product of two binomials. To find those binomial factors, we note that the product of their first terms must be x^2. Thus, the first term of each binomial must be x.

$$\overbrace{(x \quad)(x \quad)}^{x^2}$$

Because the product of their last terms must be 6, and the sum of the products of the outer and inner terms must be $5x$, we must find two numbers whose product is 6 and whose sum is 5.

$$\overbrace{(x + ?)(x + ?)}^{6}$$
$$O + I = 5x$$

Two such numbers are $+3$ and $+2$. Thus, we have

3. $x^2 + 5x + 6 = (x + 3)(x + 2)$

This factorization can be verified by multiplying $x + 3$ and $x + 2$.

$$(x + 3)(x + 2) = x^2 + 2x + 3x + 6$$
$$= x^2 + 5x + 6 \quad \text{Combine like terms.}$$

Because of the commutative property of multiplication, the order of the factors listed in Equation 3 is not important. Equation 3 can be written as

$$x^2 + 5x + 6 = (x + 2)(x + 3)$$

EXAMPLE 1

Factor $y^2 - 7y + 12$.

Solution

If this trinomial is to be the product of two binomials, the product of their first terms must be y^2. Thus, the first term of each binomial must be y.

$$\overbrace{(y \quad)(y \quad)}^{y^2}$$

Because the product of the last terms must be $+12$, and the sum of the products of the outer and inner terms must be $-7y$, we must find two negative numbers whose product is $+12$ and whose sum is -7.

$$\overbrace{(y - ?)(y - ?)}^{12}$$
$$O + I = -7y$$

Two such numbers are -4 and -3. Hence,

$$y^2 - 7y + 12 = (y - 4)(y - 3)$$

Check by verifying that the product of $y - 4$ and $y - 3$ is $y^2 - 7y + 12$. Work Progress Check 10.

Progress Check 10

Factor $x^2 - 6x + 8$.

EXAMPLE 2

Factor $a^2 + 2a - 15$.

Solution

Because the first term is a^2, the first term of each binomial factor must be a.

$$\overbrace{(a \quad)(a \quad)}^{a^2}$$

Because the product of the last terms must be -15, and the sum of the products of the outer terms and inner terms must be $+2a$, we must find two numbers whose product is -15 and whose sum is $+2$.

$$\overbrace{(a \quad ?)(a \quad ?)}^{-15}$$
$$O + I = 2a$$

10. $(x - 4)(x - 2)$

Progress Check Answers

Two such numbers are +5 and −3. Hence,
$$a^2 + 2a - 15 = (a + 5)(a - 3)$$
Check by verifying that the product of $a + 5$ and $a - 3$ is $a^2 + 2a - 15$.

EXAMPLE 3

Factor $z^2 - 4z - 21$.

Solution

Because the first term is z^2, the first term of each binomial factor must be z.

$$\overbrace{(z\quad)(z\quad)}^{z^2}$$

Because the product of the last terms must be −21, and the sum of the products of the outer terms and inner terms must be −4z, we must find two numbers whose product is −21 and whose sum is −4.

$$\overbrace{(z + ?)(z + ?)}^{-21}$$
$$O + I = -4z$$

Two such numbers are −7 and +3. Hence,
$$z^2 - 4z - 21 = (z - 7)(z + 3)$$
Check by verifying that the product of $z - 7$ and $z + 3$ is $z^2 - 4z - 21$. Work Progress Check 11.

Factor $x^2 - 4x - 12$.

EXAMPLE 4

Factor $x^2 + xy - 6y^2$.

Solution

Because the first term is x^2, the first term of each binomial factor must be x.

$$\overbrace{(x\quad)(x\quad)}^{x^2}$$

Because the product of the last terms must be $-6y^2$, and the sum of the products of the outer terms and inner terms must be xy, we must find two numbers whose product is $-6y^2$ that will give a middle term of xy.

Factor $x^2 - xy - 2y^2$.

$$\overbrace{(x + ?)(x + ?)}^{-6y^2}$$
$$O + I = xy$$

Two such numbers are $3y$ and $-2y$. Hence,
$$x^2 + xy - 6y^2 = (x + 3y)(x - 2y)$$

11. $(x + 2)(x - 6)$

12. $(x + y)(x - 2y)$

Check by verifying that the product of $x + 3y$ and $x - 2y$ is $x^2 + xy - 6y^2$. Work Progress Check 12.

If the coefficient of the first term of a trinomial is -1, begin by factoring out -1.

EXAMPLE 5

Factor $-x^2 + 11x - 18$.

Solution

$$-x^2 + 11x - 18 = -(x^2 - 11x + 18) \quad \text{Factor out } -1.$$
$$= -(x - 9)(x - 2) \quad \text{Factor } x^2 - 11x + 18.$$

Check by verifying that the product of -1, $x - 9$, and $x - 2$ is $-x^2 + 11x - 18$. ∎

EXAMPLE 6

Factor $-x^2 + 2x + 15$.

Solution

$$-x^2 + 2x + 15 = -(x^2 - 2x - 15) \quad \text{Factor out } -1.$$
$$= -(x - 5)(x + 3) \quad \text{Factor } x^2 - 2x - 15.$$

Check by verifying that the product of -1, $x - 5$, and $x + 3$ is $-x^2 + 2x + 15$. Work Progress Check 13. ∎

Progress Check 13

Factor $-x^2 + xy + 2y^2$.

Prime Polynomials

Not all trinomials are factorable. To attempt to factor the trinomial $x^2 + 2x + 3$, for example, we would begin by noting that the product of the first terms of the binomial factors is x^2. Thus, the first term of each binomial must be x.

$$\overbrace{(x \quad)(x \quad)}^{x^2}$$

Because the last term of the trinomial is 3 and the middle term is $2x$, we must find two factors of 3 whose sum is 2 so that

$$\overbrace{(x + ?)(x + ?)}^{3}$$
$$\underbrace{}_{O + I = 2x}$$

Because 3 factors only as $(1)(3)$ and $(-1)(-3)$, it has no factors whose sum is 2. Thus, $x^2 + 2x + 3$ cannot be factored. It is a prime polynomial.

EXAMPLE 7

Factor $-3ax^2 + 9a - 6ax$.

---

13. $-(x + y)(x - 2y)$

Progress Check Answers

Solution

Write the trinomial in descending powers of x and factor out the common factor of $-3a$.

$$-3ax^2 + 9a - 6ax = -3ax^2 - 6ax + 9a$$
$$= -3a(x^2 + 2x - 3)$$

Finally, factor the trinomial $x^2 + 2x - 3$.

$$-3ax^2 + 9a - 6ax = -3a(x + 3)(x - 1)$$ ∎

The next example requires factoring a trinomial and factoring a difference of two squares.

EXAMPLE 8

Factor $m^2 - 2mn + n^2 - 64a^2$.

Solution

Group the first three terms together and factor the resulting trinomial to obtain

$$m^2 - 2mn + n^2 - 64a^2 = (m - n)(m - n) - 64a^2$$
$$= (m - n)^2 - (8a)^2$$

Then factor the resulting difference of two squares:

$$m^2 - 2mn + n^2 - 64a^2 = (m - n)^2 - 64a^2$$
$$= (m - n + 8a)(m - n - 8a)$$ ∎

EXAMPLE 9

Factor $x^2 + 7x + 12 + xy + 4y$.

Solution

$$x^2 + 7x + 12 + xy + 4y = (x + 4)(x + 3) + y(x + 4)$$
$$= (x + 4)[(x + 3) + y]$$
$$= (x + 4)(x + 3 + y)$$ ∎

11.4 Exercises

In Exercises 1–12, finish each factorization.

1. $x^2 + 4x + 4$, $(x + 2)($ $)$

2. $x^2 + 8x + 16$, $(x + 4)($ $)$

3. $y^2 - 12y + 36$, $($ $)(y - 6)$

4. $y^2 - 16y + 64$, $($ $)(y - 8)$

5. $a^2 + 5a + 6$, $(a + 3)($ $)$

6. $b^2 + 7b + 12$, $(b + 3)($ $)$

7. $p^2 - 7p + 10$, $($ $)(p - 2)$

8. $q^2 - 7q + 6$, $($ $)(q - 1)$

9. $t^2 + t - 6$, $(t + 3)($ $)$

10. $r^2 - 2r - 8$, $(r - 4)($ $)$

11. $x^2 + 2x - 15$, $($ $)(x - 3)$

12. $a^2 - 4a - 21$, $($ $)(a + 3)$

In Exercises 13–24, factor each perfect square trinomial.

13. $x^2 + 6x + 9$ **14.** $x^2 + 10x + 25$ **15.** $y^2 - 8y + 16$ **16.** $z^2 - 2z + 1$

17. $t^2 + 20t + 100$ **18.** $r^2 + 24r + 144$ **19.** $u^2 - 18u + 81$ **20.** $v^2 - 14v + 49$

21. $x^2 + 4xy + 4y^2$ **22.** $a^2 + 6ab + 9b^2$ **23.** $r^2 - 10rs + 25s^2$ **24.** $m^2 - 12mn + 36n^2$

In Exercises 25–52, factor each trinomial, if possible. If the trinomial is prime, so indicate. Use the FOIL method to check each result.

25. $x^2 + 3x + 2$ **26.** $y^2 + 4y + 3$ **27.** $a^2 - 4a - 5$ **28.** $b^2 + 6b - 7$

29. $z^2 + 12z + 11$ **30.** $x^2 + 7x + 10$ **31.** $t^2 - 9t + 14$ **32.** $c^2 - 9c + 8$

33. $u^2 + 10u + 15$ **34.** $v^2 + 9v + 15$ **35.** $y^2 - y - 30$ **36.** $x^2 - 3x - 40$

37. $a^2 + 6a - 16$ **38.** $x^2 + 5x - 24$ **39.** $t^2 - 5t - 50$ **40.** $a^2 - 10a - 39$

41. $r^2 - 9r - 12$ **42.** $s^2 + 11s - 26$ **43.** $y^2 + 2yz + z^2$ **44.** $r^2 - 2rs + 4s^2$

45. $x^2 + 4xy + 4y^2$ **46.** $a^2 + 10ab + 9b^2$ **47.** $m^2 + 3mn - 10n^2$ **48.** $m^2 - mn - 12n^2$

49. $a^2 - 4ab - 12b^2$ **50.** $p^2 + pq - 6q^2$ **51.** $u^2 + 2uv - 15v^2$ **52.** $m^2 + 3mn - 10n^2$

In Exercises 53–64, factor each trinomial. Factor out -1 first.

53. $-x^2 - 7x - 10$ **54.** $-x^2 + 9x - 20$ **55.** $-y^2 - 2y + 15$ **56.** $-y^2 - 3y + 18$

57. $-t^2 - 15t + 34$ **58.** $-t^2 - t + 30$ **59.** $-r^2 + 14r - 40$ **60.** $-r^2 + 14r - 45$

61. $-a^2 - 4ab - 3b^2$ **62.** $-a^2 - 6ab - 5b^2$ **63.** $-x^2 + 6xy + 7y^2$ **64.** $-x^2 - 10xy + 11y^2$

In Exercises 65–76, write each trinomial in descending powers of one variable and then factor.

65. $4 - 5x + x^2$ **66.** $y^2 + 5 + 6y$ **67.** $10y + 9 + y^2$ **68.** $x^2 - 13 - 12x$

69. $c^2 - 5 + 4c$ **70.** $b^2 - 6 - 5b$ **71.** $-r^2 + 2s^2 + rs$ **72.** $u^2 - 3v^2 + 2uv$

73. $4rx + r^2 + 3x^2$ **74.** $-a^2 + 5b^2 + 4ab$ **75.** $-3ab + a^2 + 2b^2$ **76.** $-13yz + y^2 - 14z^2$

In Exercises 77–88, completely factor each trinomial. Factor out any common monomials first (including -1 if necessary).

77. $2x^2 + 10x + 12$ **78.** $3y^2 - 21y + 18$

79. $3y^3 + 6y^2 + 3y$ **80.** $4x^4 + 16x^3 + 16x^2$

81. $-5a^2 + 25a - 30$ **82.** $-2b^2 + 20b - 18$

83. $3z^2 - 15tz + 12t^2$ **84.** $5m^2 + 45mn - 50n^2$

85. $12xy + 4x^2y - 72y$ **86.** $48xy + 6xy^2 + 96x$

87. $-4x^2y - 4x^3 + 24xy^2$ **88.** $3x^2y^3 + 3x^3y^2 - 6xy^4$

In Exercises 89–96, completely factor each expression.

89. $ax^2 + 4ax + 4a + bx + 2b$

90. $mx^2 + mx - 6m + nx - 2n$

91. $a^2 + 8a + 15 + ab + 5b$

92. $x^2 + 2xy + y^2 + 2x + 2y$

93. $a^2 + 2ab + b^2 - 4$

94. $a^2 + 6a + 9 - b^2$

95. $b^2 - y^2 - 4y - 4$

96. $c^2 - a^2 + 8a - 16$

Review Exercises

Multiply the binomials.

1. $(2x + 1)(3x + 2)$
2. $(3y - 2)(2y - 5)$
3. $(4t - 3)(2t + 3)$
4. $(3r + 5)(2r - 5)$

5. $(2m - 3n)(3m - 2n)$
6. $(4a + 3b)(4a + b)$
7. $(4u - 3v)(5u + 2v)$
8. $(5c + 2d)(5c - 3d)$

9. $(5x^2 + 2y)(3x^2 + y)$
10. $(a - 2b^2)(a + 3b^2)$
11. $(3x^2 + 5y)(x^2 - 3y)$
12. $(z + 3t^2)(z + 3t^2)$

11.5 Factoring General Trinomials

- The Trial-and-Error Method
- Using Grouping to Factor Trinomials

The Trial-and-Error Method

There are more combinations of factors to consider when we factor trinomials with lead coefficients other than 1. To factor $2x^2 - 7x + 3$, for example, we must find binomials of the form $ax + b$ and $cx + d$ such that

$$2x^2 - 7x + 3 = (ax + b)(cx + d)$$

Because the first term of $2x^2 - 7x + 3$ is $2x^2$, the first terms of the binomial factors must be $2x$ and x.

$$2x^2$$
$$(2x \quad ?)(x \quad ?)$$

Because the product of the last terms is 3 and the sum of the products of the outer terms and inner terms is $-7x$, we must find two numbers with a product of 3 that will give a middle term of $-7x$.

$$+3$$
$$(2x \quad ?)(x \quad ?)$$
$$O + I = -7x$$

Because both $(3)(1)$ and $(-3)(-1)$ give a product of 3, there are four possible combinations to consider:

$(2x + 3)(x + 1) \quad (2x + 1)(x + 3)$
$(2x - 3)(x - 1) \quad (2x - 1)(x - 3)$

Of these possibilities, only the last one gives the required middle term of $-7x$:

$$2x^2 - 7x + 3 = (2x - 1)(x - 3)$$

We can check by multiplication:
$$(2x - 1)(x - 3) = 2x^2 - 6x - x + 3$$
$$= 2x^2 - 7x + 3$$

EXAMPLE 1

Factor $3y^2 - 4y - 4$.

Solution

Because the first term is $3y^2$, the first terms of the binomial factors must be $3y$ and y.

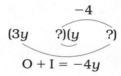

The product of the last terms must be -4, and the sum of the products of the outer terms and inner terms must be $-4y$.

$$(3y \quad ?)(y \quad ?)$$
$$O + I = -4y$$

Because $(1)(-4)$, $(-1)(4)$, and $(-2)(2)$ all give a product of -4, there are six possible combinations to consider:

$(3y + 1)(y - 4)$ $(3y - 4)(y + 1)$
$(3y - 1)(y + 4)$ $(3y + 4)(y - 1)$
$(3y - 2)(y + 2)$ $(3y + 2)(y - 2)$

Again, only the last possibility gives the required middle term of $-4y$:
$$3y^2 - 4y - 4 = (3y + 2)(y - 2)$$

Check:
$$(3y + 2)(y - 2) = 3y^2 - 6y + 2y - 4$$
$$= 3y^2 - 4y - 4$$

Work Progress Check 14.

Progress Check 14

Factor $2x^2 - 3x - 2$.

EXAMPLE 2

Factor $2x^2 + 7xy + 6y^2$.

Solution

The first terms of the two binomial factors must be $2x$ and x.

$$(2x \quad)(x \quad)$$

The product of the last terms must be $+6y^2$, and the sum of the products of the outer terms and inner terms must be $+7xy$.

$$(2x \quad ?)(x \quad ?)$$
$$O + I = 7xy$$

14. $(2x + 1)(x - 2)$

The products $(6y)(y)$, $(3y)(2y)$, $(-6y)(-y)$, and $(-3y)(-2y)$ all give a last term of $6y^2$. However, only the products $(6y)(y)$ and $(3y)(2y)$ can lead to a middle term that is preceded by a + sign. Thus there are only four possibilities to consider:

$(2x + 6y)(x + y)$ $(2x + y)(x + 6y)$
$(2x + 3y)(x + 2y)$ $(2x + 2y)(x + 3y)$

Of these possibilities, only $(2x + 3y)(x + 2y)$ gives the correct middle term of $7xy$:

$$2x^2 + 7xy + 6y^2 = (2x + 3y)(x + 2y)$$

Check:

$$(2x + 3y)(x + 2y) = 2x^2 + 4xy + 3xy + 6y^2$$
$$= 2x^2 + 7xy + 6y^2 \blacksquare$$

EXAMPLE 3

Factor $6b^2 + 7b - 20$.

Solution

This time there are many possible combinations for the first terms of the binomial factors. They are

$(b\quad)(6b\quad)$ $(6b\quad)(b\quad)$
$(3b\quad)(2b\quad)$ $(2b\quad)(3b\quad)$

There are also many combinations for the last terms of the binomial factors. We must try to find one that will (in combination with our choice of first terms) give a last term of -20 and a sum of the products of the outer terms and inner terms of $+7b$.

Begin, for example, by picking factors of b and $6b$ for the first terms and $+4$ and -5 for the last terms. The possible factorization

$(b + 4)(6b - 5)$

$O + I = 19b$

gives a middle term of $19b$, so it is incorrect.

Then try, for example, factors of $3b$ and $2b$ for the first terms and $+4$ and -5 for the last terms. The possible factorization

$(3b + 4)(2b - 5)$

$O + I = -7b$

gives a middle term of $-7b$, so it is incorrect.

The possible factorization

$(3b - 4)(2b + 5)$

does give a middle term of $+7b$ and a last term of -20, so it is correct.

$$6b^2 + 7b - 20 = (3b - 4)(2b + 5)$$

Check:

$$(3b - 4)(2b + 5) = 6b^2 + 15b - 8b - 20$$
$$= 6b^2 + 7b - 20 \blacksquare$$

EXAMPLE 4

Factor $4x^2 + 4xy - 3y^2$.

Solution

Again, there are many combinations for the first terms of the binomial factors:

$(4x\quad)(x\quad)\qquad(x\quad)(4x\quad)\qquad(2x\quad)(2x\quad)$

We must find last terms that will give a third term of $-3y^2$ and a middle term of $+4xy$. We begin by trying, for example, factors of $4x$ and x for the first terms and factors of $3y$ and $-y$ for the last terms. The possible factorization

$(4x + 3y)(x - y)$

$O + I = -xy$

gives a middle term of $-xy$, so it is incorrect.

Then try, for example, factors of $2x$ and $2x$ for the first terms and factors of $3y$ and $-y$ for the last terms. The possible factorization

$(2x + 3y)(2x - y)$

$O + I = +4xy$

gives a middle term of $+4xy$ and a last term of $-3y^2$, so it is correct:

$4x^2 + 4xy - 3y^2 = (2x + 3y)(2x - y)$

Check:

$(2x + 3y)(2x - y) = 4x^2 - 2xy + 6xy - 3y^2$
$= 4x^2 + 4xy - 3y^2$

Work Progress Check 15.

Progress Check 15

Factor $6b^2 + 5bc - 6c^2$.

EXAMPLE 5

Factor $-8x^3 + 22x^2 - 12x$.

Solution

$-8x^3 + 22x^2 - 12x = -2x(4x^2 - 11x + 6)$ Factor out the common factor of $-2x$.

$\qquad\qquad\qquad\qquad\quad = -2x(x - 2)(4x - 3)$ Factor $4x^2 - 11x + 6$.

Check:

$-2x(x - 2)(4x - 3) = -2x(4x^2 - 3x - 8x + 6)$
$= -2x(4x^2 - 11x + 6)$
$= -8x^3 + 22x^2 - 12x$

---

15. $(2b + 3c)(3b - 2c)$

Progress Check Answers

The following hints are often helpful when factoring a trinomial:

> **To factor a general trinomial, follow these steps:**
> 1. Write the trinomial in descending powers of one variable.
> 2. Factor out any greatest common factor (including -1, if necessary, to make the coefficient of the first term positive).
> 3. If the sign of the first term of the trinomial is $+$ and the sign of the third term is $+$, the signs between the terms of the binomial factors are the same as the sign of the middle term of the trinomial. If the sign of the third term is $-$, the signs between the terms of the binomial factors are opposite.
> 4. Try various combinations of first terms and last terms until one works or until you have exhausted all the possibilities. In that case, the trinomial is prime.
> 5. Check the factorization by multiplication.

EXAMPLE 6

Factor $2x^2y - 8x^3 + 3xy^2$.

Solution

Step 1

Rewrite the trinomial in descending powers of x.

$$-8x^3 + 2x^2y + 3xy^2$$

Step 2

Factor out $-x$.

$$-8x^3 + 2x^2y + 3xy^2 = -x(8x^2 - 2xy - 3y^2)$$

Step 3

Because the sign of the third term of the trinomial factor is $-$, the signs within its binomial factors must be different. Thus, the sign between the terms in one binomial must be $+$, and the sign between the terms of the other binomial must be $-$.

Step 4

Find the binomial factors of the trinomial.

$$-8x^3 + 2x^2y + 3xy^2 = -x(8x^2 - 2xy - 3y^2)$$
$$= -x(2x + y)(4x - 3y)$$

Step 5

Check:
$$-x(2x + y)(4x - 3y) = -x(8x^2 - 6xy + 4xy - 3y^2)$$
$$= -x(8x^2 - 2xy - 3y^2)$$
$$= -8x^3 + 2x^2y + 3xy^2$$
$$= 2x^2y - 8x^3 + 3xy^2$$

EXAMPLE 7

Factor $4x^2 - 4xy + y^2 - 9$.

Solution

$4x^2 - 4xy + y^2 - 9$
$= (2x - y)^2 - 9$ Factor the first three terms as a perfect square.
$= [(2x - y) + 3][(2x - y) - 3]$ Factor the difference of two squares.
$= (2x - y + 3)(2x - y - 3)$ Remove parentheses.

Check by multiplication.

Using Grouping to Factor Trinomials

We can use grouping to help factor trinomials of the form $ax^2 + bx + c$, where $c \neq 0$. For example, to factor $4x^2 - 4x - 3$, where $a = 4$, $b = -4$, and $c = -3$, we proceed as follows:

1. Find the product of a and c: $ac = 4(-3) = -12$.
2. Find two numbers whose product is -12 and whose sum is -4, the value of b. Two such numbers are 2 and -6:

$$2(-6) = -12 \quad \text{and} \quad 2 + (-6) = -4$$

3. Use 2 and -6 as coefficients of terms in the variable x and place them between $4x^3$ and -3:

$$4x^2 + 2x - 6x - 3$$

4. Factor by grouping:

$$4x^2 + 2x - 6x - 3 = 2x(2x + 1) - 3(2x + 1)$$
$$= (2x + 1)(2x - 3)$$

EXAMPLE 8

Factor $10x^2 + 13xy - 3y^2$.

Solution

In this example, the product of a and c is $ac = 10(-3) = -30$. Thus, we need two numbers whose product is -30 and whose sum is $b = 13$. Since two such numbers are -2 and 15, we place terms of $-2xy$ and $15xy$ between $10x^2$ and $-3y^2$ and factor by grouping.

$10x^2 + 13xy - 3y^2 = 10x^2 - 2xy + 15xy - 3y^2$
$= 2x(5x - y) + 3y(5x - y)$
$= (5x - y)(2x + 3y)$

11.5 Exercises

In Exercises 1–12, finish each factorization.

1. $2x^2 + 5x + 2$, $(2x + 1)($)
2. $3x^2 + 5x + 2$, $(3x + 2)($)
3. $3a^2 - 5a + 2$, $(3a - 2)($)
4. $2a^2 - 5a + 3$, $(2a - 3)($)
5. $4y^2 - 7y - 2$, $($ $)(y - 2)$
6. $5y^2 + 3y - 2$, $($ $)(y + 1)$
7. $4t^2 - 4t + 1$, $(2t - 1)($)
8. $6t^2 - 7t + 2$, $(3t - 2)($)
9. $10z^2 - 13z - 3$, $($ $)(2z - 3)$
10. $10z^2 + 17z + 3$, $($ $)(2z + 3)$
11. $12p^2 + 4p - 1$, $(6p - 1)($)
12. $14p^2 - 15p - 9$, $(7p + 3)($)

In Exercises 13–36, factor each trinomial. Check each result.

13. $2x^2 - 3x + 1$
14. $2y^2 - 7y + 3$
15. $3a^2 + 13a + 4$
16. $2b^2 + 7b + 6$
17. $4z^2 + 13z + 3$
18. $4t^2 - 4t + 1$
19. $6y^2 + 7y + 2$
20. $4x^2 + 8x + 3$
21. $6x^2 - 7x + 2$
22. $4z^2 - 9z + 2$
23. $3a^2 - 4a - 4$
24. $8u^2 - 2u - 15$
25. $2x^2 - 3x - 2$
26. $12y^2 - y - 1$
27. $2m^2 + 5m - 12$
28. $10u^2 - 13u - 3$
29. $10y^2 - 3y - 1$
30. $6m^2 + 19m + 3$
31. $12y^2 - 5y - 2$
32. $10x^2 + 21x - 10$
33. $5t^2 + 13t + 6$
34. $16y^2 + 10y + 1$
35. $16m^2 - 14m + 3$
36. $16x^2 + 16x + 3$

In Exercises 37–48, factor each trinomial.

37. $3x^2 - 4xy + y^2$
38. $2x^2 + 3xy + y^2$
39. $2u^2 + uv - 3v^2$
40. $2u^2 + 3uv - 2v^2$
41. $4a^2 - 4ab + b^2$
42. $2b^2 - 5bc + 2c^2$
43. $6r^2 + rs - 2s^2$
44. $3m^2 + 5mn + 2n^2$
45. $4x^2 + 8xy + 3y^2$
46. $4b^2 + 15bc - 4c^2$
47. $4a^2 - 15ab + 9b^2$
48. $12x^2 + 5xy - 3y^2$

In Exercises 49–68, write the terms of each trinomial in descending powers of one variable. Then factor the trinomial, if possible. If a trinomial cannot be factored, so indicate.

49. $-13x + 3x^2 - 10$
50. $-14 + 3a^2 - a$
51. $15 + 8a^2 - 26a$
52. $16 - 40a + 25a^2$
53. $12y^2 + 12 - 25y$
54. $12t^2 - 1 - 4t$
55. $3x^2 + 6 + x$
56. $25 + 2u^2 + 3u$
57. $2a^2 + 3b^2 + 5ab$
58. $11uv + 3u^2 + 6v^2$
59. $pq + 6p^2 - q^2$
60. $-11mn + 12m^2 + 2n^2$
61. $b^2 + 4a^2 + 16ab$
62. $3b^2 + 3a^2 - ab$
63. $12x^2 + 10y^2 - 23xy$
64. $5ab + 25a^2 - 2b^2$
65. $-19xy + 6x^2 + 15y^2$
66. $35r^2 - 6s^2 + rs$
67. $25a^2 - 16b^2 + 30ab$
68. $-10uv + 8u^2 - 7v^2$

In Exercises 69–92, factor completely. Remember to factor out any common factors first.

69. $4x^2 + 10x - 6$ **70.** $9x^2 + 21x - 18$ **71.** $y^3 + 13y^2 + 12y$ **72.** $2xy^2 + 8xy - 24x$

73. $6x^3 - 15x^2 - 9x$ **74.** $9y^3 + 3y^2 - 6y$ **75.** $2m^3 - m^2 - 3m$ **76.** $2a^3 + 8a^2 - 42a$

77. $6a^4 + 14a^3 - 40a^2$ **78.** $6b^5 - b^4 - 12b^3$ **79.** $30r^5 + 63r^4 - 30r^3$ **80.** $6s^5 - 26s^4 - 20s^3$

81. $4a^2 - 4ab - 8b^2$ **82.** $6x^2 + 3xy - 18y^2$ **83.** $8x^2 - 12xy - 8y^2$ **84.** $24a^2 + 14ab + 2b^2$

85. $2x^4y^2 + x^3y^3 - x^2y^4$ **86.** $2a^5b^2 + 7a^4b^3 + 6a^3b^4$

87. $-16m^3n - 20m^2n^2 - 6mn^3$ **88.** $-84x^4 - 100x^3y - 24x^2y^2$

89. $-28u^3v^3 + 26u^2v^4 - 6uv^5$ **90.** $-16x^4y^3 + 30x^3y^4 + 4x^2y^5$

91. $105x^3 - 3x^2 - 36x$ **92.** $30x^4 + 5x^3 - 200x^2$

In Exercises 93–102, factor each expression completely.

93. $4x^2 + 4xy + y^2 - 16$ **94.** $9x^2 - 6x + 1 - d^2$

95. $9 - a^2 - 4ab - 4b^2$ **96.** $25 - 9a^2 + 6ac - c^2$

97. $4x^2 + 4xy + y^2 - a^2 - 2ab - b^2$ **98.** $a^2 - 2ab + b^2 - x^2 + 2x - 1$

99. $2x^2z - 4xyz + 2y^2z - 18z^3$ **100.** $9s - r^2s + 2rs^2 - s^3$

101. $4x^2 + 4xy + y^2 + 6x + 3y$ **102.** $25 - y^2 - 2y^2 + 9y + 5$

In Exercises 103–112, use factoring by grouping to factor each trinomial.

103. $x^2 + 9x + 20$ **104.** $y^2 - 8y + 15$ **105.** $2r^2 + 9r + 10$ **106.** $2v^2 + 5v - 12$

107. $6x^2 - 7x - 5$ **108.** $2y^2 - 5y - 12$ **109.** $12t^2 + 13t - 4$ **110.** $2m^2 + 7m - 15$

111. $2x^2 - xy - 6y^2$ **112.** $2r^2 - 5rs - 3s^2$

Review Exercises

In Review Exercises 1–10, find each product.

1. $(x - 3)(x^2 + 3x + 9)$ **2.** $(x + 2)(x^2 - 2x + 4)$

3. $(y + 4)(y^2 - 4y + 16)$ **4.** $(r - 5)(r^2 + 5r + 25)$

5. $(a - b)(a^2 + ab + b^2)$ **6.** $(a + b)(a^2 - ab + b^2)$

7. $(x + 2y)(x^2 - 2xy + 4y^2)$ **8.** $(x - 2y)(x^2 + 2xy + 4y^2)$

9. $(r + s)(r^2 - rs + s^2)$ **10.** $(2y - z)(4y^2 + 2yz + z^2)$

11. The nth term, l, of an arithmetic progression is

$$l = f + (n - 1)d$$

where f is the first term and d is the common difference. Solve for n.

12. The sum, S, of n consecutive terms of an arithmetic progression is

$$S = \frac{n}{2}(f + l)$$

where f is the first term of the progression and l is the nth term. Solve for f.

11.6 Factoring the Sum and Difference of Two Cubes

■ The Sum and Difference of Two Cubes

The Sum and Difference of Two Cubes

There are formulas for factoring the sum of the cubes of two quantities and the difference of the cubes of two quantities. To discover these formulas, we find the following two products:

$$(x + y)(x^2 - xy + y^2) = (x + y)x^2 - (x + y)xy + (x + y)y^2$$
$$= x^3 + x^2y - x^2y - xy^2 + xy^2 + y^3$$
$$= x^3 + y^3$$

$$(x - y)(x^2 + xy + y^2) = (x - y)x^2 + (x - y)xy + (x - y)y^2$$
$$= x^3 - x^2y + x^2y - xy^2 + xy^2 - y^3$$
$$= x^3 - y^3$$

These results justify the formulas for factoring the **sum and difference of two cubes**.

Factoring the Sum and the Difference of Two Cubes:

$$x^3 + y^3 = (x + y)(x^2 - xy + y^2)$$
$$x^3 - y^3 = (x - y)(x^2 + xy + y^2)$$

The factorization of $x^3 + y^3$ has a first factor of $x + y$. The second factor has three terms: the square of x, the *negative* of the product of x and y, and the square of y.

The factorization of $x^3 - y^3$ has a first factor of $x - y$. The second factor has three terms: the square of x, the product of x and y, and the square of y.

EXAMPLE 1

Factor $x^3 + 8$.

Solution

$x^3 + 8$ is the sum of two cubes: the cube of x and the cube of 2.

$$x^3 + 8 = x^3 + 2^3$$

Hence, $x^3 + 8$ factors as the product of the *sum* of x and 2 and the trinomial $x^2 - 2x + 2^2$.

$$\begin{aligned}x^3 + 8 &= x^3 + 2^3 \\ &= (x + 2)(x^2 - 2x + 2^2) \\ &= (x + 2)(x^2 - 2x + 4)\end{aligned}$$

Check by multiplying:

$$\begin{aligned}(x + 2)(x^2 - 2x + 4) &= (x + 2)x^2 - (x + 2)2x + (x + 2)4 \\ &= x^3 + 2x^2 - 2x^2 - 4x + 4x + 8 \\ &= x^3 + 8\end{aligned}$$

EXAMPLE 2

Factor $a^3 - 64b^3$.

Solution

$a^3 - 64b^3$ is the difference of two cubes: the cube of a and the cube of $4b$.

$$a^3 - 64b^3 = a^3 - (4b)^3$$

Hence, its factors are the difference $a - 4b$ and the trinomial $a^2 + a(4b) + (4b)^2$.

$$\begin{aligned}a^3 - 64b^3 &= a^3 - (4b)^3 \\ &= (a - 4b)[a^2 + a(4b) + (4b)^2] \\ &= (a - 4b)(a^2 + 4ab + 16b^2)\end{aligned}$$

Check:

$$\begin{aligned}(a - 4b)&(a^2 + 4ab + 16b^2) \\ &= (a - 4b)a^2 + (a - 4b)4ab + (a - 4b)16b^2 \\ &= a^3 - 4a^2b + 4a^2b - 16ab^2 + 16ab^2 - 64b^3 \\ &= a^3 - 64b^3\end{aligned}$$

Progress Check 16 Work Progress Check 16.

Factor:

a. $8a^3 + 27$

b. $64x^3 - y^3$

EXAMPLE 3

Factor $-2t^5 + 128t^2$.

Solution

$$\begin{aligned}-2t^5 + 128t^2 &= -2t^2(t^3 - 64) & \text{Factor out } -2t^2. \\ &= -2t^2(t - 4)(t^2 + 4t + 16) & \text{Factor } t^3 - 64.\end{aligned}$$

Check by multiplication.

16. a. $(2x + 3)(4x^2 - 6x + 9)$
    b. $(4x - y)(16x^2 + 4xy + y^2)$

Progress Check Answers

EXAMPLE 4

Factor $x^6 - 64$.

Solution

The binomial $x^6 - 64$ is the difference of two squares and factors into the product of a sum and a difference.

$$x^6 - 64 = (x^3)^2 - 8^2$$
$$= (x^3 + 8)(x^3 - 8)$$

Because $x^3 + 8$ is the sum of two cubes and $x^3 - 8$ is the difference of two cubes, each of these binomials can be factored.

$$x^6 - 64 = (x^3 + 8)(x^3 - 8)$$
$$= (x + 2)(x^2 - 2x + 4)(x - 2)(x^2 + 2x + 4)$$

Check by multiplication. Work Progress Check 17.

Progress Check 17

Factor $x^6 - 7x^3 - 8$.

Progress Check Answers

17. $(x + 1)(x^2 - x + 1)(x - 2)$
    $\cdot (x^2 - 2x + 4)$

11.6 Exercises

In Exercises 1–12, finish each factorization.

1. $x^3 + 27$, $(x + 3)($ $)$
2. $a^3 - 8$, $(a - 2)($ $)$
3. $z^3 - 64$, $(z - 4)($ $)$
4. $b^3 + 1$, $(b + 1)($ $)$
5. $8t^3 + 27$, $(2t + 3)($ $)$
6. $27p^3 - 8$, $(3p - 2)($ $)$
7. $1000x^3 - y^3$, $($ $)(100x^2 + 10xy + y^2)$
8. $343q^3 + 8$, $($ $)(49q^2 - 14q + 4)$
9. $216t^3 + 125$, $(6t + 5)($ $)$
10. $p^6 - q^3$, $(p^2 - q)($ $)$
11. $x^9 - y^6$, $($ $)(x^6 + x^3y^2 + y^4)$
12. $p^3 + 8q^9$, $($ $)(p^2 - 2pq^3 + 4q^6)$

In Exercises 13–32, factor each expression.

13. $y^3 + 1$
14. $x^3 - 8$
15. $a^3 - 27$
16. $b^3 + 125$
17. $8 + x^3$
18. $27 - y^3$
19. $s^3 - t^3$
20. $8u^3 + w^3$
21. $27x^3 + y^3$
22. $x^3 - 27y^3$
23. $a^3 + 8b^3$
24. $27a^3 - b^3$
25. $64x^3 - y^3$
26. $27x^3 + 125y^3$
27. $27x^3 - 125y^3$
28. $64x^3 - 27y^3$
29. $a^6 - b^3$
30. $x^3 + y^6$
31. $x^6 - y^3$
32. $x^3 - y^9$

In Exercises 33–48, factor each expression. Factor out any greatest common factors first.

33. $2x^3 + 54$
34. $2x^3 - 2$
35. $-x^3 + 216$
36. $-x^3 - 125$
37. $64m^3x - 8n^3x$
38. $16r^4 + 128rs^3$
39. $x^4y + 216xy^4$
40. $16a^5 - 54a^2b^3$

41. $81r^4s^2 - 24rs^5$ **42.** $4m^5n + 500m^2n^4$ **43.** $125a^6b^2 + 64a^3b^5$ **44.** $216a^4b^4 - 1000ab^7$

45. $y^7z - yz^4$ **46.** $x^{10}y^2 - xy^5$ **47.** $2mp^4 + 16mpq^3$ **48.** $24m^5n - 3m^2n^4$

In Exercises 49–52, factor each expression completely. Factor a difference of two squares first.

49. $x^6 - 1$ **50.** $x^6 - y^6$ **51.** $x^{12} - y^6$ **52.** $a^{12} - 64$

In Exercises 53–66, factor each expression completely. Some exercises do not involve the sum or difference of two cubes.

53. $3(x^3 + y^3) - z(x^3 + y^3)$

54. $x(8a^3 - b^3) + 4(8a^3 - b^3)$

55. $(m^3 + 8n^3) + (m^3x + 8n^3x)$

56. $(a^3x + b^3x) - (a^3y + b^3y)$

57. $(a^4 + 27a) - (a^3b + 27b)$

58. $(x^4 + xy^3) - (x^3y + y^4)$

59. $x^2(y + z) - 4(y + z)$

60. $z^2(x + 1) - 9(x + 1)$

61. $r^2(x - a) - s^2(x - a)$

62. $pq^2(r + s) - p(r + s)$

63. $(x - 1)^2 + 2(x - 1)$

64. $(z + 3)^2 - 5(z + 3)$

65. $y^3(y^2 - 1) - 27(y^2 - 1)$

66. $z^3(y^2 - 4) + 8(y^2 - 4)$

Review Exercises

In Review Exercises 1–6, solve each equation.

1. $2x - 6 = 2$ **2.** $\dfrac{x + 11}{3} = 2$ **3.** $2(x + 5) = 4$ **4.** $\dfrac{3(x - 1)}{2} = 3$

5. $\dfrac{2(3a + 4)}{3} = 4$ **6.** $\dfrac{-3(y - 6)}{2} = -9$

In Review Exercises 7–10, simplify each expression. Write all answers without using negative exponents.

7. $\dfrac{x^2x^3}{x^5}$ **8.** $\dfrac{y^3y^{-4}}{y^3}$ **9.** $\left(\dfrac{2x^2y^3}{x^4y^2}\right)^0$ **10.** $\left(\dfrac{3x^2}{6x^3}\right)^{-4}$

11. A length of one Fermi is 1×10^{-13} centimeter, approximately the radius of a proton. Express this number in standard notation.

12. In the fourteenth century, the Black Plague killed about 25,000,000 people, which was one-quarter of the population of Europe. Express this number in scientific notation.

11.7 Summary of Factoring Techniques

■ *Identifying the Problem Type*

In this section, we discuss ways to approach a factoring problem.

Identifying the Problem Type

Suppose we wish to factor the trinomial

$$x^4y + 7x^3y - 18x^2y$$

We begin by attempting to identify the problem type. The first type to look for is **factoring out a common factor**. Because the trinomial has a common factor of x^2y, we factor it out:

$$x^4y + 7x^3y - 18x^2y = x^2y(x^2 + 7x - 18)$$

We then note that $x^2 + 7x - 18$ is a trinomial that can be factored as $(x + 9)(x - 2)$.

$$x^4y + 7x^3y - 18x^2y = x^2y(x^2 + 7x - 18)$$
$$= x^2y(x + 9)(x - 2)$$

To identify the type of factoring problem, follow these steps:

---

1. Factor out all common factors.
2. If an expression has two terms, check to see if the problem type is
   a. the **difference of two squares**: $a^2 - b^2 = (a + b)(a - b)$,
   b. the **sum of two cubes**: $a^3 + b^3 = (a + b)(a^2 - ab + b^2)$, or
   c. the **difference of two cubes**: $a^3 - b^3 = (a - b)(a^2 + ab + b^2)$.
3. If an expression has three terms, check to see if the problem type is a **perfect trinomial square**:

   $$a^2 + 2ab + b^2 = (a + b)(a + b) \quad \text{or}$$
   $$a^2 - 2ab + b^2 = (a - b)(a - b).$$

   If the trinomial is not a trinomial square, attempt to factor the trinomial as a **general trinomial**.
4. If an expression has four or more terms, try to factor the expression by **grouping**.
5. Continue factoring until each individual factor is prime.
6. Check the results by multiplying.

EXAMPLE 1

Factor $x^5y^2 - xy^6$.

Solution

Begin by factoring out the common factor of xy^2:

$$x^5y^2 - xy^6 = xy^2(x^4 - y^4)$$

Since $x^4 - y^4$ has two terms, we check to see if it is the difference of two squares, and it is. As the difference of two squares, it factors as $(x^2 + y^2)(x^2 - y^2)$.

$$x^5y^2 - xy^6 = xy^2(x^4 - y^4)$$
$$= xy^2(x^2 + y^2)(x^2 - y^2)$$

The binomial $x^2 + y^2$ is the sum of two squares and cannot be factored. However, the binomial $x^2 - y^2$ is the difference of two squares and factors as $(x + y)(x - y)$.

$$x^5y^2 - xy^6 = xy^2(x^4 - y^4)$$
$$= xy^2(x^2 + y^2)(x^2 - y^2)$$
$$= xy^2(x^2 + y^2)(x + y)(x - y)$$

Because each of the individual factors is prime, the given expression is in completely factored form. ∎

EXAMPLE 2

Factor $x^6 - x^4y^2 - x^3y^3 + xy^5$.

Solution

Begin by factoring out the common factor of x.

$$x^6 - x^4y^2 - x^3y^3 + xy^5 = x(x^5 - x^3y^2 - x^2y^3 + y^5)$$

Because $x^5 - x^3y^2 - x^2y^3 + y^5$ has four terms, we try factoring it by grouping:

$$x^6 - x^4y^2 - x^3y^3 + xy^5 = x(x^5 - x^3y^2 - x^2y^3 + y^5)$$
$$= x[x^3(x^2 - y^2) - y^3(x^2 - y^2)]$$
$$= x(x^2 - y^2)(x^3 - y^3) \quad \text{Factor out } x^2 - y^2.$$

Finally, we factor the difference of two squares and the difference of two cubes:

$$x^6 - x^4y^2 - x^3y^3 + xy^5 = x(x + y)(x - y)(x - y)(x^2 + xy + y^2)$$

Because each factor is prime, the final expression is in prime factored form. ∎

11.7 Exercises

In Exercises 1–50, factor each expression completely. If the expression is prime, so indicate.

1. $6x + 3$
2. $x^2 - 9$
3. $x^2 - 6x - 7$
4. $a^3 + b^3$
5. $6t^2 + 7t - 3$
6. $3rs^2 - 6r^2st$
7. $4x^2 - 25$
8. $ac + ad + bc + bd$
9. $t^2 - 2t + 1$
10. $6p^2 - 3p - 2$
11. $a^3 - 8$
12. $2x^2 - 32$
13. $x^2y^2 - 2x^2 - y^2 + 2$
14. $a^2c + a^2d^2 + bc + bd^2$
15. $70p^4q^3 - 35p^4q^2 + 49p^5q^2$
16. $a^2 + 2ab + b^2 - x^2 - 2xy - y^2$
17. $2ab^2 + 8ab - 24a$
18. $t^4 - 16$
19. $-8p^3q^7 - 4p^2q^3$
20. $8m^2n^3 - 24mn^4$
21. $4a^2 - 4ab + b^2 - 9$
22. $3rs + 6r^2 - 18s^2$
23. $x^2 + 7x + 1$
24. $3a^3 + 24b^3$
25. $-2x^5 + 128x^2$
26. $16 - 40z + 25z^2$
27. $14t^3 - 40t^2 + 6t^4$
28. $6x^2 + 7x - 20$

29. $a^2(x - a) - b^2(x - a)$

30. $5x^3y^3z^4 + 25x^2y^3z^2 - 35x^3y^2z^5$

31. $8p^6 - 27q^6$

32. $2c^2 - 5cd - 3d^2$

33. $125p^3 - 64y^3$

34. $8a^2x^3y - 2b^2xy$

35. $-16x^4y^2z + 24x^5y^3z^4 - 15x^2y^3z^7$

36. $2ac + 4ad + bc + 2bd$

37. $81p^4 - 16q^4$

38. $6x^2 - x - 16$

39. $4x^2 + 9y^2$

40. $30a^4 + 5a^3 - 200a^2$

41. $54x^3 + 250y^6$

42. $6a^3 + 35a^2 - 6a$

43. $10r^2 - 13r - 4$

44. $4x^2 + 4x + 1 - y^2$

45. $21t^3 - 10t^2 + t$

46. $16x^2 - 40x^3 + 25x^4$

47. $x^5 - x^3y^2 + x^2y^3 - y^5$

48. $a^3x^3 - a^3y^3 + b^3x^3 - b^3y^3$

49. $2a^2c - 2b^2c + 4a^2d - 4b^2d$

50. $3a^2x^2 + 6a^2x + 3a^2 - 6b^2x^2 - 12b^2x - 6b^2$

Review Exercises

In Review Exercises 1–8, write each expression without using parentheses.

1. $2x(x + 2)$

2. $(x - y)(x - y)$

3. $(2a)^3$

4. $3(a + b)(a - b)$

5. $(2x - 3)^2$

6. $(3a + 2b)^2$

7. $\left(\dfrac{3a^3b^{-2}}{6a^{-3}b^2}\right)^2$

8. $\left(\dfrac{8x^{-2}}{4x^3y^{-2}}\right)^{-3}$

In Review Exercises 9–10, solve each equation.

9. $2(t - 5) - t = 3(2 - t)$

10. $5 - 3(2x - 1) = 2(4 + 3x) - 24$

11. The sum of three consecutive even integers is 54. Find the smallest integer.

12. Solve $y = mx + b$ for m.

11.8 Solving Equations by Factoring

- Quadratic Equations
- Cubic Equations

Equations such as $3x + 2$ and $-2x + 7 = 0$ that contain first-degree polynomials are called **linear equations**.

Quadratic Equations

Equations such as $3x^2 + 4x - 7 = 0$ and $-4x^2 + x - 7 = 0$ that contain second-degree polynomials are called **quadratic equations**.

Definition: A **quadratic equation** is an equation of the form

$$ax^2 + bx + c = 0$$

where a, b, and c are real numbers, and $a \neq 0$.

To solve the quadratic equation

$$x^2 + 5x - 6 = 0$$

we begin by factoring the quadratic trinomial and rewriting the equation as

$$(x + 6)(x - 1) = 0$$

This equation shows that the product of two quantities is 0. However, if the product of two quantities is 0, then at least one of those quantities must be 0. This fact is called the **zero factor theorem**.

The Zero Factor Theorem: If a and b represent two real numbers and $ab = 0$, then

$$a = 0 \quad \text{or} \quad b = 0$$

By applying the zero factor theorem to the equation $(x + 6)(x - 1) = 0$, we have

$$x + 6 = 0 \quad \text{or} \quad x - 1 = 0$$

We can solve each of these linear equations to get

$$x = -6 \quad \text{or} \quad x = 1$$

To check these answers, we substitute -6 for x and 1 for x in the original equation and simplify.

For $x = -6$
$$x^2 + 5x - 6 = 0$$
$$(-6)^2 + 5(-6) - 6 \stackrel{?}{=} 0$$
$$36 - 30 - 6 \stackrel{?}{=} 0$$
$$36 - 36 \stackrel{?}{=} 0$$
$$0 = 0$$

For $x = 1$
$$x^2 + 5x - 6 = 0$$
$$(1)^2 + 5(1) - 6 \stackrel{?}{=} 0$$
$$1 + 5 - 6 \stackrel{?}{=} 0$$
$$6 - 6 \stackrel{?}{=} 0$$
$$0 = 0$$

Both solutions check.

EXAMPLE 1

Solve the equation $2x^2 + 3x = 2$.

Solution

Write the equation in the form $ax^2 + bx + c = 0$. Then solve for x as follows.

$$2x^2 + 3x = 2$$
$$2x^2 + 3x - 2 = 0 \qquad \text{Add } -2 \text{ to both sides.}$$
$$(2x - 1)(x + 2) = 0 \qquad \text{Factor } 2x^2 + 3x - 2.$$
$$2x - 1 = 0 \quad \text{or} \quad x + 2 = 0 \qquad \text{Set each factor equal to 0.}$$
$$2x = 1 \qquad \qquad x = -2 \qquad \text{Solve each linear equation.}$$
$$x = \frac{1}{2}$$

Check:

For $x = \frac{1}{2}$:
$$2x^2 + 3x = 2$$
$$2\left(\frac{1}{2}\right)^2 + 3\left(\frac{1}{2}\right) \stackrel{?}{=} 2$$
$$2\left(\frac{1}{4}\right) + \frac{3}{2} \stackrel{?}{=} 2$$
$$\frac{1}{2} + \frac{3}{2} \stackrel{?}{=} 2$$
$$2 = 2$$

For $x = -2$:
$$2x^2 + 3x = 2$$
$$2(-2)^2 + 3(-2) \stackrel{?}{=} 2$$
$$2(4) - 6 \stackrel{?}{=} 2$$
$$8 - 6 \stackrel{?}{=} 2$$
$$2 = 2$$

Both solutions check. Work Progress Check 18.

Quadratic equations such as $3x^2 + 6x = 0$ and $4x^2 - 36 = 0$ are called **incomplete quadratic equations**, because they are missing a term. Many incomplete quadratic equations can be solved by factoring.

Progress Check 18

Solve: $x^2 + x = 6$.

EXAMPLE 2

Solve the equation $3x^2 + 6x = 0$.

Solution

$$3x^2 + 6x = 0$$
$$3x(x + 2) = 0 \qquad \text{Factor out } 3x.$$
$$3x = 0 \quad \text{or} \quad x + 2 = 0 \qquad \text{Set each factor equal to 0.}$$
$$x = 0 \qquad \qquad x = -2 \qquad \text{Solve each linear equation.}$$

Both solutions check.

EXAMPLE 3

Solve the equation $4x^2 - 36 = 0$.

Progress Check Answers

18. $2, -3$

Solution

$$4x^2 - 36 = 0$$
$$x^2 - 9 = 0 \quad \text{Divide both sides by 4.}$$
$$(x + 3)(x - 3) = 0 \quad \text{Factor } x^2 - 9.$$
$$x + 3 = 0 \quad \text{or} \quad x - 3 = 0 \quad \text{Set each factor equal to 0.}$$
$$x = -3 \quad \qquad x = 3 \quad \text{Solve each linear equation.}$$

Both solutions check. Work Progress Check 19.

Progress Check 19

Solve:

a. $x^2 - 4x = 0$

b. $x^2 - 36 = 0$

EXAMPLE 4

Solve the equation $x(2x - 13) = -15$.

Solution

$$x(2x - 13) = -15$$
$$2x^2 - 13x = -15 \quad \text{Remove parentheses.}$$
$$2x^2 - 13x + 15 = 0 \quad \text{Add 15 to both sides to get 0 on the right-hand side of the equation.}$$
$$(2x - 3)(x - 5) = 0 \quad \text{Factor } 2x^2 - 13x + 15.$$
$$2x - 3 = 0 \quad \text{or} \quad x - 5 = 0 \quad \text{Set each factor equal to 0.}$$
$$2x = 3 \quad \qquad x = 5 \quad \text{Solve each linear equation.}$$
$$x = \frac{3}{2}$$

Both solutions check.

EXAMPLE 5

Solve the equation $(x - 2)(x^2 - 7x + 6) = 0$.

Solution

$$(x - 2)(x^2 - 7x + 6) = 0$$
$$(x - 2)(x - 6)(x - 1) = 0 \quad \text{Factor } x^2 - 7x + 6.$$

If the product of these three quantities is 0, then at least one of the quantities must be 0. Hence,

$$x - 2 = 0 \quad \text{or} \quad x - 6 = 0 \quad \text{or} \quad x - 1 = 0 \quad \text{Set each factor equal to 0.}$$
$$x = 2 \qquad x = 6 \qquad x = 1 \quad \text{Solve each linear equation.}$$

All three solutions check.

Cubic Equations

If the polynomial in a polynomial equation is of third degree, the equation is called a **cubic equation**.

19. a. 0, 4 b. 6, −6

Progress Check Answers

EXAMPLE 6

Solve the equation $6x^3 + 12x = 17x^2$.

Solution

$$6x^3 + 12x = 17x^2$$
$$6x^3 - 17x^2 + 12x = 0 \quad \text{Add } -17x^2 \text{ to both sides.}$$
$$x(6x^2 - 17x + 12) = 0 \quad \text{Factor out } x.$$
$$x(2x - 3)(3x - 4) = 0 \quad \text{Factor } 6x^2 - 17x + 12.$$

$x = 0$ or $2x - 3 = 0$ or $3x - 4 = 0$ Set each factor equal to 0.

$x = 0$ $2x = 3$ $3x = 4$ Solve the linear equations.

$x = \dfrac{3}{2}$ $x = \dfrac{4}{3}$

Verify that all three solutions check. Work Progress Check 20.

Progress Check 20

Solve $x^3 - x^2 - 12x = 0$.

20. $0, 4, -3$

11.8 Exercises

In Exercises 1–12, solve each equation by setting each factor equal to 0 and solving the resulting linear equations.

1. $(x - 2)(x + 3) = 0$ **2.** $(x - 3)(x - 2) = 0$ **3.** $(x - 4)(x + 1) = 0$ **4.** $(x + 5)(x + 2) = 0$

5. $(2x - 5)(3x + 6) = 0$ **6.** $(3x - 4)(x + 1) = 0$

7. $(x - 1)(x + 2)(x - 3) = 0$ **8.** $(x + 2)(x + 3)(x - 4) = 0$

9. $(2x + 4)(3x - 12)(x + 7) = 0$ **10.** $(3x - 5)(x + 6)(2x - 1) = 0$

11. $(x - 4)(x + 6)(2x - 3)(3x - 2) = 0$ **12.** $(2x - 4)(3x + 5)(4x - 6)(5x + 10) = 0$

In Exercises 13–36, solve each equation. You may have to rearrange some terms.

13. $x^2 - 13x + 12 = 0$ **14.** $x^2 + 7x + 6 = 0$ **15.** $x^2 - 2x - 15 = 0$ **16.** $x^2 - x - 20 = 0$

17. $6x^2 - x - 2 = 0$ **18.** $2x^2 - 5x - 12 = 0$ **19.** $12x^2 + 5x = 2$ **20.** $4x^2 + 9x = 9$

21. $2x^2 + 6x = 0$ **22.** $3x^2 - 9x = 0$ **23.** $4x^2 - 32x = 0$ **24.** $5x^2 + 125x = 0$

25. $4x^2 - 5 = x$ **26.** $5x^2 - 3 = 14x$ **27.** $x^2 - 16 = 0$ **28.** $x^2 - 25 = 0$

29. $x^2 - 49 = 0$ **30.** $x^2 - 64 = 0$ **31.** $6x^2 - 36x = 0$ **32.** $7x^2 - 63x = 0$

33. $5x^2 - 23x - 10 = 0$ **34.** $6x^2 - 11x + 5 = 0$ **35.** $7x^2 + 19x = 6$ **36.** $16x^2 - 3 = -2x$

In Exercises 37–48, solve each equation.

37. $a(a + 8) = -15$ **38.** $a(1 + 2a) = 6$ **39.** $2(y - 4) = -y^2$ **40.** $-3(y - 6) = y^2$

41. $2x(3x + 10) = -6$ **42.** $2x^2 = 2(x + 2)$ **43.** $x^2 + 7x = x - 9$ **44.** $x(x + 10) = 2(x - 8)$

45. $2y(y + 2) = 3(y + 1)$ **46.** $2z(z - 3) = 3 - z$ **47.** $(b - 4)^2 = 1$ **48.** $(x + 3)^2 = 9$

In Exercises 49–60, solve each equation.

49. $(x - 1)(x^2 + 5x + 6) = 0$

50. $(x - 2)(x^2 - 8x + 7) = 0$

51. $2x^3 - 4x^2 - 6x = 0$

52. $6x^3 + 22x^2 + 12x = 0$

53. $y^3 - 16y = 0$

54. $(y^2 + 6y)(y - 2) = 0$

55. $(3a^2 - 9a)(2a + 1) = 0$

56. $5b^3 - 125b = 0$

57. $21z^3 + z = 10z^2$

58. $6t^3 + 35t^2 = 6t$

59. $(x^2 - 9)(9x^2 - 4) = 0$

60. $(x^2 + 4x + 4)(4x^2 - 25) = 0$

Review Exercises

In Review Exercises 1–6, factor each expression completely.

1. $x^2 + 3x$

2. $x^3 - 4x^2$

3. $y^2 - 9$

4. $t^2 - 64$

5. $x^2 + 13x + 12$

6. $x^2 - 11x - 12$

7. One side of a square is s inches long. Find an expression that represents its perimeter.

8. One side of a square is s inches long. Find an expression that represents its area.

9. One side of a rectangle with a perimeter of 24 centimeters is 4 centimeters longer than the other. Find its dimensions.

10. Find an expression that represents the area of a rectangle if one side has a length of $(x + 2)$ inches and another side has a length of $(x + 3)$ inches.

11. The annual interest Jill earns on her investment of $15,000 is $540 less than the annual interest Carol earns on her investment of $21,000. Both are receiving interest at the same annual rate. Find the rate.

12. David bought 35 stamps. He bought three times as many 22-cent stamps as 17-cent stamps, and as many 14-cent stamps as 22-cent stamps. How many of each stamp did he buy?

11.9 Applications

- Number Problems
- Falling Object Problems
- Geometry Problems

Number Problems

EXAMPLE 1

One negative number is 5 less than another, and their product is 84. Find the numbers.

Solution

Let x represent the larger number. Then $x - 5$ represents the smaller number. Because their product is 84, form the equation $x(x - 5) = 84$ and solve it.

$$x(x - 5) = 84$$
$$x^2 - 5x = 84 \quad \text{Remove parentheses.}$$
$$x^2 - 5x - 84 = 0 \quad \text{Add } -84 \text{ to both sides.}$$
$$(x - 12)(x + 7) = 0 \quad \text{Factor.}$$
$$x - 12 = 0 \quad \text{or} \quad x + 7 = 0 \quad \text{Set each factor equal to 0.}$$
$$x = 12 \quad | \quad x = -7 \quad \text{Solve each linear equation.}$$

Because we need two negative numbers, discard the result $x = 12$. The two numbers are

$$x = -7 \quad \text{and} \quad x - 5 = -7 - 5$$
$$= -12$$

Check: The number -12 is 5 less than -7, and the product of -12 and -7 is 84. ∎

Falling Object Problems

EXAMPLE 2

If an object is thrown straight up into the air with an initial velocity of 112 feet per second, its height after t seconds is given by the formula

$$h = 112t - 16t^2$$

where h represents the height of the object in feet. After this object has been thrown, in how many seconds will it hit the ground?

Solution

When the object hits the ground, its height will be 0. Hence, set h equal to 0 and solve for t.

$$h = 112t - 16t^2$$
$$0 = 112t - 16t^2$$
$$0 = 16t(7 - t) \quad \text{Factor out } 16t.$$
$$16t = 0 \quad \text{or} \quad 7 - t = 0 \quad \text{Set each factor equal to 0.}$$
$$t = 0 \quad | \quad t = 7 \quad \text{Solve each linear equation.}$$

When $t = 0$, the object's height above the ground is 0 feet, because it has just been released. When $t = 7$, the height is again 0 feet. The object has hit the ground. The solution is 7 seconds. ∎

Geometry Problems

EXAMPLE 3

Assume that the rectangle in Figure 11-1 has an area of 52 square centimeters and that its length is 1 centimeter more than 3 times its width. Find the perimeter of the rectangle.

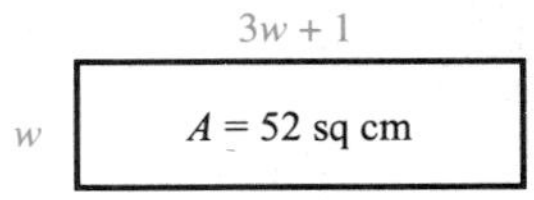

$3w + 1$

w | $A = 52$ sq cm

Figure 11-1

Solution

Let w represent the width of the rectangle. Then $3w + 1$ represents its length. Because the area is 52 square centimeters, substitute 52 for A

and $3w + 1$ for l in $A = lw$, the formula for the area of a rectangle, and solve for w.

$$A = lw$$
$$52 = (3w + 1)w$$
$$52 = 3w^2 + w \qquad \text{Remove parentheses.}$$
$$0 = 3w^2 + w - 52 \qquad \text{Add } -52 \text{ to both sides.}$$
$$0 = (3w + 13)(w - 4) \qquad \text{Factor.}$$

$3w + 13 = 0$	or	$w - 4 = 0$	Set each factor equal to 0.
$3w = -13$		$w = 4$	Solve each linear equation.
$w = -\dfrac{13}{3}$			

Because the length of a rectangle cannot be negative, discard the result $w = -\frac{13}{3}$. Hence, the width of the rectangle is 4, and the length is given by

$$3w + 1 = 3(4) + 1$$
$$= 12 + 1$$
$$= 13$$

The dimensions of the rectangle are 4 centimeters by 13 centimeters. To find the perimeter, we substitute 13 for l and 4 for w in $P = 2l + 2w$, the formula for the perimeter of a rectangle.

$$P = 2l + 2w$$
$$= 2(13) + 2(4)$$
$$= 26 + 8$$
$$= 34$$

The perimeter of the rectangle is 34 centimeters.

Check: A rectangle with dimensions of 13 centimeters by 4 centimeters does have an area of 52 square centimeters, and the length is 1 centimeter more than 3 times the width. A rectangle with these dimensions has a perimeter of 34 centimeters. ∎

EXAMPLE 4

Assume that the triangle in Figure 11-2 has an area of 10 square centimeters and that its height is 3 centimeters less than twice the length of its base. Find the length of the base and the height of the triangle.

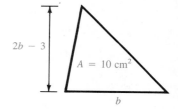

Figure 11-2

Solution

Let b represent the length of the base of the triangle. Then $2b - 3$ represents the height. Because the area is 10 square centimeters, substitute 10 for A along with $2b - 3$ for h in $A = \frac{1}{2}bh$, the formula for the area of a triangle, and solve for b.

$$A = \tfrac{1}{2}bh$$
$$10 = \tfrac{1}{2}b(2b - 3)$$
$$20 = b(2b - 3) \quad \text{Multiply both sides by 2.}$$
$$20 = 2b^2 - 3b \quad \text{Remove parentheses.}$$
$$0 = 2b^2 - 3b - 20 \quad \text{Add } -20 \text{ to both sides.}$$
$$0 = (2b + 5)(b - 4) \quad \text{Factor.}$$

$2b + 5 = 0$ or $b - 4 = 0$ Set both factors equal to 0.
$2b = -5$ $b = 4$ Solve each linear equation.
$b = -\tfrac{5}{2}$

Because a triangle cannot have a negative number for the length of its base, we must discard the result $b = -\tfrac{5}{2}$. Hence, the length of the base of the triangle is 4 centimeters. Its height is $2(4) - 3$, or 5 centimeters.

Check: If the base of a triangle has a length of 4 centimeters and the height of the triangle is 5 centimeters, its height is 3 centimeters less than twice the length of its base. Its area is 10 centimeters.

$$A = \tfrac{1}{2}bh$$
$$= \tfrac{1}{2}(4)(5)$$
$$= 2(5)$$
$$= 10$$

11.9 Exercises

1. One positive number is 2 more than another. Their product is 35. Find the numbers.

2. One positive number is 5 less than 4 times another. Their product is 21. Find the numbers.

3. If 4 is added to the square of a composite number, the result is 5 less than 10 times that number. Find the number.

4. If 3 times the square of a certain natural number is added to the number itself, the result is 14. Find the number.

In Exercises 5–8, assume that the height of the object thrown straight up is given by the formula $h = vt - 16t^2$, where h is the height after t seconds, and v is the velocity with which the object was thrown.

5. In how many seconds will an object hit the ground if it was thrown with a velocity of 144 feet per second?

6. In how many seconds will an object hit the ground if it was thrown with a velocity of 160 feet per second?

7. If an object was thrown with a velocity of 220 feet per second, at what times will the object be at a height of 600 feet?

8. If an object was thrown with a velocity of 128 feet per second, how many seconds will it take for the object to reach a height of 192 feet?

9. The length of a rectangle is 1 meter more than twice its width. Its area is 36 square meters. Find the dimensions of the rectangle.

10. The length of a rectangle is 2 inches less than 3 times its width. Its area is 21 square inches. Find the dimensions of the rectangle.

11. A room containing 143 square feet is 2 feet longer than it is wide. Find its perimeter.

12. The length of a rectangle is 2 centimeters longer than its width. If the length remains the same, but the width is doubled, the area is 48 square centimeters. Find the perimeter of the rectangle.

13. The length of the base of a triangle is 2 meters more than twice its height. The area is 30 square meters. Find the height and the length of the base of the triangle.

14. The height of a triangle is 2 inches less than 5 times the length of its base. The area is 36 square inches. Find the length of the base and the height of the triangle.

15. The base of a triangle is numerically 3 less than its area, and the height is numerically 6 less than its area. Find the area of the triangle.

16. The length of the base and the height of a triangle are numerically equal. Their sum is 6 less than the number of units in the area of the triangle. Find the area of the triangle.

17. The formula for the area of a parallelogram is $A = bh$. The area of the parallelogram in Illustration 1 is 200 square centimeters. If its base is twice its height, how long is the base?

18. The formula for the area of a trapezoid is $A = \dfrac{h(B + b)}{2}$. The area of the trapezoid in Illustration 2 is 24 square centimeters. Find the height of the trapezoid if one base is 8 centimeters and the other has the same length as the height.

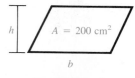

Illustration 1

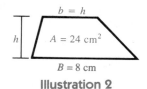

Illustration 2

19. The volume of a rectangular solid is given by the formula $V = lwh$, where l is the length, w is the width, and h is the height. The volume of the rectangular solid in Illustration 3 is 210 cubic centimeters. Find the width of the rectangular solid if its length is 10 centimeters and its height is 1 centimeter longer than twice its width.

20. The volume of a pyramid is given by the formula $V = \dfrac{Bh}{3}$, where B is the area of its base and h is its height. The volume of the pyramid in Illustration 4 is 192 cubic centimeters. Find the dimensions of its rectangular base if one edge of the base is 2 centimeters longer than the other and the height of the pyramid is 12 centimeters.

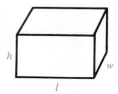

Illustration 3

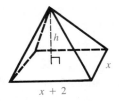

Illustration 4

21. The volume of a pyramid is 84 cubic centimeters. Its height is 9 centimeters, and one side of its rectangular base is 3 centimeters shorter than the other. Find the dimensions of its base. (See Exercise 20.)

22. The volume of a rectangular solid is 72 cubic centimeters. Its height is 4 centimeters, and its width is 3 centimeters shorter than its length. Find the sum of its length and width. (See Exercise 19.)

Review Exercises

In Review Exercises 1–10, solve each inequality and graph the solution on a number line.

1. $x - 3 > 5$
2. $x + 4 \leq 3$
3. $-3x - 5 \geq 4$
4. $2x - 3 < 7$
5. $\dfrac{3(x - 1)}{4} < 12$
6. $\dfrac{-2(x + 3)}{3} \geq 9$
7. $-2 < x \leq 4$
8. $-5 \leq x + 1 < 0$
9. $x + 1 > 2x + 3 > x - 4$
10. $1 < -7x + 8 < 15$

11. The efficiency, E, of a Carnot engine is given by the equation
$$E = 1 - \dfrac{T_2}{T_1}$$
Solve the equation for T_2.

12. Radioactive tracers are used in nuclear medicine. The *effective half-life*, H, of a radioactive material in an organism is given by the formula
$$H = \dfrac{RB}{R + B}$$
where R is the radioactive half-life, and B is the half-life of the tracer. Solve for R.

Mathematics in Geometry

Illustration 1

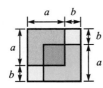

Illustration 2

The shaded area in Illustration 1 illustrates factoring the difference of two squares. Because each side of the large square is $a + b$, the area of the large square is $(a + b)^2$. Because each side of the small center square is $a - b$, the area of the small square is $(a - b)^2$. The area of the shaded region is the difference of these two areas, or

$$(a + b)^2 - (a - b)^2$$

This expression is the difference of two squares. It factors and simplifies as follows:

$$(a + b)^2 - (a - b)^2 = [(a + b) + (a - b)][(a + b) - (a - b)]$$
$$= (a + b + a - b)(a + b - a + b)$$
$$= (2a)(2b)$$
$$= 4ab$$

It is also clear from Illustration 1 that the area of the shaded region is $4ab$. Can you explain how Illustration 2 illustrates geometrically that

$$(a + b)^2 + (a - b)^2 = 2(a^2 + b^2)$$

Chapter Summary

Key Words

cubic equation *(11.8)*
difference of two cubes *(11.6)*
difference of two squares *(11.3)*
factor *(11.1)*
factoring by grouping *(11.2)*

factoring out the greatest common factor *(11.1)*
general trinomial *(11.7)*
greatest common factor (or divisor) *(11.1)*
incomplete quadratic equation *(11.8)*

linear equation *(11.8)*
perfect square trinomials *(11.4)*
prime factor *(11.1)*
prime-factored form *(11.1)*
prime number *(11.1)*
prime polynomial *(11.3)*

quadratic equation *(11.8)*
sum of two cubes *(11.6)*
sum of two squares *(11.3)*
zero factor theorem *(11.8)*

Key Ideas

(11.1) A natural number is in prime-factored form if it is written as the product of prime number factors.

The greatest common factor of several monomials is found by taking each common prime factor and variable factor the fewest number of times that it appears in any one monomial.

To factor a polynomial, first factor out all common factors.

(11.3) To factor the difference of two squares, use the pattern
$$x^2 - y^2 = (x + y)(x - y)$$

(11.6) **The sum and the difference of two cubes** factor according to the patterns
$$x^3 + y^3 = (x + y)(x^2 - xy + y^2)$$
$$x^3 - y^3 = (x - y)(x^2 + xy + y^2)$$

(11.2) If a polynomial has four or more terms, consider factoring it by grouping.

(11.4–11.5) Factor trinomials by trying these steps:

1. Write the trinomial with the exponents of one variable in descending order.
2. Factor out any greatest common factor (including -1 if that is necessary to make the coefficient of the first term positive).
3. If the sign of the third term of the trinomial is $+$, the signs between the terms of each binomial factor are the same as the sign of the trinomial's second term. If the sign of the third term is $-$, the signs between the terms of the binomials are opposite.
4. Try combinations of first terms and last terms until you find one that works or until you have exhausted all possibilities. In that case, the trinomial is prime.
5. Check the factorization by multiplication.

(11.8) **Zero factor theorem.** If a and b represent two real numbers and if $ab = 0$, then
$$a = 0 \quad \text{or} \quad b = 0$$

Chapter 11 Review Exercises

[11.1] In Review Exercises 1–8, find the prime factorization of each number.

1. 35
2. 45
3. 96
4. 102
5. 87
6. 99
7. 2050
8. 4096

[11.1–11.7] In Review Exercises 9–52, factor each expression completely.

9. $3x + 9y$
10. $5ax^2 + 15a$
11. $7x^2 + 14x$
12. $3x^2 - 3x$
13. $2x^3 + 4x^2 - 8x$
14. $ax + ay - az$
15. $ax + ay - a$
16. $x^2yz + xy^2z$
17. $5a^2 + 5ab^2 + 10acd - 15a$
18. $7axy + 21x^2y - 35x^3y + 7xy^2$
19. $(x + y)a + (x + y)b$
20. $(x + y)^2 + (x + y)$
21. $2x^2(x + 2) + 6x(x + 2)$
22. $3x(y + z) - 9x(y + z)^2$
23. $3p + 9q + ap + 3aq$
24. $ar - 2as + 7r - 14s$

25. $x^2 + ax + bx + ab$

26. $xy + 2x - 2y - 4$

27. $3x^2y - xy^2 - 6xy + 2y^2$

28. $5x^2 + 10x - 15xy - 30y$

29. $x^2 - 9$
30. $x^2y^2 - 16$
31. $(x + 2)^2 - y^2$
32. $z^2 - (x + y)^2$

33. $6x^2y - 24y^3$
34. $(x + y)^2 - z^2$
35. $x^2 + 10x + 21$
36. $x^2 + 4x - 21$

37. $x^2 + 2x - 24$
38. $x^2 - 4x - 12$
39. $2x^2 - 5x - 3$
40. $3x^2 - 14x - 5$

41. $6x^2 + 7x - 3$
42. $6x^2 + 3x - 3$
43. $6x^3 + 17x^2 - 3x$
44. $4x^3 - 5x^2 - 6x$

45. $x^2 + 2ax + a^2 - y^2$

46. $ax^2 + 4ax + 3a - bx - b$

47. $xa + yb + ya + xb$

48. $2a^2x + 2abx + a^3 + a^2b$

49. $c^3 - 27$

50. $d^3 + 8$

51. $2x^3 + 54$

52. $2ab^4 - 2ab$

[11.8] In Review Exercises 53–68, solve each equation.

53. $x^2 + 2x = 0$
54. $2x^2 - 6x = 0$
55. $x^2 - 9 = 0$
56. $x^2 - 25 = 0$

57. $a^2 - 7a + 12 = 0$
58. $x^2 - 2x - 15 = 0$
59. $2x - x^2 + 24 = 0$
60. $16 + x^2 - 10x = 0$

61. $2x^2 - 5x - 3 = 0$
62. $2x^2 + x - 3 = 0$
63. $4x^2 = 1$
64. $9x^2 = 4$

65. $x^3 - 7x^2 + 12x = 0$
66. $x^3 + 5x^2 + 6x = 0$
67. $2x^3 + 5x^2 = 3x$
68. $3x^3 - 2x = x^2$

[11.9] In Review Exercises 69–76, solve each word problem.

69. The sum of two numbers is 12, and their product is 35. Find the numbers.

70. Two positive numbers differ by 12, and their product is 45. Find the numbers.

71. A rectangle is 2 feet longer than it is wide, and its area is 48 square feet. Find its dimensions.

72. If 3 times the square of a positive number is added to 5 times the number, the result is 2. Find the number.

73. A rectangle is 3 feet longer than twice its width, and its area is 27 square feet. Find its dimensions.

74. The base of a triangle is 3 centimeters longer than twice its height. Its area is 45 square centimeters. How long is the base?

75. The area of a square is numerically equal to its perimeter. How long is a side?

76. A rectangle is 3 feet longer than it is wide. Its area is numerically equal to its perimeter. Find its dimensions.

Chapter 11 Test

1. Find the prime factorization of 196.

In Problems 2–4, factor out the greatest common factor.

2. $5x^3y^2z^3 + 10x^2y^3z^4$

3. $60ab^2c^3 + 30a^3b^2c - 25a$

4. $3x^2(a+b) - 6xy(a+b)$

In Problems 5–21, factor each expression completely.

5. $ax + ay + bx + by$

6. $x^2 - 25$

7. $3a^2 - 27b^2$

8. $x^4 - 81y^4$

9. $x^2 + 4x + 3$

10. $x^2 - 9x - 22$

11. $x^2 + 10xy + 9y^2$

12. $6x^2 - 30xy + 24y^2$

13. $3x^2 + 13x + 4$

14. $2a^2 + 5a - 12$

15. $2x^2 + 3xy - 2y^2$

16. $12 - 25x + 12x^2$ _____

17. $12a^2 + 6ab - 36b^2$ _____

18. $x^3 - 64$ _____

19. $216 + 8a^3$ _____

20. $x^9 z^3 - y^3 z^6$ _____

21. $16r^3 + 128s^3$ _____

In Problems 22–25, solve each equation.

22. $x^2 + 3x = 0$ _____

23. $2x^2 + 5x + 3 = 0$ _____

24. $9y^2 - 81 = 0$ _____

25. $-3(y - 6) + 2 = y^2 + 2$ _____

26. One positive number is 10 more than another. Their product is 39. Find their sum. _____

27. An object is fired straight up into the air with a velocity of 192 feet per second. In how many seconds will it hit the ground if its height above the ground is given by the formula $h = vt - 16t^2$, where v is the velocity and t is time? _____

28. The base of a triangle with an area of 40 square meters is 2 meters longer than its height. Find the base of the triangle. _____

12 Rational Expressions

Mathematics for Fun

Many number tricks and puzzles can be explained using the principles of mathematics. For example, try this with a friend:

1. Pick a number.
2. Multiply it by 3.
3. Add 6.
4. Subtract your original number.
5. Divide by 2.
6. Subtract your original number again.
7. Your answer is 3.

Whatever number your friend begins with, the answer will always be 3. The topics of this chapter will help explain why this trick works.

We have seen that expressions such as $\frac{4}{5}$, and $\frac{-3}{4}$ are called **arithmetic fractions** or **rational numbers**. Expressions such as

$$\frac{x}{x+2} \quad \text{and} \quad \frac{5a^2 + b^2}{3a - b}$$

are called **rational expressions**. They are the fractions of algebra.

12.1 The Basic Properties of Fractions

- Simplifying Fractions
- Factoring Out −1
- Terms Cannot Be Divided Out

Any number that can be written in the form $\frac{a}{b}$, where a and b are integers and $b \neq 0$, is a rational number. The number 0.5, for example, is rational because it can be written in the form $\frac{5}{10}$ or $\frac{1}{2}$. Symbols such as $\frac{6}{0}$ and $\frac{0}{0}$ with zeros in the **denominator** are undefined expressions.

The set of numbers that can replace a variable in a fraction to produce a meaningful expression is called the **domain** of the variable.

EXAMPLE 1

Which numbers are not in the domain of x?   **a.** $\dfrac{3}{x+2}$ **b.** $\dfrac{3x+2}{5x-4}$

c. $\dfrac{x-3}{x^2-9}$

Solution

a. The denominator is 0 when $x + 2 = 0$. Since -2 is the solution of this equation, -2 is not in the domain of x.

b. The denominator is 0 when $5x - 4 = 0$. Since $\frac{4}{5}$ is the solution of this equation, $\frac{4}{5}$ is not in the domain of x.

c. The denominator is 0 when $x^2 - 9 = 0$. Solve this equation for x.

$$x^2 - 9 = 0$$
$$(x+3)(x-3) = 0 \qquad \text{Factor } x^2 - 9.$$
$$x + 3 = 0 \quad \text{or} \quad x - 3 = 0 \qquad \text{Set each factor equal to 0.}$$
$$x = -3 \qquad \qquad x = 3$$

Since 3 and -3 are the solutions of this equation, these numbers are not in the domain of x. ∎

There are three signs associated with every fraction: the sign of the fraction, the sign of the **numerator**, and the sign of the denominator.

$$\text{Sign of the fraction} \longrightarrow -\dfrac{+12}{-4} \longleftarrow \text{Sign of the numerator}$$
$$\longleftarrow \text{Sign of the denominator}$$

Any two of these signs can be changed without changing the value of the fraction. (If no sign is indicated, a + sign is understood.) For example,

$$-\dfrac{+12}{-4} = -\dfrac{-12}{+4} = +\dfrac{-12}{-4} = +\dfrac{+12}{+4} = +3$$

In general, we have

$$\boxed{\;\dfrac{a}{b} = \dfrac{-a}{-b} = -\dfrac{a}{-b} = -\dfrac{-a}{b} \quad \text{and} \quad -\dfrac{a}{b} = \dfrac{-a}{b} = \dfrac{a}{-b} = -\dfrac{-a}{-b}\;}$$

Simplifying Fractions

Fractions can be simplified by dividing out common factors in their numerators and denominators. For example,

$$\dfrac{18}{30} = \dfrac{3 \cdot 6}{5 \cdot 6} = \dfrac{3 \cdot \cancel{6}}{5 \cdot \cancel{6}} = \dfrac{3}{5} \quad \text{and} \quad -\dfrac{6}{15} = -\dfrac{3 \cdot 2}{3 \cdot 5} = -\dfrac{\cancel{3} \cdot 2}{\cancel{3} \cdot 5} = -\dfrac{2}{5}$$

To simplify the fraction $\dfrac{ac}{bc}$, we can divide out the common factor of c to obtain

$$\frac{ac}{bc} = \frac{a\cancel{c}}{b\cancel{c}} = \frac{a}{b} \qquad (b \neq 0,\ c \neq 0)$$

This fact establishes the fundamental property of fractions.

> **The Fundamental Property of Fractions:** If a is a real number and b and c are nonzero real numbers, then
>
> $$\frac{ac}{bc} = \frac{a}{b}$$

When all common factors have been divided out, we say that the fraction has been **expressed in lowest terms**. To **simplify a fraction** means to write it in lowest terms.

EXAMPLE 2

Simplify $\dfrac{21x^2y}{14xy^2}$ $(x \neq 0,\ y \neq 0)$.

Solution

$$\frac{21x^2y}{14xy^2} = \frac{3 \cdot 7 \cdot x \cdot x \cdot y}{2 \cdot 7 \cdot x \cdot y \cdot y} \qquad \text{Factor the numerator and denominator.}$$

$$= \frac{3 \cdot \cancel{7} \cdot \cancel{x} \cdot x \cdot \cancel{y}}{2 \cdot \cancel{7} \cdot \cancel{x} \cdot y \cdot \cancel{y}} \qquad \text{Divide out the common factors of 7, } x, \text{ and } y.$$

$$= \frac{3x}{2y}$$

Note that this fraction can be simplified by using exponents:

$$\frac{21x^2y}{14xy^2} = \frac{3 \cdot 7}{2 \cdot 7} x^{2-1} y^{1-2} = \frac{3}{2} xy^{-1} = \frac{3}{2} \cdot \frac{x}{y} = \frac{3x}{2y}$$

Progress Check 1

Simplify each fraction. Assume there are no divisions by 0.

a. $\dfrac{12a^2b^3}{18a^3b^2}$

b. $\dfrac{x^2 - 3x}{3x - 9}$

EXAMPLE 3

Write the fraction $\dfrac{x^2 + 3x}{3x + 9}$ $(x \neq -3)$ in lowest terms.

Solution

$$\frac{x^2 + 3x}{3x + 9} = \frac{x(x + 3)}{3(x + 3)} \qquad \text{Factor the numerator and the denominator.}$$

$$= \frac{x\cancel{(x + 3)}}{3\cancel{(x + 3)}} \qquad \text{Divide out the common factor of } x + 3.$$

$$= \frac{x}{3}$$

Work Progress Check 1.

1. a. $\dfrac{2b}{3a}$ **b.** $\dfrac{x}{3}$

Progress Check Answers

Any number divided by the number 1 remains unchanged. For example,

$$\frac{37}{1} = 37 \quad \text{and} \quad \frac{5x}{1} = 5x$$

In general, we have

$$\frac{a}{1} = a, \quad \text{for any real number } a$$

EXAMPLE 4

Simplify $\dfrac{x^3 + x^2}{x + 1}$ $(x \neq -1)$.

Solution

$$\frac{x^3 + x^2}{x + 1} = \frac{x^2(x + 1)}{x + 1} \quad \text{Factor the numerator.}$$

$$= \frac{x^2\cancel{(x + 1)}}{\cancel{x + 1}} \quad \text{Divide out the common factor of } x + 1.$$

$$= \frac{x^2}{1}$$

$$= x^2 \quad \text{Denominators of 1 need not be written.} \quad \blacksquare$$

Factoring Out −1

Often we need to factor −1 from the numerator before we can divide out any common factors.

EXAMPLE 5

Simplify **a.** $\dfrac{x - y}{y - x}$ $(y \neq x)$ and **b.** $\dfrac{2a - 1}{1 - 2a}$ $(a \neq \tfrac{1}{2})$.

Solution

Rearrange the terms in each numerator, factor out −1, and proceed as follows:

a. $\dfrac{x - y}{y - x} = \dfrac{-y + x}{y - x}$

$= \dfrac{-(y - x)}{y - x}$

$= \dfrac{-\cancel{(y - x)}}{\cancel{y - x}}$

$= -1$

b. $\dfrac{2a - 1}{1 - 2a} = \dfrac{-1 + 2a}{1 - 2a}$

$= \dfrac{-(1 - 2a)}{1 - 2a}$

$= \dfrac{-\cancel{(1 - 2a)}}{\cancel{1 - 2a}}$

$= -1 \quad \blacksquare$

Binomials such as $x - y$ and $y - x$ are **negatives** of each other because their sum is 0. The results of Example 5 illustrate this important fact.

> The quotient of any nonzero expression and its negative is -1.

Work Progress Check 2.

Progress Check 2

Simplify: $\dfrac{1 - 3x}{3x - 1}$ $(x \neq \frac{1}{3})$.

EXAMPLE 6

Simplify $\dfrac{x^2 + 13x + 12}{x^2 - 144}$ $(x \neq 12,\ x \neq -12)$.

Solution

$$\dfrac{x^2 + 13x + 12}{x^2 - 144} = \dfrac{(x + 1)(x + 12)}{(x + 12)(x - 12)} \quad \text{Factor the numerator and denominator.}$$

$$= \dfrac{(x + 1)\cancel{(x + 12)}}{\cancel{(x + 12)}(x - 12)} \quad \text{Divide out the common factor of } x + 12.$$

$$= \dfrac{x + 1}{x - 12} \quad (x \neq 12)$$

Terms Cannot Be Divided Out

Only *factors* common to the *entire numerator* and the *entire denominator* can be divided out. *Terms* common to both the numerator and denominator *cannot* be divided out. For example, consider the correct simplification

$$\dfrac{5 + 8}{5} = \dfrac{13}{5}$$

It is incorrect to divide out the common *term* of 5 in the above simplification. Doing so gives an incorrect answer.

$$\dfrac{5 + 8}{5} \neq \dfrac{\cancel{5} + 8}{\cancel{5}} \neq \dfrac{1 + 8}{1} = 9$$

EXAMPLE 7

Express the fraction $\dfrac{5(x + 3) - 5}{7(x + 3) - 7}$ in lowest terms. Assume no divisions by 0.

Solution

Do not divide out the binomials $x + 3$, because $x + 3$ is not a *factor* of the entire numerator, nor is it a *factor* of the entire denominator. Instead,

2. -1

remove parentheses, simplify, factor the numerator and the denominator separately, and then divide out any common factors.

$$\frac{5(x+3)-5}{7(x+3)-7} = \frac{5x+15-5}{7x+21-7}$$ Remove parentheses.

$$= \frac{5x+10}{7x+14}$$ Combine terms.

$$= \frac{5(x+2)}{7(x+2)}$$ Factor the numerator and denominator.

$$= \frac{5\cancel{(x+2)}}{7\cancel{(x+2)}}$$ Divide out the common factor of $x+2$.

$$= \frac{5}{7}$$ ■

EXAMPLE 8

Simplify $\dfrac{x(x+3)-3(x-1)}{x^2+3}$. Assume no divisions by 0.

Solution

$$\frac{x(x+3)-3(x-1)}{x^2+3} = \frac{x^2+3x-3x+3}{x^2+3}$$ Remove parentheses in the numerator.

$$= \frac{x^2+3}{x^2+3}$$ Combine terms in the numerator.

$$= \frac{\cancel{x^2+3}}{\cancel{x^2+3}}$$ Divide out the common factor of x^2+3.

$$= 1$$ ■

Work Progress Check 3.

EXAMPLE 9

Simplify $\dfrac{xy+2x+3y+6}{x^2+x-6}$. Assume no divisions by 0.

Solution

$$\frac{xy+2x+3y+6}{x^2+x-6} = \frac{x(y+2)+3(y+2)}{(x-2)(x+3)}$$ Begin to factor the numerator by grouping, and factor the denominator.

$$= \frac{(y+2)(x+3)}{(x-2)(x+3)}$$ Factor the numerator.

$$= \frac{(y+2)\cancel{(x+3)}}{(x-2)\cancel{(x+3)}}$$ Divide out the common factor of $x+3$.

$$= \frac{y+2}{x-2}$$ ■

Progress Check 3

Simplify:

$$\frac{x(2x+1)-x(x-3)+4}{x^2-4}$$

$(x \neq 2, x \neq -2)$.

---

3. $\dfrac{x+2}{x-2}$

Progress Check Answers

Many fractions do not simplify. Such fractions are already in lowest terms.

EXAMPLE 10

Simplify $\dfrac{x^2 + x - 2}{x^2 + x}$. Assume no divisions by 0.

Solution

$$\dfrac{x^2 + x - 2}{x^2 + x} = \dfrac{(x + 2)(x - 1)}{x(x + 1)} \qquad \text{Factor the numerator and the denominator.}$$

Because there are no factors common to the numerator and denominator, this fraction is already in lowest terms.

12.1 Exercises

In Exercises 1–20, express each fraction in lowest terms.

1. $\dfrac{8}{10}$
2. $\dfrac{16}{20}$
3. $\dfrac{28}{35}$
4. $\dfrac{14}{20}$

5. $\dfrac{8}{52}$
6. $\dfrac{15}{21}$
7. $\dfrac{10}{45}$
8. $\dfrac{21}{35}$

9. $\dfrac{-18}{54}$
10. $\dfrac{16}{40}$
11. $\dfrac{4x}{2}$
12. $\dfrac{2x}{4}$

13. $\dfrac{-6x}{18}$
14. $\dfrac{-25y}{5}$
15. $\dfrac{45a}{9}$
16. $\dfrac{48y}{16}$

17. $\dfrac{5z}{7+3}$
18. $\dfrac{(3-18)k}{25}$
19. $\dfrac{(3+4)a}{24-3}$
20. $\dfrac{x+x}{2}$

In Exercises 21–36, find the numbers that are not in the domain of the variable.

21. $\dfrac{4}{5x}$
22. $\dfrac{6}{11y}$
23. $\dfrac{x}{x-4}$
24. $\dfrac{9x}{x+5}$

25. $\dfrac{4a}{2a-5}$
26. $\dfrac{6t^2}{5t+9}$
27. $\dfrac{7x-2}{7x+2}$
28. $\dfrac{5x+2}{5x-2}$

29. $\dfrac{3a}{a^2-4}$
30. $\dfrac{4r^4}{r^2-16}$
31. $\dfrac{6x-5}{x^2-25}$
32. $\dfrac{7x-1}{x^2-36}$

33. $\dfrac{b^3}{b^2-4b-5}$
34. $\dfrac{4b^2+1}{b^2+4b+3}$
35. $\dfrac{5x+4}{x^2+7x+6}$
36. $\dfrac{x^2-x+1}{x^2-6x-7}$

In Exercises 37–84, express each fraction in lowest terms. If a fraction is already in lowest terms, so indicate. Assume that no variable has a value that will make a denominator equal to zero.

37. $\dfrac{2x}{3x}$
38. $\dfrac{5y}{7y}$
39. $\dfrac{6x^2}{4x^2}$
40. $\dfrac{9xy}{6xy}$

41. $\dfrac{2x^2}{3y}$
42. $\dfrac{5y^2}{2y^2}$
43. $\dfrac{15x^2y}{5xy^2}$
44. $\dfrac{12xz}{4xz^2}$

45. $\dfrac{28x}{32y}$
46. $\dfrac{14xz^2}{7x^2z^2}$
47. $\dfrac{x+3}{3x+9}$
48. $\dfrac{2x+14}{x+7}$

49. $\dfrac{5x+35}{x+7}$
50. $\dfrac{x-9}{3x-27}$
51. $\dfrac{x^2+3x}{2x+6}$
52. $\dfrac{xz-2x}{yz-2y}$

53. $\dfrac{15x-3x^2}{25y-5xy}$
54. $\dfrac{3y+xy}{3x+xy}$
55. $\dfrac{6a-6b+6c}{9a-9b+9c}$
56. $\dfrac{3a-3b-6}{2a-2b-4}$

57. $\dfrac{x-7}{7-x}$
58. $\dfrac{d-c}{c-d}$
59. $\dfrac{6x-3y}{3y-6x}$
60. $\dfrac{2c-4d}{4d-2c}$

61. $\dfrac{a+b-c}{c-a-b}$
62. $\dfrac{x-y-z}{z+y-x}$
63. $\dfrac{x^2+3x+2}{x^2+x-2}$
64. $\dfrac{x^2+x-6}{x^2-x-2}$

65. $\dfrac{x^2-8x+15}{x^2-x-6}$
66. $\dfrac{x^2-6x-7}{x^2+8x+7}$
67. $\dfrac{2x^2-8x}{x^2-6x+8}$
68. $\dfrac{3y^2-15y}{y^2-3y-10}$

69. $\dfrac{xy+2x^2}{2xy+y^2}$
70. $\dfrac{3x+3y}{x^2+xy}$
71. $\dfrac{x^2+3x+2}{x^3+x^2}$
72. $\dfrac{6x^2-13x+6}{3x^2+x-2}$

73. $\dfrac{x^2-8x+16}{x^2-16}$
74. $\dfrac{3x+15}{x^2-25}$
75. $\dfrac{2x^2-8}{x^2-3x+2}$
76. $\dfrac{3x^2-27}{x^2+3x-18}$

77. $\dfrac{x^2-2x-15}{x^2+2x-15}$
78. $\dfrac{x^2+4x-77}{x^2-4x-21}$
79. $\dfrac{x^2-3(2x-3)}{9-x^2}$
80. $\dfrac{x(x-8)+16}{16-x^2}$

81. $\dfrac{4(x+3)+4}{3(x+2)+6}$
82. $\dfrac{4+2(x-5)}{3x-5(x-2)}$
83. $\dfrac{(2x+3)-(x+6)}{x^2-9}$
84. $\dfrac{2(x+3)-(x+2)}{x^2+5x+4}$

In Exercises 85–92, simplify each fraction. Assume that no variable has a value that would make a denominator equal to 0. In each exercise, you will have to factor a sum or difference of two cubes or factor by grouping.

85. $\dfrac{x^3+1}{x^2-x+1}$
86. $\dfrac{x^3-1}{x^2+x+1}$
87. $\dfrac{2a^3-16}{2a^2+4a+8}$
88. $\dfrac{3y^3+81}{y^2-3y+9}$

89. $\dfrac{ab+b+2a+2}{ab+a+b+1}$
90. $\dfrac{xy+2y+3x+6}{x^2+5x+6}$
91. $\dfrac{xy+3y+3x+9}{x^2-9}$
92. $\dfrac{ab+b^2+2a+2b}{a^2+2a+ab+2b}$

Review Exercises

In Review Exercises 1–6, state the following properties of real numbers.

1. The closure property for addition
2. The commutative property of multiplication
3. The associative property of addition
4. The distributive property
5. The identity property for multiplication
6. The inverse property for addition
7. What is the additive identity?
8. What is the multiplicative identity?
9. Find the additive inverse of -10.
10. Find the multiplicative inverse of -10.
11. What number does not have a multiplicative inverse?
12. Find the additive inverse of 0.

12.2 Multiplying Fractions

To multiply fractions, we multiply their numerators and multiply their denominators. For example, to find the product of $\frac{4}{7}$ and $\frac{3}{5}$, we proceed as follows:

$$\frac{4}{7} \cdot \frac{3}{5} = \frac{4 \cdot 3}{7 \cdot 5} = \frac{12}{35}$$

In general, we have the following rule.

> **The Rule for Multiplying Fractions:** If a and c are real numbers and b and d are nonzero real numbers, then
> $$\frac{a}{b} \cdot \frac{c}{d} = \frac{ac}{bd}$$

From now on we will assume there are no divisions by 0.

EXAMPLE 1

Perform each multiplication: **a.** $\frac{1}{3} \cdot \frac{2}{5}$, **b.** $\frac{7}{9} \cdot \frac{-5}{3x}$, **c.** $\frac{x^2}{2} \cdot \frac{3}{y^2}$, and **d.** $\frac{t+1}{t} \cdot \frac{t-1}{t-2}$.

Solution

a. $\frac{1}{3} \cdot \frac{2}{5} = \frac{1 \cdot 2}{3 \cdot 5} = \frac{2}{15}$

b. $\frac{7}{9} \cdot \frac{-5}{3x} = \frac{7(-5)}{9 \cdot 3x} = \frac{-35}{27x}$

c. $\frac{x^2}{2} \cdot \frac{3}{y^2} = \frac{x^2 \cdot 3}{2 \cdot y^2} = \frac{3x^2}{2y^2}$

d. $\frac{t+1}{t} \cdot \frac{t-1}{t-2} = \frac{(t+1)(t-1)}{t(t-2)}$

EXAMPLE 2

Multiply: $\frac{35x^2 y}{7y^2 z} \cdot \frac{z}{5xy}$.

Solution

$$\frac{35x^2 y}{7y^2 z} \cdot \frac{z}{5xy} = \frac{5 \cdot 7 \cdot x \cdot x \cdot y \cdot z}{7 \cdot y \cdot y \cdot z \cdot 5 \cdot x \cdot y}$$ Multiply the fractions and factor where possible.

$$= \frac{\cancel{5} \cdot \cancel{7} \cdot \cancel{x} \cdot x \cdot \cancel{y} \cdot \cancel{z}}{\cancel{7} \cdot \cancel{y} \cdot y \cdot \cancel{z} \cdot \cancel{5} \cdot \cancel{x} \cdot y}$$ Divide out all common factors.

$$= \frac{x}{y^2}$$

EXAMPLE 3

Multiply: $\dfrac{x^2 - x}{2x + 4} \cdot \dfrac{x + 2}{x}$.

Solution

$$\dfrac{x^2 - x}{2x + 4} \cdot \dfrac{x + 2}{x} = \dfrac{x(x - 1)(x + 2)}{2(x + 2)x} \quad \text{Multiply the fractions and factor where possible.}$$

$$= \dfrac{\cancel{x}(x - 1)\cancel{(x + 2)}}{2\cancel{(x + 2)}\cancel{x}} \quad \text{Divide out all common factors.}$$

$$= \dfrac{x - 1}{2}$$

Work Progress Check 4.

Progress Check 4

Multiply: $\dfrac{x^2 + x}{2x + 6} \cdot \dfrac{x + 3}{2x + 2}$.

EXAMPLE 4

Multiply: $\dfrac{x^2 - 3x}{x^2 - x - 6} \cdot \dfrac{x^2 + x - 2}{x^2 - x}$.

Solution

$$\dfrac{x^2 - 3x}{x^2 - x - 6} \cdot \dfrac{x^2 + x - 2}{x^2 - x}$$

$$= \dfrac{x(x - 3)(x + 2)(x - 1)}{(x + 2)(x - 3)x(x - 1)} \quad \text{Multiply the fractions and factor where possible.}$$

$$= \dfrac{\cancel{x}\cancel{(x - 3)}\cancel{(x + 2)}\cancel{(x - 1)}}{\cancel{(x + 2)}\cancel{(x - 3)}\cancel{x}\cancel{(x - 1)}} \quad \text{Divide out all common factors.}$$

$$= 1$$

EXAMPLE 5

Multiply: $\dfrac{x^2 + x}{x^2 + 8x + 7} \cdot (x + 7)$.

Progress Check 5

Multiply:

$\dfrac{x^2 + 3x + 2}{x^2 - 4} \cdot \dfrac{x^2 - 4x + 4}{2x - 4}$.

Solution

$$\dfrac{x^2 + x}{x^2 + 8x + 7} \cdot (x + 7) = \dfrac{x^2 + x}{x^2 + 8x + 7} \cdot \dfrac{x + 7}{1} \quad \text{Write } x + 7 \text{ as a fraction with a denominator of 1.}$$

$$= \dfrac{x(x + 1)(x + 7)}{(x + 1)(x + 7)1} \quad \text{Multiply the fractions and factor where possible.}$$

$$= \dfrac{x\cancel{(x + 1)}\cancel{(x + 7)}}{1\cancel{(x + 1)}\cancel{(x + 7)}} \quad \text{Divide out all common factors.}$$

$$= x$$

Work Progress Check 5.

4. $\dfrac{x}{4}$

5. $\dfrac{x + 1}{2}$

Progress Check Answers

502   Chapter 12   Rational Expressions

EXAMPLE 6

Multiply: $\dfrac{x^2 + 2x}{xy - 2y} \cdot \dfrac{x + 1}{x^2 - 4} \cdot \dfrac{x - 2}{x^2 + x}$.

Solution

$$\dfrac{x^2 + 2x}{xy - 2y} \cdot \dfrac{x + 1}{x^2 - 4} \cdot \dfrac{x - 2}{x^2 + x}$$

$$= \dfrac{x(x + 2)(x + 1)(x - 2)}{y(x - 2)(x + 2)(x - 2)x(x + 1)} \qquad \text{Multiply the fractions and factor where possible.}$$

$$= \dfrac{\cancel{x}(\cancel{x + 2})(\cancel{x + 1})(\cancel{x - 2})}{y(\cancel{x - 2})(\cancel{x + 2})(x - 2)\cancel{x}(\cancel{x + 1})} \qquad \text{Divide out all common factors.}$$

$$= \dfrac{1}{y(x - 2)}$$

12.2 Exercises

In Exercises 1–68, perform the indicated multiplications. Simplify answers if possible.

1. $\dfrac{2}{3} \cdot \dfrac{4}{5}$
2. $\dfrac{1}{2} \cdot \dfrac{3}{5}$
3. $\dfrac{5}{7} \cdot \dfrac{9}{13}$
4. $\dfrac{2}{7} \cdot \dfrac{5}{11}$

5. $\dfrac{2}{3} \cdot \dfrac{3}{5}$
6. $\dfrac{3}{7} \cdot \dfrac{7}{5}$
7. $\dfrac{-3}{7} \cdot \dfrac{14}{9}$
8. $\dfrac{-6}{9} \cdot \dfrac{15}{35}$

9. $\dfrac{25}{35} \cdot \dfrac{21}{55}$
10. $\dfrac{27}{24} \cdot \dfrac{56}{35}$
11. $\dfrac{-21}{18} \cdot \dfrac{-45}{14}$
12. $\dfrac{-33}{7} \cdot \dfrac{-5}{55}$

13. $\dfrac{2}{3} \cdot \dfrac{15}{2} \cdot \dfrac{1}{7}$
14. $\dfrac{2}{5} \cdot \dfrac{10}{9} \cdot \dfrac{3}{2}$
15. $\dfrac{3x}{y} \cdot \dfrac{y}{2}$
16. $\dfrac{2y}{z} \cdot \dfrac{z}{3}$

17. $\dfrac{5y}{7} \cdot \dfrac{7x}{5z}$
18. $\dfrac{4x}{3y} \cdot \dfrac{3y}{7x}$
19. $\dfrac{3y}{4x} \cdot \dfrac{2x}{5}$
20. $\dfrac{5x}{14y} \cdot \dfrac{7y}{x}$

21. $\dfrac{7z}{9z} \cdot \dfrac{4z}{2z}$
22. $\dfrac{8z}{2x} \cdot \dfrac{16x}{3x}$
23. $\dfrac{13x^2}{7x} \cdot \dfrac{28}{2x}$
24. $\dfrac{z^2}{z} \cdot \dfrac{5x}{5}$

25. $\dfrac{2x^2y}{3xy} \cdot \dfrac{3xy^2}{2}$
26. $\dfrac{2x^2z}{z} \cdot \dfrac{5x}{z}$
27. $\dfrac{8x^2y^2}{4x^2} \cdot \dfrac{2xy}{2y}$
28. $\dfrac{9x^2y}{3x} \cdot \dfrac{3xy}{3y}$

29. $\dfrac{-2xy}{x^2} \cdot \dfrac{3xy}{2}$
30. $\dfrac{-3x}{x^2} \cdot \dfrac{2xz}{3}$
31. $\dfrac{ab^2}{a^2b} \cdot \dfrac{b^2c^2}{abc} \cdot \dfrac{abc^2}{a^3c^2}$
32. $\dfrac{x^3y}{z} \cdot \dfrac{xz^3}{x^2y^2} \cdot \dfrac{yz}{xyz}$

33. $\dfrac{10r^2st^3}{6rs^2} \cdot \dfrac{3r^3t}{2rst} \cdot \dfrac{2s^3t^4}{5s^2t^3}$
34. $\dfrac{3a^3b}{25cd^3} \cdot \dfrac{-5cd^2}{6ab} \cdot \dfrac{10abc^2}{2bc^2d}$

35. $\dfrac{z + 7}{7} \cdot \dfrac{z + 2}{z}$
36. $\dfrac{a - 3}{a} \cdot \dfrac{a + 3}{5}$
37. $\dfrac{x - 2}{2} \cdot \dfrac{2x}{x - 2}$
38. $\dfrac{y + 3}{y} \cdot \dfrac{3y}{y + 3}$

39. $\dfrac{x + 5}{5} \cdot \dfrac{x}{x + 5}$
40. $\dfrac{y - 9}{y + 9} \cdot \dfrac{y}{9}$
41. $\dfrac{(x + 1)^2}{x + 1} \cdot \dfrac{x + 2}{x + 1}$
42. $\dfrac{(y - 3)^2}{y - 3} \cdot \dfrac{y - 3}{y - 3}$

43. $\dfrac{2x+6}{x+3} \cdot \dfrac{3}{4x}$

44. $\dfrac{3y-9}{y-3} \cdot \dfrac{y}{3y^2}$

45. $\dfrac{x^2-x}{x} \cdot \dfrac{3x-6}{3x-3}$

46. $\dfrac{5z-10}{z+2} \cdot \dfrac{3}{3z-6}$

47. $\dfrac{7y-14}{y-2} \cdot \dfrac{x^2}{7x}$

48. $\dfrac{y^2+3y}{9} \cdot \dfrac{3x}{y+3}$

49. $\dfrac{x^2+x-6}{5x} \cdot \dfrac{5x-10}{x+3}$

50. $\dfrac{z^2+4z-5}{5z-5} \cdot \dfrac{5z}{z+5}$

51. $\dfrac{m^2-2m-3}{2m+4} \cdot \dfrac{m^2-4}{m^2+3m+2}$

52. $\dfrac{p^2-p-6}{3p-9} \cdot \dfrac{p^2-9}{p^2+6p+9}$

53. $\dfrac{x^2+7xy+12y^2}{x^2+2xy-8y^2} \cdot \dfrac{x^2-xy-2y^2}{x^2+4xy+3y^2}$

54. $\dfrac{m^2+9mn+20n^2}{m^2-25n^2} \cdot \dfrac{m^2-9mn+20n^2}{m^2-16n^2}$

55. $\dfrac{3r^2+15rs+18s^2}{6r^2-24s^2} \cdot \dfrac{2r-4s}{3r+9s}$

56. $\dfrac{2u^2+8u}{2u+8} \cdot \dfrac{4u^2+8uv+4v^2}{u^2+5uv+4v^2}$

57. $\dfrac{abc^2}{a+1} \cdot \dfrac{c}{a^2b^2} \cdot \dfrac{a^2+a}{ac}$

58. $\dfrac{x^3yz^2}{4x+8} \cdot \dfrac{x^2-4}{2x^2y^2z^2} \cdot \dfrac{8yz}{x-2}$

59. $\dfrac{3x^2+5x+2}{x^2-9} \cdot \dfrac{x-3}{x^2-4} \cdot \dfrac{x^2+5x+6}{6x+4}$

60. $\dfrac{x^2-25}{3x+6} \cdot \dfrac{x^2+x-2}{2x+10} \cdot \dfrac{6x}{3x^2-18x+15}$

61. $\dfrac{x^2+5x+6}{x^2} \cdot \dfrac{x^2-2x}{x^2-9} \cdot \dfrac{x^2-3x}{x^2-4}$

62. $\dfrac{x^2-1}{1-x} \cdot \dfrac{4x}{x+x^2} \cdot \dfrac{x^2+2x+1}{2x+2}$

63. $\dfrac{x^2+4x}{xz} \cdot \dfrac{z^2+z}{x^2-16} \cdot \dfrac{z+3}{z^2+4z+3}$

64. $(x+1) \cdot \dfrac{x^3-1}{x^3+1} \cdot \dfrac{x^2-x+1}{x^2-1}$

65. $\dfrac{x^3+8}{x^3-8} \cdot \dfrac{x-2}{x^2-4} \cdot (x^2+2x+4)$

66. $(4x^3-16x) \cdot \dfrac{1}{12x^2+24x} \cdot (3x+12)$

67. $\dfrac{x^2+x-6}{5x^2+7x+2} \cdot \dfrac{5x+2}{-x^2-x+6} \cdot \dfrac{x^2+2x+1}{x^2+4x+3}$

68. $\dfrac{x^2-3x-4}{3x^2-2x-1} \cdot \dfrac{3x+1}{x^2-6x+8} \cdot \dfrac{x^2-3x+2}{x^2+x}$

In Exercises 69–72, perform the indicated multiplications. You will need to factor by grouping and factor a sum or difference of two cubes to simplify the answers.

69. $\dfrac{ax+bx+ay+by}{x^3-y^3} \cdot \dfrac{x^2+xy+y^2}{ax+bx}$

70. $\dfrac{a^2-ab+b^2}{a^3+b^3} \cdot \dfrac{ac+ad+bc+bd}{c^2-d^2}$

71. $\dfrac{x^2-y^2}{y^2-xy} \cdot \dfrac{yx^3-y^4}{ax+ay+bx+by}$

72. $\dfrac{xw-xz+wy-yz}{x^2+2xy+y^2} \cdot \dfrac{x^3-y^3}{z^2-w^2}$

Review Exercises

In Review Exercises 1–6, simplify each expression. Write all answers without using negative exponents.

1. $2x^3y^2(-3x^2y^4z)$

2. $\dfrac{8x^4y^5}{-2x^3y^2}$

3. $(3y)^{-4}$

4. $(a^{-2}a)^{-3}$

5. $\dfrac{x^{3m}}{x^{4m}}$

6. $(3x^2y^3)^0$

7. Write the number 93,000,000 in scientific notation.

8. Write the number 0.00567 in scientific notation.

9. In a mathematics class, there were 12 more women than men. The total number of people in the class was 34. How many women were in the class?

10. The area of a triangle is 48 square meters. If its height is 8 meters, find the length of its base.

11. Chuck invests $35,000 in each of two accounts, with one paying annual interest at a rate 1% greater than the other. His annual income from the two accounts is $7000. What are the annual rates of these accounts?

12. Diane has 3 more quarters than dimes, and twice as many nickels as quarters. She has 41 coins in all. How many of each does she have?

12.3 Dividing Fractions

Division by a nonzero number is equivalent to multiplying by the reciprocal of that number. Thus, to divide two fractions, we invert the **divisor** (the fraction following the ÷ sign) and multiply. For example, to divide $\frac{4}{7}$ by $\frac{3}{5}$, we proceed as follows:

$$\frac{4}{7} \div \frac{3}{5} = \frac{4}{7} \cdot \frac{5}{3} = \frac{20}{21}$$

In general, we have the following rule.

> **The Rule for Dividing Fractions:** If a is a real number and b, c, and d are nonzero real numbers, then
>
> $$\frac{a}{b} \div \frac{c}{d} = \frac{a}{b} \cdot \frac{d}{c}$$

EXAMPLE 1

Perform the divisions **a.** $\frac{7}{13} \div \frac{21}{26}$ and **b.** $\frac{-9x}{35y} \div \frac{15x^2}{14}$.

Solution

a. $\frac{7}{13} \div \frac{21}{26} = \frac{7}{13} \cdot \frac{26}{21}$ Invert the divisor and multiply.

$= \frac{7 \cdot 2 \cdot 13}{13 \cdot 3 \cdot 7}$ Multiply the fractions and factor where possible.

$= \frac{\overset{1}{7} \cdot 2 \cdot \overset{1}{13}}{\underset{1}{13} \cdot 3 \cdot \underset{1}{7}}$ Divide out common factors.

$= \frac{2}{3}$

Section 12.3 Dividing Fractions 505

b. $\dfrac{-9x}{35y} \div \dfrac{15x^2}{14} = \dfrac{-9x}{35y} \cdot \dfrac{14}{15x^2}$ Invert the divisor and multiply.

$= \dfrac{-3 \cdot 3 \cdot x \cdot 2 \cdot 7}{5 \cdot 7 \cdot y \cdot 3 \cdot 5 \cdot x \cdot x}$ Multiply the fractions and factor where possible.

$= \dfrac{-3 \cdot \cancel{3} \cdot \cancel{x} \cdot 2 \cdot \cancel{7}}{5 \cdot \cancel{7} \cdot y \cdot \cancel{3} \cdot 5 \cdot \cancel{x} \cdot x}$ Divide out all common factors.

$= \dfrac{-6}{25xy}$ Multiply the remaining factors. ■

EXAMPLE 2

Perform the division: $\dfrac{x^2 + x}{3x - 15} \div \dfrac{x^2 + 2x + 1}{6x - 30}$.

Solution

$\dfrac{x^2 + x}{3x - 15} \div \dfrac{x^2 + 2x + 1}{6x - 30} = \dfrac{x^2 + x}{3x - 15} \cdot \dfrac{6x - 30}{x^2 + 2x + 1}$ Invert the divisor and multiply.

$= \dfrac{x(x + 1) \cdot 2 \cdot 3(x - 5)}{3(x - 5)(x + 1)(x + 1)}$ Multiply the fractions and factor.

$= \dfrac{x(\cancel{x + 1}) \cdot 2 \cdot \cancel{3}(\cancel{x - 5})}{\cancel{3}(\cancel{x - 5})(\cancel{x + 1})(x + 1)}$ Divide out all common factors.

$= \dfrac{2x}{x + 1}$

Work Progress Check 6.   ■

EXAMPLE 3

Perform the division: $\dfrac{2x^2 - 3x - 2}{2x + 1} \div (4 - x^2)$.

Solution

$\dfrac{2x^2 - 3x - 2}{2x + 1} \div (4 - x^2)$

$= \dfrac{2x^2 - 3x - 2}{2x + 1} \div \dfrac{4 - x^2}{1}$ Write $4 - x^2$ as a fraction with a denominator of 1.

$= \dfrac{2x^2 - 3x - 2}{2x + 1} \cdot \dfrac{1}{4 - x^2}$ Invert the divisor and multiply.

$= \dfrac{(2x + 1)(x - 2) \cdot 1}{(2x + 1)(2 + x)(2 - x)}$ Multiply the fractions and factor where possible.

$= \dfrac{\cancel{(2x + 1)}\overset{-1}{\cancel{(x - 2)}} \cdot 1}{\cancel{(2x + 1)}(2 + x)\cancel{(2 - x)}}$ Divide out all common factors.

$= \dfrac{-1}{2 + x}$

$= -\dfrac{1}{2 + x}$ ■

Progress Check 6

Divide:
$\dfrac{x^2 + 3x + 2}{x^2 - 9} \div \dfrac{x^2 - 1}{x^2 + 2x - 3}$.

6. $\dfrac{x + 2}{x - 3}$

Progress Check Answers

Unless parentheses indicate otherwise, multiplications and divisions are to be performed in order from left to right.

EXAMPLE 4

Simplify the expression: $\dfrac{x^2 - x - 6}{x - 2} \div \dfrac{x^2 - 4x}{x^2 - x - 2} \cdot \dfrac{x - 4}{x^2 + x}$.

Solution

Since no parentheses indicate otherwise, the operation of division is performed first.

$$\dfrac{x^2 - x - 6}{x - 2} \div \dfrac{x^2 - 4x}{x^2 - x - 2} \cdot \dfrac{x - 4}{x^2 + x}$$

$$= \dfrac{x^2 - x - 6}{x - 2} \cdot \dfrac{x^2 - x - 2}{x^2 - 4x} \cdot \dfrac{x - 4}{x^2 + x}$$ Invert the divisor and multiply.

$$= \dfrac{(x + 2)(x - 3)(x + 1)(x - 2)(x - 4)}{(x - 2)x(x - 4)x(x + 1)}$$ Multiply the fractions and factor.

$$= \dfrac{(x + 2)(x - 3)\cancel{(x + 1)}\cancel{(x - 2)}\cancel{(x - 4)}}{\cancel{(x - 2)}x\cancel{(x - 4)}x\cancel{(x + 1)}}$$ Divide out all common factors.

$$= \dfrac{(x + 2)(x - 3)}{x^2}$$

Work Progress Check 7.

EXAMPLE 5

Simplify the expression: $\dfrac{x^2 + 6x + 9}{x^2 - 2x}\left(\dfrac{x^2 - 4}{x^2 + 3x} \div \dfrac{x + 2}{x}\right)$.

Solution

Do the division within the parentheses first.

$$\dfrac{x^2 + 6x + 9}{x^2 - 2x}\left(\dfrac{x^2 - 4}{x^2 + 3x} \div \dfrac{x + 2}{x}\right)$$

$$= \dfrac{x^2 + 6x + 9}{x^2 - 2x}\left(\dfrac{x^2 - 4}{x^2 + 3x} \cdot \dfrac{x}{x + 2}\right)$$ Invert the divisor and multiply.

$$= \dfrac{(x + 3)(x + 3)(x - 2)(x + 2)x}{x(x - 2)x(x + 3)(x + 2)}$$ Multiply the fractions and factor where possible.

$$= \dfrac{\cancel{(x + 3)}(x + 3)\cancel{(x - 2)}\cancel{(x + 2)}\cancel{x}}{\cancel{x}\cancel{(x - 2)}x\cancel{(x + 3)}\cancel{(x + 2)}}$$ Divide out all common factors.

$$= \dfrac{x + 3}{x}$$

Progress Check 7

Simplify:
$\dfrac{x^2 + x}{2x + 8} \div \dfrac{x^2 + 3x + 2}{5x + 15} \cdot \dfrac{x^2 + 6x + 8}{x^2 + 3x}$.

---

7. $\dfrac{5}{2}$

Progress Check Answers

12.3 Exercises

In Exercises 1–44, perform each division. Simplify answers when possible.

1. $\dfrac{1}{3} \div \dfrac{1}{2}$
2. $\dfrac{3}{4} \div \dfrac{1}{3}$
3. $\dfrac{1}{5} \div \dfrac{2}{3}$
4. $\dfrac{1}{7} \div \dfrac{2}{5}$

5. $\dfrac{2}{5} \div \dfrac{1}{3}$
6. $\dfrac{3}{7} \div \dfrac{8}{11}$
7. $\dfrac{8}{5} \div \dfrac{7}{2}$
8. $\dfrac{9}{19} \div \dfrac{4}{7}$

9. $\dfrac{21}{14} \div \dfrac{5}{2}$
10. $\dfrac{14}{3} \div \dfrac{10}{3}$
11. $\dfrac{6}{5} \div \dfrac{6}{7}$
12. $\dfrac{6}{5} \div \dfrac{14}{5}$

13. $\dfrac{35}{2} \div \dfrac{15}{2}$
14. $\dfrac{6}{14} \div \dfrac{10}{35}$
15. $\dfrac{x}{2} \div \dfrac{1}{3}$
16. $\dfrac{y}{3} \div \dfrac{1}{2}$

17. $\dfrac{2}{y} \div \dfrac{4}{3}$
18. $\dfrac{3}{a} \div \dfrac{a}{9}$
19. $\dfrac{3x}{2} \div \dfrac{x}{2}$
20. $\dfrac{y}{6} \div \dfrac{2}{3y}$

21. $\dfrac{3x}{y} \div \dfrac{2x}{4}$
22. $\dfrac{3y}{8} \div \dfrac{2y}{4y}$
23. $\dfrac{4x}{3x} \div \dfrac{2y}{9y}$
24. $\dfrac{14}{7y} \div \dfrac{10}{5z}$

25. $\dfrac{x^2}{3} \div \dfrac{2x}{4}$
26. $\dfrac{z^2}{z} \div \dfrac{z}{3z}$
27. $\dfrac{y^2}{5z} \div \dfrac{3z}{2z}$
28. $\dfrac{xy}{x^2} \div \dfrac{y^2}{5}$

29. $\dfrac{x^2y}{3xy} \div \dfrac{xy^2}{6y}$
30. $\dfrac{2xz}{z} \div \dfrac{4x^2}{z^2}$
31. $\dfrac{x+2}{3x} \div \dfrac{x+2}{2}$
32. $\dfrac{z-3}{3z} \div \dfrac{z+3}{z}$

33. $\dfrac{(z-2)^2}{3z^2} \div \dfrac{z-2}{6z}$
34. $\dfrac{(x+7)^2}{x+7} \div \dfrac{(x-3)^2}{x+7}$

35. $\dfrac{(z-7)^2}{z+2} \div \dfrac{z(z-7)}{5z^2}$
36. $\dfrac{y(y+2)}{y^2(y-3)} \div \dfrac{y^2(y+2)}{(y-3)^2}$

37. $\dfrac{x^2-4}{3x+6} \div \dfrac{x-2}{x+2}$
38. $\dfrac{x^2-9}{5x+15} \div \dfrac{x-3}{x+3}$

39. $\dfrac{x^2-1}{3x-3} \div \dfrac{x+1}{3}$
40. $\dfrac{x^2-16}{x-4} \div \dfrac{3x+12}{x}$

41. $\dfrac{5x^2+13x-6}{x+3} \div \dfrac{5x^2-17x+6}{x-2}$
42. $\dfrac{x^2-x-6}{2x^2+9x+10} \div \dfrac{x^2-25}{2x^2+15x+25}$

43. $\dfrac{2x^2+8x-42}{x-3} \div \dfrac{2x^2+14x}{x^2+5x}$
44. $\dfrac{x^2-2x-35}{3x^2+27x} \div \dfrac{x^2+7x+10}{6x^2+12x}$

In Exercises 45–68, perform the indicated operations. In the absence of grouping symbols, multiplications and divisions are performed as they are encountered from left to right.

45. $\dfrac{2}{3} \cdot \dfrac{15}{5} \div \dfrac{10}{5}$
46. $\dfrac{6}{5} \div \dfrac{3}{5} \cdot \dfrac{5}{15}$
47. $\dfrac{6}{7} \div \dfrac{5}{2} \cdot \dfrac{5}{4}$
48. $\dfrac{15}{7} \div \dfrac{5}{2} \div \dfrac{4}{2}$

49. $\dfrac{x}{3} \cdot \dfrac{9}{4} \div \dfrac{x^2}{6}$
50. $\dfrac{y^2}{2} \div \dfrac{4}{y} \cdot \dfrac{y^2}{8}$
51. $\dfrac{x^2}{18} \div \dfrac{x^3}{6} \div \dfrac{12}{x^2}$
52. $\dfrac{y^3}{3y} \cdot \dfrac{3y^2}{4} \div \dfrac{15}{20}$

53. $\dfrac{x^2-1}{x^2-9} \cdot \dfrac{x+3}{x+2} \div \dfrac{5}{x+2}$

54. $\dfrac{2}{3x-3} \div \dfrac{2x+2}{x-1} \cdot \dfrac{5}{x+1}$

55. $\dfrac{x^2-4}{2x+6} \div \dfrac{x+2}{4} \cdot \dfrac{x+3}{x-2}$

56. $\dfrac{x^2-5x}{x+1} \cdot \dfrac{x+1}{x^2+3x} \div \dfrac{x-5}{x-3}$

57. $\dfrac{x-x^2}{x^2-4}\left(\dfrac{2x+4}{x+2} \div \dfrac{5}{x+2}\right)$

58. $\dfrac{2}{3x-3} \div \left(\dfrac{2x+2}{x-1} \cdot \dfrac{5}{x+1}\right)$

59. $\dfrac{y^2}{x+1} \cdot \dfrac{x^2+2x+1}{x^2-1} \div \dfrac{3y}{xy-y}$

60. $\dfrac{x^2-y^2}{x^4-x^3} \div \dfrac{x-y}{x^2} \div \dfrac{x^2+2xy+y^2}{x+y}$

61. $\dfrac{x^2+x-6}{x^2-4} \cdot \dfrac{x^2+2x}{x-2} \div \dfrac{x^2+3x}{x+2}$

62. $\dfrac{x^2-x-6}{x^2+6x-7} \cdot \dfrac{x^2+x-2}{x^2+2x} \div \dfrac{x^2+7x}{x^2-3x}$

63. $(a+2b) \div \left(\dfrac{a^2+4ab+4b^2}{a+b} \div \dfrac{a^2+7ab+10b^2}{a^2+6ab+5b^2}\right)$

64. $(ab-2b^2) \div \left(\dfrac{a^2-ab}{b-a} \cdot \dfrac{a^2-b^2}{a^3-3a^2b+2ab^2}\right)$

65. $\dfrac{x^2+2x-3}{x^2+x} \cdot \dfrac{x^2}{x^2-1} \div (x^2+3x)$

66. $\dfrac{x^2-6x+5}{x+2} \div (x^2+3x-4) \cdot \dfrac{x^2+5x+6}{x^2-2x-15}$

67. $\dfrac{x^2+4x+3}{x^2-y^2} \div \dfrac{xy+y}{xy-x^2} \cdot \dfrac{x^2y+2xy^2+y^3}{x^2+3x}$

68. $\dfrac{a^2-b^2}{a^2-a-2} \cdot \dfrac{a^2-2a-3}{b-a} \div \dfrac{a^2+ab}{a^2-2a}$

In Exercises 69–72, perform the indicated divisions. You will need to factor by grouping and factor a sum or difference of two cubes to simplify the answers.

69. $\dfrac{ab+4a+2b+8}{b^2+4b+16} \div \dfrac{b^2-16}{b^3-64}$

70. $\dfrac{r^3-s^3}{r^2-s^2} \div \dfrac{r^2+rs+s^2}{mr+ms+nr+ns}$

71. $\dfrac{p^3-p^2q+pq^2}{mp-mq+np-nq} \div \dfrac{q^3+p^3}{q^2-p^2}$

72. $\dfrac{s^3-r^3}{r^2+rs+s^2} \div \dfrac{pr-ps-qr+qs}{q^2-p^2}$

Review Exercises

In Review Exercises 1–8, perform the indicated operations and simplify.

1. $-4(y^3-4y^2+3y-2)+6(-2y^2+4)-4(-2y^3-y)$

2. $6(3a^3+2a^2+3)-(-2a^2+4a-2)+5(-2a^3-a^2+2a-3)$

3. $(2r-3)(r^2+5)$

4. $(3x+4y)^2$

5. $(3m+2)(-2m+1)(m-1)$

6. $(2p-q)(p+q)^2$

7. $y-5\overline{)5y^3-3y^2+4y-1}$

8. $x+4\overline{)6x^3+5-4x}$

9. On a test, the lowest score was half the highest score. Find the highest score if the sum of the two scores was 126.

10. A desk is twice as long as it is wide. To protect the finish, a rectangular piece of glass is used to cover its top. Find the dimensions of the desk if the area of the glass is 18 square feet.

11. Receipts from the sale of 35,750 football tickets totaled $109,500. Regular admission tickets sold for $3.50 each, while each student admission ticket cost $1.00. How many of each were sold?

12. One type of candy is worth $4.50 per pound, and another is worth $3.00 per pound. How many pounds of each type are needed to make 150 pounds of a mixture worth $3.90 per pound?

12.4 Adding and Subtracting Fractions with Like Denominators

- Adding Fractions with Like Denominators
- Subtracting Fractions with Like Denominators

Adding Fractions with Like Denominators

To add fractions with like denominators, we add the numerators and keep the denominator. For example, to add $\frac{2}{7}$ and $\frac{3}{7}$, we proceed as follows:

$$\frac{2}{7} + \frac{3}{7} = \frac{2+3}{7} = \frac{5}{7}$$

In general, we have the following rule.

> **Adding Fractions with Like Denominators:** If a, b, and d represent real numbers, and $d \neq 0$, then
>
> $$\frac{a}{d} + \frac{b}{d} = \frac{a+b}{d}$$

EXAMPLE 1

a. $\dfrac{5}{9} + \dfrac{2}{9} = \dfrac{5+2}{9} = \dfrac{7}{9}$

b. $\dfrac{8}{41} + \dfrac{21}{41} = \dfrac{8+21}{41} = \dfrac{29}{41}$

c. $\dfrac{x}{7} + \dfrac{y}{7} = \dfrac{x+y}{7}$

d. $\dfrac{x}{7} + \dfrac{3x}{7} = \dfrac{x+3x}{7} = \dfrac{4x}{7}$

e. $\dfrac{3x+y}{5x} + \dfrac{x+y}{5x}$
$= \dfrac{3x+y+x+y}{5x}$
$= \dfrac{4x+2y}{5x}$

f. $\dfrac{3x}{7y} + \dfrac{4x}{7y} = \dfrac{3x+4x}{7y}$
$= \dfrac{7x}{7y}$
$= \dfrac{x}{y}$

EXAMPLE 2

Add: $\dfrac{3x+21}{5x+10} + \dfrac{8x+1}{5x+10}$.

Solution

$$\frac{3x + 21}{5x + 10} + \frac{8x + 1}{5x + 10} = \frac{3x + 21 + 8x + 1}{5x + 10}$$ Add the numerators and keep the denominator.

$$= \frac{11x + 22}{5x + 10}$$ Combine like terms.

$$= \frac{11(\cancel{x + 2})}{5(\cancel{x + 2})}$$ Factor and divide out the common factor of $x + 2$.

$$= \frac{11}{5}$$

Work Progress Check 8.

Progress Check 8

Add: $\dfrac{x}{2x + 1} + \dfrac{x + 1}{2x + 1}$.

Subtracting Fractions with Like Denominators

To subtract fractions with like denominators, we subtract the numerators and keep the denominator.

> **Subtracting Fractions with Like Denominators:** If a, b, and d represent real numbers, and $d \neq 0$, then
>
> $$\frac{a}{d} - \frac{b}{d} = \frac{a - b}{d}$$

EXAMPLE 3

Perform each subtraction and simplify.   **a.** $\dfrac{5x}{3} - \dfrac{2x}{3}$ and   **b.** $\dfrac{5x + 1}{x - 3} - \dfrac{4x - 2}{x - 3}$.

Solution

a. $\dfrac{5x}{3} - \dfrac{2x}{3} = \dfrac{5x - 2x}{3}$ Subtract the numerators and keep the denominator.

$\phantom{\dfrac{5x}{3} - \dfrac{2x}{3}} = \dfrac{3x}{3}$ Combine like terms.

$\phantom{\dfrac{5x}{3} - \dfrac{2x}{3}} = \dfrac{x}{1}$ Simplify the fraction.

$\phantom{\dfrac{5x}{3} - \dfrac{2x}{3}} = x$ Denominators of 1 need not be written.

b. $\dfrac{5x + 1}{x - 3} - \dfrac{4x - 2}{x - 3} = \dfrac{(5x + 1) - (4x - 2)}{x - 3}$ Subtract the numerators and keep the denominator.

$\phantom{\dfrac{5x + 1}{x - 3} - \dfrac{4x - 2}{x - 3}} = \dfrac{5x + 1 - 4x + 2}{x - 3}$ Remove parentheses.

$\phantom{\dfrac{5x + 1}{x - 3} - \dfrac{4x - 2}{x - 3}} = \dfrac{x + 3}{x - 3}$ Combine like terms.

Work Progress Check 9.

Progress Check 9

Subtract: $\dfrac{5x + 8}{2x + 3} - \dfrac{x + 2}{2x + 3}$.

8. 1
9. 2

Progress Check Answers

Section 12.4 Adding and Subtracting Fractions with Like Denominators

EXAMPLE 4

Perform the indicated operations: $\dfrac{3x+1}{x-7} - \dfrac{5x+2}{x-7} + \dfrac{2x+1}{x-7}$.

Solution

$$\dfrac{3x+1}{x-7} - \dfrac{5x+2}{x-7} + \dfrac{2x+1}{x-7}$$

$$= \dfrac{(3x+1) - (5x+2) + (2x+1)}{x-7}$$ Combine the numerators and keep the denominator.

$$= \dfrac{3x+1-5x-2+2x+1}{x-7}$$ Remove parentheses.

$$= \dfrac{0}{x-7}$$ Combine like terms.

$$= 0$$ Simplify.

12.4 Exercises

In Exercises 1–30, perform each addition. Write all answers in lowest terms.

1. $\dfrac{1}{3} + \dfrac{1}{3}$
2. $\dfrac{3}{4} + \dfrac{3}{4}$
3. $\dfrac{1}{5} + \dfrac{2}{5}$
4. $\dfrac{3}{7} + \dfrac{2}{7}$

5. $\dfrac{2}{9} + \dfrac{1}{9}$
6. $\dfrac{5}{7} + \dfrac{9}{7}$
7. $\dfrac{8}{7} + \dfrac{6}{7}$
8. $\dfrac{9}{11} + \dfrac{2}{11}$

9. $\dfrac{21}{14} + \dfrac{7}{14}$
10. $\dfrac{14}{3} + \dfrac{10}{3}$
11. $\dfrac{6}{7} + \dfrac{6}{7}$
12. $\dfrac{6}{5} + \dfrac{14}{5}$

13. $\dfrac{35}{8} + \dfrac{15}{8}$
14. $\dfrac{6}{14} + \dfrac{10}{14}$
15. $\dfrac{14x}{11} + \dfrac{30x}{11}$
16. $\dfrac{6a}{10} + \dfrac{28a}{10}$

17. $\dfrac{-77y}{126} + \dfrac{-7y}{126}$
18. $\dfrac{-39a}{15} + \dfrac{-21a}{15}$
19. $\dfrac{15z}{22} + \dfrac{-15z}{22}$
20. $\dfrac{-30rs}{21} + \dfrac{30rs}{21}$

21. $\dfrac{2x}{y} + \dfrac{2x}{y}$
22. $\dfrac{3y}{5} + \dfrac{2y}{5}$
23. $\dfrac{4y}{3x} + \dfrac{2y}{3x}$
24. $\dfrac{4}{7y} + \dfrac{10}{7y}$

25. $\dfrac{x^2}{4y} + \dfrac{x^2}{4y}$
26. $\dfrac{r^2}{r} + \dfrac{r^2}{r}$
27. $\dfrac{y+2}{5z} + \dfrac{y+4}{5z}$
28. $\dfrac{x+3}{x^2} + \dfrac{x+5}{x^2}$

29. $\dfrac{3x-5}{x-2} + \dfrac{6x-13}{x-2}$
30. $\dfrac{8x-7}{x+3} + \dfrac{2x+37}{x+3}$

In Exercises 31–60, perform each subtraction. Simplify answers if possible.

31. $\dfrac{5}{7} - \dfrac{4}{7}$
32. $\dfrac{5}{9} - \dfrac{3}{9}$
33. $\dfrac{4}{3} - \dfrac{8}{3}$
34. $\dfrac{7}{11} - \dfrac{4}{11}$

35. $\dfrac{17}{13} - \dfrac{15}{13}$
36. $\dfrac{18}{31} - \dfrac{18}{31}$
37. $\dfrac{21}{23} - \dfrac{45}{23}$
38. $\dfrac{35}{72} - \dfrac{44}{72}$

39. $\dfrac{39}{37} - \dfrac{2}{37}$

40. $\dfrac{35}{99} - \dfrac{13}{99}$

41. $\dfrac{-47}{123} - \dfrac{4}{123}$

42. $\dfrac{-23}{17} - \dfrac{11}{17}$

43. $\dfrac{15}{21} - \left(\dfrac{-15}{21}\right)$

44. $\dfrac{-37}{25} - \left(\dfrac{-22}{25}\right)$

45. $\dfrac{2x}{y} - \dfrac{x}{y}$

46. $\dfrac{7y}{5} - \dfrac{4y}{5}$

47. $\dfrac{9y}{3x} - \dfrac{6y}{3x}$

48. $\dfrac{24}{7y} - \dfrac{10}{7y}$

49. $\dfrac{3x^2}{4x} - \dfrac{x^2}{4x}$

50. $\dfrac{5r^2}{2r} - \dfrac{r^2}{2r}$

51. $\dfrac{y+2}{5z} - \dfrac{y+4}{5z}$

52. $\dfrac{x+3}{x^2} - \dfrac{x+5}{x^2}$

53. $\dfrac{6x-5}{3xy} - \dfrac{3x-5}{3xy}$

54. $\dfrac{7x+7}{5y} - \dfrac{2x+7}{5y}$

55. $\dfrac{y+2}{2z} - \dfrac{y+4}{2z}$

56. $\dfrac{2x-3}{x^2} - \dfrac{x-3}{x^2}$

57. $\dfrac{5x+5}{3xy} - \dfrac{2x-4}{3xy}$

58. $\dfrac{8x-7}{2y} - \dfrac{2x+7}{2y}$

59. $\dfrac{3y-2}{y+3} - \dfrac{2y-5}{y+3}$

60. $\dfrac{5x+8}{x+5} - \dfrac{3x-2}{x+5}$

In Exercises 61–76, perform the indicated operations. Simplify answers if possible.

61. $\dfrac{3}{7} - \dfrac{5}{7} + \dfrac{2}{7}$

62. $\dfrac{3}{4} - \dfrac{5}{4} + \dfrac{8}{4}$

63. $\dfrac{3}{5} - \dfrac{2}{5} + \dfrac{7}{5}$

64. $\dfrac{5}{11} - \dfrac{8}{11} + \dfrac{14}{11}$

65. $\dfrac{13x}{15} + \dfrac{12x}{15} - \dfrac{5x}{15}$

66. $\dfrac{13y}{32} + \dfrac{13y}{32} - \dfrac{10y}{32}$

67. $\dfrac{x}{3y} + \dfrac{2x}{3y} - \dfrac{x}{3y}$

68. $\dfrac{5y}{8x} + \dfrac{4y}{8x} - \dfrac{y}{8x}$

69. $\dfrac{3x}{y+2} - \dfrac{3y}{y+2} + \dfrac{x+y}{y+2}$

70. $\dfrac{3y}{x-5} + \dfrac{x}{x-5} - \dfrac{y-x}{x-5}$

71. $\dfrac{x+1}{x-2} - \dfrac{2(x-3)}{x-2} + \dfrac{3(x+1)}{x-2}$

72. $\dfrac{x^2-4}{x+2} + \dfrac{2(x^2-9)}{x+2} - \dfrac{3(x^2-5)}{x+2}$

73. $\dfrac{3xy}{x-y} - \dfrac{x(3y-x)}{x-y} - \dfrac{x(x-y)}{x-y}$

74. $\dfrac{x^2+4x+1}{(x-1)^2} - \dfrac{x(x+1)}{(x-1)^2} - \dfrac{x}{(x-1)^2}$

75. $\dfrac{2(2a+b)}{(a-b)^2} - \dfrac{2(2b+a)}{(a-b)^2} + \dfrac{3(b-a)}{(a-b)^2}$

76. $\dfrac{2(x-2)}{2x-3} + \dfrac{2(2x+1)}{2x-3} - \dfrac{2(5x-4)}{2x-3}$

Review Exercises

In Review Exercises 1–6, write each number in prime-factored form.

1. 49
2. 64
3. 136
4. 242
5. 102
6. 315

In Review Exercises 7–12, factor each expression.

7. $x^2 - 2x - 15$
8. $3x^2 - 5xy - 2y^2$
9. $2x^2 - 8$
10. $3y^2 - 27$
11. $ax + ay - 5x - 5y$
12. $xy - x^2 + y - x$

12.5 Adding and Subtracting Fractions with Unlike Denominators

- Adding Fractions with Unlike Denominators
- The Least Common Denominator
- Subtracting Fractions with Unlike Denominators

To add fractions with unlike denominators, we change them to fractions with like denominators. To do so, we **build** each fraction by multiplying numerator and denominator by some nonzero number to get a **common denominator**.

For example, to add the fractions $\frac{4}{7}$ and $\frac{3}{5}$, we proceed as follows:

$$\frac{4}{7} + \frac{3}{5} = \frac{4 \cdot 5}{7 \cdot 5} + \frac{3 \cdot 7}{5 \cdot 7}$$

Multiply both the numerator and the denominator of the first fraction by 5 and those of the second fraction by 7 to get a common denominator.

$$= \frac{20}{35} + \frac{21}{35}$$

Do the indicated multiplications.

$$= \frac{41}{35}$$

Add the fractions.

EXAMPLE 1

Change **a.** $\frac{1}{2}$, **b.** $\frac{3}{5}$, and **c.** $\frac{7}{10}$, into fractions with a common denominator of 30.

Solution

a. $\frac{1}{2} = \frac{1 \cdot 15}{2 \cdot 15} = \frac{15}{30}$ **b.** $\frac{3}{5} = \frac{3 \cdot 6}{5 \cdot 6} = \frac{18}{30}$

c. $\frac{7}{10} = \frac{7 \cdot 3}{10 \cdot 3} = \frac{21}{30}$

Work Progress Check 10.

Adding Fractions with Unlike Denominators

To add fractions with unlike denominators, we express the fractions in equivalent forms with the same denominator, and add them as in the previous section.

EXAMPLE 2

Add: $\frac{5}{14} + \frac{2}{21}$.

Solution

Build the fractions so that each has a denominator of 42.

$$\frac{5}{14} + \frac{2}{21} = \frac{5 \cdot 3}{14 \cdot 3} + \frac{2 \cdot 2}{21 \cdot 2}$$

Multiply both the numerator and denominator of the first fraction by 3 and those of the second fraction by 2.

$$= \frac{15}{42} + \frac{4}{42}$$

Do the indicated multiplications.

$$= \frac{19}{42}$$

Add the fractions.

Progress Check 10

Change $\frac{2}{3}$ into a fraction with a denominator of 15.

10. $\frac{10}{15}$

Progress Check Answers

The Least Common Denominator

In Example 2, the number 42 was used as the common denominator because it is divisible by both 14 and 21. The number 42 is also the *smallest* number that is divisible by both 14 and 21. The smallest number that can be divided by each of the denominators of a set of fractions is called the **least** (or **lowest**) **common denominator (LCD)** of those fractions.

We have seen that there is a process to find the LCD of a set of fractions.

Finding the Least Common Denominator (LCD):

1. Write down each of the different denominators that appear in the given fractions.
2. Factor each of these denominators completely.
3. Form a product using each of the different factors obtained in Step 2. In the product, use each different factor the *greatest* number of times it appears in any *single* denominator. The product is the least common denominator.

EXAMPLE 3

Several fractions have denominators of 18, 24, and 36. Find the LCD.

Solution

First, we write down and factor each denominator into products of prime numbers.

$$18 = 2 \cdot 3 \cdot 3$$
$$24 = 2 \cdot 2 \cdot 2 \cdot 3$$
$$36 = 2 \cdot 2 \cdot 3 \cdot 3$$

Then we form a product with the factors of 2 and 3. We use each of these factors the greatest number of times it appears in any single denominator. That is, we use the factor 2 three times because 2 appears three times as a factor of 24. We use the factor of 3 twice because it occurs twice as a factor of 18 and 36. Thus, the least common denominator of fractions with denominators of 24, 18, and 36 is

$$\text{LCD} = 2 \cdot 2 \cdot 2 \cdot 3 \cdot 3$$
$$= 8 \cdot 9$$
$$= 72$$

Work Progress Check 11.

Progress Check 11

Find the LCD of $\frac{5}{18}$, $\frac{7}{26}$, and $\frac{11}{117}$.

11. 234

EXAMPLE 4

Add: $\frac{1}{24} + \frac{5}{18} + \frac{7}{36}$.

Solution

Each of the fractions $\frac{1}{24}$, $\frac{5}{18}$, and $\frac{7}{36}$ can be written as a fraction with an LCD of $2 \cdot 2 \cdot 2 \cdot 3 \cdot 3$, or 72.

$$\frac{1}{24} + \frac{5}{18} + \frac{7}{36} = \frac{1}{2 \cdot 2 \cdot 2 \cdot 3} + \frac{5}{2 \cdot 3 \cdot 3} + \frac{7}{2 \cdot 2 \cdot 3 \cdot 3} \qquad \text{Factor each denominator.}$$

In each of these fractions, we multiply both the numerator and the denominator by whatever is necessary to build the denominator to the required $2 \cdot 2 \cdot 2 \cdot 3 \cdot 3$.

$$= \frac{1 \cdot \mathbf{3}}{2 \cdot 2 \cdot 2 \cdot 3 \cdot \mathbf{3}} + \frac{5 \cdot \mathbf{2 \cdot 2}}{2 \cdot 3 \cdot 3 \cdot \mathbf{2 \cdot 2}} + \frac{7 \cdot \mathbf{2}}{2 \cdot 2 \cdot 3 \cdot 3 \cdot \mathbf{2}} \qquad \text{Build each fraction.}$$

$$= \frac{3 + 20 + 14}{72} \qquad \text{Do the indicated multiplications and add the fractions.}$$

$$= \frac{37}{72} \qquad \text{Combine like terms.}$$

EXAMPLE 5

Add: $\dfrac{1}{x} + \dfrac{x}{y}$.

Solution

$$\frac{1}{x} + \frac{x}{y} = \frac{(1)y}{(x)y} + \frac{x(x)}{x(y)} \qquad \text{Build the fractions to get the common denominator of } xy.$$

$$= \frac{y}{xy} + \frac{x^2}{xy} \qquad \text{Do the indicated multiplication.}$$

$$= \frac{y + x^2}{xy} \qquad \text{Add the fractions.}$$

Work Progress Check 12.

Progress Check 12

Add: $\dfrac{x}{y} + \dfrac{y}{2}$.

Subtracting Fractions with Unlike Denominators

EXAMPLE 6

Subtract: $\dfrac{3}{x^2 y} - \dfrac{1}{xy^2}$.

Solution

Factor each denominator to find the least common denominator.

$$x^2 y = x \cdot x \cdot y$$
$$xy^2 = x \cdot y \cdot y$$

In these denominators, the factor x occurs at most twice, and the factor y occurs at most twice. Thus, the LCD is

$$\text{LCD} = x \cdot x \cdot y \cdot y = x^2 y^2$$

Build each given fraction into a new fraction with a denominator of $x^2 y^2$.

$$\frac{3}{x^2 y} - \frac{1}{xy^2} = \frac{3 \cdot y}{x \cdot x \cdot y \cdot y} - \frac{1 \cdot x}{x \cdot y \cdot y \cdot x} \qquad \text{Factor each denominator and build each fraction.}$$

$$= \frac{3y - x}{x^2 y^2} \qquad \text{Perform the multiplications and subtract the fractions.}$$

Progress Check Answers

12. $\dfrac{2x + y^2}{2y}$

EXAMPLE 7

Subtract: $\dfrac{x}{x+1} - \dfrac{3}{x}$.

Solution

$\dfrac{x}{x+1} - \dfrac{3}{x} = \dfrac{x(x)}{(x+1)x} - \dfrac{3(x+1)}{x(x+1)}$ Build the fractions to get the common denominator of $(x+1)x$.

$= \dfrac{x^2}{(x+1)x} - \dfrac{3x+3}{x(x+1)}$ Perform the multiplications in the numerators.

$= \dfrac{x^2 - 3x - 3}{x(x+1)}$ Subtract the numerators and keep the common denominator.

Work Progress Check 13.

Progress Check 13

Subtract: $\dfrac{x}{x-1} - \dfrac{2}{x}$.

EXAMPLE 8

Perform the indicated operations: $\dfrac{3}{x^2-y^2} + \dfrac{2}{x-y} - \dfrac{1}{x+y}$.

Solution

Find the least common denominator.

$\left.\begin{array}{r} x^2 - y^2 = (x-y)(x+y) \\ x - y = x - y \\ x + y = x + y \end{array}\right\}$ Factor each denominator, where possible.

The least common denominator is $(x-y)(x+y)$. Build each given fraction into a new fraction with that common denominator.

$\dfrac{3}{x^2-y^2} + \dfrac{2}{x-y} - \dfrac{1}{x+y}$

$= \dfrac{3}{(x-y)(x+y)} + \dfrac{2}{x-y} - \dfrac{1}{x+y}$ Factor.

$= \dfrac{3}{(x-y)(x+y)} + \dfrac{2(x+y)}{(x-y)(x+y)} - \dfrac{1(x-y)}{(x+y)(x-y)}$ Build the fractions to get a common denominator.

$= \dfrac{3 + 2(x+y) - (x-y)}{(x-y)(x+y)}$ Add the fractions.

$= \dfrac{3 + 2x + 2y - x + y}{(x-y)(x+y)}$ Remove the parentheses in the numerator.

$= \dfrac{3 + x + 3y}{(x-y)(x+y)}$ Combine like terms.

EXAMPLE 9

Perform the subtraction $\dfrac{a}{a-1} - \dfrac{2}{a^2-1}$ and simplify.

---

13. $\dfrac{x^2 - 2x + 2}{x(x-1)}$

Progress Check Answers

Section 12.5 Adding and Subtracting Fractions with Unlike Denominators

Solution

$$\frac{a}{a-1} - \frac{2}{a^2-1}$$

$$= \frac{a(a+1)}{(a-1)(a+1)} - \frac{2}{(a+1)(a-1)} \quad \text{Factor } a^2 - 1 \text{ and build the first fraction.}$$

$$= \frac{a^2+a}{(a-1)(a+1)} - \frac{2}{(a+1)(a-1)} \quad \text{Remove parentheses in the numerator.}$$

$$= \frac{a^2+a-2}{(a-1)(a+1)} \quad \text{Subtract the fractions.}$$

$$= \frac{(a+2)\cancel{(a-1)}}{\cancel{(a-1)}(a+1)} \quad \text{Factor.}$$

$$= \frac{a+2}{a+1} \quad \text{Divide out the common factor of } a-1.$$

EXAMPLE 10

Perform the addition $\dfrac{m+1}{2m+6} + \dfrac{4-m^2}{2m^2+2m-12}$.

Solution

$$\frac{m+1}{2m+6} + \frac{4-m^2}{2m^2+2m-12}$$

$$= \frac{m+1}{2(m+3)} + \frac{4-m^2}{2(m+3)(m-2)} \quad \text{Factor } 2m+6 \text{ and } 2m^2 + 2m - 12.$$

$$= \frac{(m+1)(m-2)}{2(m+3)(m-2)} + \frac{4-m^2}{2(m+3)(m-2)} \quad \text{Write each fraction with a denominator of } 2(m+3)(m-2).$$

Add the fractions by adding the numerators and keeping the denominator.

$$\frac{m+1}{2m+6} + \frac{4-m^2}{2m^2+2m-12}$$

$$= \frac{(m+1)(m-2)+4-m^2}{2(m+3)(m-2)}$$

$$= \frac{m^2-m-2+4-m^2}{2(m+3)(m-2)}$$

$$= \frac{-m+2}{2(m+3)(m-2)}$$

$$= \frac{-\cancel{(m-2)}}{2(m+3)\cancel{(m-2)}} \quad \text{Factor } -1 \text{ from } -m+2.$$

$$= \frac{-1}{2(m+3)} \quad \text{Divide out the common factor of } m-2.$$

12.5 Exercises

In Exercises 1–20, build each fraction into an equivalent fraction with the indicated denominator.

1. $\dfrac{2}{3}$; 6
2. $\dfrac{3}{4}$; 12
3. $\dfrac{25}{4}$; 20
4. $\dfrac{19}{21}$; 42

5. $\dfrac{2}{x}$; x^2
6. $\dfrac{3}{y}$; y^2
7. $\dfrac{5}{y}$; xy
8. $\dfrac{3}{x}$; xy

9. $\dfrac{8}{x}$; $x^2 y$
10. $\dfrac{7}{y}$; xy^2
11. $\dfrac{3x}{x+1}$; $(x+1)^2$
12. $\dfrac{5y}{y-2}$; $(y-2)^2$

13. $\dfrac{2y}{x}$; $x^2 + x$
14. $\dfrac{3x}{y}$; $y^2 - y$
15. $\dfrac{z}{z-1}$; $z^2 - 1$
16. $\dfrac{y}{y+2}$; $y^2 - 4$

17. $\dfrac{x+2}{x-2}$; $x^2 - 4$
18. $\dfrac{x-3}{x+3}$; $x^2 - 9$
19. $\dfrac{2}{x+1}$; $x^2 + 3x + 2$
20. $\dfrac{3}{x-1}$; $x^2 + x - 2$

In Exercises 21–38, several denominators are given. Find the least common denominator.

21. 15, 12
22. 18, 24
23. 14, 21, 42
24. 12, 15, 10
25. $2x$, $6x$
26. $3y$, $9y$
27. $x^2 y$, $x^2 y^2$, xy^2
28. xy, x^2, y^2
29. $3x$, $6y$, $9xy$
30. $2x^2$, $6y$, $3xy$
31. $x^2 - 1$, $x + 1$
32. $y^2 - 9$, $y - 3$
33. $x^2 + 6x$, $x + 6$, x
34. $xy^2 - xy$, xy, $y - 1$
35. $x^2 - x - 2$, $(x-2)^2$
36. $x^2 + 2x - 3$, $(x+3)^2$
37. $x^2 - 4x - 5$, $x^2 - 25$
38. $x^2 - x - 6$, $x^2 - 9$

In Exercises 39–82, perform the indicated operations. Simplify answers if possible.

39. $\dfrac{1}{2} + \dfrac{2}{3}$
40. $\dfrac{3}{4} + \dfrac{1}{2}$
41. $\dfrac{2}{3} - \dfrac{5}{6}$
42. $\dfrac{4}{9} - \dfrac{2}{3}$

43. $\dfrac{2y}{9} + \dfrac{y}{3}$
44. $\dfrac{3x}{8} + \dfrac{3x}{4}$
45. $\dfrac{8a}{15} - \dfrac{5a}{12}$
46. $\dfrac{2b}{15} - \dfrac{2b}{5}$

47. $\dfrac{21x}{14} - \dfrac{5x}{21}$
48. $\dfrac{7y}{6} + \dfrac{10y}{9}$
49. $\dfrac{4x}{3} + \dfrac{2x}{y}$
50. $\dfrac{2y}{5x} - \dfrac{y}{2}$

51. $\dfrac{2}{x} - 3x$
52. $14 + \dfrac{10}{y^2}$
53. $\dfrac{x^2}{2y^2} + \dfrac{x^2}{3xy}$
54. $\dfrac{r^2}{2rs} + \dfrac{r^2}{6s^2}$

55. $\dfrac{y+2}{5y} + \dfrac{y+4}{15y}$
56. $\dfrac{x+3}{x^2} + \dfrac{x+5}{2x}$
57. $\dfrac{x+5}{xy} - \dfrac{x-1}{x^2 y}$
58. $\dfrac{y-7}{y^2} - \dfrac{y+7}{2y}$

59. $\dfrac{x}{x+1} + \dfrac{x-1}{x}$
60. $\dfrac{3x}{xy} + \dfrac{x+1}{y-1}$
61. $\dfrac{3}{x-2} - (x-1)$
62. $a + 1 - \dfrac{3}{a+3}$

63. $\dfrac{x-1}{x} + \dfrac{y+1}{y}$
64. $\dfrac{a+2}{b} + \dfrac{b-2}{a}$
65. $\dfrac{x}{x-2} + \dfrac{4+2x}{x^2-4}$
66. $\dfrac{y}{y+3} - \dfrac{2y-6}{y^2-9}$

67. $\dfrac{x+1}{x-1} + \dfrac{x-1}{x+1}$ 68. $\dfrac{2x}{x+2} + \dfrac{x+1}{x-3}$ 69. $\dfrac{2x+2}{x-2} - \dfrac{2x}{x+2}$ 70. $\dfrac{y+3}{y-1} - \dfrac{y+4}{y+1}$

71. $\dfrac{x}{(x-2)^2} + \dfrac{x-4}{(x+2)(x-2)}$ 72. $\dfrac{a-2}{(a+3)^2} - \dfrac{a}{a-3}$

73. $\dfrac{2x}{x^2-3x+2} + \dfrac{2x}{x-1} - \dfrac{x}{x-2}$ 74. $\dfrac{4a}{a-2} - \dfrac{3a}{a-3} + \dfrac{4a}{a^2-5a+6}$

75. $\dfrac{2x}{x-1} + \dfrac{3x}{x+1} - \dfrac{x+3}{x^2-1}$ 76. $\dfrac{a}{a-1} - \dfrac{2}{a+2} + \dfrac{3(a-2)}{a^2+a-2}$

77. $-2 - \dfrac{y+1}{y-3} + \dfrac{3(y-2)}{y}$ 78. $\dfrac{3(a-b)}{a+b} + \dfrac{2(a+b)}{a-b} - 1$

79. $\dfrac{x+1}{2x+4} - \dfrac{x^2}{2x^2-8}$ 80. $\dfrac{x+1}{x+2} - \dfrac{x^2+1}{x^2-x-6}$

81. $\dfrac{x-1}{x+2} + \dfrac{x}{3-x} + \dfrac{9x+3}{x^2-x-6}$ 82. $\dfrac{x+1}{x-3} - \dfrac{x^2+9x}{x^2-2x-3} - \dfrac{5}{3-x}$

In Exercises 83–84, show that each formula is true.

83. $\dfrac{a}{b} + \dfrac{c}{d} = \dfrac{ad+bc}{bd}$ 84. $\dfrac{a}{b} - \dfrac{c}{d} = \dfrac{ad-bc}{bd}$

Review Exercises

In Review Exercises 1–6, let $a = -2$, $b = -3$, and $c = 4$. Simplify each expression.

1. $c + ab$ 2. $ac - bc$ 3. $a^2 + bc$ 4. $ab^2 - ac^2$

5. $\dfrac{ac+bc}{a+b+c}$ 6. $\dfrac{-4a+3b+6c}{abc-1}$

7. Define a prime number.

8. Define a natural number.

9. Define a composite number.

10. Define an integer.

In Review Exercises 11–12, write each statement as an algebraic expression.

11. The quotient obtained when the number x is divided by a number that is 3 greater than the product of x and y.

12. The product of three consecutive integers of which n is the smallest.

12.6 Complex Fractions

- **Methods for Simplifying Complex Fractions**
- **Reviewing Negative Exponents**

Fractions such as

$$\frac{\frac{1}{3}}{\frac{4}{4}}, \quad \frac{\frac{5}{3}}{\frac{2}{9}}, \quad \frac{x+\frac{1}{2}}{3-x}, \quad \text{and} \quad \frac{\frac{x+1}{2}}{x+\frac{1}{x}}$$

which have fractions in the numerators or denominators, are called **complex fractions**. Fortunately, they can often be simplified. For example, to simplify

$$\frac{\frac{1}{3}}{4}$$

we multiply both the numerator and denominator of the fraction by 3 to eliminate the denominator of the fraction in the numerator, and simplify:

$$\frac{\frac{1}{3}}{4} = \frac{\frac{1}{3} \cdot 3}{4 \cdot 3} = \frac{\frac{1}{3} \cdot \frac{3}{1}}{12} = \frac{\frac{3}{3}}{12} = \frac{1}{12}$$

Methods for Simplifying Complex Fractions

We will discuss the following two methods for simplifying complex fractions.

Methods for Simplifying a Complex Fraction:

Method 1
Write the numerator and denominator of the complex fraction as single fractions. Then perform the indicated division of the two fractions and simplify.

Method 2
Multiply both the numerator and denominator of the complex fraction by the LCD of all of the fractions that appear in that numerator and denominator. Then simplify.

To simplify the complex fraction $\dfrac{\frac{3}{5}+1}{2-\frac{1}{5}}$, for example, we can use Method 1 and proceed as follows:

$$\dfrac{\frac{3}{5}+1}{2-\frac{1}{5}} = \dfrac{\frac{3}{5}+\frac{5}{5}}{\frac{10}{5}-\frac{1}{5}}$$ Change 1 to $\frac{5}{5}$ and 2 to $\frac{10}{5}$.

$$= \dfrac{\frac{8}{5}}{\frac{9}{5}}$$ Add the fractions in the numerator and subtract the fractions in the denominator.

$$= \frac{8}{5} \div \frac{9}{5}$$ Express the complex fraction as an equivalent division problem.

$$= \frac{8}{5} \cdot \frac{5}{9}$$ Invert the divisor and multiply.

$$= \frac{8 \cdot 5}{5 \cdot 9}$$ Multiply the fractions.

$$= \frac{8}{9}$$ Simplify.

To use Method 2, we proceed as follows:

$$\dfrac{\frac{3}{5}+1}{2-\frac{1}{5}} = \dfrac{5\left(\frac{3}{5}+1\right)}{5\left(2-\frac{1}{5}\right)}$$ Multiply both the numerator and denominator by 5, the LCD of $\frac{3}{5}$ and $\frac{1}{5}$.

$$= \dfrac{5 \cdot \frac{3}{5} + 5 \cdot 1}{5 \cdot 2 - 5 \cdot \frac{1}{5}}$$ Remove parentheses.

$$= \dfrac{3+5}{10-1}$$ Simplify.

$$= \frac{8}{9}$$ Simplify.

In this case, Method 2 is easier than Method 1. With practice we will be able to see which method is best to use in any given situation.

EXAMPLE 1

Simplify $\dfrac{\frac{x}{3}}{\frac{y}{3}}$.

Solution

Method 1	Method 2
$\dfrac{\frac{x}{3}}{\frac{y}{3}} = \dfrac{x}{3} \div \dfrac{y}{3}$	$\dfrac{\frac{x}{3}}{\frac{y}{3}} = \dfrac{\left(\frac{x}{3}\right)3}{\left(\frac{y}{3}\right)3}$
$= \dfrac{x}{3} \cdot \dfrac{3}{y}$	$= \dfrac{\frac{x}{1}}{\frac{y}{1}}$
$= \dfrac{3x}{3y}$	$= \dfrac{x}{y}$
$= \dfrac{x}{y}$	

Work Progress Check 14.

Progress Check 14

Simplify: $\dfrac{\frac{a}{b}}{\frac{b}{3}}$.

EXAMPLE 2

Simplify $\dfrac{\frac{x}{x+1}}{\frac{y}{x}}$.

Solution

Method 1	Method 2
$\dfrac{\frac{x}{x+1}}{\frac{y}{x}} = \dfrac{x}{x+1} \div \dfrac{y}{x}$	$\dfrac{\frac{x}{x+1}}{\frac{y}{x}} = \dfrac{\left(\frac{x}{x+1}\right)x(x+1)}{\left(\frac{y}{x}\right)x(x+1)}$
$= \dfrac{x}{x+1} \cdot \dfrac{x}{y}$	$= \dfrac{\frac{x^2}{1}}{\frac{y(x+1)}{1}}$
$= \dfrac{x^2}{y(x+1)}$	$= \dfrac{x^2}{y(x+1)}$

EXAMPLE 3

Simplify $\dfrac{1 + \frac{1}{x}}{1 - \frac{1}{x}}$.

14. $\dfrac{3a}{b^2}$

Progress Check Answers

Solution

Method 1

$$\frac{1+\frac{1}{x}}{1-\frac{1}{x}} = \frac{\frac{x}{x}+\frac{1}{x}}{\frac{x}{x}-\frac{1}{x}}$$

$$= \frac{\frac{x+1}{x}}{\frac{x-1}{x}}$$

$$= \frac{x+1}{x} \div \frac{x-1}{x}$$

$$= \frac{x+1}{x} \cdot \frac{x}{x-1}$$

$$= \frac{x+1}{x-1}$$

Method 2

$$\frac{1+\frac{1}{x}}{1-\frac{1}{x}} = \frac{\left(1+\frac{1}{x}\right)x}{\left(1-\frac{1}{x}\right)x}$$

$$= \frac{x+1}{x-1}$$

Progress Check 15

Simplify: $\dfrac{2-\frac{1}{x}}{1+\frac{2}{x}}$.

Work Progress Check 15.

EXAMPLE 4

Simplify the complex fraction $\dfrac{1}{1+\frac{1}{x+1}}$.

Solution

Use Method 2.

$$\frac{1}{1+\frac{1}{x+1}} = \frac{1(x+1)}{\left(1+\frac{1}{x+1}\right)(x+1)} \quad \text{Multiply numerator and denominator by } x+1.$$

$$= \frac{x+1}{1(x+1)+1} \quad \text{Simplify.}$$

$$= \frac{x+1}{x+2} \quad \text{Simplify.}$$

Reviewing Negative Exponents

EXAMPLE 5

Simplify the fraction $\dfrac{x^{-1}+y^{-2}}{x^{-2}-y^{-1}}$.

---

15. $\dfrac{2x-1}{x+2}$

Progress Check Answers

Solution

Write the fraction in complex fraction form and simplify:

$$\frac{x^{-1}+y^{-2}}{x^{-2}-y^{-1}} = \frac{\frac{1}{x}+\frac{1}{y^2}}{\frac{1}{x^2}-\frac{1}{y}}$$

$$= \frac{x^2y^2\left(\frac{1}{x}+\frac{1}{y^2}\right)}{x^2y^2\left(\frac{1}{x^2}-\frac{1}{y}\right)} \quad \text{Multiply numerator and denominator by } x^2y^2.$$

$$= \frac{xy^2+x^2}{y^2-x^2y} \quad \text{Remove parentheses.}$$

$$= \frac{x(y^2+x^2)}{y(y-x^2)} \quad \text{Attempt to simplify the fraction by factoring the numerator and denominator.}$$

The result cannot be further simplified. ∎

12.6 Exercises

In Exercises 1–34, simplify each complex fraction.

1. $\dfrac{\frac{2}{3}}{\frac{3}{4}}$

2. $\dfrac{\frac{3}{5}}{\frac{2}{7}}$

3. $\dfrac{\frac{4}{5}}{\frac{32}{15}}$

4. $\dfrac{\frac{7}{8}}{\frac{49}{4}}$

5. $\dfrac{\frac{2}{3}+1}{\frac{1}{3}+1}$

6. $\dfrac{\frac{3}{5}-2}{\frac{2}{5}-2}$

7. $\dfrac{\frac{1}{2}+\frac{3}{4}}{\frac{3}{2}+\frac{1}{4}}$

8. $\dfrac{\frac{2}{3}-\frac{5}{2}}{\frac{2}{3}-\frac{3}{2}}$

9. $\dfrac{\frac{x}{y}}{\frac{1}{x}}$

10. $\dfrac{\frac{y}{x}}{\frac{x}{xy}}$

11. $\dfrac{\frac{5t^2}{9x^2}}{\frac{3t}{x^2t}}$

12. $\dfrac{\frac{5w^2}{4tz}}{\frac{15wt}{z^2}}$

13. $\dfrac{\frac{1}{x}-3}{\frac{5}{x}+2}$

14. $\dfrac{\frac{1}{y}+3}{\frac{3}{y}-2}$

15. $\dfrac{\frac{2}{x}+2}{\frac{4}{x}+2}$

16. $\dfrac{\frac{3}{x}-3}{\frac{9}{x}-3}$

17. $\dfrac{\frac{3y}{x}-y}{y-\frac{y}{x}}$

18. $\dfrac{\frac{y}{x}+3y}{y+\frac{2y}{x}}$

19. $\dfrac{\frac{1}{x+1}}{1+\frac{1}{x+1}}$

20. $\dfrac{\frac{1}{x-1}}{1-\frac{1}{x-1}}$

21. $\dfrac{\frac{x}{x+2}}{\frac{x}{x+2}+x}$

22. $\dfrac{\frac{2}{x-2}}{\frac{2}{x-2}-1}$

23. $\dfrac{1}{\frac{1}{x}+\frac{1}{y}}$

24. $\dfrac{1}{\frac{b}{a}-\frac{a}{b}}$

25. $\dfrac{\dfrac{2}{x}}{\dfrac{2}{y} - \dfrac{4}{x}}$ 26. $\dfrac{\dfrac{2y}{3}}{\dfrac{2y}{3} - \dfrac{8}{y}}$ 27. $\dfrac{3 + \dfrac{3}{x-1}}{3 - \dfrac{3}{x}}$ 28. $\dfrac{2 - \dfrac{2}{x+1}}{2 + \dfrac{2}{x}}$

29. $\dfrac{\dfrac{3}{x} + \dfrac{4}{x+1}}{\dfrac{2}{x+1} - \dfrac{3}{x}}$ 30. $\dfrac{\dfrac{5}{y-3} - \dfrac{2}{y}}{\dfrac{1}{y} + \dfrac{2}{y-3}}$ 31. $\dfrac{\dfrac{2}{x} - \dfrac{3}{x+1}}{\dfrac{2}{x+1} - \dfrac{3}{x}}$ 32. $\dfrac{\dfrac{5}{y} + \dfrac{4}{y+1}}{\dfrac{4}{y} - \dfrac{5}{y+1}}$

33. $\dfrac{\dfrac{1}{y^2+y} - \dfrac{1}{xy+x}}{\dfrac{1}{xy+x} - \dfrac{1}{y^2+y}}$ 34. $\dfrac{\dfrac{2}{b^2-1} - \dfrac{3}{ab-a}}{\dfrac{3}{ab-a} - \dfrac{2}{b^2-1}}$

In Exercises 35–44, write each expression without using negative exponents. Then simplify the resulting complex fraction.

35. $\dfrac{x^{-2}}{y^{-1}}$ 36. $\dfrac{a^{-4}}{b^{-2}}$ 37. $\dfrac{1 + x^{-1}}{x^{-1} - 1}$ 38. $\dfrac{y^{-2} + 1}{y^{-2} - 1}$

39. $\dfrac{a^{-2} + a}{a + 1}$ 40. $\dfrac{t - t^{-2}}{1 - t^{-1}}$ 41. $\dfrac{2x^{-1} + 4x^{-2}}{2x^{-2} + x^{-1}}$ 42. $\dfrac{x^{-2} - 3x^{-3}}{3x^{-2} - 9x^{-3}}$

43. $\dfrac{1 - 25y^{-2}}{1 + 10y^{-1} + 25y^{-2}}$ 44. $\dfrac{1 - 9x^{-2}}{1 - 6x^{-1} + 9x^{-2}}$

Review Exercises

In Review Exercises 1–2, identify the base and the exponent in each expression.

1. -3^4 2. $(2x)^5$

In Review Exercises 3–4, write each expression without using exponents.

3. ab^4 4. $(-3y)^5$

In Review Exercises 5–8, write each expression as an expression involving only one exponent.

5. $t^3 t^4 t^2$ 6. $(a^0 a^2)^3$ 7. $-2r(r^3)^2$ 8. $(s^3)^2 (s^4)^0$

In Review Exercises 9–12, write each expression without using parentheses or negative exponents.

9. $\left(\dfrac{3r}{4r^3}\right)^4$ 10. $\left(\dfrac{12y^{-3}}{3y^2}\right)^{-2}$ 11. $\left(\dfrac{6r^{-2}}{2r^3}\right)^{-2}$ 12. $\left(\dfrac{4x^3}{5x^{-3}}\right)^{-2}$

12.7 Solving Equations That Contain Fractions

- Extraneous Solutions
- Equations with Two Solutions

Equations with fractions often occur in application problems. For example, the equation

$$\frac{1}{r} = \frac{1}{r_1} + \frac{1}{r_2}$$

is used in electronics to calculate the combined resistance r of two resistors wired in parallel.

To solve equations with fractions, it is usually best to eliminate those fractions. To do so, we multiply both sides of the equation by the LCD of the fractions in the equation. For example, to solve the equation $\frac{x}{3} + 1 = \frac{x}{6}$, we multiply both sides of the equation by 6:

$$\frac{x}{3} + 1 = \frac{x}{6}$$

$$6\left(\frac{x}{3} + 1\right) = 6\left(\frac{x}{6}\right)$$

We then remove parentheses, simplify, and solve the resulting equation for x.

$$6 \cdot \frac{x}{3} + 6 \cdot 1 = 6 \cdot \frac{x}{6}$$

$$2x + 6 = x$$

$$x + 6 = 0 \qquad \text{Add } -x \text{ to both sides.}$$

$$x = -6 \qquad \text{Add } -6 \text{ to both sides.}$$

Check:

$$\frac{x}{3} + 1 = \frac{x}{6}$$

$$\frac{-6}{3} + 1 \stackrel{?}{=} \frac{-6}{6} \qquad \text{Replace } x \text{ with } -6.$$

$$-2 + 1 \stackrel{?}{=} -1 \qquad \text{Simplify.}$$

$$-1 = -1$$

EXAMPLE 1

Solve the equation $\frac{4}{x} + 1 = \frac{6}{x}$.

Solution

$$\frac{4}{x} + 1 = \frac{6}{x}$$

$$x\left(\frac{4}{x} + 1\right) = x\left(\frac{6}{x}\right) \qquad \text{Multiply both sides by } x.$$

$$x \cdot \frac{4}{x} + x \cdot 1 = x \cdot \frac{6}{x} \qquad \text{Remove parentheses.}$$

$$4 + x = 6 \qquad \text{Simplify.}$$

$$x = 2 \qquad \text{Add } -4 \text{ to both sides.}$$

Check:

$$\frac{4}{x} + 1 = \frac{6}{x}$$

$$\frac{4}{2} + 1 \stackrel{?}{=} \frac{6}{2} \quad \text{Replace } x \text{ with 2.}$$

$$2 + 1 \stackrel{?}{=} 3 \quad \text{Simplify.}$$

$$3 = 3$$

Extraneous Solutions

If we multiply both sides of an equation by an expression that involves a variable, we might obtain false solutions, called **extraneous solutions**. To eliminate these false solutions, we *must* check the apparent solutions.

EXAMPLE 2

Solve the equation $\dfrac{x+3}{x-1} = \dfrac{4}{x-1}$.

Solution

Multiply both sides of the equation by $x - 1$, the least common denominator of the fractions contained in the equation.

$$\frac{x+3}{x-1} = \frac{4}{x-1}$$

$$(x-1)\frac{x+3}{x-1} = (x-1)\frac{4}{x-1} \quad \text{Multiply both sides by } x - 1.$$

$$x + 3 = 4 \quad \text{Simplify.}$$

$$x = 1 \quad \text{Add } -3 \text{ to both sides.}$$

Because both sides of the equation were multiplied by an expression containing a variable, we must check the apparent solution.

$$\frac{x+3}{x-1} = \frac{4}{x-1}$$

$$\frac{1+3}{1-1} \stackrel{?}{=} \frac{4}{1-1} \quad \text{Replace } x \text{ with 1.}$$

$$\frac{4}{0} \neq \frac{4}{0} \quad \text{Simplify.}$$

Because 0's appear in the denominators of fractions, the fractions are undefined. Thus, 1 is an extraneous solution and must be discarded. The equation has no solutions.

Note that 1 is not in the domain of the variable. Work Progress Check 16.

Progress Check 16

Solve: $\dfrac{2x}{x+2} = \dfrac{-4}{x+2}$.

EXAMPLE 3

Solve the equation $\dfrac{3x+1}{x+1} - 2 = \dfrac{3(x-3)}{x+1}$.

16. -2 is extraneous

Progress Check Answers

Solution

$$\frac{3x+1}{x+1} - 2 = \frac{3(x-3)}{x+1}$$

$$(x+1)\left(\frac{3x+1}{x+1} - 2\right) = (x+1)\left[\frac{3(x-3)}{x+1}\right]$$ Multiply both sides by $x+1$.

$$3x + 1 - 2(x+1) = 3(x-3)$$ Use the distributive property to remove parentheses.

$$3x + 1 - 2x - 2 = 3x - 9$$ Remove parentheses.

$$x - 1 = 3x - 9$$ Combine like terms.

$$-2x = -8$$ Add $-3x$ and 1 to both sides.

$$x = 4$$ Divide both sides by -2.

Check:

$$\frac{3x+1}{x+1} - 2 = \frac{3(x-3)}{x+1}$$

$$\frac{3(4)+1}{4+1} - 2 \stackrel{?}{=} \frac{3(4-3)}{4+1}$$

$$\frac{13}{5} - \frac{10}{5} \stackrel{?}{=} \frac{3(1)}{5}$$

$$\frac{3}{5} = \frac{3}{5}$$

EXAMPLE 4

Solve the equation $\dfrac{x+2}{x+3} + \dfrac{1}{x^2+2x-3} = 1$.

Solution

$$\frac{x+2}{x+3} + \frac{1}{x^2+2x-3} = 1$$

$$\frac{x+2}{x+3} + \frac{1}{(x+3)(x-1)} = 1$$ Factor $x^2 + 2x - 3$.

$$(x+3)(x-1)\left[\frac{x+2}{x+3} + \frac{1}{(x+3)(x-1)}\right] = (x+3)(x-1)1$$ Multiply both sides by $(x+3)(x-1)$.

$$(x+3)(x-1)\frac{x+2}{x+3} + (x+3)(x-1)\frac{1}{(x+3)(x-1)} = (x+3)(x-1)1$$ Remove brackets.

$$(x-1)(x+2) + 1 = (x+3)(x-1)$$ Simplify.

$$x^2 + x - 2 + 1 = x^2 + 2x - 3$$ Remove parentheses.

$$x - 2 + 1 = 2x - 3$$ Add $-x^2$ to both sides.

$$x - 1 = 2x - 3$$ Combine like terms.

$$-x - 1 = -3$$ Add $-2x$ to both sides.

$$-x = -2$$ Add 1 to both sides.

$$x = 2$$ Divide both sides by -1.

Verify that 2 is a solution of the given equation. Work Progress Check 17.

Progress Check 17

Solve: $\dfrac{2a}{a-2} + \dfrac{5}{a-1} = 2$.

---

17. $\frac{14}{9}$

Progress Check Answers

Equations with Two Solutions

EXAMPLE 5

Solve the equation $\dfrac{4}{5} + y = \dfrac{4y - 50}{5y - 25}$.

Solution

$$\dfrac{4}{5} + y = \dfrac{4y - 50}{5y - 25}$$

$$\dfrac{4}{5} + y = \dfrac{4y - 50}{5(y - 5)} \qquad \text{Factor } 5y - 25.$$

$$5(y - 5)\left[\dfrac{4}{5} + y\right] = 5(y - 5)\left[\dfrac{4y - 50}{5(y - 5)}\right] \qquad \text{Multiply both sides by } 5(y - 5).$$

$$4(y - 5) + 5y(y - 5) = 4y - 50 \qquad \text{Remove brackets.}$$

$$4y - 20 + 5y^2 - 25y = 4y - 50 \qquad \text{Remove parentheses.}$$

$$5y^2 - 25y - 20 = -50 \qquad \text{Add } -4y \text{ to both sides and rearrange terms.}$$

$$5y^2 - 25y + 30 = 0 \qquad \text{Add 50 to both sides.}$$

$$y^2 - 5y + 6 = 0 \qquad \text{Divide both sides by 5.}$$

$$(y - 3)(y - 2) = 0 \qquad \text{Factor } y^2 - 5y + 6.$$

$$y - 3 = 0 \quad \text{or} \quad y - 2 = 0 \qquad \text{Set each factor equal to 0.}$$

$$y = 3 \qquad\qquad y = 2$$

Verify that 3 and 2 both satisfy the original equation. Work Progress Check 18.

Progress Check 18

Solve: $\dfrac{x - 4}{x - 3} + \dfrac{x - 2}{x - 3} = x - 3$.

18. 5; 3 is extraneous

Progress Check Answers

12.7 Exercises

In Exercises 1–58, solve each equation and check the solution. If an equation has no solution, so indicate:

1. $\dfrac{x}{2} + 4 = \dfrac{3x}{2}$

2. $\dfrac{y}{3} + 6 = \dfrac{4y}{3}$

3. $\dfrac{2y}{5} - 8 = \dfrac{4y}{5}$

4. $\dfrac{3x}{4} - 6 = \dfrac{x}{4}$

5. $\dfrac{x}{3} + 1 = \dfrac{x}{2}$

6. $\dfrac{x}{2} - 3 = \dfrac{x}{5}$

7. $\dfrac{x}{5} - \dfrac{x}{3} = -8$

8. $\dfrac{2}{3} + \dfrac{x}{4} = 7$

9. $\dfrac{3a}{2} + \dfrac{a}{3} = -22$

10. $\dfrac{x}{2} + x = \dfrac{9}{2}$

11. $\dfrac{x - 3}{3} + 2x = -1$

12. $\dfrac{x + 2}{2} - 3x = x + 8$

13. $\dfrac{z - 3}{2} = z + 2$

14. $\dfrac{b + 2}{3} = b - 2$

15. $\dfrac{5(x + 1)}{8} = x + 1$

16. $\dfrac{3(x - 1)}{2} + 2 = x$

17. $\dfrac{c - 4}{4} = \dfrac{c + 4}{8}$

18. $\dfrac{t + 3}{2} = \dfrac{t - 3}{3}$

19. $\dfrac{x + 1}{3} + \dfrac{x - 1}{5} = \dfrac{2}{15}$

20. $\dfrac{y - 5}{7} + \dfrac{y - 7}{5} = \dfrac{-2}{5}$

21. $\dfrac{3x - 1}{6} - \dfrac{x + 3}{2} = \dfrac{3x + 4}{3}$

22. $\dfrac{2x + 3}{3} + \dfrac{3x - 4}{6} = \dfrac{x - 2}{2}$

23. $\dfrac{3}{x} + 2 = 3$

24. $\dfrac{2}{x} + 9 = 11$

25. $\dfrac{5}{a} - \dfrac{4}{a} = 8 + \dfrac{1}{a}$

26. $\dfrac{11}{b} + \dfrac{13}{b} = 12$

27. $\dfrac{2}{y+1} + 5 = \dfrac{12}{y+1}$ 28. $\dfrac{1}{t-3} = \dfrac{-2}{t-3} + 1$ 29. $\dfrac{1}{x-1} + \dfrac{3}{x-1} = 1$ 30. $\dfrac{3}{p+6} - 2 = \dfrac{7}{p+6}$

31. $\dfrac{a^2}{a+2} - \dfrac{4}{a+2} = a$ 32. $\dfrac{z^2}{z+1} + 2 = \dfrac{1}{z+1}$ 33. $\dfrac{x}{x-5} - \dfrac{5}{x-5} = 3$ 34. $\dfrac{3}{y-2} + 1 = \dfrac{3}{y-2}$

35. $\dfrac{3r}{2} - \dfrac{3}{r} = \dfrac{3r}{2} + 3$ 36. $\dfrac{2p}{3} - \dfrac{1}{p} = \dfrac{2p-1}{3}$ 37. $\dfrac{1}{3} + \dfrac{2}{x-3} = 1$ 38. $\dfrac{3}{5} + \dfrac{7}{x+2} = 2$

39. $\dfrac{u}{u-1} + \dfrac{1}{u} = \dfrac{u^2+1}{u^2-u}$ 40. $\dfrac{v}{v+2} + \dfrac{1}{v-1} = 1$

41. $\dfrac{3}{x-2} + \dfrac{1}{x} = \dfrac{2(3x+2)}{x^2-2x}$ 42. $\dfrac{5}{x} + \dfrac{3}{x+2} = \dfrac{-6}{x(x+2)}$

43. $\dfrac{7}{q^2-q-2} + \dfrac{1}{q+1} = \dfrac{3}{q-2}$ 44. $\dfrac{-5}{s^2+s-2} + \dfrac{3}{s+2} = \dfrac{1}{s-1}$

45. $\dfrac{3y}{3y-6} + \dfrac{8}{y^2-4} = \dfrac{2y}{2y+4}$ 46. $\dfrac{x-3}{4x-4} + \dfrac{1}{9} = \dfrac{x-5}{6x-6}$

47. $y + \dfrac{2}{3} = \dfrac{2y-12}{3y-9}$ 48. $y + \dfrac{3}{4} = \dfrac{3y-50}{4y-24}$

49. $\dfrac{5}{4y+12} - \dfrac{3}{4} = \dfrac{5}{4y+12} - \dfrac{y}{4}$ 50. $\dfrac{3}{5x-20} + \dfrac{4}{5} = \dfrac{3}{5x-20} - \dfrac{x}{5}$

51. $\dfrac{x}{x-1} - \dfrac{12}{x^2-x} = \dfrac{-1}{x-1}$ 52. $1 - \dfrac{3}{b} = \dfrac{-8b}{b^2+3b}$

53. $\dfrac{z-4}{z-3} = \dfrac{z+2}{z+1}$ 54. $\dfrac{a+2}{a+8} = \dfrac{a-3}{a-2}$

55. $\dfrac{n}{n^2-9} + \dfrac{n+8}{n+3} = \dfrac{n-8}{n-3}$ 56. $\dfrac{x-3}{x-2} - \dfrac{1}{x} = \dfrac{x-3}{x}$

57. $\dfrac{b+2}{b+3} + 1 = \dfrac{-7}{b-5}$ 58. $\dfrac{1}{a^2-4} = \dfrac{2a}{a+2} - 1$

Review Exercises

In Review Exercises 1–4, factor each sum or difference of two cubes.

1. $a^3 + 125$ 2. $y^3 - 27$ 3. $8x^3 - 125y^6$ 4. $27a^6 + 1000b^3$

In Review Exercises 5–10, factor each expression by grouping.

5. $ab + 3a + 2b + 6$ 6. $yz + z^2 + y + z$

7. $mr + ms + nr + ns$ 8. $ac + bc - ad - bd$

9. $2a + 2b - a^2 - ab$ 10. $xa - x + ya - y$

11. Multiply: $(3rs^2 - 2)(2rs^2 + 5)$ 12. Divide: $y + 3 \overline{)2y - 4y^2 + 3y^3}$

12.8 Applications of Equations That Contain Fractions

- Electronics Problems
- Number Problems
- Shared Work Problems
- Uniform Motion Problems
- Investment Problems

Electronics Problems

EXAMPLE 1

In electronics, the formula $\frac{1}{r} = \frac{1}{r_1} + \frac{1}{r_2}$ is used to calculate the combined resistance r of two resistors wired in parallel. The symbols r_1 and r_2 use **subscripts** to denote two different resistances. Read r_1 as "r sub one" and read r_2 as "r sub two." If the combined resistance r is 4 ohms and the resistance of one of the resistors, r_1, is 12 ohms, find the resistance of the other resistor, r_2.

Solution

Substitute 4 for r and 12 for r_1 in the formula and solve for r_2.

$$\frac{1}{r} = \frac{1}{r_1} + \frac{1}{r_2}$$

$$\frac{1}{4} = \frac{1}{12} + \frac{1}{r_2}$$

$$12r_2\left(\frac{1}{4}\right) = 12r_2\left(\frac{1}{12} + \frac{1}{r_2}\right) \quad \text{Multiply both sides by } 12r_2, \text{ the LCD of } \frac{1}{4}, \frac{1}{12}, \text{ and } \frac{1}{r_2}.$$

$$3r_2 = r_2 + 12 \quad \text{Remove parentheses and simplify.}$$

$$2r_2 = 12 \quad \text{Add } -r_2 \text{ to both sides.}$$

$$r_2 = 6 \quad \text{Divide both sides by 2.}$$

The resistance of the second resistor, r_2, is 6 ohms. ■

EXAMPLE 2

Solve the formula $\frac{1}{r} = \frac{1}{r_1} + \frac{1}{r_2}$ for r.

Solution

$$\frac{1}{r} = \frac{1}{r_1} + \frac{1}{r_2}$$

$$rr_1r_2\left(\frac{1}{r}\right) = rr_1r_2\left(\frac{1}{r_1} + \frac{1}{r_2}\right) \quad \text{Multiply both sides by } rr_1r_2, \text{ the LCD of all of the fractions.}$$

$$\frac{rr_1r_2}{r} = \frac{rr_1r_2}{r_1} + \frac{rr_1r_2}{r_2} \quad \text{Remove parentheses.}$$

$$r_1r_2 = rr_2 + rr_1 \quad \text{Simplify.}$$

$$r_1r_2 = r(r_2 + r_1) \quad \text{Factor out an } r.$$

$$\frac{r_1r_2}{(r_2 + r_1)} = r \quad \text{Divide both sides by } r_2 + r_1.$$

or

$$r = \frac{r_1r_2}{(r_2 + r_1)}$$

■

Number Problems

EXAMPLE 3

If the same number is added to both the numerator and denominator of the fraction $\frac{3}{5}$, the result is $\frac{4}{5}$. Find the number.

Solution

Let n represent the number, add n to both the numerator and denominator of $\frac{3}{5}$, and set the result equal to $\frac{4}{5}$. Then solve the equation for n.

$$\frac{3+n}{5+n} = \frac{4}{5}$$

$$5(5+n)\frac{3+n}{5+n} = 5(5+n)\frac{4}{5} \quad \text{Multiply both sides by } 5(5+n).$$

$$5(3+n) = (5+n)4 \quad \text{Simplify.}$$

$$15 + 5n = 20 + 4n \quad \text{Remove parentheses.}$$

$$5n = 5 + 4n \quad \text{Add } -15 \text{ to both sides.}$$

$$n = 5 \quad \text{Add } -4n \text{ to both sides.}$$

The number is 5.

Check: Add 5 to both the numerator and denominator of $\frac{3}{5}$ and get

$$\frac{3+5}{5+5} = \frac{8}{10} = \frac{4}{5}$$

Shared Work Problems

EXAMPLE 4

An inlet pipe can fill an oil tank in 7 days, and a second inlet pipe can fill the tank in 9 days. If both pipes are used, how long will it take to fill the tank?

Analysis

If we add what the first inlet pipe can do in 1 day to what the second inlet pipe can do in 1 day, the sum is what they can do together in 1 day. Since the first inlet pipe can fill the tank in 7 days, it can do $\frac{1}{7}$ of the job in 1 day. Since it takes the second inlet pipe 9 days, it can do $\frac{1}{9}$ of the job in 1 day. Since it takes x days for both inlet pipes to fill the tank, together they can do $\frac{1}{x}$ of the job in 1 day.

Solution

Let x represent the number of days it will take to fill the tank if both inlet pipes are used. Then form the equation.

What the first inlet pipe can do in 1 day	+	what the second inlet pipe can do in 1 day	=	what they can do together in 1 day.
$\frac{1}{7}$	+	$\frac{1}{9}$	=	$\frac{1}{x}$

$$63x\left(\frac{1}{7} + \frac{1}{9}\right) = 63x\left(\frac{1}{x}\right) \quad \text{Multiply both sides by } 63x.$$

$$9x + 7x = 63 \quad \text{Remove parentheses and simplify.}$$

$$16x = 63 \quad \text{Combine terms.}$$

$$x = \frac{63}{16} \quad \text{Divide both sides by 16.}$$

It will take $\frac{63}{16}$, or $3\frac{15}{16}$ days for both inlet pipes to fill the tank.

Check: In $\frac{63}{16}$ days, the first inlet pipe does $\frac{1}{7}\left(\frac{63}{16}\right)$ of the total job and the second inlet pipe does $\frac{1}{9}\left(\frac{63}{16}\right)$ of the total job. The sum of these efforts, $\frac{9}{16} + \frac{7}{16}$, is equal to one complete job. ∎

Uniform Motion Problems

EXAMPLE 5

Tom can jog 10 miles in the same amount of time that his wife, Gail, can jog 12 miles. If Gail can run 1 mile per hour faster than Tom, how fast can Gail run?

Analysis

This is a uniform motion problem based on the formula $d = rt$, where d is the distance traveled, r is the rate, and t is the time. If we solve the formula for t, we obtain

$$t = \frac{d}{r}$$

Since Tom can run 10 miles at some unknown rate r, it will take him $\frac{10}{r}$ hours. Since Gail can run 12 miles at a rate of $r + 1$ miles per hour, it will take her $\frac{12}{r+1}$ hours. We organize the information in the chart shown in Figure 12-1.

	d	$=$	r	$\cdot$	t
Gail	12		$r+1$		$\frac{12}{r+1}$
Tom	10		r		$\frac{10}{r}$

Figure 12-1

Because the times are given to be equal, we know that $\frac{12}{r+1} = \frac{10}{r}$.

Solution

Let r be the rate that Tom can run.
Then $r + 1$ is the rate that Gail can run.

We can form the equation

$$\boxed{\text{The time it takes Gail to run 12 miles}} = \boxed{\text{the time it takes Tom to run 10 miles.}}$$

$$\frac{12}{r+1} = \frac{10}{r}$$

$$r(r+1)\frac{12}{r+1} = r(r+1)\frac{10}{r} \quad \text{Multiply both sides by } r(r+1).$$

$$12r = 10(r+1) \quad \text{Simplify.}$$

$$12r = 10r + 10 \quad \text{Remove parentheses.}$$

$$2r = 10 \quad \text{Add } -10r \text{ to both sides.}$$

$$r = 5 \quad \text{Divide both sides by 2.}$$

Thus, Tom can run 5 miles per hour. Since Gail can run 1 mile per hour faster, she can run 6 miles per hour. Verify that these results check. ∎

Investment Problems

EXAMPLE 6

At a bank, a sum of money invested for one year will earn $96 interest. If invested in bonds, that same money would earn $108, because the interest rate paid by the bonds is 1% greater than that paid by the bank. Find the bank's rate.

Analysis

The interest paid by either investment is the product of the principal (the amount invested) and the interest rate. If we let r represent the bank's rate of interest, then $r + 0.01$ represents the rate paid by the bonds. For each investment, the principal is the interest paid divided by the interest rate. See Figure 12-2.

	interest	=	principal	·	rate
bank	96		$\frac{96}{r}$		r
bonds	108		$\frac{108}{r+0.01}$		$r + 0.01$

Figure 12-2

Because the principal that would be invested in either account is the same, we can set up and solve the following equation:

$$\frac{96}{r} = \frac{108}{r + 0.01}$$

Solution

Solve the equation as follows:

$$\frac{96}{r} = \frac{108}{r + 0.01}$$

$$r(r + 0.01) \cdot \frac{96}{r} = \frac{108}{r + 0.01} \cdot r(r + 0.01) \qquad \text{Multiply both sides by } r(r + .01).$$

$$96(r + 0.01) = 108r$$

$$96r + 0.96 = 108r \qquad \text{Remove parentheses.}$$

$$0.96 = 12r \qquad \text{Subtract } 96r \text{ from both sides.}$$

$$0.08 = r \qquad \text{Divide both sides by 12.}$$

The bank's interest rate is 0.08, or 8%. The bonds pay 9% interest, a rate 1% greater than that paid by the bank. Verify that these rates check.

12.8 Exercises

1. In the equation
$$\frac{1}{a} + \frac{1}{b} = 1$$
$a = 10$. Find b.

2. In the equation
$$\frac{1}{a} + \frac{1}{b} = 1$$
$b = 7$. Find a.

3. Solve the formula in Exercise 1 for a.

4. Solve the formula in Exercise 1 for b.

5. In optics, the focal length f of a lens is given by the formula
$$\frac{1}{f} = \frac{1}{d_1} + \frac{1}{d_2}$$
where d_1 is the distance from the object to the lens and d_2 is the distance from the lens to the image. Find d_2 if $f = 8$ meters and $d_1 = 12$ meters.

6. Solve the formula in Exercise 5 for f.

7. Solve the formula in Exercise 5 for d_1.

8. Solve the formula in Exercise 5 for d_2.

9. If the denominator of the fraction $\frac{3}{4}$ is increased by a number and the numerator of the fraction is doubled, the result is 1. Find the number.

10. If a number is added to the numerator of the fraction $\frac{7}{8}$ and the same number is subtracted from the denominator, the result is 2. Find the number.

11. If a number is added to the numerator of the fraction $\frac{3}{4}$ and twice as much is added to the denominator, the result is $\frac{4}{7}$. Find the number.

12. If a number is added to the numerator of the fraction $\frac{5}{7}$ and twice as much is subtracted from the denominator, the result is 8. Find the number.

13. The sum of a number and its reciprocal is $\frac{13}{6}$. Find the numbers.

14. The sum of the reciprocals of two consecutive even integers is $\frac{7}{24}$. Find the integers.

15. An inlet pipe can fill an empty swimming pool in 5 hours, and another inlet pipe can fill the pool in 4 hours. How long will it take both pipes to fill the pool?

16. One inlet pipe can fill an empty pool in 4 hours, and a drain can empty the pool in 8 hours. How long will it take the pipe to fill the pool if the drain is left open?

17. A homeowner estimates that it will take 7 days to roof his house. A professional roofer estimates that he could roof the house in 4 days. How long will it take if the homeowner helps the roofer?

18. A pond is filled by two inlet pipes. One pipe can fill the pond in 15 days, and the other pipe can fill the pond in 21 days. However, evaporation can empty the pond in 36 days. How long will it take the two inlet pipes to fill an empty pond?

19. Juan can bicycle 28 miles in the same time he can walk 8 miles. If he can ride 10 miles per hour faster than he can walk, how fast can he walk?

20. A plane can fly 300 miles in the same time it takes a car to go 120 miles. If the car travels 90 miles per hour slower than the plane, how fast is the plane?

21. A boat that can travel 18 miles per hour in still water can travel 22 miles downstream in the same amount of time that it can travel 14 miles upstream. Find the speed of the current in the stream.

22. A plane can fly 300 miles downwind in the same amount of time it can travel 210 miles upwind. Find the velocity of the wind if the plane can fly at 255 miles per hour in still air.

23. Two certificates of deposit pay interest at rates that differ by 1%. Money invested for one year in the first CD earns $175 interest. The same principal invested in the other CD earns $200. Find the two rates of interest.

24. Two bond funds pay interest at rates that differ by 2%. Money invested for one year in the first fund earns $315 interest. The same amount invested in the other fund earns $385. Find the lower rate of interest.

25. Several employees bought a $35 gift for their boss. If there had been two more employees, everyone's cost would have been $2 less. How many workers contributed to the gift?

26. A dealer bought some radios for a total of $1200. She gave away 6 radios, sold each of the rest for $10 more than she paid for each radio, and broke even. How many did she buy?

27. A college bookstore can purchase several calculators for a total cost of $120. If each calculator cost $1 less, the bookstore could purchase 10 additional calculators at the same total cost. How many calculators can be purchased at the regular price?

28. A repairman purchased several furnace-blower motors for a total cost of $210. If his cost per motor had been $5 less, he could have purchased 1 additional motor. How many motors did he buy at the regular rate?

Review Exercises

In Review Exercises 1–8, solve each equation.

1. $x^2 - 5x - 6 = 0$
2. $x^2 - 25 = 0$
3. $(t + 2)(t^2 + 7t + 12) = 0$
4. $2(y - 4) = -y^2$

5. $y^3 - y^2 = 0$
6. $5a^3 - 125a = 0$
7. $(x^2 - 1)(x^2 - 4) = 0$
8. $6t^3 + 35t^2 = 6t$

9. A room containing 168 square feet is 2 feet longer than it is wide. Find its perimeter.

10. The base of a triangle is 4 meters longer than twice its height. Its area is 48 square meters. Find the height and the length of the base of the triangle.

11. On a trip from Edens to Grandville, Tina averaged 30 miles per hour. She returned at 50 miles per hour, and the round trip took her 8 hours. How far is Grandville from Edens?

12. On the expressway, Rick could average 60 miles per hour, but on the back roads, he could average only 45 miles per hour. If his trip of 180 miles took $3\frac{1}{2}$ hours, how far did he travel on the expressway?

Mathematics for Fun

To show that the number trick from the first page of this chapter will produce the result 3 regardless of what number we start with, we will begin with the number x. The result of following each instruction appears in the right column.

1. Pick a number. x
2. Multiply it by 3. $3x$
3. Add 6. $3x + 6$
4. Subtract your original number. $3x + 6 - x$
5. Divide by 2. $\dfrac{3x + 6 - x}{2}$
6. Subtract your original number again. $\dfrac{3x + 6 - x}{2} - x$

To show that the result is 3, we simplify the final expression, as follows.

$$\dfrac{3x + 6 - x}{2} - x = \dfrac{2x + 6}{2} - x \quad \text{Combine terms in the fraction's numerator.}$$

$$= \dfrac{2(x + 3)}{2} - x \quad \text{Factor the fraction's numerator.}$$

$$= x + 3 - x \quad \text{Simplify the fraction by removing the common factor of 2.}$$

$$= 3 \quad \text{Combine terms.}$$

Because the result is 3 regardless of the beginning number, the trick is guaranteed to work. Now it is your turn. Try to invent another number trick that always produces a result that is 5 greater than the beginning number.

Chapter Summary

Key Words

arithmetic fraction (12.1, 12.5)
common denominator (12.5)
complex fractions (12.6)
denominator (12.1)
divisor (12.3)
domain (12.1)
extraneous solutions (12.7)
fraction (12.1)
least common denominator (12.5)
lowest terms (12.1)
negatives (12.1)
numerator (12.1)
rational expression (12.1)
rational number (12.1)
simplifying a fraction (12.1)
subscripts (12.8)

Key Ideas

(12.1) If b and c are not equal to zero, then $\dfrac{a}{b} = \dfrac{a \cdot c}{b \cdot c}$.

$\dfrac{a}{1} = a; \quad \dfrac{a}{0}$ is not defined.

(12.2) $\dfrac{a}{b} \cdot \dfrac{c}{d} = \dfrac{a \cdot c}{b \cdot d}$

(12.3) $\dfrac{a}{b} \div \dfrac{c}{d} = \dfrac{a}{b} \cdot \dfrac{d}{c}$

(12.4) $\dfrac{a}{d} + \dfrac{b}{d} = \dfrac{a + b}{d} \quad \dfrac{a}{d} - \dfrac{b}{d} = \dfrac{a - b}{d}$

If $d \neq 0$, then $\dfrac{0}{d} = 0$.

(12.5) To add or subtract fractions with unlike denominators, first find the least common denominator of those fractions. Then express each of the fractions in equivalent form, with the same common denominator. Finally, add or subtract the fractions.

(12.7) To solve an equation that contains fractions, transform it into another equation without fractions. Do so by multiplying both sides by the LCD of the fractions. Check all solutions.

(12.6) To simplify a complex fraction, use either of these methods:

1. Write the numerator and the denominator of the complex fraction as single fractions; then simplify.
2. Multiply both the numerator and the denominator of the complex fraction by the LCD of the fractions that appear in the numerator and the denominator; then simplify.

Chapter 12 Review Exercises

[12.1] In Review Exercises 1–4, write each fraction in lowest terms.

1. $\dfrac{10}{25}$
2. $\dfrac{-12}{18}$
3. $\dfrac{-51}{153}$
4. $\dfrac{105}{45}$

In Review Exercises 5–8, find the numbers that are not in the domain of the variable.

5. $\dfrac{3}{x-5}$
6. $\dfrac{3x+1}{2x-5}$
7. $\dfrac{2a+5}{a^2-49}$
8. $\dfrac{3b^2-4}{b^2-5b+6}$

In Review Exercises 9–18, simplify each fraction, if possible. Assume no divisions by zero.

9. $\dfrac{3x^2}{6x^3}$
10. $\dfrac{5xy^2}{2x^2y^2}$
11. $\dfrac{x^2}{x^2+x}$
12. $\dfrac{x+2}{x^2+2x}$
13. $\dfrac{6xy}{3xy}$
14. $\dfrac{8x^2y}{2x(4xy)}$
15. $\dfrac{x^2+4x+3}{x^2-4x-5}$
16. $\dfrac{x^2-x-56}{x^2-5x-24}$
17. $\dfrac{2x^2-16x}{2x^2-18x+16}$
18. $\dfrac{x^2+x-2}{x^2-x-2}$

[12.2] In Review Exercises 19–22, perform each multiplication and simplify.

19. $\dfrac{3xy}{2x} \cdot \dfrac{4x}{2y^2}$
20. $\dfrac{3x}{x^2-x} \cdot \dfrac{2x-2}{x^2}$
21. $\dfrac{x^2+3x+2}{x^2+2x} \cdot \dfrac{x}{x+1}$
22. $\dfrac{x^3-y^3}{x^2+xy+y^2} \cdot \dfrac{x}{x^2-y^2}$

[12.3] In Review Exercises 23–26, perform the indicated division and simplify.

23. $\dfrac{3x^2}{5x^2y} \div \dfrac{6x}{15xy^2}$
24. $\dfrac{x^2+5x}{x^2+4x-5} \div \dfrac{x^2}{x-1}$
25. $\dfrac{x^2-x-6}{2x-1} \div \dfrac{x^2-2x-3}{2x^2+x-1}$
26. $\dfrac{x^3+125}{x+5} \div \dfrac{x^2-5x+25}{x^2+10x+25}$

[12.4–12.5] In Review Exercises 27–32, several denominators are given. Find the least common denominator.

27. 4, 8

28. 35, 14

29. $3x^2y$, xy^2

30. $2x + 1$, $2x^2 + x$

31. $x + 2$, $x - 3$

32. $x^2 + 4x + 3$, $x^2 + x$

In Review Exercises 33–40, perform the indicated operations. Simplify all answers.

33. $\dfrac{x}{x+y} + \dfrac{y}{x+y}$

34. $\dfrac{3x}{x-7} - \dfrac{x-2}{x-7}$

35. $\dfrac{x}{x-1} + \dfrac{1}{x}$

36. $\dfrac{1}{7} - \dfrac{1}{x}$

37. $\dfrac{3}{x+1} - \dfrac{2}{x}$

38. $\dfrac{x+2}{2x} - \dfrac{2-x}{x^2}$

39. $\dfrac{x}{x+2} + \dfrac{3}{x} - \dfrac{4}{x^2 + 2x}$

40. $\dfrac{2}{x-1} - \dfrac{3}{x+1} + \dfrac{x-5}{x^2 - 1}$

[12.6] In Review Exercises 41–46, simplify each complex fraction.

41. $\dfrac{\frac{3}{2}}{\frac{2}{3}}$

42. $\dfrac{\frac{3}{2} + 1}{\frac{2}{3} + 1}$

43. $\dfrac{\frac{1}{x} + 1}{\frac{1}{x} - 1}$

44. $\dfrac{1 + \frac{3}{x}}{2 - \frac{1}{x^2}}$

45. $\dfrac{\frac{2}{x-1} + \frac{x-1}{x+1}}{\frac{1}{x^2 - 1}}$

46. $\dfrac{\frac{a}{b} + c}{\frac{b}{a} + c}$

[12.7] In Review Exercises 47–52, solve each equation. Check all answers.

47. $\dfrac{3}{x} = \dfrac{2}{x-1}$

48. $\dfrac{5}{x+4} = \dfrac{3}{x+2}$

49. $\dfrac{2}{3x} + \dfrac{1}{x} = \dfrac{5}{9}$

50. $\dfrac{2x}{x+4} = \dfrac{3}{x-1}$

51. $\dfrac{2}{x-1} + \dfrac{3}{x+4} = \dfrac{-5}{x^2 + 3x - 4}$

52. $\dfrac{4}{x+2} - \dfrac{3}{x+3} = \dfrac{6}{x^2 + 5x + 6}$

[12.8] In Review Exercises 53–56, solve each word problem. Check all answers.

53. Solve the equation $\dfrac{1}{x} - \dfrac{1}{y} = 1$ for x.

54. If Luiz can paint a house in 14 days and Desi can paint the house in 10 days, how long will it take if they work together?

55. Tony can bicycle 30 miles in the same time he can jog 10 miles. If he can ride 10 miles per hour faster than he can jog, how fast can he jog?

56. A plane can fly 400 miles downwind in the same amount of time it can travel 320 miles upwind. Find the velocity of the wind if the plane can fly at 360 miles per hour in still air.

Chapter 12 Test

1. Simplify: $\dfrac{48x^2y}{54xy^2}$.

2. List the numbers that are not in the domain of x: $\dfrac{3x^2 - 2x + 1}{x^2 - 100}$.

3. Simplify: $\dfrac{2x^2 - x - 3}{4x^2 - 9}$.

4. Simplify: $\dfrac{3(x + 2) - 3}{2x - 4 - (x - 5)}$.

5. Multiply and simplify: $\dfrac{12x^2y}{15xyz} \cdot \dfrac{25y^2z}{16xt}$.

6. Multiply and simplify: $\dfrac{x^2 + 3x + 2}{3x + 9} \cdot \dfrac{x + 3}{x^2 - 4}$.

7. Divide and simplify: $\dfrac{8x^2y}{25xt} \div \dfrac{16x^2y^3}{30xyt^3}$.

8. Divide and simplify: $\dfrac{x^2 - x}{3x^2 + 6x} \div \dfrac{3x - 3}{3x^3 + 6x^2}$.

9. Simplify: $\dfrac{x^2 + xy}{x - y} \cdot \dfrac{x^2 - y^2}{x^2 - 2x} \div \dfrac{x^2 + 2xy + y^2}{x^2 - 4}$.

10. Add: $\dfrac{5x - 4}{x - 1} + \dfrac{5x + 3}{x - 1}$.

11. Subtract: $\dfrac{3y + 7}{2y + 3} - \dfrac{3(y - 2)}{2y + 3}$.

12. Add: $\dfrac{x + 1}{x} + \dfrac{x - 1}{x + 1}$.

13. Subtract: $\dfrac{5x}{x - 2} - 3$.

14. Simplify: $\dfrac{\dfrac{8x^2}{xy^3}}{\dfrac{4y^3}{x^2y^3}}$.

15. Simplify: $\dfrac{1 + \dfrac{y}{x}}{\dfrac{y}{x} - 1}$.

16. Solve for x: $\dfrac{x}{10} - \dfrac{1}{2} = \dfrac{x}{5}$.

17. Solve for x: $3x - \dfrac{2(x+3)}{3} = 16 - \dfrac{x+2}{2}$.

18. Solve for x: $\dfrac{7}{x+4} - \dfrac{1}{2} = \dfrac{3}{x+4}$.

19. If George could pick up all the trash on a strip of highway in 9 hours and Maria could pick up the trash in only 7 hours, how long would it take them if they work together?

20. A boat can motor 28 miles downstream in the same amount of time it can motor 18 miles upstream. Find the speed of the current if the boat can motor at 23 miles per hour in still water.

21. Maria invests equal amounts in two accounts. One pays 1% more than the other. At the end of one year, one investment paid $240 in interest, and the other paid $280. At what rates did Maria invest?

13 Graphing Linear Equations and Inequalities

Mathematics in Medicine

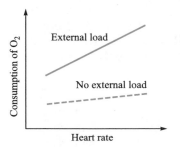

The heart pumps blood throughout the body, supplying nutrients and oxygen to all tissues. The heart muscle itself needs nutrients and oxygen, and its needs increase with increased heart activity.

Experiments have determined the oxygen requirements of an isolated animal heart, attached to measuring devices and kept beating in the laboratory. When the heart had to work extra hard because of an external load applied by the researchers, its oxygen needs increased more rapidly with increasing heart rate. To visualize the effects, researchers used a *graph*, like that in the illustration. The data of the experiment produced the two straight-line graphs.

The graph shows visually that oxygen needs were greater and increased more rapidly for a heart under load. A measure of that rate of increase is called the *slope* of the line. Graphing lines and determining their slopes are two topics of this chapter.

In this chapter, we will discuss equations and inequalities that have two or more variables.

13.1 Graphing Linear Equations

- Graphing Linear Equations
- The Intercept Method of Graphing a Line
- Graphing Lines Parallel to the x- and y-Axes

The equation $3x + 2 = 5$ has one variable, x, and its only solution is 1. This solution can be graphed (or plotted) on the number line as in Figure 13-1.

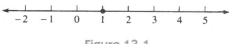

Figure 13-1

An equation such as $x + 2y = 5$, however, has two variables, x and y. The solutions of such equations are pairs of numbers. The pair $x = 1$ and $y = 2$ is a solution, because the equation is satisfied if we substitute 1 for x and 2 for y:

$x + 2y = 5$
$1 + 2(2) = 5$ Substitute 1 for x and 2 for y.
$1 + 4 = 5$
$5 = 5$

The pair $x = 5$ and $y = 0$ is also a solution, because it satisfies the equation:

$$x + 2y = 5$$
$$5 + 2(0) = 5 \quad \text{Substitute 5 for } x \text{ and 0 for } y.$$
$$5 + 0 = 5$$
$$5 = 5$$

Solutions of equations with two variables can be plotted on a **rectangular coordinate system**, sometimes called a **Cartesian coordinate system** after the 17th-century French mathematician René Descartes. The rectangular coordinate system consists of two number lines, called the ***x*-axis** and the ***y*-axis**, drawn at right angles to each other as in Figure 13-2**(a)**. The two axes intersect at a point called the **origin**, which is the 0 point on each axis. The positive direction on the *x*-axis is to the right, and the positive direction on the *y*-axis is upward. The two axes divide the coordinate system into four regions, called **quadrants**, which are numbered in a counterclockwise direction as shown in Figure 13-2**(a)**.

To plot the pair $x = 3$ and $y = 2$, we start at the origin. Because 3 is the value of x, we move 3 units to the right along the *x*-axis, as in Figure 13-2**(b)**. Because 2 is the corresponding value of y, we then move 2 units upward in the positive y direction. This locates point A in the figure. Point A has an ***x*-coordinate** of 3 and a ***y*-coordinate** of 2. This information is denoted concisely by the pair (3, 2), called the **coordinates** of point A. Point A is the **graph of the pair** (3, 2).

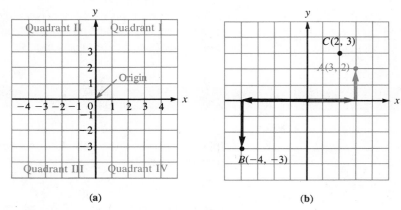

Figure 13-2

To plot the pair $(-4, -3)$, we start at the origin, move 4 units to the left along the *x*-axis, and then move 3 units down. This locates point B in the figure. Point C has coordinates of (2, 3).

Point A in Figure 13-2**(b)** with coordinates of (3, 2) is not the same as point C with coordinates (2, 3). For this reason, the pair of coordinates (x, y) of a point is often called an **ordered pair**. The *x*-coordinate is the first number in the ordered pair, and the *y*-coordinate is the second number.

EXAMPLE 1

Plot the points **a.** $A(-1, 2)$, **b.** $B(0, 0)$, **c.** $C(5, 0)$, **d.** $D(-\frac{5}{2}, -3)$, and **e.** $E(3, -2)$.

Solution

a. The point $A(-1, 2)$ has an x-coordinate of -1 and a y-coordinate of 2. To plot point A, we start at the origin, move 1 unit to the *left* and then 2 units *up*. (See Figure 13-3.) Point A lies in quadrant II.

b. To plot point $B(0, 0)$, we start at the origin, move 0 units to the *right* and 0 units *up*. Because there is no movement, point B is the origin.

c. To plot point $C(5, 0)$, we start at the origin, move 5 units to the *right* and 0 units *up*. The point C does not lie in any quadrant. It lies on the x-axis, 5 units to the right of the origin.

d. To plot point $D(-\frac{5}{2}, -3)$, we start at the origin, move $\frac{5}{2}$ units ($\frac{5}{2}$ units is $2\frac{1}{2}$ units) to the *left* and then 3 units *down*. The point D lies in quadrant III.

e. To plot point $E(3, -2)$, we start at the origin, move 3 units to the *right* and then 2 units *down*. The point E lies in quadrant IV.

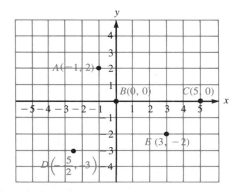

Figure 13-3

Graphing Linear Equations

To find some ordered pairs (x, y) that satisfy the equation $y = 5 - x$, we pick numbers at random, substitute them for x, and calculate the corresponding values of y. If we pick $x = 1$, we have

$y = 5 - x$
$y = 5 - 1$ Substitute 1 for x.
$y = 4$

The ordered pair $(1, 4)$ satisfies the equation. If $x = 2$, we have

$y = 5 - x$
$y = 5 - 2$ Substitute 2 for x.
$y = 3$

A second solution of the equation is $(2, 3)$. If $x = 5$, we have

$y = 5 - x$
$y = 5 - 5$ Substitute 5 for x.
$y = 0$

A third solution of the equation is $(5, 0)$. If $x = -1$, we have

$y = 5 - x$
$y = 5 - (-1)$ Substitute -1 for x.
$y = 6$

A fourth solution is $(-1, 6)$. As a final example, if $x = 6$, we have

$y = 5 - x$
$y = 5 - 6$ Substitute 6 for x.
$y = -1$

A fifth solution is $(6, -1)$. The graphs of the ordered pairs $(1, 4)$, $(2, 3)$, $(5, 0)$, $(-1, 6)$, and $(6, -1)$ appear in Figure 13-4.

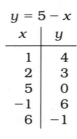
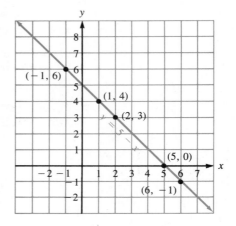

Figure 13-4

The five points lie on the line that appears in Figure 13-4. This line, called the **graph** of $y = 5 - x$, passes through many other points than the five that we have plotted. The coordinates of *every* point on this line determine a solution of $y = 5 - x$. For example, the line passes through the point $(4, 1)$, and the pair $x = 4$ and $y = 1$ is a solution because these numbers satisfy the equation:

$y = 5 - x$
$1 = 5 - 4$ Substitute 4 for x and 1 for y.
$1 = 1$

An equation, such as $y = 5 - x$, whose graph is a line is called a **linear equation in two variables**. Any point on that line has coordinates that satisfy the equation, and the graph of any pair (x, y) that satisfies the equation is a point on the line.

Although only two points are needed to draw the graph of a linear equation, it is wise to plot a third point as a check. If the three points do not lie on a line, then at least one of the points is in error.

Procedure for Graphing Linear Equations in the Variables x and y:

1. Determine two pairs (x, y) that satisfy the equation. To do so, pick arbitrary numbers for x and then solve the equation for the corresponding values of y. A third point provides a check.
2. Plot each resulting pair (x, y) on a rectangular coordinate system. If they do not appear to be on a line, check your calculations.
3. Draw the line that passes through the points.

EXAMPLE 2

Graph the equation $y = 3x - 4$.

Solution

We substitute numbers for x, and find the corresponding values of y. For example, we let $x = 1$ and find y.

$y = 3x - 4$
$y = 3(1) - 4$ Substitute 1 for x.
$y = 3 - 4$
$y = -1$

The pair $(1, -1)$ is a solution. To find another solution, we let $x = 2$ and find y.

$y = 3x - 4$
$y = 3(2) - 4$ Substitute 2 for x.
$y = 6 - 4$
$y = 2$

A second solution is $(2, 2)$. To find a third solution, we let $x = 3$ and find y.

$y = 3x - 4$
$y = 3(3) - 4$ Substitute 3 for x.
$y = 9 - 4$
$y = 5$

A third solution is $(3, 5)$. We plot these points as in Figure 13-5 and join them with a line. This line is the graph of the equation $y = 3x - 4$.

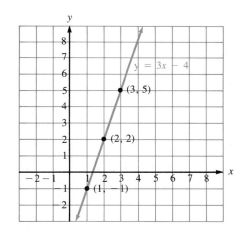

$y = 3x - 4$

x	y
1	-1
2	2
3	5

Figure 13-5

EXAMPLE 3

Graph $y - 4 = \dfrac{1}{2}(x - 8)$.

Solution

It is easier to find pairs (x, y) if we first solve the equation for y.

$$y - 4 = \frac{1}{2}(x - 8)$$

$y - 4 = \frac{1}{2}x - 4$ Use the distributive property to remove parentheses.

$y = \frac{1}{2}x$ Add 4 to both sides.

If $x = 0$, then $y = 0$. Thus, $(0, 0)$ is a solution. If $x = 2$, then $y = 1$. Thus, $(2, 1)$ is a second solution. Finally, if $x = -4$, then $y = -2$, and $(-4, -2)$ is a third solution. The graph of the equation appears in Figure 13-6.

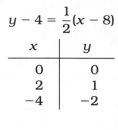
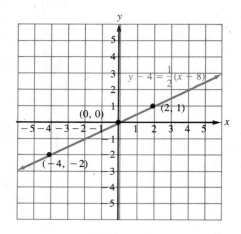

Figure 13-6

The Intercept Method of Graphing a Line

The y-coordinate of the point at which a line crosses the y-axis is called the **y-intercept** of the line. If we substitute 0 for x, the value determined for y is the y-intercept.

Likewise, the **x-intercept** of a line is the x-coordinate of the point at which the line crosses the x-axis. If we substitute 0 for y, the number determined for x is the x-intercept.

Plotting these two points and drawing a line through them is called the **intercept method of graphing a line**, a useful method for graphing linear equations written in **general form**.

General Form of the Equation of a Line: If A, B, and C are real numbers and A and B are not both 0, then the equation

$$Ax + By = C$$

is called the **general form of the equation of a line**.

Whenever possible, we will write the general form $Ax + By = C$ so that A, B, and C are integers and $A \geq 0$.

EXAMPLE 4

Use the intercept method to graph the equation $3x + 2y = 6$.

Solution

To find the y-intercept, we let $x = 0$ and solve for y.

$3x + 2y = 6$
$3(0) + 2y = 6$ Substitute 0 for x.
$2y = 6$ Simplify.
$y = 3$ Divide both sides by 2.

The y-intercept is 3, and the pair $(0, 3)$ is a solution of the equation.
To find the x-intercept, we let $y = 0$ and solve for x.

$3x + 2y = 6$
$3x + 2(0) = 6$ Substitute 0 for y.
$3x = 6$ Simplify.
$x = 2$ Divide both sides by 3.

The x-intercept is 2, and the pair $(2, 0)$ is a solution of the equation.
To find one more point, we let $x = 4$ and find the corresponding value of y:

$3x + 2y = 6$
$3(4) + 2y = 6$ Substitute 4 for x.
$12 + 2y = 6$ Simplify.
$2y = -6$ Add -12 to both sides.
$y = -3$ Divide both sides by 2.

Thus, the point $(4, -3)$ also lies on the graph of the line. We plot these three points and join them with a line. The graph of $3x + 2y = 6$ appears in Figure 13-7.

$3x + 2y = 6$

x	y
0	3
2	0
4	-3

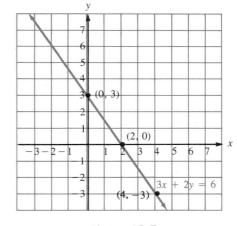

Figure 13-7

Progress Check 1

Find the x- and y-intercepts of $3x + 2y = 24$.

Work Progress Check 1.

1. 8, 12

Progress Check Answers

Graphing Lines Parallel to the x- and y-Axes

EXAMPLE 5

Graph **a.** $y = 3$ and **b.** $x = -2$.

Solution

a. Write $y = 3$ in general form as $0x + y = 3$. Because the coefficient of x is 0, the numbers assigned to x have no effect on y. The value of y is always 3. For example, if $x = -3$, then

$$0x + y = 3$$
$$0(-3) + y = 3 \quad \text{Substitute } -3 \text{ for } x.$$
$$0 + y = 3$$
$$y = 3$$

A table of values in Figure 13-8(a) gives several ordered pairs that satisfy the equation $y = 3$. After plotting these pairs (x, y) and joining them with a line, we see that the graph of $y = 3$ is a horizontal line, parallel to the x-axis and intersecting the y-axis at 3. The y-intercept of the line is 3, and the line has no x-intercept.

b. Write $x = -2$ in general form as $x + 0y = -2$. Because the coefficient of y is 0, different values of y will have no effect on x. The value of x is always -2. A table of values and the graph appear in Figure 13-8(b).

The graph of $x = -2$ is a vertical line parallel to the y-axis, intersecting the x-axis at -2. The x-intercept of the line is -2, and the line has no y-intercept.

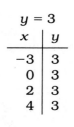
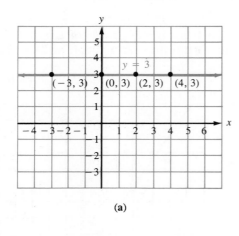

(a)

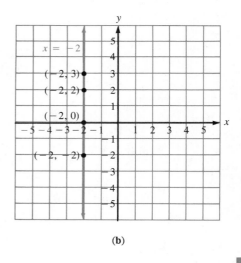

(b)

Figure 13-8

From the results of Example 5, we can conclude the following facts.

Equations of Lines Parallel to the Coordinate Axes:

The equation $x = a$ represents a vertical line that intersects the x-axis at a. If $a = 0$, the line is the y-axis.

The equation $y = b$ represents a horizontal line that intersects the y-axis at b. If $b = 0$, the line is the x-axis.

13.1 Exercises

In Exercises 1–8, plot each point. Tell in which quadrant each point lies.

1. $A(2, 5)$
2. $B(5, 2)$
3. $C(-3, 1)$
4. $D(1, -3)$
5. $E(-2, -3)$
6. $F(-3, -2)$
7. $G(3, -2)$
8. $H(-4, 5)$

In Exercises 9–16, plot each point. Indicate in which quadrant each point lies.

9. $A(-3, 5)$
10. $B(-5, 3)$
11. $C(3, -5)$
12. $D(5, -3)$
13. $E(-\frac{3}{2}, -4)$
14. $F(-5, \frac{9}{2})$
15. $G(\frac{5}{2}, \frac{7}{2})$
16. $H(\frac{7}{2}, -\frac{7}{2})$

In Exercises 17–24, plot each point. Indicate on which axis each point lies.

17. $A(0, 7)$
18. $B(0, -2)$
19. $C(2, 0)$
20. $D(-3, 0)$
21. $E(-5, 0)$
22. $F(0, -5)$
23. $G(0, 0)$
24. $H(-6, 0)$

In Exercises 25–32, refer to Illustration 1 and find the coordinates of each point.

25. A
26. B
27. C
28. D
29. E
30. F
31. G
32. H

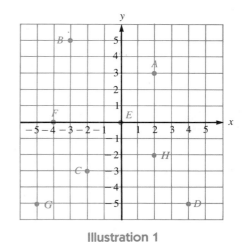

Illustration 1

In Exercises 33–34, Carlos charges $12 per hour in his part-time job tutoring mathematics.

33. Find Carlos' fee for working 1 hour, 2 hours, 3 hours, and 5 hours.

34. Let y represent Carlos' fee for working x hours. Plot the pairs (x, y) that you calculated in Exercise 33. Do the points lie on a line?

In Exercises 35–36, Wendy charges $10 for materials and $40 per hour for installing computer systems.

35. Find Wendy's fee for working 1 hour, 2 hours, 3 hours, and $5\frac{1}{2}$ hours.

36. Let y represent Wendy's total fee for working x hours. Plot the pairs (x, y) that you calculated in Exercise 35. Do the points lie on a line?

In Exercises 37–38, a person's maximum heart rate for safe aerobic exercise is approximately 220 minus that person's age.

37. Find the maximum heart rate for persons 20 years old, 40 years old, and 60 years old.

38. Let y represent the maximum heart rate for a person x years old. Write an equation that relates x and y, and graph it.

In Exercises 39–40, plot the points.

39. The accompanying table shows the approximate length l (in feet) of the skid mark caused by a car traveling at various speeds r (in mph). What conclusions can you draw?

r	10	20	30	40	50
l	5	20	45	80	125

40. The accompanying table shows the percentage P and the karat rating for several pieces of gold. What conclusions can you draw?

P	42	58	67	75	83
k	10	14	16	18	20

In Exercises 41–52, complete the table of solutions for each equation, and then graph the equation.

41. $y = x + 2$

x	y
3	
1	
−2	

42. $y = x - 3$

x	y
3	
1	
−3	

43. $y = x - 4$

x	y
5	
4	
−1	

44. $y = x + 1$

x	y
5	
1	
−1	

45. $y = -2x$

x	y
2	
1	
−3	

46. $y = 3x$

x	y
2	
0	
−2	

47. $y = \dfrac{x}{2}$

x	y
1	
−1	
−4	

48. $y = -\dfrac{x}{3}$

x	y
1	
−1	
−3	

49. $y = 2x - 1$

x	y
3	
−1	
−2	

50. $y = 3x + 1$

x	y
−2	
0	
1	

51. $y = \dfrac{x}{2} - 2$

x	y
8	
0	
−2	

52. $y = \dfrac{x}{3} - 3$

x	y
6	
0	
−3	

In Exercises 53–68, write each equation in general form if necessary. Then graph it using the intercept method.

53. $x + y = 7$

54. $x + y = -2$

55. $x - y = 7$

56. $x - y = -2$

57. $2x + y = 5$

58. $3x + y = -1$

59. $2x + 3y = 12$

60. $3x - 2y = 6$

61. $3x + 12 = 4y$

62. $2x + 12 = 9y$

63. $5x + 10 = -2y$

64. $8 - 3y = 2x$

65. $2(x + 2) - y = 4$

66. $3(y + 1) - x = 4$

67. $3(2 - y) + 2(x + 1) = 4$

68. $2(x + 2) - (y - 2) = 4$

In Exercises 69–80, graph each equation. You may have to simplify the equation first.

69. $y = -5$ **70.** $x = 4$ **71.** $x = 5$ **72.** $y = -5$

73. $y = 0$ **74.** $x = 0$ **75.** $2x = 5$ **76.** $3y = 7$

77. $3(x + 2) + x = 4$ **78.** $2(y + 3) - y = 5$

79. $3(y - 2) + 2 = y$ **80.** $4(x + 2) - 2(x - 1) = 6$

81. Find the equation of the horizontal line that passes through the point $(3, -2)$.

82. Find the equation of the vertical line that passes through the point $(3, -2)$.

83. Find the equation of the x-axis.

84. Find the equation of the y-axis.

If points $P(a, b)$ and $Q(c, d)$ are two points on a rectangular coordinate system and point M is midway between them, then point M is called the **midpoint** of the line segment joining P and Q. (See Illustration 2.) To find the coordinates of the midpoint M of a line segment PQ, we find the average of the x-coordinates and the average of the y-coordinates of P and Q:

$$x\text{-coordinate of } M = \frac{a + c}{2} \quad \text{and} \quad y\text{-coordinate of } M = \frac{b + d}{2}$$

In Exercises 85–90, find the coordinates of the midpoint of the line segment with the given endpoints.

85. $P(5, 3)$ and $Q(7, 9)$

86. $R(5, 6)$ and $S(7, 10)$

87. $R(2, -7)$ and $S(-3, 12)$

88. $P(-8, 12)$ and $Q(3, -9)$

89. $A(4, 6)$ and $B(10, 6)$

90. $A(8, -6)$ and the origin

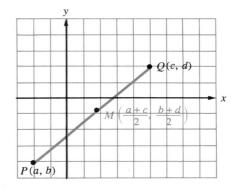

Illustration 2

Review Exercises

In Review Exercises 1–2, write each expression without using absolute value symbols.

1. $|-8 - (-3)|$

2. $-|-(-3)|$

In Review Exercises 3–4, let $x = -6$, $y = -3$, and $z = 4$. Evaluate each expression.

3. $\dfrac{2z^2 + 2x}{x + z}$

4. $\dfrac{z^2 - 2yz + y^2}{z - x + y}$

In Review Exercises 5–6, solve each equation.

5. $\dfrac{5(y-2)}{3} = -(6+y)$

6. $7(y+14) = 35(y-2)$

7. Solve the equation $A = \dfrac{1}{2}bh$ for h.

8. Solve the equation $S = 180(n-2)$ for n.

9. The sum of three consecutive even integers is 72. Find the integers.

10. The product of two consecutive integers is 7 greater than the square of the smaller. Find the integers.

In Review Exercises 11–12, perform each division.

11. $\dfrac{8x^2y^3 + 12x^3y^2}{6x^2y^2}$

12. $2x + 4 \overline{)6x^2 + 8x - 8}$

13.2 The Slope of a Line

- Slope as an Aid in Graphing
- Slope-Intercept Form of the Equation of a Line
- An Application of Slope
- Linear Depreciation

A measure of how rapidly a line rises or falls is called the **slope** of the line.

In Figure 13-9, a line passes through the points $P(1, 2)$ and $Q(3, 6)$. Moving along the line from P to Q causes the value of y to change from $y = 2$ to $y = 6$, an increase of $6 - 2$, or 4 units. In that same move, the value of x increases $3 - 1$, or 2 units. The slope of the line is the ratio of the change in y to the change in x.

Slope of line PQ

$= \dfrac{\text{change in the } y \text{ values}}{\text{change in the } x \text{ values}}$

$= \dfrac{6-2}{3-1}$

$= \dfrac{4}{2}$

$= 2$

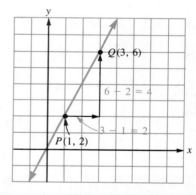

Figure 13-9

As a point on the line in Figure 13-10 moves from P to Q, its y-coordinate changes by the amount $y_2 - y_1$ (called the **rise**), while its x-coordinate changes by $x_2 - x_1$ (called the **run**). The slope of the line is the rise divided by the run.

When computing the slope of the line PQ in Figure 13-10, it doesn't matter which point is (x_1, y_1) and which is (x_2, y_2).

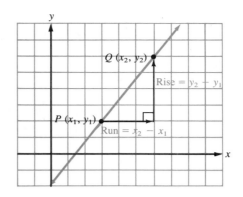

Figure 13-10

Thus, we have the following definition.

Definition: If $P(x_1, y_1)$ and $Q(x_2, y_2)$ are two points on a nonvertical line, the slope m of line PQ is given by the formula

$$m = \frac{\text{rise}}{\text{run}} = \frac{y_2 - y_1}{x_2 - x_1}$$

EXAMPLE 1

Find the slope of the line passing through $(-3, 2)$ and $(2, -5)$ and draw its graph.

Solution

We can let (x_1, y_1) be $(-3, 2)$ and let (x_2, y_2) be $(2, -5)$. Then

$$x_1 = -3 \quad \text{and} \quad x_2 = 2$$
$$y_1 = 2 \qquad\qquad y_2 = -5$$

To find the slope, we substitute these values into the formula for slope and simplify.

$$\text{Slope} = \frac{y_2 - y_1}{x_2 - x_1}$$

$$= \frac{-5 - 2}{2 - (-3)} \qquad \text{Substitute } -5 \text{ for } y_2, 2 \text{ for } y_1, 2 \text{ for } x_2, \text{ and } -3 \text{ for } x_1.$$

$$= -\frac{7}{5} \qquad \text{Simplify.}$$

The slope of the line passing through the points $(-3, 2)$ and $(2, -5)$ is $-\frac{7}{5}$.

We would obtain the same result if we had let $(x_1, y_1) = (2, -5)$ and $(x_2, y_2) = (-3, 2)$. The graph of this line appears in Figure 13-11.

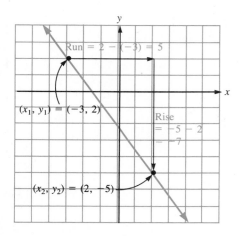

Figure 13-11

Progress Check 2

Find the slope of the line passing through $P(-4, 2)$ and $Q(6, -4)$.

Work Progress Check 2.

In the previous example, the slope of the line was negative because, as the value of x increased 5 units, the y value *decreased* 7 units. Whenever increasing values of x produce decreasing values of y, the slope of the line is negative, and the graph of the line drops as it moves to the right. If the line neither rises nor falls (if the line is horizontal), its slope is 0. See Figure 13-12.

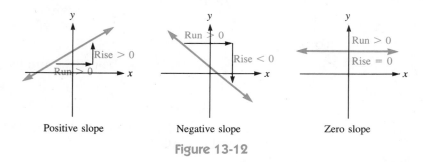

Figure 13-12

Slope as an Aid in Graphing

We can graph a line if we know the coordinates of one point on the line and its slope.

For example, to graph the line with a slope of 3 that passes through $P(2, 4)$, we plot the point $P(2, 4)$ as in Figure 13-13. Because the slope is 3, the line rises 3 units for every 1 unit it moves to the right. Thus, we can find a second point on the line by starting at $P(2, 4)$ and moving 1 unit to the right and 3 units up. This brings us to point Q with coordinates $(3, 7)$. We then draw the line passing through P and Q.

2. $-\frac{3}{5}$

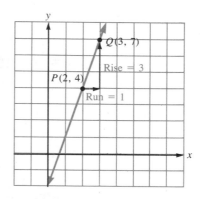

Figure 13-13

Slope-Intercept Form of the Equation of a Line

We can compute the slope of a line from its equation by finding the coordinates of two points on the line and using the definition of slope.

For example, to find the slope of the line with equation $y = 2x + 7$, we find the coordinates of two points on the line. One of these is the y-intercept, found by substituting 0 for x in the equation and solving for y.

$y = 2x + 7$
$ = 2(\mathbf{0}) + 7$ Substitute 0 for x.
$ = 7$ Simplify.

The y-intercept of the line is 7, and the line passes through the point $(0, 7)$.

To find the coordinates of a second point on the line, we replace x with some number other than 0, and find y. We will let $x = 1$.

$y = 2x + 7$
$ = 2(\mathbf{1}) + 7$ Substitute 1 for x.
$ = 9$ Simplify.

Thus, the line passes through the point $(1, 9)$.

To find the slope of the line, we let (x_1, y_1) be the point $(0, 7)$ and (x_2, y_2) be the point $(1, 9)$ and substitute into the formula for slope:

$$\text{Slope of the line} = \frac{y_2 - y_1}{x_2 - x_1} = \frac{9 - 7}{1 - 0} = \frac{2}{1} = 2$$

The slope of the line is 2, which is the same as the coefficient of x in $y = 2x + 7$. The y-intercept of the line is 7, the same as the constant in $y = 2x + 7$. This illustrates the following fact:

The Slope-Intercept Form of the Equation of a Line: If a linear equation is written in the form

$$y = mx + b$$

where m and b are constants, then the graph of that equation is a line with slope m and with y-intercept b.

EXAMPLE 2

a. Find the slope and the y-intercept of the line determined by $3x + 5y - 15 = 0$, and   **b.** graph the line.

Solution

a. We can write the equation in slope-intercept form by solving it for y:

$$3x + 5y - 15 = 0$$
$$5y = -3x + 15 \qquad \text{Add } -3x \text{ and } 15 \text{ to both sides.}$$
$$y = -\frac{3}{5}x + \frac{15}{5} \qquad \text{Divide both sides by 5.}$$
$$y = -\frac{3}{5}x + 3$$

This equation is in the form $y = mx + b$, with $m = -\frac{3}{5}$ and $b = 3$. The slope of the line is $-\frac{3}{5}$, and its y-intercept is 3.

b. We begin by graphing the y-intercept, the point $(0, 3)$, as in Figure 13-14. Then we use the slope to find a second point on the line. Because the slope is $\frac{\text{rise}}{\text{run}} = \frac{-3}{5}$, we can find a second point by starting at the y-intercept, moving 5 units to the right, and then moving 3 units *down*. This brings us to the point $(5, 0)$.

We plot the point $(5, 0)$ and draw the line passing through both points. The complete graph appears in Figure 13-14.

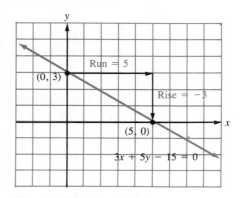

Figure 13-14

Progress Check 3

Find the slope and y-intercept of the line determined by $5x - 3y - 30 = 0$.

Work Progress Check 3.

We can find the slope and the y-intercept of a line such as $y = 3$ by writing the equation in the equivalent form

$$y = 0x + 3$$

We see that the slope of the line is 0, and its y-intercept is 3. Because its slope is 0, the line is horizontal. Its graph appears in Figure 13-15**(a)**.

The graph of an equation such as $x = 3$ is also a line. Because the equation $x = 3$ does not contain the variable y, we cannot find the slope or y-intercept by writing the equation in the form $y = mx + b$. Instead, we will find two points on the line and try to determine the slope by using the definition for slope. The points $(3, 1)$ and $(3, 2)$ lie on the given line, because *any* point with an x-coordinate of 3 lies on the line. Thus,

$$m = \frac{y_2 - y_1}{x_2 - x_1}$$
$$= \frac{2 - 1}{3 - 3} \qquad \text{Substitute 2 for } y_2, 1 \text{ for } y_1, 3 \text{ for } x_2, \text{ and 3 for } x_1.$$
$$= \frac{1}{0}$$

3. $\frac{5}{3}, -10$

Progress Check Answers

Because division by zero is not defined, the fraction $\frac{1}{0}$ has no meaning. The slope of the line $x = 3$ is undefined. The graph is the vertical line in Figure 13-15**(b)**. It has no y-intercept, and its x-intercept is 3.

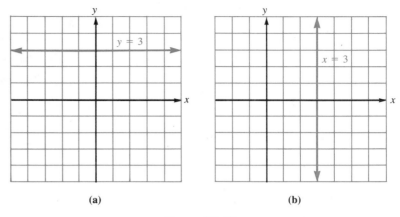

Figure 13-15

The previous discussion suggests the following facts.

All horizontal lines (lines with equations of the form $y = b$) have a slope of zero.

All vertical lines (lines with equations of the form $x = a$) have undefined slope.

An Application of Slope

The equation of a line can describe the relationship between the price of a product and the number of units that can be sold. For example, the owners of a bicycle store know that for the stock they have on hand, the number of bicycles sold is related to the price charged according to the equation of a line. If x is the price, in dollars, of one bicycle, and y is the number of bicycles sold, the equation is

$$y = 500 - \frac{3}{5}x$$

To find the number of bicycles sold at a unit price of $100, for example, we let $x = 100$ in the equation $y = 500 - \frac{3}{5}x$ and determine y.

$y = 500 - \frac{3}{5}x$

$y = 500 - \frac{3}{5}(\mathbf{100})$ Substitute 100 for x.

$ = 500 - 60$

$ = 440$

At a price of $100, 440 bicycles will be sold.

Increasing the price will decrease sales. For example, if we substitute $x = 150$ into the equation, we will find that $y = 410$. Thus, at a price of $150, the owners would expect to sell 410 bicycles. Similarly, at a price of $200, they would expect to sell 380 bicycles.

To graph this equation, we plot several points (x, y) that satisfy the equation and join them with a line. We have seen that the pairs $(100, 440)$, $(150, 410)$, and $(200, 380)$ satisfy the equation. We plot these pairs and graph the equation as in Figure 13-16.

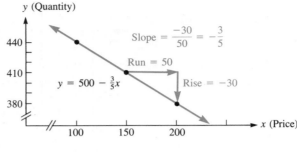

Figure 13-16

The slope of this line is $-\frac{3}{5}$, which is the coefficient of x in the equation $y = 500 - \frac{3}{5}x$. This slope represents the ratio of a change in the number of bicycles sold to the change in the price of a bicycle. The fraction $-\frac{3}{5}$ is equal to $-\frac{30}{50}$. Refer to Figure 13-16 to see that for each increase of $50 in the cost of a bicycle, the store expects to sell 30 fewer bicycles.

In economics, the graph of this equation is called a **demand curve**, and its slope is called the **marginal demand**.

Linear Depreciation

For tax purposes, accountants use **linear depreciation** to estimate the decreasing value of aging equipment.

EXAMPLE 3

An insurance company buys a $12,500 computer system with an estimated useful life of 6 years. After x years of use, the value y of the computer is linearly depreciated according to the equation

$$y = 12{,}500 - 2000x$$

a. Determine the value of the computer after $3\frac{1}{2}$ years.

b. Determine the economic meaning of the y-intercept of the graph of the equation.

c. Determine the economic meaning of the slope of the graph of the equation.

d. Determine the value of the computer at the end of its useful life.

Solution

a. To determine the computer's value after $3\frac{1}{2}$ years, we substitute 3.5 for x in the equation and calculate y.

$$y = 12{,}500 - 2000x$$
$$y = 12{,}500 - 2000(3.5)$$
$$= 12{,}500 - 7000$$
$$= 5500$$

When the computer is $3\frac{1}{2}$ years old, its value will be $5500.

b. The y-intercept of a graph is the value of y found by letting $x = 0$. In this example, the y-intercept is the value of a zero-year-old computer—that is, the computer's original cost. Because the y-intercept is 12,500, the computer cost $12,500 when it was new.

c. Each year, the value of the computer decreases by $2000, because the slope of the line is $-2,000$. The slope of the depreciation line is called the **annual depreciation rate**.

d. To determine the computer's value at the end of its 6-year useful life, we substitute 6 for x in the equation and calculate y.

$$y = 12,500 - 2000x$$
$$y = 12,500 - 2000(6)$$
$$= 12,500 - 12,000$$
$$= 500$$

At the end of its useful life, the computer is worth $500. This is known as the **salvage value** of the equipment.

13.2 Exercises

In Exercises 1–16, find the slope of the line passing through the two given points. If the slope is not defined, write "undefined."

1. (1, 3) (2, 4)   **2.** (1, 4) (2, 3)   **3.** (2, 5) (3, 4)   **4.** (3, 6) (5, 2)

5. (2, 6) (3, 7)   **6.** (5, 8) (2, 9)   **7.** (0, −5) (4, 3)   **8.** (3, −2) (3, 5)

9. (2, 3) (−3, 2)   **10.** (−7, 3) (−3, 7)   **11.** (−5, −7) (−4, −7)   **12.** (−6, −8) (−5, −8)

13. $\left(-2, \frac{1}{2}\right) \left(0, -\frac{3}{2}\right)$ **14.** $\left(\frac{2}{3}, -9\right) \left(\frac{5}{3}, 0\right)$ **15.** $\left(\frac{5}{7}, \frac{1}{2}\right) \left(-\frac{2}{7}, 0\right)$ **16.** $\left(0, \frac{7}{2}\right) \left(-\frac{5}{2}, 0\right)$

In Exercises 17–28, graph the line that passes through the given point and has the given slope.

17. (0, 3), slope 1   **18.** (3, 2), slope 3   **19.** (−3, 2), slope 4   **20.** (−1, 0), slope 2

21. (1, −3), slope −1   **22.** (1, −3), slope −2   **23.** (3, 5), slope −4   **24.** (0, 0), slope −5

25. (0, 0), slope $\frac{1}{2}$ **26.** (2, 3), slope $-\frac{1}{2}$ **27.** (−1, 3), slope $-\frac{5}{3}$ **28.** (0, 0), slope $\frac{3}{5}$

In Exercises 29–40, indicate the slope and the y-intercept of the line defined by each equation. Then use the slope and y-intercept to graph the line.

29. $y = 3x + 3$ **30.** $y = 4x - 5$ **31.** $y = 5x + 1$ **32.** $y = -3x + 2$

33. $y = -3x$ **34.** $y = -3x + 5$ **35.** $y = 3x - 2$ **36.** $y = -5x + 1$

37. $y = \frac{x}{3}$ **38.** $y = -\frac{x}{2} + 2$ **39.** $y = \frac{5}{3}x + \frac{1}{2}$ **40.** $y = \frac{3}{5}x - \frac{1}{2}$

In Exercises 41–52, write each equation in slope-intercept form, and then indicate the slope and the y-intercept of the line.

41. $3(x - 2) + y = 1$
42. $2(x + 2) - 3y = 2$
43. $5(x - 5) = 2y + 1$
44. $2(y + 7) - x = 3$

45. $2(y - 1) = y$
46. $y - 7(y + 5) = 6$
47. $\dfrac{2y + 7}{2} = x$
48. $\dfrac{2 - x}{4} = y$

49. $\dfrac{3(y - 5)}{2} = x + 3$
50. $\dfrac{5(3 + x)}{2} = y - 5$
51. $x = 3y + 5$
52. $x = -\dfrac{1}{5}y - \dfrac{1}{2}$

53. The slope of the road in Illustration 1 is the vertical rise divided by the horizontal run. If the vertical rise is 24 ft for a horizontal run of 1 mi, determine the slope of the road. (*Hint:* 1 mi = 5280 ft.)

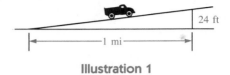

Illustration 1

54. The demand, y, for television sets depends on the sales price, x, according to the equation $y = 5500 - 6x$. Determine the impact on sales of each $1 increase in price.

55. The pitch of a roof is defined to be the vertical rise divided by the horizontal run. If the rise of the roof in Illustration 2 is 5 ft for a run of 12 ft, determine the pitch of the roof.

56. The total daily cost C to an electronics company for manufacturing x television sets is given by the equation

$$C = 1200 + 130x$$

The y-intercept of the graph of this equation is called the company's **fixed cost**, and the slope of the line is called its **marginal cost**. Determine the company's fixed and marginal costs.

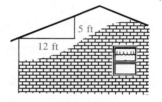

Illustration 2

57. A truck is depreciated linearly by the formula $y = 57,000 - 5500x$, where x represents its age in years. Find the value of the truck after $5\frac{1}{2}$ years.

58. Find the annual rate of depreciation of the truck in Exercise 57.

59. An office copy machine is valued annually by the linear depreciation formula $y = 6000 - 1200x$. Find the copier's original cost.

60. Find the salvage value of the copier in Exercise 59.

61. A $450 drill press will have no salvage value after 3 years. Find the annual rate of depreciation.

62. A $750 video camera has a useful life of 5 years and a salvage value of $0. Find its annual rate of depreciation.

63. Illustration 3 shows the price per share of common stock for a certain company. Find the slope between points for consecutive years. In what years did the company's stock rise most sharply? In what year did it fall most sharply?

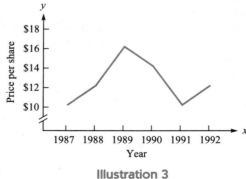

Illustration 3

64. Illustration 4 shows the sales per year of a certain company. Find the slope between points for consecutive years. In what years did the company's sales increase most sharply? In what years did sales decrease most sharply?

Illustration 4

Review Exercises

In Review Exercises 1–6, factor each expression.

1. $3x^2 - 6x$ **2.** $y^2 - 25$ **3.** $2z^2 - 5z - 3$ **4.** $9t^2 - 15t + 6$

5. $9u^2 + 24u + 16$ **6.** $x^4 - 16y^4$

In Review Exercises 7–10, solve each equation.

7. $4y^2 + 8y = 0$ **8.** $r^2 - 36 = 0$ **9.** $x^2 - 7x + 6 = 0$ **10.** $12s^2 + 13s = 4$

In Review Exercises 11–12, write each statement as an algebraic expression.

11. The product obtained when the sum of a and b is multiplied by a number that is 3 greater than z.

12. The product of three consecutive even integers, where x represents the largest integer.

13.3 Writing Equations of Lines

- Point-Slope Form of the Equation of a Line
- Parallel and Perpendicular Lines

We now consider how to determine a linear equation from its graph.

To begin, we recall that the linear equation $y = mx + b$ is written in slope-intercept form, that m is the slope of its straight-line graph, and that b is its y-intercept. If we know both the slope and the y-intercept of a line, we can always write its equation. For example, to write the equation of the line with slope 7 and y-intercept -2, we simply substitute 7 for m and -2 for b in the slope-intercept form of the equation of a line and simplify.

$y = mx + b$
$y = 7x + (-2)$ Substitute 7 for m and -2 for b.
$y = 7x - 2$

Thus, the equation of the line with slope 7 and y-intercept -2 is $y = 7x - 2$.

EXAMPLE 1

Find the equation of the line with slope $-\frac{3}{2}$ and y-intercept 3. Express the equation in general form.

Solution

$y = mx + b$	Use the slope-intercept form.
$y = -\frac{3}{2}x + 3$	Substitute $-\frac{3}{2}$ for m and 3 for b.
$2y = -3x + 6$	Multiply both sides by 2.
$3x + 2y = 6$	Add $3x$ to both sides.

The equation $3x + 2y = 6$ is written in the form $Ax + By = C$, which is the general form of the equation of a line. ∎

Point-Slope Form of the Equation of a Line

Suppose we know the slope of a line, but instead of its y-intercept we know the coordinates of a second point on the line. It is still possible to find the equation of the line. For example, we will find the equation of the line that has a slope of 3 and passes through the point $P(2, 1)$. See Figure 13-17.

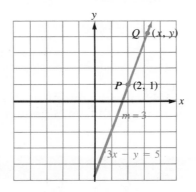

Figure 13-17

We know that $P(2, 1)$ are the coordinates of one point on the line. We can then let $Q(x, y)$ be the coordinates of some other point on the line. To find the slope of the line that passes through the points $P(2, 1)$ and $Q(x, y)$, we use the definition of slope.

$$m = \frac{y_2 - y_1}{x_2 - x_1}$$

1. $m = \dfrac{y - 1}{x - 2}$ Substitute y for y_2, 1 for y_1, x for x_2, and 2 for x_1.

However, we are given that the slope m of the line is 3. Thus, we have

$3 = \dfrac{y - 1}{x - 2}$	Substitute 3 for m in Equation 1.
$3(x - 2) = y - 1$	Multiply both sides by $x - 2$.
$3x - 6 = y - 1$	Use the distributive property to remove parentheses.
$3x - y = 5$	Add $-y$ and 6 to both sides.

The equation of the required line, written in general form, is $3x - y = 5$.

To find the equation of a line passing through a known point with a known slope, we refer to Figure 13-18, in which the slope of the line is m and the known point is (x_1, y_1). If (x, y) is any other point on the line, the slope m is

$$m = \frac{y - y_1}{x - x_1}$$

If we multiply both sides of this equation by $x - x_1$, we have

$$m(x - x_1) = y - y_1$$

or

$$y - y_1 = m(x - x_1)$$

This equation is called the **point-slope form** of the equation of a line.

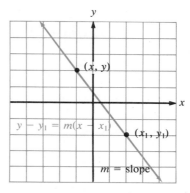

Figure 13-18

The Point-Slope Form of the Equation of a Line: If a line with slope m passes through the point (x_1, y_1), the equation of the line is

$$y - y_1 = m(x - x_1)$$

EXAMPLE 2

A line has a slope of $\frac{3}{4}$ and passes through the point $(-1, \frac{1}{2})$. Write the equation of the line in general form and graph it.

Solution

We substitute $\frac{3}{4}$ for m, -1 for x_1, and $\frac{1}{2}$ for y_1 in the point-slope form of a linear equation and simplify.

$$y - y_1 = m(x - x_1)$$

$$y - \frac{1}{2} = \frac{3}{4}[x - (-1)]$$

$4y - 2 = 3(x + 1)$ Multiply both sides by 4 and simplify.
$4y - 2 = 3x + 3$ Remove parentheses.
$-5 = 3x - 4y$ Add -3 and $-4y$ to both sides.

In general form, the equation of the required line is $3x - 4y = -5$. Work Progress Check 4.

Progress Check 4

Write the equation of a line passing through $(-2, 4)$ with a slope of $\frac{2}{3}$. Write the equation in slope-intercept form.

4. $y = \frac{2}{3}x + \frac{16}{3}$

Progress Check Answers

Section 13.3 Writing Equations of Lines

Parallel and Perpendicular Lines

Because all horizontal lines are parallel, all lines with a slope of 0 are parallel. Likewise, because all vertical lines are parallel, all lines with undefined slopes are parallel. Because the slope is a measure of how rapidly a line rises or falls, it is reasonable to assume that lines with equal slopes are parallel.

> **Two lines with the same slope are parallel.**

Thus, the equations $y = \mathbf{5}x + 7$ and $y = \mathbf{5}x - 8$ represent parallel lines, because the slope of each line is $\mathbf{5}$.

EXAMPLE 3

Find the equation of the line parallel to the line $y = -3x + 4$ that passes through the point $(5, -2)$. Write the result in general form, and graph both equations.

Solution

Because the given equation is written in slope-intercept form, we see that its slope is -3 and its y-intercept is 4.

To graph the equation, we first plot the y-intercept $(0, 4)$. Because the slope of the line is -3, the line must drop 3 units for every 1 unit it moves to the right. Beginning at the y-intercept, we move 1 unit to the right and then 3 units down. This locates the point $(1, 1)$ on the line. The line through $(0, 4)$ and $(1, 1)$ is the graph of $y = -3x + 4$. See Figure 13-19.

Because the required line is parallel to the given line, it will have the same slope as the given line. Thus, the slope of the required line is also -3, and it passes through the point $(5, -2)$. We substitute these values into the point-slope form of the equation of a line and simplify.

$$y - y_1 = m(x - x_1)$$
$$y - (\mathbf{-2}) = \mathbf{-3}(x - \mathbf{5}) \quad \text{Substitute } -3 \text{ for } m, 5 \text{ for } x_1, \text{ and } -2 \text{ for } y_1.$$
$$y + 2 = -3x + 15 \quad \text{Remove parentheses.}$$
$$3x + y = 13 \quad \text{Add } 3x \text{ and } -2 \text{ to both sides.}$$

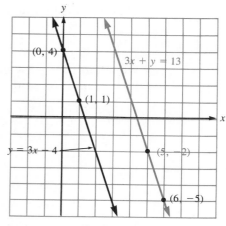

Figure 13-19

Progress Check 5

Find the equation of the line parallel to the line $y = 5x - 2$ that passes through $P(-5, 2)$.

Progress Check 6

Find the equation of the line perpendicular to the line $y = 5x - 2$ that passes through $P(-5, 2)$.

5. $y = 5x + 27$
6. $y = -\frac{1}{5}x + 1$

Progress Check Answers

In general form, the equation of the required line is $3x + y = 13$. To graph this line, we plot the point $(5, -2)$. Because the slope is -3, we move 1 unit to the right and then 3 units down to locate the point $(6, -5)$. The graph of the equation is the line through $(5, -2)$ and $(6, -5)$. See Figure 13-19. Work Progress Check 5.

Lines that meet at right angles are called **perpendicular lines**. For example, a vertical line is perpendicular to a horizontal line. Slopes of other perpendicular lines are related by the following fact.

> **Two lines whose slopes have a product of -1 are perpendicular.**

Two numbers that have a product of -1 are called **negative reciprocals** of each other. The numbers 3 and $-\frac{1}{3}$, for example, are negative reciprocals, because their product is -1:

$$3\left(-\frac{1}{3}\right) = -1$$

EXAMPLE 4

One line has a slope of 5 and another has a slope of $-\frac{1}{5}$. Determine whether the lines are parallel, perpendicular, or neither.

Solution

Because the slopes of the lines are not equal ($5 \neq -\frac{1}{5}$), the lines are not parallel. To determine whether the lines are perpendicular, find the product of their slopes. Because the product of their slopes is $5(-\frac{1}{5})$, or -1, the lines are perpendicular. Work Progress Check 6.

EXAMPLE 5

Determine whether the graphs of the following equations are parallel, perpendicular, or neither:

a. $y = 2x - 5$ **b.** $y = 2(3 - x)$

Solution

The slope of the line $y = 2x - 5$ is 2, the coefficient of x. To find the slope of the second line, solve its equation for y.

$y = 2(3 - x)$
$y = 6 - 2x$ Remove parentheses.
$y = -2x + 6$

The slope of the second line is -2, which is the coefficient of x in the equation $y = -2x + 6$. Because the slopes (2 and -2) of the two lines are not equal, the lines are not parallel. Because the product of the slopes is not -1, the lines are not perpendicular either. Thus, the lines are neither parallel nor perpendicular.

Section 13.3 Writing Equations of Lines 567

EXAMPLE 6

Find the equation of the line perpendicular to the graph of $y = \frac{1}{3}x + 5$, and passing through $P(2, -1)$. Write the result in general form and graph both equations.

Solution

The slope of the line with equation $y = \frac{1}{3}x + 5$ is $\frac{1}{3}$. Because the required line is to be perpendicular to the given line, its slope will be the negative reciprocal of $\frac{1}{3}$.

$$\text{Slope of the required line} = -\frac{1}{\frac{1}{3}} = -(1)\left(\frac{3}{1}\right) = -3$$

We use the point-slope form to find the equation of the line with a slope of -3 and passing through the point $(2, -1)$:

$y - y_1 = m(x - x_1)$
$y - (-1) = -3(x - 2)$ Substitute -3 for m, 2 for x_1, and -1 for y_1.
$y + 1 = -3x + 6$ Remove parentheses.
$3x + y = 5$ Add $3x$ and -1 to both sides.

Written in general form, the equation of the required line is $3x + y = 5$. The graphs of both lines appear in Figure 13-20.

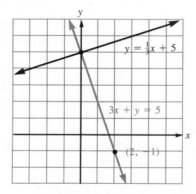

Figure 13-20

The various forms of the equation of a line are summarized as follows:

Forms of a Line:

$Ax + By = C$ **General form** of a linear equation.
                A and B cannot both be zero.

$y = mx + b$ **Slope-intercept form** of a linear equation.
                The slope is m, and the y-intercept is b.

$y - y_1 = m(x - x_1)$ **Point-slope form** of a linear equation.
                The slope is m, and the line passes through (x_1, y_1).

$y = b$ Horizontal line.
                The slope is 0, and the y-intercept is b.

$x = a$ Vertical line.
                The slope is undefined, and the x-intercept is a.

EXAMPLE 7

Each month, Kia's water bill includes a fixed service charge, plus an additional per-gallon charge for water used. The bill is related to the number of gallons used by the equation of a line. If Kia is billed $12 for using 1000 gallons and $16 for using 1800 gallons, find her bill for 2000 gallons.

Solution

Let x represent the number of gallons of water used and let y be the amount of Kia's monthly bill. To write the equation of the line, we find the slope of the line and a point on the line. Then we substitute these values into the point-slope form of the equation of a line.

When Kia uses 1000 gallons, her bill is $12. Thus, if $x = 1000$, then $y = 12$, and the point $P(x, y) = P(1000, 12)$ lies on the line.

Similarly, when $x = 1800$, then $y = 16$, and the line passes through the point $Q(x, y) = Q(1800, 16)$. The graph of the line appears in Figure 13-21. To find m, the slope of the line, we find the slope of the line PQ:

$$m = \frac{y_2 - y_1}{x_2 - x_1}$$
$$= \frac{16 - 12}{1800 - 1000}$$
$$= \frac{4}{800}$$
$$= \frac{1}{200}$$

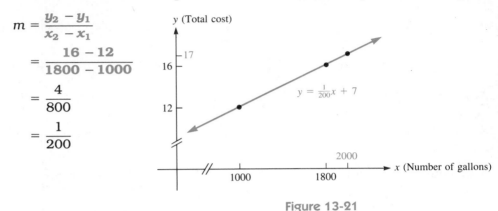

Figure 13-21

Thus, the slope of the line is $\frac{1}{200}$.

To find the equation of the line, we substitute $m = \frac{1}{200}$ and the coordinates of $P(1000, 12)$ into the point-slope form of the equation of a line:

$$y - y_1 = m(x - x_1)$$
$$y - 12 = \frac{1}{200}(x - 1000)$$
$$y - 12 = \frac{1}{200}x - \frac{1}{200}(1000)$$
$$y - 12 = \frac{1}{200}x - 5$$
$$y = \frac{1}{200}x + 7$$

To find Kia's bill for 2000 gallons, we substitute 2000 for x in the billing formula and find y:

$$y = \frac{1}{200}(2000) + 7$$
$$y = 10 + 7$$
$$y = 17$$

Her bill for using 2000 gallons of water is $17.

13.3 Exercises

In Exercises 1–12, use the slope-intercept form to find the equation of each line with the given properties. Write the equation in general form.

1. Slope 4, y-intercept 5
2. Slope 2, y-intercept -3
3. Slope -7, y-intercept -2
4. Slope -3, y-intercept 10
5. Slope $-\frac{1}{2}$, y-intercept $-\frac{1}{2}$
6. Slope $\frac{2}{3}$, y-intercept -5
7. Slope $-\frac{3}{5}$, y-intercept $-\frac{2}{5}$
8. Slope $-\frac{5}{7}$, y-intercept $-\frac{3}{7}$
9. Slope 0, y-intercept 3
10. Slope 3, y-intercept 0
11. Slope 0, y-intercept 0
12. Slope $-\frac{4}{5}$, y-intercept 0

In Exercises 13–24, use the point-slope form to find the equation of each line with the given properties. Write the equation in general form.

13. Slope 2, passing through (1, 1)
14. Slope 3, passing through (2, 3)
15. Slope 5, passing through (1, −1)
16. Slope −3, passing through (0, −3)
17. Slope −2, passing through (0, 0)
18. Slope $\frac{1}{2}$, passing through (2, 3)
19. Slope $-\frac{1}{2}$, passing through (−2, 4)
20. Slope $-\frac{2}{3}$, passing through (0, 3)
21. Slope $\frac{7}{5}$, passing through $\left(3, \frac{2}{5}\right)$
22. Slope $-\frac{3}{7}$, passing through $\left(1, \frac{5}{7}\right)$
23. Slope 0, passing through (739, 3)
24. Slope 0, passing through (0, 0)

In Exercises 25–32, determine whether the lines with the given slopes are parallel, perpendicular, or neither.

25. Slopes of 3 and −3
26. Slopes of 2 and $\frac{1}{2}$
27. Slopes of 4 and $\frac{8}{2}$
28. Slopes of $-\frac{3}{2}$ and $\frac{2}{3}$
29. Slopes of 3 and $-\frac{1}{3}$
30. Slopes of 0.5 and $\frac{1}{2}$
31. Slopes of 1.5 and $\frac{3}{2}$
32. Slopes of 1.5 and −1.5

In Exercises 33–42, indicate whether the given pairs of lines are parallel, perpendicular, or neither.

33. $y = 3x + 2$ and $y = 3x - \frac{1}{2}$
34. $y = 2x + 5$ and $y = 2x - 7$
35. $y = \frac{1}{3}x + 1$ and $y = 3x - 1$
36. $y = 5x + 5$ and $y = -\frac{1}{5}x + 5$

37. $y = x + \dfrac{7}{2}$ and $y = -x + \dfrac{2}{7}$

38. $y = 5$ and $x = 5$

39. $y = -\dfrac{3}{4}x$ and $y = \dfrac{4}{3}x - \dfrac{1}{2}$

40. $y = 7x + 1$ and $y = -\dfrac{1}{7}x - 7$

41. $2x + 3y = 5$ and $y = -\dfrac{2}{3}x - 5$

42. $3x - y = 8$ and $x + 3y = 8$

In Exercises 43–62, write the equation of each line with the given properties in general form, when possible.

43. Slope $\dfrac{2}{15}$, passing through $(15, 10)$

44. Slope $\dfrac{3}{11}$, y-intercept $\dfrac{1}{11}$

45. Slope $-\dfrac{4}{9}$, y-intercept $-\dfrac{2}{9}$

46. Slope $-\dfrac{13}{5}$, passing through $\left(3, -\dfrac{2}{5}\right)$

47. Parallel to the line $y = 5x - 8$, and passing through $(0, 0)$

48. Parallel to the line $y = -2(7 - x)$, passing through $(0, 5)$

49. Parallel to the line $y = 5(x + 3)$, with y-intercept -4

50. Parallel to the line $y = -6(x - 2)$, with y-intercept 5

51. Parallel to the line $5x - 2y = 3$, and passing through $(2, 1)$

52. Parallel to the line $3x + 4y = -9$, with y-intercept -3

53. Passing through $(2, 5)$ and $(3, 7)$ (*Hint:* First, find the slope.)

54. Passing through $(-3, 0)$ and $(3, -1)$ (*Hint:* First, find the slope.)

55. Passing through $(-5, 2)$ and $(7, 2)$ (*Hint:* First, find the slope.)

56. Passing through $(-5, 2)$ and $(-5, 7)$ (*Hint:* First, try to find the slope.)

57. Passing through $(2, 5)$ and perpendicular to a line with slope $\dfrac{1}{5}$

58. Passing through $(2, 1)$, and perpendicular to a line with slope -3

59. Passing through the origin, and perpendicular to the line $y = 5x + 11$

60. Passing through the point $(3, -5)$, and perpendicular to the line $y = -\dfrac{1}{2}x - 7$

61. Having a y-intercept of 3, and perpendicular to the line $y = \dfrac{1}{5}x + 12$

62. Having a y-intercept of $-\dfrac{1}{2}$, and perpendicular to the line $y = -\dfrac{1}{3}x$

63. Seamless aluminum rain gutter can be installed for a fixed charge, plus an additional per-foot cost. If an installation of 200 feet costs $350 and 250 feet costs $425, find the cost to install 500 feet of gutter.

64. Fahrenheit temperature, F, is related to the Celsius temperature, C, by an equation of the form

$$F = mC + b$$

If water freezes at 0° Celsius and 32° Fahrenheit, and water boils at 100° Celsius and 212° Fahrenheit, determine m and b.

65. The accounting department uses the linear method to depreciate a word-processing system. After 3 years, the system is worth $2000. After 4 years, it is worth nothing. What was the purchase price of the system?

66. A company estimates the useful life of a lathe to be 10 years. After 3 years, it is worth $330. After 9 years, it is worth $90. What is the salvage value of the lathe?

Review Exercises

In Review Exercises 1–6, perform each operation.

1. $\dfrac{x^2 - 25}{x^2 + 10x + 25} \cdot \dfrac{x^2 + 6x + 5}{x^2 - 5x}$

2. $\dfrac{x^2 - x - 2}{x^2 + 4x + 3} \div \dfrac{x^2 - 2x}{x + 3}$

3. $\dfrac{3 + x}{x + 1} + \dfrac{3 - x}{x + 1}$

4. $\dfrac{2}{x} + \dfrac{x}{3}$

5. $\dfrac{x + 1}{x - 1} - \dfrac{x - 1}{x + 1}$

6. $\dfrac{x + 3}{x - 1} - \dfrac{x + 4}{x + 1}$

In Review Exercises 7–10, write each number in scientific notation.

7. 73,000,000,000

8. 0.0000000245

9. 37.2×10^{-2}

10. 0.0043×10^5

11. Write the expression $(x - 4)^3$ without using parentheses.

12. Solve the equation $\dfrac{5(x - 4)}{2} = \dfrac{4(x - 5)}{3}$.

13.4 Graphing Inequalities

It is also possible to graph *inequalities* such as $y \geq x - 5$. These graphs contain only those points whose coordinates satisfy the given inequality. The graphs of these inequalities are areas bounded by lines, called **half-planes**.

EXAMPLE 1

Graph the inequality $y \geq x - 5$.

Solution

Because the symbol $\geq$ allows the possibility that the two sides are equal, we graph the equation $y = x - 5$ as in Figure 13-22(a).

Because the symbol $\geq$ also allows the possibility that the left-hand side is greater than the right-hand side, the coordinates of points other than those indicated by the graph in Figure 13-22(a) satisfy the inequality. For example, the coordinates of the origin satisfy the inequality. We verify this by letting x and y be 0 in the inequality:

$y \geq x - 5$
$0 \geq 0 - 5$
$0 \geq -5$

Because 0 is greater than or equal to -5, the coordinates of the origin satisfy the inequality. In fact, the coordinates of *every* point on the same side of the line as the origin satisfy the inequality. The graph of the inequality $y \geq x - 5$ is the half-plane shown in Figure 13-22(b). Because the boundary line $y = x - 5$ is included, it is drawn with a solid line.

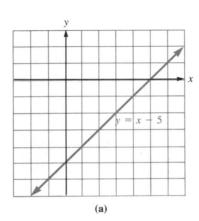
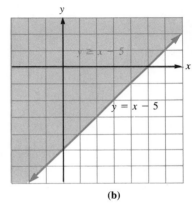

(a)   (b)

Figure 13-22

EXAMPLE 2

Graph $2(x - 3) - (x - y) \leq -1$.

Solution

We begin by simplifying the inequality:

$2(x - 3) - (x - y) \leq -1$
$2x - 6 - x + y \leq -1$ Remove parentheses.
$x - 6 + y \leq -1$ Combine terms.
$x + y \leq 5$ Add 6 to both sides.

We first graph the equation $x + y = 5$, as in Figure 13-23**(a)**.

The symbol $\leq$ allows the possibility that the left-hand side of $x + y \leq 5$ is less than the right-hand side. Again, the coordinates of the origin satisfy the inequality. We verify this by letting x and y be 0 in the given inequality.

$x + y \leq 5$
$\mathbf{0} + \mathbf{0} \leq 5$ Substitute 0 for x and for y.
$0 \leq 5$

Thus, the coordinates of the origin satisfy the original inequality. In fact, the coordinates of *every* point on the same side of the line as the origin satisfy the inequality. The graph of $x + y \leq 5$ is the half-plane that appears in color in Figure 13-23**(b)**. Because the boundary line $x + y = 5$ is included, it is drawn as a solid line.

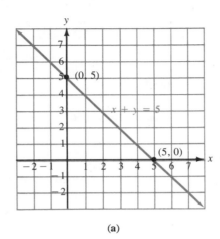
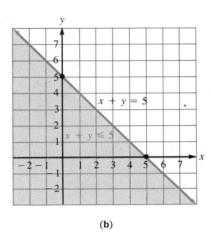

(a)   (b)

Figure 13-23

EXAMPLE 3

Graph the inequality $y > 2x$.

Solution

Although the symbol $>$ does not allow the possibility that y and $2x$ are equal, we begin by graphing the line $y = 2x$ anyway. Draw the graph as a broken line to indicate that the points on the line are not part of the solution. See Figure 13-24(a).

$y = 2x$	
x	y
-1	-2
0	0
3	6

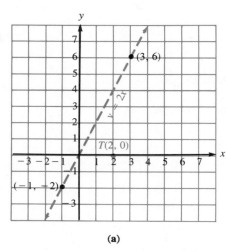
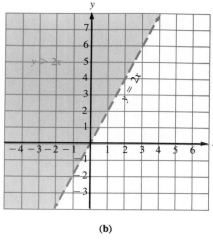

(a)   (b)

Figure 13-24

We substitute into $y > 2x$ the coordinates of any point on one side of the line. Point $T(2, 0)$, for example, is obviously *below* the line $y = 2x$. See Figure 13-24(a). To see if point $T(2, 0)$ satisfies the original inequality, we substitute 2 for x and 0 for y:

$y > 2x$
$0 > 2(2)$ Substitute 2 for x and 0 for y.
$0 > 4$

The inequality $0 > 4$ is a false statement, so the coordinates of point T do not satisfy the inequality. Thus, point T is not on the side of the line we wish to shade. The coordinates of points on the other side of the line are the ones that do satisfy the inequality. Therefore, we must shade the other side of the line. The graph of the solution set of the inequality $y > 2x$ appears in Figure 13-24(b). ■

EXAMPLE 4

Graph the inequalities   **a.** $x + 2y < 6$ and   **b.** $y \geq 0$.

Solution

a. We find the boundary line by graphing $x + 2y = 6$. This boundary is drawn as a broken line to show that the line is not part of the solution. Then we choose a point that is not on the boundary line and see if its coordinates satisfy the original inequality. The origin is a convenient choice.

$$x + 2y < 6$$
$$0 + 2(0) < 6 \quad \text{Substitute 0 for } x \text{ and for } y.$$
$$0 < 6$$

Because $0 < 6$ is a true statement, we shade the side of the line that includes the origin. The graph of $x + 2y < 6$ appears in Figure 13-25(a).

b. We first find the boundary line by graphing $y = 0$. This equation represents the x-axis. This boundary is drawn as a solid line to show that the line is part of the solution. Then we choose a point that is not on the boundary line and see if its coordinates satisfy the original inequality. The point $T(0, 1)$ is above the x-axis and is a convenient choice.

$$y \geq 0$$
$$1 \geq 0 \quad \text{Substitute 1 for } y.$$

Because $1 \geq 0$ is a true statement, we shade the side of the line that includes the point T. This is the half-plane above the x-axis. The graph of $y \geq 0$ appears in Figure 13-25(b).

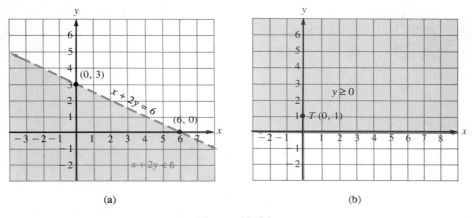

Figure 13-25

The inequalities in Examples 1–5 are called **linear inequalities**. The following is a summary of the procedure for graphing linear inequalities.

Procedure for Graphing a Linear Inequality in Two Variables:

1. Graph the boundary line of the region. If the inequality symbol is either $\leq$ or $\geq$, draw the boundary line as a solid line. If the inequality symbol is $<$ or $>$, draw the boundary line as a broken line.
2. Pick a point that is on one side of the boundary line. (Use the origin if possible.) Replace x and y with the coordinates of that point. If the inequality is satisfied, shade the side that contains that point. If the inequality is not satisfied, shade the other side.

13.4 Exercises

In Exercises 1–8, the boundary of a graph has been drawn. Complete the graph by shading the correct side of the boundary.

1. $y \leq x + 2$ **2.** $y > x - 3$ **3.** $y > 2x - 4$ **4.** $y \leq -x + 1$

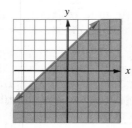 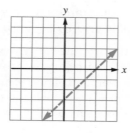 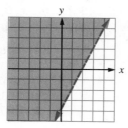 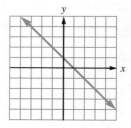

5. $x - 2y \leq 4$ **6.** $3x + 2y > 12$ **7.** $y \leq 4x$ **8.** $y + 2x < 0$

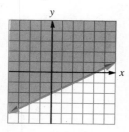 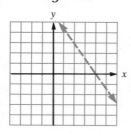 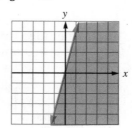 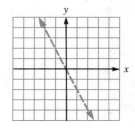

In Exercises 9–24, graph each linear inequality.

9. $y \geq 3 - x$ **10.** $y < 2 - x$ **11.** $y < 2 - 3x$ **12.** $y \geq 5 - 2x$

13. $y \geq 2x$ **14.** $y < 3x$ **15.** $2y - x < 8$ **16.** $y + 9x \geq 3$

17. $y - x \geq 0$ **18.** $y + x < 0$ **19.** $2x + y > 2$ **20.** $3x - 2y > 6$

21. $3x - 4y > 12$ **22.** $4x + 3y \leq 12$ **23.** $5x + 4y \geq 20$ **24.** $7x - 2y < 21$

In Exercises 25–40, simplify each inequality and construct its graph.

25. $3(x + y) + x < 6$

26. $2(x - y) - y \geq 4$

27. $4x - 3(x + 2y) \geq -6y$

28. $3y + 2(x + y) < 5y$

29. $7(x + 2y) - 2(x - 3y) < 50$

30. $3(6x + 5y) - 5(3x + 4y) \geq 15$

31. $5(x - 2y) > 5x$

32. $3(y - x) \leq x - 3$

33. $x(x + 2) \leq x^2 + 3x + 1$

34. $y(y - 5) > y^2 - 7$

35. $x^2 + 3y \leq x(x + 2) - 1$

36. $x + y(y - 3) < (y + 1)(y - 2)$

37. $3x + 7 \leq 5y + 7$

38. $5x \leq x + 5(y + x)$

39. $(x + 1)(x - 2) + y^2 < x^2 + y(y - 2)$

40. $(x + 2)(x - 2) + y^2 \geq (y - 2)(y + 1) + x^2$

Review Exercises

1. List the prime numbers between 40 and 50.
2. State the associative property of addition.
3. State the commutative property of multiplication.
4. What is the additive identity element?
5. What is the multiplicative identity element?
6. What is the multiplicative inverse of $\frac{5}{3}$?

In Review Exercises 7–12, let $P(x) = 3x^2 - 4x + 3$. Find each quantity.

7. $P(2)$
8. $P(-3)$
9. $P(-x)$
10. $P(x + 1)$
11. $P(x^2)$
12. $P(y^2)$

13.5 Relations and Functions

- Relations and Functions
- Function Notation
- Domain and Range
- Graphs of Functions
- Reading Domain and Range from a Graph

We have seen that equations and inequalities determine ordered pairs of numbers. Each ordered pair indicates a correspondence between two numbers. The equation $y = -x + 2$, for example, determines ordered pairs (x, y) in which each value of y corresponds to a number x. If $x = 3$ in this equation, the corresponding value of y is $-3 + 2$, or -1, and the ordered pair $(3, -1)$ indicates this correspondence. The graph of *all* such ordered pairs is the line in Figure 13-26**(a)**.

The inequality $y \leq -x + 2$ also determines a set of ordered pairs. For this inequality, however, to each number x there correspond *many* values of y. Because the boundary of the graph of $y \leq -x + 2$ is the line $y = -x + 2$, the inequality determines the ordered pair $(3, -1)$. The inequality also determines many other pairs with an x-coordinate of 3: pairs such as $(3, -2)$, $(3, -3\frac{1}{2})$, and $(3, -5)$. Thus, to the number $x = 3$ there correspond *several* values of y. The graph of the inequality appears in Figure 13-26**(b)**.

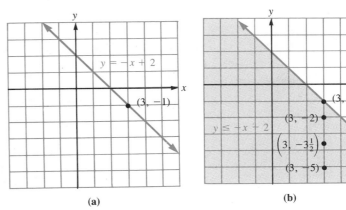

Figure 13-26

Relations and Functions

Because ordered pairs show that certain values of x and y are related, any set of ordered pairs is called a **relation**. The set of ordered pairs shown in either part of Figure 3-26 is a relation. However, the relations shown in

Figure 13-26 differ in an important way. Several ordered pairs in Part **(b)** have the same number x, but different values of y. The relation contains, for example, the pairs (3, −1) and (3, −2). In Part **(a)**, however, *exactly one* value of y corresponds to each number x. The relation of Part **(a)** is called a **function**.

> **Definition:** A **function** is a correspondence that assigns to each number x exactly one value y. The variable x is called the **independent variable**, and y is called the **dependent variable**.

EXAMPLE 1

Does the equation $y = x^2$ define a function?

Solution

To be a function, *each* number x must determine a *single* value y. To find y in this example, a number x is squared. For each x, squaring will give a single result. Thus, for each x, a single value y is produced. The equation $y = x^2$ does define a function. Work Progress Check 7. ∎

Progress Check 7

Does $y = \dfrac{x+2}{3}$ define a function?

EXAMPLE 2

Does the equation $y^2 = x$ define a function?

Solution

To be a function, *each* number x must determine a *single* value y. Let $x = 9$, for example. The variable y could be 3 or −3, because $3^2 = 9$ and $(-3)^2 = 9$. Thus, the relation contains the two ordered pairs $(9, 3)$ and $(9, -3)$ that have the same number x, but different values of y. Because there are two values of y determined when $x = 9$, the equation does not represent a function. Work Progress Check 8. ∎

Progress Check 8

Does $|y| = \dfrac{x+2}{3}$ define a function?

Function Notation

The concept of function is so important that it has a special notation.

> **Function Notation:** The notation
> $$y = f(x)$$
> denotes that y is a function of x.

The notation $f(x)$ is read as "f of x." It does not read as, nor does it mean, "f times x."

The notation $y = f(x)$ is similar to the notation $y = P(x)$ we have used for polynomials. It is used to denote the values of y that correspond to individual numbers x. If $y = f(x)$, the value of y that is determined by $x = 3$ is denoted by $f(3)$. Similarly, $f(-1)$ represents the value of y that corresponds to $x = -1$.

7. yes

8. no; if $x = 4$, y could be 2 or −2

Progress Check Answers

EXAMPLE 3

Let $f(x) = 3x + 1$. Find   **a.** $f(3)$,   **b.** $f(-1)$,   **c.** $f(0)$,   and   **d.** $f(r)$.

Solution

a. Replace x with 3:
$$f(x) = 3x + 1$$
$$f(3) = 3(3) + 1$$
$$= 9 + 1$$
$$= 10$$

b. Replace x with -1:
$$f(x) = 3x + 1$$
$$f(-1) = 3(-1) + 1$$
$$= -3 + 1$$
$$= -2$$

c. Replace x with 0:
$$f(x) = 3x + 1$$
$$f(0) = 3(0) + 1$$
$$= 0 + 1$$
$$= 1$$

d. Replace x with r:
$$f(x) = 3x + 1$$
$$f(r) = 3r + 1$$

Work Progress Check 9.

Progress Check 9

Let $f(x) = -\frac{1}{5}x + 10$. Find $f(5)$, $f(-25)$, and $f(5r)$.

The letter f is used in the notation $y = f(x)$ to represent the word *function*, but other letters can be used. The notations $y = g(x)$ and $y = h(x)$ also denote functions involving the variable x.

EXAMPLE 4

Let $g(x) = x^2 + 2x$. Calculate   **a.** $g(\frac{2}{5})$,   **b.** $g(s)$,   and   **c.** $g(s - 1)$.

Solution

a. Replace x with $\frac{2}{5}$:
$$g(x) = x^2 + 2x$$
$$g\left(\frac{2}{5}\right) = \left(\frac{2}{5}\right)^2 + 2\left(\frac{2}{5}\right)$$
$$= \frac{4}{25} + \frac{4}{5}$$
$$= \frac{24}{25}$$

b. Replace x with s:
$$g(x) = x^2 + 2x$$
$$g(s) = s^2 + 2s$$

c. Replace x with $s - 1$:
$$g(x) = x^2 + 2x$$
$$g(s - 1) = (s - 1)^2 + 2(s - 1)$$
$$= (s^2 - 2s + 1) + 2s - 2$$
$$= s^2 - 1$$

Work Progress Check 10.

Progress Check 10

Let $h(x) = 5x - 3$. Calculate $h\left(\dfrac{x + 3}{5}\right)$.

9. 9, 15, $-r + 10$

10. $h\left(\dfrac{x + 3}{5}\right) = x$

Progress Check Answers

EXAMPLE 5

Let $f(x) = 3x + 2$. Calculate   **a.** $f(3) + f(2)$ and   **b.** $f(a) - f(b)$.

Solution

a. First calculate $f(3)$.

$$f(x) = 3x + 2$$
$$f(3) = 3(3) + 2$$
$$= 9 + 2$$
$$= 11$$

Then calculate $f(2)$.

$$f(x) = 3x + 2$$
$$f(2) = 3(2) + 2$$
$$= 6 + 2$$
$$= 8$$

Finally, add the results to obtain

$$f(3) + f(2) = 11 + 8$$
$$= 19$$

b. First calculate $f(a)$.

$$f(x) = 3x + 2$$
$$f(a) = 3a + 2$$

Then calculate $f(b)$.

$$f(x) = 3x + 2$$
$$f(b) = 3b + 2$$

Finally, subtract the results to obtain

$$f(a) - f(b)$$
$$= (3a + 2) - (3b + 2)$$
$$= 3a + 2 - 3b - 2$$
$$= 3a - 3b$$

Domain and Range

In sets of ordered pairs (x, y), the set of all possible replacements for x is called the **domain** of the relation, and the set of all possible values of y is called the **range**. In the relations shown in Figure 13-26, both x and y can be any real number. The domain and the range of both relations is the set of real numbers.

The equation $y = x^2$ determines a function. Because any real number can be squared, the domain of the function is the set of real numbers. Because the result of squaring a real number cannot be negative, the range of the function is the set of nonnegative real numbers.

The equation

$$y = \frac{1}{x - 2}$$

also defines a function. Because the denominator of this fraction cannot be 0, x cannot be 2. Thus, the domain of this function consists of all real numbers except 2. To determine the range of the function, we note that $\frac{1}{x-2}$ cannot be 0, because the numerator can never be 0. The range of the function is the set of all real numbers except 0.

EXAMPLE 6

Find the domain and the range of   **a.** $f(x) = x^2 + 2$ and   **b.** $f(x) = \frac{1}{x^2 - 4}$.

Solution

a. Because 2 can be added to the result of squaring any real number, the domain of the function $f(x) = x^2 + 2$ is the set of real numbers.

Because x^2 is never less than 0, the value $x^2 + 2$ can never be less than 2. The range of the function is the set of real numbers 2 or greater.

b. The domain of the function $f(x) = \dfrac{1}{x^2 - 4}$ does not include any numbers x that would make the denominator of the fraction equal to 0. To find the numbers x that are *excluded* from the domain of the function, solve the equation $x^2 - 4 = 0$.

$$x^2 - 4 = 0$$
$$(x + 2)(x - 2) = 0$$
$$x + 2 = 0 \quad \text{or} \quad x - 2 = 0$$
$$x = -2 \quad \quad \quad x = 2$$

Thus, the domain of the function is the set of all real numbers except 2 and -2.

Because the numerator of the fraction $\dfrac{1}{x^2 - 4}$ is not 0, the fraction cannot equal 0. Thus, the range of the function is the set of all real numbers except 0.

Work Progress Check 11. ∎

Progress Check 11

Find the domain of the function

Graphs of Functions

The equation $y = 3x - 2$ defines a function. This equation also determines a graph, the line that appears in Figure 13-27**(a)**. This line is the **graph of the function**.

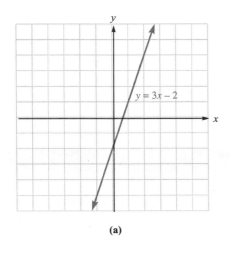
(a)

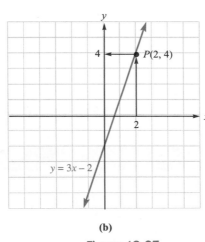
(b)

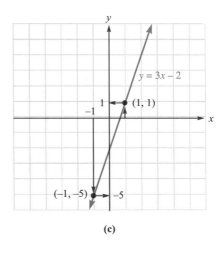
(c)

Figure 13-27

The function given by $y = 3x - 2$ defines a correspondence between numbers x and values of y. The graph of the function provides a picture of this correspondence. To see how, we draw a vertical and a horizontal line through any point on the graph, such as $P(2, 4)$. These lines intersect the x-axis at $x = 2$ and the y-axis at $y = 4$. Thus, the point $P(2, 4)$ associates the number 2 on the x-axis with the corresponding value $y = 4$ on the y-axis. Figure 13-27**(b)** shows this correspondence with a vertical arrow leaving the point 2 on the x-axis, taking a sharp turn at the point $P(2, 4)$ on the graph, and pointing to 4, the value of y that corresponds to $x = 2$,

11. Real numbers except 3 and -4

Progress Check Answers

as in Figure 13-27(b). In Figure 13-27(c), arrows show that the number $x = 1$ determines the value $y = 1$, and that $x = -1$ determines the value $y = -5$.

Reading Domain and Range from a Graph

A graph provides a way to visualize the domain and the range of a function. The set of all numbers x from which a vertical arrow can reach the graph is the domain of the function. The set of all values on the y-axis to which a horizontal arrow can point is the range of the function. As shown in Figure 13-28(a), vertical arrows can start anywhere on the x-axis and reach the graph of the function $f(x) = 3x - 2$. Thus, the domain of the function is the set of all real numbers, represented by the entire x-axis. The range of the function is also the set of real numbers, represented by the entire y-axis.

The equation $y = |x|$ also determines a function. To determine its graph, we plot several ordered pairs (x, y) that satisfy the equation, and join them as in Figure 13-28(b). Because vertical arrows can start anywhere on the x-axis, the domain is the set of real numbers. Because horizontal arrows can point only to values of y that are 0 or greater, the range is the set of nonnegative real numbers.

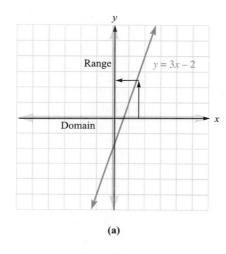

$y = |x|$

x	y
-4	4
-1	1
0	0
1	1
4	4

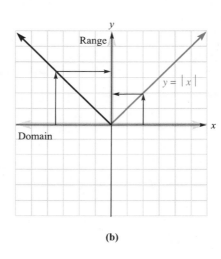

(a)   (b)

Figure 13-28

13.5 Exercises

In Exercises 1–14, indicate whether the equation or inequality determines y to be a function of x. If it does not, indicate some numbers x for which there is more than one corresponding value of y.

1. $y = x$
2. $y = 2x$
3. $y = x + 3$
4. $y = 2x - 1$
5. $y = 3x^2$
6. $y^2 = x + 1$
7. $y = 3 + 7x$
8. $y = 3 - 2x$
9. $y \leq x$
10. $y > x$
11. $x + y = 2$
12. $x - y = 3$
13. $x^2 = y^2$
14. $x^2 = 4y$

In Exercises 15–22, find a. $f(3)$, b. $f(0)$, and c. $f(-1)$.

15. $f(x) = 3x$ **16.** $f(x) = -4x$ **17.** $f(x) = 2x - 3$ **18.** $f(x) = 3x - 5$

19. $f(x) = 7 + 5x$ **20.** $f(x) = 3 + 3x$ **21.** $f(x) = 9 - 2x$ **22.** $f(x) = 12 + 3x$

In Exercises 23–30, find a. $f(1)$, b. $f(2)$, and c. $f(3)$.

23. $f(x) = x^2$ **24.** $f(x) = x^2 - 2$ **25.** $f(x) = x^3 - 1$ **26.** $f(x) = x^3$

27. $f(x) = (x + 1)^2$ **28.** $f(x) = (x - 3)^2$ **29.** $f(x) = 2x^2 - x$ **30.** $f(x) = 5x^2 + 2x - 1$

In Exercises 31–38, find a. $f(2)$, b. $f(1)$, and c. $f(-2)$.

31. $f(x) = |x| + 2$ **32.** $f(x) = |x| - 5$ **33.** $f(x) = x^2 - 2$ **34.** $f(x) = x^2 + 3$

35. $f(x) = \dfrac{1}{x + 3}$ **36.** $f(x) = \dfrac{3}{x - 4}$ **37.** $f(x) = \dfrac{x}{x - 3}$ **38.** $f(x) = \dfrac{x}{x^2 + 2}$

In Exercises 39–46, find a. $g(w)$ and b. $g(w + 1)$.

39. $g(x) = 2x$ **40.** $g(x) = -3x$ **41.** $g(x) = 3x - 5$ **42.** $g(x) = 2x - 7$

43. $g(x) = x^2 + x$ **44.** $g(x) = x^2 - 2x$ **45.** $g(x) = x^2 - 1$ **46.** $g(x) = |x - 1|$

In Exercises 47–54, let $f(x) = 2x + 1$. Then calculate the value requested.

47. $f(3) + f(2)$ **48.** $f(1) - f(-1)$ **49.** $f(b) - f(a)$ **50.** $f(b) + f(a)$

51. $f(b) - 1$ **52.** $f(b) - f(1)$ **53.** $f(0) + f(-\tfrac{1}{2})$ **54.** $f(a) + f(2a)$

In Exercises 55–62, find the domain and the range of each function.

55. $y = 3x - 2$ **56.** $y = 7 - 3x$ **57.** $y = x^2 + 3$ **58.** $y = 2x^2$

59. $y = \dfrac{1}{x}$ **60.** $y = \dfrac{1}{x + 2}$ **61.** $f(x) = \dfrac{1}{x^2 - 9}$ **62.** $f(x) = \dfrac{1}{x^2 - 5x + 6}$

In Exercises 63–66, graph each function defined by the equation $y = f(x)$.

63. $f(x) = 3x - 1$ **64.** $f(x) = 5x - 1$ **65.** $f(x) = \dfrac{x + 1}{2}$ **66.** $f(x) = 3 - 2x$

In Exercises 67–70, determine the domain and the range from the graph.

67. 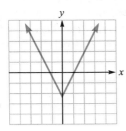 **68.** 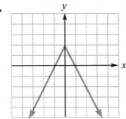 **69.** 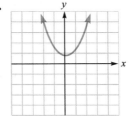 **70.**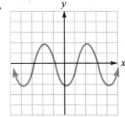

Section 13.5 Exercises 583

Review Exercises

In Review Exercises 1–6, solve each equation.

1. $\dfrac{y+2}{2} = 4(y+2)$

2. $\dfrac{3z-1}{6} - \dfrac{3z+4}{3} = \dfrac{z+3}{2}$

3. $\dfrac{2x+1}{9} = \dfrac{x}{27}$

4. $\dfrac{y+4}{5} = \dfrac{3y-6}{3}$

5. $\dfrac{2}{x-3} - 1 = -\dfrac{1}{3}$

6. $\dfrac{5}{x} + \dfrac{6}{x^2+2x} = \dfrac{-3}{x+2}$

In Review Exercises 7–12, solve each inequality and graph the solution set on a number line.

7. $3(x-2) + 3x < 24$

8. $\dfrac{4(y+3)}{2} - 3(y-4) \geqslant 6$

9. $x + 5 \leqslant 2x - 4 < 10 + x$

10. $\dfrac{-3(x-1)}{4} \geqslant -x + 4$

11. $3x^2 + 5x - 1 \leqslant (3x+2)(x-1)$

12. $(x-5)(x+1) \geqslant x^2 - 3x + 2$

13.6 Variation

- Direct Variation
- Inverse Variation
- Joint Variation
- Combined Variation
- Reading Variation from a Graph

We now introduce some terminology that scientists use to describe special functions.

Direct Variation

The first type is called **direct variation**.

> **Definition:** The words "*y* varies directly with *x*" mean that
>
> $y = kx$
>
> for some constant k. The constant k is called the **constant of variation**.

If a force is applied to a spring, the spring will stretch. The greater the force, the more it will stretch. This fact is called **Hooke's Law**, and it says that the *distance a spring will stretch varies directly with the force applied.* If d represents distance and f represents force, this relationship can be expressed as

1. $d = kf$ k is determined by the stiffness of the spring.

where k is the constant of variation. If a spring stretches 5 inches when a weight of 2 pounds is attached, the constant of variation can be computed by substituting 5 for d and 2 for f in Equation 1 and solving for k.

$$d = kf$$
$$5 = k(2)$$
$$\frac{5}{2} = k$$

To find the distance that the spring will stretch when a weight of 6 pounds is attached, we substitute $\frac{5}{2}$ for k and 6 for f in Equation 1 and evaluate d.

$$d = kf$$
$$d = \frac{5}{2}(6)$$
$$d = 15$$

The spring will stretch 15 inches when a weight of 6 pounds is attached.

EXAMPLE 1

At a constant speed, the distance traveled varies directly with the time. If Adrian can drive 105 miles in 3 hours, how far could he drive in 5 hours?

Solution

Let d represent the distance traveled, and let t represent the time. Translate the words "distance varies directly with time" into the equation

1. $d = kt$

To find k, we substitute 105 for d and 3 for t in Equation 1 and solve for k.

$$d = kt$$
$$105 = k(3)$$
$$35 = k \quad \text{Divide both sides by 3.}$$

We now substitute 35 for k into Equation 1.

2. $d = 35t$

To find the distance traveled in 5 hours, substitute $t = 5$ into Equation 2.

$$d = 35t$$
$$d = 35(5)$$
$$d = 175$$

In 5 hours, Adrian can travel 175 miles. Work Progress Check 12. ∎

Inverse Variation

Definition: The words "y varies inversely with x" mean that

$$y = \frac{k}{x}$$

for some constant k. The constant k is called the **constant of variation**.

Progress Check 12

The distance s that an object will fall varies directly with the square of the time t that it falls. If an object falls 64 feet in 2 seconds, how far will it fall in 4 seconds?

12. 256 feet

Section 13.6 Variation

Under a constant temperature, the volume occupied by a gas varies inversely with its pressure. If V represents volume and p represents pressure, this relationship is expressed by the formula

2. $V = \dfrac{k}{p}$

EXAMPLE 2

A gas occupies a volume of 15 cubic inches when it is placed under 4 pounds per square inch of pressure. Find how much pressure is needed to compress the gas into a volume of 10 cubic inches.

Solution

To find the constant of variation, we substitute 15 for V and 4 for p in Equation 2 and solve for k.

$V = \dfrac{k}{p}$

$15 = \dfrac{k}{4}$

$60 = k$ Multiply both sides by 4.

To find the pressure needed to compress the gas into a volume of 10 cubic inches, we substitute 60 for k and 10 for V in Equation 2 and solve for p.

$V = \dfrac{k}{p}$

$10 = \dfrac{60}{p}$

$10p = 60$ Multiply both sides by p.

$p = 6$ Divide both sides by 10.

It will take a pressure of 6 pounds per square inch to compress the gas into a volume of 10 cubic inches. Work Progress Check 13. ■

Joint Variation

Definition: The words "y varies jointly with x and z" mean that

$y = kxz$

for some constant k. The constant k is called the **constant of variation**.

The area A of a rectangle depends on its length l and its width w by the formula

$A = lw$

We can say that the area of a rectangle varies jointly with its length and its width. In this example, the constant of variation, k, is 1.

Progress Check 13

The electrical resistance of a wire varies inversely with the square of its diameter. If the resistance of a length of wire having a diameter of 0.1 inch is 0.5 ohm, find the resistance of a wire of equal length with a diameter of 0.2 inch.

13. 0.125 ohm

Progress Check Answers

EXAMPLE 3

The area of a triangle varies jointly with the length of its base and its height. If a triangle with an area of 63 square inches has a base of 18 inches and a height of 7 inches, find the area of a triangle with a base of 12 inches and a height of 10 inches.

Solution

Let A represent the area of a triangle, let b represent the length of the base, and let h represent the height. Translate the words "area varies jointly with the length of the base and the height" into the formula

1. $A = kbh$

We are given that $A = 63$ when $b = 18$ and $h = 7$. To determine k, we substitute these values into Equation 1 and solve for k:

$$A = kbh$$
$$63 = k(18)(7)$$
$$63 = k(126)$$
$$\frac{63}{126} = k$$
$$\frac{1}{2} = k$$

Thus, $k = \frac{1}{2}$, and the complete formula for finding the area of any triangle is

2. $A = \frac{1}{2}bh$

To find the area of a triangle with a base of 12 inches and a height of 10 inches, we substitute 12 for b and 10 for h in Equation 2 and calculate A:

$$A = \frac{1}{2}bh$$
$$A = \frac{1}{2}(12)(10)$$
$$A = 60$$

The area of the triangle is 60 square inches. ■

Combined Variation

Many applications involve a combination of direct and inverse variation. Such problems involve **combined variation**.

EXAMPLE 4

The pressure of a fixed amount of gas varies directly with its temperature and inversely with its volume. A sample of gas at a pressure of 1 atmosphere occupies a volume of 3 cubic meters when its temperature is 273 degrees Kelvin (about 0° Celsius). The gas is heated to 364° K and compressed to 1 cubic meter. Find the pressure of the gas.

Solution

Let P represent the pressure of the gas, let T be its temperature, and let V be the volume. Then the words "the pressure varies directly with temperature and inversely with volume" translate into the equation

1. $P = \dfrac{kT}{V}$

To determine the constant of variation, k, we substitute the given information into Equation 1 and solve for k. Let $P = 1$, $T = 273$, and $V = 3$:

$$P = \dfrac{kT}{V}$$
$$1 = \dfrac{k \cdot 273}{3}$$
$$1 = 91k$$
$$k = \dfrac{1}{91}$$

Thus, $k = \dfrac{1}{91}$, and the complete formula is $P = \dfrac{\tfrac{1}{91}T}{V}$ or

2. $P = \dfrac{T}{91V}$

To find the pressure of the gas under the new conditions, we substitute 364 for T and 1 for V into Equation 2 and calculate P:

$$P = \dfrac{T}{91V}$$
$$P = \dfrac{364}{91 \cdot 1}$$
$$= 4$$

The pressure of the heated and compressed gas is now 4 atmospheres. ∎

Reading Variation from a Graph

Direct variation can be described by the straight-line graph of the equation $y = kx$, where k is a constant. Because $y = 0$ whenever $x = 0$, the line graph representing direct variation always passes through the origin.

To draw the graph of an equation describing direct variation, we first find k and then substitute numbers for x to find the corresponding values of y. We plot the resulting ordered pairs and draw the line connecting the points.

For example, to graph the equation $y = kx$ when the constant of variation is 2, we graph the equation $y = 2x$, as shown in Figure 13-29**(a)**.

Inverse variation can be described by the curved graph of the equation $y = \dfrac{k}{x}$, where k is a constant. To draw the curve, called a **hyperbola**, we first find k and then substitute numbers for x to find the corresponding values of y. We plot the resulting ordered pairs and draw a smooth curve connecting the points.

For example, the equation $y = \dfrac{4}{x}$ describes inverse variation with a constant of variation of 4. To graph the equation, we construct the table of values shown in Figure 13-29**(b)**, plot the points, and connect them with a smooth curve. The resulting hyperbola is a curve whose ends come close to but never touch either the x- or the y-axis.

$y = 2x$	
x	y
−2	−4
−1	−2
0	0
1	2

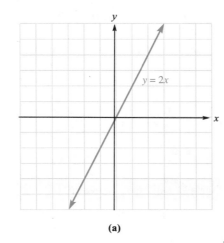

(a)

$y = \dfrac{4}{x}$	
x	y
−4	−1
−2	−2
−1	−4
1	4
2	2
4	1

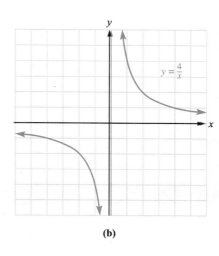

(b)

Figure 13-29

EXAMPLE 5

Tell whether each graph indicates direct or inverse variation, and then find the constant of variation.

a.
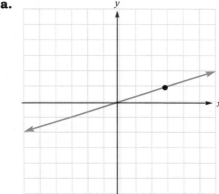

b.
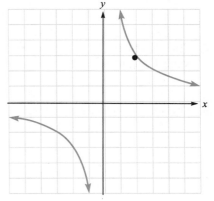

Solution

a. Because the graph shown in Part **(a)** is a straight line passing through the origin, the graph represents direct variation. By using the coordinate grid, we can determine that the coordinates of point A are (3, 1). We can then substitute 3 for x and 1 for y in the equation $y = kx$ and find k:

$$y = kx$$
$$1 = k(3)$$
$$\frac{1}{3} = k$$

The constant of variation is $\frac{1}{3}$. It is interesting to note that the constant of variation is the slope of the line.

b. Because the graph shown in Part **(b)** is a hyperbola that approaches but never touches either the x- or the y- axis, the graph represents inverse variation. By using the coordinate grid, we can find that the coordinates of point A are $(2, 3)$. We can then substitute 2 for x and 3 for y in the equation $y = \frac{k}{x}$ and find k:

$$y = \frac{k}{x}$$

$$3 = \frac{k}{2}$$

$$6 = k$$

The constant of variation is 6.

13.6 Exercises

In Exercises 1–10, express each sentence as a formula.

1. The distance d a car can travel while moving at a constant speed varies directly with n, the number of gallons of gasoline it consumes.

2. A farmer's harvest h varies directly with a, the number of acres planted.

3. For a fixed area, the length l of a rectangle varies inversely with its width w.

4. The value v of a car varies inversely with its age, a.

5. The area A of a circle varies directly with the square of its radius, r.

6. The distance s that a body falls varies directly with the square of the time t.

7. The distance D traveled varies jointly with the speed s and the time t.

8. The interest I on a savings account varies jointly with the interest rate r and the time t that the money is left on deposit.

9. The current I varies directly with the voltage V and inversely with the resistance R.

10. The force of gravity F varies directly with the product of the masses m_1 and m_2, and inversely with the square of the distance d between them.

In Exercises 11–28, assume that all variables represent positive numbers.

11. Assume that y varies directly with x. If $y = 10$ when $x = 2$, find y when $x = 7$.

12. Assume that A varies directly with z. If $A = 30$ when $z = 5$, find A when $z = 9$.

13. Assume that r varies directly with s. If $r = 21$ when $s = 6$, find r when $s = 12$.

14. Assume that d varies directly with t. If $d = 15$ when $t = 3$, find t when $d = 3$.

15. Assume that s varies directly with t^2. If $s = 12$ when $t = 2$, find s when $t = 10$.

16. Assume that y varies directly with x^3. If $y = 16$ when $x = 2$, find y when $x = 3$.

17. Assume that y varies inversely with x. If $y = 8$ when $x = 1$, find y when $x = 8$.

18. Assume that V varies inversely with p. If $V = 30$ when $p = 5$, find V when $p = 6$.

19. Assume that r varies inversely with s. If $r = 40$ when $s = 10$, find r when $s = 15$.

20. Assume that J varies inversely with v. If $J = 90$ when $v = 5$, find J when $v = 45$.

21. Assume that y varies inversely with x^2. If $y = 6$ when $x = 4$, find y when $x = 2$.

22. Assume that i varies inversely with d^2. If $i = 6$ when $d = 3$, find i when $d = \frac{1}{2}$.

23. Assume that y varies jointly with r and s. If $y = 4$ when $r = 2$ and $s = 6$, find y when $r = 3$ and $s = 4$.

24. Assume that A varies jointly with x and y. If $A = 18$ when $x = 3$ and $y = 3$, find A when $x = 7$ and $y = 9$.

25. Assume that D varies jointly with p and q. If $D = 20$ when p and q are both 5, find D when p and q are both 10.

26. Assume that z varies jointly with r and the square of s. If $z = 24$ when r and s are both 2, find z when $r = 3$ and $s = 4$.

27. Assume that y varies directly with a and inversely with b. If $y = 1$ when $a = 2$ and $b = 10$, find y when $a = 7$ and $b = 14$.

28. Assume that y varies directly with the square of x and inversely with z. If $y = 1$ when $x = 2$ and $z = 10$, find y when $x = 4$ and $z = 5$.

29. The distance traveled by an object in free fall varies directly with the square of the time that it falls. If the object falls 256 feet in 4 seconds, how far will it fall in 6 seconds?

30. The distance that a car can travel without refueling varies directly with the number of gallons of gasoline in the tank. If a car can go 360 miles on 12 gallons of gas, how far could the car go on 7 gallons?

31. For a fixed rate and principal, the interest earned in a bank account paying simple interest varies directly with the length of time the principal is left on deposit. If an investment of $5000 earns $700 in 2 years, how much will it earn in 7 years?

32. The force of gravity acting on an object varies directly with the mass of the object. The force on a mass of 5 kilograms is 49 Newtons. (A Newton is a unit of force.) What is the force acting on a mass of 12 kilograms?

33. The time it takes a car to travel a certain distance varies inversely with its rate of speed. If a certain trip takes 3 hours when the driver travels at 50 miles per hour, how long will the trip take when the driver travels at 60 miles per hour?

34. For a fixed area, the length of a rectangle varies inversely with its width. A rectangle has a width of 12 feet and a length of 20 feet. If the length is increased to 24 feet, find the width of the rectangle.

35. If the temperature of a gas is constant, the occupied volume varies inversely with the pressure. If a gas occupies a volume of 40 cubic meters under a pressure of 8 atmospheres, what will the volume be if the pressure is changed to 6 atmospheres?

36. Assume that the value of a machine varies inversely with the machine's age. If a drill press is worth $300 when it is 2 years old, find its worth when it is 6 years old.

37. The interest earned on a fixed amount of money varies jointly with the annual interest rate and the time that the money is left on deposit. If an account earns $120 at 8% annual interest when left on deposit for 2 years, find the interest that would be earned in 3 years, if the annual rate were 12%.

38. The total cost of a quantity of identical items varies jointly with the unit cost and the quantity. If the unit cost is tripled and the quantity is doubled, by what factor is the total cost multiplied?

39. The current in a circuit varies directly with the voltage and inversely with the resistance. If a current of 4 amperes flows when 36 volts is applied to a 9-ohm resistance, find the current if the voltage is 42 volts and the resistance is 11 ohms.

40. The deflection of a beam is inversely proportional to its width and the cube of its depth. If the deflection is 2 inches when the width is 4 inches and the depth is 3 inches, find the deflection if the width is 3 inches and the depth is 4 inches.

In Exercises 41–48, tell whether each graph represents direct variation or inverse variation. If it represents neither, so indicate.

41.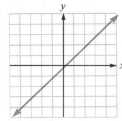
42.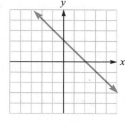
43.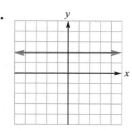
44.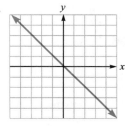

45.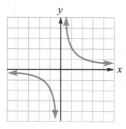
46.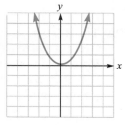
47.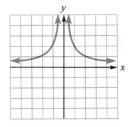
48.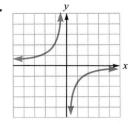

Review Exercises

In Review Exercises 1–4, remove parentheses and simplify each expression.

1. $2(x + 4) + 3(2x - 1)$
2. $-3(3x + 5) - 2(2x + 4)$
3. $3x(x^2 - 2) - 6x^2(x - 1)$
4. $-5a^2(a + 1) - 3a(a^2 + 4a - 3)$

In Review Exercises 5–8, simplify each fraction.

5. $\dfrac{y^2 + 2 + 3y}{y^3 + y^2}$
6. $\dfrac{a^2 - 9}{18 - 3a - a^2}$
7. $\dfrac{\dfrac{1}{t} + 1}{1 - \dfrac{1}{t}}$
8. $\dfrac{\dfrac{1}{r} - \dfrac{1}{s}}{\dfrac{1}{r} + \dfrac{1}{s}}$

In Review Exercises 9–10, perform each operation and simplify, if possible.

9. $\dfrac{x^2 + 5x + 6}{x + 3} \cdot \dfrac{x^2 + 2x - 8}{x^2 - 4}$
10. $\dfrac{3x + 1}{x + 2} + \dfrac{2x}{x + 1}$

11. If one pair of opposite sides of a square are each increased by 8 inches, and the other sides are each decreased by 4 inches, the area remains unchanged. Find the dimensions of the original square.

12. The perimeter of a triangle is 19 feet. The first side is twice as long as the second side, and the third side is 3 feet greater than the second. Find the length of each side.

Mathematics in Medicine

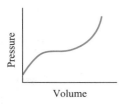

Not all graphs are straight lines. Experiments have determined that bladder pressure is related to the amount of urine in the bladder, and that the graph of the relationship is not a straight line, but rather the curve shown in the illustration. The graph indicates that the pressure increases very little over a wide volume range. After a point, however, small increases in volume cause substantial increases in pressure, causing discomfort.

It is reasonable that increasing volume causes increasing pressure. The rate of such increase is still the slope of the graph, but because the graph rises more rapidly at some points than at others, the slope changes. The study of such changing slopes is one topic of *calculus*.

Chapter Summary

Key Words

annual depreciation rate *(13.2)*
Cartesian coordinate system *(13.1)*
combined variation *(13.6)*
constant of variation *(13.6)*
demand curve *(13.2)*
dependent variable *(13.5)*
direct variation *(13.6)*
domain *(13.5)*
function *(13.5)*
function notation *(13.5)*
general form of a linear equation *(13.1)*
graph *(13.1)*
graph of a function *(13.5)*
graph of a line *(13.1)*
graph of a pair *(13.1)*
half-plane *(13.4)*
Hooke's law *(13.6)*
hyperbola *(13.6)*
independent variable *(13.5)*
intercept method of graphing *(13.1)*
inverse variation *(13.6)*
joint variation *(13.6)*
linear depreciation *(13.2)*
linear equation *(13.1)*
linear inequality *(13.4)*
marginal demand *(13.2)*
negative reciprocals *(13.3)*
ordered pair *(13.1)*
origin *(13.1)*
parallel lines *(13.3)*
perpendicular lines *(13.3)*
point-slope form of a linear equation *(13.3)*
quadrant *(13.1)*
range *(13.5)*
rectangular coordinate system *(13.1)*
relation *(13.5)*
rise *(13.2)*
run *(13.2)*
salvage value *(13.2)*
slope *(13.2)*
slope-intercept form of a linear equation *(13.2)*
x-axis *(13.1)*
x-coordinate *(13.1)*
x-intercept *(13.1)*
y-axis *(13.1)*
y-coordinate *(13.1)*
y-intercept *(13.1)*

Key Ideas

(13.1) Ordered pairs of numbers are associated with points in a rectangular coordinate system.

To graph an equation in the variables x and y, choose several numbers for x (at least three), calculate the corresponding value of y, plot the points (x, y), and draw the line that passes through them.

General form of a linear equation:
$Ax + By = C$, where A and B are not both zero

To graph a linear equation by the intercept method, plot the points corresponding to the x- and y-intercepts and draw the line through them. Plot a third point as a check.

The equation $x = a$ represents the y-axis or a line parallel to the y-axis.

The equation $y = b$ represents the x-axis or a line parallel to the x-axis.

(13.2) The slope of a line passing through (x_1, y_1) and (x_2, y_2) is given by the formula

$$\text{Slope} = \frac{y_2 - y_1}{x_2 - x_1} \qquad (x_2 \neq x_1)$$

The graph of the equation $y = mx + b$ is a line with a slope of m and y-intercept of b.

Slope-intercept form of a linear equation:
$y = mx + b$

Horizontal lines have a slope of zero.

The slope of a vertical line is undefined.

(13.3) **Point-slope form of a linear equation:**
$$y - y_1 = m(x - x_1)$$
Nonvertical parallel lines have the same slope.

The product of the slopes of perpendicular lines is -1, provided neither line is vertical.

(13.4) To graph an inequality in variables x and y, first graph the boundary and then use a convenient point not on that boundary to determine which side of the line to shade.

(13.5) A relation is any set of ordered pairs (x, y).

The set of all possible x values is the domain of the relation, and the set of all possible y values is the range.

A function is a correspondence that assigns to each number x exactly one value y. To indicate such a correspondence, the notation $y = f(x)$ is used.

The domain and range of a function can be determined from its graph.

(13.6) A formula of the form $y = kx$ represents direct variation.

A formula of the form $y = \dfrac{k}{x}$ represents inverse variation.

A formula of the form $y = kxz$ represents joint variation.

Direct and inverse variation are used together in combined variation.

Chapter 13 Review Exercises

[13.1] In Review Exercises 1–6, plot each point on a rectangular coordinate system.

1. $A(1, 3)$
2. $B(1, -3)$
3. $C(-3, 1)$
4. $D(-3, -1)$
5. $E(0, 5)$
6. $F(-5, 0)$

In Review Exercises 7–14, find the coordinates of each indicated point in Illustration 1.

7. A
8. B
9. C
10. D
11. E
12. F
13. G
14. H

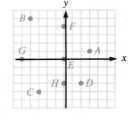

Illustration 1

In Review Exercises 15–22, graph each equation on a rectangular coordinate system.

15. $y = x - 5$
16. $y = 2x + 1$
17. $y = \dfrac{x}{2} + 2$
18. $y = 3$
19. $x + y = 4$
20. $x - y = -3$
21. $3x + 5y = 15$
22. $7x - 4y = 28$

[13.2] In Review Exercises 23–26, find the slope of the line passing through the two given points. If the slope is undefined, write "undefined slope."

23. $(1, 4)$ $(2, 3)$
24. $(-1, 3)$ $(3, -2)$
25. $(-1, -1)$ $(-3, 0)$
26. $(-8, 2)$ $(3, 2)$

In Review Exercises 27–30, graph the line that passes through the given point and has the given slope.

27. $(-1, 4)$, slope 2
28. $(1, -2)$, slope -2
29. $\left(0, \dfrac{1}{2}\right)$, slope $\dfrac{3}{2}$
30. $(-3, 0)$, slope $-\dfrac{5}{2}$

In Review Exercises 31–34, find the slope and the y-intercept of the line defined by the given equation, and then graph the equation. If the line has no defined slope, write "undefined slope."

31. $y = 5x + 2$
32. $y = -\dfrac{x}{2} + 4$
33. $y + 3 = 0$
34. $x + 3y = 1$

[13.3] In Review Exercises 35–38, use the slope-intercept form of a linear equation to find the equation of each line with the given properties. Write the equation in general form.

35. Slope -3, y-intercept 2
36. Slope 0, y-intercept -7
37. Slope 7, y-intercept 0
38. Slope $\dfrac{1}{2}$, y-intercept $-\dfrac{3}{2}$

In Review Exercises 39–42, use the point-slope form of a linear equation to find the equation of each line with the given properties. Write the equation in general form.

39. Slope 3, passing through $(0, 0)$
40. Slope $-\dfrac{1}{3}$, passing through $\left(1, \dfrac{2}{3}\right)$
41. Slope $\dfrac{1}{9}$, passing through $(-27, -2)$
42. Slope $-\dfrac{3}{5}$, passing through $\left(1, -\dfrac{1}{5}\right)$

In Review Exercises 43–46, determine if lines with the given slopes are parallel, perpendicular, or neither.

43. 5 and $\dfrac{1}{5}$
44. $\dfrac{2}{4}$ and 0.5
45. -5 and $\dfrac{1}{5}$
46. 0.25 and $\dfrac{1}{4}$

In Review Exercises 47–50, determine whether the graphs of the given equations are parallel, perpendicular, or neither.

47. $3x = y$
$x = 3y$

48. $3x = y$
$x = -3y$

49. $x + 2y = y - x$
$2x + y = 3$

50. $3x + 2y = 7$
$2x - 3y = 8$

In Review Exercises 51–54, find the equation of each line with the given properties. Write the equation in general form.

51. Parallel to $y = 7x - 18$ and passing through $(2, 5)$

52. Parallel to $3x + 2y = 7$ and passing through $(-3, 5)$

53. Perpendicular to $2x - 5y = 12$ and passing through the origin

54. Perpendicular to $y = \dfrac{x}{3} + 17$ and having a y-intercept of -4

[13.4] In Review Exercises 55–58, graph each linear inequality.

55. $y \leq 3x + 1$
56. $y > x - 5$
57. $2x - 3y \geq 6$
58. $2y + 3(x - y) < 5y$

[13.5] In Review Exercises 59–62, indicate whether the equation determines y as a function of x. If it does not, indicate some number x for which there are two corresponding values of y.

59. $y = 2x$
60. $y = 5x^2$
61. $|y| = x$
62. $y^2 = 4x^2$

In Review Exercises 63–70, let $f(x) = x^2 - x + 1$. Then find the value indicated.

63. $f(0)$ **64.** $f(-2)$ **65.** $f(3)$ **66.** $f(w)$

67. $f(1) - f(-1)$ **68.** $f(2) + f(-2)$ **69.** $f(a) + f(0)$ **70.** $f(1) - f(a)$

In Review Exercises 71–74, determine the domain and range of each function.

71. $y = 3x$ **72.** $y = \dfrac{5}{x}$ **73.** $f(x) = \dfrac{1}{x^2 - 25}$ **74.** $f(x) = \dfrac{1}{x^2 - 3x + 2}$

In Review Exercises 75–76, determine the domain and range of the function from its graph.

75.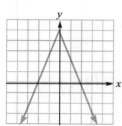

76.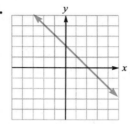

[13.6] In Review Exercises 77–80, express each variation as an equation. Then find the value requested.

77. Assume that s varies directly with the square of t. Find s when $t = 10$ if $s = 64$ when $t = 4$.

78. Assume that l varies inversely with w. Find the constant of variation if $l = 30$ when $w = 20$.

79. Assume that R varies jointly with b and c. If $R = 72$ when $b = 4$ and $c = 24$, find R when $b = 6$ and $c = 18$.

80. Assume that s varies directly with w and inversely with the square of m. If $s = \frac{7}{4}$ when w and m are both 4, find s when $w = 5$ and $m = 7$.

In Review Exercises 81–82, tell whether each graph represents direct or inverse variation.

81.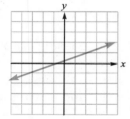

82.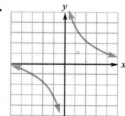

Name _____   Section _____

Chapter 13 Test

In Problems 1–4, graph each equation.

1. $y = \dfrac{x}{2} + 1$ _____

2. $2(x + 1) - y = 4$ _____

3. $x = 1$ _____

4. $2y = 8$ _____

5. Find the slope of the line passing through $(0, 0)$ and $(6, 8)$.   _____

6. Find the slope of the line passing through $(-1, 3)$ and $(3, -1)$.   _____

7. Find the slope of the line determined by $2x + y = 3$.   _____

8. Find the y-intercept of the line determined by $2y - 7(x + 5) = 7$.   _____

9. Find the slope of the line parallel to the line determined by $y = \dfrac{3}{5}x + 3$.   _____

10. Find the slope of a line perpendicular to a line with a slope of 2.   _____

11. In general form, write the equation of a line with a slope of $\dfrac{1}{2}$ and a y-intercept of 3.   _____

12. In general form, write the equation of a line with a slope of 7 that passes through the point $(-2, 5)$.   _____

13. Write the equation of a line parallel to the y-axis that passes through $(-3, 17)$.   _____

14. Write the equation of a line passing through $(3, -5)$ and perpendicular to the line with the equation $y = \dfrac{1}{3}x + 11$.   _____

In Problems 15–16, graph each inequality.

15. $y \geq x + 2$

16. $x < 3$

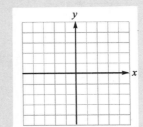
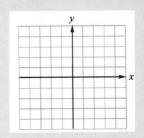

17. Does the equation $x = 7y - 8$ determine y to be a function of x?

18. Does the equation $y^2 = x$ determine y to be a function of x?

In Problems 19–22, assume that $f(x) = 3x + 2$.

19. Find $f(3)$.

20. Find $f(-2) + f(0)$.

21. Find $f(a) - f(b)$.

22. Find $f(t^2)$.

23. Find the domain of the function $y = \dfrac{3}{x^2 + x - 20}$.

24. Find the range of the function $y = 3x^2 + 2$.

25. Find the domain of the function whose graph appears in Illustration 1.

26. Find the range of the function whose graph appears in Illustration 1.

27. If y varies directly with x and $y = 32$ when $x = 8$, find x when $y = 4$.

28. If i varies inversely with the square of d, find the constant of variation if $i = 100$ when $d = 2$.

29. Does the graph in Illustration 2 represent direct variation?

30. Does the graph in Illustration 3 represent inverse variation?

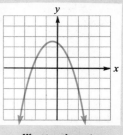

Illustration 1

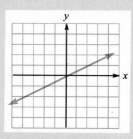

Illustration 2

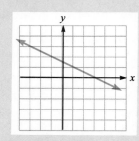

Illustration 3

14 Solving Systems of Equations and Inequalities

Mathematics in Electronics

The technology of FM radio transmission allows programming on separate channels, called *subcarriers*, that hitch a ride on the main transmission. To transmit stereo FM, it would seem reasonable to transmit the left channel and the right channel separately—the left as the main program, for example, and the right on a subcarrier. In practice, it is not so simple, because a receiver without stereo capability would receive only one channel.

A technique called *multiplexing*, which is based on an idea of this chapter, provides both channels on a monaural receiver, as well as full stereo on better receivers.

We have considered equations such as $x + y = 3$ that contain two variables. Because there are infinitely many pairs of numbers whose sum is 3, there are infinitely many ordered pairs (x, y) that will satisfy this equation. Some of these pairs are

$x + y = 3$

x	y
0	3
1	2
2	1
3	0

Likewise, there are infinitely many ordered pairs (x, y) that will satisfy the equation $3x - y = 1$. Some of these pairs are

$3x - y = 1$

x	y
0	-1
1	2
2	5
3	8

Although infinitely many ordered pairs satisfy each of these equations, only the pair $(1, 2)$ satisfies both equations at the same time. The pair of equations

$$\begin{cases} x + y = 3 \\ 3x - y = 1 \end{cases}$$

is called a **system of equations**. Because the ordered pair (1, 2) satisfies both equations simultaneously, it is called a **simultaneous solution**, or just a **solution of the system of equations**. We will discuss three methods for finding the simultaneous solution of a system of two equations, each containing two variables.

14.1 Solving Systems of Equations by Graphing

- Inconsistent Systems
- Dependent Equations

To use the **graphing method** to solve the system

$$\begin{cases} x + y = 3 \\ 3x - y = 1 \end{cases}$$

we graph both equations on a single set of coordinate axes as in Figure 14-1. Although there are infinitely many pairs of real numbers (x, y) that satisfy the equation $x + y = 3$, and infinitely many pairs of real numbers (x, y) that satisfy the equation $3x - y = 1$, only the coordinates of the point where their graphs intersect satisfy both equations simultaneously. Thus, the solution of this system is $x = 1$ and $y = 2$, or just $(1, 2)$.

$x + y = 3$	
x	y
0	3
1	2
2	1
3	0

$3x - y = 1$	
x	y
0	-1
1	2
2	5
3	8

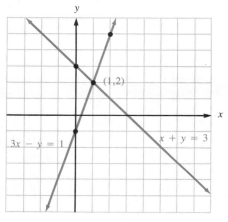

Figure 14-1

To check this solution, we substitute 1 for x and 2 for y in each equation and verify that the pair (1, 2) satisfies each equation.

$x + y = 3$	$3x - y = 1$
$1 + 2 \stackrel{?}{=} 3$	$3(1) - 2 \stackrel{?}{=} 1$
$3 = 3$	$3 - 2 \stackrel{?}{=} 1$
	$1 = 1$

The graphs in this system are different lines. When this is so, the equations of the system are called **independent equations**. The system in this example has a solution. When a system has at least one solution, it is called a **consistent system of equations**.

EXAMPLE 1

Use the graphing method to solve the system $\begin{cases} 2x + 3y = 2 \\ 3x = 2y + 16 \end{cases}$.

Solution

We begin by graphing both equations on a single set of coordinate axes, as in Figure 14-2.

Although there are infinitely many pairs (x, y) that satisfy $2x + 3y = 2$ and infinitely many pairs (x, y) that satisfy $3x = 2y + 16$, only the coordinates of the point where the graphs intersect satisfy both equations simultaneously. Thus, the solution is $x = 4$ and $y = -2$, or just $(4, -2)$.

$2x + 3y = 2$

x	y
1	0
-2	2
4	-2

$3x = 2y + 16$

x	y
6	1
0	-8
4	-2

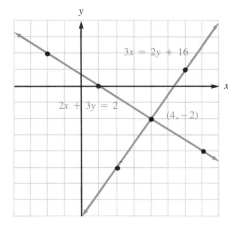

Figure 14-2

To check this solution, we substitute 4 for x and -2 for y in each equation and verify that the pair $(4, -2)$ satisfies each equation.

$$2x + 3y = 2 \qquad\qquad 3x = 2y + 16$$
$$2(4) + 3(-2) \stackrel{?}{=} 2 \qquad\qquad 3(4) \stackrel{?}{=} 2(-2) + 16$$
$$8 - 6 \stackrel{?}{=} 2 \qquad\qquad 12 \stackrel{?}{=} -4 + 16$$
$$2 = 2 \qquad\qquad 12 = 12$$

The equations in this system are *independent equations* (their graphs are different lines), and the system is a *consistent system of equations* (it has a solution). Work Progress Check 1.

Inconsistent Systems

In the next example, the graphs of the equations are different lines. Thus, the equations of the system are independent equations. However, the system does not have a solution. Such a system is called an **inconsistent system of equations**.

EXAMPLE 2

Solve the system $\begin{cases} 2x + y = -6 \\ 4x + 2y = 8 \end{cases}$.

Solution

Graph both equations on one set of coordinate axes as in Figure 14-3.

Progress Check 1

Solve $\begin{cases} 2x + y = 4 \\ 4x - y = 2 \end{cases}$ by graphing.

1.
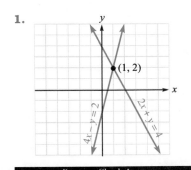

Progress Check Answers

2x + y = −6		4x + 2y = 8	
x	y	x	y
−3	0	2	0
0	6	0	4
−2	−2	1	2

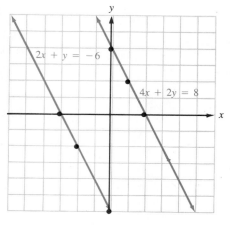

Figure 14-3

The lines in Figure 14-3 are parallel. We can show this is true by writing each equation in slope-intercept form and observing that the coefficients of x are equal.

$$2x + y = -6 \qquad\qquad 4x + 2y = 8$$
$$y = -2x - 6 \qquad\qquad 2y = -4x + 8$$
$$\qquad\qquad\qquad\qquad y = -2x + 4$$

Since parallel lines do not intersect, there is no simultaneous solution to this system. ■

Dependent Equations

In the next example, the graphs of two lines will coincide (are the same line). In this case, the equations of the system are called **dependent equations**. Because the system has at least one solution, the system is consistent.

EXAMPLE 3

Solve the system $\begin{cases} y - 2x = 4 \\ 4x + 8 = 2y \end{cases}$.

Solution

Graph both equations on one set of coordinate axes as in Figure 14-4.

y − 2x = 4		4x + 8 = 2y	
x	y	x	y
0	4	0	4
−2	0	−2	0
−3	−2	−3	−2

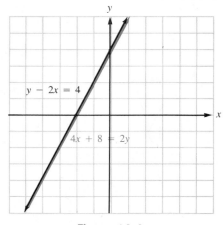

Figure 14-4

Progress Check 2

a. Solve $\begin{cases} y = 2x + 1 \\ 4x - 2y = 4 \end{cases}$ by graphing.

b. Solve $\begin{cases} y = 2x - 1 \\ 2 = 4x - 2y \end{cases}$ by graphing.

The lines in Figure 14-4 coincide. Because these lines intersect at infinitely many points, there are an unlimited number of solutions. Any pair (x, y) that satisfies one of the equations satisfies the other also. Some possible solutions are $(0, 4)$, $(1, 6)$, and $(2, 8)$, because each of these ordered pairs satisfies each equation. Work Progress Check 2. ■

The possibilities that can occur when two equations, each with two variables, are graphed are summarized as follows:

Possible graph	If the	then
	lines are different and intersect,	the equations are independent, and the system is consistent. One solution exists.
	lines are different and parallel,	the equations are independent, and the system is inconsistent. No solutions exist.
	lines coincide,	the equations are dependent, and the system is consistent. An unlimited number of solutions exist.

EXAMPLE 4

Solve the system $\begin{cases} \dfrac{2}{3}x - \dfrac{1}{2}y = 1 \\ \dfrac{1}{2}x + \dfrac{1}{3}y = 5 \end{cases}$.

Solution

Multiply both sides of the first equation by 6 to clear it of fractions.

$$\frac{2}{3}x - \frac{1}{2}y = 1$$

$$6\left(\frac{2}{3}x - \frac{1}{2}y\right) = 6(1)$$

1. $4x - 3y = 6$

Then multiply both sides of the second given equation by 6 to clear it of fractions.

$$\frac{1}{2}x + \frac{1}{3}y = 5$$

$$6\left(\frac{1}{2}x + \frac{1}{3}y\right) = 6(5)$$

2. $3x + 2y = 30$

Equations 1 and 2 form the following equivalent system, which has the same solutions as the original system.

$$\begin{cases} 4x - 3y = 6 \\ 3x + 2y = 30 \end{cases}$$

Graph each equation as in Figure 14-5 and find that their point of intersection is $(6, 6)$. Thus, the solution of the given system is $x = 6$ and $y = 6$, or just $(6, 6)$.

2. a.
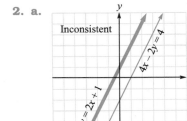

b.
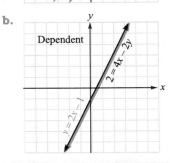

Progress Check Answers

4x − 3y = 6		3x + 2y = 30	
x	y	x	y
0	−2	10	0
3	2	8	3
6	6	6	6

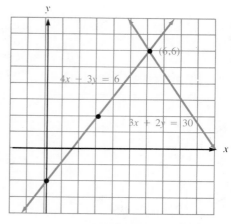

Figure 14-5

To verify that the pair (6, 6) satisfies each equation of the original system, substitute 6 for x and 6 for y in each of the original equations and simplify.

$$\frac{2}{3}x - \frac{1}{2}y = 1 \qquad\qquad \frac{1}{2}x + \frac{1}{3}y = 5$$

$$\frac{2}{3}(6) - \frac{1}{2}(6) \stackrel{?}{=} 1 \qquad\qquad \frac{1}{2}(6) + \frac{1}{3}(6) \stackrel{?}{=} 5$$

$$4 - 3 \stackrel{?}{=} 1 \qquad\qquad 3 + 2 \stackrel{?}{=} 5$$

$$1 = 1 \qquad\qquad 5 = 5$$

The equations in this system are independent, and the system itself is consistent. ∎

14.1 Exercises

In Exercises 1–12, determine whether the given ordered pair is a solution for the given system of equations.

1. $(1, 1)$; $\begin{cases} x + y = 2 \\ 2x - y = 1 \end{cases}$

2. $(1, 3)$; $\begin{cases} 2x + y = 5 \\ 3x - y = 0 \end{cases}$

3. $(3, -2)$; $\begin{cases} 2x + y = 4 \\ x + y = 1 \end{cases}$

4. $(-2, 4)$; $\begin{cases} 2x + 2y = 4 \\ x + 3y = 10 \end{cases}$

5. $(4, 5)$; $\begin{cases} 2x - 3y = -7 \\ 4x - 5y = 25 \end{cases}$

6. $(2, 3)$; $\begin{cases} 3x - 2y = 0 \\ 5x - 3y = -1 \end{cases}$

7. $(-2, -3)$; $\begin{cases} 4x + 5y = -23 \\ -3x + 2y = 0 \end{cases}$

8. $(-5, 1)$; $\begin{cases} -2x + 7y = 17 \\ 3x - 4y = -19 \end{cases}$

9. $\left(\frac{1}{2}, 3\right)$; $\begin{cases} 2x + y = 4 \\ 4x - 3y = 11 \end{cases}$

10. $\left(2, \frac{1}{3}\right)$; $\begin{cases} x - 3y = 1 \\ -2x + 6y = -6 \end{cases}$

11. $\left(-\frac{2}{5}, \frac{1}{4}\right)$; $\begin{cases} 5x - 4y = -6 \\ 8y = 10x + 12 \end{cases}$

12. $\left(-\frac{1}{3}, \frac{3}{4}\right)$; $\begin{cases} 3x + 4y = 2 \\ 12y = 3(2 - 3x) \end{cases}$

In Exercises 13–24, use the graphing method to solve each system of equations. Write each answer as an ordered pair where possible. If a system is inconsistent, or if the equations of a system are dependent, so indicate.

13. $\begin{cases} x + y = 2 \\ x - y = 0 \end{cases}$

14. $\begin{cases} x + y = 4 \\ x - y = 0 \end{cases}$

15. $\begin{cases} x + y = 2 \\ x - y = 4 \end{cases}$

16. $\begin{cases} x + y = 1 \\ x - y = -5 \end{cases}$

17. $\begin{cases} 3x + 2y = -8 \\ 2x - 3y = -1 \end{cases}$

18. $\begin{cases} x + 4y = -2 \\ x + y = -5 \end{cases}$

19. $\begin{cases} 4x - 2y = 8 \\ y = 2x - 4 \end{cases}$

20. $\begin{cases} 3x - 6y = 18 \\ x = 2y + 3 \end{cases}$

21. $\begin{cases} 2x - 3y = -18 \\ 3x + 2y = -1 \end{cases}$

22. $\begin{cases} -x + 3y = -11 \\ 3x - y = 17 \end{cases}$

23. $\begin{cases} 4x = 3(4 - y) \\ 2y = 4(3 - x) \end{cases}$

24. $\begin{cases} 2x = 3(2 - y) \\ 3y = 2(3 - x) \end{cases}$

In Exercises 25–32, use the graphing method to solve each system of equations. Write each answer as an ordered pair.

25. $\begin{cases} x + 2y = -4 \\ x - \frac{1}{2}y = 6 \end{cases}$

26. $\begin{cases} \frac{2}{3}x - y = -3 \\ 3x + y = 3 \end{cases}$

27. $\begin{cases} -\frac{3}{4}x + y = 3 \\ \frac{1}{4}x + y = -1 \end{cases}$

28. $\begin{cases} \frac{1}{3}x + y = 7 \\ \frac{2}{3}x - y = -4 \end{cases}$

29. $\begin{cases} \frac{1}{2}x + \frac{1}{4}y = 0 \\ \frac{1}{4}x - \frac{3}{8}y = -2 \end{cases}$

30. $\begin{cases} \frac{1}{2}x + \frac{2}{3}y = -5 \\ \frac{3}{2}x - y = 3 \end{cases}$

31. $\begin{cases} \frac{1}{3}x - \frac{1}{2}y = \frac{1}{6} \\ \frac{2}{5}x + \frac{1}{2}y = \frac{13}{10} \end{cases}$

32. $\begin{cases} \frac{3}{4}x + \frac{2}{3}y = -\frac{19}{6} \\ y - x = -\frac{4x}{3} \end{cases}$

Review Exercises

In Review Exercises 1–4, write each expression as a single exponential expression.

1. $x^3 x^4 x^5$

2. $\dfrac{y^7}{y^2 y^3}$

3. $\dfrac{(a^3)^2}{(a^2)^3}$

4. $\dfrac{(t^2 t^3)^3}{(t^2 t)^4}$

In Review Exercises 5–10, perform the indicated operations and simplify.

5. $2(x^2 + 3) + 3(2x^2 - 3x + 2)$

6. $-4(2y^2 - 3y + 2) - 2(3y^2 + 4y - 3)$

7. $3x^2 y(-4xy^2 - 3x^2 y^2)$

8. $(2z^2 - 5)(3z^2 + 2)$

9. $\dfrac{6a^2 b^2 + 8ab^2 - 10a^2 b}{2ab}$

10. $3x - 1 \overline{) 6x^2 + 7x - 3}$

11. The volume V of a circular cylinder varies directly with its height, h. If $V = 27$ when $h = 8$, find V when $h = \frac{4}{3}$.

12. The volume V of a circular cone varies directly with the square of r, the radius of its base. If $V = 27$ when $r = 3$, find V when $r = 6$.

14.2 Solving Systems of Equations by Substitution

- The Substitution Method
- Inconsistent Systems
- Dependent Equations

The graphing method for solving systems of equations does not always provide exact solutions. For example, if the solution to a system is $x = \frac{11}{97}$ and $y = \frac{13}{97}$, the graphs of the equations do not intersect at a point where we can read the solutions exactly. Fortunately, other methods exist that will determine exact solutions.

The Substitution Method

To solve the system
$$\begin{cases} y = 3x - 2 \\ 2x + y = 8 \end{cases}$$

by the **substitution method**, we note that $y = 3x - 2$. Thus, we can substitute $3x - 2$ for y in the equation $2x + y = 8$ to get

$$2x + y = 8$$
$$2x + (3x - 2) = 8$$

The equation that results has only one variable. We solve it as follows:

$2x + (3x - 2) = 8$	
$2x + 3x - 2 = 8$	Use the distributive property to remove parentheses.
$5x - 2 = 8$	Combine like terms.
$5x = 10$	Add 2 to both sides.
$x = 2$	Divide both sides by 5.

We can find the value of y by substituting 2 for x in either equation of the given system. Because the equation $y = 3x - 2$ is already solved for y, it is easiest to substitute in this equation.

$$y = 3x - 2$$
$$= 3(2) - 2$$
$$= 6 - 2$$
$$= 4$$

The solution to the given system is $x = 2$ and $y = 4$, or just $(2, 4)$.

Check:

$y = 3x - 2$	$2x + y = 8$
$4 \stackrel{?}{=} 3(2) - 2$	$2(2) + 4 \stackrel{?}{=} 8$
$4 \stackrel{?}{=} 6 - 2$	$4 + 4 \stackrel{?}{=} 8$
$4 = 4$	$8 = 8$

Because the pair $x = 2$ and $y = 4$ is a solution, the line graphs of the equations of the system would intersect at the point $(2, 4)$.

EXAMPLE 1

Use the method of substitution to solve the system $\begin{cases} 2x + y = -5 \\ 3x + 5y = -4 \end{cases}$.

Solution

First solve one equation for one of its variables. Because the term y in the first equation has a coefficient of 1, we solve the first equation for y. We then substitute this value of y into the second equation and solve it for x.

$$\begin{cases} 2x + y = -5 \\ 3x + 5y = -4 \end{cases} \rightarrow y = -5 - 2x$$

$$3x + 5(-5 - 2x) = -4$$

$3x - 25 - 10x = -4$	Remove parentheses.
$-7x - 25 = -4$	Combine like terms.
$-7x = 21$	Add 25 to both sides.
$x = -3$	Divide both sides by -7.

We find y by substituting -3 for x in the equation $y = -5 - 2x$.

$$y = -5 - 2x$$
$$= -5 - 2(-3)$$
$$= -5 + 6$$
$$= 1$$

The solution to the given system is $x = -3$ and $y = 1$, or just $(-3, 1)$.

Check:

$$2x + y = -5 \qquad\qquad 3x + 5y = -4$$
$$2(-3) + 1 \stackrel{?}{=} -5 \qquad\qquad 3(-3) + 5(1) \stackrel{?}{=} -4$$
$$-6 + 1 \stackrel{?}{=} -5 \qquad\qquad -9 + 5 \stackrel{?}{=} -4$$
$$-5 = -5 \qquad\qquad -4 = -4$$

Work Progress Check 3.

✏️ Progress Check 3

Solve the system $\begin{cases} 3x - 2y = 8 \\ x = 2y + 4 \end{cases}$ by substitution.

EXAMPLE 2

Use the substitution method to solve the system $\begin{cases} 2x + 3y = 5 \\ 3x + 2y = 0 \end{cases}$.

Solution

Solve the second equation for x and proceed as follows:

$$\begin{cases} 2x + 3y = 5 \\ 3x + 2y = 0 \end{cases} \rightarrow 3x = -2y$$
$$x = \frac{-2y}{3}$$

$$2\left(\frac{-2y}{3}\right) + 3y = 5$$

$\dfrac{-4y}{3} + 3y = 5$	Remove parentheses.
$3\left(\dfrac{-4y}{3}\right) + 3(3y) = 3(5)$	Multiply both sides by 3.
$-4y + 9y = 15$	Remove parentheses.
$5y = 15$	Combine like terms.
$y = 3$	Divide both sides by 5.

3. $(2, -1)$

Progress Check Answers

We find x by substituting 3 for y in the equation $x = \dfrac{-2y}{3}$.

$$x = \dfrac{-2y}{3}$$
$$= \dfrac{-2(3)}{3}$$
$$= -2$$

The solution to this system is $(-2, 3)$. Check this solution in each equation. ∎

Inconsistent Systems

EXAMPLE 3

Use the substitution method to solve the system $\begin{cases} x = 4(3 - y) \\ 2x = 4(3 - 2y) \end{cases}$.

Solution

Substitute $4(3 - y)$ for x in the second equation and solve for y.

$$\begin{cases} x = \boxed{4(3 - y)} \\ 2x = 4(3 - 2y) \end{cases}$$

$$2 \cdot 4(3 - y) = 4(3 - 2y)$$
$$8(3 - y) = 4(3 - 2y) \quad \text{Simplify.}$$
$$24 - 8y = 12 - 8y \quad \text{Remove parentheses.}$$
$$24 = 12 \quad \text{Add } 8y \text{ to both sides.}$$

The impossible result that $24 = 12$ shows that the equations in this system are independent, but that the system is inconsistent. If each equation in this system were graphed, the graphs would be parallel lines. The system has no solution. ∎

Dependent Equations

EXAMPLE 4

Use the substitution method to solve the system $\begin{cases} 3x = 4(6 - y) \\ 4y + 3x = 24 \end{cases}$.

Solution

We substitute $4(6 - y)$ for $3x$ in the second equation and proceed as follows:

$$4y + \mathbf{3x} = 24$$
$$4y + \mathbf{4(6 - y)} = 24$$
$$4y + 24 - 4y = 24 \quad \text{Remove parentheses.}$$
$$24 = 24 \quad \text{Combine like terms.}$$

Although $24 = 24$ is true, we have not found a value for y. This result shows that the equations of the system are dependent. If each equation

Progress Check 4

a. Solve the system
$\begin{cases} x + y = -2(y - 2) \\ 3y = 6 - x \end{cases}$

b. Solve the system
$\begin{cases} x + y = -2(y - 2) \\ 3y = 4 - x \end{cases}$

---

were graphed, the same line would result. Hence, this system has an unlimited number of solutions. Some of them are (8, 0), (0, 6), and (4, 3), because each of these ordered pairs satisfies both equations. Work Progress Check 4.

EXAMPLE 5

Use the substitution method to solve the system $\begin{cases} 3(x - y) = 5 \\ x + 3 = -\dfrac{5}{2}y \end{cases}$.

Solution

Begin by writing each equation in general form:

	$3(x - y) = 5$	First equation.
1.	$3x - 3y = 5$	Remove parentheses.
	$x + 3 = -\dfrac{5}{2}y$	Second equation.
	$2x + 6 = -5y$	Multiply both sides by 2.
2.	$2x + 5y = -6$	Add $5y$ and -6 to both sides.

Equations 1 and 2 form a new system that can be solved as follows:

1. $\quad 3x - 3y = 5 \quad \rightarrow \quad 3x = 5 + 3y$

$$x = \dfrac{5 + 3y}{3}$$

2. $\quad 2x + 5y = -6$

$$2\left(\dfrac{5 + 3y}{3}\right) + 5y = -6$$

$2(5 + 3y) + 15y = -18$	Multiply both sides by 3.
$10 + 6y + 15y = -18$	Remove parentheses.
$10 + 21y = -18$	Combine like terms.
$21y = -28$	Add -10 to both sides.
$y = \dfrac{-28}{21}$	Divide both sides by 21.
$y = -\dfrac{4}{3}$	Simplify $\dfrac{-28}{21}$.

To find x, we substitute $-\dfrac{4}{3}$ for y in the equation $x = \dfrac{5 + 3y}{3}$ and simplify.

$$x = \dfrac{5 + 3y}{3}$$
$$= \dfrac{5 + 3\left(-\frac{4}{3}\right)}{3}$$
$$= \dfrac{5 - 4}{3}$$
$$= \dfrac{1}{3}$$

The solution is $\left(\dfrac{1}{3}, -\dfrac{4}{3}\right)$. Check this solution in each equation.

---

4. a. inconsistent
b. dependent equations

Progress Check Answers

14.2 Exercises

Use the substitution method to solve each system of equations. Write each answer as an ordered pair, where possible. If a system is inconsistent or if the equations of a system are dependent, so indicate.

1. $\begin{cases} y = 2x \\ x + y = 6 \end{cases}$
2. $\begin{cases} y = 3x \\ x + y = 4 \end{cases}$
3. $\begin{cases} y = 2x - 6 \\ 2x + y = 6 \end{cases}$
4. $\begin{cases} y = 2x - 9 \\ x + 3y = 8 \end{cases}$

5. $\begin{cases} y = 2x + 5 \\ x + 2y = -5 \end{cases}$
6. $\begin{cases} y = -2x \\ 3x + 2y = -1 \end{cases}$
7. $\begin{cases} 2a + 4b = -24 \\ a = 20 - 2b \end{cases}$
8. $\begin{cases} 3a + 6b = -15 \\ a = -2b - 5 \end{cases}$

9. $\begin{cases} 2a = 3b - 13 \\ b = 2a + 7 \end{cases}$
10. $\begin{cases} a = 3b - 1 \\ b = 2a + 2 \end{cases}$
11. $\begin{cases} r + 3s = 9 \\ 3r + 2s = 13 \end{cases}$
12. $\begin{cases} x - 2y = 2 \\ 2x + 3y = 11 \end{cases}$

13. $\begin{cases} 4x + 5y = 2 \\ 3x - y = 11 \end{cases}$
14. $\begin{cases} 5u + 3v = 5 \\ 4u - v = 4 \end{cases}$
15. $\begin{cases} 2x + y = 0 \\ 3x + 2y = 1 \end{cases}$
16. $\begin{cases} 3x - y = 7 \\ 2x + 3y = 1 \end{cases}$

17. $\begin{cases} 3x + 4y = -7 \\ 2y - x = -1 \end{cases}$
18. $\begin{cases} 4x + 5y = -2 \\ x + 2y = -2 \end{cases}$
19. $\begin{cases} 9x = 3y + 12 \\ 4 = 3x - y \end{cases}$
20. $\begin{cases} 8y = 15 - 4x \\ x + 2y = 4 \end{cases}$

21. $\begin{cases} 2x + 3y = 5 \\ 3x + 2y = 5 \end{cases}$
22. $\begin{cases} 3x - 2y = -1 \\ 2x + 3y = -5 \end{cases}$
23. $\begin{cases} 2x + 5y = -2 \\ 4x + 3y = 10 \end{cases}$
24. $\begin{cases} 3x + 4y = -6 \\ 2x - 3y = -4 \end{cases}$

25. $\begin{cases} 2x - 3y = -3 \\ 3x + 5y = -14 \end{cases}$
26. $\begin{cases} 4x - 5y = -12 \\ 5x - 2y = 2 \end{cases}$
27. $\begin{cases} 7x - 2y = -1 \\ -5x + 2y = -1 \end{cases}$
28. $\begin{cases} -8x + 3y = 22 \\ 4x + 3y = -2 \end{cases}$

29. $\begin{cases} 2a + 3b = 2 \\ 8a - 3b = 3 \end{cases}$
30. $\begin{cases} 3a - 2b = 0 \\ 9a + 4b = 5 \end{cases}$

31. $\begin{cases} y - x = 3x \\ 2(x + y) = 14 - y \end{cases}$
32. $\begin{cases} y + x = 2x + 2 \\ 2(3x - 2y) = 21 - y \end{cases}$

33. $\begin{cases} 3(x - 1) + 3 = 8 + 2y \\ 2(x + 1) = 4 + 3y \end{cases}$
34. $\begin{cases} 4(x - 2) = 19 - 5y \\ 3(x + 1) - 2y = 2y \end{cases}$

35. $\begin{cases} 6a = 5(3 + b + a) - a \\ 3(a - b) + 4b = 5(1 + b) \end{cases}$
36. $\begin{cases} 5(x + 1) + 7 = 7(y + 1) \\ 5(y + 1) = 6(1 + x) + 5 \end{cases}$

37. $\begin{cases} \dfrac{1}{2}x + \dfrac{1}{2}y = -1 \\ \dfrac{1}{3}x - \dfrac{1}{2}y = -4 \end{cases}$
38. $\begin{cases} \dfrac{2}{3}y + \dfrac{1}{5}z = 1 \\ \dfrac{1}{3}y - \dfrac{2}{5}z = 3 \end{cases}$

39. $\begin{cases} 5x = \dfrac{1}{2}y - 1 \\ \dfrac{1}{4}y = 10x - 1 \end{cases}$
40. $\begin{cases} \dfrac{2}{3}x = 1 - 2y \\ 2(5y - x) + 11 = 0 \end{cases}$

41. $\begin{cases} \dfrac{6x - 1}{3} - \dfrac{5}{3} = \dfrac{3y + 1}{2} \\ \dfrac{1 + 5y}{4} + \dfrac{x + 3}{4} = \dfrac{17}{2} \end{cases}$
42. $\begin{cases} \dfrac{5x - 2}{4} + \dfrac{1}{2} = \dfrac{3y + 2}{2} \\ \dfrac{7y + 3}{3} = \dfrac{x}{2} + \dfrac{7}{3} \end{cases}$

Review Exercises

In Review Exercises 1–6, factor each expression completely.

1. $8x^2y^2 - 32xy^2z + 16xyz^2$
2. $(x-y)a - (x-y)b$
3. $a^6 - 25$
4. $b^4 - 625$
5. $r^2 + 2rs - 15s^2$
6. $4m^2 - 15mn + 9n^2$

In Review Exercises 7–10, simplify each fraction.

7. $\dfrac{21ab^2c^3}{14abc^4}$
8. $\dfrac{x^2 + xy}{x^2 - y^2}$
9. $\dfrac{-2x + 2y + 2z}{4x - 4y - 4z}$
10. $\dfrac{2t^2 + t - 3}{2t^2 + 7t + 6}$

11. If the voltage is constant, then the current I passing through a resistance R varies inversely with R. If $I = 17$ when $R = 2$, find I when $R = 17$.

12. If the power dissipated by a resistance R is constant, then R is inversely proportional to the square of the current I. If $R = 9$ when $I = 2$, find R when $I = 3$.

14.3 Solving Systems of Equations by Addition

- The Addition Method
- Inconsistent Systems
- Dependent Equations

The Addition Method

To use the **addition method** to solve

$$\begin{cases} x + y = 8 \\ x - y = -2 \end{cases}$$

we add the left-hand sides and the right-hand sides of the equations to eliminate y. The resulting equation can then be solved for x.

$$\begin{aligned} x + y &= 8 \\ x - y &= -2 \\ \hline 2x &= 6 \end{aligned}$$

We can now solve the equation $2x = 6$ for x.

$2x = 6$

$x = 3$ Divide both sides by 2.

To find the value of y, we multiply the first equation of the system by -1 to obtain the system

$$\begin{cases} -x - y = -8 \\ x - y = -2 \end{cases}$$

When we add these equations, the terms involving x are eliminated, and we can solve the resulting equation for y.

$$\begin{aligned} -x - y &= -8 \\ x - y &= -2 \\ \hline -2y &= -10 \\ y &= 5 \end{aligned}$$ Divide both sides by -2.

Thus, the solution of this system of equations is $(3, 5)$.

We could have found the value of y by substituting 3 for x in either of the equations of the system and solving the resulting equation for y. For example, substituting 3 for x in the equation $x + y = 8$ and solving for y gives

$$x + y = 8$$
$$3 + y = 8$$
$$y = 5 \quad \text{Add } -3 \text{ to both sides.}$$

Check by verifying that the pair $(3, 5)$ satisfies each equation of the system.

EXAMPLE 1

Use the addition method to solve the system $\begin{cases} 3y = 14 + x \\ x + 22 = 5y \end{cases}$.

Solution

Write the equations in the form $\begin{cases} -x + 3y = 14 \\ x - 5y = -22 \end{cases}$.

Add the equations to eliminate the terms involving x and solve for y.

$$-x + 3y = 14$$
$$\underline{x - 5y = -22}$$
$$-2y = -8$$
$$y = 4 \quad \text{Divide both sides by } -2.$$

To find the value of x, we substitute 4 for y in either equation of the system. For example, we can substitute 4 for y in $-x + 3y = 14$ and solve for x.

$$-x + 3y = 14$$
$$-x + 3(4) = 14$$
$$-x + 12 = 14 \quad \text{Simplify.}$$
$$-x = 2 \quad \text{Add } -12 \text{ to both sides.}$$
$$x = -2 \quad \text{Divide both sides by } -1.$$

The solution to this system is $(-2, 4)$.

Check this solution by verifying that the pair $(-2, 4)$ satisfies each equation of the system. Work Progress Check 5. ∎

Progress Check 5

Solve $\begin{cases} 2x + y = -2 \\ 3x - y = -8 \end{cases}$ by addition.

EXAMPLE 2

Use the addition method to solve the system $\begin{cases} 2x - 5y = 10 \\ 3x - 2y = -7 \end{cases}$.

Solution

In this system, each equation must be adjusted so that one of the variables will be eliminated when the equations are added. To eliminate the x variable, for example, multiply the first equation by 3 and the second equation by -2. This gives the system

$$\begin{cases} 3(2x - 5y) = 3(10) \\ -2(3x - 2y) = -2(-7) \end{cases} \quad \text{or} \quad \begin{cases} 6x - 15y = 30 \\ -6x + 4y = 14 \end{cases}$$

5. $(-2, 2)$

Progress Check Answers

When these equations are added, the terms involving the variable x are eliminated.

$$6x - 15y = 30$$
$$-6x + 4y = 14$$
$$\overline{-11y = 44}$$
$$y = -4 \quad \text{Divide both sides by } -11.$$

To find x, we substitute -4 for y in the equation $2x - 5y = 10$:

$$2x - 5y = 10$$
$$2x - 5(-4) = 10$$
$$2x + 20 = 10 \quad \text{Simplify.}$$
$$2x = -10 \quad \text{Add } -20 \text{ to both sides.}$$
$$x = -5 \quad \text{Divide both sides by 2.}$$

The solution to this system is $(-5, -4)$. Check this solution. Work Progress Check 6. ∎

Progress Check 6

Solve $\begin{cases} 2x + 3y = 7 \\ 3x - 2y = -9 \end{cases}$ by addition.

Inconsistent Systems

EXAMPLE 3

Use the addition method to solve the system $\begin{cases} x - \dfrac{2}{3}y = \dfrac{8}{3} \\ -\dfrac{3}{2}x + y = -6 \end{cases}$

Solution

We multiply both sides of the first equation by 3 and both sides of the second equation by 2 to clear the equations of fractions. This gives the system

$$\begin{cases} 3x - 2y = 8 \\ -3x + 2y = -12 \end{cases}$$

Add these equations to eliminate the x variable.

$$3x - 2y = 8$$
$$-3x + 2y = -12$$
$$\overline{0 = -4}$$

In this case, *both* variables are eliminated, and the *false* result $0 = -4$ is obtained. This indicates that the equations of the system are independent but the system itself is inconsistent. The given system has no solutions. ∎

Dependent Equations

EXAMPLE 4

Use the addition method to solve the system $\begin{cases} x - \dfrac{5}{2}y = \dfrac{19}{2} \\ -\dfrac{2}{5}x + y = -\dfrac{19}{5} \end{cases}$

6. $(-1, 3)$

Progress Check Answers

Solution

We multiply both sides of the first equation by 2 and both sides of the second equation by 5 to clear the equations of fractions. This gives the system

$$\begin{cases} 2x - 5y = 19 \\ -2x + 5y = -19 \end{cases}$$

Add these equations to eliminate a variable.

$$\begin{aligned} 2x - 5y &= 19 \\ -2x + 5y &= -19 \\ \hline 0 &= 0 \end{aligned}$$

As in Example 3, both x and y were eliminated. However, this time the *true* result $0 = 0$ was obtained. This shows that the equations are dependent and the system has an unlimited number of solutions. Any ordered pair that satisfies one of the equations satisfies the other also. Some solutions are $(2, -3)$, $(12, 1)$, and $(0, -\frac{19}{5})$. Work Progress Check 7. ∎

Progress Check 7

a. Solve $\begin{cases} 2x + 3y = 12 \\ y = -\frac{2}{3}x + 5 \end{cases}$ by addition.

b. Solve $\begin{cases} 3x + 2y = 12 \\ y = -\frac{3}{2}x + 6 \end{cases}$ by addition.

EXAMPLE 5

Use the addition method to solve the system $\begin{cases} \frac{5}{6}x + \frac{2}{3}y = \frac{7}{6} \\ \frac{10}{7}x - \frac{4}{9}y = \frac{17}{21} \end{cases}$.

Solution

To clear the equations of fractions, we multiply both sides of the first equation by 6 and both sides of the second equation by 63. This gives the system

1. $\begin{cases} 5x + 4y = 7 \\ 90x - 28y = 51 \end{cases}$
2.

We solve for x by eliminating the terms involving y. To do so, we multiply Equation 1 by 7 and add the result to Equation 2.

$$\begin{aligned} 35x + 28y &= 49 \\ 90x - 28y &= 51 \\ \hline 125x &= 100 \end{aligned}$$

$$x = \frac{100}{125} \qquad \text{Divide both sides by 125.}$$

$$x = \frac{4}{5} \qquad \text{Simplify.}$$

To solve for y, we substitute $\frac{4}{5}$ for x in Equation 1 and simplify.

$$5x + 4y = 7$$
$$5\left(\frac{4}{5}\right) + 4y = 7$$
$$4 + 4y = 7 \qquad \text{Simplify.}$$
$$4y = 3 \qquad \text{Add } -4 \text{ to both sides.}$$
$$y = \frac{3}{4} \qquad \text{Divide both sides by 4.}$$

The solution of this system is $(\frac{4}{5}, \frac{3}{4})$. Check this solution. ∎

7. a. inconsistent
   b. dependent equations

Progress Check Answers

14.3 Exercises

In Exercises 1–12, use the addition method to solve each system of equations. Write each answer as an ordered pair.

1. $\begin{cases} x + y = 5 \\ x - y = -3 \end{cases}$
2. $\begin{cases} x - y = 1 \\ x + y = 7 \end{cases}$
3. $\begin{cases} x - y = -5 \\ x + y = 1 \end{cases}$
4. $\begin{cases} x + y = 1 \\ x - y = 5 \end{cases}$

5. $\begin{cases} 2x + y = -1 \\ -2x + y = 3 \end{cases}$
6. $\begin{cases} 3x + y = -6 \\ x - y = -2 \end{cases}$
7. $\begin{cases} 2x - 3y = -11 \\ 3x + 3y = 21 \end{cases}$
8. $\begin{cases} 3x - 2y = 16 \\ -3x + 8y = -10 \end{cases}$

9. $\begin{cases} 2x + y = -2 \\ -2x - 3y = -6 \end{cases}$
10. $\begin{cases} 3x + 4y = 8 \\ 5x - 4y = 24 \end{cases}$
11. $\begin{cases} 4x + 3y = 24 \\ 4x - 3y = -24 \end{cases}$
12. $\begin{cases} 5x - 4y = 8 \\ -5x - 4y = 8 \end{cases}$

In Exercises 13–42, use the addition method to solve each system of equations. Write each answer as an ordered pair, where possible. If the equations of a system are dependent or if a system is inconsistent, so indicate.

13. $\begin{cases} x + y = 5 \\ x + 2y = 8 \end{cases}$
14. $\begin{cases} x + 2y = 0 \\ x - y = -3 \end{cases}$
15. $\begin{cases} 2x + y = 4 \\ 2x + 3y = 0 \end{cases}$
16. $\begin{cases} 2x + 5y = -13 \\ 2x - 3y = -5 \end{cases}$

17. $\begin{cases} 3x + 29 = 5y \\ 4y - 34 = -3x \end{cases}$
18. $\begin{cases} 3x - 16 = 5y \\ 33 - 5y = 4x \end{cases}$
19. $\begin{cases} 2x = 3(y - 2) \\ 2(x + 4) = 3y \end{cases}$
20. $\begin{cases} 3(x - 2) = 4y \\ 2(2y + 3) = 3x \end{cases}$

21. $\begin{cases} -2(x + 1) = 3(y - 2) \\ 3(y + 2) = 6 - 2(x - 2) \end{cases}$
22. $\begin{cases} 5(x - 1) = 8 - 3(y + 2) \\ 4(x + 2) - 7 = 3(2 - y) \end{cases}$

23. $\begin{cases} 4(x + 1) = 17 - 3(y - 1) \\ 2(x + 2) + 3(y - 1) = 9 \end{cases}$
24. $\begin{cases} 3(x + 3) + 2(y - 4) = 5 \\ 3(x - 1) = -2(y + 2) \end{cases}$

25. $\begin{cases} 2x + y = 10 \\ x + 2y = 10 \end{cases}$
26. $\begin{cases} 3x + 2y = 0 \\ 2x - 3y = -13 \end{cases}$
27. $\begin{cases} 2x - y = 16 \\ 3x + 2y = 3 \end{cases}$
28. $\begin{cases} 3x + 4y = -17 \\ 4x - 3y = -6 \end{cases}$

29. $\begin{cases} 4x + 5y = -20 \\ 5x - 4y = -25 \end{cases}$
30. $\begin{cases} 3x - 5y = 4 \\ 7x + 3y = 68 \end{cases}$
31. $\begin{cases} 6x = -3y \\ 5y = 2x + 12 \end{cases}$
32. $\begin{cases} 3y = 4x \\ 5x = 4y - 2 \end{cases}$

33. $\begin{cases} 4(2x - y) = 18 \\ 3(x - 3) = 2y - 1 \end{cases}$
34. $\begin{cases} 2(2x + 3y) = 5 \\ 8x = 3(1 + 3y) \end{cases}$
35. $\begin{cases} \dfrac{3}{5}x + \dfrac{4}{5}y = 1 \\ -\dfrac{1}{4}x + \dfrac{3}{8}y = 1 \end{cases}$
36. $\begin{cases} \dfrac{1}{2}x - \dfrac{1}{4}y = 1 \\ \dfrac{1}{3}x + y = 3 \end{cases}$

37. $\begin{cases} \dfrac{3}{5}x + y = 1 \\ \dfrac{4}{5}x - y = -1 \end{cases}$
38. $\begin{cases} \dfrac{1}{2}x + \dfrac{4}{7}y = -1 \\ 5x - \dfrac{4}{5}y = -10 \end{cases}$

39. $\begin{cases} \dfrac{x}{2} - \dfrac{y}{3} = -2 \\ \dfrac{2x - 3}{2} + \dfrac{6y + 1}{3} = \dfrac{17}{6} \end{cases}$
40. $\begin{cases} \dfrac{x + 2}{4} + \dfrac{y - 1}{3} = \dfrac{1}{12} \\ \dfrac{x + 4}{5} - \dfrac{y - 2}{2} = \dfrac{5}{2} \end{cases}$

41. $\begin{cases} \dfrac{x - 3}{2} + \dfrac{y + 5}{3} = \dfrac{11}{6} \\ \dfrac{x + 3}{3} - \dfrac{5}{12} = \dfrac{y + 3}{4} \end{cases}$
42. $\begin{cases} \dfrac{x + 2}{3} = \dfrac{3 - y}{2} \\ \dfrac{x + 3}{2} = \dfrac{2 - y}{3} \end{cases}$

Review Exercises

In Review Exercises 1–6, solve each equation.

1. $8(3x - 5) - 12 = 4(2x + 3)$
2. $3y + \dfrac{y + 2}{2} = \dfrac{2(y + 3)}{3} + 16$
3. $3z^2 - 5z + 2 = 0$
4. $3t^2 + 4 = 8t$
5. $10y^2 + 21y = 10$
6. $x(9x - 24) + 12 = 0$
7. Solve $P = 2l + 2w$ for w.
8. Solve $A = p + prt$ for p.

In Review Exercises 9–12, solve each word problem.

9. Find three consecutive integers whose sum is 318.
10. The length of a rectangle is 6 feet greater than its width, and its perimeter is 72 feet. Find its dimensions.
11. Manuel had $14,000 to invest, some at 7% annual interest and the rest at 10%. How much did he invest at the lower rate if his annual income from the two investments is $1280?
12. In a triangle, one angle is twice the second angle, and the third is three times the second. How large is each angle? (*Hint:* There are 180° in the sum of the angles of a triangle.)

14.4 Applications of Systems of Equations

- *Agriculture Problem*
- *Plumbing Problem*
- *Geometry Problem*
- *Coin Problem*
- *Investment Problem*
- *Uniform Motion Problem*
- *Medical Problem*

We now consider applications that involve systems of equations. We will use the following steps to solve a problem with two unknown quantities and two variables.

1. Read the problem and analyze the facts. A sketch, chart, or diagram may help you visualize the facts of the problem.
2. Pick different variables to represent each of the unknown quantities, and write a sentence stating what each variable represents.
3. Find two equations involving each of the two variables. This will give a system of two equations in two variables.
4. Solve the system using the most convenient method.
5. State the solution or solutions.
6. Check the answer in the words of the problem.

Agriculture Problem

EXAMPLE 1

A farmer raises wheat and soybeans on 215 acres. He plants wheat on 31 more acres of land than he plants soybeans. How many acres of each does he plant?

Solution

Let w represent the number of acres of wheat planted and s equal the number of acres of soybeans planted. Because the total land used by both crops is 215 acres, we have the equation

$$w + s = 215$$

Because the area devoted to wheat exceeds that devoted to soybeans by 31 acres, we also have the equation

$$w - s = 31$$

We can now solve the system

1. $\begin{cases} w + s = 215 \\ w - s = 31 \end{cases}$
2.

by using the addition method.

$$\begin{aligned} w + s &= 215 \\ w - s &= 31 \\ \hline 2w &= 246 \\ w &= 123 \quad \text{Divide both sides by 2.} \end{aligned}$$

To find the value of s, we substitute 123 for w in one of the original equations of the system, such as Equation 1, and solve for s.

$$\begin{aligned} w + s &= 215 \\ \mathbf{123} + s &= 215 \quad \text{Substitute 123 for } w. \\ s &= 92 \quad \text{Add } -123 \text{ to both sides.} \end{aligned}$$

Thus, the farmer plants 123 acres of wheat and 92 acres of soybeans.

Check: The total acreage planted is $123 + 92$, or 215 acres. The area devoted to wheat is 31 acres greater than that used for soybeans, because $123 - 92 = 31$. ∎

Plumbing Problem

EXAMPLE 2

A plumber wants to cut a 19-foot pipe into two pieces. The longer piece is to be 1 foot longer than twice the shorter piece. Find the length of each piece.

Solution

Let s represent the length of the shorter piece and l represent the length of the longer piece (see Figure 14-6).

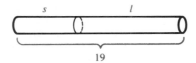

Figure 14-6

Because the length of pipe is 19 feet long, we have

$$s + l = 19$$

Because the longer piece is 1 foot longer than twice the shorter piece, we have

$$l = 2s + 1$$

We can use the substitution method to solve this system.

1. $\begin{cases} s + l = 19 \\ l = 2s + 1 \end{cases}$
2.

$$s + 2s + 1 = 19$$
$$3s + 1 = 19 \quad \text{Combine like terms.}$$
$$3s = 18 \quad \text{Add } -1 \text{ to both sides.}$$
$$s = 6 \quad \text{Divide both sides by 3.}$$

The shorter piece should be 6 feet long.

To find the length of the longer piece, we substitute 6 for s in Equation 1 and solve for l.

$$s + l = 19$$
$$6 + l = 19$$
$$l = 13 \quad \text{Add } -6 \text{ to both sides.}$$

Thus, the longer piece should be 13 feet long.

Check: The sum of 13 and 6 is 19. 13 is 1 more than twice 6.

Geometry Problem

EXAMPLE 3

Tom has 150 feet of fencing to enclose a rectangular garden. The length of the garden is 5 feet less than three times its width. Find the area of the garden.

Solution

Let l represent the length of the garden and w represent the width (see Figure 14-7). Because the perimeter (the distance around the garden) is 150 feet, the sum of two lengths and two widths must equal 150. Thus,

$$2l + 2w = 150$$

Because the length is 5 feet less than 3 times the width,

$$l = 3w - 5$$

We can use the substitution method to solve the system.

1. $\begin{cases} 2l + 2w = 150 \\ l = 3w - 5 \end{cases}$
2.

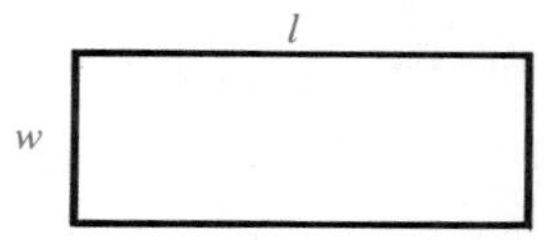

$$2(3w - 5) + 2w = 150$$
$$6w - 10 + 2w = 150 \quad \text{Remove parentheses.}$$
$$8w - 10 = 150 \quad \text{Combine like terms.}$$
$$8w = 160 \quad \text{Add 10 to both sides.}$$
$$w = 20 \quad \text{Divide both sides by 8.}$$

Figure 14-7

Thus, the width is 20 feet.
To find the length, we substitute 20 for w in Equation 2 and simplify.

$$l = 3w - 5$$
$$= 3(20) - 5$$
$$= 60 - 5$$
$$= 55$$

Because the dimensions of the rectangle are 55 feet by 20 feet, and the area of a rectangle is given by the formula

$$A = l \cdot w \qquad \text{Area = length times width.}$$

we have

$$A = 55 \cdot 20$$
$$= 1100$$

The garden covers an area of 1100 square feet.

Check:

$$P = 2l + 2w$$
$$= 2(55) + 2(20) \qquad \text{Substitute 55 for } l \text{ and 20 for } w.$$
$$= 110 + 40$$
$$= 150$$

It is also true that 55 feet is 5 feet less than 3 times 20 feet. ∎

Coin Problem

EXAMPLE 4

Nancy has 1056 coins, consisting of nickels and dimes. The coins are worth $84.55. How many of each coin does she have?

Solution

Let n represent the number of nickels and d represent the number of dimes. Because there are 1056 coins in all, we can form the equation

$$n + d = 1056$$

Because n nickels are worth $0.05n$ and d dimes are worth $0.10d$, the sum of $0.05n$ and $0.10d$ represents the total value of the coins. Thus, we can form the equation

$$0.05n + 0.10d = 84.55$$

We now have the system

1. $\begin{cases} n + d = 1056 \\ 0.05n + 0.10d = 84.55 \end{cases}$
2.

Multiply Equation 2 by 100 to eliminate the decimal fractions. Then multiply Equation 1 by -5 to get a system that can be solved by addition.

$$-5n - 5d = -5280$$
$$\underline{5n + 10d = 8455}$$
$$5d = 3175$$
$$d = 635 \qquad \text{Divide both sides by 5.}$$

To find the number of nickels, substitute 635 for d in Equation 1 and simplify.

$$n + d = 1056$$
$$n + 635 = 1056$$
$$n = 421 \qquad \text{Add } -635 \text{ to both sides.}$$

Nancy has 421 nickels and 635 dimes.

Check: $421 + 635 = 1056$

The value of 421 nickels is	$21.05
The value of 635 dimes is	$63.50
The total value is	$84.55

Investment Problem

EXAMPLE 5

Jeffrey invested some money at 8% annual interest, and Grant invested some at 10%. The first year's income on their combined investment of $15,000 is $1340. How much did each invest?

Solution

Let x represent the amount of money invested by Jeffrey and y represent the amount of money invested by Grant. Because their total investment was $15,000,

$$x + y = 15{,}000$$

The income on the x dollars invested at 8% is $0.08x$. The income on the y dollars invested at 10% is $0.10y$. The combined income is $1340. Hence, we can form the equation

$$0.08x + 0.10y = 1340$$

to get the system

1. $\begin{cases} x + y = 15{,}000 \\ 0.08x + 0.10y = 1340 \end{cases}$
2.

We use the addition method and proceed as follows.

$$-8x - 8y = -120{,}000 \qquad \text{Multiply both sides of Equation 1 by } -8.$$
$$\underline{8x + 10y = 134{,}000} \qquad \text{Multiply both sides of Equation 2 by 100.}$$
$$2y = 14{,}000$$
$$y = 7000 \qquad \text{Divide both sides by 2.}$$

To find x, we substitute 7000 for y in Equation 1 and simplify.

$$x + y = 15{,}000$$
$$x + 7000 = 15{,}000$$
$$x = 8000 \qquad \text{Add } -7000 \text{ to both sides.}$$

Jeffrey invested $8000, and Grant invested $7000.

Check:

$$\$8000 + \$7000 = \$15{,}000$$
$$0.08(\$8000) = \$640$$
$$0.10(\$7000) = \$700$$

The combined interest is $1340.

Uniform Motion Problem

EXAMPLE 6

A boat travels 30 kilometers downstream in 3 hours and makes the return trip in 5 hours. Find the speed of the boat in still water.

Solution

Let s represent the speed of the boat in still water, and let c represent the speed of the current. Then the speed of the boat going downstream is $s + c$. The speed of the boat going upstream is $s - c$. Organize the information as in Figure 14-8.

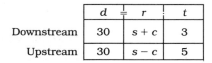

	d	r	t
Downstream	30	$s + c$	3
Upstream	30	$s - c$	5

Figure 14-8

Because $d = r \cdot t$, the information in the table gives two equations in two variables.

$$\begin{cases} 30 = 3(s + c) \\ 30 = 5(s - c) \end{cases}$$

After removing parentheses and rearranging terms, we have

1. $\quad \begin{cases} 3s + 3c = 30 \\ 5s - 5c = 30 \end{cases}$
2.

To solve this system by addition, we multiply Equation 1 by 5, Equation 2 by 3, add the equations, and solve for s.

$$15s + 15c = 150$$
$$15s - 15c = 90$$
$$\overline{30s = 240}$$
$$s = 8 \qquad \text{Divide both sides by 30.}$$

The speed of the boat in still water is 8 kilometers per hour. Check the result. ∎

Medical Problem

EXAMPLE 7

A laboratory technician has a batch of solution that is 40% alcohol and a second batch that is 60% alcohol. She wants to make 8 liters of solution that is 55% alcohol. How many liters of each batch should she use?

Solution

Let x represent the number of liters to be used from batch 1 and y represent the number of liters to be used from batch 2. Organize the information of the problem as in Figure 14-9.

	Fractional part that is alcohol	Number of liters of solution	Number of liters of alcohol
Batch 1	0.40	x	$0.40x$
Batch 2	0.60	y	$0.60y$
Mixture	0.55	8	$0.55(8)$

Figure 14-9

The information in Figure 14-9 provides two equations.

$$\begin{cases} x + y = 8 & \text{(1)} \\ 0.40x + 0.60y = 0.55(8) & \text{(2)} \end{cases}$$

1. The number of liters of batch 1 plus the number of liters of batch 2 equals the total number of liters in the mixture.

2. The amount of alcohol in batch 1 plus the amount of alcohol in batch 2 equals the amount of alcohol in the mixure.

We can use addition to solve this system.

$$-40x - 40y = -320 \quad \text{Multiply both sides of Equation 1 by } -40.$$
$$40x + 60y = 440 \quad \text{Multiply both sides of Equation 2 by 100.}$$
$$\overline{20y = 120}$$
$$y = 6 \quad \text{Divide both sides by 20.}$$

To find x, we substitute 6 for y in Equation 1 and simplify.

$$x + y = 8$$
$$x + 6 = 8$$
$$x = 2 \quad \text{Add } -6 \text{ to both sides.}$$

The technician should use 2 liters of the 40% solution and 6 liters of the 60% solution. Check the result. ■

14.4 Exercises

Use two equations in two variables to solve each word problem.

1. One number is twice another. Their sum is 96. Find the numbers.

2. If the sum of two numbers is 38 and the difference is 12, what are the numbers?

3. Three times a certain number plus another number is 29, but the first number plus twice the second number is 18. Find the numbers.

4. Twice a certain number plus another number is 21, but the first number plus 3 times the second number is 33. Find the numbers.

5. Eight cans of paint and three paintbrushes cost $135. How much does each cost if six cans of paint and two brushes cost $100?

6. One catcher's mitt and 10 outfielder's gloves cost $239.50. One catcher's mitt and 5 outfielder's gloves cost $134.50. How much does each cost?

7. Two bottles of contact lens cleaner and three bottles of soaking solution cost $29.40, and three bottles of cleaner and two bottles of soaking solution cost $28.60. Find the cost of each.

8. Two pairs of shoes and four pairs of socks cost $109, and three pairs of shoes and five pairs of socks cost $160. Find the cost of a pair of socks.

9. Jerry wants to cut a 25-foot pole into two pieces. He wants one piece to be 5 feet longer than the other. How long should each piece be?

10. A carpenter wants to cut a 20-foot pole into two pieces so that one piece is 4 times as long as the other. How long should each piece be?

11. The perimeter of a rectangle is 110 feet. The length is 5 feet longer than its width. Find its dimensions.

12. A rectangle is 3 times as long as it is wide. Its perimeter is 80 centimeters. Find its dimensions.

13. A rectangle has a length that is 2 feet more than twice its width. Its perimeter is 34 feet. Find its area.

14. The perimeter of a rectangle is 50 meters. Its width is two-thirds its length. Find its area.

15. A girl has 80 coins that are quarters and dimes. The coins are worth $14. How many quarters and how many dimes does she have?

16. A girl has some dimes and some nickels. There are 25 coins in all, and they are worth $1.75. How many of each type of coin does she have?

17. David has equal numbers of dimes and quarters worth $3.50. How many of each does he have?

18. A girl has twice as many nickels as dimes. Her coins are worth $12.40. How many coins does she have?

19. Bill invested some money at 5% annual interest, and Janette invested some at 7%. If their combined interest was $310 on a total investment of $5000, how much did each invest?

20. Peter invested some money at 6% annual interest, and Martha invested some at twice that rate. If their combined investment was $6000 and their combined interest was $540, how much money did each invest?

21. Students can buy tickets to a basketball game for $1. However, the admission for nonstudents is $2. If 350 tickets are sold and the total receipts are $450, how many student tickets and how many nonstudent tickets are sold?

22. General-admission movie tickets cost $4, but senior citizens are admitted for $3. If the receipts were $720 for a total audience of 190 people, how many senior citizens attended?

23. A boat can travel 24 miles downstream in 2 hours and can make the return trip in 3 hours. Find the speed of the boat in still water.

24. With the wind, a plane can fly 3000 miles in 5 hours. Against the same wind, the trip takes 6 hours. What is the airspeed of the plane (the speed in still air)?

25. An airplane can fly downwind a distance of 600 miles in 2 hours. However, the return trip against the same wind takes 3 hours. What is the speed of the wind?

26. It takes a motorboat 4 hours to travel 56 miles down a river, but it takes 3 hours longer to make the return trip. Find the speed of the current.

27. A chemist has a solution that is 40% alcohol and another solution that is 55% alcohol. How much of each must she use to make 15 liters of a solution that is 50% alcohol?

28. A nurse has a solution that is 25% alcohol and another that is 50% alcohol. How much of each must he use to make 20 liters of a solution that is 40% alcohol?

29. A merchant wants to mix peanuts worth $3 per pound with cashews worth $6 per pound to get 48 pounds of mixed nuts to sell at $4 per pound. How many pounds of peanuts and how many pounds of cashews must the merchant use?

30. A merchant wants to mix peanuts worth $3 per pound with jelly beans worth $1.50 per pound to make 30 pounds of a mixture worth $2.10 per pound. How many pounds of each should he use?

31. Lisa wants to buy gifts for seven friends. On some she will only spend $7, but on each of her best friends she will spend $9. Lisa spends $53, total. How many of each gift did she buy?

32. Three soft drinks and a sandwich cost $4.15. One soft drink and three sandwiches cost $6.45. Find the cost of each.

33. An electronics store put two types of car radio on sale. One model costs $87, and the other costs $119. During the sale, the receipts for the 25 radios sold were $2495. How many of each model were sold?

34. At a restaurant, ice cream cones cost $.90, and sundaes cost $1.65. One day, the receipts for a total of 148 cones and sundaes were $180.45. Determine how many cones and how many sundaes were sold.

35. An investment of $950 at one rate of interest and $1200 at a higher rate together generate an annual income of $205.50. The investment rates differ by 1%. Find the two rates of interest. (*Hint:* Treat 1% as 0.01.)

36. A man drives for a while at 45 miles per hour. Realizing that he is running late, he increases his speed to 60 miles per hour and completes his 405-mile trip in 8 hours. How long does he travel at each speed?

Review Exercises

In Review Exercises 1–4, graph each equation.

1. $y = 2x - 4$ **2.** $2x + y = 6$ **3.** $2x = 3y + 1$ **4.** $\dfrac{x}{2} - \dfrac{y}{3} = 2$

In Review Exercises 5–8, graph each inequality.

5. $2x - y \leq 4$ **6.** $x + 3y > -2$ **7.** $2x + 3y > 0$ **8.** $\dfrac{1}{3}x + 2y \geq 1$

In Review Exercises 9–12, graph each expression on a rectangular coordinate system.

9. $y = 5$ **10.** $x < 2$ **11.** $y \geq -3$ **12.** $x = -1$

14.5 Solving Systems of Inequalities

We now consider **systems of inequalities**. To solve such systems, we will

> 1. Graph each inequality in the system on the same coordinate axes.
> 2. Find the region where the graphs overlap.

EXAMPLE 1

Graph the solution set of the system $\begin{cases} x + y \geq 1 \\ x - y \geq 1 \end{cases}$.

Solution

We graph each inequality on the same set of coordinate axes as in Figure 14-10.

$x + y = 1$		$x - y = 1$	
x	y	x	y
0	1	0	−1
1	0	1	0
2	−1	2	1

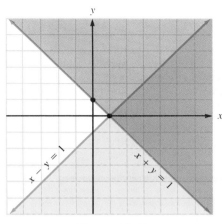

Figure 14-10

The graph of $x + y \geq 1$ includes the graph of the equation $x + y = 1$ and all points above it. Because the boundary line is included, we draw it with a solid line. The graph of the inequality $x - y \geq 1$ includes the graph of the equation $x - y = 1$ and all points below it. Because the boundary line is included, we draw it with a solid line also.

The area shaded twice represents the set of simultaneous solutions of the given system of inequalities. Any point in the doubly shaded region has coordinates that will satisfy both of the inequalities of the system. ■

EXAMPLE 2

Graph the solution set of the system $\begin{cases} 2x + y < 4 \\ -2x + y > 2 \end{cases}$.

Solution

We graph each inequality on the same set of coordinate axes, as in Figure 14-11.

$2x + y = 4$		$-2x + y = 2$	
x	y	x	y
0	4	−1	0
1	2	0	2
2	0	2	6

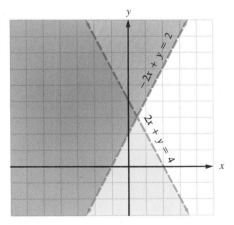

Figure 14-11

The graph of $2x + y < 4$ includes all points below the line $2x + y = 4$. Because the boundary line itself is *not* included, we draw it as a broken line. The graph of $-2x + y > 2$ includes all points above the line $-2x + y = 2$. Because the boundary line itself is again *not* included, we draw it as a broken line.

Progress Check 8

Solve $\begin{cases} 2x + y < 4 \\ -2x + y < 2 \end{cases}$ by graphing.

The area shaded twice represents the set of simultaneous solutions of the given system of inequalities. Any point in the doubly shaded region has coordinates that will satisfy both inequalities of the system. Work Progress Check 8.

EXAMPLE 3

Graph the solution set of the system $\begin{cases} x \leq 2 \\ y > 3 \end{cases}$.

Solution

We graph each inequality on the same set of coordinate axes, as in Figure 14-12.

$x = 2$

x	y
2	0
2	2
2	4

$y = 3$

x	y
0	3
1	3
4	3

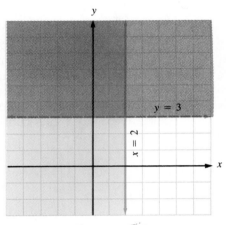

Figure 14-12

The graph of $x \leq 2$ includes all points to the left of the line $x = 2$. Because the boundary line *is* included, we draw it as a solid line. The graph $y > 3$ includes all points above the line $y = 3$. Because the boundary line is *not* included, draw it as a broken line.

The area that is shaded twice represents the set of simultaneous solutions of the given system of inequalities. Any point in the doubly shaded region has coordinates that will satisfy both inequalities of the system.

EXAMPLE 4

Graph the solution set of the system $\begin{cases} y < 3x - 1 \\ y \geq 3x + 1 \end{cases}$.

Solution

We graph each inequality, as shown in Figure 14-13. The graph of $y < 3x - 1$ includes all of the points below the broken line $y = 3x - 1$. The graph of $y \geq 3x + 1$ includes all of the points on and above the solid line $y = 3x + 1$. Because the graphs of these inequalities do not intersect, the solution set is empty. There are no solutions.

8.
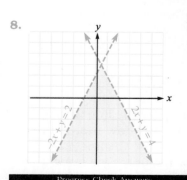

Progress Check Answers

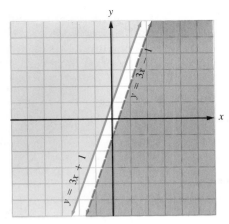

Figure 14-13

14.5 Exercises

In Exercises 1–26, use the method of graphing to find the solution set of each system of inequalities. If a system has no solutions, so indicate.

1. $\begin{cases} x + 2y \leq 3 \\ 2x - y \geq 1 \end{cases}$
2. $\begin{cases} 2x + y \geq 3 \\ x - 2y \leq -1 \end{cases}$
3. $\begin{cases} x + y < -1 \\ x - y > -1 \end{cases}$
4. $\begin{cases} x + y > 2 \\ x - y < -2 \end{cases}$

5. $\begin{cases} 2x - y < 4 \\ x + y \geq -1 \end{cases}$
6. $\begin{cases} x - y \geq 5 \\ x + 2y < -4 \end{cases}$
7. $\begin{cases} x > 2 \\ y \leq 3 \end{cases}$
8. $\begin{cases} x \geq -1 \\ y > -2 \end{cases}$

9. $\begin{cases} 2x - 3y < 0 \\ y > x - 1 \end{cases}$
10. $\begin{cases} 3x - y \geq -1 \\ y \geq 3x + 1 \end{cases}$
11. $\begin{cases} x + y < 1 \\ x + y > 3 \end{cases}$
12. $\begin{cases} x + y > 2 \\ x + y < 4 \end{cases}$

13. $\begin{cases} x > 0 \\ y > 0 \end{cases}$
14. $\begin{cases} x \leq 0 \\ y < 0 \end{cases}$
15. $\begin{cases} 3x + 4y > -7 \\ 2x - 3y \geq 1 \end{cases}$
16. $\begin{cases} 3x + y \leq 1 \\ 4x - y > -8 \end{cases}$

17. $\begin{cases} x < 3y - 1 \\ y \geq 2x - 3 \end{cases}$
18. $\begin{cases} y \geq x + 2 \\ x \leq y - 2 \end{cases}$
19. $\begin{cases} 2x + y < 7 \\ y > 2(1 - x) \end{cases}$
20. $\begin{cases} 2x + y \geq 6 \\ y \leq 2(2x - 3) \end{cases}$

21. $\begin{cases} 2x - 4y > -6 \\ 3x + y \geq 5 \end{cases}$
22. $\begin{cases} 2x - 3y < 0 \\ 2x + 3y \geq 12 \end{cases}$
23. $\begin{cases} 3x - y \leq -4 \\ 3y > -2(x + 5) \end{cases}$
24. $\begin{cases} 3x + y < -2 \\ y > 3(1 - x) \end{cases}$

25. $\begin{cases} \dfrac{x}{2} + \dfrac{y}{3} \geq 2 \\ \dfrac{x}{2} - \dfrac{y}{2} < -1 \end{cases}$
26. $\begin{cases} \dfrac{x}{3} - \dfrac{y}{2} < -3 \\ \dfrac{x}{3} + \dfrac{y}{2} > -1 \end{cases}$

In Exercises 27–30, use the graphing method to find the region that satisfies all of the inequalities of the system.

27. $\begin{cases} x \geq 0 \\ y \geq 0 \\ x + y \leq 3 \end{cases}$
28. $\begin{cases} x - y \leq 6 \\ x + 2y \leq 6 \\ x \geq 0 \end{cases}$
29. $\begin{cases} x \geq 0 \\ y \geq 0 \\ x \leq 5 \\ y \leq x \end{cases}$
30. $\begin{cases} x \geq 0 \\ y \geq 0 \\ y \leq 2 + x \\ y \geq 4x - 2 \end{cases}$

Review Exercises

In Review Exercises 1–8, simplify each expression. Write each answer without using negative exponents.

1. $x^5 x^2$
2. $\dfrac{y^6}{y^2}$
3. $(a^2)^5$
4. a^{-3}
5. $(z^3 z^{-2})^{-3}$
6. $\dfrac{t^2 t^{-1}}{t^3}$
7. $\left(\dfrac{3m^2}{n^3}\right)^4$
8. $\left(\dfrac{y^4}{y^{-2}}\right)^3$

In Review Exercises 9–10, simplify each complex fraction.

9. $\dfrac{\frac{x^2}{y^3 z}}{\frac{x}{yz^2}}$

10. $\dfrac{\frac{x}{y} + \frac{y}{x}}{\frac{y}{x} - \frac{x}{y}}$

In Review Exercises 11–12, factor each expression.

11. $15x^2 - 27x$
12. $18x^3 - 50x$

14.6 Solving Systems of Three Equations in Three Variables

A solution to the system of equations

$$\begin{cases} ax + by = e \\ cx + dy = f \end{cases}$$

is an ordered pair of real numbers (x, y) that satisfies both of the given equations simultaneously. Likewise, a solution to the system

$$\begin{cases} ax + by + cz = j \\ dx + ey + fz = k \\ gx + hy + iz = l \end{cases}$$

is an ordered triple of numbers (x, y, z) that satisfies each of the three given equations simultaneously.

The graph of an equation in three variables of the form $ax + by + cz = j$ is a flat surface called a **plane**. A system of three equations in three variables is consistent or inconsistent depending on how the three planes corresponding to the three equations intersect. The drawings in Figure 14-14 illustrate some of the possibilities.

Example 1 discusses a consistent system of three equations in three variables. Example 2 discusses a system that is inconsistent.

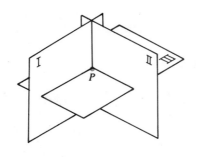
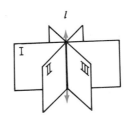
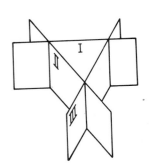

The three planes intersect at a single point P: One solution.

The three planes have a line l in common: An infinite number of solutions.

The three planes have no point in common: No solutions.

(a)   (b)   (c)

Figure 14-14

EXAMPLE 1

Solve the system $\begin{cases} 2x + y + 4z = 12 \\ x + 2y + 2z = 9 \\ 3x - 3y - 2z = 1 \end{cases}$.

Solution

We are given the following system of equations in three variables:

1. $\begin{cases} 2x + y + 4z = 12 \\ x + 2y + 2z = 9 \\ 3x - 3y - 2z = 1 \end{cases}$
2.
3.

We use the addition method to eliminate the variable z and thereby obtain a system of two equations in two variables. If Equations 2 and 3 are added, the variable z is eliminated:

2. $x + 2y + 2z = 9$
3. $3x - 3y - 2z = 1$
4. $4x - y = 10$

We now pick a different pair of equations and eliminate the variable z again. If each side of Equation 3 is multiplied by 2 and the resulting equation is added to Equation 1, the variable z is eliminated again:

1. $2x + y + 4z = 12$
     $6x - 6y - 4z = 2$
5. $8x - 5y = 14$

Equations 4 and 5 form a system of two equations in two variables:

4. $\begin{cases} 4x - y = 10 \\ 8x - 5y = 14 \end{cases}$
5.

To solve this system, we multiply Equation 4 by -5, add the resulting equation to Equation 5 to eliminate the variable y, and solve for x:

     $-20x + 5y = -50$
5. $8x - 5y = 14$
6. $-12x = -36$
            $x = 3$ Divide both sides by -12.

To find the variable y, we substitute 3 for x in an equation containing the variables x and y, such as Equation 5, and solve for y:

5.
$$8x - 5y = 14$$
$$8(3) - 5y = 14$$
$$24 - 5y = 14 \quad \text{Simplify.}$$
$$-5y = -10 \quad \text{Add } -24 \text{ to both sides.}$$
$$y = 2 \quad \text{Divide both sides by } -5.$$

To find the variable z, we substitute 3 for x and 2 for y in an equation that contains the variables x, y, and z, such as Equation 1, and solve for z:

1.
$$2x + y + 4z = 12$$
$$2(3) + 2 + 4z = 12$$
$$8 + 4z = 12 \quad \text{Simplify.}$$
$$4z = 4 \quad \text{Add } -8 \text{ to both sides.}$$
$$z = 1 \quad \text{Divide both sides by 4.}$$

The solution of the system is $(x, y, z) = (3, 2, 1)$. Verify that these values satisfy each of the equations in the system. ∎

EXAMPLE 2

Solve the system $\begin{cases} 2x + y - 3z = -3 \\ 3x - 2y + 4z = 2 \\ 4x + 2y - 6z = -7 \end{cases}$.

Solution

We are given the following system of equations in three variables:

1. $\begin{cases} 2x + y - 3z = -3 \\ 3x - 2y + 4z = 2 \\ 4x + 2y - 6z = -7 \end{cases}$
2.
3.

We begin by multiplying Equation 1 by 2 and adding the resulting equation to Equation 2 to eliminate the variable y:

$$4x + 2y - 6z = -6$$
2. $\quad 3x - 2y + 4z = 2$
$$\overline{}$$
4. $\quad 7x - 2z = -4$

We now add Equations 2 and 3 to eliminate the variable y again:

2. $\quad 3x - 2y + 4z = 2$
3. $\quad 4x + 2y - 6z = -7$
$$\overline{}$$
5. $\quad 7x - 2z = -5$

Equations 4 and 5 form the system

4. $\begin{cases} 7x - 2z = -4 \\ 7x - 2z = -5 \end{cases}$
5.

Because no values of x and z can cause $7x - 2z$ to equal both -4 and -5 at the same time, this system must be inconsistent. Thus, the original system has no solutions either; the original system is inconsistent. ∎

EXAMPLE 3

The sum of three integers is 2. The third integer is 2 greater than the second and 17 greater than the first. Find the three integers.

Solution

Let a, b, and c represent the three integers. Because their sum is 2, we have

1. $\quad a + b + c = 2$

Because the third integer is 2 greater than the second, we have $c = b + 2$, or

2. $\quad -b + c = 2$

Because the third integer is 17 greater than the first, we have $c = a + 17$ or

3. $\quad -a + c = 17$

These three equations form a system of three equations in three variables:

1.
2.
3.
$$\begin{cases} a + b + c = 2 \\ -b + c = 2 \\ -a + c = 17 \end{cases}$$

We add Equations 1 and 2 to get Equation 4:

4. $\quad a + 2c = 4$

Equations 3 and 4 form a system of two equations in two variables:

3.
4.
$$\begin{cases} -a + c = 17 \\ a + 2c = 4 \end{cases}$$

We add Equations 3 and 4 to get

$3c = 21$

$c = 7$

We substitute 7 for c in Equation 4 to find a:

4. $\quad a + 2c = 4$

$\phantom{\textbf{4.} \quad} a + 2(7) = 4$

$\phantom{\textbf{4.} \quad} a + 14 = 4 \qquad$ Simplify.

$\phantom{\textbf{4.} \quad\quad\quad\quad} a = -10 \qquad$ Add -14 to both sides.

We substitute 7 for c in Equation 2 to find b:

2. $\quad -b + c = 2$

$\phantom{\textbf{2.} \quad} -b + 7 = 2$

$\phantom{\textbf{2.} \quad\quad\quad} -b = -5 \qquad$ Add -7 to both sides.

$\phantom{\textbf{2.} \quad\quad\quad\quad} b = 5 \qquad$ Divide both sides by -1.

Thus, the three integers are -10, 5, and 7. Note that these three integers have a sum of 2, that 7 is 2 greater than 5, and that 7 is 17 greater than -10. ∎

14.6 Exercises

In Exercises 1–12, solve each system of equations. If a system of equations is inconsistent, or if the equations are dependent, so indicate.

1. $\begin{cases} x + y + z = 4 \\ 2x + y - z = 1 \\ 2x - 3y + z = 1 \end{cases}$

2. $\begin{cases} x + y + z = 4 \\ x - y + z = 2 \\ x - y - z = 0 \end{cases}$

3. $\begin{cases} 2x + 2y + 3z = 10 \\ 3x + y - z = 0 \\ x + y + 2z = 6 \end{cases}$

4. $\begin{cases} x - y + z = 4 \\ x + 2y - z = -1 \\ x + y - 3z = -2 \end{cases}$

5. $\begin{cases} x + y + 2z = 7 \\ x + 2y + z = 8 \\ 2x + y + z = 9 \end{cases}$

6. $\begin{cases} x + 2y + 2z = 10 \\ 2x + y + 2z = 9 \\ 2x + 2y + z = 11 \end{cases}$

7. $\begin{cases} 2x + y - z = 1 \\ x + 2y + 2z = 2 \\ 4x + 5y + 3z = 3 \end{cases}$

8. $\begin{cases} 4x + 3z = 4 \\ 2y - 6z = -1 \\ 8x + 4y + 3z = 9 \end{cases}$

9. $\begin{cases} 2x + 3y + 4z = 6 \\ 2x - 3y - 4z = -4 \\ 4x + 6y + 8z = 12 \end{cases}$

10. $\begin{cases} x - 3y + 4z = 2 \\ 2x + y + 2z = 3 \\ 4x - 5y + 10z = 7 \end{cases}$

11. $\begin{cases} x + \frac{1}{3}y + z = 13 \\ \frac{1}{2}x - y + \frac{1}{3}z = -2 \\ x + \frac{1}{2}y - \frac{1}{3}z = 2 \end{cases}$

12. $\begin{cases} x - \frac{1}{5}y - z = 9 \\ \frac{1}{4}x + \frac{1}{5}y - \frac{1}{2}z = 5 \\ 2x + y + \frac{1}{6}z = 12 \end{cases}$

In Exercises 13–22, solve each word problem.

13. The sum of three numbers is 18. The third number is four times the second, and the second number is 6 more than the first. Find the numbers.

14. The sum of three numbers is 48. If the first number is doubled, the sum is 60. If the second number is doubled, the sum is 63. Find the numbers.

15. Three numbers have a sum of 30. The third number is 8 less than the sum of the first and second, and the second number is one-half the sum of the first and third. Find the numbers.

16. The sum of the three angles in any triangle is 180°. In triangle ABC, angle A is 100° less than the sum of angles B and C, and angle C is 40° less than twice angle B. Find each angle.

17. A collection of 17 nickels, dimes, and quarters has a value of $1.50. There are twice as many nickels as dimes. How many of each kind are there?

18. A unit of food contains 1 gram of fat, 1 gram of carbohydrate, and 2 grams of protein. A second contains 2 grams of fat, 1 gram of carbohydrate, and 1 gram of protein. A third contains 2 grams of fat, 1 gram of carbohydrate, and 2 grams of protein. How many units of each must be used to provide exactly 11 grams of fat, 6 grams of carbohydrate, and 10 grams of protein?

19. A factory manufactures three types of footballs at a monthly cost of $2425 for 1125 footballs. The manufacturing costs for the three types of footballs are $4, $3, and $2. These footballs sell for $16, $12, and $10, respectively. How many of each type are manufactured if the monthly profit is $9275? (*Hint:* Profit = income − cost.)

20. A retailer purchased 105 radios from sources A, B, and C. Five fewer units were purchased from C than from A and B combined. If twice as many had been purchased from A, the total would have been 130. Find the number purchased from each source.

21. Tickets for a concert cost $5, $3, and $2. Twice as many $5 tickets were sold as $2 tickets. The receipts for 750 tickets were $2625. How many of each price ticket were sold?

22. The owner of a candy store wants to mix some peanuts worth $3 per pound, some cashews worth $9 per pound, and some Brazil nuts worth $9 per pound to get 50 pounds of a mixture that will sell for $6 per pound. She used 15 fewer pounds of cashews than peanuts. How many pounds of each did she use?

Review Exercises

1. If $3x - 2 = 5x + 8$, find the value of $3x - 1$.

2. If $5(y + 3) = 4y - 7$, find the value of $2y + 5$.

3. If $\dfrac{3 - x}{2} = \dfrac{2 - x}{3}$, find the value of $x + 1$.

4. If $h + 6.7 = 3(h - 2.9)$, find the value of $h + 0.3$.

5. If $2(x - 5y) = 5(x - 3y)$, find the value of x in terms of y.

6. If $2(x - 5y) = 5(x - 3y)$, find the value of y in terms of x.

7. A line with a slope of 3 and y-intercept of -5 passes through $(a, 1)$. Find a.

8. A line passes through $(2, 3)$, $(5, y)$, and $(-7, 3)$. Find y.

9. Find the slope of the line that passes through $(-3, 5)$ and $(3, -5)$.

10. Find the y-intercept of the line that passes through $(-3, 5)$ and $(3, -5)$.

Mathematics in Electronics

To accommodate monaural radios, the main programming of a stereo broadcast must contain both the left and the right channels of the stereo signal. If M represents the main programming, L represents the left stereo channel, and R the right, then $M = L + R$. The subcarrier signal, S, carries the *difference* of the left and right signals: $S = L - R$. The multiplexing circuitry electronically solves this system of equations:

$$\begin{cases} L + R = M \\ L - R = S \end{cases}$$

Electronically adding the signals corresponds to adding the equations, to give $M + S = 2L$, from which the left signal can be recovered. The difference of the signals is $M - S = 2R$, which provides the right signal. The left and right signals are amplified separately and sent to separate speakers.

Chapter Summary

Key Words

addition method (14.3)
consistent system of equations (14.1)
dependent equations (14.1)
graphing method (14.1)
inconsistent system of equations (14.1)
independent equations (14.1)
plane (14.6)
simultaneous solution (14.1)
solution of a system of equations (14.1)
substitution method (14.2)
system of equations (14.1)
system of inequalities (14.5)

Key Ideas

(14.1) To solve a system of equations graphically, carefully graph each equation of the system. If the lines intersect, the coordinates of the point of intersection give the solution of the system.

(14.2) To solve a system of equations by substitution, solve one of the equations of the system for one of its variables, substitute the resulting expression into the other equation, and solve for the other variable.

(14.3) To solve a system of equations by addition, first multiply one or both of the equations by suitable constants, if necessary, to eliminate one of the variables when the equations are added. The equation that results can be solved for its single variable. Then substitute the value obtained back into one of the original equations and solve for the other variable.

(14.4) Systems of equations are useful in solving many different types of word problems.

(14.5) To graph a system of inequalities, first graph the individual inequalities of the system. The final solution, if one exists, is that region where all the individual graphs intersect.

(14.6) To solve a system of three equations in three variables, eliminate one variable from two of the equations. From another pair, eliminate the same variable. Solve the resulting system of two equations in two variables using the methods of Sections 14.2 or 14.3.

Chapter 14 Review Exercises

[14.1] In Review Exercises 1–4, determine whether the given ordered pair is a solution of the given system of equations.

1. $(1, 5)$ $\begin{cases} 3x - y = -2 \\ 2x + 3y = 17 \end{cases}$

2. $(-2, 4)$ $\begin{cases} 5x + 3y = 2 \\ -3x + 2y = 16 \end{cases}$

3. $\left(14, \dfrac{1}{2}\right)$ $\begin{cases} 2x + 4y = 30 \\ \dfrac{x}{4} - y = 3 \end{cases}$

4. $\left(\dfrac{7}{2}, -\dfrac{2}{3}\right)$ $\begin{cases} 4x - 6y = 18 \\ \dfrac{x}{3} + \dfrac{y}{2} = \dfrac{5}{6} \end{cases}$

In Review Exercises 5–8, use the graphing method to solve each system. If the system is inconsistent, or if the equations are dependent, so indicate.

5. $\begin{cases} x + y = 7 \\ 2x - y = 5 \end{cases}$

6. $\begin{cases} \dfrac{x}{3} + \dfrac{y}{5} = -1 \\ x - 3y = -3 \end{cases}$

7. $\begin{cases} 3x + 6y = 6 \\ x + 2y = 2 \end{cases}$

8. $\begin{cases} 6x + 3y = 12 \\ 2x + y = 2 \end{cases}$

[14.2] In Review Exercises 9–12, use the substitution method to solve each system of equations.

9. $\begin{cases} x = 3y + 5 \\ 5x - 4y = 3 \end{cases}$

10. $\begin{cases} 3x - \dfrac{2y}{5} = 2(x - 2) \\ 2x - 3 = 3 - 2y \end{cases}$

11. $\begin{cases} 8x + 5y = 3 \\ 5x - 8y = 13 \end{cases}$

12. $\begin{cases} 6(x + 2) = y - 1 \\ 5(y - 1) = x + 2 \end{cases}$

[14.3] In Review Exercises 13–20, use the addition method to solve each system of equations. If the equations of a system are dependent, or if the system is inconsistent, so indicate.

13. $\begin{cases} 2x + y = 1 \\ 5x - y = 20 \end{cases}$

14. $\begin{cases} x + 8y = 7 \\ x - 4y = 1 \end{cases}$

15. $\begin{cases} 5x + y = 2 \\ 3x + 2y = 11 \end{cases}$

16. $\begin{cases} x + y = 3 \\ 3x = 2 - y \end{cases}$

17. $\begin{cases} 11x + 3y = 27 \\ 8x + 4y = 36 \end{cases}$

18. $\begin{cases} 9x + 3y = 5 \\ 3x = 4 - y \end{cases}$

19. $\begin{cases} 9x + 3y = 5 \\ 3x + y = \dfrac{5}{3} \end{cases}$

20. $\begin{cases} \dfrac{x}{3} + \dfrac{y + 2}{2} = 1 \\ \dfrac{x + 8}{8} + \dfrac{y - 3}{3} = 0 \end{cases}$

[14.4] In Review Exercises 21–26, use a system of equations to solve each word problem.

21. One number is 5 times another, and their sum is 18. Find the numbers.

22. The length of a rectangle is 3 times its width. The perimeter is 24 feet. Find the dimensions of the rectangle.

23. A grapefruit costs 15 cents more than an orange. Together, they cost 85 cents. Find the cost of each.

24. A man's electric bill for January was $23 less than his gas bill. These two utilities cost him a total of $109. Find the amount of each bill.

25. Two gallons of milk and 3 dozen eggs cost $6.80. Three gallons of milk and 2 dozen eggs cost $7.35. How much does each gallon of milk and each dozen eggs cost?

26. Carlos invested part of $3000 in a 10% certificate account and the rest in a 6% passbook account. The total annual interest from both accounts is $270. How much did he invest in each account?

[14.5] In Review Exercises 27–30, solve each system of inequalities.

27. $\begin{cases} 5x + 3y < 15 \\ 3x - y > 3 \end{cases}$

28. $\begin{cases} 5x - 3y \geq 5 \\ 3x + 2y \geq 3 \end{cases}$

29. $\begin{cases} x \geq 3y \\ y < 3x \end{cases}$

30. $\begin{cases} x > 0 \\ x \leq 3 \end{cases}$

[14.6] In Review Exercises 31–34, solve each system of equations.

31. $\begin{cases} x - 2y - 2z = 7 \\ 2x - y + z = -1 \\ x + 5y + 3z = -8 \end{cases}$

32. $\begin{cases} x + 2z = 4 \\ y - 3z = -4 \\ 2x + 3y + z = 2 \end{cases}$

33. $\begin{cases} x + y - z = 1 \\ 3x - y + z = 11 \\ x - 2y + 2z = 7 \end{cases}$

34. $\begin{cases} 2x - y + z = 1 \\ x + 2y + 2z = 2 \\ x - 3y + 2z = 2 \end{cases}$

Chapter 14 Test

In Problems 1–2, determine whether the given ordered pair is a solution of the given system.

1. $(2, -3)$; $\begin{cases} 3x - 2y = 12 \\ 2x + 3y = -5 \end{cases}$

2. $(-2, -1)$; $\begin{cases} 4x + y = -9 \\ 2x - 3y = -7 \end{cases}$

In Problems 3–4, solve each system by graphing.

3. $\begin{cases} 3x + y = 7 \\ x - 2y = 0 \end{cases}$

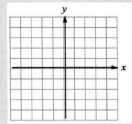

4. $\begin{cases} x + \dfrac{y}{2} = 1 \\ y = 1 - 3x \end{cases}$

In Problems 5–6, solve each system by substitution.

5. $\begin{cases} y = x - 1 \\ 2x + y = -7 \end{cases}$

6. $\begin{cases} \dfrac{x}{6} + \dfrac{y}{10} = 3 \\ \dfrac{5x}{16} - \dfrac{3y}{16} = \dfrac{15}{8} \end{cases}$

In Problems 7–8, solve each system by addition.

7. $\begin{cases} 3x - y = 2 \\ 2x + y = 8 \end{cases}$

8. $\begin{cases} 4x + 3 = -3y \\ \dfrac{-x}{7} + \dfrac{4y}{21} = 1 \end{cases}$

In Problems 9–10, state whether each system is consistent or inconsistent.

9. $\begin{cases} 2x + 3(y - 2) = 0 \\ -3y = 2(x - 4) \end{cases}$ _____

10. $\begin{cases} \dfrac{x}{3} + y - 4 = 0 \\ -3y = x - 12 \end{cases}$ _____

In Problems 11–12, use a system of equations in two variables to solve each word problem.

11. The sum of two numbers is -18. One number is 2 greater than 3 times the other. Find the product of the numbers. _____

12. A woman invested some money at 8% and some at 9%. The interest on the combined investment of $10,000 was $840. How much was invested at 9%? _____

In Problems 13–14, solve each system of inequalities by graphing.

13. $\begin{cases} x + y < 3 \\ x - y > 1 \end{cases}$

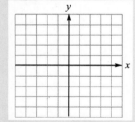

14. $\begin{cases} 2x + 3y \leq 6 \\ x \geq 2 \end{cases}$

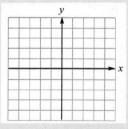

In Problems 15–16, solve each system of equations.

15. $\begin{cases} 2x + y - 3z = 5 \\ x + 2y - z = 7 \\ x + y + 5z = 4 \end{cases}$ _____

16. $\begin{cases} 3x + 2y - 2z = 1 \\ x + 2y - 3z = 5 \\ -x + y - 3z = 7 \end{cases}$ _____

15 Roots and Radical Expressions

Mathematics in Racing

The America's Cup is more than a race between sleek yachts and skillful crews. Each boat is the result of the design efforts of many scientists, including naval architects, oceanographers, aerospace engineers, and computer experts.

The new International America's Cup Class (IACC) yachts are longer, lighter, and faster than those raced previously. The 200-page document outlining the rules contains a formula that limits the yacht's length L, its sail area A, and its displacement D:

$$\frac{L + 1.25\sqrt{A} - 9.8\sqrt[3]{D}}{0.388} \leq 42$$

In 1988, an America's Cup entry, New Zealand's 20.95-meter sloop *Challenge*, had a 277.3-square-meter sail and a 17.56-cubic-meter displacement. Did the *Challenge* satisfy the IACC rules?

In this chapter, we will study *roots* of numbers, denoted by such symbols as $\sqrt{A}$ and $\sqrt[3]{D}$, and learn how to perform the calculations of the IACC formula.

The product $b \cdot b$ is called the **square of b** and is usually denoted as b^2. For example:

The square of 3 is 9, because $3^2 = 9$.
The square of -3 is 9, because $(-3)^2 = 9$.
The square of 12 is 144, because $12^2 = 144$.
The square of -12 is 144, because $(-12)^2 = 144$.
The square of 0 is 0, because $0^2 = 0$.

We now reverse the squaring process and find **square roots** of numbers.

15.1 Radicals

- Using Calculators to Find Square Roots
- Using Tables to Find Square Roots
- Perfect Integer Squares
- Imaginary Numbers
- Cube Roots

The number 4 is called a **square root of 16** because $4^2 = 16$. Other examples are:

3 is a square root of 9, because $3^2 = 9$.
-3 is a square root of 9, because $(-3)^2 = 9$.
12 is a square root of 144, because $12^2 = 144$.
-12 is a square root of 144, because $(-12)^2 = 144$.
0 is a square root of 0, because $0^2 = 0$.

In general, a number b is a **square root of a** if the square of b is equal to a.

> **Definition:** The number b is a **square root of a** if $b^2 = a$.

Positive numbers have two square roots, one that is positive and one that is negative. Zero is the only number with a single square root, which is 0. The symbol $\sqrt{}$, called a **radical sign**, represents the *positive* square root of a number. The expression under a radical sign is called a **radicand**.

> **Definition:** If a is positive, the **principal square root of a**, denoted by $\sqrt{a}$, is the positive square root of a.
>
> The principal square root of 0 is 0: $\sqrt{0} = 0$.

The principal square root of a positive number is positive also. Although 8 and -8 are both square roots of 64, only 8 is the principal square root. The symbol $\sqrt{64}$ represents 8. To designate -8, we place a $-$ sign in front of the radical. Thus,

$$\sqrt{64} = 8 \quad \text{and} \quad -\sqrt{64} = -8$$

EXAMPLE 1

a. $\sqrt{0} = 0$ **b.** $\sqrt{1} = 1$
c. $\sqrt{49} = 7$ **d.** $\sqrt{121} = 11$
e. $-\sqrt{4} = -2$ **f.** $-\sqrt{81} = -9$
g. $\sqrt{225} = 15$ **h.** $\sqrt{169} = 13$
i. $-\sqrt{625} = -25$

Work Progress Check 1.

Square roots of many numbers are hard to compute by hand. However, we can approximate the value of a number such as $\sqrt{7}$ with a calculator or with a table of square roots.

Using Calculators to Find Square Roots

To find the principal square root of 7 with a calculator, we enter 7 and press the $\boxed{\sqrt{}}$ key. The approximate value of $\sqrt{7}$ will appear on the calculator's display.

$$\sqrt{7} \approx 2.6457513 \qquad \text{Read} \approx \text{as "is approximately equal to."}$$

Progress Check 1

Find each square root:

a. $\sqrt{900}$.

b. $-\sqrt{256}$.

1. **a.** 30 **b.** -16

Using Tables to Find Square Roots

To find the principal square root of 7, we use the table of square roots in Appendix I. In the left column, headed by n, we locate 7. The column headed $\sqrt{n}$ contains the approximate value of $\sqrt{7}$.

$$\sqrt{7} \approx 2.646$$

Perfect Integer Squares

Any number that is the square of an integer is called a **perfect integer square**:

1, 4, 9, 16, 25, 36, 49, 64, 81, 100, 121, 144, 169, 196, 225, 256, 289, 324, 361, 400

are the first twenty perfect integer squares. The square roots of perfect square integers are **rational numbers**. For example, $\sqrt{256} = 16$, and 16 is a rational number.

Imaginary Numbers

Some numbers, such as $\sqrt{7}$, are not rational. Numbers that are not square roots of perfect integer squares are called **irrational numbers**. Recall that the union of the set of rational numbers and the set of irrational numbers is the set of **real numbers**.

Square roots with negative radicands, such as $\sqrt{-4}$, are not real numbers, because the square of no real number is negative. The number $\sqrt{-4}$ is an example of an **imaginary number**. We emphasize this important fact:

> **The square root of a negative number is an imaginary number.**

In this chapter, we assume that *all radicands under the square root symbol are positive numbers or 0*. Thus, all square roots will be real numbers.

Cube Roots

The **cube root of a** is any number whose *cube is a*.

> **Definition:** The **cube root of a** is denoted as $\sqrt[3]{a}$, and
> $$\sqrt[3]{a} = b \quad \text{if} \quad b^3 = a$$

EXAMPLE 2

a. $\sqrt[3]{8} = 2$ because $2^3 = 8$.   **b.** $\sqrt[3]{27} = 3$ because $3^3 = 27$.

c. $\sqrt[3]{-8} = -2$ because $(-2)^3 = -8$. The cube root of a negative number is a negative number. However, the *square root* of a negative number is imaginary.

EXAMPLE 3

a. $\sqrt[3]{0} = 0$
b. $\sqrt[3]{-1} = -1$
c. $\sqrt[3]{64} = 4$
d. $\sqrt[3]{-64} = -4$
e. $\sqrt[3]{125} = 5$
f. $\sqrt[3]{-125} = -5$

Work Progress Check 2.

Progress Check 2

Find each cube root:

a. $\sqrt[3]{1000}$.

b. $\sqrt[3]{-216}$.

There are also fourth roots, fifth roots, sixth roots, and so on. In general,

> **Definition:** The **nth root of a** is denoted by $\sqrt[n]{a}$, and
> $$\sqrt[n]{a} = b \quad \text{if} \quad b^n = a$$
> The number n is called the **index** of the radical. If n is even, then both a and $\sqrt[n]{a}$ must be either positive numbers or 0.

In the square root symbol $\sqrt{}$, the unwritten index is understood to be 2.

EXAMPLE 4

a. $\sqrt[4]{81} = 3$ because $3^4 = 81$.

b. $\sqrt[5]{32} = 2$ because $2^5 = 32$.

c. $\sqrt[5]{-32} = -2$ because $(-2)^5 = -32$.

d. $\sqrt[4]{-81}$ is an imaginary number because no number raised to the fourth power can equal -81.

We can also find the square roots of certain quantities that contain variables, provided we know that these variables represent positive numbers or 0.

EXAMPLE 5

Assume that each variable represents a positive number and find each root.

a. $\sqrt{x^2} = x$ because $x^2 = x^2$.

b. $\sqrt{x^4} = x^2$ because $(x^2)^2 = x^4$.

c. $\sqrt{x^4y^2} = x^2y$ because $(x^2y)^2 = x^4y^2$.

d. $\sqrt[3]{x^6y^3} = x^2y$ because $(x^2y)^3 = x^6y^3$.

e. $\sqrt[4]{x^{12}y^8} = x^3y^2$ because $(x^3y^2)^4 = x^{12}y^8$.

Work Progress Check 3.

Progress Check 3

Find each root:

a. $\sqrt{x^6y^4}$.

b. $\sqrt[3]{x^9y^{12}}$.

---

2. a. 10 b. -6

3. a. x^3y^2 b. x^3y^4

Progress Check Answers

EXAMPLE 6

Assume that each variable represents a positive number and find each root.

a. $\sqrt{(x+1)^2} = x+1$ because $(x+1)^2 = (x+1)^2$.

b. $\sqrt{x^2 + 4x + 4} = \sqrt{(x+2)^2}$ Factor $x^2 + 4x + 4$.
$= x + 2$

c. $\sqrt{x^2 + 2xy + y^2} = \sqrt{(x+y)^2}$ Factor $x^2 + 2xy + y^2$.
$= x + y$

15.1 Exercises

In Exercises 1–24, find the value of each expression.

1. $\sqrt{9}$
2. $\sqrt{16}$
3. $\sqrt{49}$
4. $\sqrt{100}$

5. $\sqrt{36}$
6. $\sqrt{4}$
7. $\sqrt{81}$
8. $\sqrt{121}$

9. $-\sqrt{25}$
10. $-\sqrt{49}$
11. $-\sqrt{81}$
12. $-\sqrt{36}$

13. $\sqrt{196}$
14. $\sqrt{169}$
15. $\sqrt{256}$
16. $\sqrt{225}$

17. $-\sqrt{289}$
18. $\sqrt{400}$
19. $\sqrt{10,000}$
20. $-\sqrt{2500}$

21. $\sqrt{324}$
22. $-\sqrt{625}$
23. $-\sqrt{3600}$
24. $\sqrt{1600}$

 In Exercises 25–36, use a calculator or a table of square roots to find each square root to three decimal places.

25. $\sqrt{2}$
26. $\sqrt{3}$
27. $\sqrt{5}$
28. $\sqrt{10}$

29. $\sqrt{6}$
30. $\sqrt{8}$
31. $\sqrt{11}$
32. $\sqrt{17}$

33. $\sqrt{23}$
34. $\sqrt{53}$
35. $\sqrt{95}$
36. $\sqrt{99}$

In Exercises 37–48, use a calculator to find each square root to three decimal places.

37. $\sqrt{6428}$
38. $\sqrt{4444}$
39. $-\sqrt{9876}$
40. $-\sqrt{3619}$

41. $\sqrt{21.35}$
42. $\sqrt{13.78}$
43. $\sqrt{0.3588}$
44. $\sqrt{0.9999}$

45. $\sqrt{0.9925}$
46. $\sqrt{0.12345}$
47. $-\sqrt{0.8372}$
48. $-\sqrt{0.4279}$

In Exercises 49–56, indicate whether each number is rational, irrational, or imaginary.

49. $\sqrt{9}$
50. $\sqrt{17}$
51. $\sqrt{49}$
52. $\sqrt{-49}$

53. $-\sqrt{5}$
54. $\sqrt{0}$
55. $\sqrt{-100}$
56. $-\sqrt{225}$

In Exercises 57–72, find the value of each expression.

57. $\sqrt[3]{1}$
58. $\sqrt[3]{8}$
59. $\sqrt[3]{27}$
60. $\sqrt[3]{0}$

61. $\sqrt[3]{-8}$
62. $\sqrt[3]{-1}$
63. $\sqrt[3]{-64}$
64. $\sqrt[3]{-27}$

65. $\sqrt[3]{125}$
66. $\sqrt[3]{1000}$
67. $-\sqrt[3]{-1}$
68. $-\sqrt[3]{-27}$

69. $-\sqrt[3]{64}$
70. $-\sqrt[3]{343}$
71. $\sqrt[3]{729}$
72. $\sqrt[3]{512}$

In Exercises 73–80, find the value of each expression.

73. $\sqrt[4]{16}$
74. $\sqrt[4]{81}$
75. $-\sqrt[5]{32}$
76. $-\sqrt[5]{243}$

77. $\sqrt[6]{1}$
78. $\sqrt[6]{0}$
79. $\sqrt[5]{-32}$
80. $\sqrt[7]{-1}$

In Exercises 81–104, write each expression without using a radical sign. Assume that all variables represent positive numbers.

81. $\sqrt{x^2y^2}$
82. $\sqrt{x^2y^4}$
83. $\sqrt{x^4z^4}$
84. $\sqrt{y^6z^8}$

85. $-\sqrt{x^4y^2}$
86. $-\sqrt{x^6y^4}$
87. $\sqrt{4z^2}$
88. $\sqrt{9t^6}$

89. $-\sqrt{9x^4y^2}$
90. $-\sqrt{16x^2y^4}$
91. $\sqrt{x^2y^2z^2}$
92. $\sqrt{x^4y^6z^8}$

93. $-\sqrt{x^2y^2z^4}$
94. $-\sqrt{a^8b^6c^2}$
95. $-\sqrt{25x^4z^{12}}$
96. $-\sqrt{100a^6b^4}$

97. $\sqrt{36z^{36}}$
98. $\sqrt{64y^{64}}$
99. $-\sqrt{2^4z^2}$
100. $-\sqrt{3^6x^8y^2}$

101. $\sqrt[3]{27y^3z^6}$
102. $\sqrt[3]{64x^3y^6z^9}$
103. $\sqrt[3]{-8p^6q^3}$
104. $\sqrt[3]{-r^{12}s^3t^6}$

In Exercises 105–114, write each expression without using a radical sign. Assume that all variables represent positive numbers.

105. $\sqrt{(x+9)^2}$
106. $\sqrt{(x+6)^2}$
107. $\sqrt{(x+11)^2}$
108. $\sqrt{(x+13)^2}$

109. $\sqrt{x^2+6x+9}$
110. $\sqrt{x^2+10x+25}$
111. $\sqrt{x^2+20x+100}$
112. $\sqrt{x^2+16x+64}$

113. $\sqrt{x^2+14xy+49y^2}$
114. $\sqrt{x^2+50xy+625y^2}$

Review Exercises

In Review Exercises 1–4, let $P(x) = -2x^2 + 3x - 5$. Evaluate each quantity.

1. $P(0)$
2. $P(4)$
3. $P(-1)$
4. $P(2t)$

In Review Exercises 5–8, let $y = f(x) = 2x^2 - x - 1$. Evaluate each quantity.

5. $f(0)$
6. $f(2)$
7. $f(-2)$
8. $f(-t)$

9. Express the words "y varies directly with x" as an equation.

10. Express the words "y varies inversely with x" as an equation.

11. The number of feet, s, that a body falls from rest in t seconds varies directly with the square of t. If the body falls 64 feet in 2 seconds, how long will it take for it to fall 256 feet?

12. For a constant voltage, the resistance, R, in a circuit varies inversely with the amperage, I. If the resistance is 3 ohms when the current is 6 amperes, what will the amperage be when the resistance is 9 ohms?

15.2 Simplifying Radical Expressions

- The Multiplication Property of Radicals
- Simplifying Radicals
- The Division Property of Radicals
- Simplifying Cube Roots

The Multiplication Property of Radicals

We introduce the multiplication property of radicals with the following examples:

$$\sqrt{4 \cdot 25} = \sqrt{100} \qquad \sqrt{4}\sqrt{25} = 2 \cdot 5$$
$$= 10 \qquad \qquad\qquad = 10$$

In each case the answer is 10. Thus, $\sqrt{4 \cdot 25} = \sqrt{4} \cdot \sqrt{25}$. Likewise,

$$\sqrt{9 \cdot 16} = \sqrt{144} \qquad \sqrt{9}\sqrt{16} = 3 \cdot 4$$
$$= 12 \qquad \qquad\qquad = 12$$

In each case, the answer is 12. Thus, $\sqrt{9 \cdot 16} = \sqrt{9} \cdot \sqrt{16}$. These results suggest the **multiplication property of radicals**.

> **The Multiplication Property of Radicals:** If $a \geq 0$ and $b \geq 0$, then
>
> $$\sqrt{ab} = \sqrt{a}\sqrt{b}$$

The multiplication property of radicals points out that the *square root of the product of two nonnegative numbers is equal to the product of their square roots*.

Simplifying Radicals

The multiplication property of radicals can be used to simplify radicals such as $\sqrt{12}$:

$$\sqrt{12} = \sqrt{4 \cdot 3} \qquad \text{Factor 12 as } 4 \cdot 3.$$
$$= \sqrt{4} \cdot \sqrt{3} \qquad \text{Use the multiplication property of radicals.}$$
$$= 2\sqrt{3} \qquad \text{Write } \sqrt{4} \text{ as 2.}$$

To simplify radicals, we must remember the perfect square integers:

1, 4, 9, 16, 25, 36, 49, 64, 81, 100, 121, 144, 169, 196, 225, 256, 289, 324, 361

Expressions with variables such as x^4y^2 are also perfect squares, because they can be written as the square of a quantity:

$$x^4y^2 = (x^2y)^2$$

A square root radical is in **simplified form** when the radicand has no perfect square factors. The radical $\sqrt{10}$ is in simplified form because 10 factors as $2 \cdot 5$, and neither 2 nor 5 is a perfect square. However, $\sqrt{12}$ is not in simplified form, because 12 factors as $4 \cdot 3$, and 4 is a perfect square.

EXAMPLE 1

Simplify $\sqrt{72x^3}$. Assume that $x \geq 0$.

Solution

We factor $72x^3$ into two factors, one of which is the greatest perfect square that divides $72x^3$. Because the greatest perfect square that divides $72x^3$ is $36x^2$, such a factorization is $72x^3 = 36x^2 \cdot 2x$.

$\sqrt{72x^3} = \sqrt{36x^2}\sqrt{2x}$ Use the multiplication property of radicals to write $\sqrt{72x^3}$ as $\sqrt{36x^2}\sqrt{2x}$.

$= 6x\sqrt{2x}$ $\sqrt{36x^2} = 6x$ ∎

EXAMPLE 2

Simplify $\sqrt{45x^2y^3}$. Assume that $x \geq 0$ and $y \geq 0$.

Solution

We look for the greatest perfect square that divides $45x^2y^3$. Because

9 is the greatest perfect square that divides 45,
x^2 is the greatest perfect square that divides x^2, and
y^2 is the greatest perfect square that divides y^3,

the quantity $9x^2y^2$ is the greatest perfect square that divides $45x^2y^3$.

$\sqrt{45x^2y^3} = \sqrt{9x^2y^2}\sqrt{5y}$ Use the multiplication property of radicals to write $\sqrt{45x^2y^3}$ as $\sqrt{9x^2y^2}\sqrt{5y}$.

$= 3xy\sqrt{5y}$ $\sqrt{9x^2y^2} = 3xy$

Work Progress Check 4.   ∎

EXAMPLE 3

Simplify $3a\sqrt{288a^5b^7}$. Assume that $a \geq 0$ and $b \geq 0$.

Progress Check 4

Simplify:

a. $\sqrt{40x^3y^2}$.

b. $\sqrt{98x^5y^3}$.

4. **a.** $2xy\sqrt{10x}$ **b.** $7x^2y\sqrt{2xy}$

Progress Check Answers

Solution

We look for the greatest perfect square that divides $288a^5b^7$. Because

144 is the greatest perfect square that divides 288,
a^4 is the greatest perfect square that divides a^5, and
b^6 is the greatest perfect square that divides b^7,

the quantity $144a^4b^6$ is the greatest perfect square that divides $288a^5b^7$.

$$3a\sqrt{288a^5b^7} = 3a\sqrt{144a^4b^6}\sqrt{2ab}$$

Use the multiplication property of radicals to write $\sqrt{288a^5b^7}$ as $\sqrt{144a^4b^6}\sqrt{2ab}$.

$$= 3a\left(12a^2b^3\sqrt{2ab}\right)$$

$\sqrt{144a^4b^6} = 12a^2b^3$

$$= 36a^3b^3\sqrt{2ab}$$

■

The Division Property of Radicals

To find the division property of radicals, we consider these examples.

$$\sqrt{\frac{100}{25}} = \sqrt{4} \qquad \frac{\sqrt{100}}{\sqrt{25}} = \frac{10}{5}$$
$$= 2 \qquad\qquad\qquad = 2$$

In each case, the answer is 2. Thus, $\sqrt{\dfrac{100}{25}} = \dfrac{\sqrt{100}}{\sqrt{25}}$. Likewise,

$$\sqrt{\frac{36}{4}} = \sqrt{9} \qquad \frac{\sqrt{36}}{\sqrt{4}} = \frac{6}{2}$$
$$= 3 \qquad\qquad\qquad = 3$$

In each case, the answer is 3. Thus, $\sqrt{\dfrac{36}{4}} = \dfrac{\sqrt{36}}{\sqrt{4}}$. These results suggest the **division property of radicals**.

The Division Property of Radicals: If a and b are nonnegative numbers and $b \neq 0$, then

$$\sqrt{\frac{a}{b}} = \frac{\sqrt{a}}{\sqrt{b}}$$

The division property of radicals points out that the *square root of the quotient of two numbers is the quotient of their square roots.*

The division property of radicals can be used to simplify radicals such as $\sqrt{\dfrac{59}{49}}$:

$$\sqrt{\frac{59}{49}} = \frac{\sqrt{59}}{\sqrt{49}}$$
$$= \frac{\sqrt{59}}{7} \qquad \sqrt{49} = 7$$

EXAMPLE 4

Simplify $\sqrt{\dfrac{108}{25}}$.

Solution

$\sqrt{\dfrac{108}{25}} = \dfrac{\sqrt{108}}{\sqrt{25}}$ Use the division property of radicals.

$= \dfrac{\sqrt{36 \cdot 3}}{5}$ Factor 108 using the factorization involving 36, the largest perfect square factor of 108, and write $\sqrt{25}$ as 5.

$= \dfrac{\sqrt{36}\sqrt{3}}{5}$ Use the multiplication property of radicals.

$= \dfrac{6\sqrt{3}}{5}$ $\sqrt{36} = 6$

Progress Check 5

Simplify: $\sqrt{\dfrac{72}{49}}$.

Work Progress Check 5.

EXAMPLE 5

Simplify $\sqrt{\dfrac{44x^3}{9xy^2}}$. Assume that $x > 0$ and $y > 0$.

Solution

$\sqrt{\dfrac{44x^3}{9xy^2}} = \sqrt{\dfrac{44x^2}{9y^2}}$ Simplify the fraction by dividing out the common factor of x.

$= \dfrac{\sqrt{44x^2}}{\sqrt{9y^2}}$ Use the division property of radicals.

$= \dfrac{\sqrt{4x^2}\sqrt{11}}{\sqrt{9y^2}}$ $4x^2$ is the greatest perfect square factor of $44x^2$.

$= \dfrac{2x\sqrt{11}}{3y}$ $\sqrt{4x^2} = 2x$ and $\sqrt{9y^2} = 3y$.

Simplifying Cube Roots

The multiplication and division properties of radicals extend to cube roots and higher. To simplify a cube root, recall the first ten perfect cube integers:

1, 8, 27, 64, 125, 216, 343, 512, 729, 1000

Expressions containing variables such as x^6y^3 are also perfect cubes, because they can be written as the cube of a quantity:

$x^6y^3 = (x^2y)^3$

EXAMPLE 6

Simplify   **a.** $\sqrt[3]{16x^3y^4}$ and   **b.** $\sqrt[3]{\dfrac{64n^4}{27m^3}}$.

Progress Check Answers

5. $\dfrac{6\sqrt{2}}{7}$

Solution

a. Try to factor $16x^3y^4$ into two factors, one of which is the greatest perfect cube that divides $16x^3y^4$. Such a factorization is $16x^3y^4 = 8x^3y^3 \cdot 2y$.

$\sqrt[3]{16x^3y^4} = \sqrt[3]{8x^3y^3} \sqrt[3]{2y}$ Use the multiplication property of radicals and write $\sqrt[3]{16x^3y^4}$ as $\sqrt[3]{8x^3y^3}\sqrt[3]{2y}$.

$= 2xy\sqrt[3]{2y}$ $\sqrt[3]{8x^3y^3} = 2xy$

b. $\sqrt[3]{\dfrac{64n^4}{27m^3}} = \dfrac{\sqrt[3]{64n^4}}{\sqrt[3]{27m^3}}$ Use the division property of radicals.

$= \dfrac{\sqrt[3]{64n^3}\sqrt[3]{n}}{3m}$ Use the multiplication property of radicals and write $\sqrt[3]{27m^3}$ as $3m$.

$= \dfrac{4n\sqrt[3]{n}}{3m}$

Work Progress Check 6.

Progress Check 6

Simplify:

a. $\sqrt[3]{24x^4y^3}$.

b. $\sqrt[3]{\dfrac{54x^3}{8y^3}}$.

Warning: It is important to note that $\sqrt{a+b} \neq \sqrt{a} + \sqrt{b}$.

To illustrate the warning, we consider this correct simplification:

$\sqrt{9+16} = \sqrt{25} = 5$

However, it is incorrect to write

$\sqrt{9+16} = \sqrt{9} + \sqrt{16} = 3 + 4 = 7$

Likewise, $\sqrt{25-16} = \sqrt{9} = 3$. It is incorrect to write

$\sqrt{25-16} = \sqrt{25} - \sqrt{16} = 5 - 4 = 1$

6. a. $2xy\sqrt[3]{3x}$ **b.** $\dfrac{3x\sqrt[3]{2}}{2y}$

Progress Check Answers

15.2 Exercises

In Exercises 1–48, simplify each radical. Assume that all variables represent positive numbers.

1. $\sqrt{20}$
2. $\sqrt{18}$
3. $\sqrt{50}$
4. $\sqrt{75}$

5. $\sqrt{45}$
6. $\sqrt{54}$
7. $\sqrt{98}$
8. $\sqrt{27}$

9. $\sqrt{48}$
10. $\sqrt{128}$
11. $\sqrt{200}$
12. $\sqrt{300}$

13. $\sqrt{192}$
14. $\sqrt{250}$
15. $\sqrt{88}$
16. $\sqrt{275}$

17. $\sqrt{324}$
18. $\sqrt{405}$
19. $\sqrt{147}$
20. $\sqrt{722}$

21. $\sqrt{180}$
22. $\sqrt{320}$
23. $\sqrt{432}$
24. $\sqrt{720}$

25. $4\sqrt{288}$
26. $2\sqrt{800}$
27. $-7\sqrt{1000}$
28. $-3\sqrt{252}$

29. $2\sqrt{245}$
30. $3\sqrt{196}$
31. $-5\sqrt{162}$
32. $-4\sqrt{243}$

33. $\sqrt{25x}$
34. $\sqrt{36y}$
35. $\sqrt{a^2b}$
36. $\sqrt{rs^2}$
37. $\sqrt{9x^2y}$
38. $\sqrt{16xy^2}$
39. $8x^2y\sqrt{50x^2y^2}$
40. $3x^5y\sqrt{75x^3y^2}$
41. $12x\sqrt{16x^2y^3}$
42. $-4x^5y^3\sqrt{36x^3y^3}$
43. $-3xyz\sqrt{18x^3y^5}$
44. $15xy^2\sqrt{72x^2y^3}$
45. $\frac{3}{4}\sqrt{192a^3b^5}$
46. $-\frac{2}{9}\sqrt{162r^3s^3t}$
47. $-\frac{2}{5}\sqrt{80mn^2}$
48. $\frac{5}{6}\sqrt{180ab^2c}$

In Exercises 49–64, write each quotient as the quotient of two radicals and simplify.

49. $\sqrt{\frac{25}{9}}$
50. $\sqrt{\frac{36}{49}}$
51. $\sqrt{\frac{81}{64}}$
52. $\sqrt{\frac{121}{144}}$
53. $\sqrt{\frac{26}{25}}$
54. $\sqrt{\frac{17}{169}}$
55. $\sqrt{\frac{20}{49}}$
56. $\sqrt{\frac{50}{9}}$
57. $\sqrt{\frac{48}{81}}$
58. $\sqrt{\frac{27}{64}}$
59. $\sqrt{\frac{32}{25}}$
60. $\sqrt{\frac{75}{16}}$
61. $\sqrt{\frac{125}{121}}$
62. $\sqrt{\frac{250}{49}}$
63. $\sqrt{\frac{245}{36}}$
64. $\sqrt{\frac{500}{81}}$

In Exercises 65–72, simplify each expression. Assume that all variables represent positive numbers.

65. $\sqrt{\frac{72x^3}{y^2}}$
66. $\sqrt{\frac{108a^3b^2}{c^2d^4}}$
67. $\sqrt{\frac{125m^2n^5}{64n}}$
68. $\sqrt{\frac{72p^5q^7}{16pq^3}}$
69. $\sqrt{\frac{128m^3n^5}{36mn^7}}$
70. $\sqrt{\frac{75p^3q^2}{9p^5q^4}}$
71. $\sqrt{\frac{12r^7s^6t}{81r^5s^2t}}$
72. $\sqrt{\frac{36m^2n^9}{100mn^3}}$

In Exercises 73–86, simplify each cube root.

73. $\sqrt[3]{8x^3}$
74. $\sqrt[3]{27x^3y^3}$
75. $\sqrt[3]{-64x^5}$
76. $\sqrt[3]{-16x^4y^3}$
77. $\sqrt[3]{54x^3y^4z^6}$
78. $\sqrt[3]{-24x^5y^5z^4}$
79. $\sqrt[3]{-81x^2y^3z^4}$
80. $\sqrt[3]{1600xy^2z^3}$
81. $\sqrt[3]{\frac{27m^3}{8n^6}}$
82. $\sqrt[3]{\frac{125t^9}{27s^6}}$
83. $\sqrt[3]{\frac{16r^4s^5}{1000t^3}}$
84. $\sqrt[3]{\frac{54m^4n^3}{r^3s^6}}$
85. $\sqrt[3]{\frac{250a^3b^4}{16b}}$
86. $\sqrt[3]{\frac{81p^5q^3}{1000p^2q^6}}$

Review Exercises

In Review Exercises 1–4, simplify each fraction.

1. $\frac{5xy^2z^3}{10x^2y^2z^4}$
2. $\frac{35a^3b^2c}{63a^2b^3c^2}$
3. $\frac{a^2-a-2}{a^2+a-6}$
4. $\frac{y^2+3y-18}{y^2-9}$

In Review Exercises 5–10, perform the indicated operations and simplify the result, if possible.

5. $\dfrac{t^2 - t - 6}{t^2 - 3t} \cdot \dfrac{t^2 - t}{t^2 + t - 2}$

6. $\dfrac{x + 3}{x^2 + x - 6} \div \dfrac{x - 2}{x^2 - 5x + 6}$

7. $\dfrac{2r}{r + 3} + \dfrac{6}{r + 3}$

8. $\dfrac{2(u - 4)}{u + 1} - \dfrac{2(u + 4)}{u + 1}$

9. $\dfrac{3a}{2b} + \dfrac{3b}{5a} - \dfrac{ab}{2c}$

10. $\dfrac{2}{y^2 - 1} - \dfrac{y}{y - 1}$

11. The number of n straight lines that can join P points is given by the formula

$$n = \dfrac{P(P - 1)}{2}$$

Find P when n is 21.

12. The height, h, that an object will reach in t seconds when it is thrown upward at 128 feet per second is given by the formula

$$h = 128t - 16t^2$$

At what times will the height of the object be 112 feet?

15.3 Adding and Subtracting Radical Expressions

- Like Radicals
- Combining Square Roots
- Combining Cube Roots

Like Radicals

It is often possible to combine terms that have **like radicals**.

> **Definition:** Radicals are called **like radicals** if they have the same index and the same radicand.

Because terms such as $3\sqrt{2}$ and $5\sqrt{2}$ have like radicals, they are considered to be like terms. Such terms can be combined.

$3\sqrt{2} + 5\sqrt{2} = (3 + 5)\sqrt{2}$ Use the distributive property.
$= 8\sqrt{2}$

Likewise,

$2x\sqrt{3y} - x\sqrt{3y} = (2x - x)\sqrt{3y}$ Use the distributive property.
$= x\sqrt{3y}$

However, the terms in the expression

$3\sqrt{2} + 5\sqrt{7}$

cannot be combined, because $3\sqrt{2}$ and $5\sqrt{7}$ do not have like radicals.

Combining Square Roots

Unlike radical expressions, such as $3\sqrt{2}$ and $5\sqrt{8}$, can often be combined after they are simplified.

EXAMPLE 1

Simplify $3\sqrt{2} + 5\sqrt{8}$.

Solution

The radical $\sqrt{8}$ is not in simplified form, because 8 has a perfect square factor. We simplify the radical and combine like terms as follows.

$3\sqrt{2} + 5\sqrt{8} = 3\sqrt{2} + 5\sqrt{4 \cdot 2}$	Factor 8. Look for perfect square factors.
$= 3\sqrt{2} + 5\sqrt{4}\sqrt{2}$	Use the multiplication property of radicals.
$= 3\sqrt{2} + 5(2)\sqrt{2}$	Simplify.
$= 3\sqrt{2} + 10\sqrt{2}$	Simplify.
$= 13\sqrt{2}$	Combine like terms.

EXAMPLE 2

Simplify $\sqrt{20} + \sqrt{45} + 3\sqrt{5}$.

Solution

$\sqrt{20} + \sqrt{45} + 3\sqrt{5} = \sqrt{4 \cdot 5} + \sqrt{9 \cdot 5} + 3\sqrt{5}$	Factor.
$= \sqrt{4}\sqrt{5} + \sqrt{9}\sqrt{5} + 3\sqrt{5}$	Use the multiplication property of radicals.
$= 2\sqrt{5} + 3\sqrt{5} + 3\sqrt{5}$	Simplify.
$= 8\sqrt{5}$	Combine like terms.

Work Progress Check 7.

Progress Check 7

Simplify: $\sqrt{18} + \sqrt{32} - \sqrt{50}$.

EXAMPLE 3

Simplify $\sqrt{8x^2y} + \sqrt{18x^2y}$.

Solution

$\sqrt{8x^2y} + \sqrt{18x^2y} = \sqrt{4 \cdot 2x^2y} + \sqrt{9 \cdot 2x^2y}$	Factor.
$= \sqrt{4x^2}\sqrt{2y} + \sqrt{9x^2}\sqrt{2y}$	Use the multiplication property of radicals.
$= 2x\sqrt{2y} + 3x\sqrt{2y}$	Simplify.
$= 5x\sqrt{2y}$	Combine like terms.

EXAMPLE 4

Simplify $\sqrt{28x^2y} - 2\sqrt{63y^3}$.

---

7. $2\sqrt{2}$

Progress Check Answers

Solution

$$\sqrt{28x^2y} - 2\sqrt{63y^3} = \sqrt{4 \cdot 7x^2y} - 2\sqrt{9 \cdot 7y^2y}$$ Factor.
$$= \sqrt{4x^2}\sqrt{7y} - 2\sqrt{9y^2}\sqrt{7y}$$ Use the multiplication property of radicals.
$$= 2x\sqrt{7y} - 2 \cdot 3y\sqrt{7y}$$ Simplify.
$$= 2x\sqrt{7y} - 6y\sqrt{7y}$$ Simplify.
$$= (2x - 6y)\sqrt{7y}$$ Factor out $\sqrt{7y}$.

Because the variables of the coefficients $2x$ and $6y$ are not the same, the expression does not simplify any further. Work Progress Check 8.

Progress Check 8

Simplify: $\sqrt{128x^3y} - \sqrt{27xy^3}$.

EXAMPLE 5

Simplify $\sqrt{27xy} + \sqrt{20xy}$.

Solution

$$\sqrt{27xy} + \sqrt{20xy} = \sqrt{9 \cdot 3xy} + \sqrt{4 \cdot 5xy}$$ Factor.
$$= \sqrt{9}\sqrt{3xy} + \sqrt{4}\sqrt{5xy}$$ Use the multiplication property of radicals.
$$= 3\sqrt{3xy} + 2\sqrt{5xy}$$ Simplify.

Because these terms do not have like radicals, we cannot simplify any further.

EXAMPLE 6

Simplify $\sqrt{8x} + \sqrt{3y} - \sqrt{50x} + \sqrt{27y}$.

Solution

$$\sqrt{8x} + \sqrt{3y} - \sqrt{50x} + \sqrt{27y} = \sqrt{4 \cdot 2x} + \sqrt{3y} - \sqrt{25 \cdot 2x} + \sqrt{9 \cdot 3y}$$
$$= \sqrt{4}\sqrt{2x} + \sqrt{3y} - \sqrt{25}\sqrt{2x} + \sqrt{9}\sqrt{3y}$$
$$= 2\sqrt{2x} + \sqrt{3y} - 5\sqrt{2x} + 3\sqrt{3y}$$
$$= -3\sqrt{2x} + 4\sqrt{3y}$$

Combining Cube Roots

We can also combine like terms containing like radicals for higher roots.

Progress Check 9

Simplify: $\sqrt[3]{108a^4} + a\sqrt[3]{32a}$.

EXAMPLE 7

Simplify $\sqrt[3]{81x^4} - x\sqrt[3]{24x}$.

Solution

$$\sqrt[3]{81x^4} - x\sqrt[3]{24x} = \sqrt[3]{27x^3 \cdot 3x} - x\sqrt[3]{8 \cdot 3x}$$
$$= \sqrt[3]{27x^3}\sqrt[3]{3x} - x\sqrt[3]{8}\sqrt[3]{3x}$$
$$= 3x\sqrt[3]{3x} - 2x\sqrt[3]{3x}$$
$$= x\sqrt[3]{3x}$$

Work Progress Check 9.

8. $8x\sqrt{2xy} - 3y\sqrt{3xy}$

9. $5a\sqrt[3]{4a}$

15.3 Exercises

In Exercises 1–24, find each indicated sum.

1. $\sqrt{12} + \sqrt{27}$
2. $\sqrt{20} + \sqrt{45}$
3. $\sqrt{48} + \sqrt{75}$
4. $\sqrt{48} + \sqrt{108}$
5. $\sqrt{45} + \sqrt{80}$
6. $\sqrt{80} + \sqrt{125}$
7. $\sqrt{125} + \sqrt{245}$
8. $\sqrt{36} + \sqrt{196}$
9. $\sqrt{20} + \sqrt{180}$
10. $\sqrt{80} + \sqrt{245}$
11. $\sqrt{160} + \sqrt{360}$
12. $\sqrt{12} + \sqrt{147}$
13. $3\sqrt{45} + 4\sqrt{245}$
14. $2\sqrt{28} + 7\sqrt{63}$
15. $2\sqrt{28} + 2\sqrt{112}$
16. $4\sqrt{63} + 6\sqrt{112}$
17. $5\sqrt{32} + 3\sqrt{72}$
18. $3\sqrt{72} + 2\sqrt{128}$
19. $3\sqrt{98} + 8\sqrt{128}$
20. $5\sqrt{90} + 7\sqrt{250}$
21. $\sqrt{20} + \sqrt{45} + \sqrt{80}$
22. $\sqrt{48} + \sqrt{27} + \sqrt{75}$
23. $\sqrt{24} + \sqrt{150} + \sqrt{240}$
24. $\sqrt{28} + \sqrt{63} + \sqrt{112}$

In Exercises 25–44, find each indicated difference. If the terms cannot be combined, so indicate.

25. $\sqrt{18} - \sqrt{8}$
26. $\sqrt{32} - \sqrt{18}$
27. $\sqrt{9} - \sqrt{50}$
28. $\sqrt{50} - \sqrt{32}$
29. $\sqrt{72} - \sqrt{32}$
30. $\sqrt{98} - \sqrt{72}$
31. $\sqrt{12} - \sqrt{48}$
32. $\sqrt{48} - \sqrt{75}$
33. $\sqrt{108} - \sqrt{75}$
34. $\sqrt{147} - \sqrt{48}$
35. $\sqrt{1000} - \sqrt{360}$
36. $\sqrt{180} - \sqrt{125}$
37. $2\sqrt{80} - 3\sqrt{125}$
38. $3\sqrt{245} - 2\sqrt{180}$
39. $8\sqrt{96} - 5\sqrt{24}$
40. $3\sqrt{216} - 3\sqrt{150}$
41. $\sqrt{288} - 3\sqrt{200}$
42. $\sqrt{392} - 2\sqrt{128}$
43. $5\sqrt{250} - 3\sqrt{160}$
44. $4\sqrt{490} - 3\sqrt{360}$

In Exercises 45–56, simplify each expression.

45. $\sqrt{12} + \sqrt{18} - \sqrt{27}$
46. $\sqrt{8} - \sqrt{50} + \sqrt{72}$
47. $\sqrt{200} - \sqrt{75} + \sqrt{48}$
48. $\sqrt{20} + \sqrt{80} - \sqrt{125}$
49. $\sqrt{24} - \sqrt{150} - \sqrt{54}$
50. $\sqrt{98} - \sqrt{300} + \sqrt{800}$
51. $\sqrt{200} + \sqrt{300} - \sqrt{75}$
52. $\sqrt{175} + \sqrt{125} - \sqrt{28}$
53. $\sqrt{48} - \sqrt{8} + \sqrt{27} - \sqrt{32}$
54. $\sqrt{162} + \sqrt{50} - \sqrt{75} - \sqrt{108}$
55. $\sqrt{147} + \sqrt{216} - \sqrt{108} - \sqrt{27}$
56. $\sqrt{180} - \sqrt{112} + \sqrt{45} - \sqrt{700}$

In Exercises 57–70, simplify each expression. If the terms cannot be combined, so indicate. Assume that all variables represent positive numbers.

57. $\sqrt{2x^2} + \sqrt{8x^2}$
58. $\sqrt{3y^2} - \sqrt{12y^2}$
59. $\sqrt{2x^3} + \sqrt{8x^3}$
60. $\sqrt{3y^3} - \sqrt{12y^3}$
61. $\sqrt{18x^2y} - \sqrt{27x^2y}$
62. $\sqrt{49xy} + \sqrt{xy}$
63. $\sqrt{32x^5} - \sqrt{18x^5}$
64. $\sqrt{27xy^3} - \sqrt{48xy^3}$

Chapter 15 Roots and Radical Expressions

65. $3\sqrt{54x^2} + 5\sqrt{24x^2}$

66. $3\sqrt{24x^4y^3} + 2\sqrt{54x^4y^3}$

67. $y\sqrt{490y} - 2\sqrt{360y^3}$

68. $3\sqrt{20x} + 2\sqrt{63y}$

69. $\sqrt{20x^3y} + \sqrt{45x^5y^3} - \sqrt{80x^7y^5}$

70. $x\sqrt{48xy^2} - y\sqrt{27x^3} + \sqrt{75x^3y^2}$

In Exercises 71–82, simplify each expression.

71. $\sqrt[3]{16} + \sqrt[3]{54}$

72. $\sqrt[3]{24} - \sqrt[3]{81}$

73. $\sqrt[3]{81} - \sqrt[3]{24}$

74. $\sqrt[3]{32} + \sqrt[3]{108}$

75. $\sqrt[3]{40} + \sqrt[3]{125}$

76. $\sqrt[3]{3000} - \sqrt[3]{192}$

77. $\sqrt[3]{x^4} - \sqrt[3]{x^7}$

78. $\sqrt[3]{8x^5} + \sqrt[3]{27x^8}$

79. $\sqrt[3]{192x^4y^5} - \sqrt[3]{24x^4y^5}$

80. $\sqrt[3]{24a^5b^4} + \sqrt[3]{81a^5b^4}$

81. $\sqrt[3]{135x^7y^4} - x\sqrt[3]{40x^4y^4}$

82. $a\sqrt[3]{7ab^5} + b\sqrt[3]{56a^4b^2}$

Review Exercises

In Review Exercises 1–4, express each phrase as a ratio in lowest terms.

1. 3 to 8

2. 6 ounces to 18 ounces

3. 5 inches to 3 feet

4. 3 months to 10 years

In Review Exercises 5–10, solve each proportion.

5. $\dfrac{a-2}{8} = \dfrac{a+10}{24}$

6. $\dfrac{6}{t+12} = \dfrac{18}{4t}$

7. $\dfrac{-2}{x+14} = \dfrac{6}{x-6}$

8. $\dfrac{y-4}{4} = \dfrac{y+2}{12}$

9. $\dfrac{z+3}{4} = \dfrac{9}{2z}$

10. $\dfrac{s}{s+3} = \dfrac{2s}{s+2}$

11. The total surface area, A, of the cylindrical solid in Illustration 1 is given approximately by the formula

$$A = \dfrac{44}{7}(r^2 + rh)$$

Solve for h.

12. The total surface area, A, of the closed box in Illustration 2 is given by the formula

$$A = 2lw + 2wd + 2ld$$

Solve the formula for d.

Illustration 1

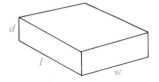

Illustration 2

15.4 Multiplying and Dividing Radical Expressions

- Multiplying Monomials by Monomials
- Multiplying Polynomials by Monomials
- Multiplying Binomials by Binomials
- Dividing Radical Expressions
- Rationalizing the Denominator

The definition of square root implies that $\sqrt{5}$ is the number whose square is 5:

$$(\sqrt{5})^2 = 5 \quad \text{and} \quad \sqrt{5}\sqrt{5} = 5$$

In general, we have

$$(\sqrt{x})^2 = x \quad \text{and} \quad \sqrt{x}\sqrt{x} = x$$

Because of the multiplication property of radicals, the *product of the square roots of two nonnegative numbers is equal to the square root of the product of those numbers*. For example,

$$\sqrt{2}\sqrt{8} = \sqrt{2 \cdot 8} = \sqrt{16} = 4$$
$$\sqrt{3}\sqrt{27} = \sqrt{3 \cdot 27} = \sqrt{81} = 9$$

and

$$\sqrt{x}\sqrt{x^3} = \sqrt{x \cdot x^3} = \sqrt{x^4} = x^2$$

Likewise, the *product of the cube roots of two numbers is equal to the cube root of the product of those numbers*. For example,

$$\sqrt[3]{2}\sqrt[3]{4} = \sqrt[3]{2 \cdot 4} = \sqrt[3]{8} = 2$$
$$\sqrt[3]{4}\sqrt[3]{16} = \sqrt[3]{4 \cdot 16} = \sqrt[3]{64} = 4$$

and

$$\sqrt[3]{3x^2}\sqrt[3]{9x} = \sqrt[3]{3x^2 \cdot 9x} = \sqrt[3]{27x^3} = 3x$$

Multiplying Monomials by Monomials

To multiply monomials with like radicals, we multiply the coefficients and multiply the radicals separately. We then simplify the result, if possible.

EXAMPLE 1

Multiply **a.** $3\sqrt{6}$ by $4\sqrt{3}$ and **b.** $-2\sqrt[3]{7}$ by $6\sqrt[3]{49}$.

Solution

a. $3\sqrt{6} \cdot 4\sqrt{3} = 3(4)\sqrt{6}\sqrt{3}$
$= 12\sqrt{18}$
$= 12\sqrt{9}\sqrt{2}$
$= 12(3)\sqrt{2}$
$= 36\sqrt{2}$

b. $-2\sqrt[3]{7} \cdot 6\sqrt[3]{49} = -2(6)\sqrt[3]{7}\sqrt[3]{49}$
$= -12\sqrt[3]{7 \cdot 49}$
$= -12\sqrt[3]{343}$
$= -12(7)$
$= -84$

Work Progress Check 10.

Multiplying Polynomials by Monomials

To multiply a polynomial by a monomial, we use the distributive property to remove parentheses and then combine like terms, if possible.

Progress Check 10

Multiply:

a. $2\sqrt{5} \cdot 3\sqrt{3}$.

b. $-4\sqrt[3]{5} \cdot 2\sqrt[3]{5}$.

10. **a.** $6\sqrt{15}$ **b.** $-8\sqrt[3]{25}$

Progress Check Answers

EXAMPLE 2

Multiply   **a.** $\sqrt{2}(\sqrt{6} + \sqrt{8})$ and   **b.** $\sqrt[3]{3}(\sqrt[3]{9} - 2)$.

Solution

a. $\sqrt{2}(\sqrt{6} + \sqrt{8}) = \sqrt{2}\sqrt{6} + \sqrt{2}\sqrt{8}$ Use the distributive property to remove parentheses.

$\qquad = \sqrt{12} + \sqrt{16}$ Use the multiplication property of radicals.

$\qquad = \sqrt{4 \cdot 3} + \sqrt{16}$ Factor 12.

$\qquad = \sqrt{4}\sqrt{3} + \sqrt{16}$ Use the multiplication property of radicals.

$\qquad = 2\sqrt{3} + 4$ Simplify.

b. $\sqrt[3]{3}(\sqrt[3]{9} - 2) = \sqrt[3]{3}\sqrt[3]{9} - 2\sqrt[3]{3}$ Use the distributive property to remove parentheses.

$\qquad = \sqrt[3]{27} - 2\sqrt[3]{3}$ Use the multiplication property of radicals.

$\qquad = 3 - 2\sqrt[3]{3}$ Simplify.

Work Progress Check 11.

Progress Check 11

Multiply:

a. $\sqrt{3}(\sqrt{3} + \sqrt{2})$.

b. $\sqrt[3]{2}(\sqrt[3]{4} + \sqrt[3]{2})$.

Multiplying Binomials by Binomials

The FOIL method is used to multiply one binomial by another.

EXAMPLE 3

Multiply and simplify $(\sqrt{3} + \sqrt{2})(\sqrt{3} - \sqrt{2})$.

Solution

Find the product by the FOIL method.

$(\sqrt{3} + \sqrt{2})(\sqrt{3} - \sqrt{2})$

$= \sqrt{3}\sqrt{3} - \sqrt{3}\sqrt{2} + \sqrt{2}\sqrt{3} - \sqrt{2}\sqrt{2}$ Use the FOIL method.

$= \sqrt{9} - \sqrt{4}$

$= 3 - 2$ Combine like terms and simplify.

$= 1$ Simplify.

EXAMPLE 4

Multiply and simplify $(\sqrt{3x} + 1)(\sqrt{3x} + 2)$.

Solution

$(\sqrt{3x} + 1)(\sqrt{3x} + 2)$

$= \sqrt{3x}\sqrt{3x} + 2\sqrt{3x} + \sqrt{3x} + 2$ Use the FOIL method.

$= 3x + 3\sqrt{3x} + 2$ Combine like terms and simplify.

---

11. a. $3 + \sqrt{6}$ **b.** $2 + \sqrt[3]{4}$

Progress Check Answers

EXAMPLE 5

Multiply and simplify $(\sqrt[3]{4x} - 3)(\sqrt[3]{2x^2} + 1)$.

Solution

$(\sqrt[3]{4x} - 3)(\sqrt[3]{2x^2} + 1)$

$= \sqrt[3]{4x} \sqrt[3]{2x^2} + 1\sqrt[3]{4x} - 3\sqrt[3]{2x^2} - 3$ Use the FOIL method.

$= \sqrt[3]{8x^3} + \sqrt[3]{4x} - 3\sqrt[3]{2x^2} - 3$ Use the multiplication property of radicals.

$= 2x + \sqrt[3]{4x} - 3\sqrt[3]{2x^2} - 3$ Simplify.

Work Progress Check 12.

Progress Check 12

Multiply: $(\sqrt{2a} - 1)(\sqrt{2a} + 2)$.

Dividing Radical Expressions

We divide radical expressions such as $\sqrt{108}$ by $\sqrt{36}$ as follows:

$\dfrac{\sqrt{108}}{\sqrt{36}} = \sqrt{\dfrac{108}{36}}$ Use the division property of radicals.

$= \sqrt{3}$ $\dfrac{108}{36} = \dfrac{36 \cdot 3}{36} = 3$

EXAMPLE 6

Simplify $\dfrac{\sqrt{22a^2b^7}}{\sqrt{99a^4b^3}}$.

Solution

$\dfrac{\sqrt{22a^2b^7}}{\sqrt{99a^4b^3}} = \sqrt{\dfrac{22a^2b^7}{99a^4b^3}}$ Use the division property of radicals.

$= \sqrt{\dfrac{2b^4}{9a^2}}$ Simplify the radicand.

$= \dfrac{\sqrt{2b^4}}{\sqrt{9a^2}}$ Use the division property of radicals.

$= \dfrac{\sqrt{b^4}\sqrt{2}}{\sqrt{9a^2}}$ Use the multiplication property of radicals.

$= \dfrac{b^2\sqrt{2}}{3a}$ Simplify.

Rationalizing the Denominator

It is impossible to do the long division indicated by the fraction

$$\dfrac{1}{\sqrt{2}} \quad \text{or} \quad \dfrac{1}{1.4142135\ldots}$$

because of the unending decimal in the denominator. However, if the radical were not present in the denominator, the division would be possible.

12. $2a + \sqrt{2a} - 2$

We can eliminate the radical in the denominator by multiplying both the numerator and the denominator by $\sqrt{2}$.

$$\frac{1}{\sqrt{2}} = \frac{1\sqrt{2}}{\sqrt{2}\sqrt{2}} \quad \text{Multiply both numerator and denominator by } \sqrt{2}.$$

$$= \frac{\sqrt{2}}{2} \quad \text{Simplify.}$$

The division indicated by the fraction

$$\frac{1.4142135\ldots}{2}$$

can be carried out easily to any desired number of decimal places.

The process of removing radicals from the denominator of a fraction is called **rationalizing the denominator**.

EXAMPLE 7

Rationalize the denominator of the fractions **a.** $\dfrac{3}{\sqrt{3}}$ and **b.** $\dfrac{2}{\sqrt[3]{3}}$.

Solution

a. $\dfrac{3}{\sqrt{3}} = \dfrac{3\sqrt{3}}{\sqrt{3}\sqrt{3}}$ Multiply both the numerator and denominator by $\sqrt{3}$.

$\phantom{\dfrac{3}{\sqrt{3}}} = \dfrac{3\sqrt{3}}{3}$ $\sqrt{3}\sqrt{3} = 3$

$\phantom{\dfrac{3}{\sqrt{3}}} = \sqrt{3}$ Divide out the common factor of 3.

b. Because $\sqrt[3]{3} \cdot \sqrt[3]{9} = \sqrt[3]{27}$, and 27 is a perfect integer cube, we multiply the numerator and the denominator by $\sqrt[3]{9}$ and simplify.

$$\frac{2}{\sqrt[3]{3}} = \frac{2\sqrt[3]{9}}{\sqrt[3]{3}\sqrt[3]{9}}$$

$$= \frac{2\sqrt[3]{9}}{\sqrt[3]{27}}$$

$$= \frac{2\sqrt[3]{9}}{3}$$

Work Progress Check 13.

If a fraction has radicals in its denominator, it is not in simplified form. The following examples show how to simplify such fractions.

EXAMPLE 8

Divide 5 by $\sqrt{20}$ by simplifying the fraction $\dfrac{5}{\sqrt{20}}$.

Progress Check 13

Rationalize the denominator:

a. $\dfrac{3}{\sqrt{2}}$.

b. $\dfrac{3}{\sqrt[3]{2}}$.

---

13. **a.** $\dfrac{3\sqrt{2}}{2}$ **b.** $\dfrac{3\sqrt[3]{4}}{2}$

Progress Check Answers

Solution

To rationalize the denominator, it is not necessary to multiply the numerator and the denominator by $\sqrt{20}$. It is sufficient to multiply by $\sqrt{5}$ because $5 \cdot 20$ is 100, which is a perfect integer square.

$\dfrac{5}{\sqrt{20}} = \dfrac{5\sqrt{5}}{\sqrt{20}\sqrt{5}}$ Multiply both the numerator and denominator by $\sqrt{5}$.

$= \dfrac{5\sqrt{5}}{\sqrt{100}}$ Use the multiplication property of radicals.

$= \dfrac{5\sqrt{5}}{10}$ $\sqrt{100} = 10$

$= \dfrac{\sqrt{5}}{2}$ Simplify the fraction.

EXAMPLE 9

Simplify the fraction $\dfrac{\sqrt{72x^5}}{\sqrt{45}}$.

Solution

$\dfrac{\sqrt{72x^5}}{\sqrt{45}} = \dfrac{\sqrt{36 \cdot x^4 \cdot 2x}}{\sqrt{9 \cdot 5}}$ Factor the radicands, looking for perfect square factors.

$= \dfrac{\sqrt{36} \cdot \sqrt{x^4} \cdot \sqrt{2x}}{\sqrt{9} \cdot \sqrt{5}}$ Use the multiplication property of radicals.

$= \dfrac{6x^2\sqrt{2x}}{3\sqrt{5}}$ Simplify.

$= \dfrac{6x^2\sqrt{2x}\,\sqrt{5}}{3\sqrt{5}\,\sqrt{5}}$ Multiply both the numerator and denominator by $\sqrt{5}$ to rationalize the denominator.

$= \dfrac{2x^2\sqrt{10x}}{5}$ Simplify.

EXAMPLE 10

Simplify the fraction $\sqrt{\dfrac{3x^3y^2}{27xy^3}}$.

Progress Check 14

Simplify: $\sqrt{\dfrac{2x^5y^2}{32x^3y^3}}$.

14. $\dfrac{x\sqrt{y}}{4y}$

Progress Check Answers

Solution

$\sqrt{\dfrac{3x^3y^2}{27xy^3}} = \sqrt{\dfrac{x^2}{9y}}$ Simplify the fraction within the radical.

$= \sqrt{\dfrac{x^2 \cdot y}{9y \cdot y}}$ Multiply both the numerator and denominator by y.

$= \dfrac{\sqrt{x^2y}}{\sqrt{9y^2}}$ Use the division property of radicals.

$= \dfrac{x\sqrt{y}}{3y}$ Simplify.

Work Progress Check 14.

EXAMPLE 11

Simplify the fraction $\dfrac{2}{\sqrt{3}-1}$.

Solution

Since the denominator is a *binomial*, multiplying the numerator and the denominator by $\sqrt{3}$ will not work. Instead, we multiply the numerator and the denominator by the binomial $\sqrt{3}+1$, since the product of $(\sqrt{3}+1)$ and $(\sqrt{3}-1)$ is free of radicals.

$$(\sqrt{3}+1)(\sqrt{3}-1) = \sqrt{3}\sqrt{3} - \sqrt{3} + \sqrt{3} - 1 = 3 - 1 = 2$$

Radical expressions such as $\sqrt{3}+1$ and $\sqrt{3}-1$ are called **conjugates** of each other.

$\dfrac{2}{\sqrt{3}-1} = \dfrac{2(\sqrt{3}+1)}{(\sqrt{3}-1)(\sqrt{3}+1)}$ Multiply both the numerator and denominator by $\sqrt{3}+1$.

$= \dfrac{2(\sqrt{3}+1)}{3-1}$ Multiply the binomials in the denominator.

$= \dfrac{2(\sqrt{3}+1)}{2}$ Simplify.

$= \sqrt{3}+1$ Divide out the common factor of 2.

Work Progress Check 15.

Progress Check 15

Simplify: $\dfrac{2}{\sqrt{3}+1}$.

EXAMPLE 12

Simplify the fraction $\dfrac{10\sqrt{7}}{\sqrt{7x}+\sqrt{2x}}$.

Solution

We multiply both the numerator and the denominator by $\sqrt{7x} - \sqrt{2x}$, which is the conjugate of the denominator.

$\dfrac{10\sqrt{7}}{\sqrt{7x}+\sqrt{2x}} = \dfrac{10\sqrt{7}(\sqrt{7x}-\sqrt{2x})}{(\sqrt{7x}+\sqrt{2x})(\sqrt{7x}-\sqrt{2x})}$ Multiply both the numerator and the denominator by $\sqrt{7x}-\sqrt{2x}$.

$= \dfrac{10\sqrt{7}(\sqrt{7x}-\sqrt{2x})}{7x-2x}$ Multiply the binomials.

$= \dfrac{10\sqrt{7}(\sqrt{7x}-\sqrt{2x})}{5x}$ Simplify.

$= \dfrac{2\sqrt{7}(\sqrt{7x}-\sqrt{2x})}{x}$ Simplify the fraction.

$= \dfrac{14\sqrt{x} - 2\sqrt{14x}}{x}$ Remove parentheses.

Progress Check Answers

15. $\sqrt{3}-1$

15.4 Exercises

In Exercises 1–62, perform each indicated multiplication. Assume that all variables represent positive numbers.

1. $\sqrt{3}\sqrt{3}$
2. $\sqrt{7}\sqrt{7}$
3. $\sqrt{2}\sqrt{8}$
4. $\sqrt{27}\sqrt{3}$
5. $\sqrt{16}\sqrt{4}$
6. $\sqrt{32}\sqrt{2}$
7. $\sqrt[3]{8}\sqrt[3]{8}$
8. $\sqrt[3]{4}\sqrt[3]{250}$
9. $\sqrt{x^3}\sqrt{x^3}$
10. $\sqrt{a^7}\sqrt{a^3}$
11. $\sqrt{b^8}\sqrt{b^6}$
12. $\sqrt{y^4}\sqrt{y^8}$
13. $(2\sqrt{5})(2\sqrt{3})$
14. $(4\sqrt{3})(2\sqrt{2})$
15. $(-5\sqrt{6})(4\sqrt{3})$
16. $(6\sqrt{3})(-7\sqrt{3})$
17. $(2\sqrt[3]{4})(3\sqrt[3]{16})$
18. $(-3\sqrt[3]{100})(\sqrt[3]{10})$
19. $(4\sqrt{x})(-2\sqrt{x})$
20. $(3\sqrt{y})(15\sqrt{y})$
21. $(-14\sqrt{50x})(-5\sqrt{20x})$
22. $(12\sqrt{24y})(-16\sqrt{2y})$
23. $\sqrt{8x}\sqrt{2x^3y}$
24. $\sqrt{27y}\sqrt{3y^3}$
25. $\sqrt{2}(\sqrt{2}+1)$
26. $\sqrt{3}(\sqrt{3}-2)$
27. $\sqrt{3}(\sqrt{27}-1)$
28. $\sqrt{2}(\sqrt{8}-1)$
29. $\sqrt{7}(\sqrt{7}-3)$
30. $\sqrt{5}(\sqrt{5}+2)$
31. $\sqrt{5}(3-\sqrt{5})$
32. $\sqrt{7}(2+\sqrt{7})$
33. $\sqrt{3}(\sqrt{6}+1)$
34. $\sqrt{2}(\sqrt{6}-2)$
35. $\sqrt[3]{7}(\sqrt[3]{49}-2)$
36. $\sqrt[3]{5}(\sqrt[3]{25}+3)$
37. $\sqrt{x}(\sqrt{3x}-2)$
38. $\sqrt{y}(\sqrt{y}+5)$
39. $2\sqrt{x}(\sqrt{9x}+3)$
40. $3\sqrt{z}(\sqrt{4z}-\sqrt{z})$
41. $3\sqrt{x}(2+\sqrt{x})$
42. $5\sqrt{y}(5-\sqrt{5y})$
43. $\sqrt{21x}(\sqrt{3x}+\sqrt{2x})$
44. $\sqrt{35y}(\sqrt{7y}-\sqrt{5y})$
45. $(\sqrt{2}+1)(\sqrt{2}-1)$
46. $(\sqrt{3}-1)(\sqrt{3}+1)$
47. $(\sqrt{5}+2)(\sqrt{5}-2)$
48. $(\sqrt{7}+5)(\sqrt{7}-5)$
49. $(\sqrt[3]{2}+1)(\sqrt[3]{2}+1)$
50. $(\sqrt[3]{5}-2)(\sqrt[3]{5}-2)$
51. $(\sqrt{7}-x)(\sqrt{7}+x)$
52. $(\sqrt{2}-\sqrt{x})(\sqrt{x}+\sqrt{2})$
53. $(\sqrt{2}-\sqrt{x})^2$
54. $(\sqrt{a}+\sqrt{3})^2$
55. $(\sqrt{6x}+\sqrt{7})(\sqrt{6x}-\sqrt{7})$
56. $(\sqrt{8y}+\sqrt{2z})(\sqrt{8y}-\sqrt{2z})$
57. $(\sqrt{2x}+3)(\sqrt{8x}-6)$
58. $(\sqrt{5y}-3)(\sqrt{20y}+6)$
59. $(\sqrt{8xy}+1)(\sqrt{8xy}+1)$
60. $(\sqrt{5x}+3\sqrt{y})(\sqrt{5x}-3\sqrt{y})$
61. $(\sqrt{16x}-\sqrt{x})(\sqrt{16x}-\sqrt{x})$
62. $(\sqrt{9xz}+2\sqrt{xz})(\sqrt{25xz}-\sqrt{xz})$

In Exercises 63–74, simplify each expression. Assume that all variables represent positive numbers.

63. $\dfrac{\sqrt{12x^3}}{\sqrt{27x}}$
64. $\dfrac{\sqrt{32}}{\sqrt{98x^2}}$
65. $\dfrac{\sqrt{18xy^2}}{\sqrt{25x}}$
66. $\dfrac{\sqrt{27y^3}}{\sqrt{75x^2y}}$
67. $\dfrac{\sqrt{196xy^3}}{\sqrt{49x^3y}}$
68. $\dfrac{\sqrt{50xyz^4}}{\sqrt{98xyz^2}}$
69. $\dfrac{\sqrt[3]{16x^6}}{\sqrt[3]{54x^3}}$
70. $\dfrac{\sqrt[3]{128a^6b^3}}{\sqrt[3]{16a^3b^6}}$
71. $\dfrac{\sqrt{3x^2y^3}}{\sqrt{27x}}$
72. $\dfrac{\sqrt{44x^2y^5}}{\sqrt{99x^4y}}$
73. $\dfrac{\sqrt{5x}\sqrt{10y^2}}{\sqrt{x^3y}}$
74. $\dfrac{\sqrt{7y}\sqrt{14x}}{\sqrt{8xy}}$

In Exercises 75–106, perform each indicated division by rationalizing a denominator and simplifying. Assume that all variables represent positive numbers.

75. $\dfrac{1}{\sqrt{3}}$ 76. $\dfrac{1}{\sqrt{5}}$ 77. $\dfrac{2}{\sqrt{7}}$ 78. $\dfrac{3}{\sqrt{11}}$

79. $\dfrac{5}{\sqrt[3]{5}}$ 80. $\dfrac{7}{\sqrt[3]{7}}$ 81. $\dfrac{9}{\sqrt{27}}$ 82. $\dfrac{4}{\sqrt{20}}$

83. $\dfrac{3}{\sqrt{32}}$ 84. $\dfrac{5}{\sqrt{18}}$ 85. $\dfrac{4}{\sqrt[3]{4}}$ 86. $\dfrac{7}{\sqrt[3]{10}}$

87. $\dfrac{\sqrt{5}}{\sqrt{3}}$ 88. $\dfrac{\sqrt{3}}{\sqrt{5}}$ 89. $\dfrac{10}{\sqrt{x}}$ 90. $\dfrac{12}{\sqrt{y}}$

91. $\dfrac{\sqrt{9}}{\sqrt{2x}}$ 92. $\dfrac{\sqrt{4}}{\sqrt{3z}}$ 93. $\dfrac{\sqrt{2x}}{\sqrt{9y}}$ 94. $\dfrac{\sqrt{3xy}}{\sqrt{4x}}$

95. $\dfrac{2\sqrt{3}}{\sqrt{8x^2}}$ 96. $\dfrac{3\sqrt{5}}{\sqrt{27y^2}}$ 97. $\dfrac{5\sqrt{6x}}{\sqrt{50}}$ 98. $\dfrac{8\sqrt{10y}}{\sqrt{40}}$

99. $\dfrac{\sqrt[3]{5}}{\sqrt[3]{2}}$ 100. $\dfrac{\sqrt[3]{2}}{\sqrt[3]{5}}$ 101. $\dfrac{\sqrt[3]{2x^2}}{\sqrt[3]{2x}}$ 102. $\dfrac{\sqrt[3]{3y^4}}{\sqrt[3]{3y}}$

103. $\dfrac{2}{\sqrt[3]{4x^2y}}$ 104. $\dfrac{3}{\sqrt[3]{9xy^2}}$ 105. $\dfrac{-5}{\sqrt[3]{25a^2b^2}}$ 106. $\dfrac{-4}{\sqrt[3]{4ab^2c^2}}$

In Exercises 107–130, perform each indicated division by rationalizing a denominator and simplifying. Assume that all variables represent positive numbers.

107. $\dfrac{3}{\sqrt{3}-1}$ 108. $\dfrac{3}{\sqrt{5}-2}$ 109. $\dfrac{3}{\sqrt{7}+2}$ 110. $\dfrac{5}{\sqrt{8}+3}$

111. $\dfrac{12}{3-\sqrt{3}}$ 112. $\dfrac{10}{5-\sqrt{5}}$ 113. $\dfrac{\sqrt{2}}{\sqrt{2}+1}$ 114. $\dfrac{\sqrt{3}}{\sqrt{3}-1}$

115. $\dfrac{-\sqrt{3}}{\sqrt{3}+1}$ 116. $\dfrac{-\sqrt{2}}{\sqrt{2}-1}$ 117. $\dfrac{5}{\sqrt{3}+\sqrt{2}}$ 118. $\dfrac{3}{\sqrt{3}-\sqrt{2}}$

119. $\dfrac{\sqrt{8}}{\sqrt{5}-\sqrt{3}}$ 120. $\dfrac{\sqrt{32}}{\sqrt{7}-\sqrt{3}}$ 121. $\dfrac{\sqrt{3}-\sqrt{2}}{\sqrt{3}+\sqrt{2}}$ 122. $\dfrac{\sqrt{5}+\sqrt{3}}{\sqrt{5}-\sqrt{3}}$

123. $\dfrac{\sqrt{3x}-1}{\sqrt{3x}+1}$ 124. $\dfrac{\sqrt{5x}+3}{\sqrt{5x}-3}$ 125. $\dfrac{\sqrt{2x}+5}{\sqrt{2x}+3}$ 126. $\dfrac{\sqrt{3y}+3}{\sqrt{3y}-2}$

127. $\dfrac{1-\sqrt{2z}}{\sqrt{2z}+1}$ 128. $\dfrac{1+\sqrt{3z}}{\sqrt{3z}-1}$ 129. $\dfrac{y-\sqrt{15}}{y+\sqrt{15}}$ 130. $\dfrac{x+\sqrt{17}}{x-\sqrt{17}}$

Review Exercises

In Review Exercises 1–6, factor each polynomial.

1. $x^2 - 4x - 21$
2. $y^2 + 6y - 27$
3. $6x^2y - 15xy$
4. $2x^3y^2 - 6x^2y$
5. $x^3 + 8$
6. $y^2 - 9$

In Review Exercises 7–10, solve each equation.

7. $x^2 - 13x + 30 = 0$
8. $2x^2 + x = 1$
9. $2x^2 - 8 = 0$
10. $3x^2 = 15x$

11. A 16-foot rope is to be cut into two pieces, one three times as long as the other. Find the length of each piece.

12. George gave twice as much money to charity as Bob, and Sally gave $6 more than Bob. The total amount given was $506. How much did George give?

15.5 Solving Equations Containing Radicals

- The Squaring Property of Equality
- Extraneous Solutions

The Squaring Property of Equality

To solve equations with radical expressions, we use the **squaring property of equality**.

> **The Squaring Property of Equality:**
> If $a = b$, then $a^2 = b^2$.

The equation $x = 2$ has only one solution, the number 2. If we apply the squaring property and square both sides of $x = 2$, we get the equation $x^2 = 4$. This new equation has *two* solutions, the numbers 2 and -2. Thus, squaring both sides leads to another equation with more solutions than the first. However, no solutions were lost when both sides were squared. The original solution of 2 is still a solution of the squared equation.

To solve equations such as $\sqrt{x + 2} = 3$, we must square both sides to eliminate the radical. Sometimes this will produce an equation with more solutions than the original equation. Therefore, we *must* check every solution of the squared equation in the *original* equation.

EXAMPLE 1

Solve the equation $\sqrt{x + 2} = 3$.

Solution

We square both sides to eliminate the radical and proceed as follows:

$$\sqrt{x+2} = 3$$
$$(\sqrt{x+2})^2 = 3^2 \quad \text{Square both sides.}$$
$$x + 2 = 9 \quad \text{Simplify.}$$
$$x = 7 \quad \text{Add } -2 \text{ to both sides.}$$

Check:
$$\sqrt{x+2} = 3$$
$$\sqrt{7+2} \stackrel{?}{=} 3 \quad \text{Replace } x \text{ with 7.}$$
$$\sqrt{9} \stackrel{?}{=} 3$$
$$3 = 3$$

The solution checks. Since no solutions are lost in this process, 7 is the only solution of the original equation. ■

Extraneous Solutions

EXAMPLE 2

Solve the equation $\sqrt{x+1} + 5 = 3$.

Solution

We isolate the radical on one side of the equation and proceed as follows:

$$\sqrt{x+1} + 5 = 3$$
$$\sqrt{x+1} = -2 \quad \text{Add } -5 \text{ to both sides.}$$
$$(\sqrt{x+1})^2 = (-2)^2 \quad \text{Square both sides.}$$
$$x + 1 = 4 \quad \text{Simplify.}$$
$$x = 3 \quad \text{Add } -1 \text{ to both sides.}$$

Check:
$$\sqrt{x+1} + 5 = 3$$
$$\sqrt{3+1} + 5 \stackrel{?}{=} 3 \quad \text{Replace } x \text{ with 3.}$$
$$\sqrt{4} + 5 \stackrel{?}{=} 3$$
$$2 + 5 \stackrel{?}{=} 3$$
$$7 = 3$$

Since $7 = 3$ is *false*, the number 3 is *not* a solution of the given equation. Thus, the original equation has *no* solution. Work Progress Check 16. ■

Example 2 shows that squaring both sides of an equation can lead to **extraneous solutions**. These solutions do not satisfy the original equation and must be discarded.

Follow these steps to solve an equation with radical expressions.

Progress Check 16

Solve: $\sqrt{x+2} - 5 = 4$.

16. 79

> 1. Whenever possible, rearrange the terms so that no more than one radical appears on either side of the equation.
> 2. Square both sides of the equation and solve the resulting equation.
> 3. Check the solution in the original equation. This step is required.

EXAMPLE 3

Solve the equation $\sqrt{x + 12} = 3\sqrt{x + 4}$.

Solution

We square both sides to eliminate the radicals and proceed as follows:

$$\sqrt{x + 12} = 3\sqrt{x + 4}$$
$$(\sqrt{x + 12})^2 = (3\sqrt{x + 4})^2 \quad \text{Square both sides.}$$
$$x + 12 = 9(x + 4) \quad \text{Simplify.}$$
$$x + 12 = 9x + 36 \quad \text{Remove parentheses.}$$
$$-24 = 8x \quad \text{Add } -x \text{ and } -36 \text{ to both sides.}$$
$$-3 = x \quad \text{Divide both sides by 8.}$$

Check:

$$\sqrt{x + 12} = 3\sqrt{x + 4}$$
$$\sqrt{-3 + 12} \stackrel{?}{=} 3\sqrt{-3 + 4} \quad \text{Replace } x \text{ with } -3.$$
$$\sqrt{9} \stackrel{?}{=} 3\sqrt{1}$$
$$3 = 3$$

EXAMPLE 4

Solve the equation $x = \sqrt{2x + 10} - 1$.

Solution

$$x = \sqrt{2x + 10} - 1$$
$$x + 1 = \sqrt{2x + 10} \quad \text{Add 1 to both sides.}$$
$$(x + 1)^2 = (\sqrt{2x + 10})^2 \quad \text{Square both sides.}$$
$$x^2 + 2x + 1 = 2x + 10 \quad \text{Remove parentheses.}$$
$$x^2 - 9 = 0 \quad \text{Add } -2x \text{ and } -10 \text{ to both sides.}$$
$$(x - 3)(x + 3) = 0 \quad \text{Factor.}$$
$$x - 3 = 0 \quad \text{or} \quad x + 3 = 0 \quad \text{Set each factor equal to 0.}$$
$$x = 3 \quad \quad x = -3 \quad \text{Solve each linear equation.}$$

Check:

For $x = 3$	For $x = -3$
$x = \sqrt{2x + 10} - 1$	$x = \sqrt{2x + 10} - 1$
$3 \stackrel{?}{=} \sqrt{2(3) + 10} - 1$	$-3 \stackrel{?}{=} \sqrt{2(-3) + 10} - 1$
$3 \stackrel{?}{=} \sqrt{16} - 1$	$-3 \stackrel{?}{=} \sqrt{4} - 1$
$3 \stackrel{?}{=} 4 - 1$	$-3 \stackrel{?}{=} 2 - 1$
$3 = 3$	$-3 \stackrel{?}{=} 1$

The given equation has only one solution, $x = 3$. The other value, $x = -3$, does not satisfy the equation. It is an extraneous solution. Work Progress Check 17.

Progress Check 17

Solve: $\sqrt{a} = a - 6$.

EXAMPLE 5

The square root of the sum of 3 and a number is 3 greater than the number. Find the number.

Solution

Let x represent the number. Then $3 + x$ represents the sum of 3 and the number. We translate the words of the problem into the following equation.

The square root of the sum of 3 and a number	is	3 greater than the number.
$\sqrt{3 + x}$	=	$3 + x$

$$\sqrt{3 + x} = 3 + x$$

$3 + x = 9 + 6x + x^2$	Square both sides.
$0 = x^2 + 5x + 6$	Add -3 and $-x$ to both sides.
$0 = (x + 2)(x + 3)$	Factor $x^2 + 5x + 6$.
$x + 2 = 0$ or $x + 3 = 0$	Set each factor equal to 0.
$x = -2$ $\quad$ $x = -3$	Solve each linear equation.

The number can be either -2 or -3.

Check: The square root of 3 plus -2 equals 1, and 1 is 3 greater than -2. Thus, -2 checks. The square root of 3 plus -3 equals 0, and 0 is 3 greater than -3. Thus, -3 checks.

17. 9; -4 is extraneous
Progress Check Answers

15.5 Exercises

In Exercises 1–44, solve each equation. Check all solutions. If an equation has no solutions, so indicate.

1. $\sqrt{x} = 3$
2. $\sqrt{x} = 5$
3. $\sqrt{x} = 7$
4. $\sqrt{x} = 2$
5. $\sqrt{x} = -4$
6. $\sqrt{x} = -1$
7. $\sqrt{x + 3} = 2$
8. $\sqrt{x - 2} = 3$
9. $\sqrt{x - 5} = 5$
10. $\sqrt{x + 8} = 12$
11. $\sqrt{3 - x} = -2$
12. $\sqrt{5 - x} = 10$
13. $\sqrt{6 + 2x} = 4$
14. $\sqrt{7 + 3x} = -4$
15. $\sqrt{5x - 5} = 5$
16. $\sqrt{6x + 19} = 7$
17. $\sqrt{4x - 3} = 3$
18. $\sqrt{11x - 2} = 3$
19. $\sqrt{13x + 14} = 1$
20. $\sqrt{8x + 9} = 1$

21. $\sqrt{x+3} + 5 = 12$
22. $\sqrt{x-5} - 3 = 4$
23. $\sqrt{2x+10} + 3 = 5$
24. $\sqrt{3x+4} + 7 = 12$

25. $\sqrt{5x+9} + 4 = 7$
26. $\sqrt{9x+25} - 2 = 3$
27. $\sqrt{7-5x} + 4 = 3$
28. $\sqrt{7+6x} - 4 = -3$

29. $\sqrt{3x+3} = 3\sqrt{x-1}$
30. $2\sqrt{4x+5} = 5\sqrt{x+4}$

31. $2\sqrt{3x+4} = \sqrt{5x+9}$
32. $\sqrt{10-3x} = \sqrt{2x+20}$

33. $\sqrt{3x+6} = 2\sqrt{2x-11}$
34. $2\sqrt{9x+16} = \sqrt{3x+64}$

35. $\sqrt{x+1} = x - 1$
36. $\sqrt{x+4} = x - 2$
37. $\sqrt{x+1} = x + 1$
38. $\sqrt{x+9} = x + 7$

39. $\sqrt{7x+2} - 2x = 0$
40. $\sqrt{3x+3} + 5 = x$
41. $x - 1 = \sqrt{x-1}$
42. $x - 2 = \sqrt{x+10}$

43. $x - 3 = \sqrt{3-x}$
44. $x - 4 = \sqrt{x-4}$

45. The square root of the sum of a certain number and 8 is 13. Find the number.

46. The square root of the sum of a certain number and 12 is 15. Find the number.

47. The square root of 3 less than a certain number is 5. Find the number.

48. The square root of 5 less than a certain number is 0. Find the number.

49. The square root of 7 more than a number is the square root of 14. Find the number.

50. The square root of twice a number is equal to the square root of 8 more than the number. Find the number.

51. The square root of twice a number is equal to the square root of 18 more than that number. Find the number.

52. Three times the square root of 2 greater than a number is equal to the square root of 4 more than 11 times that number. Find the number.

53. The square root of 4 more than a number is equal to 2 less than the number. Find the number.

54. Five times the square root of 1 less than a number is equal to 3 more than the number. Find the number. (*Hint:* There are two answers.)

55. What numbers are equal to their own square roots?

56. The sum of a number and its square root is equal to 0. Find the number.

57. The square root of the sum of two consecutive integers is 1 less than the smaller integer. Find the smallest integer.

58. The square root of the sum of two consecutive odd integers is 1 less than 3 times the smaller integer. Find the integers.

59. Can the sum of 4 and the principal square root of a number ever equal 0? Explain.

60. Can the difference of 4 and the principal square root of a number ever equal 0? Explain.

61. Einstein's Theory of Relativity predicts that an object moving at speed v will be shortened in the direction of its motion by a factor f, given by

$$f = \sqrt{1 - \frac{v^2}{c^2}}$$

where c is the speed of light. Solve this equation for v^2.

62. Einstein's Theory of Relativity predicts that a clock moving at speed v will run slower by a factor f, given by

$$f = \frac{1}{\sqrt{1 - \frac{v^2}{c^2}}}$$

where c is the speed of light. Solve this equation for v^2.

Review Exercises

In Review Exercises 1–6, perform the indicated operations.

1. $3x^2y^3(-2xy^4)$
2. $-2a^4b^2c(5a^3bc^4)$
3. $(x+4)(x-3)$
4. $(2x-3)(3x+4)$

5. $(3x+4)^2$
6. $(x+1)(x+2)(x+3)$

In Review Exercises 7–10, solve each system of equations.

7. $\begin{cases} x+y = 5 \\ x-y = -1 \end{cases}$
8. $\begin{cases} 2x+y = 0 \\ x+3y = 5 \end{cases}$
9. $\begin{cases} 2x+3y = 0 \\ 3x-2y = 13 \end{cases}$
10. $\begin{cases} 3x-4y = 11 \\ 4x+y = -17 \end{cases}$

11. A woman has $2.75 in nickels, dimes, and quarters. She has twice as many dimes as nickels and 4 fewer quarters than dimes. How many nickels does she have?

12. John invested equal amounts in each of three accounts paying 6%, 7%, and 8% annual interest. His annual income from these investments is $105. How much did he invest at each rate?

15.6 Rational Exponents

- Rational Exponents with Numerators of 1
- Rational Exponents with Numerators Other Than 1
- Simplifying Expressions with Rational Exponents

Rational Exponents with Numerators of 1

We can raise many bases to fractional powers. To give meaning to rational (fractional) exponents, we consider the number $\sqrt{7}$. Because $\sqrt{7}$ is the positive number whose square is 7, we have

$$(\sqrt{7})^2 = 7$$

We now consider the symbol $7^{1/2}$. If we demand that fractional exponents obey the same rules as integral exponents, then the square of $7^{1/2}$ must be 7, because

$$(7^{1/2})^2 = 7^{(1/2)2}$$
$$= 7^1$$
$$= 7$$

Since the square of $7^{1/2}$ is 7, and the square of $\sqrt{7}$ is 7, we can define $7^{1/2}$ to be $\sqrt{7}$. Similarly, we define

$7^{1/3}$ to be $\sqrt[3]{7}$

$7^{1/7}$ to be $\sqrt[7]{7}$

and so on. In general, we define the **rational exponent** $\frac{1}{n}$ as follows:

Definition: If n is a positive integer greater than 1 and $\sqrt[n]{x}$ is a real number, then

$$x^{1/n} = \sqrt[n]{x}$$

EXAMPLE 1

Simplify **a.** $64^{1/2}$, **b.** $64^{1/3}$, **c.** $(-64)^{1/3}$, and **d.** $64^{1/6}$.

Solution

a. $64^{1/2} = \sqrt{64} = 8$ **b.** $64^{1/3} = \sqrt[3]{64} = 4$

c. $(-64)^{1/3} = \sqrt[3]{-64} = -4$ **d.** $64^{1/6} = \sqrt[6]{64} = 2$

Work Progress Check 18.

Progress Check 18

Simplify: $125^{1/3}$.

Rational Exponents with Numerators Other Than 1

The definition of $x^{1/n}$ can be extended to fractional exponents for which the numerator is not 1. For example, because $4^{3/2}$ can be written as $(4^{1/2})^3$, we have

$$4^{3/2} = (4^{1/2})^3 = (\sqrt{4})^3 = 2^3 = 8$$

Similarly, because $4^{3/2}$ can be written as $(4^3)^{1/2}$, we have

$$4^{3/2} = (4^3)^{1/2} = 64^{1/2} = \sqrt{64} = 8$$

In general, $x^{m/n}$ can be written as $(x^{1/n})^m$ or as $(x^m)^{1/n}$. Since $(x^{1/n})^m = (\sqrt[n]{x})^m$ and $(x^m)^{1/n} = \sqrt[n]{x^m}$, we make the following definition.

> **Definition:** If m and n are positive integers, x is nonnegative, and the fraction m/n cannot be simplified, then
> $$x^{m/n} = \sqrt[n]{x^m} = (\sqrt[n]{x})^m$$

EXAMPLE 2

Simplify **a.** $8^{2/3}$ and **b.** $(-27)^{4/3}$.

Solution

a. $8^{2/3} = (\sqrt[3]{8})^2 = 2^2 = 4$ or $8^{2/3} = \sqrt[3]{8^2} = \sqrt[3]{64} = 4$

b. $(-27)^{4/3} = (\sqrt[3]{-27})^4 = (-3)^4 = 81$ or
$(-27)^{4/3} = \sqrt[3]{(-27)^4} = \sqrt[3]{531,441} = 81$

The work in Example 2 suggests that in order to avoid large numbers, it is usually easier to take the root of the base first.

EXAMPLE 3

Simplify **a.** $125^{4/3}$, **b.** $9^{5/2}$, and **c.** $-25^{3/2}$.

---

18. 5

Progress Check Answers

Progress Check 19

Simplify:

a. $27^{2/3}$.

b. $36^{3/2}$.

Solution

a. $125^{4/3} = (\sqrt[3]{125})^4 = (5)^4 = 625$ b. $9^{5/2} = (\sqrt{9})^5 = (3)^5 = 243$

c. $-25^{3/2} = -(\sqrt{25})^3 = -(5)^3 = -125$

Work Progress Check 19.

Because of the definition of $x^{1/n}$, the familiar rules of exponents are valid for rational exponents. The following example illustrates the use of each rule.

EXAMPLE 4

a. $4^{2/5} 4^{1/5} = 4^{2/5 + 1/5} = 4^{3/5}$ $x^m x^n = x^{m+n}$

b. $(5^{2/3})^{1/2} = 5^{(2/3)(1/2)} = 5^{1/3}$ $(x^m)^n = x^{mn}$

c. $(3x)^{2/3} = 3^{2/3} x^{2/3}$ $(xy)^m = x^m y^m$

d. $\dfrac{4^{3/5}}{4^{2/5}} = 4^{3/5 - 2/5} = 4^{1/5}$ $\dfrac{x^m}{x^n} = x^{m-n}$

e. $\left(\dfrac{3}{2}\right)^{2/5} = \dfrac{3^{2/5}}{2^{2/5}}$ $\left(\dfrac{x}{y}\right)^n = \dfrac{x^n}{y^n}$

f. $4^{-2/3} = \dfrac{1}{4^{2/3}}$ $x^{-n} = \dfrac{1}{x^n}$

g. $5^0 = 1$ $x^0 = 1$

Simplifying Expressions with Rational Exponents

We can use the rules of exponents to simplify expressions with rational exponents.

EXAMPLE 5

Simplify **a.** $64^{-2/3}$, **b.** $(x^2)^{1/2}$, **c.** $(x^6 y^4)^{1/2}$, and **d.** $(27x^{12})^{-1/3}$. Assume that all variables represent positive numbers.

Solution

a. $64^{-2/3} = \dfrac{1}{64^{2/3}}$ b. $(x^2)^{1/2} = x^{2(1/2)}$

$= \dfrac{1}{(64^{1/3})^2}$ $= x^1$

$= \dfrac{1}{4^2}$ $= x$

$= \dfrac{1}{16}$

c. $(x^6 y^4)^{1/2} = x^{6(1/2)} y^{4(1/2)}$ d. $(27x^{12})^{-1/3} = \dfrac{1}{(27x^{12})^{1/3}}$

$= x^3 y^2$ $= \dfrac{1}{27^{1/3} x^{12(1/3)}}$

$= \dfrac{1}{3x^4}$

Progress Check 20

Simplify:

a. $27^{-2/3}$.

b. $(a^8 b^4)^{3/2}$.

---

19. a. 9   b. 216
20. a. $\dfrac{1}{9}$ b. $a^{12} b^6$

Progress Check Answers

Work Progress Check 20.

15.6 Exercises

In Exercises 1–24, simplify each expression.

1. $81^{1/2}$
2. $100^{1/2}$
3. $-144^{1/2}$
4. $-400^{1/2}$
5. $\left(\dfrac{1}{4}\right)^{1/2}$
6. $\left(\dfrac{1}{25}\right)^{1/2}$
7. $\left(\dfrac{4}{49}\right)^{1/2}$
8. $\left(\dfrac{9}{64}\right)^{1/2}$
9. $27^{1/3}$
10. $8^{1/3}$
11. $-125^{1/3}$
12. $-1000^{1/3}$
13. $(-8)^{1/3}$
14. $(-125)^{1/3}$
15. $\left(\dfrac{1}{64}\right)^{1/3}$
16. $\left(\dfrac{1}{1000}\right)^{1/3}$
17. $\left(\dfrac{27}{64}\right)^{1/3}$
18. $\left(\dfrac{64}{125}\right)^{1/3}$
19. $16^{1/4}$
20. $81^{1/4}$
21. $32^{1/5}$
22. $-32^{1/5}$
23. $-243^{1/5}$
24. $\left(-\dfrac{1}{32}\right)^{1/5}$

In Exercises 25–44, simplify each expression.

25. $81^{3/2}$
26. $16^{3/2}$
27. $25^{3/2}$
28. $4^{5/2}$
29. $125^{2/3}$
30. $8^{4/3}$
31. $1000^{2/3}$
32. $27^{2/3}$
33. $(-8)^{2/3}$
34. $(-125)^{2/3}$
35. $(32)^{3/5}$
36. $-243^{3/5}$
37. $81^{3/4}$
38. $256^{3/4}$
39. $(-32)^{3/5}$
40. $243^{2/5}$
41. $\left(\dfrac{8}{27}\right)^{2/3}$
42. $\left(\dfrac{27}{64}\right)^{2/3}$
43. $\left(\dfrac{16}{625}\right)^{-3/4}$
44. $\left(\dfrac{49}{64}\right)^{-3/2}$

In Exercises 45–68, simplify each expression. Write each answer without using negative exponents.

45. $6^{3/5} 6^{2/5}$
46. $3^{4/7} 3^{3/7}$
47. $5^{2/3} 5^{4/3}$
48. $2^{7/8} 2^{9/8}$
49. $(7^{2/5})^{5/2}$
50. $(8^{1/3})^3$
51. $(5^{2/7})^7$
52. $(3^{3/8})^8$
53. $\dfrac{8^{3/2}}{8^{1/2}}$
54. $\dfrac{11^{9/7}}{11^{2/7}}$
55. $\dfrac{5^{11/3}}{5^{2/3}}$
56. $\dfrac{27^{13/15}}{27^{8/15}}$
57. $(2^{1/2} 3^{1/2})^2$
58. $(3^{2/3} 5^{1/3})^3$
59. $(4^{3/4} 3^{1/4})^4$
60. $(2^{1/5} 3^{2/5})^5$
61. $4^{-1/2}$
62. $8^{-1/3}$
63. $27^{-2/3}$
64. $36^{-3/2}$
65. $16^{-3/2}$
66. $100^{-5/2}$
67. $(-27)^{-4/3}$
68. $(-8)^{-4/3}$

In Exercises 69–88, simplify each expression. Assume that all variables represent positive numbers.

69. $(x^{1/2})^2$
70. $(x^9)^{1/3}$
71. $(x^{12})^{1/6}$
72. $(x^{18})^{1/9}$
73. $(x^{18})^{2/9}$
74. $(x^{12})^{3/4}$
75. $x^{5/6} x^{7/6}$
76. $x^{2/3} x^{7/3}$

672 Chapter 15 Roots and Radical Expressions

77. $y^{4/7}y^{10/7}$

78. $y^{5/11}y^{6/11}$

79. $\dfrac{x^{3/5}}{x^{1/5}}$

80. $\dfrac{x^{4/3}}{x^{1/3}}$

81. $\dfrac{x^{1/7}x^{3/7}}{x^{2/7}}$

82. $\dfrac{x^{5/6}x^{5/6}}{x^{7/6}}$

83. $\left(\dfrac{x^{3/5}}{x^{2/5}}\right)^5$

84. $\left(\dfrac{x^{2/9}}{x^{1/9}}\right)^9$

85. $\left(\dfrac{y^{2/7}y^{3/7}}{y^{4/7}}\right)^{49}$

86. $\left(\dfrac{z^{3/5}z^{6/5}}{z^{2/5}}\right)^5$

87. $\left(\dfrac{y^{5/6}y^{7/6}}{y^{1/3}y}\right)^3$

88. $\left(\dfrac{t^{4/9}t^{5/9}}{t^{1/9}t^{2/9}}\right)^9$

In Exercises 89–100, simplify each expression. Assume that all variables represent positive numbers.

89. $x^{2/3}x^{3/4}$

90. $a^{3/5}a^{1/2}$

91. $(b^{1/2})^{3/5}$

92. $(x^{2/5})^{4/7}$

93. $\dfrac{t^{2/3}}{t^{2/5}}$

94. $\dfrac{p^{3/4}}{p^{1/3}}$

95. $\dfrac{x^{4/5}x^{1/3}}{x^{2/15}}$

96. $\dfrac{y^{2/3}y^{3/5}}{y^{1/5}}$

97. $\dfrac{a^{2/5}a^{1/5}}{a^{-1/3}}$

98. $\dfrac{q^{3/4}q^{4/5}}{q^{-2/3}}$

99. $\dfrac{12b^{-1/3}b^{-3/4}}{4b^{3/5}}$

100. $\dfrac{4c^{-3/4}c^{1/6}}{8c^{1/4}}$

Review Exercises

In Review Exercises 1–4, factor each expression.

1. $3x^3y^2z^4 - 6xyz^5 + 15x^2yz^2$

2. $3z^2 - 15tz + 12t^2$

3. $a^4 - b^4$

4. $30r^4 - 200r^2 + 5r^3$

In Review Exercises 5–6, solve each equation.

5. $\dfrac{x-5}{7} + \dfrac{2}{5} = \dfrac{7-x}{5}$

6. $\dfrac{t}{t+2} - 1 = \dfrac{1}{1-t}$

7. Find the slope m of the line represented by the equation $2x - 3y = 18$.

8. Write the equation of the line with slope -4 and y-intercept 7.

In Review Exercises 9–10, solve each system of equations.

9. $\begin{cases} 2x + y = 4 \\ 4x - 3y = 13 \end{cases}$

10. $\begin{cases} 3x + 4y = 2 \\ 12y = 3(2 - 3x) \end{cases}$

11. The **harmonic mean** of two numbers is equal to the quotient obtained when twice their product is divided by their sum. Find two consecutive positive integers with harmonic mean of $\frac{4}{3}$.

12. The **geometric mean** of two numbers is the square root of their product. Find two consecutive positive integers with geometric mean of $3\sqrt{10}$.

15.7 The Distance Formula

- The Pythagorean Theorem
- The Distance Formula

A triangle with a 90° angle is called a **right triangle**. The longest side of a right triangle is called the **hypotenuse**, which is the side opposite the right angle. The remaining two sides are called **legs**. Side c is the hypotenuse, and sides a and b are the legs of the triangle shown in Figure 15-1.

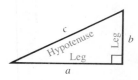

Figure 15-1

The Pythagorean Theorem

A theorem from geometry, called the **Pythagorean theorem**, relates the lengths of the sides of any right triangle.

> **The Pythagorean Theorem:** If the length of the hypotenuse of a right triangle is c and the lengths of the two legs are a and b, then
> $$c^2 = a^2 + b^2$$

The Pythagorean theorem is useful because equal positive numbers have equal positive square roots. This property is called the **square root property of equality**.

> **The Square Root Property of Equality:** If a and b are positive numbers, then
> If $a = b$, then $\sqrt{a} = \sqrt{b}$.

Because the lengths of the sides of a triangle are positive, we can use the square root property of equality and the Pythagorean theorem to calculate the lengths of the sides of a right triangle.

EXAMPLE 1

In the right triangle of Figure 15-1, let $a = 9$ feet and $b = 12$ feet. Find the length of the hypotenuse c.

Solution

We can use the Pythagorean theorem, with $a = 9$ and $b = 12$.

$$c^2 = a^2 + b^2$$
$$c^2 = 9^2 + 12^2 \quad \text{Let } a = 9 \text{ and } b = 12.$$
$$c^2 = 81 + 144 \quad \text{Simplify.}$$
$$c^2 = 225 \quad \text{Simplify.}$$
$$\sqrt{c^2} = \sqrt{225} \quad \text{Use the square root property of equality.}$$
$$c = 15 \quad \text{Simplify.}$$

The hypotenuse of the triangle is 15 feet.

EXAMPLE 2

In the right triangle of Figure 15-1, let $a = 5$ feet and the hypotenuse $c = 13$ feet. Find the length of leg b.

Solution

We can use the Pythagorean theorem, with $a = 5$ and $c = 13$.

$$c^2 = a^2 + b^2$$
$$13^2 = 5^2 + b^2 \quad \text{Let } a = 5 \text{ and } c = 13.$$
$$169 = 25 + b^2 \quad \text{Simplify.}$$
$$169 - 25 = b^2 \quad \text{Add } -25 \text{ to both sides.}$$
$$144 = b^2 \quad \text{Simplify.}$$
$$\sqrt{144} = \sqrt{b^2} \quad \text{Use the square root property of equality.}$$
$$12 = b \quad \text{Simplify.}$$

The length of leg b is 12 feet. Work Progress Check 21.

Progress Check 21

If two legs of a right triangle are 7 feet and 24 feet, find the length of the hypotenuse.

EXAMPLE 3

A 26-foot ladder rests against the side of a building. The base of the ladder is 10 feet from the wall. How far up the side of the building does the ladder reach?

Solution

The wall, the ground, and the ladder form a right triangle, as in Figure 15-2. The square of the hypotenuse is equal to the sum of the squares of the two legs. The hypotenuse is 26 feet, and one of the legs is the base-to-wall distance of 10 feet. Let d represent the other leg, which is the distance that the ladder reaches up the wall.

The hypotenuse squared	is	one leg squared	+	the other leg squared.
26^2	=	10^2	+	d^2

$$26^2 = 10^2 + d^2$$
$$676 = 100 + d^2 \quad \text{Simplify.}$$
$$676 - 100 = d^2 \quad \text{Add } -100 \text{ to both sides.}$$
$$576 = d^2 \quad \text{Simplify.}$$
$$\sqrt{576} = \sqrt{d^2} \quad \text{Use the square root property of equality.}$$
$$24 = d \quad \text{Simplify.}$$

Figure 15-2

The ladder will reach 24 feet up the side of the building.

EXAMPLE 4

The gable end of a roof is an isosceles right triangle with a span of 48 feet (see Figure 15-3). Find the distance from the eave to the peak.

21. 25 feet

Solution

In Figure 15-3, the equal sides of the isosceles triangle are the two legs of the right triangle, and the span of 48 feet is the length of the hypotenuse. Let x represent the length of each of the legs, the distance from eave to peak.

Figure 15-3

The hypotenuse squared	is	one leg squared	+	the other leg squared.
48^2	=	x^2	+	x^2

$48^2 = x^2 + x^2$

$2304 = 2x^2$ Simplify and combine like terms.

$1152 = x^2$ Divide both sides by 2.

$\sqrt{1152} = \sqrt{x^2}$ Use the square root property of equality.

$34 \approx x$ Use a calculator to find the approximate value of $\sqrt{1152}$.

The eave-to-peak distance of the roof is approximately 34 feet. ∎

The Distance Formula

We can derive a formula for finding the distance between two points $P(x_1, y_1)$ and $Q(x_2, y_2)$ plotted on a rectangular coordinate system. The distance d between points P and Q is the length of the hypotenuse of the triangle in Figure 15-4. The two legs have lengths $x_2 - x_1$ and $y_2 - y_1$.

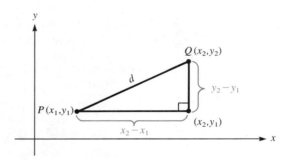

Figure 15-4

By the Pythagorean theorem, the square of the hypotenuse of this triangle is equal to the sum of the squares of the two legs. Thus, we have

$$d^2 = (x_2 - x_1)^2 + (y_2 - y_1)^2$$

To solve for d, we can use the square root property of equality to get

$$d = \sqrt{(x_2 - x_1)^2 + (y_2 - y_1)^2}$$

The result is called the **distance formula**.

The Distance Formula: The distance d between points $P(x_1, y_1)$ and $Q(x_2, y_2)$ is given by the formula

$$d = \sqrt{(x_2 - x_1)^2 + (y_2 - y_1)^2}$$

EXAMPLE 5

Find the distance between the two points $P(1, 5)$ and $Q(4, 9)$.

Solution

Let $(x_1, y_1) = (1, 5)$ and let $(x_2, y_2) = (4, 9)$. In other words, we let

$x_1 = 1, \quad y_1 = 5, \quad x_2 = 4,$ and $y_2 = 9$

and substitute these values into the distance formula and simplify.

$d = \sqrt{(x_2 - x_1)^2 + (y_2 - y_1)^2}$
$= \sqrt{(4 - 1)^2 + (9 - 5)^2}$
$= \sqrt{3^2 + 4^2}$
$= \sqrt{9 + 16}$
$= \sqrt{25}$
$= 5$

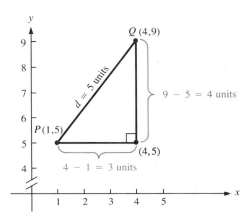

Figure 15-5

The distance between points P and Q is 5 units (see Figure 15-5).

EXAMPLE 6

Find the distance between the points $P(-3, 6)$ and $Q(2, -6)$.

Solution

Let $(x_1, y_1) = (-3, 6)$ and let $(x_2, y_2) = (2, -6)$. In other words, we let

$x_1 = -3, \quad y_1 = 6, \quad x_2 = 2,$ and $y_2 = -6$

and substitute these values into the distance formula and simplify.

$d = \sqrt{(x_2 - x_1)^2 + (y_2 - y_1)^2}$
$= \sqrt{[2 - (-3)]^2 + (-6 - 6)^2}$
$= \sqrt{(5)^2 + (-12)^2}$
$= \sqrt{25 + 144}$
$= \sqrt{169}$
$= 13$

The distance between points P and Q is 13 units. Work Progress Check 22.

Progress Check 22

Find the distance between $P(-2, 4)$ and $Q(4, -4)$.

22. 10 units

15.7 Exercises

In Exercises 1–10, refer to the right triangle of Illustration 1. Find the length of the unknown side.

1. $a = 4$ and $b = 3$. Find c.
2. $a = 6$ and $b = 8$. Find c.
3. $a = 5$ and $b = 12$. Find c.
4. $a = 15$ and $c = 17$. Find b.
5. $a = 21$ and $c = 29$. Find b.
6. $b = 16$ and $c = 34$. Find a.
7. $b = 45$ and $c = 53$. Find a.
8. $a = 7$ and $b = 1$. Find c.
9. $a = 5$ and $c = 9$. Find b.
10. $a = 1$ and $c = \sqrt{2}$. Find b.

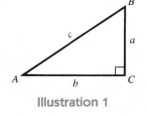

Illustration 1

In Exercises 11–20, find the distance between the given points.

11. (1, 2) (4, 6)
12. (2, 2) (5, 6)
13. (−2, 5) (6, −1)
14. (1, −8) (5, −12)
15. (−1, 4) (4, 16)
16. (−5, 7) (10, −1)
17. (−17, −3) (−23, 5)
18. (0, 0) (21, 20)
19. $\left(-\frac{1}{2}, 0\right) \left(\frac{5}{2}, -4\right)$
20. $(-2\sqrt{3}, \sqrt{3}) (\sqrt{3}, -2\sqrt{3})$

21. A 20-foot ladder reaches a window 16 feet above the ground. How far from the wall is the base of the ladder?

22. A 150-foot-tall tower is secured by three guy wires fastened at the top and to anchors 15 feet from the base of the tower. How long is each guy wire?

23. A 34-foot-long wire reaches from the top of a telephone pole to a point on the ground 16 feet from the base of the pole. How tall is the telephone pole?

24. A rectangular garden has sides of 28 and 45 feet. How long is a path that extends from one corner to the opposite corner?

25. The legs of a certain right triangle are equal, and the hypotenuse is $2\sqrt{2}$ units long. Find the length of each leg.

26. The sides of a square are 3 feet long. Find the length of each diagonal of the square.

27. The diagonal of a square is 3 feet long. Find the square's perimeter.

28. A woman drives 4.2 miles east and then 4.0 miles north. How far is she from her starting point?

29. A man drives 7 miles north, 8 miles west, and 5 miles north. How far is he from his starting point?

30. The entrance to a one-way tunnel is a rectangle with a semicircular roof. Its dimensions are given in Illustration 2. How high can a 10-foot-wide truck be, without getting stuck in the tunnel?

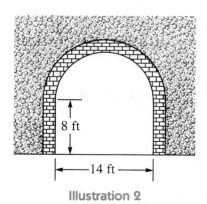

Illustration 2

678   Chapter 15   Roots and Radical Expressions

31. The square in Illustration 3 is inscribed in a circle. The sides of the square are 6 inches long. Find the area of the circle.

32. Will a square with area 40 square inches fit inside a circle with an area of 120 square inches?

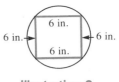

Illustration 3

Review Exercises

In Review Exercises 1–4, graph each equation.

1. $x = 3$ **2.** $y = -3$ **3.** $-2x + y = 4$ **4.** $3x - 4y = 12$

In Exercises 5–8, graph each inequality.

5. $3y \leq 9$ **6.** $-x > 3$ **7.** $4x - y > 4$ **8.** $2x - 5y \leq 10$

9. Write 7.2×10^6 in standard notation.

10. Write 0.435×10^{-6} in standard notation.

11. Two cars leave Malta at the same time, one heading east at 55 miles per hour and the other heading west at 50 miles per hour. How long will it take for the cars to be 315 miles apart?

12. Lemon drops worth $1.10 per pound are mixed with jelly beans worth $1.25 per pound to make 30 pounds of a mixture worth $1.15 per pound. How many pounds of jelly beans should be used?

15.8 Special Right Triangles

■ Isosceles Right Triangles
■ 30°-60° Right Triangles

In this section, we will discuss certain right triangles that are of special interest.

Isosceles Right Triangles

An isosceles right triangle is shown in Figure 15-6. Since the sum of the measures of the angles of the triangle must equal 180°, the sum of the base angles ∠A and ∠B must be 90°. Because the base angles of an isosceles triangle are congruent, the measure of each base angle is 45°.

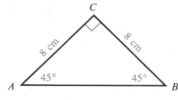

Figure 15-6

To determine the relationships between the lengths of the sides of a 45°-45°-90° triangle, we consider the isosceles right triangle shown in Figure 15-7, which has legs of x units and a hypotenuse of h units. By using the Pythagorean theorem, we have

$$h^2 = x^2 + x^2$$
$$h^2 = 2x^2$$
$$h = \sqrt{2x^2}$$
$$h = x\sqrt{2}$$

Thus, the measure of the hypotenuse of an isosceles right triangle is the measure of one of the congruent legs multiplied by $\sqrt{2}$.

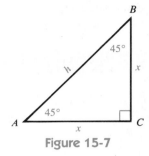

Figure 15-7

Theorem: In an isosceles right triangle, the length of the hypotenuse is equal to the length of one of the congruent legs multiplied by $\sqrt{2}$.

EXAMPLE 1

Find the length of the hypotenuse of an isosceles right triangle with one leg of length 5 centimeters.

Solution

In an isosceles right triangle, the length of the hypotenuse is always equal to the length of one of the congruent legs multiplied by $\sqrt{2}$. Because the length of a leg is 5 centimeters, the length of the hypotenuse is $5\sqrt{2}$ centimeters. ■

EXAMPLE 2

The length of the hypotenuse of an isosceles right triangle is 9 meters. Find the length of each leg.

Solution

Let the length of each of the congruent legs be x meters. Since the length of the hypotenuse is equal to the length of one of the legs multiplied by $\sqrt{2}$, we have

$$\sqrt{2}x = 9$$
$$x = \frac{9}{\sqrt{2}}$$
$$x = \frac{9\sqrt{2}}{\sqrt{2}\sqrt{2}} \quad \text{Multiply both the numerator and denominator by } \sqrt{2}.$$
$$x = \frac{9\sqrt{2}}{2}$$

The length of each leg of the triangle is $\dfrac{9\sqrt{2}}{2}$ meters. Work Progress Check 23. ■

Progress Check 23

Find the length of the hypotenuse of an isosceles right triangle with one leg 10 centimeters in length.

23. $10\sqrt{2}$ cm

30°-60° Right Triangles

Another special right triangle is the 30°-60° right triangle ABC shown in Figure 15-8.

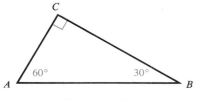

Figure 15-8

To determine the relationships between the lengths of the sides of a 30°-60° right triangle, we refer to Figure 15-9, in which $\triangle ABC \cong \triangle DBC$. If $m(\overline{AB}) = x$ units, $m(\overline{BD}) = x$ units also, since they are corresponding parts of congruent triangles. It follows that $m(\overline{AD}) = 2x$ units. Since $\triangle ADC$ is equiangular, it is also equilateral, and

$$m(\overline{AC}) = m(\overline{AD}) = 2x \text{ units}$$

Thus, in $\triangle ABC$, the length of the hypotenuse $\overline{AC}$ is twice as long as the leg opposite the 30° angle:

$$m(\overline{AC}) = 2[m(\overline{AB})]$$

To find the length of $\overline{BC}$, we use the Pythagorean theorem:

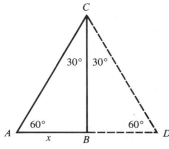

Figure 15-9

$$m(\overline{AC})^2 = m(\overline{AB})^2 + m(\overline{BC})^2$$
$$(2x)^2 = x^2 + m(\overline{BC})^2 \quad \text{Substitute } 2x \text{ for } m(\overline{AC}) \text{ and } x \text{ for } m(\overline{AB}).$$
$$4x^2 = x^2 + m(\overline{BC})^2$$
$$3x^2 = m(\overline{BC})^2$$
$$\sqrt{3}\,x = m(\overline{BC}) \quad \text{Take the square root of both sides.}$$

Thus, the leg opposite the 60° angle is $\sqrt{3}$ times as long as the leg opposite the 30° angle:

$$m(\overline{BC}) = \sqrt{3}\,[m(\overline{AB})]$$

These results are summarized in the following theorem:

Theorem: In a 30°-60° right triangle, the length of the hypotenuse is twice the length of the leg opposite the 30° angle.

The length of the leg opposite the 60° angle is equal to the length of the leg opposite the 30° angle multiplied by $\sqrt{3}$.

EXAMPLE 3

From level ground, a girl 43 feet from the base of a flagpole sights its top at an angle of 30° (see Figure 15-10). How tall is the flagpole?

Solution

The triangle in Figure 15-10 is a 30°-60° right triangle in which h represents the height of the flagpole. By the previous theorem, the leg opposite the 60° angle is $\sqrt{3}$ times the length of the leg opposite the 30° angle. Proceed as follows:

$$43 = \sqrt{3}\,h$$

$$\frac{43}{\sqrt{3}} = h$$

$$h = \frac{43\sqrt{3}}{3}$$

The flagpole is $\frac{43\sqrt{3}}{3}$ feet tall. Work Progress Check 24.

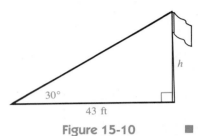

Figure 15-10

Progress Check 24

A 30°-60° right triangle has a hypotenuse that is 12 centimeters long. Find the length of the leg that is opposite the 60° angle.

EXAMPLE 4

The equilateral triangle in Figure 15-11 has a perimeter of 18 meters. Find $m(\overline{CD})$.

Solution

If the perimeter of an equilateral triangle is 18 meters, the length of one leg is 6 meters. Thus,

$$m(\overline{AC}) = 6$$

Because $\triangle ADC$ is a 30°-60° right triangle,

$$m(\overline{AC}) = 2\,[m(\overline{AD})]$$
$$6 = 2\,[m(\overline{AD})]$$
$$3 = m(\overline{AD})$$

and

$$m(\overline{CD}) = [m(\overline{AD})]\,\sqrt{3}$$
$$= 3\sqrt{3}$$

Thus, $m(\overline{CD}) = 3\sqrt{3}$ meters.

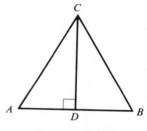

Figure 15-11

24. $6\sqrt{3}$ cm

Progress Check Answers

15.8 Exercises

In Exercises 1–10, refer to the isosceles right triangle shown in Illustration 1. Find the length of the indicated side.

1. $m(\overline{BC}) = 20$; $m(\overline{AB}) = $ _____

2. $m(\overline{AC}) = 30$; $m(\overline{AB}) = $ _____

3. $m(\overline{BC}) = 14$; $m(\overline{AC}) = $ _____

4. $m(\overline{AC}) = 12$; $m(\overline{BC}) = $ _____

5. $m(\overline{BC}) = \sqrt{3}$; $m(\overline{AB}) = $ _____

6. $m(\overline{AC}) = \sqrt{2}$; $m(\overline{AB}) = $ _____

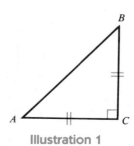

Illustration 1

7. m($\overline{AB}$) = 5; m($\overline{BC}$) = _____

8. m($\overline{AB}$) = 4; m($\overline{AC}$) = _____

9. m($\overline{AB}$) = 5$\sqrt{3}$; m($\overline{BC}$) = _____

10. m($\overline{AB}$) = 5$\sqrt{3}$; m($\overline{AC}$) = _____

In Exercises 11–20, refer to the 30°-60° right triangle in Illustration 2. Find the length of the indicated side.

11. m($\overline{AC}$) = 20; m($\overline{AB}$) = _____

12. m($\overline{BC}$) = 30$\sqrt{3}$; m($\overline{AB}$) = _____

13. m($\overline{BC}$) = 14; m($\overline{AC}$) = _____

14. m($\overline{AC}$) = 12; m($\overline{BC}$) = _____

15. m($\overline{BC}$) = $\dfrac{\sqrt{3}}{2}$; m($\overline{AB}$) = _____

16. m($\overline{AC}$) = $\dfrac{\sqrt{3}}{2}$; m($\overline{AB}$) = _____

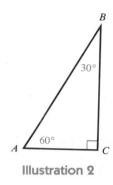

Illustration 2

17. m($\overline{AB}$) = 5; m($\overline{BC}$) = _____

18. m($\overline{AB}$) = 4; m($\overline{AC}$) = _____

19. m($\overline{AB}$) = $\dfrac{2\sqrt{3}}{3}$; m($\overline{BC}$) = _____

20. m($\overline{AB}$) = $\dfrac{2\sqrt{3}}{3}$; m($\overline{AC}$) = _____

21. Two airplanes flying at the same altitude are each 155 miles from an airport, one due west and one due south. How far are the planes from each other?

22. A boy is at camp 220 miles due east of his home. His sister is directly northeast of home and is also directly north of her brother. How far is the girl from home?

23. The guy wires to the top of a 230-foot tower make angles of 60° with the ground. How far from the base of the tower are the wires anchored?

24. A circus tightrope walker ascends to a platform by walking 212 feet up a wire that makes an angle of 30° with the horizontal. How high is the platform?

25. The hypotenuse of a 30°-60° right triangle is 4 inches. Find the length of each leg.

26. The hypotenuse of a 30°-60° right triangle is 8 inches. Find its perimeter.

27. The perimeter of a 30°-60° right triangle is (6 + 2$\sqrt{3}$) meters. How long is each leg?

28. The hypotenuse of an isosceles right triangle is 12 meters. Find the length of each leg.

29. The hypotenuse of an isosceles right triangle is 10 centimeters. Find its perimeter.

In Exercises 30–34, refer to Illustration 3.

30. If the perimeter of $\triangle ABC$ = 21 inches, find m($\overline{CD}$).

31. If m($\overline{CD}$) = 5$\sqrt{3}$ centimeters, find the perimeter of $\triangle ABC$.

32. If m($\overline{CD}$) = 8 centimeters, find m($\overline{BC}$).

33. If m($\overline{CD}$) = 14 meters, find the perimeter of $\triangle DBC$.

34. If the perimeter of $\triangle ABC$ is 36 meters, find m($\overline{CD}$).

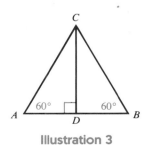

Illustration 3

Review Exercises

In Review Exercises 1–4, write each expression without using parentheses or negative exponents.

1. $\left(\dfrac{3b^2}{5b}\right)^3$
2. $\left(\dfrac{16x^{-4}}{4x^{-2}}\right)^3$
3. $\left(\dfrac{6m^{-3}}{12m^4}\right)^{-3}$
4. $\left(\dfrac{12y^3}{4y^{-2}}\right)^{-5}$

In Review Exercises 5–10, perform each multiplication.

5. $(3x + 4)(2x - 5)$
6. $(3a - 4b)(2a + 5b)$
7. $(3m + 5n)(2m - 3n)$
8. $(3x - 4y)^2$
9. $(3r - 2)(2r^2 - 3r + 2)$
10. $(2a + 3b)(a^2 - ab + b^2)$

Mathematics in Racing

To determine whether or not New Zealand's *Challenge* satisfied the IACC rules, substitute its dimensions into the official formula. Substitute 20.95 for L, 277.3 for A, and 17.56 for D and simplify. Use a calculator to verify the following steps:

$$\dfrac{L + 1.25\sqrt{A} - 9.8\sqrt[3]{D}}{0.388} \leq 42$$

$$\dfrac{20.95 + 1.25\sqrt{277.3} - 9.8\sqrt[3]{17.56}}{0.388} \leq 42$$

$$\dfrac{20.95 + 1.25(16.65) - 9.8(2.6)}{0.388} \leq 42$$

$$\dfrac{16.283}{0.388} \leq 42$$

$$41.96 \leq 42$$

Yes, New Zealand's *Challenge* was acceptable under the IACC rules. Even so, New Zealand lost the 1988 America's Cup trophy to the Americans.

Chapter Summary

Key Words

conjugate (15.4)
cube root (15.1)
distance formula (15.7)
division property
 of radicals (15.2)
extraneous
 solution (15.5)
hypotenuse (15.7)
imaginary number (15.1)
index (15.1)

irrational number (15.1)
leg of a right
 triangle (15.7)
like radicals (15.3)
multiplication property
 of radicals (15.2)
nth root of a
 number (15.1)
perfect integer
 square (15.1)

principal square
 root (15.1)
Pythagorean
 theorem (15.7)
radical sign (15.1)
radicand (15.1)
rational exponent (15.6)
rational number (15.1)
rationalizing the
 denominator (15.4)

real number (15.1)
right triangle (15.7)
simplified form of
 a radical (15.2)
square root (15.1)
square root property
 of equality (15.7)
squaring property of
 equality (15.5)

Key Ideas

(15.1) The number b is a square root of a if $b^2 = a$.

If a is a positive number, then the principal square root of a, denoted by $\sqrt{a}$, is the positive square root of a. The principal square root of 0 is 0.

If a is a positive integer and not a perfect square, then $\sqrt{a}$ is an irrational number.

The square root of a negative number is not a real number.

The cube root of a is denoted by $\sqrt[3]{a}$, and $\sqrt[3]{a} = b$ if $b^3 = a$.

The nth root of a is denoted by $\sqrt[n]{a}$, and $\sqrt[n]{a} = b$ if $b^n = a$.

(15.2) If a and b are nonnegative numbers, then
$$\sqrt{ab} = \sqrt{a}\sqrt{b}$$
and if $b \neq 0$, then
$$\sqrt{\frac{a}{b}} = \frac{\sqrt{a}}{\sqrt{b}}$$

To simplify an expression involving square roots, use the multiplication and division properties of radicals to remove perfect square factors from the radicands.

(15.3) Radical expressions can be added or subtracted if they contain like radicals. Often radicals can be converted to like radicals and then added.

(15.4) If a square root appears as a monomial in the denominator of a fraction, rationalize the denominator by multiplying both the numerator and the denominator of the fraction by some appropriate square root.

If the denominator of a fraction contains radicals within a binomial, multiply numerator and denominator by the conjugate of the denominator.

(15.5) If $a = b$, then $a^2 = b^2$.

To solve an equation that involves square roots, rearrange the terms of the equation so that no more than one radical appears on one side. Then square both sides of the equation and solve the resulting equation. Finally, *check the solution.*

(15.6) $x^{1/n} = \sqrt[n]{x}$ $x^{m/n} = \sqrt[n]{x^m} = (\sqrt[n]{x})^m$

(15.7) **The Pythagorean theorem.** In any right triangle, the sum of the squares of the two legs is equal to the square of the hypotenuse.

Let a and b be positive numbers. If $a = b$, then $\sqrt{a} = \sqrt{b}$.

The distance formula. The distance between the points (x_1, y_1) and (x_2, y_2) is given by the formula
$$d = \sqrt{(x_2 - x_1)^2 + (y_2 - y_1)^2}$$

(15.8) In an isosceles right triangle, the length of the hypotenuse is equal to the length of one of the congruent legs multiplied by $\sqrt{2}$.

In a 30°-60° right triangle, the length of the hypotenuse is twice the length of the leg opposite the 30° angle. The length of the leg opposite the 60° angle is equal to the length of the leg opposite the 30° angle multiplied by $\sqrt{3}$.

Chapter 15 Review Exercises

[15.1] In Review Exercises 1–12, find the value of each expression.

1. $\sqrt{25}$
2. $\sqrt{64}$
3. $-\sqrt{144}$
4. $-\sqrt{289}$
5. $\sqrt{256}$
6. $-\sqrt{64}$
7. $\sqrt{169}$
8. $-\sqrt{225}$
9. $-\sqrt[3]{-27}$
10. $-\sqrt[3]{125}$
11. $\sqrt[4]{81}$
12. $\sqrt[5]{32}$

In Review Exercises 13–16, use a calculator to find each value to three decimal places.

13. $\sqrt{21}$
14. $-\sqrt{15}$
15. $-\sqrt{57.3}$
16. $\sqrt{751.9}$

[15.2] In Review Exercises 17–36, simplify each expression. Assume that all variables represent positive numbers.

17. $\sqrt{32}$
18. $\sqrt{50}$
19. $\sqrt{500}$
20. $\sqrt{112}$
21. $\sqrt{80x^2}$
22. $\sqrt{63y^2}$
23. $-\sqrt{250t^3}$
24. $-\sqrt{700z^5}$
25. $\sqrt{200x^2y}$
26. $\sqrt{75y^2z}$
27. $\sqrt[3]{8x^2y^3}$
28. $\sqrt[3]{250x^4y^3}$
29. $\sqrt{\dfrac{16}{25}}$
30. $\sqrt{\dfrac{100}{49}}$
31. $\sqrt[3]{\dfrac{1000}{27}}$
32. $\sqrt[3]{\dfrac{16}{64}}$
33. $\sqrt{\dfrac{60}{49}}$
34. $\sqrt{\dfrac{80}{225}}$
35. $\sqrt{\dfrac{242x^4}{169x^2}}$
36. $\sqrt{\dfrac{450a^6}{196a^2}}$

[15.3–15.4] In Review Exercises 37–60, perform the indicated operations. Assume that all variables represent positive numbers.

37. $\sqrt{2} + \sqrt{8} - \sqrt{18}$
38. $\sqrt{3} + \sqrt{27} - \sqrt{12}$
39. $3\sqrt{5} + 5\sqrt{45}$
40. $5\sqrt{28} - 3\sqrt{63}$
41. $3\sqrt{2x^2y} + 2x\sqrt{2y}$
42. $3y\sqrt{5xy^3} - y^2\sqrt{20xy}$
43. $\sqrt[3]{16} + \sqrt[3]{54}$
44. $\sqrt[3]{2000x^3} - \sqrt[3]{128x^3}$
45. $(3\sqrt{2})(-2\sqrt{3})$
46. $(-5\sqrt{x})(-2\sqrt{x})$
47. $(3\sqrt{3x})(4\sqrt{6x})$
48. $(-2\sqrt{27y^3})(y\sqrt{2y})$
49. $(\sqrt[3]{4})(2\sqrt[3]{4})$
50. $(-2\sqrt[3]{32x^2})(3\sqrt[3]{2x^2})$
51. $\sqrt{2}(\sqrt{8} - \sqrt{18})$
52. $\sqrt{6y}(\sqrt{2y} + \sqrt{75})$
53. $(\sqrt{3} + \sqrt{5})(\sqrt{3} - \sqrt{5})$
54. $(\sqrt{15} + 3x)(\sqrt{15} + 3x)$
55. $(\sqrt[3]{3} + 2)(\sqrt[3]{3} - 1)$
56. $(\sqrt[3]{5} - 1)(\sqrt[3]{5} + 1)$
57. $(3\sqrt{5} + 2)^2$
58. $(2\sqrt{3} - 1)^2$
59. $(\sqrt{x} - \sqrt{2})^2$
60. $(\sqrt{7} - \sqrt{a})^2$

In Review Exercises 61–68, rationalize each denominator. Assume that all variables represent positive numbers.

61. $\dfrac{1}{\sqrt{7}}$
62. $\dfrac{3}{\sqrt{18}}$
63. $\dfrac{8}{\sqrt[3]{16}}$
64. $\dfrac{10}{\sqrt[3]{32}}$
65. $\dfrac{\sqrt{7}}{\sqrt{35}}$
66. $\dfrac{3}{\sqrt{3} - 1}$
67. $\dfrac{2\sqrt{5}}{\sqrt{5} + \sqrt{3}}$
68. $\dfrac{\sqrt{7x} + \sqrt{x}}{\sqrt{7x} - \sqrt{x}}$

[15.5] In Review Exercises 69–76, solve each equation. Check all solutions. If an equation has no solutions, so indicate.

69. $\sqrt{x+3} = 3$

70. $\sqrt{2x+10} = 2$

71. $\sqrt{3x+4} = -2\sqrt{x}$

72. $\sqrt{2(x+4)} - \sqrt{4x} = 0$

73. $\sqrt{x+5} = x - 1$

74. $\sqrt{2x+9} = x - 3$

75. $\sqrt{2x+5} - 1 = x$

76. $\sqrt{4a+13} + 2 = a$

[15.6] In Review Exercises 77–92, simplify each expression. Write each answer without using negative exponents.

77. $49^{1/2}$

78. $(-1000)^{1/3}$

79. $36^{3/2}$

80. $\left(\dfrac{4}{9}\right)^{5/2}$

81. $8^{2/3} 8^{4/3}$

82. $\dfrac{5^{17/7}}{5^{3/7}}$

83. $\dfrac{x^{4/5} x^{3/5}}{(x^{1/5})^2}$

84. $\left(\dfrac{r^{1/3} r^{2/3}}{r^{4/3}}\right)^3$

85. $6^{5/3} 6^{-2/3}$

86. $\dfrac{5^{2/3}}{5^{-1/3}}$

87. $\dfrac{x^{2/5} x^{1/5}}{x^{-2/5}}$

88. $(a^4 b^8)^{-1/2}$

89. $x^{1/3} x^{2/5}$

90. $\dfrac{t^{3/4}}{t^{2/3}}$

91. $\dfrac{x^{-4/5} x^{1/3}}{x^{1/3}}$

92. $\dfrac{r^{1/4} r^{1/3}}{r^{5/6}}$

[15.7] In Review Exercises 93–96, refer to the right triangle of Illustration 1. Find the length of the unknown side.

93. $a = 21$ and $b = 28$. Find c.

94. $a = 25$ and $c = 65$. Find b.

95. $a = 1$ and $c = \sqrt{2}$. Find b.

96. $b = 5$ and $c = 7$. Find a.

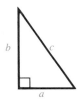

Illustration 1

In Review Exercises 97–100, find the distance between the given points.

97. $(-7, 12)\ (-4, 8)$

98. $(-15, -3)\ (-10, -15)$

99. $(1, 1)\ (-1, 1)$

100. $(-10, 11)\ (10, -10)$

101. A window frame is 32 inches by 60 inches. How long is the frame's diagonal?

102. A 53-foot wire runs from the top of a building to a point 28 feet from the base of the building. How tall is the building?

[15.8] In Review Exercises 103–104, solve each problem.

103. Find the area of a square with a diagonal 7 inches long.

104. A 78-foot guy wire supporting a tower makes an angle of 60° with the ground. How far from the base of the tower is the wire attached?

Chapter 15 Test

Assume that all variables represent positive numbers.

In Problems 1–4, write each expression without a radical sign.

1. $\sqrt{100}$

2. $-\sqrt{400}$

3. $\sqrt[3]{-27}$

4. $\sqrt{3x}\sqrt{27x}$

In Problems 5–10, simplify each expression.

5. $\sqrt{8x^2}$

6. $\sqrt{54x^3y}$

7. $\sqrt{\dfrac{320}{10}}$

8. $\sqrt{\dfrac{18x^2y^3}{2xy}}$

9. $\sqrt[3]{x^6y^6}$

10. $\sqrt[4]{\dfrac{16x^8}{y^4}}$

In Problems 11–15, perform each operation. Give all answers in simplified form.

11. $\sqrt{12} + \sqrt{27}$

12. $\sqrt{8x^3} - x\sqrt{18x}$

13. $(-2\sqrt{8x})(3\sqrt{12x})$

14. $\sqrt{3}(\sqrt{8} + \sqrt{6})$

15. $(\sqrt{2} + \sqrt{3})(\sqrt{2} - \sqrt{3})$

In Problems 16–19, rationalize each denominator.

16. $\dfrac{2}{\sqrt{2}}$

17. $\sqrt{\dfrac{3xy^3}{48x^2}}$

18. $\dfrac{2}{\sqrt{5} - 2}$

19. $\dfrac{\sqrt{3x}}{\sqrt{x} + 2}$

In Problems 20–24, solve each equation.

20. $\sqrt{x} + 3 = 9$

21. $\sqrt{x - 2} - 2 = 6$

22. $\sqrt{3x + 9} = 2\sqrt{x + 1}$

23. $3\sqrt{x - 3} = \sqrt{2x + 8}$

24. $\sqrt{3x + 1} = x - 1$

In Problems 25–30, simplify each expression. Write all answers without using negative exponents. Assume that all variables represent positive numbers.

25. $121^{1/2}$

26. $27^{-4/3}$

27. $(y^{15})^{2/5}$

28. $\left(\dfrac{a^{5/3}a^{4/3}}{(a^{1/3})^2 a^{2/3}}\right)^6$

29. $p^{2/3}p^{3/4}$

30. $\dfrac{x^{2/3}x^{-4/5}}{x^{2/15}}$

31. Find the length of the hypotenuse of a right triangle with legs of 5 inches and 12 inches.

32. Find the distance between the points (1, 4) and (7, 12).

33. Find the distance between the points (−2, −3) and (−5, 1).

34. A 26-foot ladder reaches a point on a wall 24 feet above the ground. How far from the wall is the ladder's base?

In Problems 35–36, refer to Illustration 1 and find each measure.

35. m($\overline{BC}$)

36. m($\overline{AB}$)

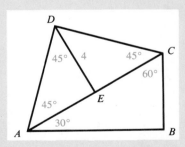
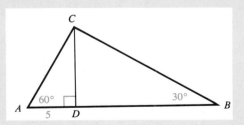

In Problems 37–38, refer to Illustration 2 and find each measure.

37. m($\overline{CB}$)

38. m($\overline{AB}$)

16 Quadratic Equations

Mathematics in Chemistry

The research of chemists has produced stronger and more durable materials, as well as processes that help save energy and reduce pollution. DuPont says it well—*Better things for better living—through chemistry*. Mathematics is important in a chemist's work. Here is a typical problem.

A weak acid (0.1 M concentration) is dissolved in water. Some of its molecules break into positively charged cations (the free hydrogen ions) and free negatively charged anions. Some cations and anions rejoin to form the acid, and some of the acid dissociates into free cations and anions. Chemists have determined that this acid solution satisfies the **equilibrium equation**

$$\frac{x^2}{0.1 - x} = 0.0004$$

where x is the concentration of free hydrogen ions.

This is a quadratic equation, but it cannot easily be solved by factoring. In this chapter, we will learn how to solve any quadratic equation and determine the concentration of free hydrogen ions.

In this chapter, we develop techniques for solving quadratic equations—equations of the form $ax^2 + bx + c = 0$, where a, b, and c are real numbers and $a \neq 0$.

16.1 Solving Equations of the Form $x^2 = c$

- Incomplete Quadratic Equations
- The Square Root Method

Previously, we have solved quadratic equations by the **factoring method**.

> **The Factoring Method:** To solve an equation by factoring:
> 1. Write the equation in $ax^2 + bx + c = 0$ form (called **quadratic form**).
> 2. Factor the polynomial on the left-hand side of the equation.
> 3. Use the zero factor theorem to set each factor equal to 0.
> 4. Solve each resulting linear equation.

EXAMPLE 1

Use the factoring method to solve a. $6x^2 - 3x = 0$ and b. $6x^2 - x - 2 = 0$.

Solution

a. $6x^2 - 3x = 0$

$3x(2x - 1) = 0$ Factor out $3x$.

$3x = 0$ or $2x - 1 = 0$ Set each factor equal to 0.

$x = 0$ $2x = 1$ Solve each linear equation.

$x = \dfrac{1}{2}$

Check: For $x = 0$

$6x^2 - 3x = 0$

$6(0)^2 - 3(0) \stackrel{?}{=} 0$

$6(0) - 0 \stackrel{?}{=} 0$

$0 - 0 \stackrel{?}{=} 0$

$0 = 0$

For $x = \dfrac{1}{2}$

$6x^2 - 3x = 0$

$6\left(\dfrac{1}{2}\right)^2 - 3\left(\dfrac{1}{2}\right) \stackrel{?}{=} 0$

$6\left(\dfrac{1}{4}\right) - \dfrac{3}{2} \stackrel{?}{=} 0$

$\dfrac{3}{2} - \dfrac{3}{2} \stackrel{?}{=} 0$

$0 = 0$

b. $6x^2 - x - 2 = 0$

$(3x - 2)(2x + 1) = 0$ Factor the trinomial.

$3x - 2 = 0$ or $2x + 1 = 0$ Set each factor equal to 0.

$3x = 2$ $2x = -1$ Solve each linear equation.

$x = \dfrac{2}{3}$ $x = -\dfrac{1}{2}$

Check: For $x = \dfrac{2}{3}$

$6x^2 - x - 2 = 0$

$6\left(\dfrac{2}{3}\right)^2 - \dfrac{2}{3} - 2 \stackrel{?}{=} 0$

$6\left(\dfrac{4}{9}\right) - \dfrac{6}{9} - \dfrac{18}{9} \stackrel{?}{=} 0$

$\dfrac{24}{9} - \dfrac{24}{9} \stackrel{?}{=} 0$

$0 = 0$

For $x = -\dfrac{1}{2}$

$6x^2 - x - 2 = 0$

$6\left(-\dfrac{1}{2}\right)^2 - \left(-\dfrac{1}{2}\right) - 2 \stackrel{?}{=} 0$

$6\left(\dfrac{1}{4}\right) + \dfrac{2}{4} - \dfrac{8}{4} \stackrel{?}{=} 0$

$\dfrac{6}{4} - \dfrac{6}{4} \stackrel{?}{=} 0$

$0 = 0$

Work Progress Check 1.

Progress Check 1

Solve by factoring:

a. $4x^2 - 36 = 0$.

b. $6x^2 - 7x - 3 = 0$.

1. a. $3, -3$ b. $\dfrac{3}{2}, -\dfrac{1}{3}$

Progress Check Answers

Incomplete Quadratic Equations

Since the second-degree polynomial in many quadratic equations cannot be factored easily, we must develop other methods for solving such quadratic equations. We begin by discussing a method for solving **incomplete quadratic equations** of the form $x^2 = c$, where c is a nonnegative number.

If $x^2 = c$, then x must be a number whose square is c. There are two such numbers: $\sqrt{c}$ and $-\sqrt{c}$. Hence, there are *two* solutions for the equation. They are

$$x = \sqrt{c} \quad \text{and} \quad x = -\sqrt{c}$$

Both of these solutions check, because $(\sqrt{c})^2 = c$ and $(-\sqrt{c})^2 = c$.

It is common to write the previous result using double-sign notation. The equation $x = \pm\sqrt{c}$ (read as "x equals plus or minus $\sqrt{c}$") means that either $x = \sqrt{c}$ or $x = -\sqrt{c}$.

The Square Root Property: If $c > 0$, the quadratic equation $x^2 = c$ has two real solutions. They are

$$x = \sqrt{c} \quad \text{or} \quad x = -\sqrt{c}$$

The Square Root Method

The method used in Example 2 is called the **square root method**.

EXAMPLE 2

Solve the equation $x^2 = 16$.

Solution

The equation $x^2 = 16$ has two solutions. They are

$$x = \sqrt{16} \quad \text{or} \quad x = -\sqrt{16}$$
$$\quad = 4 \qquad\qquad\qquad = -4$$

The solution can also be written as $x = \pm 4$.

Check: For $x = 4$:
$$x^2 = 16$$
$$4^2 \stackrel{?}{=} 16$$
$$16 = 16$$

For $x = -4$:
$$x^2 = 16$$
$$(-4)^2 \stackrel{?}{=} 16$$
$$16 = 16$$

We note that Example 2 can also be solved by factoring:

$$x^2 = 16$$
$$x^2 - 16 = 0$$
$$(x + 4)(x - 4) = 0$$
$$x + 4 = 0 \quad \text{or} \quad x - 4 = 0$$
$$x = -4 \qquad\qquad\qquad x = 4$$

EXAMPLE 3

Solve $3x^2 - 9 = 0$.

Solution

Solve this equation by the square root method.

$$3x^2 - 9 = 0$$
$$3x^2 = 9 \quad \text{Add 9 to both sides.}$$
$$x^2 = 3 \quad \text{Divide both sides by 3.}$$

This incomplete quadratic equation has two solutions. They are

$$x = \sqrt{3} \quad \text{or} \quad x = -\sqrt{3}$$

These solutions can be written as $x = \pm\sqrt{3}$.

Check:

For $x = \sqrt{3}$
$$3x^2 - 9 = 0$$
$$3(\sqrt{3})^2 - 9 \stackrel{?}{=} 0$$
$$3(3) - 9 \stackrel{?}{=} 0$$
$$9 - 9 \stackrel{?}{=} 0$$
$$0 = 0$$

For $x = -\sqrt{3}$
$$3x^2 - 9 = 0$$
$$3(-\sqrt{3})^2 - 9 \stackrel{?}{=} 0$$
$$3(3) - 9 \stackrel{?}{=} 0$$
$$9 - 9 \stackrel{?}{=} 0$$
$$0 = 0$$

Work Progress Check 2.

Progress Check 2

Solve: $4x^2 - 24 = 0$.

EXAMPLE 4

Solve $(x + 1)^2 - 9 = 0$.

Solution

$$(x + 1)^2 - 9 = 0$$
$$(x + 1)^2 = 9 \quad \text{Add 9 to both sides.}$$

The two solutions are

$$x + 1 = \sqrt{9} \quad \text{or} \quad x + 1 = -\sqrt{9}$$
$$x + 1 = +3 \quad \quad\quad x + 1 = -3$$
$$x = 2 \quad\quad\quad\quad x = -4$$

Both solutions check.

EXAMPLE 5

Solve $(x - 2)^2 - 18 = 0$.

Progress Check 3

Solve: $(x - 3)^2 = 27$.

Solution

$$(x - 2)^2 - 18 = 0$$
$$(x - 2)^2 = 18 \quad \text{Add 18 to both sides.}$$

The two solutions are

$$x - 2 = \sqrt{18} \quad \text{or} \quad x - 2 = -\sqrt{18}$$
$$x = 2 + \sqrt{18} \quad\quad\quad x = 2 - \sqrt{18}$$
$$x = 2 + 3\sqrt{2} \quad\quad\quad x = 2 - 3\sqrt{2} \quad \sqrt{18} = \sqrt{9}\sqrt{2} = 3\sqrt{2}$$

Work Progress Check 3.

2. $\sqrt{6}, -\sqrt{6}$

3. $3 + 3\sqrt{3}, 3 - 3\sqrt{3}$

Progress Check Answers

EXAMPLE 6

Solve $3x^2 - 4 = 2(x^2 + 2)$.

Solution

$$3x^2 - 4 = 2(x^2 + 2)$$
$$3x^2 - 4 = 2x^2 + 4 \quad \text{Remove parentheses.}$$
$$3x^2 = 2x^2 + 8 \quad \text{Add 4 to both sides.}$$
$$x^2 = 8 \quad \text{Add } -2x^2 \text{ to both sides.}$$

The two solutions are

$$x = \sqrt{8} \quad \text{or} \quad x = -\sqrt{8}$$
$$= 2\sqrt{2} \quad \quad\quad = -2\sqrt{2} \quad \text{Simplify the radical.}$$

16.1 Exercises

In Exercises 1–12, use the factoring method to solve each equation.

1. $x^2 - 9 = 0$
2. $x^2 + x = 0$
3. $3x^2 + 9x = 0$
4. $2x^2 - 8 = 0$
5. $x^2 - 5x + 6 = 0$
6. $x^2 + 7x + 12 = 0$
7. $3x^2 + x - 2 = 0$
8. $2x^2 - x - 6 = 0$
9. $6x^2 + 11x + 3 = 0$
10. $5x^2 + 13x - 6 = 0$
11. $10x^2 + x - 2 = 0$
12. $6x^2 + 37x + 6 = 0$

In Exercises 13–24, use the square root method to solve each equation.

13. $x^2 = 1$
14. $x^2 = 4$
15. $x^2 = 9$
16. $x^2 = 32$
17. $x^2 = 20$
18. $x^2 = 0$
19. $3x^2 = 27$
20. $4x^2 = 64$
21. $4x^2 = 16$
22. $5x^2 = 125$
23. $x^2 = a^2$
24. $x^2 = 4b^2$

In Exercises 25–34, use the square root method to solve each equation for x.

25. $(x + 1)^2 = 25$
26. $(x - 1)^2 = 49$
27. $(x + 2)^2 = 81$
28. $(x + 3)^2 = 16$
29. $(x - 2)^2 = 8$
30. $(x + 2)^2 = 50$
31. $(x - a)^2 = 4a^2$
32. $(x + y)^2 = 9y^2$
33. $(x + b)^2 = 16c^2$
34. $(x - c)^2 = 25b^2$

In Exercises 35–40, use the square root method to solve each equation. (*Hint:* Factor the perfect trinomial square first.)

35. $x^2 + 4x + 4 = 4$
36. $x^2 - 6x + 9 = 9$
37. $9x^2 - 12x + 4 = 16$
38. $4x^2 - 20x + 25 = 36$
39. $4x^2 + 4x + 1 = 20$
40. $9x^2 + 12x + 4 = 12$

In Exercises 41–46, solve each equation.

41. $6(x^2 - 1) = 4(x^2 + 3)$

42. $5(x^2 - 2) = 2(x^2 + 1)$

43. $8(x^2 - 6) = 4(x^2 + 13)$

44. $8(x^2 - 1) = 5(x^2 + 10) + 50$

45. $5(x + 1)^2 = (x + 1)^2 + 32$

46. $6(x - 4)^2 = 4(x - 4)^2 + 36$

Review Exercises

In Review Exercises 1–10, write each expression without using parentheses.

1. $(y - 1)^2$
2. $(z + 2)^2$
3. $(x + y)^2$
4. $(a - b)^2$
5. $(2r - s)^2$
6. $(m + 3n)^2$
7. $(3a + 2b)^2$
8. $(2p - 5q)^2$
9. $(5r - 8s)^2$
10. $(7y + 9z)^2$

11. Some robbers leave the scene of a crime by car and proceed out of town at 60 miles per hour. One-half hour later, the police follow in a helicopter traveling at 120 miles per hour. How long will it take the police to overtake the robbers?

12. A grocer mixes 40 pounds of a 60 cents-per-pound candy with 60 pounds of a 45 cents-per-pound candy. How much should he charge per pound for the mixture?

16.2 Completing the Square

- Completing the Square
- Using Completing the Square to Solve Quadratic Equations

If a polynomial in a quadratic equation does not factor easily, we can use the method of **completing the square**. In this method, the left-hand side of an equation such as $x^2 - 4x - 12 = 0$ is rewritten so that it becomes a perfect trinomial square. The resulting equation can be solved by the square root method.

Completing the Square

The method of completing the square depends on perfect trinomial squares. In boldface is a list of some perfect trinomial squares with lead coefficients of 1:

$$\mathbf{x^2 + 2x + 1} = (x + 1)^2 \qquad \mathbf{x^2 - 2x + 1} = (x - 1)^2$$
$$\mathbf{x^2 + 4x + 4} = (x + 2)^2 \qquad \mathbf{x^2 - 4x + 4} = (x - 2)^2$$
$$\mathbf{x^2 + 6x + 9} = (x + 3)^2 \qquad \mathbf{x^2 - 6x + 9} = (x - 3)^2$$

These trinomials are perfect trinomial squares because each one factors as the square of a binomial. In each of these trinomials, the third term is the square of one-half of the coefficient of x in the second term. For example, the third term in $x^2 - 6x + 9$ is

$$\left[\frac{1}{2}(-6)\right]^2 = (-3)^2 = 9$$

This suggests that to make $x^2 + 12x$ a perfect trinomial square, we should take one-half of 12, square it, and add it to $x^2 + 12x$.

$$x^2 + 12x + \left[\frac{1}{2}(12)\right]^2 = x^2 + 12x + (6)^2$$
$$= x^2 + 12x + 36$$

Because $x^2 + 12x + 36 = (x + 6)^2$, it is a perfect trinomial square.

EXAMPLE 1

Add the square of one-half of the coefficient of x to make **a.** $x^2 + 4x$, **b.** $x^2 - 6x$, and **c.** $x^2 - 5x$ into perfect trinomial squares.

Solution

a. $x^2 + 4x + \left[\frac{1}{2}(4)\right]^2 = x^2 + 4x + (2)^2$
$$= x^2 + 4x + 4$$

Note that $x^2 + 4x + 4 = (x + 2)^2$.

b. $x^2 - 6x + \left[\frac{1}{2}(-6)\right]^2 = x^2 - 6x + (-3)^2$
$$= x^2 - 6x + 9$$

Note that $x^2 - 6x + 9 = (x - 3)^2$.

c. $x^2 - 5x + \left[\frac{1}{2}(-5)\right]^2 = x^2 - 5x + \left(-\frac{5}{2}\right)^2$
$$= x^2 - 5x + \frac{25}{4}$$

Note that $x^2 - 5x + \frac{25}{4} = \left(x - \frac{5}{2}\right)^2$.

Work Progress Check 4.

Progress Check 4

Find what number must be added to each binomial to make it a trinomial square:

a. $x^2 + 8x$.

b. $x - 7x$.

Using Completing the Square to Solve Quadratic Equations

EXAMPLE 2

Use completing the square to solve $x^2 - 4x - 12 = 0$.

Solution

Because the coefficient of x^2 is 1, we can proceed as follows:

$$x^2 - 4x - 12 = 0$$
$$x^2 - 4x = 12 \quad \text{Add 12 to both sides.}$$

4. **a.** 16 **b.** $\frac{49}{4}$

Progress Check Answers

Section 16.2 Completing the Square   699

We complete the square on x to make the left-hand side a perfect trinomial square.

$$x^2 - 4x + \left[\frac{1}{2}(-4)\right]^2 = 12 + \left[\frac{1}{2}(-4)\right]^2 \quad \text{Add } \left[\frac{1}{2}(-4)\right]^2 \text{ to both sides.}$$

$$x^2 - 4x + 4 = 12 + 4 \quad \text{Simplify.}$$

$$(x - 2)^2 = 16 \quad \text{Factor } x^2 - 4x + 4 \text{ and simplify.}$$

$$x - 2 = \pm\sqrt{16} \quad \text{Solve the quadratic equation for } x - 2.$$

$$x = 2 \pm 4 \quad \text{Add 2 to both sides and simplify.}$$

Thus, there are two solutions: either

$$x = 2 + 4 \quad \text{or} \quad x = 2 - 4$$
$$= 6 \qquad \qquad = -2$$

Check both solutions. Note that this equation can be solved by factoring. Work Progress Check 5.

Progress Check 5

Solve by completing the square: $x^2 - 4x - 12 = 0$.

EXAMPLE 3

Use completing the square to solve the equation $4x^2 + 4x - 3 = 0$.

Solution

We divide both sides by 4 to make the coefficient of x^2 equal to 1 and proceed as follows.

$$4x^2 + 4x - 3 = 0$$

$$x^2 + x - \frac{3}{4} = 0 \quad \text{Divide both sides by 4.}$$

$$x^2 + x = \frac{3}{4} \quad \text{Add } \tfrac{3}{4} \text{ to both sides.}$$

$$x^2 + x + \left(\frac{1}{2}\right)^2 = \frac{3}{4} + \left(\frac{1}{2}\right)^2 \quad \text{Add } (\tfrac{1}{2})^2 \text{ to both sides to complete the square.}$$

$$\left(x + \frac{1}{2}\right)^2 = 1 \quad \text{Factor and simplify.}$$

$$x + \frac{1}{2} = \pm 1 \quad \text{Solve the quadratic equation for } x + \tfrac{1}{2}.$$

$$x = -\frac{1}{2} \pm 1 \quad \text{Add } -\tfrac{1}{2} \text{ to both sides.}$$

$$x = -\frac{1}{2} + 1 \quad \text{or} \quad x = -\frac{1}{2} - 1$$
$$= \frac{1}{2} \qquad \qquad = -\frac{3}{2}.$$

Check both solutions. This equation can also be solved by factoring. Work Progress Check 6.

Progress Check 6

Solve by completing the square: $2x^2 - x - 6 = 0$.

5. 6, −2

6. 2, $-\frac{3}{2}$

Progress Check Answers

700   Chapter 16   Quadratic Equations

EXAMPLE 4

Use completing the square to solve $2x^2 - 5x - 3 = 0$.

Solution

We divide both sides by 2 to make the coefficient of x^2 equal to 1 and proceed as follows.

$$2x^2 - 5x - 3 = 0$$

$$x^2 - \frac{5}{2}x - \frac{3}{2} = 0 \qquad \text{Divide both sides by 2.}$$

$$x^2 - \frac{5}{2}x = \frac{3}{2} \qquad \text{Add } \tfrac{3}{2} \text{ to both sides.}$$

$$x^2 - \frac{5}{2}x + \left[\frac{1}{2}\left(-\frac{5}{2}\right)\right]^2 = \frac{3}{2} + \left[\frac{1}{2}\left(-\frac{5}{2}\right)\right]^2 \qquad \text{Add } \left[\tfrac{1}{2}\left(-\tfrac{5}{2}\right)\right]^2 \text{ to both sides to complete the square.}$$

$$x^2 - \frac{5}{2}x + \frac{25}{16} = \frac{3}{2} + \frac{25}{16} \qquad \text{Simplify.}$$

$$\left(x - \frac{5}{4}\right)^2 = \frac{24}{16} + \frac{25}{16} \qquad \text{Factor on the left-hand side and get a common denominator on the right-hand side.}$$

$$\left(x - \frac{5}{4}\right)^2 = \frac{49}{16} \qquad \text{Add the fractions.}$$

$$x - \frac{5}{4} = \pm\frac{7}{4} \qquad \text{Solve the quadratic equation for } x - \tfrac{5}{4}.$$

$$x = \frac{5}{4} \pm \frac{7}{4} \qquad \text{Add } \tfrac{5}{4} \text{ to both sides.}$$

$$x = \frac{5}{4} + \frac{7}{4} \quad \text{or} \quad x = \frac{5}{4} - \frac{7}{4}$$

$$= \frac{12}{4} \qquad\qquad\qquad = -\frac{2}{4}$$

$$= 3 \qquad\qquad\qquad\quad = -\frac{1}{2}$$

Check both solutions. Note that this equation can be solved by factoring. ■

EXAMPLE 5

Use completing the square to solve the equation $2x^2 + 4x - 2 = 0$.

Solution

$$2x^2 + 4x - 2 = 0$$

$$x^2 + 2x - 1 = 0 \qquad \text{Divide both sides by 2.}$$

$$x^2 + 2x = 1 \qquad \text{Add 1 to both sides.}$$

$$x^2 + 2x + (1)^2 = 1 + (1)^2 \qquad \text{Add } (1)^2 \text{ to both sides to complete the square.}$$

$$(x + 1)^2 = 2 \qquad \text{Factor and simplify.}$$

$$x + 1 = \pm\sqrt{2} \qquad \text{Solve the quadratic equation for } x + 1.$$

$$x = -1 \pm \sqrt{2} \qquad \text{Add } -1 \text{ to both sides.}$$

$$x = -1 + \sqrt{2} \quad \text{or} \quad x = -1 - \sqrt{2}$$

Both solutions check. Note that this equation cannot be solved by factoring. Work Progress Check 7. ■

Progress Check 7

Solve by completing the square:
$2x^2 + 8x - 2 = 0$.

7. $-2 + \sqrt{5}, -2 - \sqrt{5}$

Progress Check Answers

To solve an equation by completing the square, follow these steps.

> 1. Make sure that the coefficient of x^2 is 1. If it is not, make it 1 by dividing both sides of the equation by the coefficient of x^2.
> 2. If necessary, add a number to both sides of the equation so that the constant term is on the right-hand side of the equation.
> 3. Complete the square.
>   a. Identify the coefficient of x.
>   b. Take half the coefficient of x.
>   c. Square half the coefficient of x.
>   d. Add that square to both sides of the equation.
> 4. Factor the trinomial square and combine terms.
> 5. Solve the resulting incomplete quadratic equation.
> 6. Check each solution.

16.2 Exercises

In Exercises 1–12, complete the square to make each binomial into a perfect trinomial square. Factor each trinomial answer to show that it is the square of a binomial.

1. $x^2 + 4x$
2. $x^2 + 6x$
3. $x^2 - 10x$
4. $x^2 - 8x$
5. $x^2 + 11x$
6. $x^2 + 21x$
7. $a^2 - 3a$
8. $b^2 - 13b$
9. $b^2 + \frac{2}{3}b$
10. $a^2 + \frac{8}{5}a$
11. $c^2 - \frac{5}{2}c$
12. $c^2 - \frac{11}{3}c$

In Exercises 13–30, solve each quadratic equation by completing the square. In some equations, you may have to rearrange some terms.

13. $x^2 + 6x + 8 = 0$
14. $x^2 + 8x + 12 = 0$
15. $x^2 - 8x + 12 = 0$
16. $x^2 - 4x + 3 = 0$
17. $x^2 - 2x - 15 = 0$
18. $x^2 - 2x - 8 = 0$
19. $x^2 - 7x + 12 = 0$
20. $x^2 - 7x + 10 = 0$
21. $x^2 + 5x - 6 = 0$
22. $x^2 = 14 - 5x$
23. $2x^2 = 4 - 2x$
24. $3x^2 + 9x + 6 = 0$
25. $3x^2 + 48 = -24x$
26. $3x^2 = 3x + 6$
27. $2x^2 = 3x + 2$
28. $3x^2 = 2 - 5x$
29. $4x^2 = 2 - 7x$
30. $2x^2 = 5x + 3$

In Exercises 31–38, use completing the square to solve each equation.

31. $x^2 + 4x + 1 = 0$
32. $x^2 + 6x + 2 = 0$
33. $x^2 - 2x - 4 = 0$
34. $x^2 - 4x - 2 = 0$
35. $x^2 = 4x + 3$
36. $x^2 = 6x - 3$
37. $2x^2 = 2 - 4x$
38. $3x^2 = 12 - 6x$

In Exercises 39–44, write each equation in quadratic form and solve it by completing the square.

39. $2x(x + 3) = 8$
40. $3x(x - 2) = 9$
41. $6(x^2 - 1) = 5x$
42. $2(3x^2 - 2) = 5x$
43. $x(x + 3) - \frac{1}{2} = -2$
44. $x[(x - 2) + 3] = 3\left(x - \frac{2}{9}\right)$

702   Chapter 16   Quadratic Equations

Review Exercises

In Review Exercises 1–6, solve each equation.

1. $\dfrac{3t(2t+1)}{2} + 6 = 3t^2$
2. $\dfrac{2(x+2)}{4} - 4x = 8$
3. $20r^2 - 11r - 3 = 0$
4. $\dfrac{2}{3x} - \dfrac{5}{9} = -\dfrac{1}{x}$
5. $\sqrt{x+12} = \sqrt{3x}$
6. $\dfrac{1}{2}\sqrt{3(t+2)} = \sqrt{2t-11}$

In Review Exercises 7–10, simplify each radical expression.

7. $\sqrt{80}$
8. $12\sqrt{x^3 y^2}$
9. $\dfrac{x}{\sqrt{7x}}$
10. $\dfrac{\sqrt{x}+2}{\sqrt{x}-2}$

11. The bus for a biology field trip costs $195. If two more students had signed up for the trip, it would have cost each student $2 less. How many students took the trip?

12. The team members will share equally in the $120 cost of new equipment. When two more join the team, each member's share decreases by $2. How much is each paying now?

16.3 The Quadratic Formula

Using the Quadratic Formula to Solve Quadratic Equations

The method of completing the square can be used to solve the **general quadratic equation** $ax^2 + bx + c = 0$, with $a \ne 0$.

$$ax^2 + bx + c = 0$$
$$\dfrac{ax^2}{a} + \dfrac{bx}{a} + \dfrac{c}{a} = \dfrac{0}{a} \qquad \text{Divide both sides by } a.$$
$$x^2 + \dfrac{b}{a}x + \dfrac{c}{a} = 0 \qquad \text{Simplify.}$$
$$x^2 + \dfrac{b}{a}x \phantom{+ \dfrac{c}{a}} = -\dfrac{c}{a} \qquad \text{Add } -\dfrac{c}{a} \text{ to both sides.}$$

We then complete the square on x by adding $\left(\dfrac{1}{2}\cdot\dfrac{b}{a}\right)^2$, or $\dfrac{b^2}{4a^2}$, to both sides:

$$x^2 + \dfrac{b}{a}x + \dfrac{b^2}{4a^2} = \dfrac{b^2}{4a^2} - \dfrac{c}{a}$$

We now factor the trinomial on the left-hand side and add the fractions on the right-hand side to obtain

$$\left(x + \dfrac{b}{2a}\right)^2 = \dfrac{b^2}{4a^2} - \dfrac{4ac}{4aa}$$
$$\left(x + \dfrac{b}{2a}\right)^2 = \dfrac{b^2 - 4ac}{4a^2}$$

We solve this equation by the square root method. Its solutions are

$$x + \frac{b}{2a} = \sqrt{\frac{b^2 - 4ac}{4a^2}} \quad \text{and} \quad x + \frac{b}{2a} = -\sqrt{\frac{b^2 - 4ac}{4a^2}}$$

$$x + \frac{b}{2a} = \frac{\sqrt{b^2 - 4ac}}{\sqrt{4a^2}} \qquad\qquad x + \frac{b}{2a} = -\frac{\sqrt{b^2 - 4ac}}{\sqrt{4a^2}}$$

$$x = -\frac{b}{2a} + \frac{\sqrt{b^2 - 4ac}}{2a} \qquad\qquad x = -\frac{b}{2a} - \frac{\sqrt{b^2 - 4ac}}{2a}$$

$$x = \frac{-b + \sqrt{b^2 - 4ac}}{2a} \qquad\qquad x = \frac{-b - \sqrt{b^2 - 4ac}}{2a}$$

These results are often written as one expression, called the **quadratic formula**.

The Quadratic Formula: The solutions of the general quadratic equation $ax^2 + bx + c = 0$, where $a \neq 0$, are

$$x = \frac{-b \pm \sqrt{b^2 - 4ac}}{2a}$$

Using the Quadratic Formula to Solve Quadratic Equations

EXAMPLE 1

Use the quadratic formula to solve $x^2 + 5x + 6 = 0$.

Solution

In this example, $a = 1$, $b = 5$, and $c = 6$.

$$x = \frac{-b \pm \sqrt{b^2 - 4ac}}{2a}$$

$$= \frac{-5 \pm \sqrt{5^2 - 4(1)(6)}}{2(1)} \quad \text{Substitute 1 for } a, 5 \text{ for } b, \text{ and } 6 \text{ for } c.$$

$$= \frac{-5 \pm \sqrt{25 - 24}}{2}$$

$$= \frac{-5 \pm \sqrt{1}}{2}$$

$$= \frac{-5 \pm 1}{2}$$

Thus,

$$x = \frac{-5 + 1}{2} \quad \text{and} \quad x = \frac{-5 - 1}{2}$$

$$= \frac{-4}{2} \qquad\qquad\qquad = \frac{-6}{2}$$

$$= -2 \qquad\qquad\qquad\quad = -3$$

The solutions of the equation are -2 and -3. Check both solutions. Work Progress Check 8.

Progress Check 8

Use the quadratic formula to solve $2x^2 + 7x + 6 = 0$.

8. $-2, -\frac{3}{2}$

EXAMPLE 2

Use the quadratic formula to solve $2x^2 = 5x + 3$.

Solution

We begin by writing the given equation in quadratic form.

$$2x^2 = 5x + 3$$

$2x^2 - 5x - 3 = 0$ Add $-5x$ and -3 to both sides.

In this example, $a = 2$, $b = -5$, and $c = -3$.

$$\begin{aligned} x &= \frac{-b \pm \sqrt{b^2 - 4ac}}{2a} \\ &= \frac{-(-5) \pm \sqrt{(-5)^2 - 4(2)(-3)}}{2(2)} \quad \text{Substitute 2 for } a, -5 \text{ for } b, \text{ and } -3 \text{ for } c. \\ &= \frac{5 \pm \sqrt{25 + 24}}{4} \\ &= \frac{5 \pm \sqrt{49}}{4} \\ &= \frac{5 \pm 7}{4} \end{aligned}$$

Thus,

$$x = \frac{5 + 7}{4} \quad \text{or} \quad x = \frac{5 - 7}{4}$$
$$= \frac{12}{4} \quad\quad\quad\quad = \frac{-2}{4}$$
$$= 3 \quad\quad\quad\quad\quad = -\frac{1}{2}$$

The solutions of the equation are 3 and $-\frac{1}{2}$. Check both solutions. ∎

EXAMPLE 3

Use the quadratic formula to solve $3x^2 = 2x + 4$.

Solution

We begin by writing the given equation in quadratic form.

$$3x^2 = 2x + 4$$

$3x^2 - 2x - 4 = 0$ Add $-2x$ and -4 to both sides.

In this example, $a = 3$, $b = -2$, and $c = -4$.

$$x = \frac{-b \pm \sqrt{b^2 - 4ac}}{2a}$$

$$= \frac{-(-2) \pm \sqrt{(-2)^2 - 4(3)(-4)}}{2(3)} \quad \text{Substitute 3 for } a, -2 \text{ for } b, \text{ and } -4 \text{ for } c.$$

$$= \frac{2 \pm \sqrt{4 + 48}}{6}$$

$$= \frac{2 \pm \sqrt{52}}{6}$$

$$= \frac{2 \pm 2\sqrt{13}}{6} \quad \sqrt{52} = \sqrt{4 \cdot 13} = \sqrt{4}\sqrt{13} = 2\sqrt{13}$$

$$= \frac{2(1 \pm \sqrt{13})}{6} \quad \text{Factor out 2.}$$

$$= \frac{1 \pm \sqrt{13}}{3} \quad \text{Divide out the common factor of 2.}$$

Thus

$$x = \frac{1}{3} + \frac{\sqrt{13}}{3} \quad \text{or} \quad x = \frac{1}{3} - \frac{\sqrt{13}}{3}$$

Both solutions check. Work Progress Check 9.

Progress Check 9

Use the quadratic formula to solve $x^2 - 6x - 3 = 0$.

Progress Check Answers

9. $3 \pm 2\sqrt{3}$

16.3 Exercises

In Exercises 1–12, write each equation in quadratic form, if necessary. Then determine the values of a, b, and c for each quadratic equation. Do not solve the equation.

1. $x^2 + 4x + 3 = 0$
2. $x^2 - x - 4 = 0$
3. $3x^2 - 2x + 7 = 0$
4. $4x^2 + 7x - 3 = 0$
5. $4y^2 = 2y - 1$
6. $2x = 3x^2 + 4$
7. $x(3x - 5) = 2$
8. $y(5y + 10) = 8$
9. $7(x^2 + 3) = -14x$
10. $5(a^2 + 5) = -4a$
11. $(2a + 3)(a - 2) = (a + 1)(a - 1)$
12. $(3a + 2)(a - 1) = (2a + 7)(a - 1)$

In Exercises 13–36, use the quadratic formula to solve each equation.

13. $x^2 - 5x + 6 = 0$
14. $x^2 + 5x + 4 = 0$
15. $x^2 + 7x + 12 = 0$
16. $x^2 - x - 12 = 0$
17. $2x^2 - x - 1 = 0$
18. $2x^2 + 3x - 2 = 0$
19. $3x^2 + 5x + 2 = 0$
20. $3x^2 - 4x + 1 = 0$
21. $4x^2 + 4x - 3 = 0$
22. $4x^2 + 3x - 1 = 0$
23. $5x^2 - 8x - 4 = 0$
24. $6x^2 - 8x + 2 = 0$
25. $x^2 + 3x + 1 = 0$
26. $x^2 + 3x - 2 = 0$
27. $x^2 + 5x - 3 = 0$
28. $x^2 + 5x + 3 = 0$
29. $2x^2 + x - 5 = 0$
30. $3x^2 - x - 1 = 0$
31. $x^2 + 4x + 1 = 0$
32. $x^2 + 8x + 1 = 0$
33. $x^2 + 2x - 1 = 0$
34. $x^2 + 2x - 2 = 0$
35. $3x^2 - 6x - 2 = 0$
36. $3x^2 + 8x + 2 = 0$

In Exercises 37–38, use these facts: The two solutions of $ax^2 + bx + c = 0$ (with $a \neq 0$) are

$$x_1 = \frac{-b + \sqrt{b^2 - 4ac}}{2a} \quad \text{and} \quad x_2 = \frac{-b - \sqrt{b^2 - 4ac}}{2a}$$

37. Show that $x_1 + x_2 = -\frac{b}{a}$.

38. Show that $x_1 x_2 = \frac{c}{a}$.

Review Exercises

In Review Exercises 1–2, solve each equation for the indicated variable.

1. $A = p + prt$, for r

2. $F = \dfrac{GMm}{d^2}$, for M

In Review Exercises 3–4, graph each equation.

3. $y = -2x + 1$

4. $x = 3y - 2$

In Review Exercises 5–6, write the equation of the line with the given properties in general form.

5. Slope of $\dfrac{3}{5}$ and passing through $(0, 12)$

6. Passing through $(6, 8)$ and the origin

In Review Exercises 7–8, graph each inequality.

7. $y < 2x$

8. $2x + y \geq 4$

In Review Exercises 9–10, let $f(x) = -x^2 + 7x - 3$. Calculate each value.

9. $f(0)$

10. $f(-3)$

11. The distance between two locations, measured in nautical miles, is directly proportional to that distance measured in statute miles. If 4.6 statute miles equal 4 nautical miles, how many nautical miles are in 230 statute miles?

12. The illumination I measured at a distance d from a light source varies inversely as the square of the distance. If $I = 300$ when $d = 4$, what is I when $d = 6$?

16.4 Complex Numbers

- Imaginary Numbers
- Complex Numbers
- Adding and Subtracting Complex Numbers
- Multiplying Complex Numbers
- Dividing Complex Numbers
- The Absolute Value of a Complex Number

The solutions of many quadratic equations are not real numbers.

EXAMPLE 1

Solve $x^2 + x + 1 = 0$.

Solution

We use the quadratic formula, with $a = 1$, $b = 1$, and $c = 1$.

$$x = \dfrac{-b \pm \sqrt{b^2 - 4ac}}{2a}$$

$$= \dfrac{-1 \pm \sqrt{1^2 - 4(1)(1)}}{2(1)} \quad \text{Substitute 1 for } a, \text{ 1 for } b, \text{ and 1 for } c.$$

$$= \frac{-1 \pm \sqrt{1-4}}{2}$$

$$= \frac{-1 \pm \sqrt{-3}}{2}$$

$$x = \frac{-1 + \sqrt{-3}}{2} \quad \text{or} \quad x = \frac{-1 - \sqrt{-3}}{2}$$

■

Imaginary Numbers

Each solution in Example 1 involves the nonreal number $\sqrt{-3}$. For years, mathematicians believed that numbers like these were nonsense. Even the great English mathematician Sir Isaac Newton (1642–1727) called them impossible. In the seventeenth century, they were named **imaginary numbers** by René Descartes (1596–1650).

Imaginary numbers have important applications, such as describing the behavior of alternating current in electronics.

The letter i is used to denote the imaginary number $\sqrt{-1}$. Because i represents the square root of -1, it follows that

$$i^2 = -1$$

The powers of the imaginary number i produce an interesting pattern:

$i = \sqrt{-1} = i$ $\quad\quad$ $i^5 = i^4 \cdot i = 1 \cdot i = i$
$i^2 = \sqrt{-1}\sqrt{-1} = -1$ $\quad\quad$ $i^6 = i^4 \cdot i^2 = 1(-1) = -1$
$i^3 = i^2 \cdot i = -1 \cdot i = -i$ $\quad\quad$ $i^7 = i^4 \cdot i^3 = 1(-i) = -i$
$i^4 = i^2 \cdot i^2 = (-1)(-1) = 1$ $\quad\quad$ $i^8 = i^4 \cdot i^4 = (1)(1) = 1$

The pattern continues: $i, -1, -i, 1, \ldots$.

If we assume that multiplication of imaginary numbers is commutative and associative, then

$$(2i)^2 = 2^2 i^2 = 4(-1) = -4$$

Because $(2i)^2 = -4$, it follows that $2i$ is a square root of -4, and we write

$$\sqrt{-4} = 2i$$

Note that this result could have been obtained by the following process:

$$\sqrt{-4} = \sqrt{4(-1)} = \sqrt{4}\sqrt{-1} = 2i$$

Similarly, we have

$$\sqrt{-\frac{1}{9}} = \sqrt{\frac{1}{9}(-1)} = \sqrt{\frac{1}{9}}\sqrt{-1} = \frac{1}{3}i$$

and

$$\sqrt{\frac{-100}{49}} = \sqrt{\frac{100}{49}(-1)} = \frac{\sqrt{100}}{\sqrt{49}}\sqrt{-1} = \frac{10}{7}i$$

The previous examples illustrate the following rule.

Properties of Radicals: If at least one of a and b is a nonnegative real number and if there are no divisions by 0, then

$$\sqrt{ab} = \sqrt{a}\sqrt{b} \quad \text{and} \quad \sqrt{\frac{a}{b}} = \frac{\sqrt{a}}{\sqrt{b}}$$

Complex Numbers

Imaginary numbers such as $\sqrt{-3}$, $\sqrt{-1}$, and $\sqrt{-9}$ form a subset of a broader set of numbers called **complex numbers**.

> **Definition:** A **complex number** is any number that can be written in the form $a + bi$, where a and b are real numbers and $i = \sqrt{-1}$.
> The number a is called the **real part** and the number b is called the **imaginary part** of the complex number $a + bi$.

If $b = 0$, the complex number $a + bi$ is a real number. If $b \neq 0$ and $a = 0$, the complex number $0 + bi$ (or just bi) is an imaginary number. Some examples of complex numbers are

$$0, \quad 3, \quad 5i, \quad 3 - 7i, \quad -5 + 12i$$

> **Definition:** The complex numbers $a + bi$ and $c + di$ are equal if and only if $a = c$ and $b = d$.

EXAMPLE 2

a. $2 + 3i = \sqrt{4} + \dfrac{6}{2}i$, because $2 = \sqrt{4}$ and $3 = \dfrac{6}{2}$.

b. $4 - 5i = \dfrac{12}{3} - \sqrt{25}\,i$, because $4 = \dfrac{12}{3}$ and $-5 = -\sqrt{25}$.

c. $x + yi = 4 + 7i$ if and only if $x = 4$ and $y = 7$.

Adding and Subtracting Complex Numbers

> **Definition:** Complex numbers are added and subtracted as if they were binomials:
> $$(a + bi) + (c + di) = (a + c) + (b + d)i$$

Progress Check 10

Add:

a. $(3 + 5i) + (4 - 3i)$.

b. $(3 + 5i) - (4 - 3i)$.

EXAMPLE 3

a. $(8 + 4i) + (12 + 8i)$
$= 8 + 4i + 12 + 8i$
$= 20 + 12i$

b. $(7 - 4i) + (9 + 2i)$
$= 7 - 4i + 9 + 2i$
$= 16 - 2i$

c. $(-6 + i) - (3 - 4i)$
$= -6 + i - 3 + 4i$
$= -9 + 5i$

d. $(2 - 4i) - (-4 + 3i)$
$= 2 - 4i + 4 - 3i$
$= 6 - 7i$

Work Progress Check 10.

10. **a.** $7 + 2i$ **b.** $-1 + 8i$

Progress Check Answers

Multiplying Complex Numbers

To multiply a complex number by an imaginary number, we use the distributive property to remove parentheses and then simplify. For example,

$$-5i(4 - 8i) = -5i(4) - (-5i)(8i)$$
$$= -20i + 40i^2$$
$$= -40 - 20i \qquad \text{Remember that } i^2 = -1.$$

To multiply two complex numbers, we use the following definition.

> **Definition:** Complex numbers are multiplied as if they were binomials, with $i^2 = -1$:
> $$(a + bi)(c + di) = ac + adi + bci + bdi^2$$
> $$= (ac - bd) + (ad + bc)i$$

EXAMPLE 4

a. $(2 + 3i)(3 - 2i)$
$= 6 - 4i + 9i - 6i^2$
$= 6 + 5i + 6$
$= 12 + 5i$

b. $(3 + i)(1 + 2i)$
$= 3 + 6i + i + 2i^2$
$= 3 + 7i - 2$
$= 1 + 7i$

c. $(-4 + 2i)(2 + i)$
$= -8 - 4i + 4i + 2i^2$
$= -8 - 2$
$= -10$

d. $(3 + 2i)^2 = (3 + 2i)(3 + 2i)$
$= 9 + 6i + 6i + 4i^2$
$= 9 + 12i - 4$
$= 5 + 12i$

Work Progress Check 11.

Progress Check 11

Multiply: $(2 + 4i)(3 - 2i)$.

The next example shows how to write several complex numbers in $a + bi$ form. It is common practice to accept the form $a - bi$ as a substitute for $a + (-b)i$.

EXAMPLE 5

a. $7 = 7 + 0i$
b. $3i = 0 + 3i$
c. $4 - \sqrt{-16} = 4 - \sqrt{-1(16)} = 4 - \sqrt{16}\sqrt{-1} = 4 - 4i$
d. $5 + \sqrt{-11} = 5 + \sqrt{-1(11)} = 5 + \sqrt{11}\sqrt{-1} = 5 + \sqrt{11}i$
e. $2i^2 + 4i^3 = 2(-1) + 4(-i) = -2 - 4i$

Progress Check Answers

11. $14 + 8i$

f. $\dfrac{3}{2i} = \dfrac{3}{2i} \cdot \dfrac{i}{i} = \dfrac{3i}{2i^2} = \dfrac{3i}{2(-1)} = \dfrac{3i}{-2} = 0 - \dfrac{3}{2}i$

g. $-\dfrac{5}{i} = -\dfrac{5}{i} \cdot \dfrac{i^3}{i^3} = -\dfrac{5(-i)}{1} = 5i = 0 + 5i$

Progress Check 12

Write $5 - \sqrt{-25}$ in $a + bi$ form.

Work Progress Check 12.

Dividing Complex Numbers

We rationalize denominators to write complex numbers such as

$$\dfrac{1}{3+i}, \quad \dfrac{3-i}{2+i}, \quad \text{and} \quad \dfrac{5+i}{5-i}$$

in $a + bi$ form. To this end, we make the following definition.

> **Definition:** The complex numbers $a + bi$ and $a - bi$ are called **complex conjugates** of each other.

For example,

$3 + 4i$ and $3 - 4i$ are complex conjugates
$5 - 7i$ and $5 + 7i$ are complex conjugates
$8 + 17i$ and $8 - 17i$ are complex conjugates

EXAMPLE 6

Find the product of the complex number $3 + i$ and its complex conjugate.

Solution

The complex conjugate of $3 + i$ is $3 - i$, and we find the product of these binomials:

$$\begin{aligned}(3 + i)(3 - i) &= 9 - 3i + 3i - i^2 \\ &= 9 - i^2 \\ &= 9 - (-1) \qquad \text{Because } i^2 = -1. \\ &= 10\end{aligned}$$

In general, the product of the complex number $a + bi$ and its complex conjugate $a - bi$ is the real number $a^2 + b^2$, as the following work shows:

$$\begin{aligned}(a + bi)(a - bi) &= a^2 - abi + abi - b^2 i^2 \\ &= a^2 - b^2(-1) \\ &= a^2 + b^2\end{aligned}$$

Thus, we have

$$(a + bi)(a - bi) = a^2 + b^2$$

12. $5 - 5i$

EXAMPLE 7

Write $\dfrac{1}{3+i}$ in $a + bi$ form.

Solution

We can rationalize the denominator by multiplying the numerator and the denominator by the complex conjugate of the denominator and simplifying.

$$\dfrac{1}{3+i} = \dfrac{1}{3+i} \cdot \dfrac{3-i}{3-i}$$
$$= \dfrac{3-i}{9 - 3i + 3i - i^2}$$
$$= \dfrac{3-i}{9 - (-1)}$$
$$= \dfrac{3-i}{10}$$
$$= \dfrac{3}{10} - \dfrac{1}{10}i$$

Work Progress Check 13.

Progress Check 13

Write $\dfrac{1}{2-i}$ in $a + bi$ form.

EXAMPLE 8

Write $\dfrac{3-i}{2+i}$ in $a + bi$ form.

Solution

$$\dfrac{3-i}{2+i} = \dfrac{3-i}{2+i} \cdot \dfrac{2-i}{2-i}$$ Multiply numerator and denominator by $2 - i$.
$$= \dfrac{6 - 3i - 2i + i^2}{4 - 2i + 2i - i^2}$$
$$= \dfrac{5 - 5i}{4 - (-1)}$$
$$= \dfrac{5(1 - i)}{5}$$ Factor out 5 in the numerator.
$$= 1 - i$$ Simplify.

EXAMPLE 9

Divide $5 + i$ by $5 - i$ and express the quotient in $a + bi$ form.

13. $\frac{2}{5} + \frac{1}{5}i$

Solution

This division can be expressed by the fraction $\frac{5+i}{5-i}$. To express this quotient in $a+bi$ form, we rationalize the denominator.

$\frac{5+i}{5-i} = \frac{5+i}{5-i} \cdot \frac{5+i}{5+i}$ Multiply the numerator and denominator by $5+i$.

$= \frac{25 + 5i + 5i + i^2}{25 + 5i - 5i - i^2}$

$= \frac{25 + 10i - 1}{25 - (-1)}$

$= \frac{24 + 10i}{26}$

$= \frac{2(12 + 5i)}{26}$ Factor out 2 in the numerator.

$= \frac{12 + 5i}{13}$ Simplify.

$= \frac{12}{13} + \frac{5}{13}i$

Progress Check 14

Divide: $\frac{5-i}{5+i}$.

Work Progress Check 14.

Complex numbers are not always in $a + bi$ form. To avoid mistakes, always put complex numbers in $a + bi$ form before doing any arithmetic involving them.

EXAMPLE 10

Write $\frac{4 + \sqrt{-16}}{2 + \sqrt{-4}}$ in $a + bi$ form.

Solution

$\frac{4 + \sqrt{-16}}{2 + \sqrt{-4}} = \frac{4 + 4i}{2 + 2i}$ Write each complex number in $a + bi$ form.

$= \frac{2(2 + 2i)}{2 + 2i}$ Factor out 2 in the numerator.

$= 2 + 0i$ Simplify.

The Absolute Value of a Complex Number

Definition: The **absolute value** of $a + bi$ is $\sqrt{a^2 + b^2}$. In symbols,

$$|a + bi| = \sqrt{a^2 + b^2}$$

14. $\frac{12}{13} - \frac{5}{13}i$

Progress Check Answers

EXAMPLE 11

a. $|3 + 4i| = \sqrt{3^2 + 4^2} = \sqrt{9 + 16} = \sqrt{25} = 5$

b. $|5 - 12i| = \sqrt{5^2 + (-12)^2} = \sqrt{25 + 144} = \sqrt{169} = 13$

c. $|1 + i| = \sqrt{1^2 + 1^2} = \sqrt{1 + 1} = \sqrt{2}$

d. $|a + 0i| = \sqrt{a^2 + 0^2} = \sqrt{a^2} = |a|$

In each part, we see that the absolute value of a complex number is a nonnegative real number. We also see that the result of Part **d** is consistent with the definition of the absolute value of a real number. ∎

16.4 Exercises

In Exercises 1–10, solve each quadratic equation. Write all roots in *bi* or *a + bi* form.

1. $x^2 + 9 = 0$
2. $x^2 + 16 = 0$
3. $3x^2 = -16$
4. $2x^2 = -25$
5. $x^2 + 2x + 2 = 0$
6. $x^2 + 3x + 3 = 0$
7. $2x^2 + x + 1 = 0$
8. $3x^2 + 2x + 1 = 0$
9. $3x^2 - 4x + 2 = 0$
10. $2x^2 - 3x + 2 = 0$

In Exercises 11–18, simplify each expression.

11. i^{21}
12. i^{19}
13. i^{27}
14. i^{22}
15. i^{100}
16. i^{42}
17. i^{97}
18. i^{200}

In Exercises 19–60, express each number in *a + bi* form, if necessary, and perform the indicated operations. Give all answers in *a + bi* form.

19. $(3 + 4i) + (5 - 6i)$
20. $(5 + 3i) - (6 - 9i)$
21. $(7 - 3i) - (4 + 2i)$
22. $(8 + 3i) + (-7 - 2i)$
23. $(8 + \sqrt{-25}) + (7 + \sqrt{-4})$
24. $(-7 + \sqrt{-81}) - (-2 - \sqrt{-64})$
25. $(-8 - \sqrt{-3}) - (7 - \sqrt{-27})$
26. $(2 + \sqrt{-8}) + (-3 - \sqrt{-2})$
27. $3i(2 - i)$
28. $-4i(3 + 4i)$
29. $(2 + 3i)(3 - i)$
30. $(4 - i)(2 + i)$
31. $(2 - 4i)(3 + 2i)$
32. $(3 - 2i)(4 - 3i)$
33. $(2 + \sqrt{-2})(3 - \sqrt{-2})$
34. $(5 + \sqrt{-3})(2 - \sqrt{-3})$
35. $(-2 - \sqrt{-16})(1 + \sqrt{-4})$
36. $(-3 - \sqrt{-81})(-2 + \sqrt{-9})$
37. $(2 + \sqrt{-3})(3 - \sqrt{-2})$
38. $(1 + \sqrt{-5})(2 - \sqrt{-3})$
39. $(8 - \sqrt{-5})(-2 - \sqrt{-7})$
40. $(-1 + \sqrt{-6})(2 - \sqrt{-3})$

41. $\dfrac{1}{i}$ **42.** $\dfrac{1}{i^3}$ **43.** $\dfrac{4}{5i^3}$ **44.** $\dfrac{3}{2i}$

45. $\dfrac{3i}{8\sqrt{-9}}$ **46.** $\dfrac{5i^3}{2\sqrt{-4}}$ **47.** $\dfrac{-3}{5i^5}$ **48.** $\dfrac{-4}{6i^7}$

49. $\dfrac{-6}{\sqrt{-32}}$ **50.** $\dfrac{5}{\sqrt{-125}}$ **51.** $\dfrac{3}{5+i}$ **52.** $\dfrac{-2}{2-i}$

53. $\dfrac{-12}{7-\sqrt{-1}}$ **54.** $\dfrac{4}{3+\sqrt{-1}}$ **55.** $\dfrac{5i}{6+2i}$ **56.** $\dfrac{-4i}{2-6i}$

57. $\dfrac{3-2i}{3+2i}$ **58.** $\dfrac{2+3i}{2-3i}$ **59.** $\dfrac{3+\sqrt{-2}}{2+\sqrt{-5}}$ **60.** $\dfrac{2-\sqrt{-5}}{3+\sqrt{-7}}$

In Exercises 61–70, find each indicated value.

61. $|6+8i|$ **62.** $|12+5i|$ **63.** $|12-5i|$ **64.** $|3-4i|$

65. $|5+7i|$ **66.** $|6-5i|$ **67.** $|4+\sqrt{-2}|$ **68.** $|3+\sqrt{-3}|$

69. $|8+\sqrt{-5}|$ **70.** $|7-\sqrt{-6}|$

Review Exercises

In Review Exercises 1–6, factor each polynomial.

1. $3x^2 - 27$ **2.** $2x^2 - 8$ **3.** $2x^2 + x - 1$ **4.** $3x^2 + 7x + 4$

5. $-x^2 - 4x + 21$ **6.** $-x^2 - x + 6$

In Review Exercises 7–12, find the volume of each figure.

7.

8.

9.

10.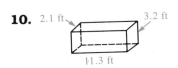

11.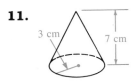

12.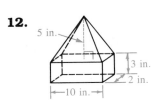

Section 16.4 Exercises 715

16.5 Graphing Quadratic Functions

- Parabolas
- Finding the Coordinates of the Vertex of a Parabola

Parabolas

We now graph equations of the form $y = ax^2 + bx + c$, where $a \neq 0$. These graphs are curves called **parabolas**.

EXAMPLE 1

Graph $y = x^2$.

Solution

To find several ordered pairs (x, y) that satisfy the equation, we pick several numbers x and find their corresponding values y. We begin by letting x be 3.

$y = x^2$
$y = 3^2$ Substitute 3 for x.
$y = 9$

The ordered pair $(3, 9)$ and several others that satisfy the equation appear in the table of values shown in Figure 16-1.

To graph the equation, we plot each ordered pair given in the table and draw a smooth curve that passes through each of the plotted points. The resulting curve, called a **parabola**, is the graph of $y = x^2$. The lowest point on the graph is called the **vertex of the parabola**. The vertex is the point $(0, 0)$.

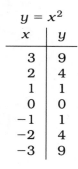
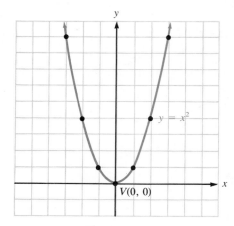

Figure 16-1

EXAMPLE 2

Graph $y = x^2 - 4x + 4$ and find its vertex.

Solution

We construct a table of values as in Figure 16-2, plot the points, and join them with a smooth curve.

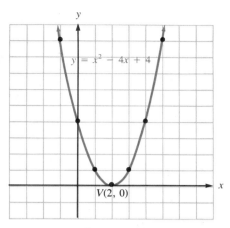

$y = x^2 - 4x + 4$

x	y
−1	9
0	4
1	1
2	0
3	1
4	4
5	9

Figure 16-2

Again, the graph is a parabola that opens upward. As before, the vertex is the lowest point on the graph, the point (2, 0). ∎

EXAMPLE 3

Graph $y = -x^2 + 2x - 1$ and find its vertex.

Solution

We construct a table of values as in Figure 16-3, plot the points, and draw the graph.

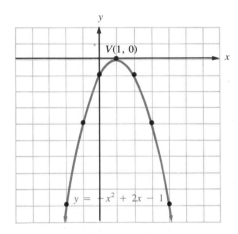

$y = -x^2 + 2x - 1$

x	y
−2	−9
−1	−4
0	−1
1	0
2	−1
3	−4
4	−9

Figure 16-3

Since this parabola opens downward, its vertex is its highest point, the point (1, 0). ∎

The results of these first three examples suggest the following fact.

> The graph of the equation $y = ax^2 + bx + c$ is a parabola. It opens upward if $a > 0$, and it opens downward if $a < 0$.

Finding the Coordinates of the Vertex of a Parabola

It is easier to graph a parabola when we know the coordinates of its vertex. We can find the coordinates of the vertex of the graph of an equation such as

$$y = x^2 - 6x + 8$$

if we complete the square in the following way.

$y = x^2 - 6x + 8$

$y = x^2 - 6x + 9 - 9 + 8$ Add 9 to complete the square on $x^2 - 6x$ and subtract 9.

$y = (x - 3)^2 - 1$ Factor $x^2 - 6x + 9$ and combine like terms.

Since $a = 1$ in $y = x^2 - 6x + 8$ and $1 > 0$, the graph is a parabola that opens upward. Thus, its vertex will be its lowest point, and this occurs when y is its smallest possible value.

Because $(x - 3)^2$ is always nonnegative, the smallest value of y will occur when $(x - 3)^2 = 0$ or when $x = 3$.

To find the corresponding value of y, we substitute 3 for x in $y = (x - 3)^2 - 1$ and simplify.

$y = (x - 3)^2 - 1$

$y = (3 - 3)^2 - 1$ Substitute 3 for x.

$y = 0^2 - 1$ Simplify.

$y = -1$ Simplify.

The vertex of the graph of $y = x^2 - 6x + 8$, or $y = (x - 3)^2 - 1$, is the point $(3, -1)$.

A generalization of this discussion leads to the following fact.

The graph of an equation of the form

$$y = a(x - h)^2 + k$$

is a parabola with its vertex at the point with coordinates (h, k). The parabola opens upward if $a > 0$, and it opens downward if $a < 0$.

EXAMPLE 4

Find the vertex of the parabola determined by $y = -4(x - 3)^2 - 2$. Will the parabola open upward or downward?

Solution

In $y = a(x - h)^2 + k$, the coordinates of the vertex are given by the ordered pair (h, k). In $y = -4(x - 3)^2 - 2$, 3 takes the place of h, -2 takes the place of k, and -4 takes the place of a. Thus, the vertex is the point $(h, k) = (3, -2)$. Because $a = -4$ and $-4 < 0$, the parabola opens downward. ∎

EXAMPLE 5

Find the vertex of the parabola given by $y = 5(x + 1)^2 + 4$. Will the parabola open upward or downward?

Solution

The equation
$$y = 5(x + 1)^2 + 4$$
is equivalent to the equation
$$y = 5[x - (-1)]^2 + 4$$
In this equation, $h = -1$, $k = 4$, and $a = 5$. Thus the vertex is the point $(h, k) = (-1, 4)$. Because $a = 5$ and $5 > 0$, the parabola opens upward. Work Progress Check 15. ∎

Progress Check 15

a. Find the vertex of $y = -3(x - 2)^2 + 4$.

b. Does the parabola open upward or downward?

EXAMPLE 6

Find the vertex of the parabola given by $y = 2x^2 + 8x + 2$ and graph it.

Solution

To complete the square, the coefficient of the term involving x^2 must be 1. To make this so, we factor 2 out of the binomial $2x^2 + 8x$.

$y = 2x^2 + 8x + 2$
$= 2(x^2 + 4x) + 2$ Factor 2 out of $2x^2 + 8x$.
$= 2(x^2 + 4x + 4 - 4) + 2$ Complete the square on $x^2 + 4x$.
$= 2\left[(x + 2)^2 - 4\right] + 2$ Factor $x^2 + 4x + 4$.
$= 2(x + 2)^2 - 2 \cdot 4 + 2$ Distribute the multiplication by 2.
$= 2(x + 2)^2 - 6$ Simplify and combine terms.

or
$$y = 2[x - (-2)]^2 + (-6)$$

Because $h = -2$ and $k = -6$, the vertex of the parabola is the point with coordinates $(h, k) = (-2, -6)$.

Because $a = 2$, the parabola opens upward. Pick some numbers x on either side of $x = -2$ and construct a table of values such as the one in Figure 16-4.

To find the y-intercept of the graph, we substitute 0 for x in the given equation and determine that $y = 2$. Thus, the y-intercept of the graph is 2.

Find and plot some other ordered pairs, then draw the parabola.

$y = 2x^2 + 8x + 2$

x	y
0	2
-1	-4
-2	-6
-3	-4
-4	2

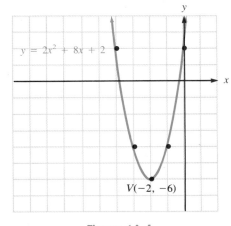

Figure 16-4

15. a. $(2, 4)$ **b.** downward

Much can be learned about the graph of $y = ax^2 + bx + c$ from the coefficients a, b, and c:

- The y-intercept of the graph is the value of y attained when $x = 0$: the y-intercept is $y = c$.
- The x-intercepts (if any) of the graph are those values of x that cause y to be 0. To find them, we can solve $ax^2 + bx + c = 0$.
- Finally, by using the methods of Example 6, we could complete the square on the right-hand side of $y = ax^2 + bx + c$ and determine the coordinates of the vertex of the parabola.

We summarize these results as follows.

Graphing the Parabola $y = ax^2 + bx + c$:

The x-coordinate of the vertex of the parabola $y = ax^2 + bx + c$ is

$$x = -\frac{b}{2a}$$

To find the y-coordinate of the vertex, substitute $-\frac{b}{2a}$ for x into the equation $y = ax^2 + bx + c$ and solve for y.

The y-intercept is $y = c$.

The x-intercepts (if any) are the roots of the quadratic equation

$$ax^2 + bx + c = 0$$

EXAMPLE 7

Graph $y = x^2 - 2x - 3$.

Solution

The equation is written in the form $y = ax^2 + bx + c$, with $a = 1$, $b = -2$, and $c = -3$. Because $a > 0$, the parabola opens upward. To find the x-coordinate of the vertex, we substitute the values for a and b into the formula $x = -\frac{b}{2a}$.

$$x = -\frac{b}{2a}$$

$$x = -\frac{-2}{2(1)} \quad \text{Substitute 1 for } a \text{ and } -2 \text{ for } b.$$

$$= 1$$

The x-coordinate of the vertex is $x = 1$. To find the y-coordinate, we substitute $x = 1$ into the equation of the parabola and solve for y.

$$y = x^2 - 2x - 3$$
$$y = 1^2 - 2 \cdot 1^2 - 3$$
$$= 1 - 2 - 3$$
$$= -4$$

The vertex of the parabola is the point $(1, -4)$.

To graph the parabola, we find several other points with coordinates that satisfy the equation. One easy point to find is the y-intercept of the graph. It is the value $y = c$, attained when $x = 0$. Thus, the parabola passes through the point $(0, -3)$.

To find the x-intercepts of the graph, set y equal to 0 and solve the resulting quadratic equation:

$$y = x^2 - 2x - 3$$
$$0 = x^2 - 2x - 3$$
$$0 = (x - 3)(x + 1) \quad \text{Factor.}$$
$$x - 3 = 0 \quad \text{or} \quad x + 1 = 0 \quad \text{Set each factor equal to 0.}$$
$$x = 3 \quad \quad \quad x = -1$$

Thus, the x-intercepts of the graph are 3 and -1. The graph passes through the points $(3, 0)$ and $(-1, 0)$. The graph appears in Figure 16-5.

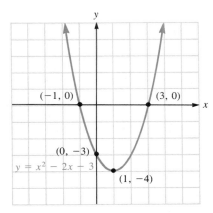

Figure 16-5

16.5 Exercises

In Exercises 1–12, graph each equation.

1. $y = x^2 + 1$
2. $y = x^2 - 4$
3. $y = -x^2$
4. $y = -(x - 1)^2$
5. $y = x^2 + x$
6. $y = x^2 - 2x$
7. $y = -x^2 - 4x$
8. $y = -x^2 + 2x$
9. $y = x^2 + 4x + 4$
10. $y = x^2 - 6x + 9$
11. $y = x^2 - 4x + 6$
12. $y = x^2 + 2x - 3$

In Exercises 13–24, find the vertex of the graph of each equation. Do not draw the graph.

13. $y = -3(x - 2)^2 + 4$
14. $y = 4(x - 3)^2 + 2$
15. $y = 5(x + 1)^2 - 5$
16. $y = 4(x + 3)^2 + 1$
17. $y = (x - 1)^2$
18. $y = -(x - 2)^2$
19. $y = -7x^2 + 4$
20. $y = 5x^2 - 2$
21. $y = x^2 + 2x + 5$
22. $y = x^2 + 4x + 1$
23. $y = x^2 - 6x - 12$
24. $y = x^2 - 8x - 20$

In Exercises 25–36, complete the square, if necessary, to determine the vertex of the graph of each equation. Then graph the equation.

25. $y = x^2 - 4x + 4$
26. $y = x^2 + 6x + 9$
27. $y = -x^2 - 2x - 1$
28. $y = -x^2 + 2x - 1$
29. $y = x^2 + 2x - 3$
30. $y = x^2 + 6x + 5$
31. $y = -x^2 - 6x - 7$
32. $y = -x^2 + 8x - 14$
33. $y = 2x^2 + 8x + 6$
34. $y = 3x^2 - 12x + 9$
35. $y = -3x^2 + 6x - 2$
36. $y = -2x^2 - 4x + 2$

In Exercises 37–42, determine the vertex and the x- and y-intercepts of the graph of each equation. Then graph each equation.

37. $y = x^2 - x - 2$ **38.** $y = x^2 - 6x + 8$ **39.** $y = -x^2 + 2x + 3$ **40.** $y = -x^2 + 5x - 4$

41. $y = 2x^2 + 3x - 2$ **42.** $y = 3x^2 - 7x + 2$

Review Exercises

In Review Exercises 1–10, simplify each expression.

1. $\sqrt{8}$ **2.** $\sqrt[3]{125}$ **3.** $\sqrt{24}$ **4.** $\sqrt[3]{128}$

5. $\sqrt{12} + \sqrt{27}$ **6.** $3\sqrt{6y}(-4\sqrt{3y})$ **7.** $(\sqrt{3} + 1)(\sqrt{3} - 1)$ **8.** $\dfrac{1}{\sqrt{5}}$

9. $\dfrac{2}{\sqrt{5} - 1}$ **10.** $\dfrac{x + \sqrt{3}}{\sqrt{3} - \sqrt{2}}$

11. The base of a 41-foot ladder is 9 feet from a vertical wall. How far up the wall does the ladder reach? See Illustration 1.

12. A rectangular garden is 20 feet wide by 21 feet long. How long is a diagonal path joining opposite corners? See Illustration 2.

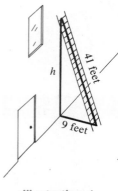

Illustration 1

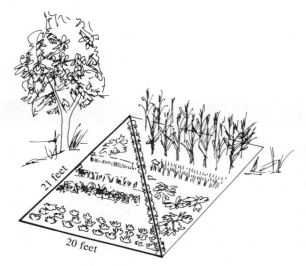

Illustration 2

Mathematics in Chemistry

To determine the concentration of free hydrogen ions in a 0.1 molar solution of a weak acid, a chemist would write the equilibrium equation in standard quadratic form and solve it using the quadratic formula:

$$\dfrac{x^2}{0.1 - x} = 0.0004$$

$x^2 = 0.0004(0.1 - x)$ Multiply both sides by $0.1 - x$.

$x^2 = 0.00004 - 0.0004x$ Remove parentheses.

$x^2 + 0.0004x - 0.00004 = 0$ Put the equation in $ax^2 + bx + c = 0$ form.

Substitute into the quadratic formula with $a = 1$, $b = 0.0004$, and $c = -0.00004$. You need to find only the positive solution.

$$x = \frac{-b + \sqrt{b^2 - 4ac}}{2a}$$

$$x = \frac{-0.0004 + \sqrt{(0.0004)^2 - 4(1)(-0.00004)}}{2(1)}$$

$$x = \frac{-0.0004 + 0.01266}{2}$$

$$x = 0.00613$$

At equilibrium, the concentration of free hydrogen ions is approximately 6.13×10^{-3} M.

Chapter Summary

Key Words

absolute value (16.4)
completing the square (16.2)
complex conjugates (16.4)
complex number (16.4)
factoring method (16.1)
general quadratic equation (16.3)
imaginary number (16.4)
imaginary part of a complex number (16.4)
incomplete quadratic equation (16.1)
parabola (16.5)
quadratic form (16.1)
quadratic formula (16.3)
real part of a complex number (16.4)
square root method (16.1)
vertex of a parabola (16.5)

Key Ideas

(16.1) The two solutions of the incomplete quadratic equation $x^2 = c$ are $+\sqrt{c}$ and $-\sqrt{c}$.

To make a binomial $x^2 + 2bx$ into a perfect trinomial square, add the square of one-half of the coefficient of x:

$$x^2 + 2bx + b^2 = (x + b)^2$$

(16.3) The solutions of $ax^2 + bx + c = 0$, $a \neq 0$, can be found by using the quadratic formula

$$x = \frac{-b \pm \sqrt{b^2 - 4ac}}{2a}$$

(16.4) **Properties of complex numbers:** If a, b, c, and d are real numbers, and given that $i^2 = -1$, then

$a + bi = c + di$ if and only if
$\quad a = c$ and $b = d$
$(a + bi) + (c + di) = (a + c) + (b + d)i$
$(a + bi)(c + di) = (ac - bd) + (ad + bc)i$
$|a + bi| = \sqrt{a^2 + b^2}$

(16.2) To solve a quadratic equation by completing the square, follow these steps:

1. If necessary, divide both sides of the equation by the coefficient of x^2 to make its coefficient 1.
2. If necessary, get the constant on the right-hand side of the equation.
3. Complete the square.
4. Solve the resulting incomplete quadratic equation.
5. Check each solution.

(16.5) The graph of $y = ax^2 + bx + c$ is a parabola. It opens upward if $a > 0$, and it opens downward if $a < 0$.

The graph of $y = a(x - h)^2 + k$ is a parabola with its vertex at the point (h, k). The parabola opens upward if $a > 0$, and it opens downward if $a < 0$.

The x-coordinate of the vertex of the parabola $y = ax^2 + bx + c$ is $x = -\frac{b}{2a}$.

To find the y-coordinate of the vertex, substitute $-\frac{b}{2a}$ for x in the equation of the parabola, and determine y.

Chapter 16 Review Exercises

[16.1] In Review Exercises 1–6, use the square root method to solve each incomplete quadratic equation.

1. $x^2 = 25$
2. $x^2 = 36$
3. $2x^2 = 18$
4. $4x^2 = 9$
5. $x^2 = 8$
6. $x^2 = 75$

In Review Exercises 7–12, use the square root method to solve each equation.

7. $(x - 1)^2 = 25$
8. $(x + 3)^2 = 36$
9. $2(x + 1)^2 = 18$
10. $4(x - 2)^2 = 9$
11. $(x - 8)^2 = 8$
12. $(x + 5)^2 = 75$

In Review Exercises 13–16, solve each equation. Note that each trinomial is a perfect square.

13. $x^2 + 2x + 1 = 9$
14. $x^2 - 6x + 9 = 4$
15. $x^2 - 8x + 16 = 20$
16. $x^2 - 2x + 1 = 18$

[16.2] In Review Exercises 17–24, solve each quadratic equation by completing the square.

17. $x^2 + 5x - 14 = 0$
18. $x^2 - 8x + 15 = 0$
19. $x^2 + 4x - 77 = 0$
20. $x^2 - 2x - 1 = 0$
21. $x^2 + 4x - 3 = 0$
22. $x^2 - 6x + 4 = 0$
23. $2x^2 + 5x - 3 = 0$
24. $2x^2 - 2x - 1 = 0$

[16.3] In Review Exercises 25–32, use the quadratic formula to solve each quadratic equation.

25. $x^2 - 2x - 15 = 0$
26. $x^2 - 6x - 7 = 0$
27. $x^2 - 15x + 26 = 0$
28. $2x^2 - 7x + 3 = 0$
29. $6x^2 - 7x - 3 = 0$
30. $x^2 + 4x + 1 = 0$
31. $x^2 - 6x + 7 = 0$
32. $x^2 + 3x = 0$

[16.4] In Review Exercises 33–48, perform the indicated operations. Give all answers in $a + bi$ form.

33. $(5 + 4i) + (7 - 12i)$
34. $(-6 - 40i) - (-8 + 28i)$
35. $7i(-3 + 4i)$
36. $6i(2 + i)$
37. $(5 - 3i)(-6 + 2i)$
38. $(2 + \sqrt{128}i)(3 - \sqrt{98}i)$
39. $\dfrac{3}{4i}$
40. $\dfrac{-2}{5i^3}$
41. $\dfrac{6}{2 + i}$
42. $\dfrac{7}{3 - i}$
43. $\dfrac{4 + i}{4 - i}$
44. $\dfrac{3 - i}{3 + i}$
45. $\dfrac{3}{5 + \sqrt{-4}}$
46. $\dfrac{2}{3 - \sqrt{-9}}$
47. $|9 + 12i|$
48. $|24 - 10i|$

[16.5] In Review Exercises 49–52, find the vertex of the graph of each equation. Do not draw the graph.

49. $y = 5(x - 6)^2 + 7$
50. $y = 3(x + 3)^2 - 5$
51. $y = 2x^2 - 4x + 7$
52. $y = -3x^2 + 18x - 11$

In Review Exercises 53–54, graph each equation.

53. $y = x^2 + 8x + 10$
54. $y = -2x^2 - 4x - 6$

Name _____   Section _____

Chapter 16 Test

1. Solve by factoring: $6x^2 + x - 1 = 0$.   _____

2. Use the square root method to solve $x^2 = 16$.   _____

3. Solve $(x - 2)^2 = 3$.   _____

4. Solve $2(x^2 - 6) = x^2 + x - 6$.   _____

In Problems 5–6, find the number required to complete the square.

5. $x^2 + 14x$ _____

6. $x^2 - 7x$ _____

7. Use the method of completing the square to solve $3a^2 + 6a - 12 = 0$.   _____

8. Write the quadratic formula.   _____

9. Use the quadratic formula to solve $x^2 + 3x - 10 = 0$.   _____

10. Use the quadratic formula to solve $2x^2 - 5x = 12$.   _____

11. Solve $2x^2 + 5x + 1 = 2$.   _____

12. Write the number $\sqrt{-49}$ in bi form.   _____

13. Add: $(3 + 4i) + (-2 + 5i)$.   _____

14. Subtract: $(4 - 3i) - (-5 + 2i)$.   _____

15. Multiply: $(-2 - 5i)(-3 + 2i)$.   _____

16. Rationalize the denominator: $\dfrac{-2}{3 + i}$.   _____

17. Rationalize the denominator: $\dfrac{2+i}{2-i}$.

18. Evaluate: $|3-4i|$.

19. Find the vertex of the parabola determined by the equation $y = -4(x+5)^2 - 4$.

20. Graph the equation $y = x^2 + 4x + 2$.

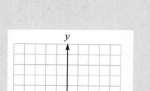

Roots and Powers

n	n^2	$\sqrt{n}$	n^3	$\sqrt[3]{n}$	n	n^2	$\sqrt{n}$	n^3	$\sqrt[3]{n}$
1	1	1.000	1	1.000	51	2,601	7.141	132,651	3.708
2	4	1.414	8	1.260	52	2,704	7.211	140,608	3.733
3	9	1.732	27	1.442	53	2,809	7.280	148,877	3.756
4	16	2.000	64	1.587	54	2,916	7.348	157,464	3.780
5	25	2.236	125	1.710	55	3,025	7.416	166,375	3.803
6	36	2.449	216	1.817	56	3,136	7.483	175,616	3.826
7	49	2.646	343	1.913	57	3,249	7.550	185,193	3.849
8	64	2.828	512	2.000	58	3,364	7.616	195,112	3.871
9	81	3.000	729	2.080	59	3,481	7.681	205,379	3.893
10	100	3.162	1,000	2.154	60	3,600	7.746	216,000	3.915
11	121	3.317	1,331	2.224	61	3,721	7.810	226,981	3.936
12	144	3.464	1,728	2.289	62	3,844	7.874	238,328	3.958
13	169	3.606	2,197	2.351	63	3,969	7.937	250,047	3.979
14	196	3.742	2,744	2.410	64	4,096	8.000	262,144	4.000
15	225	3.873	3,375	2.466	65	4,225	8.062	274,625	4.021
16	256	4.000	4,096	2.520	66	4,356	8.124	287,496	4.041
17	289	4.123	4,913	2.571	67	4,489	8.185	300,763	4.062
18	324	4.243	5,832	2.621	68	4,624	8.246	314,432	4.082
19	361	4.359	6,859	2.668	69	4,761	8.307	328,509	4.102
20	400	4.472	8,000	2.714	70	4,900	8.367	343,000	4.121
21	441	4.583	9,261	2.759	71	5,041	8.426	357,911	4.141
22	484	4.690	10,648	2.802	72	5,184	8.485	373,248	4.160
23	529	4.796	12,167	2.844	73	5,329	8.544	389,017	4.179
24	576	4.899	13,824	2.884	74	5,476	8.602	405,224	4.198
25	625	5.000	15,625	2.924	75	5,625	8.660	421,875	4.217
26	676	5.099	17,576	2.962	76	5,776	8.718	438,976	4.236
27	729	5.196	19,683	3.000	77	5,929	8.775	456,533	4.254
28	784	5.292	21,952	3.037	78	6,084	8.832	474,552	4.273
29	841	5.385	24,389	3.072	79	6,241	8.888	493,039	4.291
30	900	5.477	27,000	3.107	80	6,400	8.944	512,000	4.309
31	961	5.568	29,791	3.141	81	6,561	9.000	531,441	4.327
32	1,024	5.657	32,768	3.175	82	6,724	9.055	551,368	4.344
33	1,089	5.745	35,937	3.208	83	6,889	9.110	571,787	4.362
34	1,156	5.831	39,304	3.240	84	7,056	9.165	592,704	4.380
35	1,225	5.916	42,875	3.271	85	7,225	9.220	614,125	4.397
36	1,296	6.000	46,656	3.302	86	7,396	9.274	636,056	4.414
37	1,369	6.083	50,653	3.332	87	7,569	9.327	658,503	4.431
38	1,444	6.164	54,872	3.362	88	7,744	9.381	681,472	4.448
39	1,521	6.245	59,319	3.391	89	7,921	9.434	704,969	4.465
40	1,600	6.325	64,000	3.420	90	8,100	9.487	729,000	4.481
41	1,681	6.403	68,921	3.448	91	8,281	9.539	753,571	4.498
42	1,764	6.481	74,088	3.476	92	8,464	9.592	778,688	4.514
43	1,849	6.557	79,507	3.503	93	8,649	9.644	804,357	4.531
44	1,936	6.633	85,184	3.530	94	8,836	9.695	830,584	4.547
45	2,025	6.708	91,125	3.557	95	9,025	9.747	857,375	4.563
46	2,116	6.782	97,336	3.583	96	9,216	9.798	884,736	4.579
47	2,209	6.856	103,823	3.609	97	9,409	9.849	912,673	4.595
48	2,304	6.928	110,592	3.634	98	9,604	9.899	941,192	4.610
49	2,401	7.000	117,649	3.659	99	9,801	9.950	970,299	4.626
50	2,500	7.071	125,000	3.684	100	10,000	10.000	1,000,000	4.642

Answers for Selected Exercises

Exercises 1.1 (Page 5)

1. 2 hundreds + 4 tens + 5 ones; two hundred forty-five
3. 3 thousands + 6 hundreds + 9 ones; three thousand six hundred nine
5. 3 ten thousands + 2 thousands + 5 hundreds; thirty-two thousand five hundred
7. 1 hundred thousand + 4 thousands + 4 hundreds + 1 one; one-hundred four thousand four hundred one
9. 3 hundred thousands + 6 ten thousands + 3 thousands + 7 hundreds + 8 ones; three hundred sixty-three thousand seven hundred eight
11. 9 million + 4 hundred thousand; nine million four hundred thousand
13. 425 15. 2736 17. 456 19. 27,598 21. 660 23. 138 25. 863 27. 10
29. 90 31. 100 33. 250 35. 800 37. 200 39. 100 41. 3000 43. 2000
45. 1000 47. 0 49. 99,000 51. 10,000 53. 50,000 55. 500,000 57. 2,000,000
59. 2,000,000 61. 1,000,000 63. 79,590 65. 80,000 67. 5,926,000 69. 5,900,000
71. 45,049 73. 1,233,995 75. 876,543,219 77. 299,800,000 m/sec

Review Exercises (Page 6)

1. 12 3. 2 5. 7 7. 17 9. 13 11. 5

Exercises 1.2 (Page 14)

1. 38 3. 190 5. 461 7. 1305 9. 2134 11. 979 13. 1985 15. 10,000
17. 15,907 19. 1861 21. 5312 23. 43,731 25. 35 hrs 27. 88 ft 29. 538 cm
31. 220 ft 33. 3 35. 25 37. 103 39. 81 41. 35 43. 2107 45. 24
47. 118 49. 958 51. 1689 53. 10,457 55. 303 57. 6 59. 1542 61. $213
63. 15 years 65. $7 = 3 + 4$ 67. $15 = 7 + 8$ 69. $35 = 35 + 0$ 71. $12 = x + 5$
73. $8 - 3 = 5$; $8 - 5 = 3$ 75. $20 - 12 = 8$; $20 - 8 = 12$ 77. $17 - 4 = 13$; $17 - 13 = 4$
79. $20 - x = 17$; $20 - 17 = x$ 81. 9688 ft 83. 69 years

Review Exercises (Page 16)

1. 56 3. 9 5. 35 7. 2 9. 72 11. 8

Exercises 1.3 (Page 25)

1. 28 **3.** 84 **5.** 324 **7.** 2350 **9.** 6345 **11.** 48,144 **13.** 296 **15.** 299
17. 7623 **19.** 1060 **21.** 2576 **23.** 20,079 **25.** 4 **27.** 9 **29.** 16
31. 1000 **33.** 36 **35.** 162 **37.** $(3^3)(2^2)$, or 108 **39.** $132 **41.** 925¢ **43.** 406
45. 125,800 **47.** 16 **49.** 312 **51.** 1728 **53.** yes **55.** 864 in.^{2} **57.** 388 ft^2
59. 8 **61.** 3 **63.** 12 **65.** 13 **67.** 13 **69.** 23 **71.** 12 **73.** 19 **75.** 73
77. 41 **79.** 205 **81.** 210 **83.** 8 with remainder of 25 **85.** 20 with remainder of 3
87. 30 with remainder of 13 **89.** 31 with remainder of 28 **91.** 4 **93.** 4 **95.** not
97. $3 \cdot 9 = 27$ **99.** $12 \cdot 12 = 144$ **101.** $x \cdot y = 29$
103. $45 \div 9 = 5$; $45 \div 5 = 9$ **105.** $105 \div 15 = 7$; $105 \div 7 = 15$
107. $63 \div x = 7$; $63 \div 7 = x$ **109.** 32 **111.** 53

Review Exercises (Page 28)

1. 8 **3.** 46,000 **5.** 872 **7.** 238 **9.** $52 - 37 = x$; 15 **11.** $22

Exercises 1.4 (Page 32)

1. 5 **3.** 2 **5.** 11 **7.** 11 **9.** 3 **11.** 28 **13.** 64 **15.** 5 **17.** 8 **19.** 20
21. 64 **23.** 1 **25.** 11 **27.** 1 **29.** $(3 \cdot 8) + (5 \cdot 3)$ **31.** $(3 \cdot 8 + 5) \cdot 3$ **33.** $(4 + 3) \cdot (5 - 3)$
35. $(4 + 3) \cdot 5 - 3$ **37.** $3(16) = 48$; $3(7) + 3(9) = 48$ **39.** $11(3) = 33$; $11(8) - 11(5) = 33$
41. $17(4) = 68$; $17(12) - 17(8) = 68$ **43.** $19(24) = 456$; $19(17) + 19(7) = 456$
45. $7(10) = 70$; $7(24) - 7(14) = 70$ **47.** $5(30) = 150$; $5(5) + 5(25) = 150$
49. $11(35) = 385$; $11(92) - 11(57) = 385$ **51.** $151(83) = 12{,}533$; $151(31) + 151(52) = 12{,}533$
53. $32(15) = 480$; $32(25) - 32(10) = 480$ **55.** $27(54) = 1458$; $27(27) + 27(27) = 1458$
57. 52 **59.** 79° **61.** 4 **63.** 6

Review Exercises (Page 33)

1. 90 **3.** 360 **5.** 11,025 **7.** 1573 **9.** 13,013

Exercises 1.5 (Page 37)

1. yes **3.** yes **5.** yes **7.** no **9.** yes **11.** yes **13.** yes **15.** no
17. 1, 2, 3, 6, 7, 14, 21, 42 **19.** 1, 2, 3, 6, 9, 18, 27, 54 **21.** 1, 11, 121
23. 1, 2, 3, 4, 6, 8, 12, 16, 24, 32, 48, 96 **25.** 1, 2, 23, 46 **27.** 1, 23 **29.** prime
31. composite **33.** composite **35.** neither **37.** $2 \cdot 13$ **39.** $3^2 \cdot 19$ **41.** $3^3 \cdot 19$
43. $2 \cdot 3 \cdot 17$ **45.** $2 \cdot 3^2 \cdot 17$ **47.** $2 \cdot 3^2 \cdot 5 \cdot 29$ **49.** $2^3 \cdot 3^2 \cdot 5^2$ **51.** $2^4 \cdot 5 \cdot 7^2$
53. $5^2 \cdot 7 \cdot 11$ **55.** $5 \cdot 7 \cdot 11^2$
57. The sum of the divisors of 28 is $1 + 2 + 4 + 7 + 14 + 28 = 56$, and $56 = 2 \cdot 28$.
59. 79 and 97

Review Exercises (Page 38)

1. 868 **3.** 36 **5.** 1924 **7.** 37 **9.** 2592 **11.** 122

Exercises 1.6 (Page 41)

1. 3, 6, 9, 12, 15 **3.** 11, 22, 33, 44, 55 **5.** 15 **7.** 42 **9.** 66 **11.** 30 **13.** 600
15. 140 **17.** 72 **19.** 792 **21.** 70 **23.** 3150 **25.** 1 **27.** 7 **29.** 11
31. 15 **33.** 20 **35.** 14 **37.** 6 **39.** 3 **41.** 1 **43.** 5
45. $56 = 2^3 \cdot 7$, $135 = 3^3 \cdot 5$; GCD = 1 **47.** $15 = 5 \cdot 3$, $21 = 7 \cdot 3$, and $35 = 5 \cdot 7$; GCD = 1

Review Exercises (Page 42)

1. $x = 103 - 95$; $x = 8$ **3.** $x = 12 + 17$; $x = 29$ **5.** $x = 35 \div 5$; $x = 7$ **7.** $x = 15$; $x = 45$
9. $2^5 \cdot 3^2 \cdot 5$

Chapter 1 Review Exercises (Page 44)

1. 2 hundred thousands + 6 ten thousands + 2 hundreds + 6 ones **3.** 3207 **5.** 348
7. 8824 **9.** 17 **11.** 8 **13.** 2746 ft **15.** $785 **17.** $13 = 5 + 8$ **19.** $57 = x + 12$
21. $12 - 5 = 7$; $12 - 7 = 5$ **23.** $57 - x = 21$; $57 - 21 = x$ **25.** 3297 **27.** 83,044
29. 625 **31.** 567 **33.** Craig **35.** 21 **37.** 107 **39.** 19 with remainder of 6
41. 16 with remainder of 27 **43.** $81 = 3 \cdot 27$ **45.** $143 = x \cdot 11$ **47.** $40 \div 8 = 5$; $40 \div 5 = 8$
49. $493 \div 29 = x$; $493 \div x = 29$ **51.** 16 **53.** 49 **55.** 75 **57.** 38 **59.** 24 **61.** 3
63. $7(2) = 14$; $7(11) - 7(9) = 14$ **65.** $8(26) = 208$; $8(12) + 8(14) = 208$ **67.** 1, 3, 9, 27, 81
69. $2^4 \cdot 3^2 \cdot 5$ **71.** 90 **73.** 300 **75.** 3 **77.** 1

Chapter 1 Test (Page 47)

1. 0, 1, 2, 3, 4 **3.** 7507 **5.** 8100 **7.** 2168 in. **9.** $x = 29 + 17$; $x = 46$ **11.** 39
13. 123 **15.** 44 **17.** $x \cdot z$ **19.** 350 **21.** 25

Exercises 2.1 (Page 53)

1. $\frac{2}{5}$ **3.** $\frac{5}{8}$ **5.** $\frac{1}{2}$ **7.** $\frac{3}{8}$ **9.** $\frac{1}{6}$ **11.** $\frac{1}{2}$ **13.** $\frac{5}{12}$ **15.** $\frac{11}{24}$ **17.** $\frac{3}{4}$ **19.** $\frac{5}{264}$
21. $\frac{1}{4}$ **23.** 2 **25.** 1 **27.** 9 **29.** 17 **31.** equal **33.** not equal **35.** not equal
37. equal **39.** $\frac{1}{4}$ **41.** $\frac{3}{4}$ **43.** $\frac{9}{8}$ **45.** in lowest terms **47.** $\frac{55}{91}$ **49.** in lowest terms
51. $\frac{55}{39}$ **53.** $\frac{1}{2}$ **55.** $\frac{7}{25}$ **57.** $\frac{9}{25}$ **59.** $\frac{3}{4}$ pizza **61.** $\frac{1}{7}$

Review Exercises (Page 55)

1. $3^2 \cdot 5 \cdot 7$ **3.** $2^2 \cdot 3^2 \cdot 5 \cdot 17$ **5.** $2^2 \cdot 3^2 \cdot 5^4$ **7.** 33,312 **9.** 97 **11.** 376

Exercises 2.2 (Page 62)

1. $\frac{3}{10}$ **3.** $\frac{20}{99}$ **5.** $\frac{8}{5}$ **7.** $\frac{3}{2}$ **9.** 1 **11.** 10 **13.** $\frac{20}{3}$ **15.** $\frac{10}{3}$ **17.** $\frac{5}{3}$ **19.** 11
21. $\frac{1}{19}$ **23.** 23 **25.** $\frac{9}{10}$ **27.** $\frac{44}{21}$ **29.** $\frac{5}{8}$ **31.** $\frac{1}{4}$ **33.** $\frac{14}{5}$ **35.** 1 **37.** 28 **39.** $\frac{1}{5}$
41. 2 **43.** $\frac{3}{4}$ **45.** 6 **47.** 10 **49.** $\frac{18}{5}$ **51.** $\frac{45}{8}$ **53.** 1 **55.** $\frac{2}{9}$ **57.** $\frac{9}{16}$ **59.** $\frac{3}{5}$
61. $\frac{4}{9}$ ft^2 **63.** $\frac{6}{7}$ **65.** 100 **67.** $2160 **69.** 3771 **71.** 430 **73.** 72 **75.** 64
77. 120 gal **79.** 2,200,000 acres

Review Exercises (Page 64)

1. 32 **3.** 350 **5.** 105 **7.** 126 **9.** 5 **11.** 1

Exercises 2.3 (Page 70)

1. $\frac{6}{5}$ **3.** 1 **5.** 2 **7.** $\frac{3}{5}$ **9.** $\frac{2}{12}$ **11.** $\frac{9}{21}$ **13.** $\frac{3}{45}$ **15.** $\frac{21}{39}$ **17.** $\frac{60}{12}$ **19.** $\frac{45}{3}$
21. $\frac{19}{15}$ **23.** $\frac{37}{12}$ **25.** $\frac{19}{20}$ **27.** $\frac{23}{42}$ **29.** $\frac{81}{55}$ **31.** $\frac{14}{15}$ **33.** $\frac{15}{4}$ **35.** $\frac{29}{3}$ **37.** 1
39. $\frac{3}{2}$ **41.** 0 **43.** $\frac{4}{7}$ **45.** $\frac{3}{7}$ **47.** $\frac{2}{15}$ **49.** $\frac{29}{30}$ **51.** $\frac{9}{20}$ **53.** $\frac{19}{42}$ **55.** $\frac{5}{42}$ **57.** $\frac{13}{36}$
59. $\frac{11}{5}$ **61.** $\frac{2}{5}$ **63.** $\frac{35}{2}$ **65.** $\frac{9}{10}$ **67.** $\frac{23}{6}$ **69.** $\frac{11}{5}$ **71.** $\frac{3}{8}$ **73.** $\frac{48}{5}$ **75.** $\frac{7}{8}$
77. $\frac{15}{4}$ **79.** $\frac{19}{6}$ **81.** $x = \frac{2}{5} + \frac{3}{5}; x = 1$ **83.** $x = \frac{3}{10} + \frac{2}{5}; x = \frac{7}{10}$ **85.** $x = \frac{5}{9} + \frac{5}{6}; x = \frac{25}{18}$
87. $x = \frac{2}{7} + 3; x = \frac{23}{7}$ **89.** $x = \frac{2}{5} - \frac{1}{5}; x = \frac{1}{5}$ **91.** $x = \frac{1}{6} - \frac{1}{12}; x = \frac{1}{12}$ **93.** $x = \frac{2}{9} - \frac{1}{6}; x = \frac{1}{18}$
95. $x = 4 - \frac{5}{7}; x = \frac{23}{7}$ **97.** $\frac{38}{15}$ m **99.** $\frac{68}{5}$ cm **101.** $\frac{41}{30}$ bu **103.** yes **105.** $\frac{1}{2}$ **107.** $\frac{1}{4}$
109. 5 **111.** $\frac{41}{10}$ **113.** 7 **115.** $\frac{57}{2}$ **117.** $\frac{3}{2}$ **119.** 54 **121.** 9 **123.** $\frac{15}{2}$

Review Exercises (Page 72)

1. yes **3.** yes **5.** 2, 3, 5, 7 **7.** true **9.** false

Exercises 2.4 (Page 80)

1. $4 + \frac{3}{5}$ **3.** $4 + \frac{1}{11}$ **5.** $5 + \frac{3}{10}$ **7.** $4 + \frac{1}{8}$ **9.** $4\frac{3}{5}$ **11.** $4\frac{1}{11}$ **13.** $5\frac{3}{10}$ **15.** $4\frac{1}{8}$
17. $6\frac{1}{3}$ **19.** $12\frac{6}{17}$ **21.** 23 **23.** 2 **25.** $\frac{3}{2}$ **27.** $\frac{31}{8}$ **29.** $\frac{60}{11}$ **31.** $\frac{17}{3}$ **33.** $\frac{68}{9}$
35. $\frac{796}{31}$ **37.** $6\frac{1}{6}$ **39.** $9\frac{1}{10}$ **41.** $\frac{1}{6}$ **43.** $\frac{8}{9}$ **45.** $8\frac{7}{8}$ **47.** $8\frac{5}{16}$ **49.** $42\frac{3}{10}$ **51.** $\frac{57}{100}$
53. $9\frac{1}{6}$ **55.** $8\frac{1}{6}$ **57.** $6\frac{6}{25}$ **59.** $1\frac{2}{3}$ **61.** $1\frac{2}{5}$ **63.** $1\frac{9}{11}$ **65.** $55\frac{11}{15}$ **67.** 2 **69.** $\frac{36}{113}$
71. $1\frac{7}{10}$ **73.** 17 **75.** $11\frac{4}{5}$ **77.** $3\frac{2}{7}$ **79.** $\frac{1}{8}$ **81.** $5\frac{1}{3}$ **83.** $6\frac{5}{6}$ **85.** $\frac{77}{8}$ **87.** $\frac{52}{9}$
89. $5\frac{5}{6}$ gal **91.** $24\frac{3}{10}$ ft **93.** $4\frac{27}{50}$ m **95.** $16\frac{19}{20}$ ft **97.** 21 in. **99.** $5\frac{7}{8}$ in. by 11 in.
101. $32\frac{1}{2}$ in. **103.** $64\frac{5}{8}$ in.^{2} **105.** $656\frac{11}{16}$ cm^2 **107.** $313\frac{1}{4}$ cm^2 **109.** $4\frac{7}{12}$ ft^2

Review Exercises (Page 83)

1. 916 **3.** 21,432 **5.** 845 **7.** $\frac{1}{8}$ **9.** $\frac{65}{12}$ **11.** $\frac{12}{11}$

Chapter 2 Review Exercises (Page 84)

1. $\frac{5}{3}$ **3.** $\frac{2}{9}$ **5.** $\frac{1}{3}$ **7.** $\frac{343}{50}$ **9.** $\frac{15}{21}$ **11.** $\frac{49}{217}$ **13.** $\frac{10}{21}$ **15.** $\frac{13}{30}$ **17.** $x = \frac{1}{2} + \frac{3}{2}; x = 2$
19. $x = 3 - \frac{3}{7}; x = \frac{18}{7}$ **21.** $3\frac{6}{7}$ **23.** $11\frac{2}{11}$ **25.** $\frac{59}{7}$ **27.** $\frac{149}{12}$ **29.** $\frac{5}{8}$ **31.** 1 **33.** $\frac{19}{12}$
35. $3\frac{9}{11}$ **37.** yes **39.** $18\frac{1}{2}$ million

Chapter 2 Test (Page 87)

1. $\frac{1}{6}$ **3.** $\frac{1}{6}$ **5.** yes **7.** 2 **9.** $\frac{1}{9}$ **11.** $\frac{99}{189}$ **13.** $\frac{29}{30}$ **15.** $\frac{3}{13}$ **17.** $5\frac{7}{13}$ **19.** 1
21. $\frac{29}{12}$ or $2\frac{5}{12}$ ft **23.** $861\frac{1}{3}$ m

Exercises 3.1 (Page 92)

1. 2 tens + 5 ones + 4 tenths; twenty-five and four tenths
3. 1 tenth + 6 hundredths; sixteen hundredths
5. 5 ones + 3 tenths + 7 hundredths + 4 thousandths; five and three hundred seventy-four thousandths
7. 1 ten + 1 ten thousandth; ten and one ten thousandth
9. 3.256 **11.** 0.356 **13.** 70.707 **15.** 3.571 **17.** $2\frac{1}{2}$ **19.** $3\frac{2}{5}$ **21.** $\frac{3}{20}$ **23.** $\frac{1}{8}$
25. 2.23, 2.32, 2.33, 3.2, 3.22 **27.** 0.3201, 0.3210, 1.3012, 1.3102 **29.** 3.4 **31.** 0.4
33. 0.0 **35.** 0.3 **37.** 3.38 **39.** 0.44 **41.** 0.01 **43.** 0.26 **45.** 3.375 **47.** 0.439
49. 0.020 **51.** 0.256 **53.** 14 **55.** 14.2 **57.** 60 **59.** 59.9

Review Exercises   (Page 94)

1. 462   **3.** 876   **5.** 1639   **7.** 544   **9.** 12   **11.** 902   **13.** 170

Exercises 3.2   (Page 97)

1. 137.6   **3.** 211.02   **5.** 42.1   **7.** 22.36   **9.** 166.48   **11.** 171.76   **13.** 27.1
15. 93.84   **17.** 12.9   **19.** 0.357   **21.** 133.12   **23.** 2.84   **25.** 221.22   **27.** 398.16
29. 1647.3   **31.** 1724.758   **33.** 90.3   **35.** 271.57   **37.** 117.8 ft   **39.** 626.34 m
41. 233.2 m   **43.** $239.92   **45.** $116.40   **47.** 96.1 cm   **49.** $22.56   **51.** $14.37
53. 59¢

Review Exercises   (Page 98)

1. 11,729   **3.** 5664   **5.** 126,750   **7.** 38   **9.** 55   **11.** 19 with remainder of 151
13. 311.8 kwh

Exercises 3.3   (Page 104)

1. 126.15   **3.** 33.152   **5.** 5.818   **7.** 1315.13   **9.** 306.6   **11.** 53,786.112
13. 2394.18   **15.** 123,399.15   **17.** 85,623.72   **19.** 1378   **21.** 3523.1   **23.** 10.08
25. 235,100   **27.** 3.6   **29.** 5.7   **31.** 15   **33.** 18   **35.** 5.3   **37.** 14.7   **39.** 0.388
41. 0.0035781   **43.** 0.0083   **45.** 0.023791   **47.** 101.558   **49.** 37,120   **51.** 617.76
53. 43.7   **55.** 2.1   **57.** 19.01   **59.** $33.83   **61.** 252 of 100 and 200 speed; 168 of 400 speed
63. 21.2 mpg

Review Exercises   (Page 106)

1. $4\frac{1}{2}$ **3.** $7\frac{5}{7}$ **5.** $\frac{50}{9}$ **7.** $\frac{89}{9}$ **9.** $2^2 \cdot 3^2 \cdot 5^2$ **11.** $2^4 \cdot 3^2 \cdot 17$ **13.** 28

Exercises 3.4   (Page 111)

1. 98.6 m^2 **3.** 341.8 mi^2 **5.** 198.8 m^2 **7.** 670.8 in.^{2} **9.** 35.6 mm^2 **11.** 6.7 in.^{2}
13. 75.4 in.   **15.** 91.1 ft   **17.** 14.9 mm   **19.** 61.2 ft   **21.** 113.0 in.^{2} **23.** 2640.7 ft^2
25. 17.8 mm^2 **27.** 687.8 ft^2 **29.** 56 m^2 **31.** 19 m^2 **33.** 22 m^2 **35.** 6 m^2
37. 68.0 m^2 **39.** 88.4 mi^2 **41.** 24.6 in.^{2} **43.** 43.9 in.^{2}

Review Exercises   (Page 112)

1. 235.76   **3.** 4639.7   **5.** 351   **7.** 23.678   **9.** 0.000378   **11.** 0.003567

Chapter 3 Review Exercises   (Page 113)

1. 4 tens + 3 ones + 7 tenths; forty-three and seven tenths
3. 3 tens + 1 one + 1 tenth + 6 thousandths; thirty-one and one hundred six thousandths
5. 253.053   **7.** 0.253   **9.** 341.003   **11.** 34.04   **13.** 0.6   **15.** 107.8   **17.** 0.491
19. 244.4325   **21.** 2.7   **23.** 113.2 ft   **25.** 86.4 in.   **27.** 92.66 in.^{2}

Chapter 3 Test   (Page 115)

1. 2 tens + 7 ones + 5 hundredths   **3.** 3.53   **5.** 35.0302   **7.** 35.711   **9.** 56.15   **11.** 23.1
13. 0.0099999   **15.** 0.023567   **17.** 1206.4 ft   **19.** 1536.64 cm^2

Exercises 4.1   (Page 121)

1. $\frac{1}{4}$ **3.** $\frac{3}{4}$ **5.** $\frac{1}{5}$ **7.** $\frac{3}{5}$ **9.** $\frac{1}{20}$ **11.** $\frac{7}{20}$ **13.** $1\frac{7}{10}$ **15.** $2\frac{3}{4}$ **17.** $\frac{17}{10,000}$ **19.** $\frac{11}{400}$
21. $\frac{1}{200}$ **23.** $\frac{9}{400}$ **25.** 0.32   **27.** 0.98   **29.** 0.49   **31.** 1.38   **33.** 0.047   **35.** 0.487
37. 0.0057   **39.** 8.92   **41.** 0.0025   **43.** 0.0375   **45.** 0.004   **47.** 0.0012   **49.** 17%
51. 93%   **53.** 70%   **55.** 45%   **57.** 60%   **59.** 36%   **61.** 132%   **63.** 1520%
65. 396%   **67.** 1340%   **69.** 32.5%   **71.** $10\frac{2}{3}$%   **73.** 35%   **75.** 77%   **77.** 99%
79. 78.7%   **81.** 83.5%   **83.** 820%   **85.** 1492%   **87.** 8880%   **89.** 8350%   **91.** 0.82%
93. 10,000%   **95.** 0.1%

Review Exercises   (Page 123)

1. 35   **3.** 300   **5.** 1000   **7.** 3000   **9.** $\frac{7}{10}$ **11.** 1

Exercises 4.2   (Page 126)

1. 35   **3.** 300   **5.** 182   **7.** 39.1   **9.** 102.5   **11.** 300   **13.** 0.116   **15.** 1400
17. 200   **19.** 180   **21.** 150   **23.** 400   **25.** 10   **27.** 550   **29.** 380   **31.** 1375
33. 50%   **35.** 20%   **37.** 20%   **39.** 8%   **41.** 200%   **43.** 100%   **45.** $33\frac{1}{3}$%
47. 10,000%   **49.** 28   **51.** 70%   **53.** 8%   **55.** 28%   **57.** 36   **59.** 760   **61.** 18
63. $1.85   **65.** $270   **67.** 5600   **69.** 30%   **71.** 55%   **73.** 349

Review Exercises   (Page 128)

1. 390   **3.** 612.5   **5.** 308.75   **7.** 129.2

Exercises 4.3   (Page 133)

1. 50%   **3.** $33\frac{1}{3}$%   **5.** 42%   **7.** 42%   **9.** 14%   **11.** 71%   **13.** 100%   **15.** 50%
17. 20%   **19.** 42%   **21.** 88%   **23.** 1780%   **25.** $4.80   **27.** $718.20   **29.** $455.10
31. 48¢   **33.** $531.00   **35.** yes   **37.** $128   **39.** $420   **41.** $3400   **43.** $97,500
45. yes, it is within $\frac{1}{2}$ cent

Review Exercises   (Page 134)

1. 25.6   **3.** 3.72   **5.** 82.25   **7.** 8   **9.** 2.39   **11.** 143

Chapter 4 Review Exercises   (Page 136)

1. $\frac{43}{100}$ **3.** $\frac{153}{200}$ **5.** 0.47   **7.** 0.559   **9.** 74%   **11.** 250%   **13.** 47%   **15.** 9.9%
17. 27   **19.** 1.25   **21.** $666\frac{2}{3}$ **23.** 225   **25.** 20%   **27.** 20%   **29.** 400%   **31.** 80%
33. $74.16   **35.** $0.98   **37.** $624

Chapter 4 Test   (Page 137)

1. $\frac{23}{50}$ **3.** 0.68   **5.** 57%   **7.** 7.5%   **9.** 135   **11.** 700   **13.** 30%   **15.** 700%
17. 22%   **19.** $84   **21.** $2000

Exercises 5.1   (Page 146)

1. $2\frac{5}{8}$ in.   **3.** 1 in.   **5.** 11 in.   **7.** $4\frac{3}{4}$ in.   **9.** $\frac{1}{2}$ in.   **11.** $6\frac{1}{2}$ in.   **13.** 154 mm
15. 21 mm   **17.** 13 mm   **19.** 28 cm   **21.** 19 cm   **23.** 16 cm   **25.** 48 in.   **27.** 108 in.
29. 60 in.   **31.** 120 in.   **33.** 42 in.   **35.** 63 in.   **37.** 93 in.   **39.** 101 in.   **41.** $\frac{1}{12}$ ft
43. $\frac{1}{24}$ ft   **45.** $\frac{1}{2}$ ft   **47.** $\frac{2}{3}$ ft   **49.** $1\frac{1}{3}$ ft   **51.** $1\frac{1}{6}$ ft   **53.** $4\frac{2}{3}$ ft   **55.** $6\frac{1}{3}$ ft
57. 8 ft 10 in.   **59.** 10 ft 11 in.   **61.** 100 ft 10 in.   **63.** 4 ft 6 in.   **65.** 10 ft 3 in.
67. 7 ft 5 in.   **69.** 15 ft   **71.** 7 ft   **73.** $10\frac{1}{2}$ ft   **75.** $8\frac{1}{4}$ ft   **77.** 10 ft 6 in.
79. 3 ft $4\frac{1}{2}$ in.   **81.** 7 ft 0 in.   **83.** 15 ft 6 in.   **85.** 1   **87.** $\frac{1}{2}$ **89.** 1320   **91.** 9240
93. 1760   **95.** $1173\frac{1}{3}$ **97.** 300   **99.** 570   **101.** 3.1   **103.** 732   **105.** 7,680,000
107. 0.472   **109.** 4.532   **111.** 0.0325   **113.** 37.5   **115.** 125   **117.** 32.5
119. 675,000   **121.** 6.383   **123.** 0.63   **125.** 69.5   **127.** 5.689   **129.** 5.762
131. 0.000645   **133.** 0.65823

Review Exercises   (Page 149)

1. 3700   **3.** 3673.26   **5.** 0.101   **7.** 0.1

Exercises 5.2   (Page 151)

1. 360 ft^2 **3.** 900 ft^2 **5.** 4000   **7.** 450   **9.** 44%   **11.** 5   **13.** no   **15.** $8000
17. $160   **19.** $270   **21.** 210,000

Review Exercises   (Page 152)

1. 25%   **3.** 12   **5.** 92   **7.** 5.6   **9.** $33\frac{1}{3}$%

Exercises 5.3   (Page 161)

1. nuclear energy   **3.** 47%   **5.** 40%   **7.** 1980   **9.** 1970   **11.** 300 thousand metric tons
13. speeding and following too closely   **15.** reckless driving   **17.** seniors   **19.** $50
21. French and German   **23.** English   **25.** 51.4%   **27.** $7\frac{9}{13}$%   **29.** $11\frac{1}{9}$%   **31.** $4\frac{8}{13}$%
33. 65   **35.** 45   **37.** 1977–1979   **39.** 1   **41.** 1   **43.** 1 is running, 2 is stopped
45. $250   **47.** $1125   **49.** miners   **51.** 27   **53.** 112   **55.** $5.90   **57.** $10.35
59. $11,568   **61.** $2445.50

Review Exercises   (Page 166)

1. 22   **3.** $\frac{5}{2}$ **5.** 11, 13, 17, 19, 23, 29   **7.** $2^2 \cdot 73$ **9.** no   **11.** 210

Exercises 5.4 (Page 170)

1. 8 **3.** 35 **5.** 19 **7.** 9 **9.** 6 **11.** 17.5 **13.** 3 **15.** none **17.** 22.7
19. about 57¢ **21.** 50¢ **23.** 50¢ **25.** 82.5 **27.** 83 **29.** 2670 mi **31.** 89 mi
33. Both are 85. **35.** They had the same average of 56; Ginger's were more consistent.

Review Exercises (Page 171)

1. 9 **3.** 225 **5.** $\frac{1}{6}$ **7.** 6 **9.** $1\frac{9}{10}$

Chapter 5 Review Exercises (Page 172)

1. $2\frac{1}{4}$ in. **3.** 17 mm **5.** 66 **7.** $1\frac{1}{3}$ **9.** 8448 **11.** 160 **13.** 1.43 **15.** 167
17. 4.570 **19.** $49 **21.** no **23.** 7 billion **25.** 1985–1986 **27.** 830 million
29. 1987 **31.** $1659 **33.** 4.1% **35.** about 1967 **37.** about 40% **39.** 180
41. −18° **43.** mean is 33, median is 32, mode is 32

Chapter 5 Test (Page 177)

1. $2\frac{1}{4}$ in. **3.** 88 in. **5.** $7\frac{1}{3}$ ft **7.** 2 yds **9.** $1\frac{1}{2}$ mi **11.** 470 cm **13.** 0.0372 m
15. 31 dam **17.** about 19% **19.** about 290,000 **21.** 1986 **23.** A **25.** E
27. mean is 23, median is 21, mode is 22

Exercises 6.1 (Page 186)

1. fine, excellent, wonderful, etc. **3.** 1 **5.** 0 **7.** an infinite number **9.** 3 **11.** 3
13. 1 **15.** point B **17.** yes **19.** ∠G denotes many angles. **21.** true
23. false; a line has no endpoints **25.** true **27.** false; point G is the vertex of ∠DGB
29. true **31.** true **33.** true **35.** true **37.** acute **39.** obtuse **41.** right
43. straight **45.** 1 **47.** 6

Review Exercises (Page 188)

1. $\frac{11}{3}$ **3.** $\frac{23}{12}$ or $1\frac{11}{12}$ **5.** $\frac{1}{10}$ **7.** 1.08 **9.** 0.07488

Exercises 6.2 (Page 192)

1. 60° **3.** 75° **5.** There is no angle whose measure, added to 105°, gives 90°. **7.** 175°
9. true **11.** false; the common side FG is not between the angles **13.** true
15. false; they are adjacent angles **17.** true **19.** true **21.** false **23.** true **25.** true
27. true **29.** ∠1 and ∠2, ∠3 and ∠4, ∠4 and ∠5, ∠5 and ∠6, and ∠6 and ∠3 **31.** none
33. 130° **35.** 230° **37.** 100° **39.** 40° **41.** If I can see, then my eyes are open; yes.
43. If I am healthy, then I eat apples; not necessarily.

Review Exercises (Page 194)

1. 5.4 **3.** 0.8 **5.** $\frac{27}{64}$ **7.** 11 **9.** 18

Exercises 6.3   (Page 199)

1. alternate interior angles   **3.** corresponding angles
5. interior angles on the same side of the transversal
7. corresponding angles   **9.** alternate interior angles   **11.** not necessarily   **13.** parallel
15. parallel   **17.** parallel   **19.** parallel   **21.** 60°   **23.** 120°   **25.** 110°   **27.** 110°
29. no   **31.** yes   **33.** no

Review Exercises   (Page 200)

1. $\frac{1}{2}$ **3.** $\frac{3}{5}$ **5.** <   **7.** >   **9.** 15% of 5200

Exercises 6.4   (Page 205)

1. right triangle   **3.** acute triangle   **5.** acute triangle   **7.** obtuse triangle
9. isosceles triangle   **11.** scalene triangle   **13.** scalene triangle   **15.** isosceles triangle
17. 85 cm   **19.** $28\frac{1}{3}$ ft   **21.** 90°   **23.** 45°   **25.** 51°   **27.** 90.7°   **29.** 60°   **31.** 55°
33. 95°   **35.** 75°   **37.** 540°   **39.** 1440°

Review Exercises   (Page 207)

1. 7.5   **3.** $\frac{6}{55}$ **5.** 40%   **7.** 2500   **9.** 0.10625

Exercises 6.5   (Page 212)

1. DF **3.** EF **5.** $\angle B$ **7.** congruent; ASA $\cong$ ASA   **9.** not necessarily congruent
11. congruent; SAS $\cong$ SAS   **13.** SAS $\cong$ SAS   **15.** 50°   **17.** SAS $\cong$ SAS   **19.** 90°   **21.** 60°
23. 120°   **25.** 60°   **27.** 45°   **29.** 60°   **31.** 90°   **33.** 60°   **35.** 20°   **37.** 20°

Review Exercises   (Page 214)

1. $\frac{29}{15}$ or $1\frac{14}{15}$ **3.** $\frac{53}{60}$ **5.** 1.0696   **7.** $4\frac{17}{20}$ **9.** $11\frac{1}{3}$

Exercises 6.6   (Page 219)

1. rectangle   **3.** rectangle, rhombus, square   **5.** rhombus   **7.** rectangle, rhombus, square
9. 60°   **11.** 60°   **13.** 40°   **15.** 100°   **17.** 90°   **19.** 30°   **21.** 60°   **23.** 10 cm
25. 70°   **27.** 10 in.   **29.** 50°   **31.** 5 cm   **33.** yes

Review Exercises   (Page 220)

1. $\frac{13}{15}$ **3.** 0.4356   **5.** 1   **7.** 2   **9.** 23.35

Exercises 6.7   (Page 225)

1. $\overline{OA}$, $\overline{OC}$, $\overline{OB}$ **3.** $\overline{DA}$, $\overline{DC}$, $\overline{AC}$ **5.** $\angle 3$, $\angle 4$, $\angle D$ **7.** $\widehat{AD}$, $\widehat{DC}$, $\widehat{CB}$, $\widehat{BA}$
9. $\widehat{ADCB}$, $\widehat{DCBA}$, $\widehat{CBAD}$, $\widehat{BADC}$ **11.** 80°   **13.** 50°   **15.** 50°   **17.** 80°   **19.** 160°
21. 150°   **23.** 15°   **25.** 125°   **27.** 70°   **29.** 50°   **31.** 60°   **33.** 40°   **35.** 70°
37. 20°   **39.** 36°

Review Exercises   (Page 227)

1. $0.15   **3.** $13.41   **5.** yes, if he doesn't have to pay sales tax

Chapter 6 Review Exercises   (Page 229)

1. 5   **3.** point B   **5.** ∠1 and ∠2   **7.** ∠3   **9.** ∠1 and ∠2, ∠2 and ∠3, ∠3 and ∠4, ∠4 and ∠1
11. 55°   **13.** 150°   **15.** 150°   **17.** 60°   **19.** 60°   **21.** 30°   **23.** 60°
25. 65°   **27.** 25°   **29.** 720°   **31.** SAS ≅ SAS   **33.** 6 cm   **35.** 30°   **37.** 25°
39. 75°   **41.** 90°   **43.** 110°   **45.** 70°   **47.** 70°   **49.** 55°   **51.** 18 cm

Chapter 6 Test   (Page 233)

1. 4   **3.** 63°   **5.** 110°   **7.** 40°   **9.** 70°   **11.** 30°   **13.** 360°   **15.** 8 cm
17. 145°   **19.** 100°   **21.** 50°   **23.** 50°   **25.** 60°   **27.** 140°   **29.** 70°

Exercises 7.1   (Page 738)

1. $\frac{5}{7}$ **3.** $\frac{1}{2}$ **5.** $\frac{2}{3}$ **7.** $\frac{1}{3}$ **9.** $\frac{1}{5}$ **11.** $\frac{3}{7}$ **13.** $\frac{1}{18}$ **15.** $\frac{3}{4}$ **17.** $\frac{6}{5}$ **19.** $\frac{25}{1}$
21. no   **23.** yes   **25.** no   **27.** yes   **29.** 4   **31.** 6   **33.** 3   **35.** 9   **37.** $17
39. $6.50   **41.** $7\frac{1}{2}$ gal   **43.** $309   **45.** 49 ft $3\frac{1}{2}$ in.   **47.** 162   **49.** not exactly, but close

Review Exercises   (Page 240)

1. 90%   **3.** $\frac{1}{3}$ **5.** 480   **7.** $73.50   **9.** $88.70

Exercises 7.2   (Page 244)

1. true   **3.** false   **5.** true   **7.** AA ≅ AA   **9.** AA ≅ AA   **11.** $8\frac{4}{7}$ cm   **13.** $6\frac{18}{25}$ cm
15. 8 units   **17.** 144 units   **19.** 44 units   **21.** 50°   **27.** $\frac{9}{2}$ units   **29.** $37\frac{1}{2}$ ft   **31.** 60 ft
33. 48 ft   **35.** 528 ft   **37.** 174 ft

Review Exercises   (Page 246)

1. Base is 7; exponent is 6.   **3.** Base is $\frac{2}{3}$; exponent is 2.   **5.** 192   **7.** 432   **9.** 38

Exercises 7.3   (Page 250)

1. 15 in.^{2} **3.** 60 cm^2 **5.** 16 cm^2 **7.** 160 ft^2 **9.** 156 m^2 **11.** 108 ft^2 **13.** 45 cm^2
15. 144 m^2 **17.** 9 ft^2 **19.** 10,000 cm^2 **21.** $1200   **23.** $361.20   **25.** about 9.72 ft^2

Review Exercises   (Page 252)

1. $10\frac{1}{2}$ **3.** 250   **5.** 1040   **7.** 44%   **9.** 16

Exercises 7.4   (Page 255)

1. 25 in.^{2} **3.** 18 cm^2 **5.** 12 mm^2 **7.** 64 ft^2 **9.** 169 mm^2 **11.** 36 m^2 **13.** 80 ft^2
15. 144 ft^2 **17.** $192   **19.** $1658.48

Review Exercises (Page 257)

1. $1\frac{1}{3}$ **3.** 20 **5.** 79.35 **7.** 38.6

Exercises 7.5 (Page 259)

1. 8π in. **3.** 36 m **5.** 9π in.² **7.** $(60 + 9\pi)$ in.² **9.** $(117 - 16\pi)$ in.² **11.** $(40 - 4\pi)$ in.²
13. 75π ft² **15.** $\left(135 + \frac{9\pi}{4}\right)$ cm² **17.** 3.14 mi² **19.** 12.74 times **21.** It is multiplied by 4.
23. 1.59 ft

Review Exercises (Page 261)

1. $1\frac{5}{12}$ **3.** $6\frac{1}{12}$ **5.** $17\frac{17}{45}$ **7.** 80.44 **9.** $1\frac{7}{8}$

Exercises 7.6 (Page 267)

1. 94 cm²; 60 cm³ **3.** 108 m²; 48 m³ **5.** 324π in.²; 972π in.³ **7.** 216π m²; 432π m³
9. 90π cm²; 100π cm³ **11.** 360 m²; 400 m³ **13.** 576 cm³ **15.** $\frac{320\pi}{3}$ in.³ **17.** $\frac{1}{8}$ in.³
19. 197.82 ft³ **21.** 314 ft² **23.** 33,493.33 ft³ **25.** $V = lwh = s \cdot s \cdot s = s^3$ **27.** 27 ft³

Review Exercises (Page 269)

1. 36 **3.** 33 **5.** $\frac{4}{11}$ **7.** 6% of 36

Exercises 7.7 (Page 272)

1. 17 **3.** 17 **5.** 6 **7.** 3 **9.** 11 **11.** 9 **13.** [figure] **15.** [figure]

17. Maria **19.** Mercedes **21.** green, blue, yellow, red

Chapter 7 Review Exercises (Page 275)

1. $\frac{5}{12}$ **3.** no **5.** 3 **7.** $7\frac{1}{2}$ **9.** Show that m($\angle B$) = m($\angle E$) and that m($\angle ACB$) = m($\angle DCE$)
11. 18 **13.** 5 **15.** 50° **17.** 40° **19.** 90 ft **21.** 3 ft **23.** 49 in.² **25.** 180 ft²
27. 240 cm² **29.** 152.5 in.² **31.** 113.04 ft **33.** 56.25π cm² **35.** 1350 cm²
37. 254.52 in.³ **39.** 36π m³ **41.** 48π in.³ **43.** 2π in.³ **45.** 6 **47.** □
49. the sheep

Chapter 7 Test (Page 279)

1. $\frac{1}{3}$ **3.** yes **5.** 18 **7.** 6 **9.** yes **11.** 4 units **13.** 12 units **15.** $13\frac{1}{2}$ ft
17. 96 ft² **19.** 81 ft² **21.** 120 m³ **23.** 21π m² **25.** 5 **27.** △

Exercises 8.1 (Page 287)

1. 1, 2, 3, 4, 9, 18   **3.** 1, 3, 9   **5.** 0, 1, 2, 3, 4, 9, 18   **7.** 2   **9.** 2, 5   **11.** 2, 10
13. 10   **15.** 10, 15   **17.** number line with dots at 3, 4, 5, 6, 7   **19.** number line with dots at 20, 21, 22, 23, 24, 25, 26
21. number line with dots at 11, 12, 13, 14, 15, 16, 17, 18, 19   **23.** number line with dots at 0, 1, 2, 3, 4   **25.** number line with dots at 12, 13, ..., 24
27. =   **29.** <   **31.** >   **33.** =   **35.** =   **37.** =   **39.** <   **41.** $7 > 3$ **43.** $17 \leq 17$
45. $3 + 4 = 7$ **47.** $7 \geq 3$ **49.** $0 < 6$ **51.** $8 < 3 + 8$ **53.** $10 - 4 > 6 - 2$ **55.** $9 \div 3 \leq 8 \div 2$
57. 6 is the greater.   **59.** 11 is the greater.   **61.** 2 is the greater and lies to the right.
63. By adding 1, we can always obtain a larger number.
65. A natural number can either be divided by 2 or it can't.

Review Exercises (Page 288)

1. 250   **3.** 148   **5.** 16,606   **7.** 105   **9.** 726   **11.** $2480

Exercises 8.2 (Page 292)

1. $x + y$ **3.** $x(2y)$ or $2xy$ **5.** $y - x$ **7.** $\frac{y}{x}$ **9.** $\frac{x}{y} + z$ **11.** $z - xy$
13. $3xy$ **15.** $\frac{x+y}{y+z}$ **17.** $xy + \frac{y}{z}$ **19.** $c + 4$ **21.** $22t$ cents or $0.22t$ dollars   **23.** $\frac{x}{5}$ ft
25. $(3d + 5)$ dollars   **27.** 3 greater than a number x
29. the quotient obtained when x is divided by y **31.** twice the product of x and y
33. the quotient obtained when 2 is divided by the sum of x and y
35. the quotient obtained when the sum of 3 and x is divided by y
37. the quotient obtained when the product of x and y is divided by the sum of x and y
39. 10   **41.** 2   **43.** 24   **45.** 48   **47.** 1   **49.** 16   **51.** 1 term; 6   **53.** 3 terms; 1
55. 4 terms; 3   **57.** 3 terms; 4   **59.** 4 terms; 3   **61.** 19 and x **63.** 29, x, y, and z
65. 3, x, y, and z **67.** 2, 3, 3, x, and z **69.** 5, 1, 8   **71.** x and y **73.** 3, 1, and 25; 75
75. x and y

Review Exercises (Page 294)

1. odd natural number   **3.** odd natural number and prime number   **5.** $\frac{1}{5}$ **7.** $2 \cdot 3 \cdot 5 \cdot 5$
9. 8   **11.** $\frac{23}{30}$

Exercises 8.3 (Page 297)

1. 64   **3.** 36   **5.** 10,000   **7.** xx **9.** $3zzzz$ **11.** $(5t)(5t)$ or $25tt$
13. $5(2x)(2x)(2x)$ or $40xxx$ **15.** 36   **17.** 1000   **19.** 18   **21.** 216   **23.** 11
25. 3   **27.** 28   **29.** 64   **31.** 13   **33.** 16   **35.** 90   **37.** 28   **39.** 21   **41.** 56
43. 1   **45.** 1   **47.** 28   **49.** 17   **51.** 9   **53.** 8   **55.** 1   **57.** 60   **59.** 11   **61.** 1
63. $\frac{8}{9}$ **65.** 4   **67.** 4   **69.** 12   **71.** 4   **73.** 11   **75.** 24   **77.** 12   **79.** 25   **81.** 1
83. 28   **85.** 35   **87.** 1   **89.** 2   **91.** $(3 \cdot 8) + (5 \cdot 3)$ **93.** $(3 \cdot 8 + 5) \cdot 3$ **95.** $(4 + 3) \cdot (5 - 3)$
97. $(4 + 3) \cdot 5 - 3$

Review Exercises (Page 299)

1. >   **3.** ≥   **5.** 14   **7.** 24   **9.** 17   **11.** 8

Exercises 8.4 (Page 305)

1. positive, integer, rational, real
3. negative, integer, rational, real
5. positive, rational, real
7. positive, rational, real
9. negative, integer, rational, real
11. negative, rational, real

13. [number line: −4/3, 1/3, 5/2 marked between −1 and 5]
15. [number line: −9/2, −5/2, 3/2 marked between −4 and 2]
17. [number line: open circles at 2 and 5, segment between]
19. [number line: open circle at 2, ray to right]
21. [number line: open circle at −5, ray to left]
23. [number line: closed circles at −7 and −2, segment between]
25. [number line: open circles at −2 and 4, rays outward]
27. [number line: open circles at −7 and −2, segment between with ray]
29. [number line: open circles at −4, −2, 0]

31. 8 **33.** 8 **35.** 0 **37.** 9 **39.** −10 **41.** −5 **43.** $\frac{34}{15}$ **45.** 29 **47.** 38
49. 20 **51.** −2 **53.** −3 **55.** 2 **57.** 0 **59.** −3 **61.** 5 **63.** 4 **65.** 49
67. 1 **69.** 2 **71.** 5 **73.** 11

75. because every even natural number is a rational number
77. On a number line, the graphs of a number and its negative lie at equal distances from the origin.

Review Exercises (Page 306)

1. 52 **3.** 100 **5.** 12 **7.** 8 **9.** [number line with 5/3 marked] **11.** 153.86

Exercises 8.5 (Page 312)

1. 12 **3.** −10 **5.** 2 **7.** −2 **9.** −12 **11.** −13 **13.** $\frac{12}{35}$ **15.** $-\frac{1}{12}$ **17.** 1
19. 2.2 **21.** 7 **23.** −1 **25.** −7 **27.** −8 **29.** −18 **31.** $\frac{1}{5}$ **33.** 1.3 **35.** −1
37. 3 **39.** 10 **41.** −3 **43.** −1 **45.** 9 **47.** 1 **49.** 7 **51.** 4 **53.** 12
55. −17 **57.** 5 **59.** $\frac{1}{2}$ **61.** $-\frac{22}{5}$ **63.** $-8\frac{3}{4}$ **65.** −4.2 **67.** 4 **69.** −7 **71.** 10
73. 0 **75.** 8 **77.** 64 **79.** 3 **81.** 2.45 **83.** 1 **85.** 1 **87.** −3 **89.** −15
91. 4 **93.** 15 **95.** −105 **97.** 1 **99.** $-\frac{3}{5}$ **101.** 3 **103.** −1 **105.** −1
107. $\frac{7}{6}$ **109.** $8 **111.** +4 degrees **113.** +6 degrees **115.** $351.20 **117.** 0 yard gained
119. 2162 **121.** 700 shares **123.** $5.50 **125.** 18° **127.** 2000 yrs **129.** −$55

Review Exercises (Page 315)

1. $\frac{8}{9}$ **3.** 15 **5.** 13 **7.** $x + 2|y|$ **9.** $|x + 2y|$ **11.** $(36 + 10x + 25y)$ cents

Exercises 8.6 (Page 320)

1. 3 **3.** 18 **5.** 32 **7.** −32 **9.** −54 **11.** −16 **13.** 2 **15.** 1 **17.** 72
19. −24 **21.** −420 **23.** −96 **25.** 4 **27.** −9 **29.** −2 **31.** 5 **33.** −3 **35.** −8
37. −8 **39.** 6 **41.** 36 **43.** 18 **45.** 5 **47.** 7 **49.** −4 **51.** 2 **53.** −4
55. −2 **57.** 2 **59.** 1 **61.** −6 **63.** −30 **65.** 7 **67.** −10 **69.** −66 **71.** 6
73. −10 **75.** 14 **77.** −81 **79.** 88 **81.** −30 **83.** −21 **85.** $-\frac{1}{6}$ **87.** $-\frac{11}{12}$
89. $-\frac{7}{36}$ **91.** $-\frac{11}{48}$ **93.** 6° warmer **95.** −$450 **97.** 720 gal **99.** −$7 **101.** −5 hr

Review Exercises (Page 322)

1. $=$ 3. $<$ 5. $>$ 7. $>$ 9. $=$ 11. ←—○━━○—→
                                              -3 $3/2$

Exercises 8.7 (Page 327)

1. 10 3. -24 5. 144 7. 3 9. $x + y = 5 + 7 = 12; y + x = 7 + 5 = 12$
11. $3x + 2y = 3(5) + 2(7) = 15 + 14 = 29; 2y + 3x = 2(7) + 3(5) = 14 + 15 = 29$
13. $x(x + y) = 5(5 + 7) = 5(12) = 60; (x + y)x = (5 + 7)5 = (12)5 = 60$
15. $(x + y) + z = [2 + (-3)] + 1 = (-1) + 1 = 0; x + (y + z) = 2 + (-3 + 1) = 2 + (-2) = 0$
17. $(xz)y = [2(1)](-3) = -6; x(yz) = 2[(-3)1] = 2(-3) = -6$
19. $x^2(yz^2) = 2^2[-3(1)^2] = 4[-3(1)] = 4(-3) = -12; (x^2y)z^2 = [2^2(-3)]1^2 = [4(-3)]1 = (-12)1 = -12$
21. $3x + 3y$ 23. $x^2 + 3x$ 25. $-ax - bx$ 27. $4x^2 + 4x$ 29. $-5t - 10$ 31. $-2; \frac{1}{2}$
33. $-\frac{1}{3}; 3$ 35. 0; none 37. $\frac{5}{2}; -\frac{2}{5}$ 39. $0.2; -5$
41. commutative property of addition   43. commutative property of multiplication
45. distributive property   47. commutative property of addition
49. multiplicative identity property   51. additive inverse property
53. $3x + 3 \cdot 2$ 55. xy^2 57. $(y + x)z$ 59. $x(yz)$ 61. x

Review Exercises (Page 328)

1. $x + y^2$ 3. $(x + 3)^2$ 5. x 7. $-y$ 9. 15 is positive, integer, and rational
11. $-\frac{5}{2}$ is negative and rational

Chapter 8 Review Exercises (Page 330)

1. 1, 2, 3, 4, 5 3. 1, 3, 5 5. ←●—●—●—●—●—●—●—●—●—●—●→ 7. $>$ 9. $2x + 3$ 11. 7
                                  10 11 12 13 14 15 16 17 18 19 20
13. three 15. 1 17. 81 19. 22 21. $<$ 23. $>$ 25. 0 27. ←○━━○—→
                                                              -4 3
29. -5 31. -4 33. -2 35. 4 37. -6 39. 2 41. -7 43. 6 45. -8
47. closure property of addition   49. associative property of addition
51. commutative property of addition   53. commutative property of addition
55. additive inverse property

Chapter 8 Test (Page 333)

1. 31, 37, 41, 43, 47 3. ←●━━●→ 5. $5y - (x + y)$ 7. $<$ 9. 16 11. 0 13. -4
                          5 15
15. 5 17. -23 19. 5 21. distributive property 23. multiplicative inverse property

Exercises 9.1 (Page 339)

1. yes 3. no 5. yes 7. yes 9. yes 11. yes 13. no 15. no 17. no
19. yes 21. yes 23. yes 25. yes 27. 10 29. -3 31. 7 33. 4 35. 13
37. 32 39. 740 41. 0 43. $\frac{7}{6}$ 45. $\frac{7}{3}$ 47. -6 49. 5 51. 9 53. -36
55. -1 57. -62 59. 0 61. $\frac{5}{6}$ 63. $\frac{14}{3}$ 65. $-\frac{5}{2}$ 67. -10 69. 12 71. 16
73. 21 75. 74 77. -28 79. 806 81. 460 83. 0 85. $\frac{46}{35}$ 87. $\frac{47}{4}$ 89. $-\frac{46}{17}$
91. 5 93. 18 95. 62 97. $45,200 99. 3 101. $28 103. $39.50 105. $190

Review Exercises   (Page 341)

1. 15   **3.** 12   **5.** −1   **7.** 1   **9.** −18   **11.** −124

Exercises 9.2   (Page 347)

1. 1   **3.** $\frac{5}{2}$ **5.** −3   **7.** 5   **9.** −2   **11.** 1   **13.** 3   **15.** −2   **17.** 2   **19.** −5
21. 3   **23.** −2   **25.** −2   **27.** 0   **29.** −9   **31.** −33   **33.** 28   **35.** 5   **37.** 7
39. −8   **41.** 10   **43.** −6   **45.** 12   **47.** 3   **49.** $\frac{9}{2}$ **51.** −2   **53.** 6   **55.** 35
57. $-\frac{2}{3}$ **59.** 0   **61.** 6   **63.** 3   **65.** 5   **67.** 5   **69.** 7   **71.** 8   **73.** 7   **75.** 3
77. −4   **79.** 117   **81.** $2000   **83.** 18   **85.** 61   **87.** 3   **89.** 1200 gal

Review Exercises   (Page 349)

1. 50 cm   **3.** 27 cm^2 **5.** 635.664 cm^3 **7.** closure property of addition
9. commutative property of addition   **11.** additive inverse property

Exercises 9.3   (Page 354)

1. $20x$ **3.** $3x^2$ **5.** not possible   **7.** $7x + 6$ **9.** $7z - 15$ **11.** $12x + 121$
13. $-22x^2 + 46$ **15.** $x^3 - 10$ **17.** $5x^2 + 24x$ **19.** $6y + 62$ **21.** $-2x + 7y$ **23.** $18y$
25. $2x - y + 2$ **27.** $5x + 7$ **29.** $5x + 35$ **31.** −2   **33.** −3   **35.** 1   **37.** 1   **39.** 6
41. 35   **43.** −1   **45.** $\frac{3}{2}$ **47.** −2   **49.** −9   **51.** 0   **53.** −20   **55.** −41   **57.** −6
59. −9   **61.** $\frac{37}{2}$ **63.** 9   **65.** −1   **67.** 8   **69.** 2   **71.** 5   **73.** 4   **75.** −3   **77.** 1
79. −5   **81.** $\frac{8}{3}$ **83.** 0   **85.** −3   **87.** 2   **89.** identity   **91.** impossible equation
93. 16   **95.** impossible equation   **97.** identity   **99.** identity

Review Exercises   (Page 356)

1. 0   **3.** −122   **5.** 2   **7.** 2   **9.** 9   **11.** −17

Exercises 9.4   (Page 360)

1. $I = \dfrac{E}{R}$ **3.** $w = \dfrac{V}{lh}$ **5.** $b = P - a - c$ **7.** $w = \dfrac{P - 2l}{2}$ **9.** $t = \dfrac{A - P}{Pr}$
11. $r = \dfrac{C}{2\pi}$ **13.** $w = \dfrac{2Kg}{v^2}$ **15.** $R = \dfrac{P}{I^2}$ **17.** $g = \dfrac{wv^2}{2K}$ **19.** $M = \dfrac{Fd^2}{Gm}$ **21.** $d^2 = \dfrac{GMm}{F}$
23. $r = \dfrac{G}{2b} + 1$ or $r = \dfrac{G + 2b}{2b}$ **25.** $t = \dfrac{d}{r}$; 3   **27.** $t = \dfrac{i}{pr}$; 2   **29.** $c = P - a - b$; 3
31. $h = \dfrac{2K}{a + b}$; 8   **33.** $I = \dfrac{E}{R}$; 4 amp   **35.** $r = \dfrac{C}{2\pi}$; 2.71 ft   **37.** $R = \dfrac{P}{I^2}$; 13.78 ohms
39. $m = \dfrac{Fd^2}{GM}$ **41.** $D = \dfrac{L - 3.25r - 3.25R}{2}$; 6 ft

Review Exercises   (Page 361)

1. $-2x$ **3.** impossible   **5.** $7y^3 - 2x^2 + 5$ **7.** $-x - 13$ **9.** 21   **11.** 6

Exercises 9.5 (Page 365)

1. 3   **3.** 2   **5.** 11   **7.** 35   **9.** 3   **11.** 0   **13.** 4 ft and 8 ft   **15.** 5 m, 15 m, 25 m
17. 7 ft, 14 ft, 14 ft   **19.** $316   **21.** 3 first-place, 11 second-place, 22 third-place   **23.** 250
25. $10   **27.** 125 lb   **29.** 300 g

Review Exercises (Page 366)

1. $3 + 2 = 2 + 3$ **3.** $4(5 \cdot 6) = (4 \cdot 5)6$ **5.** $-(-7) = 7$ **7.** 12   **9.** 12   **11.** 0

Exercises 9.6 (Page 372)

1. 26 and 28   **3.** 39, 40, and 41   **5.** 7   **7.** 13   **9.** 19 ft   **11.** 29 m by 18 m
13. 17 in. by 39 in.   **15.** 60°   **17.** 26°   **19.** 102°   **21.** 2
23. 6 nickels, 6 dimes, 9 quarters   **25.** 9   **27.** 7   **29.** 960 pairs   **31.** 300
33. 7500 gal   **35.** A

Review Exercises (Page 374)

1. 200 cm³   **3.** $\frac{1372}{3}$ m³   **5.** 27   **7.** $7x - 6$ **9.** $-\frac{3}{2}$ **11.** x

Exercises 9.7 (Page 378)

1. $4500 at 6%; $19,500 at 7%   **3.** $3750 in each account   **5.** $2350 at 5%; $4700 at 10%
7. $2500 at 10%; $5000 at 12%   **9.** 3 hr   **11.** 6.5 hr   **13.** 7.5 hr   **15.** 500 mph   **17.** 20
19. 50 gal   **21.** 7.5 oz   **23.** 40 lb lemon drops; 60 lb jelly beans   **25.** $1.20   **27.** 80

Review Exercises (Page 379)

1. [number line graph]   **3.** [number line graph]
5. [number line graph]   **7.** [number line graph]   **9.** $1488   **11.** $3

Exercises 9.8 (Page 385)

1. [graph: open at 3]   **3.** [graph: closed at −10]   **5.** [graph: open at −1]   **7.** [graph: closed at 4]
9. [graph: open at −2]   **11.** [graph: open at −4]   **13.** [graph: closed at −1]   **15.** [graph: closed at −13]
17. [graph: open at −3]   **19.** [graph: open at −2]   **21.** [graph: closed at 2]   **23.** [graph: open at 3]
25. [graph: open at −15]   **27.** [graph: closed at 20]   **29.** [graph: open at −3]   **31.** [graph: closed at 3]
33. [graph: open at 2]   **35.** [graph: closed at 3]   **37.** [graph: open at −7]   **39.** [graph: closed at 4]
41. [graph: open 7 to 10]   **43.** [graph: open −9 to closed 3]   **45.** [graph: closed −10 to 0]   **47.** [graph: open −5 to −2]
49. [graph: closed −6 to 10]   **51.** [graph: closed 2 to 3]   **53.** [graph: closed −1 to open 2]   **55.** [graph: open −4 to 1]
57. [graph: closed −3 to −1]   **59.** [graph: open −2 to 2]   **61.** 0 ft $< s \leq$ 19 ft   **63.** 0.1 mi $\leq x \leq$ 2.5 mi
65. 3.3 mi $\leq x \leq$ 4.1 mi   **67.** 66.2° $< F <$ 71.6°
69. 37.052 in. $< C <$ 38.308 in.   **71.** 68.18 kg $< w <$ 86.36 kg
73. 140 lb $\leq c \leq$ 150 lb   **75.** 5 ft $< w <$ 9 ft

Review Exercises   (Page 387)

1. $5x^2 - 2y^2$ **3.** $-x^2 + 2xy$ **5.** $-x + 14$ **7.** $\frac{1}{3}$ **9.** 3   **11.** $12,500

Chapter 9 Review Exercises   (Page 388)

1. yes   **3.** no   **5.** yes   **7.** yes   **9.** -4 **11.** 9   **13.** 3   **15.** -1 **17.** 1
19. 2   **21.** -2 **23.** 5   **25.** 13   **27.** 5   **29.** 15   **31.** 15   **33.** $14x$ **35.** $5b$
37. impossible   **39.** $4y^2 - 6$ **41.** $9x$ **43.** 0   **45.** -7 **47.** 1   **49.** 7   **51.** -3
53. 9   **55.** 1   **57.** $R = \frac{E}{I}$ **59.** $R = \frac{P}{I^2}$ **61.** $h = \frac{V}{lw}$ **63.** $h = \frac{V}{\pi r^2}$ **65.** $G = \frac{Fd^2}{Mm}$
67. $V = \frac{T}{n} + 3$ or $V = \frac{T + 3n}{n}$ **69.** -8 **71.** $4.85   **73.** 85   **75.** 15°   **77.** 13 in.
79. 40 units on each machine; 80 total units   **81.** 20 min.   **83.** $17,500
85. ⟵○⟶ 1   **87.** ⟵●⟶ 4   **89.** ⟵●⟶ 3   **91.** ⟵○—○⟶ 6 11
93. ⟵○—●⟶ 0 9

Chapter 9 Test   (Page 391)

1. a solution   **3.** not a solution   **5.** -2 **7.** -3 **9.** -2 **11.** $6x - 15$ **13.** $3x^2 - 6x$
15. $-18x$ **17.** $t = \frac{d}{r}$ **19.** $h = \frac{A}{2\pi r}$ **21.** $v = \frac{RT}{P}$ **23.** 165   **25.** $5250   **27.** $7\frac{1}{2}$ liters
29. ⟵○⟶ -2

Exercises 10.1   (Page 398)

1. base of 4; exponent of 3   **3.** base of x; exponent of 5   **5.** base of $2y$; exponent of 3
7. base of x; exponent of 4   **9.** base of x; exponent of 1   **11.** base of x; exponent of 3
13. $5 \cdot 5 \cdot 5$ **15.** $xxxxxxx$ **17.** $-4xxxxx$ **19.** $(3t)(3t)(3t)(3t)(3t)$ **21.** 2^3 **23.** x^4
25. $(2x)^3$ **27.** $-4t^4$ **29.** 625   **31.** 13   **33.** 561   **35.** -725 **37.** x^7 **39.** x^7
41. t^3 **43.** a^{12} **45.** y^9 **47.** $12x^7$ **49.** $-4y^5$ **51.** $12x^9$ **53.** $6a^6$ **55.** 3^8
57. y^{15} **59.** a^{21} **61.** x^{25} **63.** $243z^{30}$ **65.** x^{31} **67.** r^{36} **69.** s^{33} **71.** x^3y^3
73. r^6s^4 **75.** $16a^2b^4$ **77.** $-8r^6s^9t^3$ **79.** $\frac{a^3}{b^3}$ **81.** $\frac{x^{10}}{y^{15}}$ **83.** $\frac{-32a^5}{b^5}$ **85.** a^{16}
87. $\frac{y^2}{4}$ **89.** $-\frac{8}{27}$ **91.** x^2 **93.** y^4 **95.** $3a$ **97.** ab^4 **99.** $\frac{10r^{13}s^3}{3}$ **101.** $\frac{x^{12}}{2}$

Review Exercises   (Page 400)

1. ⟵●—●⟶ -4 $5/2$ **3.** $\frac{19}{18}$ **5.** 1   **7.** three times the sum of x and y
9. the absolute value of the difference obtained when y is subtracted from x **11.** $3 + |2x|$

Exercises 10.2   (Page 403)

1. 8   **3.** 1   **5.** 1   **7.** 512   **9.** 2   **11.** 1   **13.** 1   **15.** -2 **17.** $\dfrac{1}{x^2}$ **19.** $\dfrac{1}{b^5}$
21. $\dfrac{1}{16^4}$ **23.** $\dfrac{1}{a^3b^6}$ **25.** $\dfrac{1}{y}$ **27.** $\dfrac{1}{r^6}$ **29.** y^5 **31.** 1   **33.** $\dfrac{1}{a^2b^4}$ **35.** $\dfrac{1}{x^6y^3}$
37. $\dfrac{1}{x^3}$ **39.** $\dfrac{1}{y^2}$ **41.** a^8b^{12} **43.** $-\dfrac{y^{10}}{32x^{15}}$ **45.** a^{14} **47.** $\dfrac{1}{b^{14}}$ **49.** $\dfrac{256x^{28}}{81}$
51. $\dfrac{16y^{14}}{z^{10}}$ **53.** $\dfrac{x^{14}}{128y^{28}}$ **55.** $\dfrac{16u^4v^8}{81}$ **57.** $\dfrac{1}{9a^2b^2}$ **59.** $\dfrac{27c^{15}}{8a^9b^3}$ **61.** $\dfrac{1}{512}$
63. $\dfrac{17y^{27}z^5}{x^{35}}$ **65.** x^{3m} **67.** u^{5m} **69.** y^{2m+2} **71.** y^m **73.** $\dfrac{1}{x^{3n}}$ **75.** x^{2m+2}
77. x^{8n-12} **79.** y^{4n-8}

Review Exercises   (Page 404)

1. 2   **3.** 8   **5.** 9   **7.** $s = \dfrac{f(P-L)}{i}$ or $s = \dfrac{fP - fL}{i}$ **9.** ⟵○⟶ 1/3
11. $13,000 and $28,400

Exercises 10.3   (Page 407)

1. 2.3×10^4 **3.** 1.7×10^6 **5.** 6.2×10^{-2} **7.** 5.1×10^{-6} **9.** 4.25×10^3 **11.** 2.5×10^{-3}
13. 230   **15.** 812,000   **17.** 0.00115   **19.** 0.000976   **21.** 25,000,000   **23.** 0.00051
25. 2.52×10^{13} **27.** 114,000,000 mi   **29.** 6.22×10^{-3} mi   **31.** 714,000   **33.** 30,000
35. 200,000   **37.** 1.9008×10^{11} **39.** 3.3×10^{-1} km/sec

Review Exercises   (Page 408)

1. 11, 13, 17, 19, 23, 29   **3.** 3   **5.** commutative property of addition   **7.** 6   **9.** ⟵●⟶ -7
11. 11 ft $\leq s \leq$ 15 ft

Exercises 10.4   (Page 412)

1. binomial   **3.** trinomial   **5.** monomial   **7.** binomial   **9.** trinomial   **11.** none of these
13. 4th   **15.** 3rd   **17.** 8th   **19.** 6th   **21.** 12th   **23.** zeroth   **25.** 7   **27.** -8
29. $5w - 3$ **31.** $-5y - 3$ **33.** -4 **35.** -5 **37.** $-r^2 - 4$ **39.** $-9s^2 - 4$ **41.** 3
43. 11   **45.** $b^2 + 2b + 3$ **47.** $\frac{1}{16}w^2 + \frac{1}{2}w + 3$ **49.** -1 **51.** $5u^2 - 2$ **53.** $-20z^6 - 2$
55. $5x^2y^2 - 2$ **57.** $5x + 5h - 2$ **59.** $5x + 5h - 4$ **61.** $10y + 5z - 2$ **63.** $10y + 5z - 4$

Review Exercises   (Page 413)

1. 8   **3.** ⟵●⟶ $-7/3$ **5.** 560   **7.** x^{18} **9.** y^9
11. 1,900,000,000,000,000,000,000,000,000 kg

Exercises 10.5 (Page 416)

1. like terms; $7y$ 3. unlike terms   5. like terms; $13x^3$ 7. like terms; $8x^3y^2$
9. like terms; $65t^6$ 11. unlike terms   13. $9y$ 15. $-12t^2$ 17. $16u^3$ 19. $7x^5y^2$
21. $14rst$ 23. $-6a^2bc$ 25. $15x^2$ 27. $4x^2y^2$ 29. $98x^8y^4$ 31. $7x+4$ 33. $2a+7$
35. $7x-7y$ 37. $-19x-4y$ 39. $6x^2+x-5$ 41. $7b+4$ 43. $3x+1$ 45. $5x+15$
47. $3x-3y$ 49. $5x^2-25x-20$ 51. $5x^2+x+11$ 53. $-7x^3-7x^2-x-1$
55. $2x^2y+xy+13y^2$ 57. $5x^2+6x-8$ 59. $-x^3+6x^2+x+14$ 61. $-12x^2y^2-13xy+36y^2$
63. $6x^2-2x-1$ 65. t^3+3t^2+6t-5 67. $-3x^2+5x-7$ 69. $6x-2$ 71. $-5x^2-8x-19$
73. $4y^3-12y^2+8y+8$ 75. $3a^2b^2-6ab+b^2-6ab^2$ 77. $-6x^2y^2+4xy^2z-20xy^3+2y$
79. $6x+3h-10$

Review Exercises (Page 418)

1. ⟵−5——5⟶   3. -8 5. -9 7. -3 9. $315\ \text{ft}^2$ 11. $m=\dfrac{2K}{v^2}$

Exercises 10.6 (Page 423)

1. $12x^5$ 3. $-24b^6$ 5. $6x^5y^5$ 7. $-3x^4y^7z^8$ 9. $x^{10}y^{15}$ 11. $a^5b^4c^7$ 13. $3x+12$
15. $-4t-28$ 17. $3x^2-6x$ 19. $-6x^4+2x^3$ 21. $3x^2y+3xy^2$ 23. $6x^4+8x^3-14x^2$
25. $-3x^5y^6+3x^6y^5-3x^4y^5$ 27. $-a^4bc^3-ab^4c^3+abc^6$ 29. $-6x^4y^4-6x^3y^5$
31. $a^2+9a+20$ 33. $3x^2+10x-8$ 35. $6a^2+2a-20$ 37. $6x^2-7x-5$
39. $2x^2+3x-9$ 41. $6t^2+7st-3s^2$ 43. $x^2-xy-2y^2$ 45. $-4r^2-20rs-21s^2$
47. $-6a^2-16ab-8b^2$ 49. $-12t^2+7tu-u^2$ 51. $x^2+xz+yx+yz$
53. $u^2+2tu+uv+2vt$ 55. $4x^2+11x+6$ 57. $12x^2+14xy-10y^2$ 59. x^3-1
61. $x^2+8x+16$ 63. t^2-6t+9 65. r^2-16 67. $x^2+10x+25$ 69. $4s^2+4s+1$
71. $16x^2-25$ 73. $9r^2+24rs+16s^2$ 75. $x^2-4xy+4y^2$ 77. $4a^2-12ab+9b^2$
79. $16x^2-25y^2$ 81. $2x^2-6x-8$ 83. $3a^3-3ab^2$ 85. $-6y^4z-9y^3z^2+6y^2z^3$
87. $4t^3+11t^2+18t+9$ 89. $-3x^3+25x^2y-56xy^2+16y^3$ 91. $5t^2-11t$
93. $x^2y+3xy^2+2x^2$ 95. $2x^2+xy-y^2$ 97. $8x$ 99. $5s^2-7s-9$ 101. -3
103. -8 105. -1 107. 0 109. 1 111. 6 113. 7 and 9   115. 90 ft   117. $\frac{3}{2}$ in.

Review Exercises (Page 425)

1. $-\frac{1}{2}$ 3. 4 5. distributive property   7. commutative property of multiplication   9. 0
11. 4.8×10^{18}

Exercises 10.7 (Page 428)

1. $\dfrac{1}{3}$ 3. $-\dfrac{5}{3}$ 5. $\dfrac{3}{4}$ 7. 1 9. $-\dfrac{1}{4}$ 11. $\dfrac{42}{19}$ 13. $\dfrac{x}{z}$ 15. $\dfrac{r^2}{s}$ 17. $\dfrac{2x^2}{y}$
19. $-\dfrac{3u^3}{v^2}$ 21. $\dfrac{4r}{y^2}$ 23. $-\dfrac{13}{3rs}$ 25. $\dfrac{x^4}{y^6}$ 27. a^8b^8 29. $-\dfrac{3r}{s^9}$ 31. $-\dfrac{x^3}{4y^3}$ 33. $\dfrac{125}{8b^3}$
35. $\dfrac{xy^2}{3}$ 37. a^8 39. z^3 41. $\dfrac{2}{y}+\dfrac{3}{x}$ 43. $\dfrac{1}{5y}-\dfrac{2}{5x}$ 45. $\dfrac{1}{y^2}+\dfrac{2y}{x^2}$ 47. $3a-2b$
49. $\dfrac{1}{y}-\dfrac{1}{2x}+\dfrac{2z}{xy}$ 51. $3x^2y-2x-\dfrac{1}{y}$ 53. $5x-6y+1$ 55. $\dfrac{10x^2}{y}-5x$ 57. $-\dfrac{4x}{3}+\dfrac{3x^2}{2}$
59. $xy-1$ 61. $\dfrac{x}{y}-\dfrac{11}{6}+\dfrac{y}{2x}$ 63. 2

Review Exercises (Page 429)

1. 52 **3.** 2.65×10^{-4} **5.** 1 **7.** $\dfrac{2y^2}{x}$ **9.** -2 **11.** 19 and 20

Exercises 10.8 (Page 435)

1. $x + 2$ **3.** $y + 12$ **5.** $a + b$ **7.** $3a - 2$ **9.** $b + 3$ **11.** $x - 3y$ **13.** $2x + 1$
15. $x - 7$ **17.** $3x + 2y$ **19.** $2x - y$ **21.** $x + 5y$ **23.** $x - 5y$ **25.** $x^2 + 2x - 1$
27. $2x^2 + 2x + 1$ **29.** $x^2 + xy + y^2$ **31.** $x + 1 + \dfrac{-1}{2x + 3}$ **33.** $2x + 2 + \dfrac{-3}{2x + 1}$
35. $x^2 + 2x + 1$ **37.** $x^2 + 2x - 1 + \dfrac{6}{2x + 3}$ **39.** $2x^2 + 8x + 14 + \dfrac{31}{x - 2}$ **41.** $x + 1$
43. $2x - 3$ **45.** $x^2 - x + 1$ **47.** $4a - 3b$ **49.** $5x^2 - 3x - 4$

Review Exercises (Page 436)

1. 21, 22, 24, 25, 26, 27, 28 **3.** -2 **5.** 25 **7.** -2 **9.** $8x^2 - 6x + 1$ **11.** 880 in.^{2}

Chapter 10 Review Exercises (Page 438)

1. 125 **3.** 64 **5.** 13 **7.** 162 **9.** x^5 **11.** y^{10} **13.** $2b^{12}$ **15.** $256s^3$ **17.** x^{15}
19. 9 **21.** x^4 **23.** $\dfrac{y^2}{x^2}$ **25.** x **27.** x^{10} **29.** $\dfrac{1}{x^4}$ **31.** $\dfrac{1}{s^3}$ **33.** 7.28×10^2
35. 1.36×10^{-2} **37.** 7.61×10^0 **39.** 1.2×10^{-4} **41.** 726,000 **43.** 2.68 **45.** 7.31
47. 7th **49.** 5th **51.** 11 **53.** -4 **55.** 402 **57.** 82 **59.** $7x$ **61.** $4x^2y^2$
63. $5x^2$ **65.** $8x^2 - 6x$ **67.** $9x^2 + 3x + 9$ **69.** $10x^3y^5$ **71.** $5x + 15$ **73.** $3x^4 - 5x^2$
75. $-x^2y^3 + x^3y^2$ **77.** $x^2 + 5x + 6$ **79.** $6a^2 - 6$ **81.** $2a^2 - ab - b^2$ **83.** $-9a^2 + b^2$
85. $y^2 - 4$ **87.** $y^2 - 6y + 9$ **89.** 1 **91.** 7 **93.** 1 **95.** $\dfrac{3}{2y} + \dfrac{3}{x}$ **97.** $-3a - 4b + 5c$
99. $x + 1 + \dfrac{3}{x + 2}$ **101.** $2x + 1$ **103.** $3x^2 + 2x + 1$

Chapter 10 Test (Page 441)

1. $2x^3y^4$ **3.** y^6 **5.** $32x^{15}$ **7.** 3 **9.** y^3 **11.** 2.8×10^4 **13.** 7400 **15.** binomial
17. 0 **19.** $-7x + 2y$ **21.** $5x^3 + 2x^2 + 2x - 5$ **23.** $-4x^5y$ **25.** $6x^2 - 7x - 20$
27. $2x^3 - 7x^2 + 14x - 12$ **29.** $\dfrac{y}{2x}$ **31.** -2

Exercises 11.1 (Page 447)

1. $2^2 \cdot 3$ **3.** $3 \cdot 5$ **5.** $2^3 \cdot 5$ **7.** $2 \cdot 7^2$ **9.** $3^2 \cdot 5^2$ **11.** $2^5 \cdot 3^2$ **13.** $x + 5$
15. a **17.** $1 + 2y$ **19.** $ab + 2$ **21.** $3(x + 2)$ **23.** $x(y - z)$ **25.** $t(t + 2)$
27. $2r^2(r^2 - 2)$ **29.** $a^2b^3z^2(az - 1)$ **31.** $8xy^2z^3(3xyz + 1)$ **33.** $6uvw^2(2w - 3v)$
35. $3(x + y - 2z)$ **37.** $a(b + c - d)$ **39.** $2y(2y + 4 - x)$ **41.** $3r(4r - s + 3rs^2)$
43. $abx(1 - b + x)$ **45.** $2xyx^2(2xy - 3y + 6)$ **47.** $7a^2b^2c^2(10a + 7bc - 3)$ **49.** $-(a + b)$
51. $-(2x - 5y)$ **53.** $-(2a - 3b)$ **55.** $-(3m + 4n - 1)$ **57.** $-(3xy - 2z - 5w)$
59. $-(3abc + 6abd - 9abe)$ **61.** $-3xy(x + 2y)$ **63.** $-4a^2b^2(b - 3a)$
65. $-2ab^2c(2ac - 7a + 5c)$ **67.** $-7ab(2a^5b^5 - 7ab^2 + 3)$ **69.** $-5a^2b^3c(1 - 3abc + 5a^2)$

Review Exercises (Page 448)

1. $6xy + 3x + 4y + 2$ **3.** $2ab - 2a + b - 1$ **5.** $6pr - 3pq - 2qr + q^2$ **7.** $3x + 3y + ax + ay$
9. $x^2 + 4x + 3 - xy - y$ **11.** $3x^3 - 6x - x^2y + 2y + 3x^2 - 6$

Exercises 11.2 (Page 450)

1. $3 + x$ **3.** $5 - b$ **5.** $3 - x$ **7.** $z + y + 3$ **9.** $2 + a$ **11.** $2x^2 - 3y^2$ **13.** $(x + y)(2 + b)$
15. $(x + y)(3 - a)$ **17.** $(r - 2s)(3 - x)$ **19.** $(x - 3)(x - 2)$ **21.** $2(a^2 + b)(x + y)$
23. $3(r + 3s)(x^2 - 2y^2)$ **25.** $(a + b + c)(3x - 2y)$ **27.** $7xy(r + 2s - t)(2x - 3)$
29. $(x + 1)(x + 3 - y)$ **31.** $(x + y)(2 + a)$ **33.** $(r + s)(7 - k)$ **35.** $(r + s)(x + y)$
37. $(2x + 3)(a + b)$ **39.** $(b + c)(2a + 3)$ **41.** $(x + y)(2x - 3)$ **43.** $(v - 3w)(3t + u)$
45. $(3p + q)(3m - n)$ **47.** $(m - n)(p - 1)$ **49.** $x^2(a + b)(x + 2y)$ **51.** $4a(b + 3)(a - 2)$
53. $(x^2 + 1)(x + 2)$ **55.** $y(x^2 - y)(x - 1)$ **57.** $(x + 2)(x + y + 1)$ **59.** $(m - n)(a + b + c)$
61. $(d + 3)(a - b - c)$ **63.** $(a + b + c)(x^2 - y)$ **65.** $(r - s)(2 + b)$ **67.** $(x + y)(a + b)$
69. $(a - b)(c - d)$ **71.** $r(r + s)(a - b)$ **73.** $(b + 1)(a + 3)$

Review Exercises (Page 452)

1. u^9 **3.** $\dfrac{a}{b}$ **5.** 4.5×10^{-4} **7.** $a^2 - b^2$ **9.** $9x^2 - 4y^2$ **11.** $\dfrac{x + y}{xy}$

Exercises 11.3 (Page 455)

1. $x - 3$ **3.** $t - 1$ **5.** $r + 2$ **7.** $q - 12$ **9.** $x + 20$ **11.** $a + 25$ **13.** $(x - 4)(x - 4)$
15. $(y + 7)(y - 7)$ **17.** $(2y + 7)(2y - 7)$ **19.** $(3x + y)(3x - y)$ **21.** $(5t + 6u)(5t - 6u)$
23. $(4a + 5b)(4a - 5b)$ **25.** prime **27.** prime **29.** $(7y + 15z^2)(7y - 15z^2)$
31. $8(x + 2y)(x - 2y)$ **33.** $2(a + 2y)(a - 2y)$ **35.** $3(r + 2s)(r - 2s)$ **37.** $x(x + y)(x - y)$
39. $x(2a + 3b)(2a - 3b)$ **41.** $3m(m + n)(m - n)$ **43.** $x^2(2x + y)(2x - y)$
45. $2ab(a + 11b)(a - 11b)$ **47.** $(x^2 + 9)(x + 3)(x - 3)$ **49.** $(a^2 + 4)(a + 2)(a - 2)$
51. $(a^2 + b^2)(a + b)(a - b)$ **53.** $(9r^2 + 16s^2)(3r + 4s)(3r - 4s)$ **55.** $(a^2 + b^4)(a + b^2)(a - b^2)$
57. $(x^2 + y^4)(x^2 + y^2)(x + y)(x - y)$ **59.** $2(x^2 + y^2)(x + y)(x - y)$ **61.** $b(a^2 + b^2)(a + b)(a - b)$
63. $3n(4m^2 + 9n^2)(2m + 3n)(2m - 3n)$ **65.** $3ay(a^4 + 2y^4)$ **67.** $3a^2(a^4 + b^2)(a^2 + b)(a^2 - b)$
69. $2y^2(x^4 + 4y^2)(x^2 + 2y)(x^2 - 2y)$ **71.** $a^2b^2(a^2 + b^2c^2)(a + bc)(a - bc)$
73. $a^2b^3(b^2 + 25)(b + 5)(b - 5)$ **75.** $3rs(9r^2 + 4s^2)(3r + 2s)(3r - 2s)$ **77.** $(4x - 4y + 3)(4x - 4y - 3)$
79. $(a + 3)^2(a - 3)$ **81.** $(y + 4)(y - 4)(y - 3)$ **83.** $3(x + 2)(x - 2)(x + 1)$
85. $3(m + n)(m - n)(m + a)$ **87.** $2(m + 4)(m - 4)(mn^2 + 4)$

Review Exercises (Page 456)

1. $x^2 + 12x + 36$ **3.** $a^2 - 6a + 9$ **5.** $x^2 + 9xy + 20y^2$ **7.** $m^2 + mn - 6n^2$
9. $u^2 - 8uv + 15v^2$ **11.** $p = w\left(k - h - \dfrac{v^2}{2g}\right)$

Exercises 11.4   (Page 461)

1. $x + 2$
3. $y - 6$
5. $a + 2$
7. $p - 5$
9. $t - 2$
11. $x + 5$
13. $(x + 3)(x + 3)$
15. $(y - 4)(y - 4)$
17. $(t + 10)(t + 10)$
19. $(u - 9)(u - 9)$
21. $(x + 2y)(x + 2y)$
23. $(r - 5s)(r - 5s)$
25. $(x + 2)(x + 1)$
27. $(a - 5)(a + 1)$
29. $(z + 11)(z + 1)$
31. $(t - 7)(t - 2)$
33. prime polynomial
35. $(y - 6)(y + 5)$
37. $(a + 8)(a - 2)$
39. $(t - 10)(t + 5)$
41. prime polynomial
43. $(y + z)(y + z)$
45. $(x + 2y)(x + 2y)$
47. $(m + 5n)(m - 2n)$
49. $(a - 6b)(a + 2b)$
51. $(u + 5v)(u - 3v)$
53. $-(x + 5)(x + 2)$
55. $-(y + 5)(y - 3)$
57. $-(t + 17)(t - 2)$
59. $-(r - 10)(r - 4)$
61. $-(a + 3b)(a + b)$
63. $-(x - 7y)(x + y)$
65. $(x - 4)(x - 1)$
67. $(y + 9)(y + 1)$
69. $(c + 5)(c + 1)$
71. $-(r - 2s)(r + s)$
73. $(r + 3x)(r + x)$
75. $(a - 2b)(a - b)$
77. $2(x + 3)(x + 2)$
79. $3y(y + 1)^2$
81. $-5(a - 3)(a - 2)$
83. $3(z - 4t)(z - t)$
85. $4y(x + 6)(x - 3)$
87. $-4x(x + 3y)(x - 2y)$
89. $(x + 2)(ax + 2a + b)$
91. $(a + 5)(a + 3 + b)$
93. $(x + b + 2)(a + b - 2)$
95. $(b + y + 2)(b - y - 2)$

Review Exercises   (Page 463)

1. $6x^2 + 7x + 2$
3. $8t^2 + 6t - 9$
5. $6m^2 - 13mn + 6n^2$
7. $20u^2 - 7uv - 6v^2$
9. $15x^4 + 11x^2y + 2y^2$
11. $3x^4 - 4x^2y - 15y^2$

Exercises 11.5   (Page 469)

1. $x + 2$
3. $a - 1$
5. $4y + 1$
7. $2t - 1$
9. $5z + 1$
11. $2p + 1$
13. $(2x - 1)(x - 1)$
15. $(3a + 1)(a + 4)$
17. $(z + 3)(4z + 1)$
19. $(3y + 2)(2y + 1)$
21. $(3x - 2)(2x - 1)$
23. $(3a + 2)(a - 2)$
25. $(2x + 1)(x - 2)$
27. $(2m - 3)(m + 4)$
29. $(5y + 1)(2y - 1)$
31. $(3y - 2)(4y + 1)$
33. $(5t + 3)(t + 2)$
35. $(8m - 3)(2m - 1)$
37. $(3x - y)(x - y)$
39. $(2u + 3v)(u - v)$
41. $(2a - b)(2a - b)$
43. $(3r + 2s)(2r - s)$
45. $(2x + 3y)(2x + y)$
47. $(4a - 3b)(a - 3b)$
49. $(3x + 2)(x - 5)$
51. $(2a - 5)(4a - 3)$
53. $(4y - 3)(3y - 4)$
55. prime polynomial
57. $(2a + 3b)(a + b)$
59. $(3p - q)(2p + q)$
61. prime polynomial
63. $(4x - 5y)(3x - 2y)$
65. $(3x - 5y)(2x - 3y)$
67. $(5a + 8b)(5a - 2b)$
69. $2(2x - 1)(x + 3)$
71. $y(y + 12)(y + 1)$
73. $3x(2x + 1)(x - 3)$
75. $m(2m - 3)(m + 1)$
77. $2a^2(3a - 5)(a + 4)$
79. $3r^3(5r - 2)(2r + 5)$
81. $4(a - 2b)(a + b)$
83. $4(2x + y)(x - 2y)$
85. $x^2y^2(2x - y)(x + y)$
87. $-2mn(4m + 3n)(2m + n)$
89. $-uv^3(7u - 3v)(2u - v)$
91. $3x(7x + 4)(5x - 3)$
93. $(2x + y + 4)(2x + y - 4)$
95. $(3 + a + 2b)(3 - a - 2b)$
97. $(2x + y + a + b)(2x + y - a - b)$
99. $2z(x - y + 3z)(x - y - 3z)$
101. $(2x + y)(2x + y + 3)$
103. $(x + 5)(x + 4)$
105. $(2r + 5)(r + 2)$
107. $(3x - 5)(2x + 1)$
109. $(3t + 4)(4t - 1)$
111. $(2x + 3y)(x - 2y)$

Review Exercises   (Page 470)

1. $x^3 - 27$
3. $y^3 + 64$
5. $a^3 - b^3$
7. $x^3 + 8y^3$
9. $r^3 + s^3$
11. $n = \dfrac{l - f}{d} + 1$ or $n = \dfrac{l - f + d}{d}$

Exercises 11.6 (Page 473)

1. $x^2 - 3x + 9$ 3. $z^2 + 4z + 16$ 5. $4t^2 - 6t + 9$ 7. $10x - y$ 9. $36t^2 - 30t + 25$
11. $x^3 - y^2$ 13. $(y + 1)(y^2 - y + 1)$ 15. $(a - 3)(a^2 + 3a + 9)$ 17. $(2 + x)(4 - 2x + x^2)$
19. $(s - t)(s^2 + st + t^2)$ 21. $(3x + y)(9x^2 - 3xy + y^2)$
23. $(a + 2b)(a^2 - 2ab + 4b^2)$ 25. $(4x - y)(16x^2 + 4xy + y^2)$
27. $(3x - 5y)(9x^2 + 15xy + 25y^2)$ 29. $(a^2 - b)(a^4 + a^2b + b^2)$
31. $(x^2 - y)(x^4 + x^2y + y^2)$ 33. $2(x + 3)(x^2 - 3x + 9)$
35. $-(x - 6)(x^2 + 6x + 36)$ 37. $8x(2m - n)(4m^2 + 2mn + n^2)$
39. $xy(x + 6y)(x^2 - 6xy + 36y^2)$ 41. $3rs^2(3r - 2s)(9r^2 + 6rs + 4s^2)$
43. $a^3b^2(5a + 4b)(25a^2 - 20ab + 16b^2)$ 45. $yz(y^2 - z)(y^4 + y^2z + z^2)$
47. $2mp(p + 2q)(p^2 - 2pq + 4q^2)$ 49. $(x + 1)(x^2 - x + 1)(x - 1)(x^2 + x + 1)$
51. $(x^2 + y)(x^4 - x^2y + y^2)(x^2 - y)(x^4 + x^2y + y^2)$ 53. $(x + y)(x^2 - xy + y^2)(3 - z)$
55. $(m + 2n)(m^2 - 2mn + 4n^2)(1 + x)$ 57. $(a + 3)(a^2 - 3a + 9)(a - b)$
59. $(x + 2)(x - 2)(y + z)$ 61. $(r + s)(r - s)(x - a)$
63. $(x - 1)(x + 1)$ 65. $(y + 1)(y - 1)(y - 3)(y^2 + 3y + 9)$

Review Exercises (Page 474)

1. 4 3. -3 5. $\frac{2}{3}$ 7. 1 9. 1 11. 0.0000000000001

Exercises 11.7 (Page 476)

1. $3(2x + 1)$ 3. $(x - 7)(x + 1)$ 5. $(3t - 1)(2t + 3)$ 7. $(2x + 5)(2x - 5)$ 9. $(t - 1)^2$
11. $(a - 2)(a^2 + 2a + 4)$ 13. $(y^2 - 2)(x + 1)(x - 1)$ 15. $7p^4q^2(10q - 5 + 7p)$
17. $2a(b + 6)(b - 2)$ 19. $-4p^2q^3(2pq^4 + 1)$ 21. $(2a - b + 3)(2a - b - 3)$ 23. prime polynomial
25. $-2x^2(x + 4)(x^2 - 4x + 16)$ 27. $2t^2(3t - 5)(t + 4)$ 29. $(x - a)(a + b)(a - b)$
31. $(2p^2 - 3q^2)(4p^4 + 6p^2q^2 + 9q^4)$ 33. $(5p - 4y)(25p^2 + 20py + 16y^2)$
35. $-x^2y^2z(16x^2 - 24x^3yz^3 + 15yz^6)$ 37. $(9p^2 + 4q^2)(3p + 2q)(3p - 2q)$
39. prime polynomial 41. $2(3x + 5y^2)(9x^2 - 15xy^2 + 25y^4)$
43. prime polynomial 45. $t(7t - 1)(3t - 1)$
47. $(x + y)^2(x - y)(x^2 - xy + y^2)$ 49. $2(a + b)(a - b)(c + 2d)$

Review Exercises (Page 477)

1. $2x^2 + 4x$ 3. $8a^4$ 5. $4x^2 - 12x + 9$ 7. $\dfrac{a^{12}}{4b^8}$ 9. 4 11. 16

Exercises 11.8 (Page 481)

1. $2, -3$ 3. $4, -1$ 5. $\frac{5}{2}, -2$ 7. $1, -2, 3$ 9. $-2, 4, -7$ 11. $4, -6, \frac{3}{2}, \frac{2}{3}$ 13. $12, 1$
15. $5, -3$ 17. $\frac{2}{3}, -\frac{1}{2}$ 19. $\frac{1}{4}, -\frac{2}{3}$ 21. $0, -3$ 23. $0, 8$ 25. $\frac{5}{4}, -1$ 27. $4, -4$
29. $7, -7$ 31. $0, 6$ 33. $5, -\frac{2}{5}$ 35. $\frac{2}{7}, -3$ 37. $-3, -5$ 39. $2, -4$ 41. $-3, -\frac{1}{3}$
43. $-3, -3$ 45. $1, -\frac{2}{3}$ 47. $5, 3$ 49. $1, -2, -3$ 51. $0, 3, -1$ 53. $0, 4, -4$
55. $0, 3, -\frac{1}{2}$ 57. $0, \frac{1}{3}, \frac{1}{7}$ 59. $3, -3, \frac{2}{3}, -\frac{2}{3}$

Review Exercises (Page 482)

1. $x(x + 3)$ 3. $(y + 3)(y - 3)$ 5. $(x + 12)(x + 1)$ 7. $4s$ in. 9. 4 cm by 8 cm 11. 9%

Exercises 11.9 (Page 485)

1. 5 and 7 **3.** 9 **5.** 9 sec **7.** $\frac{15}{4}$ sec and 10 sec **9.** 4 m by 9 m **11.** 48 ft
13. $h = 5$ m, $b = 12$ m **15.** 9 sq units **17.** 20 cm **19.** 3 cm **21.** 4 cm by 7 cm

Review Exercises (Page 487)

1. ←——○——→ at 8 **3.** ←——●——→ at −3 **5.** ←——○—— at 17 **7.** ——○———●—— at −2, 4
9. ——○——○—— at −7, −2 **11.** $T_1 = \dfrac{T_2}{1 - E}$

Chapter 11 Review Exercises (Page 488)

1. $5 \cdot 7$ **3.** $2^5 \cdot 3$ **5.** $3 \cdot 29$ **7.** $2 \cdot 5^2 \cdot 41$
9. $3(x + 3y)$ **11.** $7x(x + 2)$ **13.** $2x(x^2 + 2x - 4)$ **15.** $a(x + y - 1)$
17. $5a(a + b^2 + 2cd - 3)$ **19.** $(x + y)(a + b)$ **21.** $2x(x + 2)(x + 3)$ **23.** $(p + 3q)(3 + a)$
25. $(x + a)(x + b)$ **27.** $y(3x - y)(x - 2)$ **29.** $(x + 3)(x - 3)$ **31.** $(x + 2 + y)(x + 2 - y)$
33. $6y(x + 2y)(x - 2y)$ **35.** $(x + 3)(x + 7)$ **37.** $(x + 6)(x - 4)$ **39.** $(2x + 1)(x - 3)$
41. $(2x + 3)(3x - 1)$ **43.** $x(x + 3)(6x - 1)$ **45.** $(x + a + y)(x + a - y)$ **47.** $(x + y)(a + b)$
49. $(c - 3)(c^2 + 3c + 9)$ **51.** $2(x + 3)(x^2 - 3x + 9)$ **53.** $0, -2$ **55.** $3, -3$
57. $3, 4$ **59.** $-4, 6$ **61.** $3, -\frac{1}{2}$ **63.** $\frac{1}{2}, -\frac{1}{2}$
65. $0, 3, 4$ **67.** $0, \frac{1}{2}, -3$ **69.** 5 and 7 **71.** 6 ft by 8 ft
73. 3 ft by 9 ft **75.** 4 units

Chapter 11 Test (Page 491)

1. $2^2 \cdot 7^2$ **3.** $5a(12b^2c^3 + 6a^2b^2c - 5)$ **5.** $(x + y)(a + b)$ **7.** $3(a + 3b)(a - 3b)$
9. $(x + 3)(x + 1)$ **11.** $(x + 9y)(x + y)$ **13.** $(3x + 1)(x + 4)$ **15.** $(2x - y)(x + 2y)$
17. $6(2a - 3b)(a + 2b)$ **19.** $8(3 + a)(9 - 3a + a^2)$ **21.** $16(r + 2s)(r^2 - 2rs + 4s^2)$ **23.** $-1, -\frac{3}{2}$
25. $3, -6$ **27.** 12 sec

Exercises 12.1 (Page 499)

1. $\frac{4}{5}$ **3.** $\frac{4}{5}$ **5.** $\frac{2}{13}$ **7.** $\frac{2}{9}$ **9.** $-\frac{1}{3}$ **11.** $2x$ **13.** $-\frac{x}{3}$ **15.** $5a$ **17.** $\frac{x}{2}$ **19.** $\frac{a}{3}$
21. 0 **23.** 4 **25.** $\frac{5}{2}$ **27.** $-\frac{2}{7}$ **29.** $2, -2$ **31.** $5, -5$ **33.** $5, -1$ **35.** $-6, -1$
37. $\frac{2}{3}$ **39.** $\frac{3}{2}$ **41.** in lowest terms **43.** $\frac{3x}{y}$ **45.** $\frac{7x}{8y}$ **47.** $\frac{1}{3}$ **49.** 5 **51.** $\frac{x}{2}$
53. $\frac{3x}{5y}$ **55.** $\frac{2}{3}$ **57.** -1 **59.** -1 **61.** -1 **63.** $\frac{x + 1}{x - 1}$ **65.** $\frac{x - 5}{x + 2}$ **67.** $\frac{2x}{x - 2}$
69. $\frac{x}{y}$ **71.** $\frac{x + 2}{x^2}$ **73.** $\frac{x - 4}{x + 4}$ **75.** $\frac{2(x + 2)}{x - 1}$ **77.** in lowest terms **79.** $-\frac{x - 3}{x + 3}$ or $\frac{3 - x}{3 + x}$
81. $\frac{4}{3}$ **83.** $\frac{1}{x + 3}$ **85.** $x + 1$ **87.** $a - 2$ **89.** $\frac{b + 2}{b + 1}$ **91.** $\frac{y + 3}{x - 3}$

Review Exercises (Page 500)

1. For all real numbers a and b, $a + b$ is a real number.
3. For all real numbers a, b, and c, $(a + b) + c = a + (b + c)$.
5. For any real number a, $a \cdot 1 = a$. **7.** 0 **9.** 10 **11.** 0

Exercises 12.2 (Page 503)

1. $\frac{8}{15}$ 3. $\frac{45}{91}$ 5. $\frac{2}{5}$ 7. $-\frac{2}{3}$ 9. $\frac{3}{11}$ 11. $\frac{15}{4}$ 13. $\frac{5}{7}$ 15. $\frac{3x}{2}$ 17. $\frac{xy}{z}$
19. $\frac{3y}{10}$ 21. $\frac{14}{9}$ 23. 26   25. x^2y^2 27. $2xy^2$ 29. $-3y^2$ 31. $\frac{b^3c}{a^4}$ 33. $\frac{r^3t^4}{s}$
35. $\frac{(z+7)(z+2)}{7z}$ 37. x 39. $\frac{x}{5}$ 41. $\frac{3}{2x}$ 43. $x-2$ 45. $x-2$ 47. x
49. $\frac{(x-2)^2}{x}$ 51. $\frac{(m-2)(m-3)}{2(m+2)}$ 53. 1   55. $\frac{1}{3}$ 57. $\frac{c^2}{ab}$ 59. $\frac{x+1}{2(x-2)}$ 61. 1
63. $\frac{1}{x-4}$ 65. $\frac{x^2-2x+4}{x-2}$ 67. $-\frac{1}{x+3}$ 69. $\frac{x+y}{x(x-y)}$ 71. $\frac{-(x-y)(x^2+xy+y^2)}{a+b}$

Review Exercises (Page 504)

1. $-6x^5y^6z$ 3. $\frac{1}{81y^4}$ 5. $\frac{1}{x^m}$ 7. 9.3×10^7 9. 23   11. 9.5% and 10.5%

Exercises 12.3 (Page 508)

1. $\frac{2}{3}$ 3. $\frac{3}{10}$ 5. $\frac{6}{5}$ 7. $\frac{16}{35}$ 9. $\frac{3}{5}$ 11. $\frac{7}{5}$ 13. $\frac{7}{3}$ 15. $\frac{3x}{2}$ 17. $\frac{3}{2y}$ 19. 3
21. $\frac{6}{y}$ 23. 6   25. $\frac{2x}{3}$ 27. $\frac{2y^2}{15z}$ 29. $\frac{2}{y}$ 31. $\frac{2}{3x}$ 33. $\frac{2(z-2)}{z}$ 35. $\frac{5z(z-7)}{z+2}$
37. $\frac{x+2}{3}$ 39. 1   41. $\frac{x-2}{x-3}$ 43. $x+5$ 45. 1   47. $\frac{3}{7}$ 49. $\frac{9}{2x}$ 51. $\frac{x}{36}$
53. $\frac{(x+1)(x-1)}{5(x-3)}$ 55. 2   57. $\frac{2x(1-x)}{5(x-2)}$ 59. $\frac{y^2}{3}$ 61. $\frac{x+2}{x-2}$ 63. 1   65. $\frac{1}{(x+1)^2}$
67. $-x-y$ 69. $a+2$ 71. $\frac{-p}{m+n}$

Review Exercises (Page 509)

1. $4y^3 + 4y^2 - 8y + 32$ 3. $2r^3 - 3r^2 + 10r - 15$ 5. $-6m^3 + 5m^2 + 3m - 2$
7. $5y^2 + 22y + 114 + \frac{569}{y-5}$ 9. 84   11. 29,500 regular; 6250 student admissions

Exercises 12.4 (Page 512)

1. $\frac{2}{3}$ 3. $\frac{3}{5}$ 5. $\frac{1}{3}$ 7. 2   9. 2   11. $\frac{12}{7}$ 13. $\frac{25}{4}$ 15. $4x$ 17. $-\frac{2y}{3}$ 19. 0
21. $\frac{4x}{y}$ 23. $\frac{2y}{x}$ 25. $\frac{x^2}{2y}$ 27. $\frac{2y+6}{5z}$ 29. 9   31. $\frac{1}{7}$ 33. $-\frac{4}{3}$ 35. $\frac{2}{13}$ 37. $-\frac{24}{23}$
39. 1   41. $-\frac{17}{41}$ 43. $\frac{10}{7}$ 45. $\frac{x}{y}$ 47. $\frac{y}{x}$ 49. $\frac{x}{2}$ 51. $\frac{-2}{5x}$ 53. $\frac{1}{y}$ 55. $\frac{-1}{z}$
57. $\frac{x+3}{xy}$ 59. 1   61. 0   63. $\frac{8}{5}$ 65. $\frac{4x}{3}$ 67. $\frac{2x}{3y}$ 69. $\frac{4x-2y}{y+2}$ 71. $\frac{2x+10}{x-2}$
73. $\frac{xy}{x-y}$ 75. $\frac{-1}{a-b}$

Review Exercises (Page 513)

1. 7^2 3. $2^3 \cdot 17$ 5. $2 \cdot 3 \cdot 17$ 7. $(x-5)(x+3)$ 9. $2(x+2)(x-2)$ 11. $(x+y)(a-5)$

Exercises 12.5 (Page 519)

1. $\dfrac{4}{6}$ 3. $\dfrac{125}{20}$ 5. $\dfrac{2x}{x^2}$ 7. $\dfrac{5x}{xy}$ 9. $\dfrac{8xy}{x^2y}$ 11. $\dfrac{3x(x+1)}{(x+1)^2}$ 13. $\dfrac{2y(x+1)}{x^2+x}$ 15. $\dfrac{z(z+1)}{z^2-1}$
17. $\dfrac{(x+2)^2}{x^2-4}$ 19. $\dfrac{2(x+2)}{x^2+3x+2}$ 21. 60 23. 42 25. $6x$ 27. x^2y^2 29. $18xy$
31. x^2-1 33. x^2+6x 35. $(x+1)(x-2)^2$ 37. $(x+1)(x+5)(x-5)$ 39. $\dfrac{7}{6}$
41. $-\dfrac{1}{6}$ 43. $\dfrac{5y}{9}$ 45. $\dfrac{7a}{60}$ 47. $\dfrac{53x}{42}$ 49. $\dfrac{4xy+6x}{3y}$ 51. $\dfrac{2-3x^2}{x}$ 53. $\dfrac{3x^2+2xy}{6y^2}$
55. $\dfrac{4y+10}{15y}$ 57. $\dfrac{x^2+4x+1}{x^2y}$ 59. $\dfrac{2x^2-1}{x(x+1)}$ 61. $\dfrac{-x^2+3x+1}{x-2}$ 63. $\dfrac{2xy+x-y}{xy}$
65. $\dfrac{x+2}{x-2}$ 67. $\dfrac{2x^2+2}{(x-1)(x+1)}$ 69. $\dfrac{10x+4}{(x-2)(x+2)}$ 71. $\dfrac{2x^2-4x+8}{(x-2)^2(x+2)}$ 73. $\dfrac{x}{x-2}$
75. $\dfrac{5x+3}{x+1}$ 77. $\dfrac{-10y+18}{y(y-3)}$ 79. $\dfrac{-1}{2(x-2)}$ 81. $\dfrac{3}{x-3}$

Review Exercises (Page 520)

1. 10 3. -8 5. 20
7. A prime number is a natural number greater than 1 that is divisible only by itself and 1.
9. A composite number is a natural number greater than 1 that is not prime. 11. $\dfrac{x}{xy+3}$

Exercises 12.6 (Page 525)

1. $\dfrac{8}{9}$ 3. $\dfrac{3}{8}$ 5. $\dfrac{5}{4}$ 7. $\dfrac{5}{7}$ 9. $\dfrac{x^2}{y}$ 11. $\dfrac{5t^2}{27}$ 13. $\dfrac{1-3x}{5+2x}$ 15. $\dfrac{1+2}{2+x}$ 17. $\dfrac{3-x}{x-1}$
19. $\dfrac{1}{x+2}$ 21. $\dfrac{1}{x+3}$ 23. $\dfrac{xy}{y+x}$ 25. $\dfrac{y}{x-2y}$ 27. $\dfrac{x^2}{(x-1)^2}$ 29. $\dfrac{7x+3}{-x-3}$ 31. $\dfrac{x-2}{x+3}$
33. -1 35. $\dfrac{y}{x^2}$ 37. $\dfrac{x+1}{1-x}$ 39. $\dfrac{a^2-a+1}{a^2}$ 41. 2 43. $\dfrac{y-5}{y+5}$

Review Exercises (Page 526)

1. base of 3; exponent of 4 3. $abbbb$ 5. t^9 7. $-2r^7$ 9. $\dfrac{81}{256r^8}$ 11. $\dfrac{r^{10}}{9}$

Exercises 12.7 (Page 530)

1. 4 3. -20 5. 6 7. 60 9. -12 11. 0 13. -7 15. -1 17. 12
19. 0 21. -3 23. 3 25. no solution; 0 is extraneous 27. 1 29. 5
31. no solution; -2 is extraneous 33. no solution; 5 is extraneous 35. -1 37. 6
39. 2 41. -3 43. 1 45. no solution; -2 is extraneous 47. 1, 2
49. 3; -3 is extraneous 51. 3, -4 53. 1 55. 0 57. -2, 1

Review Exercises (Page 531)

1. $(a+5)(a^2-5a+25)$ 3. $(2x-5y^2)(4x^2+10xy^2+25y^4)$ 5. $(b+3)(a+2)$ 7. $(r+s)(m+n)$
9. $(2-a)(a+b)$ 11. $6r^2s^4+11rs^2-10$

Exercises 12.8 (Page 536)

1. $\frac{10}{9}$ **3.** $a = \frac{b}{b-1}$ **5.** 24 m **7.** $d_1 = \frac{fd_2}{d_2 - f}$ **9.** 2 **11.** 5 **13.** $\frac{2}{3}, \frac{3}{2}$ **15.** $2\frac{2}{9}$ hr **17.** $2\frac{6}{11}$ days **19.** 4 mph **21.** 4 mph **23.** 7% and 8% **25.** 5 **27.** 30

Review Exercises (Page 537)

1. 3, 2 **3.** $-2, -3, -4$ **5.** 0, 0, 1 **7.** 1, -1, 2, -2 **9.** 52 ft **11.** 150 m

Chapter 12 Review Exercises (Page 539)

1. $\frac{2}{5}$ **3.** $-\frac{1}{3}$ **5.** 5 **7.** 7, -7 **9.** $\frac{1}{2x}$ **11.** $\frac{x}{x+1}$ **13.** 2 **15.** $\frac{x+3}{x-5}$ **17.** $\frac{x}{x-1}$ **19.** $\frac{3x}{y}$ **21.** 1 **23.** $\frac{3y}{2}$ **25.** $x+2$ **27.** 8 **29.** $3x^2y^2$ **31.** $(x+2)(x-3)$ **33.** 1 **35.** $\frac{x^2+x-1}{x(x-1)}$ **37.** $\frac{x-2}{x(x+1)}$ **39.** $\frac{x+1}{x}$ **41.** $\frac{9}{4}$ **43.** $\frac{1+x}{1-x}$ **45.** x^2+3 **47.** 3 **49.** 3 **51.** -2 **53.** $x = \frac{y}{y+1}$ **55.** 5 mph

Chapter 12 Test (Page 541)

1. $\frac{8x}{9y}$ **3.** $\frac{x+1}{2x+3}$ **5.** $\frac{5y^2}{4t}$ **7.** $\frac{3t^2}{5y}$ **9.** $x+2$ **11.** $\frac{13}{2y+3}$ **13.** $\frac{2x+6}{x-2}$ **15.** $\frac{x+y}{y-x}$ **17.** 6 **19.** $3\frac{15}{16}$ hr **21.** 6% and 7%

Exercises 13.1 (Page 551)

1. QI **3.** QII **5.** QIII **7.** QIV **9.** QII **11.** QIV **13.** QIII **15.** QI **17.** y-axis **19.** x-axis **21.** x-axis **23.** both axes **25.** (2, 3) **27.** $(-2, -3)$ **29.** (0, 0) **31.** $(-5, -5)$ **33.** $12, $24, $36, $60 **35.** $50, $90, $130, $230 **37.** 200, 180, 160 **39.** The relationship is not linear. The lengths grow faster than the speeds.

41.

x	y
3	5
1	3
-2	0

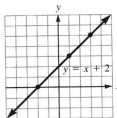

43.

x	y
5	1
4	0
-1	-5

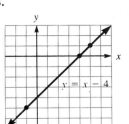

45.

x	y
2	-4
1	-2
-3	6

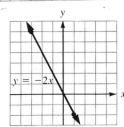

47.

x	y
1	$\frac{1}{2}$
-1	$-\frac{1}{2}$
-4	-2

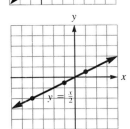

49.

x	y
3	5
−1	−3
−2	−5

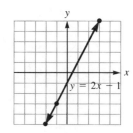

51.

x	y
8	2
0	−2
−2	−3

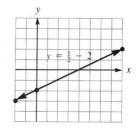

53. 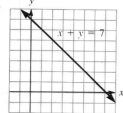   **55.** 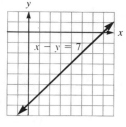   **57.** 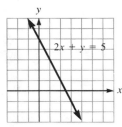   **59.**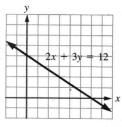

61. 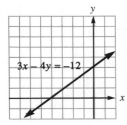   **63.** 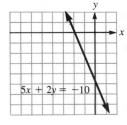   **65.** 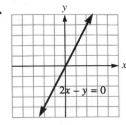   **67.**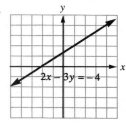

69. 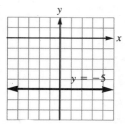   **71.** 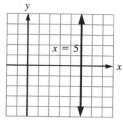   **73.** 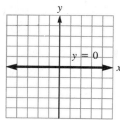   **75.**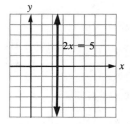

77. 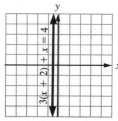   **79.** 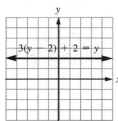   **81.** $y = -2$ **83.** $y = 0$

85. $M(6, 6)$ **87.** $M(-\frac{1}{2}, \frac{5}{2})$ **89.** $M(7, 6)$

Review Exercises   (Page 553)

1. 5   **3.** −10   **5.** −1   **7.** $h = \frac{2A}{b}$ **9.** 22, 24, and 26   **11.** $\frac{4y}{3} + 2x$

Exercises 13.2 (Page 561)

1. 1 **3.** −1 **5.** 1 **7.** 2 **9.** $\frac{1}{5}$ **11.** 0 **13.** −1 **15.** $\frac{1}{2}$

17. 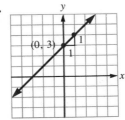 **19.** 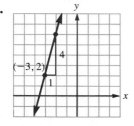 **21.** 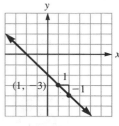 **23.**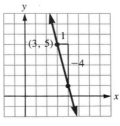

25. 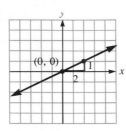 **27.** 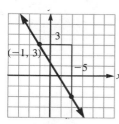 **29.** slope 3, y-int. 3 **31.** slope 5, y-int. 1

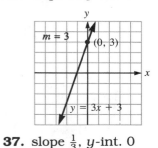

33. slope −3, y-int. 0 **35.** slope 3, y-int. −2 **37.** slope $\frac{1}{3}$, y-int. 0 **39.** slope $\frac{5}{3}$, y-int. $\frac{1}{2}$

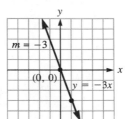 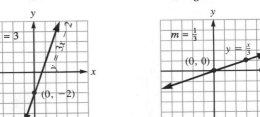 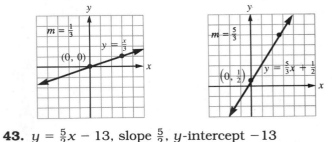

41. $y = -3x + 7$, slope −3, y-intercept 7 **43.** $y = \frac{5}{2}x - 13$, slope $\frac{5}{2}$, y-intercept −13
45. $y = 2$, slope 0, y-intercept 2 **47.** $y = x - \frac{7}{2}$, slope 1, y-intercept $-\frac{7}{2}$
49. $y = \frac{2}{3}x + 7$, slope $\frac{2}{3}$, y-intercept 7 **51.** $y = \frac{1}{3}x - \frac{5}{3}$, slope $\frac{1}{3}$, y-intercept $-\frac{5}{3}$
53. $\frac{1}{220}$ **55.** $\frac{5}{12}$ **57.** $26,750 **59.** $6000 **61.** $150
63. 2, 4, −2, −4, 2; from 1988 to 1989; from 1990 to 1991

Review Exercises (Page 563)

1. $3x(x - 2)$ **3.** $(2z + 1)(z - 3)$ **5.** $(3u + 4)(3u + 4)$ **7.** −2 **9.** 1, 6 **11.** $(a + b)(z + 3)$

Exercises 13.3 (Page 570)

1. $4x - y = -5$ **3.** $7x + y = -2$ **5.** $x + 2y = -1$ **7.** $3x + 5y = -2$ **9.** $y = 3$ **11.** $y = 0$
13. $2x - y = 1$ **15.** $5x - y = 6$ **17.** $2x + y = 0$ **19.** $x + 2y = 6$ **21.** $7x - 5y = 19$
23. $y = 3$ **25.** neither **27.** parallel **29.** perpendicular **31.** parallel **33.** parallel
35. neither **37.** perpendicular **39.** perpendicular **41.** parallel **43.** $2x - 15y = -120$
45. $4x + 9y = -2$ **47.** $5x - y = 0$ **49.** $5x - y = 4$ **51.** $5x - 2y = 8$ **53.** $2x - y = -1$
55. $y = 2$ **57.** $5x + y = 15$ **59.** $x + 5y = 0$ **61.** $5x + y = 3$ **63.** $800 **65.** $8000

Review Exercises (Page 572)

1. $\dfrac{x+1}{x}$ 3. $\dfrac{6}{x+1}$ 5. $\dfrac{4x}{(x+1)(x-1)}$ 7. 7.3×10^{10} 9. 3.2×10^{-1}
11. $x^3 - 12x^2 + 48x - 64$

Exercises 13.4 (Page 576)

1. 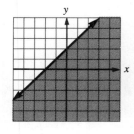   3. 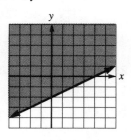   5. 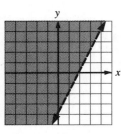   7.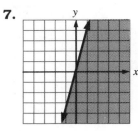

9. 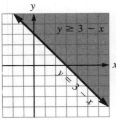   11. 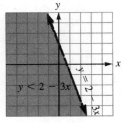   13. 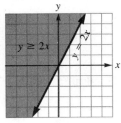   15.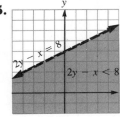

17. 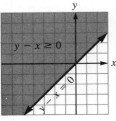   19. 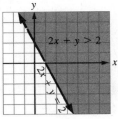   21. 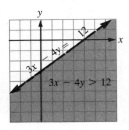   23.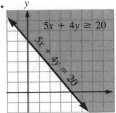

25. 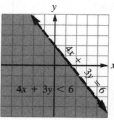   27. 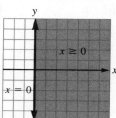   29. 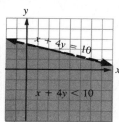   31.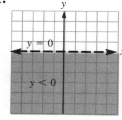

33. 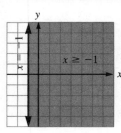   35. 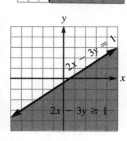   37. 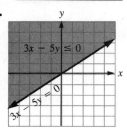   39.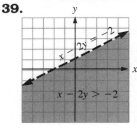

Review Exercises (Page 577)

1. 41, 43, 47   3. $ab = ba$ 5. 1   7. 7   9. $3x^2 + 4x + 3$ 11. $3x^4 - 4x^2 + 3$

Exercises 13.5 (Page 582)

1. function **3.** function **5.** function **7.** function
9. No. If $x = 2$, y can be 2 or any number less than 2. **11.** function
13. No. If $x = 2$, then $y = 2$ or -2. **15.** 9, 0, -3 **17.** 3, -3, -5 **19.** 22, 7, 2
21. 3, 9, 11 **23.** 1, 4, 9 **25.** 0, -7, 26 **27.** 4, 9, 16 **29.** 1, 6, 15 **31.** 4, 3, 4
33. 2, -1, 2 **35.** $\frac{1}{5}, \frac{1}{4}, 1$ **37.** $-2, -\frac{1}{2}, \frac{2}{5}$ **39.** $2w, 2w + 2$ **41.** $3w - 5, 3w - 2$
43. $w^2 + w, w^2 + 3w + 2$ **45.** $w^2 - 1, w^2 + 2w$ **47.** 12 **49.** $2b - 2a$ **51.** 2b **53.** 1
55. Domain is the set of real numbers; range is the set of real numbers.
57. Domain is the set of real numbers; range is the set of real numbers that are 3 or greater.
59. Domain is the set of real numbers except 0; range is the set of real numbers except 0.
61. Domain is the set of real numbers except 3 and -3; range is the set of real numbers except 0.

63. 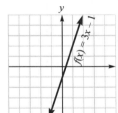 **65.** 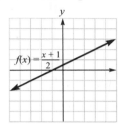 **67.** D = reals
R = reals greater than or equal to -2
69. D = reals
R = reals greater than or equal to 1

Review Exercises (Page 584)

1. -2 **3.** $-\frac{3}{5}$ **5.** 6 **7.** ←——○——→ 5 **9.** ←——●——○——→ 9 14 **11.** ←——●——→ $-1/6$

Exercises 13.6 (Page 590)

1. $d = km$ **3.** $I = \frac{k}{w}$ **5.** $A = kr^2$ **7.** $D = kst$ **9.** $I = \frac{kV}{R}$ **11.** 35 **13.** 42
15. 300 **17.** 1 **19.** $\frac{80}{3}$ **21.** 24 **23.** 4 **25.** 80 **27.** $\frac{5}{2}$ **29.** 576 ft **31.** $2450
33. $2\frac{1}{2}$ hr **35.** $53\frac{1}{3}$ m^3 **37.** $270 **39.** $3\frac{9}{11}$ amp **41.** direct variation **43.** neither
45. inverse variation **47.** neither

Review Exercises (Page 592)

1. $8x + 5$ **3.** $-3x^3 + 6x^2 - 6x$ **5.** $\frac{y + 2}{y^2}$ **7.** $\frac{1 + t}{t - 1}$ **9.** $x + 4$ **11.** 8 in. by 8 in.

Chapter 13 Review Exercises (Page 594)

1–6. 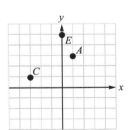 **7.** (3, 1) **9.** ($-3, -4$) **11.** (0, 0) **13.** ($-5, 0$)

15. 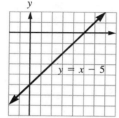 **17.** 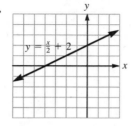 **19.** 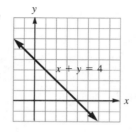 **21.**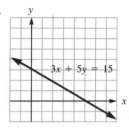

23. -1 **25.** $-\frac{1}{2}$

27. 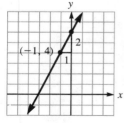 **29.** 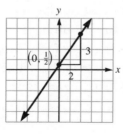 **31.** slope 5, y-int. 2 **33.** slope 0, y-int. -3

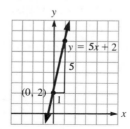 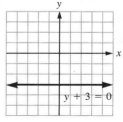

35. $3x + y = 2$ **37.** $7x - y = 0$ **39.** $3x - y = 0$ **41.** $x - 9y = -9$
43. neither **45.** perpendicular **47.** neither **49.** parallel
51. $7x - y = 9$ **53.** $5x + 2y = 0$ **55.** 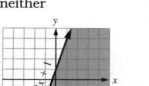 **57.**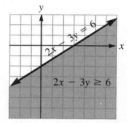

59. function **61.** No. If $x = 1$, then $y = 1$ or -1. **63.** 1 **65.** 7 **67.** -2 **69.** $a^2 - a + 2$
71. Domain is the set of real numbers; range is the set of real numbers.
73. Domain is the set of real numbers except 5 and -5; range is the set of real numbers except 0.
75. Domain is the set of real numbers; range is the set of real numbers less than or equal to 5.
77. 400 **79.** 81 **81.** direct variation

Chapter 13 Test   (Page 597)

1. 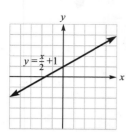 **3.** 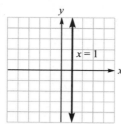 **5.** $\frac{4}{3}$ **7.** -2 **9.** $\frac{3}{5}$

11. $x - 2y = -6$ **13.** $x = -3$ **15.** 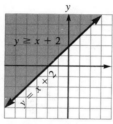 **17.** yes **19.** 11 **21.** $3a - 3b$

23. Domain is the set of real numbers except 4 and -5 **25.** Domain is the set of real numbers
27. 1 **29.** yes

Exercises 14.1 (Page 604)

1. yes **3.** yes **5.** no **7.** yes **9.** no **11.** no

13. 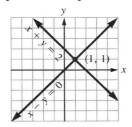 **15.** 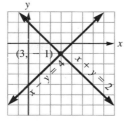 **17.** 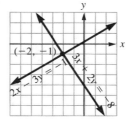 **19.**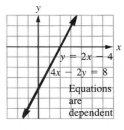

21. 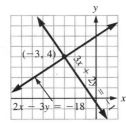 **23.** 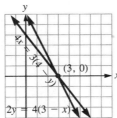 **25.** 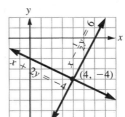 **27.**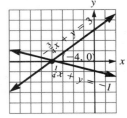

29. 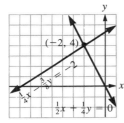 **31.**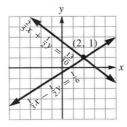

Review Exercises (Page 605)

1. x^{12} **3.** 1 **5.** $8x^2 - 9x + 12$ **7.** $-12x^3y^3 - 9x^4y^3$ **9.** $3ab + 4b - 5a$ **11.** $\frac{9}{2}$

Exercises 14.2 (Page 610)

1. $(2, 4)$ **3.** $(3, 0)$ **5.** $(-3, -1)$ **7.** inconsistent **9.** $(-2, 3)$ **11.** $(3, 2)$ **13.** $(3, -2)$
15. $(-1, 2)$ **17.** $(-1, -1)$ **19.** dependent **21.** $(1, 1)$ **23.** $(4, -2)$ **25.** $(-3, -1)$
27. $(-1, -3)$ **29.** $(\frac{1}{2}, \frac{1}{3})$ **31.** $(1, 4)$ **33.** $(4, 2)$ **35.** $(-5, -5)$ **37.** $(-6, 4)$ **39.** $(\frac{1}{5}, 4)$
41. $(5, 5)$

Review Exercises (Page 611)

1. $8xy(xy - 4yz + 2z^2)$ **3.** $(a^3 + 5)(a^3 - 5)$ **5.** $(r + 5s)(r - 3s)$ **7.** $\frac{3b}{2c}$ **9.** $-\frac{1}{2}$ **11.** 2

Exercises 14.3 (Page 615)

1. $(1, 4)$ **3.** $(-2, 3)$ **5.** $(-1, 1)$ **7.** $(2, 5)$ **9.** $(-3, 4)$ **11.** $(0, 8)$ **13.** $(2, 3)$
15. $(3, -2)$ **17.** $(2, 7)$ **19.** inconsistent **21.** dependent **23.** $(4, 0)$ **25.** $(\frac{10}{3}, \frac{10}{3})$
27. $(5, -6)$ **29.** $(-5, 0)$ **31.** $(-1, 2)$ **33.** $(1, -\frac{5}{2})$ **35.** $(-1, 2)$ **37.** $(0, 1)$ **39.** $(-2, 3)$
41. $(2, 2)$

Review Exercises (Page 616)

1. 4 **3.** $\frac{2}{3}, 1$ **5.** $-\frac{5}{2}, \frac{2}{5}$ **7.** $w = \dfrac{P - 2l}{2}$ **9.** 105, 106, 107 **11.** $4000

Exercises 14.4 (Page 622)

1. 32 and 64   3. 8 and 5   5. Paint costs $15; brush costs $5.
7. Cleaner costs $5.40; soaking solution costs $6.20.   9. 10 ft and 15 ft   11. 25 ft by 30 ft
13. 60 ft^2 15. 40 quarters, 40 dimes   17. 10 dimes, 10 quarters
19. Bill $2000, Janette $3000   21. 250 student, 100 nonstudent   23. 10 mph
25. 50 mph   27. 5 liters 40% solution, 10 liters 55% solution   29. 32 lb peanuts, 16 lb cashews
31. 5 $7 gifts, 2 $9 gifts   33. 15 $87 radios, 10 $119 radios   35. 9% and 10%

Review Exercises (Page 624)

1. 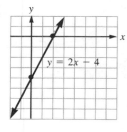   3. 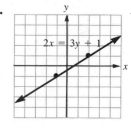   5. 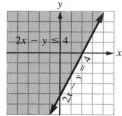   7.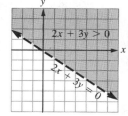

9. 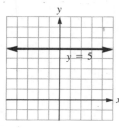   11.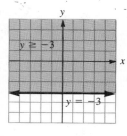

Exercises 14.5 (Page 627)

1. 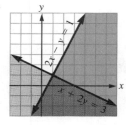   3. 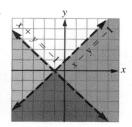   5. 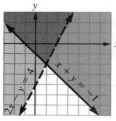   7.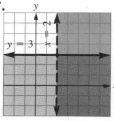

9. 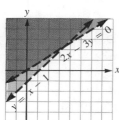   11. 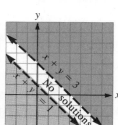   13. 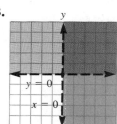   15.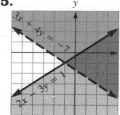

17. 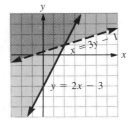 **19.** 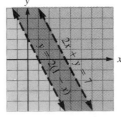 **21.** 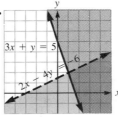 **23.**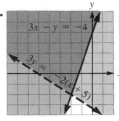

25. 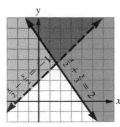 **27.** 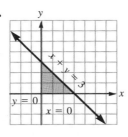 **29.**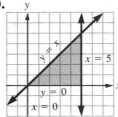

Review Exercises   (Page 628)

1. x^7 **3.** a^{10} **5.** $\dfrac{1}{z^3}$ **7.** $\dfrac{81m^8}{n^{12}}$ **9.** $\dfrac{xz}{y^2}$ **11.** $3x(5x-9)$

Exercises 14.6   (Page 632)

1. $(1,1,2)$ **3.** $(0,2,2)$ **5.** $(3,2,1)$ **7.** inconsistent   **9.** dependent; one solution is $(\tfrac{1}{2}, \tfrac{5}{3}, 0)$
11. $(2,6,9)$ **13.** $-2, 4, 16$ **15.** $9, 10, 11$ **17.** 10 nickels, 5 dimes, 2 quarters
19. 50 expensive footballs, 75 middle-priced footballs, 1000 cheap footballs
21. 250 $5 tickets, 375 $3 tickets, 125 $2 tickets

Review Exercises   (Page 633)

1. -16 **3.** 6 **5.** $\tfrac{5}{3}y$ **7.** 2 **9.** $-\tfrac{5}{3}$

Chapter 14 Review Exercises   (Page 634)

1. yes   **3.** yes   **5.** 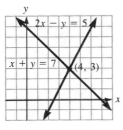   **7.** The equations are dependent.   **9.** $(-1,-2)$

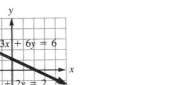

11. $(1,-1)$ **13.** $(3,-5)$ **15.** $(-1,7)$ **17.** $(0,9)$ **19.** dependent equations   **21.** 3 and 15
23. Orange costs 35¢; grapefruit costs 50¢.   **25.** Milk costs $1.69; eggs cost $1.14.
27. 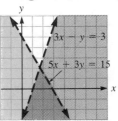   **29.**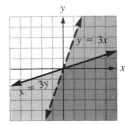

31. $(1, 0, -3)$ **33.** Dependent equations, one solution is $(3, 1, 3)$

Chapter 14 Test   (Page 637)

1. a solution
3.
5. $(-2, -3)$
7. $(2, 4)$
9. inconsistent
11. 65
13.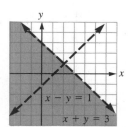
15. $(1, 3, 0)$

Exercises 15.1   (Page 643)

1. 3   3. 7   5. 6   7. 9   9. -5 11. -9 13. 14   15. 16   17. -17 19. 100
21. 18   23. -60 25. 1.414   27. 2.236   29. 2.449   31. 3.317   33. 4.796
35. 9.747   37. 80.175   39. -99.378 41. 4.621   43. 0.599   45. 0.996   47. -0.915
49. rational   51. rational   53. irrational   55. imaginary   57. 1   59. 3   61. -2
63. -4 65. 5   67. 1   69. -4 71. 9   73. 2   75. -2 77. 1   79. -2
81. xy 83. x^2z^2 85. $-x^2y$ 87. $2z$ 89. $-3x^2y$ 91. xyz 93. $-xyz^2$ 95. $5x^2z^6$
97. $6z^{18}$ 99. $-4z$ 101. $3yz^2$ 103. $-2p^2q$ 105. $x + 9$ 107. $x + 11$ 109. $x + 3$
111. $x + 10$ 113. $x + 7y$

Review Exercises   (Page 644)

1. -5 3. -10 5. -1 7. 9   9. $y = kx$ 11. 4 sec

Exercises 15.2   (Page 649)

1. $2\sqrt{5}$ 3. $5\sqrt{2}$ 5. $3\sqrt{5}$ 7. $7\sqrt{2}$ 9. $4\sqrt{3}$ 11. $10\sqrt{2}$ 13. $8\sqrt{3}$ 15. $2\sqrt{22}$
17. 18   19. $7\sqrt{3}$ 21. $6\sqrt{5}$ 23. $12\sqrt{3}$ 25. $48\sqrt{2}$ 27. $-70\sqrt{10}$ 29. $14\sqrt{5}$
31. $-45\sqrt{2}$ 33. $5\sqrt{x}$ 35. $a\sqrt{b}$ 37. $3x\sqrt{y}$ 39. $40x^3y^2\sqrt{2}$ 41. $48x^2y\sqrt{y}$
43. $-9x^2y^3z\sqrt{2xy}$ 45. $6ab^2\sqrt{3ab}$ 47. $-\dfrac{8n\sqrt{5m}}{5}$ 49. $\dfrac{5}{3}$ 51. $\dfrac{9}{8}$ 53. $\dfrac{\sqrt{26}}{5}$
55. $\dfrac{2\sqrt{5}}{7}$ 57. $\dfrac{4\sqrt{3}}{9}$ 59. $\dfrac{4\sqrt{2}}{5}$ 61. $\dfrac{5\sqrt{5}}{11}$ 63. $\dfrac{7\sqrt{5}}{6}$ 65. $\dfrac{6x\sqrt{2x}}{y}$ 67. $\dfrac{5mn^2\sqrt{5}}{8}$
69. $\dfrac{4m\sqrt{2}}{3n}$ 71. $\dfrac{2rs^2\sqrt{3}}{9}$ 73. $2x$ 75. $-4x\sqrt[3]{x^2}$ 77. $3xyz^2\sqrt[3]{2y}$ 79. $-3yz\sqrt[3]{3x^2z}$
81. $\dfrac{3m}{2n^2}$ 83. $\dfrac{rs\sqrt[3]{2rs^2}}{5t}$ 85. $\dfrac{5ab}{2}$

Review Exercises   (Page 650)

1. $\dfrac{1}{2xz}$ 3. $\dfrac{a+1}{a+3}$ 5. 1   7. 2   9. $\dfrac{15a^2c + 6b^2c - 5a^2b^2}{10abc}$ 11. 7

Exercises 15.3 (Page 654)

1. $5\sqrt{3}$ 3. $9\sqrt{3}$ 5. $7\sqrt{5}$ 7. $12\sqrt{5}$ 9. $8\sqrt{5}$ 11. $10\sqrt{10}$ 13. $37\sqrt{5}$ 15. $12\sqrt{7}$
17. $38\sqrt{2}$ 19. $85\sqrt{2}$ 21. $9\sqrt{5}$ 23. $7\sqrt{6}+4\sqrt{15}$ 25. $\sqrt{2}$ 27. $3-5\sqrt{2}$ 29. $2\sqrt{2}$
31. $-2\sqrt{3}$ 33. $\sqrt{3}$ 35. $4\sqrt{10}$ 37. $-7\sqrt{5}$ 39. $22\sqrt{6}$ 41. $-18\sqrt{2}$ 43. $13\sqrt{10}$
45. $3\sqrt{2}-\sqrt{3}$ 47. $10\sqrt{2}-\sqrt{3}$ 49. $-6\sqrt{6}$ 51. $10\sqrt{2}+5\sqrt{3}$ 53. $7\sqrt{3}-6\sqrt{2}$
55. $6\sqrt{6}-2\sqrt{3}$ 57. $3x\sqrt{2}$ 59. $3x\sqrt{2x}$ 61. $3x\sqrt{2y}-3x\sqrt{3y}$ 63. $x^2\sqrt{2x}$ 65. $19x\sqrt{6}$
67. $-5y\sqrt{10y}$ 69. $2x\sqrt{5xy}+3x^2y\sqrt{5xy}-4x^3y^2\sqrt{5xy}$ 71. $5\sqrt[3]{2}$ 73. $\sqrt[3]{3}$ 75. $2\sqrt[3]{5}+5$
77. $(1-x)x\sqrt[3]{x}$ 79. $2xy\sqrt[3]{3xy^2}$ 81. $x^2y\sqrt[3]{5xy}$

Review Exercises (Page 655)

1. $\frac{3}{8}$ 3. $\frac{5}{36}$ 5. 8 7. -9 9. $3, -6$ 11. $h=\dfrac{\frac{7A}{44}-r^2}{r}$ or $h=\dfrac{7A}{44r}-r$

Exercises 15.4 (Page 662)

1. 3 3. 4 5. 8 7. 4 9. x^3 11. b^7 13. $4\sqrt{15}$ 15. $-60\sqrt{2}$ 17. 24
19. $-8x$ 21. $700x\sqrt{10}$ 23. $4x^2\sqrt{y}$ 25. $2+\sqrt{2}$ 27. $9-\sqrt{3}$ 29. $7-3\sqrt{7}$
31. $3\sqrt{5}-5$ 33. $3\sqrt{2}+\sqrt{3}$ 35. $7-2\sqrt[3]{7}$ 37. $x\sqrt{3}-2\sqrt{x}$ 39. $6x+6\sqrt{x}$ 41. $6\sqrt{x}+3x$
43. $3x\sqrt{7}+x\sqrt{42}$ 45. 1 47. 1 49. $\sqrt[3]{4}+2\sqrt[3]{2}+1$ 51. $7-x^2$ 53. $2-2\sqrt{2x}+x$
55. $6x-7$ 57. $4x-18$ 59. $8xy+4\sqrt{2xy}+1$ 61. $9x$ 63. $\frac{2x}{3}$ 65. $\frac{3y\sqrt{2}}{5}$ 67. $\frac{2y}{x}$
69. $\frac{2x}{3}$ 71. $\frac{y\sqrt{xy}}{3}$ 73. $\frac{5\sqrt{2y}}{x}$ 75. $\frac{\sqrt{3}}{3}$ 77. $\frac{2\sqrt{7}}{7}$ 79. $\sqrt[3]{25}$ 81. $\sqrt{3}$ 83. $\frac{3\sqrt{2}}{8}$
85. $2\sqrt[3]{2}$ 87. $\frac{\sqrt{15}}{3}$ 89. $\frac{10\sqrt{x}}{x}$ 91. $\frac{3\sqrt{2x}}{2x}$ 93. $\frac{\sqrt{2xy}}{3y}$ 95. $\frac{\sqrt{6}}{2x}$ 97. $\sqrt{3x}$
99. $\frac{\sqrt[3]{20}}{2}$ 101. $\sqrt[3]{x}$ 103. $\frac{\sqrt[3]{2xy^2}}{xy}$ 105. $\frac{-\sqrt[3]{5ab}}{ab}$ 107. $\frac{3(\sqrt{3}+1)}{2}$ 109. $\sqrt{7}-2$
111. $6+2\sqrt{3}$ 113. $2-\sqrt{2}$ 115. $\frac{\sqrt{3}-3}{2}$ 117. $5\sqrt{3}-5\sqrt{2}$ 119. $\sqrt{10}+\sqrt{6}$
121. $5-2\sqrt{6}$ 123. $\frac{3x-2\sqrt{3x}+1}{3x-1}$ 125. $\frac{2x+2\sqrt{2x}-15}{2x-9}$ 127. $\frac{2\sqrt{2z}-1-2z}{2z-1}$
129. $\frac{y^2-2y\sqrt{15}+15}{y^2-15}$

Review Exercises (Page 664)

1. $(x-7)(x+3)$ 3. $3xy(2x-5)$ 5. $(x+2)(x^2-2x+4)$ 7. 3, 10 9. 2, -2
11. 4 ft and 12 ft

Exercises 15.5 (Page 667)

1. 9 3. 49 5. no solution 7. 1 9. 30 11. no solution 13. 5 15. 6 17. 3
19. -1 21. 46 23. -3 25. 0 27. no solution 29. 2 31. -1 33. 10
35. 3 37. $0, -1$ 39. 2 41. 2, 1 43. 3 45. 161 47. 28 49. 7 51. 18
53. 5 55. 0, 1 57. 4 59. no 61. $v^2 = c^2 - f^2c^2$

Review Exercises (Page 669)

1. $-6x^3y^7$ **3.** $x^2 + x - 12$ **5.** $9x^2 + 24x + 16$ **7.** $(2, 3)$ **9.** $(3, -2)$ **11.** 5

Exercises 15.6 (Page 672)

1. 9 **3.** -12 **5.** $\frac{1}{2}$ **7.** $\frac{2}{7}$ **9.** 3 **11.** -5 **13.** -2 **15.** $\frac{1}{4}$ **17.** $\frac{3}{4}$ **19.** 2
21. 2 **23.** -3 **25.** 729 **27.** 125 **29.** 25 **31.** 100 **33.** 4 **35.** 8 **37.** 27
39. -8 **41.** $\frac{4}{9}$ **43.** $\frac{125}{8}$ **45.** 6 **47.** 25 **49.** 7 **51.** 25 **53.** 8 **55.** 125
57. 6 **59.** 192 **61.** $\frac{1}{2}$ **63.** $\frac{1}{9}$ **65.** $\frac{1}{64}$ **67.** $\frac{1}{81}$ **69.** x **71.** x^2 **73.** x^4
75. x^2 **77.** y^2 **79.** $x^{2/5}$ **81.** $x^{2/7}$ **83.** x **85.** y^7 **87.** y^2 **89.** $x^{17/12}$
91. $b^{3/10}$ **93.** $t^{4/15}$ **95.** x **97.** $a^{14/15}$ **99.** $\dfrac{3}{b^{101/60}}$

Review Exercises (Page 673)

1. $3xyz^2(x^2yz^2 - 2z^3 + 5x)$ **3.** $(a^2 + b^2)(a + b)(a - b)$ **5.** 5 **7.** $\frac{2}{3}$ **9.** $(\frac{5}{2}, -1)$
11. 1 and 2

Exercises 15.7 (Page 678)

1. 5 **3.** 13 **5.** 20 **7.** 28 **9.** $2\sqrt{14}$ **11.** 5 **13.** 10 **15.** 13 **17.** 10
19. 5 **21.** 12 ft **23.** 30 ft **25.** 2 units **27.** $6\sqrt{2}$ ft **29.** $4\sqrt{13}$ mi **31.** 18π in.^{2}

Review Exercises (Page 679)

1. 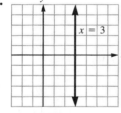 **3.** 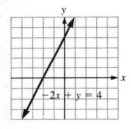 **5.** 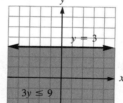 **7.**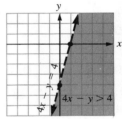

9. 7,200,000 **11.** 3 hr

Exercises 15.8 (Page 682)

1. $20\sqrt{2}$ **3.** 14 **5.** $\sqrt{6}$ **7.** $\dfrac{5\sqrt{2}}{2}$ **9.** $\dfrac{5\sqrt{6}}{2}$ **11.** 40 **13.** $\dfrac{14\sqrt{3}}{3}$ **15.** 1
17. $\dfrac{5\sqrt{3}}{2}$ **19.** 1 **21.** $155\sqrt{2}$ mi **23.** $\dfrac{230\sqrt{3}}{3}$ ft **25.** 2 in.; $2\sqrt{3}$ in. **27.** 2 m; $2\sqrt{3}$ m
29. $(10 + 10\sqrt{2})$ cm **31.** 30 cm **33.** $(14 + 14\sqrt{3})$ m

Review Exercises (Page 684)

1. $\dfrac{27b^3}{125}$ **3.** $8m^{21}$ **5.** $6x^2 - 7x - 20$ **7.** $6m^2 + mn - 15n^2$ **9.** $6r^3 - 13r^2 + 12r - 4$

Chapter 15 Review Exercises   (Page 686)

1. 5   **3.** -12 **5.** 16   **7.** 13   **9.** 3   **11.** 3   **13.** 4.583   **15.** -7.570 **17.** $4\sqrt{2}$
19. $10\sqrt{5}$ **21.** $4x\sqrt{5}$ **23.** $-5t\sqrt{10t}$ **25.** $10x\sqrt{2y}$ **27.** $2y\sqrt[3]{x^2}$ **29.** $\frac{4}{5}$ **31.** $\frac{10}{3}$
33. $\frac{2\sqrt{15}}{7}$ **35.** $\frac{11x\sqrt{2}}{13}$ **37.** 0   **39.** $18\sqrt{5}$ **41.** $5x\sqrt{2y}$ **43.** $5\sqrt[3]{2}$ **45.** $-6\sqrt{6}$
47. $36x\sqrt{2}$ **49.** $4\sqrt[3]{2}$ **51.** -2 **53.** -2 **55.** $\sqrt[3]{9}+\sqrt[3]{3}-2$ **57.** $49+12\sqrt{5}$
59. $x-2\sqrt{2x}+2$ **61.** $\frac{\sqrt{7}}{7}$ **63.** $2\sqrt[3]{4}$ **65.** $\frac{\sqrt{5}}{5}$ **67.** $5-\sqrt{15}$ **69.** 6   **71.** no solution
73. 4; -1 is extraneous   **75.** 2; -2 is extraneous   **77.** 7   **79.** 216   **81.** 64   **83.** x
85. 6   **87.** x **89.** $x^{11/15}$ **91.** $\frac{1}{x^{4/5}}$ **93.** 35   **95.** 1   **97.** 5   **99.** 2   **101.** 68 in.
103. $24\frac{1}{2}$ in.^{2}

Chapter 15 Test   (Page 689)

1. 10   **3.** -3 **5.** $2x\sqrt{2}$ **7.** $4\sqrt{2}$ **9.** x^2y^2 **11.** $5\sqrt{3}$ **13.** $-24x\sqrt{6}$ **15.** -1
17. $\frac{y\sqrt{xy}}{4x}$ **19.** $\frac{x\sqrt{3}-2\sqrt{3x}}{x-4}$ **21.** 66   **23.** 5   **25.** 11   **27.** y^6 **29.** $p^{17/12}$
31. 13 in.   **33.** 5   **35.** $10\sqrt{3}$ **37.** $32\sqrt{3}$

Exercises 16.1   (Page 697)

1. 3, -3 **3.** 0, -3 **5.** 2, 3   **7.** $-1, \frac{2}{3}$ **9.** $-\frac{1}{3}, -\frac{3}{2}$ **11.** $\frac{2}{5}, -\frac{1}{2}$ **13.** 1, -1
15. 3, -3 **17.** $2\sqrt{5}, -2\sqrt{5}$ **19.** 3, -3 **21.** 2, -2 **23.** $a, -a$ **25.** $-6, 4$
27. 7, -11 **29.** $2+2\sqrt{2}, 2-2\sqrt{2}$ **31.** $3a, -a$ **33.** $-b+4c, -b-4c$ **35.** 0, -4
37. 2, $-\frac{2}{3}$ **39.** $\frac{-1\pm 2\sqrt{5}}{2}$ **41.** 3, -3 **43.** 5, -5 **45.** $-1\pm 2\sqrt{2}$

Review Exercises   (Page 698)

1. y^2-2y+1 **3.** $x^2+2xy+y^2$ **5.** $4r^2-4rs+s^2$ **7.** $9a^2+12ab+4b^2$
9. $25r^2-80rs+64s^2$ **11.** $\frac{1}{2}$ hr

Exercises 16.2   (Page 702)

1. $x^2+4x+4, (x+2)^2$ **3.** $x^2-10x+25, (x-5)^2$ **5.** $x^2+11x+\frac{121}{4}, (x+\frac{11}{2})^2$
7. $a^2-3a+\frac{9}{4}, (a-\frac{3}{2})^2$ **9.** $b^2+\frac{2}{3}b+\frac{1}{9}, (b+\frac{1}{3})^2$ **11.** $c^2-\frac{5}{2}c+\frac{25}{16}, (c-\frac{5}{4})^2$
13. $-2, -4$ **15.** 2, 6   **17.** 5, -3 **19.** 3, 4   **21.** 1, -6 **23.** 1, -2 **25.** $-4, -4$
27. 2, $-\frac{1}{2}$ **29.** $-2, \frac{1}{4}$ **31.** $-2\pm\sqrt{3}$ **33.** $1\pm\sqrt{5}$ **35.** $2\pm\sqrt{7}$ **37.** $-1\pm\sqrt{2}$
39. 1, -4 **41.** $\frac{3}{2}, -\frac{2}{3}$ **43.** $\frac{-3\pm\sqrt{3}}{2}$

Review Exercises   (Page 703)

1. -4 **3.** $\frac{3}{4}, -\frac{1}{5}$ **5.** 6   **7.** $4\sqrt{5}$ **9.** $\frac{\sqrt{7x}}{7}$ **11.** 13

Exercises 16.3 (Page 706)

1. $a = 1, b = 4, c = 3$
3. $a = 3, b = -2, c = 7$
5. $a = 4, b = -2, c = 1$
7. $a = 3, b = -5, c = -2$
9. $a = 7, b = 14, c = 21$
11. $a = 1, b = -1, c = -5$
13. $2, 3$
15. $-3, -4$
17. $1, -\frac{1}{2}$
19. $-1, -\frac{2}{3}$
21. $\frac{1}{2}, -\frac{3}{2}$
23. $2, -\frac{2}{5}$
25. $\frac{-3 \pm \sqrt{5}}{2}$
27. $\frac{-5 \pm \sqrt{37}}{2}$
29. $\frac{-1 \pm \sqrt{41}}{4}$
31. $-2 \pm \sqrt{3}$
33. $-1 \pm \sqrt{2}$
35. $\frac{3 \pm \sqrt{15}}{3}$

Review Exercises (Page 707)

1. $r = \dfrac{A - p}{pt}$
3.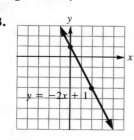
5. $3x - 5y = -60$
7.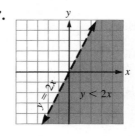
9. -3
11. 200

Exercises 16.4 (Page 714)

1. $\pm 3i$
3. $\pm \frac{4\sqrt{3}}{3} i$
5. $-1 \pm i$
7. $-\frac{1}{4} \pm \frac{\sqrt{7}}{4} i$
9. $\frac{2}{3} \pm \frac{\sqrt{2}}{3} i$
11. i
13. $-i$
15. 1
17. i
19. $8 - 2i$
21. $3 - 5i$
23. $15 + 7i$
25. $-15 + 2\sqrt{3} i$
27. $3 + 6i$
29. $9 + 7i$
31. $14 - 8i$
33. $8 + \sqrt{2} i$
35. $6 - 8i$
37. $6 + \sqrt{6} + (3\sqrt{3} - 2\sqrt{2}) i$
39. $-16 - \sqrt{35} + (2\sqrt{5} - 8\sqrt{7}) i$
41. $0 - i$
43. $0 + \frac{4}{5} i$
45. $\frac{1}{8} + 0i$
47. $0 + \frac{3}{5} i$
49. $0 + \frac{3\sqrt{2}}{4} i$
51. $\frac{15}{26} - \frac{3}{26} i$
53. $-\frac{42}{25} - \frac{6}{25} i$
55. $\frac{1}{4} + \frac{3}{4} i$
57. $\frac{5}{13} - \frac{12}{13} i$
59. $\frac{6 + \sqrt{10}}{9} + \frac{2\sqrt{2} - 3\sqrt{5}}{9} i$
61. 10
63. 13
65. $\sqrt{74}$
67. $3\sqrt{2}$
69. $\sqrt{69}$

Review Exercises (Page 715)

1. $3(x + 3)(x - 3)$
3. $(2x - 1)(x + 1)$
5. $-(x - 3)(x + 7)$
7. $637\pi \text{ ft}^3$
9. $\dfrac{10{,}976\pi}{3} \text{ m}^3$
11. $21\pi \text{ m}^3$

Exercises 16.5 (Page 721)

1.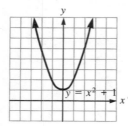
3.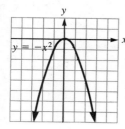
5.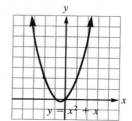
7.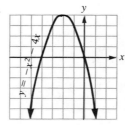

9. 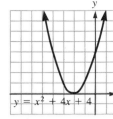 **11.** 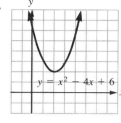 **13.** $(2, 4)$ **15.** $(-1, -5)$

17. $(1, 0)$ **19.** $(0, 4)$ **21.** $(-1, 4)$ **23.** $(3, -21)$

25. 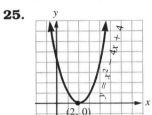 **27.** 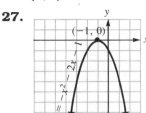 **29.** 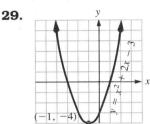 **31.**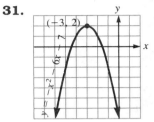

33. 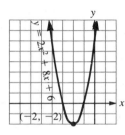 **35.** 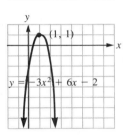 **37.** vertex: $(\frac{1}{2}, -\frac{9}{4})$
x-intercepts: $2, -1$
y-intercept: -2
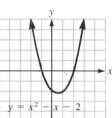

39. vertex: $(1, 4)$
x-intercepts: $3, -1$
y-intercept: 3
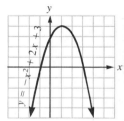

41. vertex: $(-\frac{3}{4}, -\frac{25}{8})$
x-intercepts: $-2, \frac{1}{2}$
y-intercept: -2
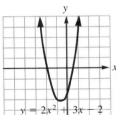

Review Exercises   (Page 722)

1. $2\sqrt{2}$ **3.** $2\sqrt{6}$ **5.** $5\sqrt{3}$ **7.** 2 **9.** $\dfrac{\sqrt{5}+1}{2}$ **11.** 40 ft

Chapter 16 Review Exercises (Page 724)

1. $5, -5$ **3.** $3, -3$ **5.** $\pm 2\sqrt{2}$ **7.** $-4, 6$ **9.** $2, -4$ **11.** $8 \pm 2\sqrt{2}$ **13.** $2, -4$
15. $4 \pm 2\sqrt{5}$ **17.** $2, -7$ **19.** $7, -11$ **21.** $-2 \pm \sqrt{7}$ **23.** $\frac{1}{2}, -3$ **25.** $5, -3$
27. $13, 2$ **29.** $\frac{3}{2}, -\frac{1}{3}$ **31.** $3 \pm \sqrt{2}$ **33.** $12 - 8i$ **35.** $-28 - 21i$ **37.** $-24 + 28i$
39. $0 - \frac{3}{4}i$ **41.** $\frac{12}{5} - \frac{6}{5}i$ **43.** $\frac{15}{17} + \frac{8}{17}i$ **45.** $\frac{15}{29} - \frac{6}{29}i$ **47.** $15 + 0i$ **49.** $(6, 7)$ **51.** $(1, 5)$
53.

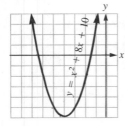

Chapter 16 Test (Page 725)

1. $\frac{1}{3}, -\frac{1}{2}$ **3.** $2 \pm \sqrt{3}$ **5.** 49 **7.** $-1 \pm \sqrt{5}$ **9.** $2, -5$ **11.** $\frac{-5 \pm \sqrt{33}}{4}$ **13.** $1 + 9i$
15. $16 + 11i$ **17.** $\frac{3}{5} + \frac{4}{5}i$ **19.** $(-5, -4)$

Index

Absolute value:
  of a complex number, 713
  of a real number, 303
Acute angle, 185
Acute triangle, 201
Addends, 7
Addition:
  associative property of, 7
  commutative property of, 7
  of complex numbers, 709
  of decimals, 94
  of fractions, 65, 510, 514
  of like terms, 350, 413
  of monomials, 413
  of polynomials, 414
  of radical expressions, 651
  of real numbers, 307, 308
  of whole numbers, 7
Addition property:
  of equality, 336
  of inequality, 381
Additive inverse, 326
Adjacent angles, 189
Ahmes, 283
Algebraic expression, 289
Algebraic terms, 291, 409
al-Khowarazmi, 283
Alternate exterior angles, 195
Alternate interior angles, 194
Altitude of a parallelogram, 248
American units of measurement, 141
Angle(s), 184
  acute, 185
  adjacent, 189
  alternate exterior, 195
  alternate interior, 194
  base, 210, 368
  bisecting, 185
  central, 222
  complementary, 188
  congruent, 185
  corresponding, 194
  inscribed, 222
  naming, 184
  obtuse, 185
  right, 185
  straight, 185
  supplementary, 188
  vertex, 184, 368
  vertical, 190
Annual depreciation rate, 561
Arc, 222

Area:
  of a circle, 109, 258
  of a parallelogram, 249
  of a polygon, 246
  of a rectangle, 20, 107, 247
  of a rectangular-solid surface, 262
  of a right-circular cone, 266
  of a right-circular cylinder, 265
  of a sphere, 263
  of a square, 107
  of a trapezoid, 254
  of a triangle, 110, 252
Arithmetic, fundamental theorem of, 36
Arithmetic mean, 30
Associative properties, 7, 17, 324
Average, 30, 167
Axes, equations of lines parallel to, 551
Axioms, 271

Bar graph, 152, 153
Base:
  of an exponential expression, 21, 294, 393
  of a parallelogram, 248
  of a percentage, 123
  of a right prism, 261
  of a trapezoid, 253
  of a triangle, 110
Base angle, of an isosceles triangle, 210, 368
Between, 182
Binomial(s), 409
  conjugate, 422
Bisecting an angle, 185
Borrowing, in subtraction, 10
Break-even analysis, 371
Building a fraction, 514

Carrying, in addition, 7
Cartesian coordinate system, 544
Center:
  of circle, 221
  of a sphere, 263
Centimeter, 144
Central angle, 222
Changing a decimal to a percent, 120
Changing a fraction to a percent, 120
Changing a percent to a decimal, 119
Changing a percent to a fraction, 118
Chord of circle, 221

Circle(s), 108, 221
  arc of, 222
  area of, 109, 258
  center of, 221
  central angle of, 222
  chord of, 221
  circumference of, 109, 257
  concentric, 260
  diameter of, 108, 221
  radius of, 108, 221
Circumference of a circle, 109, 257
Closed figure, 200
Closure property, 323
Coefficient, 291
  numerical, 291, 349
Coin problems, 369
Collinear, 182
Combined variation, 587
Combining like terms, 350, 413
Common denominator, 514, 515
  least, 515
  lowest, 515
Common divisor, greatest, 40, 444
Common factor, greatest, 444, 445
Common multiple, least, 39
Commutative properties, 7, 17, 323
Comparing decimals, 90
Complementary angles, 188
Completing the square, 698
Complex conjugates, 711
Complex fraction, 521
Complex number, 709
  absolute value of, 713
  addition of, 709
  division of, 711
  imaginary part of, 709
  multiplication of, 710
  real part of, 709
  subtraction of, 709
Composite number, 35, 284
Compound inequality, 384
Concentric circles, 260
Conclusion, of a conditional statement, 191
Conditional statement, 191
  conclusion of, 191
  converse of, 191
  hypothesis of, 191
Cone, right-circular, 266
  surface area of, 266
  volume of, 266
Congruent angles, 185

Congruent triangles, 208
Conjugate binomials, 422
Conjugate complex numbers, 711
Consistent system of equations, 600
Constant of variation, 584, 585, 586
Contradiction, 353
Converse of a conditional statement, 191
Conversion factor, 142
Coordinate axes:
    equations of lines parallel to, 551
Coordinate system, 544
Coplanar points, 183
Corresponding angles, 194
Corresponding parts of triangles, 207
Cost(s):
    fixed, 371, 562
    marginal, 562
    unit, 371
Cube root, 641
Cubes, 261
    factoring the sum and difference of, 471
    perfect integer, 648
Cubic equation, 480
Cylinder, 264
    surface area of, 265
    volume of, 265

Data, grouped, 160
Decimal(s):
    addition of, 94
    comparing, 90
    expanded notation of, 89
    multiplication of, 99
    as percent, 119, 120
    rounding, 92
    subtraction of, 95
Decimal notation, 89
Decimal point, 89
Decimeter, 144
Decrease, percent of, 129
Deductive reasoning, 181, 271
Defined terms, 271
Degree:
    of a monomial, 409
    of a polynomial, 410
Degree measure, 184
Dekameter, 144
Demand, marginal, 560
Demand curve, 560
Denominator(s), 22, 51
    common, 514
    least common, 515
    lowest common, 515
    rationalizing, 659
Dependent equations, 602, 608, 613
Dependent variable, 578
Depreciation, linear, 560
Depreciation rate, 561
Depth, 182
Descartes, René, 544
Diameter, 108, 221
Difference, 10, 289, 310
    of real numbers, 310
    of two cubes, 471
    of two squares, 453
    of whole numbers, 10
Digits, 2
Direct variation, 584
Distance formula, 677
Distributive property, 31, 325
Dividend, 21, 74, 430
Divisibility, tests for, 34
Division:
    of complex numbers, 711
    of fractions, 59, 505

Division (continued)
    long, 23
    of monomials, 426
    of polynomials by monomials, 427
    of polynomials by polynomials, 430
    by a power of ten, 104
    of radical expressions, 658
    of real numbers, 318
Division property:
    of equality, 341
    of inequality, 381
    of radicals, 647
    of whole numbers, 21
    by zero, 22, 301
Divisor, 21, 33, 74, 430, 505
    greatest common, 40, 444
Domain:
    of a function, 580
    reading from a graph, 582
    of a relation, 580
    of a variable, 493
Double inequality, 384
Double negative rule, 303

Element, of a set, 284
Ellipsis, 283
Endpoint:
    of a line segment, 183
    of a ray, 182
Equality:
    addition property of, 336
    division property of, 341
    of fractions, 52
    multiplication property of, 341
    square root property of, 674, 695
    squaring property of, 664
    subtraction property of, 337
Equal ratios, 235
Equation(s), 335
    containing radicals, 664
    cubic, 480
    dependent, 602, 608, 613
    equilibrium, 693
    false, 335
    with fractions, 527
    general quadratic, 703
    of a horizontal line, 559
    impossible, 353
    incomplete quadratic, 479, 694
    independent, 600
    of a line, 548, 557, 565
    linear, 477, 546
    literal, 356
    of a parabola, 717, 718
    quadratic, 478
    root of, 336
    solution of, 336
    system of, 600
    in two variables, 546
    of a vertical line, 559
Equiangular triangle, 201
Equilibrium equation, 693
Estimating, 8, 149
Euclid, 181
Evaluating an expression, 28
Even number, 284
Expanded notation, 2, 89
Exponential expression, 21, 294
    base of, 21, 294, 393
Exponent(s), 21, 294, 393
    negative, 401, 524
    power rule, 396, 397
    product rule, 395
    properties of, 398, 402
    quotient rule, 397

Exponent(s) (continued)
    rational, 669, 670
    zero, 400
Exterior angles:
    alternate, 195
    of a closed figure, 200
Extraneous solution, 528, 665
Extremes of a proportion, 236

Face of a rectangular solid, 261
Factor, 16, 36, 291, 443
    greatest common, 444, 445
Factoring:
    the difference of two cubes, 471
    the difference of two squares, 453
    greatest common factor, 445
    by grouping, 449, 468
    the sum of two cubes, 471
    by the trial-and-error method, 463
    trinomials, 456, 457, 463, 467
False equation, 335
Figure, 182
    closed, 200
    exterior of, 200
    interior of, 200
    plane, 261
    three-dimensional, 261
Fixed cost, 371, 562
FOIL method, 421
Foot, 141
Formula(s), 106, 356
    area of a rectangle, 107
    area of a square, 107
    distance, 677
    quadratic, 704
    simple interest, 132
Fraction(s), 22, 50, 493
    addition of, 65, 68, 510, 511, 514, 516
    building, 514
    complex, 521
    denominator of, 22, 51
    division of, 59, 505
    equality of, 52
    fundamental property of, 53, 495
    improper, 73
    lowest terms, 495
    multiplication of, 56, 501
    numerator of, 22, 51
    as percent, 118, 120
    proper, 73
    simplifying, 494, 495
Frequency polygon, 160
Function, 578
    domain of, 580
    graph of, 581
    range of, 580
Function notation, 578
Fundamental property of fractions, 53, 495
Fundamental theorem of arithmetic, 36

GCD, 40
General form:
    of the equation of a line, 548
    of the quadratic equation, 703
Geometry, 181
Graph(s):
    bar, 152, 153
    of a function, 581
    line, 156
    of linear equations, 546
    of linear inequalities, 575
    of ordered pairs, 544
    pie, 155
    of sets of real numbers, 302

Greater than, 2
Greatest common divisor, 40, 444
Greatest common factor, 444, 445
Grouped data, 160
Grouping, factoring by, 449, 468

Half plane, 572
Hectometer, 144
Height:
   of a parallelogram, 248
   slant, 266
   of a triangle, 110
Hemisphere, 265
Histogram, 159
Hooke's law, 584
Horizontal line, equation of, 559
Hyperbola, 588
Hypotenuse, 201
Hypothesis, 191

Identity, 353
   element, 326
Imaginary number, 641, 708
Imaginary part of a complex number, 709
Impossible equation, 353
Improper fraction, 73
Inch, 139, 141
Incomplete quadratic equation, 479, 694
Inconsistent system of equations, 601, 608, 613
Increase, percent of, 131
Independent equations, 600
Independent variable, 578
Index, of a radical, 642
Inductive reasoning, 269
Inequality(ies), 380
   addition property of, 381
   compound, 384
   division property of, 381
   double, 384
   multiplication property of, 381
   solution of, 380
   subtraction property of, 381
   systems of, 624
Inequality symbols, 286
Inscribed angle, 222
Integer, 300
Intercept method of graphing a line, 548
Interest, simple, 132
   rate of, 132
Interior angle(s):
   alternate, 194
   of a closed figure, 200
Inverse:
   additive, 326
   multiplicative, 326
Inverse variation, 585
Investment problems, 374, 535
Irrational number, 301, 641
Isosceles triangle, 201, 368, 679
   base angle of, 210, 368
   vertex of, 210, 368

Joint variation, 586

Kilometer, 144

LCD, 515
LCM, 39
Least common denominator, 515

Least common multiple, 39
Length, 9, 182, 247
Less than, 2
Like terms, 350, 413
Line(s), 182
   equations of, 548, 551, 559, 565
   number, 285
   parallel, 184, 566
   parallel to the coordinate axes, 551
   perpendicular, 186, 567
   point-slope form, 565
   slope of, 555
   slope-intercept form, 557
Linear depreciation, 560
Linear equation, 477
   graph of, 546
   in two variables, 546
Linear inequality, 575
Line graph, 156
Line segment, 183
Liquid mixture problems, 376
Literal equations, 356
Long division, 23
   remainder of, 23
Lowest common denominator, 515
Lowest terms, 495

Major arc, 222
Marginal cost, 562
Marginal demand, 560
Markdown, 129
Markup, 131
Mean, 167
   arithmetic, 30
   of a proportion, 236
Measure:
   of an arc of a circle, 222
   degree, 184
Measurement, 139
   American units of, 141
   metric units of, 141
Median, 168
Meter, 144
Metric prefixes, 146
Metric units of measurement, 141
Midpoint of a line segment, 553
Mile, 141
Millimeter, 144
Minor arc, 222
Minuend, 10
Mixed number, 73
Mixture problems, 376, 377
Möbius strip, 181
Mode, 169
Monomial(s), 409
   addition of, 413
   degree of, 409
   division of, 426, 427
   multiplication of, 419
   subtraction of, 414
Motion, uniform, 375, 534
Multiples, 38
   least common, 39
Multiplication:
   associative property of, 17
   commutative property of, 17
   of complex numbers, 710
   of decimals, 99
   of fractions, 56, 501
   of monomials, 419
   of polynomials, 419
   by a power of ten, 101
   of radical expressions, 645, 656
   of real numbers, 316
   of whole numbers, 16

Multiplication property(ies):
   of equality, 341
   of inequality, 381
Multiplicative inverse, 326

Naming angles, 184
Natural number, 1, 283
Negative exponent, 401, 524
Negative reciprocals, 567
Negatives, 302, 326, 497
Nonperfect square trinomials, 457
Notation:
   decimal, 89
   expanded, 2, 89
   function, 578
   scientific, 404
   standard, 2, 405
Number(s):
   complex, 709
   composite, 35, 284
   even, 284
   imaginary, 641, 708
   irrational, 301, 641
   mixed, 73
   natural, 1, 283
   odd, 284
   perfect, 38
   prime, 35, 284, 443
   rational, 301, 493
   real, 301
   whole, 2, 284
Number line, 285
Numeration, decimal, 89
Numerator, 22, 51
Numerical coefficient, 291, 349

Obtuse angle, 185
Obtuse triangle, 201
Odd number, 284
Ohm's law, 361
Operations, order of, 29, 295
Opposite, 302, 326, 497
Ordered pair, 544
Order of mathematical operations, 29, 295
Origin, 285, 544

Parabola, 716
   equation of, 717, 718
   vertex of, 716, 718
Parallel lines, 184
   slope of, 566
Parallelogram, 215
   altitude of, 248
   area of, 249
   base of, 248
   height of, 248
Parallel planes, 184
Pentagon, 201
Percent, 117
   and decimals, 119, 120
   of decrease, 129
   and fractions, 118, 120
   of increase, 131
Percentage, 123
   base of, 123
Perfect integer cube, 648
Perfect integer square, 641
Perfect number, 38
Perfect-square trinomial, 456, 698
   factoring, 456
Perimeter, 9
Perpendicular lines, 186
   slopes of, 567

Pi (π), 108, 257
Pictograph, 154
Pie graph, 155
Place value, 2
Plane(s), 182, 628
    half-, 572
    parallel, 184
Plane figure, 261
Point, 182
    decimal, 89
    of tangency, 223
Point-slope form of the equation of a line, 565
Polygon(s), 201
    area of, 246
    frequency, 160
    sum of angles in, 203
    vertex of, 201
Polynomial(s), 409
    addition of, 414
    degree of, 410
    division of, 427, 430
    multiplication of, 419
    prime, 454, 460
    subtraction of, 415
Postulate, 271
Power(s), 294, 393
    of ten, 101, 104
Power rules for exponent, 396, 397
Prefixes, metric, 146
Prime-factored form, 36, 444
Prime number, 35, 284, 443
Prime polynomials, 454, 460
Principal square root, 640
Priority rules, 295
Prism, 261
    volume of, 262
Product(s), 16
    of real numbers, 316
    special, 422
    of whole numbers, 16
Product rule for exponents, 395
Proper fractions, 73
Property(ies):
    of addition, 7
    associative, 7, 17, 324
    closure, 323
    commutative, 7, 17, 323
    distributive, 31, 325
    of equality, 336, 337, 341, 664, 674, 695
    of exponents, 398, 402
    of fractions, 53, 495
    of inequality, 381
    of multiplication, 17
    of radicals, 645, 647
    of real numbers, 323
    square-root, 674, 695
    squaring, 664
Proportion, 236
    extremes of, 236
    means of, 236
Proportional, 237
Proportionality, constant of, 584–586
Pyramid, 266
    volume of, 266
Pythagoras, 49, 83
Pythagorean theorem, 674

Quadrant, 544
Quadratic equation, 478, 703
    general, 703
    incomplete, 479, 694
    solving by completing the square, 698
    solving by the quadratic formula, 704

Quadratic form, 693
Quadrilateral, 201
Quotient, 21, 74
Quotient rule for exponents, 397

Radicals:
    addition of, 651
    division of, 647, 658
    index of, 642
    like, 651
    multiplication of, 645, 656
    simplified form of, 645
    solving equations containing, 664
    subtraction of, 651
Radical sign, 640
Radicand, 640
Radius:
    of a circle, 108, 221
    of a cylinder, 264
    of a sphere, 263
Range:
    of a function, 580
    reading from a graph, 582
    of a relation, 580
Rate:
    annual depreciation, 561
    of interest, 132
    in a percentage, 123
Ratio(s), 235
    equal, 235
Rational exponent, 669, 670
Rational expression, 493
Rationalizing the denominator, 659
Rational number, 301, 493
Ray, 182
    endpoint of, 182
Reading data from tables, 160
Reading the domain from a graph, 582
Reading the range from a graph, 582
Reading variation from a graph, 588
Real number(s), 301
    absolute value of, 303
    addition of, 307, 308
    difference of, 310
    division of, 318
    graphing sets of, 302
    multiplication of, 316
    product of, 316
    properties of, 323
    subtraction of, 310
Real part of a complex number, 709
Reasoning:
    deductive, 181, 271
    inductive, 269
Reciprocals, 58, 326
    negative, 567
Rectangle, 216
    area of, 20, 107, 247
    length of, 9, 247
    perimeter of, 9
    width of, 9, 247
Rectangular coordinate system, 544
Rectangular solid, 261
    surface area of, 262
    volume of, 262
Relation, 577
    domain of, 580
    range of, 580
Relatively prime, 42
Remainder, 23, 74, 431
Rhombus, 217
Right angle, 185
Right-circular cone, 266
Right-circular cylinder, 264
Right prism, 261
Right pyramid, 266

Right triangle, 201
    hypotenuse of, 201
    isosceles, 679
Root:
    of an equation, 336
    nth, 642
    principal square, 640
    square, 640
Rounding, 2, 92
Rules, priority, 295

Salvage value, 561
Scalene triangle, 202
Scientific notation, 404
Segment, line, 183
    midpoint of, 553
Semicircle, 221
Set, element of, 284
Shared-work problems, 533
Side(s):
    of an angle, 184
    of an equation, 335
Sign, radical, 640
Similar terms, 349
Similar triangles, 240
Simple interest, 132
Simplified Employee Pension (SEP), 335
Simplified form of a radical expression, 645
Simplifying a fraction, 494, 495
Simultaneous solution, 600
Slant height, 266
Slope(s), 555
    as aid in graphing, 556
    of a line, 555
    of parallel lines, 566
    of perpendicular lines, 567
Slope-intercept form of the equation of a line, 557
Solution:
    of an equation, 336
    extraneous, 528, 665
    of an inequality, 380
    of a system of equations, 600
Solving equations:
    linear, 477
    quadratic equations, 698, 704
    that contain fractions, 527
    that contain radicals, 664
Special products, 422
Sphere, 263
    center of, 263
    radius of, 263
    surface area of, 263
    volume of, 263
Square(s), 218, 639
    area of, 107
    difference of two, 453
    sum of two, 454
Square root, 640
Square root method for solving a quadratic equation, 695
Square root property of equality, 674, 695
Squaring property of equality, 664
Standard notation, 2, 405
Statement, conditional, 191
Straight angle, 185
Subscript, 532
Substitution method, 606
Subtraction:
    of complex numbers, 709
    of decimals, 95
    of fractions, 68, 511, 516
    of monomials, 414
    of polynomials, 415

Subtraction *(continued)*
  of radical expressions, 651
  of real numbers, 310
  of whole numbers, 10
Subtraction property:
  of equality, 337
  of inequality, 381
Subtrahend, 10
Sum, 289
  of angles in a polygon, 203
  of angles in a triangle, 203
  of two cubes, 471
  of two squares, 454
  of whole numbers, 7
Supplementary angles, 188
Surface, 182
Surface area:
  of a cone, 266
  of a cylinder, 265
  of a rectangular solid, 262
  of a sphere, 263
Symbols, inequality, 286
System of equations, 600, 628
  consistent, 600
  inconsistent, 601, 608, 613
  of inequalities, 624
  solution of, 600
  solving by addition, 611
  solving by substitution, 606
  in three variables, 628

Tables, reading data from, 160
Tangency, point of, 223
Tangent line, 223
Term(s), 349
  algebraic, 291, 409
  defined, 271
  fraction lowest, 495
  like, 349, 350, 413
  similar, 349
  undefined, 271
  unlike, 350
Tests for divisibility, 34
Theorems, 271
  Pythagorean, 674
  zero-factor, 478
Three-dimensional figures, 261

Topology, 181
Transversal, 194
Trapezoid, 253
  area of, 254
Trial-and-error method of factoring, 463
Triangle(s), 201
  acute, 201
  area of, 110, 252
  congruent, 208
  corresponding parts of, 207
  equiangular, 201
  equilateral, 201
  hypotenuse of, 201
  isosceles, 201, 368
  isosceles right, 679
  obtuse, 201
  right, 201
  scalene, 202
  similar, 240
  sum of angles in, 203
  30°-60° right, 681
  vertex of, 210, 368
Trinomial(s), 409
  factoring, 463, 468, 456, 457, 467

Undefined terms, 271
Uniform-motion problems, 375, 534
Unit conversion factor, 142
Unit cost, 371
Units of measurement, 141
Unknown, 335
Unlike denominators, 65, 68, 514, 516
Unlike terms, 350

Value:
  absolute, 303, 713
  salvage, 561
Variables, 13, 335
  dependent, 578
  domain of, 493
  independent, 578
Variation:
  combined, 587
  constant of, 584, 585, 586
  direct, 584
  inverse, 585

Variation *(continued)*
  joint, 586
  reading from a graph, 588
Vertex:
  of an angle, 184
  of a parabola, 716, 718
  of a polygon, 201
Vertex angle, 210, 368
Vertical angle, 190
Vertical line, equation of, 559
Volume:
  of a rectangular solid, 262
  of a right-circular cone, 266
  of a right-circular cylinder, 265
  of a right prism, 262
  of a right pyramid, 266
  of a sphere, 263

Whole number(s), 2, 284
  addition of, 7
  difference of, 10
  division of, 21
  multiplication of, 16
  product of, 16
  rounding, 3
  subtraction of, 10
  sum of, 7
Width, 182
  of a rectangle, 9, 247

x-axis, 544
x-coordinate, 544
x-intercept, 548

Yard, 141
y-axis, 544
y-coordinate, 544
y-intercept, 548

Zero:
  division by, 22, 301
  exponent, 400
Zero-factor theorem, 478

Rules of Exponents

If x and y are real numbers and there are no divisions by 0, then

$$x^m x^n = x^{m+n} \qquad (x^m)^n = x^{mn} \qquad (xy)^n = x^n y^n$$

$$\left(\frac{x}{y}\right)^n = \frac{x^n}{y^n} \qquad \frac{x^m}{x^n} = x^{m-n} \qquad x^0 = 1 \qquad x^{-n} = \frac{1}{x^n}$$

Rules for Expanding Binomials

$$(x + y)^2 = x^2 + 2xy + y^2 \qquad\qquad (x - y)^2 = x^2 - 2xy + y^2$$
$$(x + y)(x - y) = x^2 - y^2$$

Factoring Formulas

$$x^2 - y^2 = (x + y)(x - y) \qquad\qquad x^3 + y^3 = (x + y)(x^2 - xy + y^2)$$
$$x^2 + 2xy + y^2 = (x + y)^2 \qquad\qquad x^3 - y^3 = (x - y)(x^2 + xy + y^2)$$
$$x^2 - 2xy + y^2 = (x - y)^2$$

Properties of Fractions

If there are no divisions by 0, then

$$\frac{a}{b} = \frac{c}{d} \text{ is equivalent to } ad = bc \qquad\qquad \frac{a \cdot x}{b \cdot x} = \frac{a}{b}$$

$$\frac{a}{b} \cdot \frac{c}{d} = \frac{ac}{bd} \qquad\qquad \frac{a}{b} \div \frac{c}{d} = \frac{a}{b} \cdot \frac{d}{c} = \frac{ad}{bc}$$

$$\frac{a}{d} + \frac{b}{d} = \frac{a + b}{d} \qquad\qquad \frac{a}{d} - \frac{b}{d} = \frac{a - b}{d}$$

Slope and Equations of Lines

Midpoint formula: If $P(x_1, y_1)$ and $Q(x_2, y_2)$ are endpoints of a line segment, then the coordinates of the midpoint of $\overline{PQ}$ is

$$\left(\frac{x_1 + x_2}{2}, \frac{y_1 + y_2}{2}\right)$$

General form of the equation of a line: $Ax + By = C$

Slope of a nonvertical line: $m = \dfrac{y_2 - y_1}{x_2 - x_1} \qquad (x_2 \neq x_1)$

Slope-intercept form of the equation of a line: $y = mx + b$ (m is the slope, b is the y-intercept)

Point-slope form of the equation of a line: $y - y_1 = m(x - x_1)$ (m is the slope, x_1 and y_1 are coordinates of a known point)